DIGITAL VLSI SYSTEMS DESIGN

Digital VLSI Systems Design

A Design Manual for Implementation of Projects on FPGAs and ASICs Using Verilog

By

Dr. S. Ramachandran

Indian Institute of Technology Madras, India

Springer

A C.I.P. Catalogue record for this book is available from the Library of Congress.

Additional material to this book can be downloaded from http://extras.springer.com.

ISBN 978-1-4020-5828-8 (HB)
ISBN 978-1-4020-5829-5 (e-book)

Published by Springer,
P.O. Box 17, 3300 AA Dordrecht, The Netherlands.

www.springer.com

Printed on acid-free paper

Contents

Preface

This book deals with actual design applications rather than the technology of VLSI Systems. This book is written basically for an advanced level course in Digital VLSI Systems Design using a Hardware Design Language (HDL), Verilog. This book may be used for teaching undergraduates, graduates, and research scholars of Electrical, Electronics, Computer Science and Engineering, Embedded Systems, Measurements and Instrumentation, Applied Electronics, and interdisciplinary departments such as Biomedical, Mechanical Engineering, Information Technology, Physics, etc. This book also serves as a reference design manual for practicing engineers and researchers. Although this book is written for an advanced level course, diligent freelance readers, and consultants, especially, those who do not have a first level exposure of digital logic design, may also start using this book after a short term course or self-study on digital logic design. In order to help these readers as well as regular students, the book starts with a good review of digital systems design, which lays a solid foundation to understand the rest of this book right up to involved Project Designs unfolded gradually.

Contents of the Book

The book presents new source material and theory as well as synthesis of recent work with complete Project Designs using industry standard CAD tools and FPGA boards, enabling the serious readers to design VLSI Systems on their own. The reader is guided into systematic design step by step starting from a buffer to full-fledged designs right up to 120,000 gates. At every stage, the reader's grasp of the developing subject is challenged so that he or she may stand on his or her own feet after completing the course. Easy to learn Verilog HDL is made use of to realize the designs. A bird's eye view on what the reader is going to learn shortly is shown as follows:

- Introduction to VLSI Systems Design
- Features and architectures of latest FPGAs of leading vendors
- Detailed review of Digital Systems Design
- Introduction to Verilog Design
- Verilog coding of Combinational and Sequential Circuits
- Writing a Test Bench
- RTL Coding Guidelines
- Design Flow for VLSI Systems and Design Methodology
- Simulation using industry standard Modelsim Tool

- Synthesis using industry standard Synplify Tool
- Place and Route and Back Annotation using industry standard Xilinx Tool
- Verilog Coding of Memories and Arithmetic Circuits
- Development of Algorithms and Verification using High Level Languages such as Matlab
- Architectural Design
- VLSI Systems Design for a couple of projects as examples: PCI Arbiter and the Discrete Cosine Transform and Quantization Processor for Video com pression applications
- Complete Hardware Implementations using FPGA Board: A Traffic Light Controller and a Real Time Clock as examples
- Suggestion of Projects for Implementation on FPGAs/ASICs

Approach

The reader is taken step by step into designing of VLSI Systems using Verilog. To start with, an overview of VLSI Systems is presented. Features and architectures of the latest FPGAs of leading vendors are also presented. The design starts right from implementing a single digital gate to a massive design consuming well over 100,000 gates. Following a review of basic concepts of digital systems design, a number of design examples are illustrated using conventional digital components such as Flip-flops, Multiplexers, De-multiplexers, Decoders, ROM, Programable Array Logic, etc. With these foundations, the reader is introduced to Verilog coding of the components mentioned earlier as well as designing systems for small end applications. These designs are tested using Test Benches, also written in Verilog. All HDL codes the reader wishes to develop for various applications must conform to Register Transfer Level (RTL) Coding Guidelines, without which no chip can work satisfactorily. These guidelines are presented at length.

The sequel to the design is to simulate, synthesize, and place and route. Since our sincere interest lies in making the reader a full-fledged engineer, we use the same popular development tools used in the Industries and Research Laboratories such as Modelsim (a simulation tool of Mentor Graphics), Synplify (a synthesis tool of Synplicity Inc.), and Place and Route and Back Annotation of Xilinx Inc. Equipped with these powerful tools, the elucidation of design progresses into more and more complex designs such as Memories and Arithmetic Circuits extending into VLSI realms.

Complex Project Designs usually need the development of new Algorithms and Architectures for Optimum realization. These issues, which stimulate creative thinking and indispensable for researchers, are thoroughly analyzed and solved in this book. Armed with all the features mentioned earlier, complete Project Designs are presented. This is followed by hands-on experience in designing systems using FPGA and Input/Output Boards. Once the path is shown for designing VLSI Systems systematically, numerous Project Designs are suggested for FPGA/ASIC Implementations.

The book gives complete RTL Compliant Verilog codes for several projects, which are synthesizable and works on the hardware based on FPGA or ASIC. Most of the Verilog codes developed in this book can be readily used in new projects the reader may undertake in his or her study or career. The advantages accruing out of this strategy are two fold: One, the students are trained to suit industries/R&D Laboratories and, therefore, industries and other employers need not spend time and money to train them when they are hired. The other advantage is to provide in-plant training in industries, etc. based on this book to retrain old (as well as new) personnel in the new technologies.

Brief Description of the Topics Covered

The following is a brief description of the topics covered in each chapter of this book:

Chapter 1 presents an introduction to Digital VLSI Systems Design. The evolution of VLSI Systems over the years has been described. This chapter also outlines a number of applications for the VLSI Systems, thus motivating the reader to undertake a serious study of the subject and in turn be a contributor. This chapter shows how FPGA based system designs score over processor based systems. The features and architectures of latest FPGAs available in the market are presented.

Chapter 2 provides a detailed review of digital systems design so that the reader may understand the rest of the book without any difficulty or needing to refer to logic design fundamentals from another book. This chapter starts with the number systems, design of combinational circuits followed by designs using Programmable Logic Devices such as ROM and Programmable Array Logic. Thereafter, it deals with the design of sequential circuits. It shows how to design systems using the conventional state graphs and ASM Charts. Digital system designs are illustrated with a number of examples.

Chapter 3 presents a brief introduction of the evolution of Hardware Design Language. Verilog is introduced as a tool for realizing digital systems design. The advantages of Verilog coding over the traditional schematic circuit diagram approach are established, especially when the design of VLSI circuits crossing over 50,000 transistors mark is encountered. A number of design examples using Verilog are illustrated for both combinational and sequential circuits. These examples cater to the frequently used digital circuits in a system design, especially in industries. Only the cores of the designs are initially presented to expedite the learning process. Later on, full-fledged codes are presented so that the reader may simulate, synthesize and place and route. Register Transfer Level (or Logic) coding, vital for designing chips that work successfully, is the main emphasis of this design book and is discussed in a later chapter.

Chapter 4 shows how to write an effective test bench. The basic concept of a test bench is shown by presenting a simple design and a model test bench for testing the design exhaustively. For bigger designs, an elaborate test may prove to be difficult. In such cases, the test may be carried out for a range of inputs covering minimum, maximum, center, and few other input values applied judiciously.

In the previous chapter, designs for combinational and sequential circuits were dealt. Test benches for the same are presented in this chapter. Usually, the test bench size will be smaller than that of the design. This chapter is especially useful for verification engineers.

Chapter 5 deals with the Register Transfer Level Coding Guidelines. Every designer is vitally interested in making his or her design work when implemented as a hardware, which uses FPGA or ASIC. For successful working of a system, RTL coding techniques are inevitable. This requires a high degree of discipline or care while designing such systems. This chapter discusses RTL coding techniques in depth, which is basically adhering to synchronous design practices. RTL approach deals with the regulation of data flow, and how the data is processed using register transfer level as the primary means. Since we deal with a synchronous design, it should naturally run smoothly through various tools such as simulation, synthesis, and place and route, which tools are described at length in succeeding chapters.

Chapter 6 presents the VLSI Design flow along with the design methodologies that may be gainfully employed so that one may become conversant with various steps involved in designing a product. Several designs were considered in Chapter 3 and their test benches in Chapter 4. This was followed by RTL coding guidelines in Chapter 5. The next logical step in the design flow is the simulation, vital for testing one's design. All the Verilog designs presented in the third chapter are analyzed using waveforms in the present chapter. Industry standard Modelsim tool of Mentor Graphics is employed for the simulation. A command summary of the Modelsim tool, which serves as a quick reference while using the tool, is also furnished. Even though the functionality of a design is checked using simulation, it does not test time critical paths or furnish insights into gate delays, unless back annotated, since simulation does not map a target chip such as an FPGA, where the design will have to reside ultimately. These features are possible with synthesis tool, which is presented in the next chapter.

Chapter 7 covers the synthesis of designs using Synplify tool, widely used in industries. The salient features of synthesis are mapping of an FPGA device, logic optimization, and viewing schematic circuit diagrams of the Verilog code. The tool creates optimized Verilog file and Electronic Data Information Format (EDIF) file, which may be used for simulation and vendor specific place and route respectively. EDIF file is exported to the next tool, the Place and Route tool, for creating a bit stream of the design. Synplify tool supports all types of FPGAs. In order to learn and fix compilation errors and simulation errors in Modelsim and Synplify tools, errors are created deliberately and known correction applied. A command summary of the Synplify tool is furnished for quick reference while using the tool.

Chapter 8 covers the Place and Route (P&R) tool of Xilinx, also widely used in industries. The salient features of Xilinx P&R are the creation of a 'bit' file from EDIF file created by the Synplify tool or from source file directly, specification of user constraints such as clock speed and FPGA pins, remapping of the target FPGA device if desired, back annotation and floor planning. The back annotated file, which reflects the actual gate delays, is simulated again in Modelsim to ensure that the design is working correctly at the maximum frequency reported by the Place and Route tool. Report file generated by the P&R tool gives the maximum frequency of operation possible as well as the gate count (chip complexity) for the design. A command summary of the Xilinx P&R tool is furnished for ready reference.

Chapter 9 presents the memory design, which is one of the most important aspects of a VLSI System Design. This chapter shows the way to design various types of on-chip ROMs and RAMs, some of them unconventional, in order to meet the special requirements of a particular application. The size of memory that could be incorporated on-chip is usually limited by the order of a few tens of Kilo Bytes with the currently available FPGAs. This limitation is fast changing with the advances in the technology. In applications, where large memories are called for, external memories such as the commercially available static RAMs, ROMs, Flash RAMs, dynamic RAMs, etc. may be used. Towards this end, a controller design that interfaces with commercially available external memories is presented in this chapter. However, the access speed of external memory falls by a factor of two when compared to on-chip memory. On the other hand, on-chip memory increases the chip area consumed. Therefore, the designer must consider carefully the pros and cons before making the choice for on-chip memory or external memory in a system design.

Chapter 10 presents arithmetic circuits such as add/subtract, multiply, etc. These circuits are computationally intensive and, therefore, conventional methods are not sufficient for real time implementations on FPGA or ASIC. In order to speed up the processing considerably, we will have to base our designs on massively parallel circuits and heavy pipelining. This chapter presents arithmetic circuit designs such as signed adders and multiplier with a high degree of parallelism and pipelining for computationally intensive applications such as video compression.

Chapter 11 deals with the development of algorithms for various projects so that they are suitable for implementation on FPGAs/ASICs. Complex applications such as video codecs have involved algorithms at their core, which need to be adapted or developed depending upon how we wish to implement the system. The design methodology or strategy would depend upon whether we need to implement the system using software such as C or by a HDL such as Verilog. The viability of the project is decided by the successful working of the algorithms or the design methodology using Matlab or C, which is also discussed at length. While developing algorithms for hardware implementation, we need to keep the actual hardware in mind and subsequently, design the architecture. Only then, the algorithm can be converted into an actual working product. This chapter also covers the verification of concepts and algorithms developed earlier using a high level language such as Matlab. This must be carried out before taking up the architecture and Verilog design. Designers, in general, have a tendency to bypass this vital step and get into trouble later on, after completing Verilog codes, at the time of debugging. This is especially true in large designs, and is worthwhile to take little more trouble of coding in Matlab or C, preferably in Matlab, to save lots of trouble and time.

Chapter 12 presents the development of architectural designs. In the last chapter, development of algorithms were presented and verified for a number of applications in the field of video processing as examples. These algorithms were developed in such a way that the applications may be mapped onto an FPGA or an ASIC. The next logical step is to work out a detailed architecture keeping the actual hardware such as registers, counters, combination circuits, etc. in mind. In this chapter, the architectural designs are developed for the same applications that we had undertaken in the previous chapter.

Chapter 13 presents VLSI System Design examples for two projects, namely, PCI Arbiter and the Discrete Cosine Transform and Quantization Processor for Video compression applications. While presenting these designs, emphasis is laid on systematic design. This comprises identification of a project based on need, formulating detailed specifications, development of algorithm and verification, or proving an algorithm or concept using a high level language such as Matlab or C to establish its feasibility, development of detailed architecture based on actual hardware components, Verilog RTL coding, simulation, synthesis, place and route, and back annotation. The methodologies adopted in these designs are to use highly parallel and heavily pipelined circuits in order to increase the throughput and to be platform independent, whether an implementation uses an FPGA or an ASIC. No vendor specific modules are used and, hence, these designs are universal and can work on any FPGA or ASIC. The design methodologies presented in this book are equally applicable to other HDLs such as VHDL.

Chapter 14 presents a couple of complete hardware implementations, namely, a traffic light controller and a real time clock, as examples. These applications are based on ready made boards available such as an FPGA board and a digital input/output board. The design methodologies adopted in these designs may be extended to any other Project Design or any other FPGA or ASIC and I/O boards. These system designs are presented in a systematic manner, starting from detailed specification. The need for formulating the right type of architecture is emphasized and designed with actual hardware components in mind. The signal nomenclatures adopted in the architectures are actually used as it is in realizing their designs in Verilog conforming to the RTL coding guidelines. Simple test benches are developed and simulated using Modelsim tool to ensure the correct functioning of the designs. These are followed by running the synthesis tool and the place and route tool to get the bit stream files. The hardware for each of the designs is subsequently setup and the corresponding bit streams downloaded into the FPGA. Elaborate testing is thereafter conducted on the actual hardware to ensure the correct working of the systems designed.

Chapter 15 suggests a number of projects for the reader to design and implement on FPGAs/ASICs, category-wise. Issues involved in Digital VLSI Systems Design are discussed at length in order to aid the reader to quickly develop products. A brief introduction of embedded systems design is presented. Detailed specifications and basic architectures for a couple of applications for FPGA/ASIC implementations are furnished for the reader to make a start and gain hands-on experience in Digital VLSI Systems Design.

How to Use This Book?

The material in this book has been developed over several years as a result of teaching students of various disciplines and guiding practicing engineers and students in their projects/thesis work. Teaching/training styles vary among countries, universities, colleges, and industries and may, therefore, be suitably organized. In the light of imparting VLSI Systems Design to students and engineers of various disciplines and, research over the years, the following pattern of education is recommended:

Second Year Under Graduates of Electrical/Electronics/Computer Engineering

Specializations such as Power systems, Electronics, Communication, Power electronics, Embedded Systems, Controls, Measurements and Instrumentation, Applied Electronics, etc. are included in Electrical/Electronic Engineering. Chapter 1 presents a general view of what VLSI Systems are and Chapter 2 is a review material, which the students may study independently if they have just completed a first level digital design course. Optionally, they may be taught Chapter 2, especially if there is a semester gap after they have completed their first level digital design course. Thereafter, they may be taught Chapters 3, 4, and 6. Chapter 5, which presents RTL coding guidelines, may be deferred to the third year.

Third Year Under Graduates of Electrical/Electronics/Computer Engineering

Chapter 5, and review of Chapter 6 if there is a semester gap after they have completed their second year course mentioned earlier, may be taught to start with. Thereafter, they may be taught other tools such as synthesis and place and route from Chapters 7 and 8 respectively. Finally, Memory designs presented in Chapter 9 may be taught.

Fourth Year Under Graduates of Electrical/Electronics/Computer Engineering

After a brief review of simulation, synthesis, and place and route tools, Arithmetic circuits design (Chapter 10) may be taught. A simple Project Design, PCI arbiter, may be taught from Chapter 13. More involved design such as the DCTQ processor may be by-passed. Thereafter, hardware designs based on FPGA boards may be taught from Chapter 14. This may be followed by Chapter 15. Finally, mini projects may be assigned to students in small groups encouraging team work, yet clearly defining individual student goals. The second half of a semester may be used for the mini projects. These projects may be based on FPGA boards, if feasible. Students may be assessed by asking them to present their progress from time to time. These presentations may be strategically scheduled as follows:

First week:	Detailed specification of the project.
Second week:	Algorithmic development, if any, and Hardware Architecture.
Fifth week :	Verilog coding of the design and test bench.
Sixth week :	Simulation/synthesis/place and route results and
Eighth week :	Demonstration, documentation, and final presentation.

Many projects are suggested throughout the book, and many more may be created by the instructor and the students.

First Year Post-Graduate Students and Ph.D. Scholars of Electrical/Electronics/Computer Engineering

Chapters 1 and 2 are review materials, which the students may study independently. Chapters 3 to 5 may be taught in detail, if the students have not covered these topics in their earlier studies. A quick exposure to three industry standard tools in Chapters 6 to 8 may be made using command summary of these tools, leaving the students to gain proficiency with the tools in the laboratory. This may be followed by Chapters 9 and 10. Chapters 11 and 12 are important for developing new algorithms and their verification and the design of architecture, especially for those doing research (MS or Ph.D.). If hard pressed for time, one of the three applications in these chapters, say, the DCTQ alone need be taught. Thereafter, Chapters 13 to 15 may be taught. Individual mini projects may be assigned to students in the second half of a semester as detailed for the fourth year students earlier. The above recommendation, although appear to be too crammed for a semester study, has been actually tested for post-graduates and Ph.D. scholars of Electrical/Electronics engineering. If there is room, the entire book may be taught spread over two semesters: Chapters 3 to 10 in the first semester and the rest in the second semester.

Third Year under Graduates of Information Technology/Computer Science and Interdisciplinary Departments such as Bio-medical, Mechanical Engineering, and Post-graduate Students of Physics

Chapters 1 to 4 and Chapter 6 may be taught spread over a semester. The Chapter 5 on RTL coding guidelines may be deferred to the fourth year.

Fourth Year Under Graduates of Information Technology/Computer Science and Interdisciplinary Departments such as Biomedical, Mechanical Engineering and Post-graduate Students of Physics

Chapters 5, 9, 10, and all topics of Chapter 11 for developing new algorithms and their verification may be taught. PCI arbiter design in Chapter 13 and Traffic light controller design in Chapter 14 may also be taught. Applications from Chapter 15 can be taught finally. Individual mini projects may be assigned to students in the second half of a semester as detailed for the fourth year electrical engineering students earlier.

Each of the above recommendations is to be taught over a semester.

In-plant Industrial Training

Experience shows that in many industries, R&D laboratories, etc., there are no periodic and systematic orientation programs or training, either for new recruits or for existing personnel. As a result, with rapid technological strides, employees do not come up to the expected level of the employer. The lack of proper and regular training has also rendered the 'attrition management' increasingly difficult. These problems can be alleviated to a great extent by conducting regular orientation programs for the new recruits even if they come with experience, and periodical hands-on training based on this book to retrain old (as well as new) personnel in the new technologies. Depending upon the level of personnel, the management can form their own curriculum for orientation programs and training as per the recommendations made earlier for various categories of students using this book. Designers may be encouraged to create a library of developed codes in Verilog/VHDL/Matlab/C with proper documentation on-line so that the on-going product designs are completed quickly without reinventing the wheel, thus improving the productivity and hence profitability dramatically.

Assignments and Laboratory Work

Merely studying the text would not make the reader an adept in designing VLSI Systems. On the other hand, one is sure to become proficient if every assignment given towards the end of every chapter is sincerely solved. These must be supplemented by inventing more number of assignments and solving them. Most of the assignments, especially those presented in the second chapter, are based on industry related problems and placement questions faced by a large number of students globally. Chapters 3 to 10, 13, and 14 contain a large numbers of illustrated examples including full-fledged projects, all of which can be used as source materials for laboratory work. Chapters 6, 7, and 8 respectively present the simulator (Modelsim), synthesis (Synplify), and place and route and back annotation (Xilinx) tools and may be thoroughly studied before starting their respective laboratory work. As laboratory assignments, the problems/assignments presented in each of the Chapters 3 to 15 may be selectively allocated, depending upon the categories of students as presented earlier.

Verilog and Matlab Source Codes Supplied in CD

All Verilog codes, be they designs or test benches, illustrated in Chapters 3 to 5, 7, 9, 10, 13, and 14 are in their respective folders in the CD provided in the book. Similarly, the source codes of Matlab illustrated in Chapters 11 and 13 are in different folders. These may be copied in hard disks and tested using the tools. 'Readme' files explain the usage of various files. Command summaries for Modelsim, Synplify and Xilinx tools are also included. Also, a summary of the usage of RTL

Verilog codes is provided in separate files. Reader's technical skills may be enhanced by going through the PPT: How to make oral/written presentations?

Mini Project and Project Work

Some assignments in Chapters 7, 8, and 11 to 15 are suitable for mini projects for students of fourth year undergraduate and post-graduate students as recommended earlier. Over 100 projects are listed in Chapter 15, which may be selected for FPGA/ASIC implementation by both undergraduate and post-graduate/research students. Many mini projects may also be carved out of these projects. Many more mini projects, full-fledged projects and research projects, may be created on similar lines as illustrated in above mentioned chapters.

Solution Manual

Solution Manual for the assignments presented towards the end of each chapter is available to teachers from the publishers on a CD or on their website. The solution manual contains all source codes and reports of solved assignments in the book and presentations for quick and easy dissemination of the subject.

Acknowledgment

The author is thankful to various industries he has been associated with over decades and Indian Institute of Technology, Madras for providing excellent resources and working environment. Heart-felt thanks are due to Prof. S. Srinivasan, IITM, who has been a great source of inspiration for writing this book. Thanks are also due to numerous co-designers in industries, undergraduate to research students for their lively discussions on projects and staff for their timely help in preparing the manuscript of this book. Special thanks are due to the publishers in bringing out this book quickly, yet maintaining high quality.

S. Ramachandran

Chapter 1

Introduction to Digital VLSI Systems Design

The electronics industry has achieved a phenomenal growth over the last few decades, mainly due to the rapid advances in large scale integration technologies and system design applications. With the advent of very large scale integration (VLSI) designs, the number of applications of integrated circuits (ICs) in high-performance computing, controls, telecommunications, image and video processing, and consumer electronics has been rising at a very fast pace. The current cutting-edge technologies such as high resolution and low bit-rate video and cellular communications provide the end-users a marvelous amount of applications, processing power and portability. This trend is expected to grow rapidly, with very important implications on VLSI design and systems design.

Information technology (IT) focuses on state of the art technologies pertaining to digital information and communication. The IT sector is the fastest growing industry in recent times. With the world growing smaller day by day and business going global, the need for better devices and means for communications becomes all the more important. One of the most important characteristics of IT is its increasing need for very high processing power and bandwidth in order to handle real-time applications: video, for example. This has led to the need for faster and increasingly more efficient products to enable better telecommunications. It is this ever-growing demand in the modern world that is making many countries invest heavily in VLSI systems design. Manufacturing VLSI systems on chips is an involved process and comprises a number of activities: VLSI systems design using electronic design automation (EDA) tools; computer aided design (CAD) in the manufacture of VLSI chips; foundry activity starting from base wafer to packaged and tested ICs; and design, development, and manufacture of capital equipments for producing VLSI chips. All these activities except VLSI systems design using EDA tools are capital intensive. The latter, however, is knowledge intensive. Since applications are numerous and growing rapidly, challenging as well as interesting, VLSI system application designers are in greater demand than professionals working on chip technology.

Ever-increasing global communications has opened up a brand new vista for people interested in a career in the information technology and VLSI design including embedded systems. The advent of advanced EDA tools gives one the freedom to innovate and experiment to develop a new product that could be the next breakthrough in all spheres of research and development. A career in these sectors will give the reader access to the technical resources to work with and design better world-class products. As a product developer, the reader will be providing solutions to international markets and, therefore, technical skills need to be of the best quality. It requires continuous learning, systematic approach, and sustained efforts to realize one's dreams. It is both intellectually stimulating and exciting to be part of creating the technology of the future. This book is an earnest attempt to equip the reader completely for this challenging task and shape him or her as a highly skilled professional.

The rest of this chapter is organized as follows: In the next section, a brief introduction to the evolution of VLSI systems is presented. This is followed by presenting a short list of growing applications for VLSI systems in order to motivate the reader towards undertaking product designs seriously. In Sections 1.3 and 1.4, the advantages as well as limitations are discussed for processor based systems and embedded systems respectively. Section 1.5 presents field programmable gate arrays (FPGAs) based designs and their advantages over processor and embedded controller based designs. It discusses briefly the selection criteria of hardware. It also discusses in detail various issues involved in one of the latest applications, namely, video compression. Complete project design for video compression is presented in later chapters. An introduction to digital system design using FPGAs is presented in Section 1.6. Following this, detailed descriptions of features and architectures of various types of popular FPGAs manufactured by leading vendors are presented so that the readers may select the right type of FPGAs for their applications.

1.1 Evolution of VLSI Systems

With the advent of discrete semiconductor devices such as bipolar transistors, unijunction transistors, field effect transistors, etc., miniaturization started in full-swing, replacing bulky systems that used vacuum tubes. During 1950s computers that were made using vacuum tubes occupied an entire floor of a big building. Vacuum tubes are even now used in high power applications such as radio transmission and HAM radios. Gradually, attempts were made to integrate several circuits, be it analog or digital, in a single package. These attempts succeeded in producing both analog and digital ICs, as well as mixed signal ICs. Analog ICs offered operational amplifiers, multipliers, modulators/demodulators, etc., while digital ICs integrated AND, OR, XOR gates and so on.

Digital ICs are broadly classified according to their circuit complexity measured in terms of the number of logic gates or transistors in a single package. Chips falling under the category of small scale integration (SSI) contain up to 10 independent gates in a single package. The inputs and outputs of these gates are connected directly to the pins in the package with provision for connections to a power supply. With the advances in integration technology, more devices having a complexity of approximately 10 to 100 gates were packed in a single package. They were called medium scale integration (MSI) devices. Decoders, adders, multiplexers, de-multiplexers, encoders, comparators are examples of MSIs. Thereafter, large scale integration (LSI) devices emerged, which integrated between 100 and 1000 gates in a single package. Examples of this category include digital systems such as processors, memory chips, and programmable logic devices. Finally in late 1970s, very large scale integration devices containing thousands of gates within a single package became a reality. Personal computer chips such as 80186, 80286 of Intel are examples of this category. Since then, integration has been growing by leaps and bounds crossing 10 million gates in a single package, going into realms of ultra large scale integration (ULSI), system level integration (SLI), and system-on-chip (SOC). FPGAs fall under all the above high-end categories starting from VLSI. The foregoing classifications are summarized in the following.

Category	**Date**	**Density (gates)**
Single transistor	1959	1 device
Logic gate	1960	1
Small scale integration (SSI)	1964	Up to 10
Medium scale integration (MSI)	1967	10 – 100
Large scale integration (LSI)	1972	100 – 1000
Very large scale integration (VLSI)	1978	1000 – 10000
Ultra large scale integration (ULSI)	1989	10000 and above
SLI/SOC	Late 1990s	> 10 million

Systems were implemented in all the above categories and are still being implemented using discrete transistors for large power applications and SSIs to SOCs for progressively larger systems. In the next sub-section, we will see what applications these systems can be configured for.

1.2 Applications of VLSI Systems

VLSI system applications have become all pervasive in various walks of life like communications including internet, image and video processing, digital signal processing, instrumentation, power, automation, automobiles, avionics, robotics, health and environment, agriculture, defense, games, etc. There is hardly anyone who does not know what a cell phone is. From MP3 players, camera cell phones and GSM to Bluetooth and Ipods, everyone wants all the features squeezed into a single device as small as possible. Some of the ever-growing applications are as follows:

- Digital cameras
- Digital camcorders
- Digital camera interface
- Digital cinema
- Digital display

- Digital TV and digital cable TV
- Digitizer for analog NTSC/PAL/SECAM cameras
- Display interface

- Mobile phone
- FAX machine
- PDA

- Scanner

- Anti-lock brakes

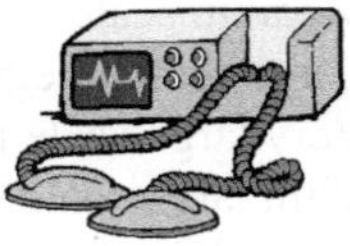

- Automatic transmission
- Cruise control
- Global positioning system for automobiles

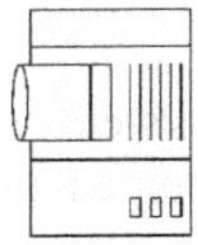

- Electro cardiograph
- Life-support systems

- MRI/CT scan
- LCD projector

- Low-cost computer
- Mobile phone personal computers

- Scan pen and PC notes taker
- Automated baggage clearance system in airports
- Avionic systems

- Flight simulator
- Instrument landing system

- Ship controls
- Driverless shuttle

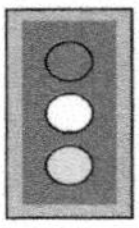

- Cruise controls
- Traffic controller
- Washing machine

- Petrol/diesel dispenser
- Demodulator for satellite communication
- Encryption/decryption
- Network card
- Network switches/routers
- Quadrature amplitude modulator (QAM) and demodulator
- Wireless LAN/WAN

More applications are presented in the final chapter, which may be taken up for implementation by the readers. They are classified into various categories, such as automotives, avionics, control system applications, medical applications, and video processing applications, to name a few. Brief descriptions are also presented. Curious readers may have a peep into those applications before going over to the next section.

VLSI systems can be designed using any of the following: 8/16/32/64 bit general-purpose processors, microcontrollers, DSPs, FPGAs, or ASICs depending upon the applications, throughput, market potential, etc. The advantages and limitations for each of these categories are discussed in the following sections.

1.3 Processor Based Systems

Designers have wide choice of selecting processors, which include general- purpose processors, microcontrollers, application specific instruction processors (ASIP), reduced instruction set computers (RISC), complex instruction set computers (CISC), digital signal processors (DSP), etc. ASIPs are optimized for specific class of applications such as telecommunications, digital signal processing, embedded controls, etc. Each of these processors has an instruction set with a specific class of applications in mind. Nevertheless, they are all processors executing instructions sequentially, differing only in performance, processing speed, effectiveness, power, cost, etc. Most of these processors fall under the category of VLSI. Over the years, a wide variety of systems have been designed with these processors. These cover an impressive spectrum: data processing systems, data acquisition systems, programmable logic controllers and numerous industrial control systems, measuring instruments, image processing systems, etc.

Selection of a processor for an application is not an easy task, especially when numerous processors are available. The designer is overwhelmed by numerous instruction sets. With a change of processor for a new project, the designer is forced to learn a new instruction set, which is as arduous as learning a new language, if the designer is not already familiar with the processor. Often, there is great confusion and struggle if instruction sets of processors have conflicting meaning such as the position of source and destination in an instruction. For example, Intel microprocessor instructions start with the destination first, followed by the source, whereas Motorola microprocessor instructions start with the source first, followed by the destination. Even in the hardware realms, process timings of processors vary widely. The associated peripheral chips also differ, necessitating major redesign of software/hardware, if processors are changed. All these introduce considerable delays in the project. In order to avert disaster, designers try to bend the new project towards their favorite processor(s). For small performance requirements, this tactic may serve the intended purpose. However, in time critical applications, the favorite processor may be a misfit.

Earlier, we discussed about the difficulties a designer undergoes while designing systems using processors, especially when the need arises to unlearn instruction set of the familiar processor and, instead, learn a new processor assembly language instructions. This difficulty may be eliminated if the designer has knowledge of a high level language such as C or C++. Of course, the designer has to learn C if he or she has no knowledge of the same. It is a well known fact to designers that C codes generate longer codes than assembly language instructions do. This, in turn, would lower the processing speed considerably and may prove to be a bottleneck in real-time applications.

1.4 Embedded Systems

General-purpose processors as well as microcontrollers are popular in embedded systems due to several good features such as low cost, good performance, etc. If hardware is already available, the designer needs to concentrate only on software development and integration of the system. Therefore, for small quantity of end products, it will be cost-effective as well as reduce the development cycle time dramatically if the embedded systems are built using bought out populated electronic cards such as STD/VME bus cards and the like. The designer may write some part of the programs in C and other parts in assembly language and link them together. Many tools are available to aid designers in product development – hardware: troubleshooters, emulators, logic analyzers, and programming units and software: assemblers, compilers, linkers, and C. These tools are a must if the embedded system designer is to bring out a working product into the market fast. Designers must try to achieve the goal of designing complete systems by treating hardware and software in a unified way. Hardware/software co-design emphasizes this unified view that enables the co-development of systems using both hardware and software, especially during synthesis [1, 2].

Processor based embedded systems are quite effective for small and medium-end applications. For medium to high-end embedded systems design, field programmable gate arrays (FPGAs) and application specific integrated circuits (ASICs) are the right choice. In the near future, FPGAs may be expected to be cost-effective even for small-end applications. The development tools that will convert ideas into reality of a working system for this category are Verilog/VHDL (hardware design language) compilers, simulation, synthesis and place and route tools and programmers. In the next section, we will discuss how FPGAs offer much higher performance than processors including DSPs and those used in embedded systems.

1.5 FPGA Based Systems

A number of software and hardware implementations have been reported for various real-time applications such as video codecs. Although software implementations are easy to realize on general-purpose microprocessors, multiprocessors, microcontrollers, or digital signal processors, their instruction sets are not well suited for fast processing of computationally intensive real time processing applications such as high resolution compression/video scaling of motion pictures, satellite communication modulator/demodulator, etc. In addition, the instructions are executed sequentially, thus slowing down the processing further. For example, one of the promising digital signal processors, which was used to implement MPEG based video codec could only process still monochrome images of resolution 512 × 512 pixels at one frame per second instead of the required frame rate of 30. In contrast to this, the hardware implementations based on FPGAs and ASICs can exploit pipelining and massively parallel processing resulting in faster and cost-effective designs.

ASIC designs are suitable if high-volume production is envisaged. However, in the research and development phase and for rapid prototyping of a new design, FPGA is the right choice. Further, FPGA implementation is cost-effective for low volume applications. In a later chapter on project designs, we will show that high resolution motion pictures of sizes 1600 × 1200 pixels can be processed (actually compressed) at 30 frames per second using FPGAs. The Verilog code developed for this application can also be implemented without any modification of the codes on an ASIC for still higher resolution pictures by over three times in the present day technology. The following sub-sections discuss briefly video compression application using FPGAs as an example.

1.5.1 FPGA Based Design: Video Compression as an Example

FPGAs offer high performance in terms of processing speed and high chip density, thus suiting every conceivable application, whether small or high end, yet remaining cost-effective. An entire VLSI system can be housed in a single FPGA device. Although many applications are possible, we will discuss the basics of video compression implementation as an example. Video compression is an application that demands high performance and high density. For example, a color motion picture of high resolution, 1600 × 1200 pixels can be compressed and transmitted or received over a serial channel at a real time processing speed of 30 frames/second using an FPGA. Complete project design for this application, in addition to many other applications, is presented in later chapters. The design methodology presented for this application is equally applicable for any other application. In the following sub-sections, we will briefly discuss the need for video compression, what standards govern the implementations, various issues involved in the design, and a review of the evolution of video compression implementations.

Need for Video Compression

Image processing applications such as high definition television, video conferencing, computer communication, etc. require large storage and high speed channels for handling huge volumes of data. For instance, one hour of color motion picture of size 1024 × 768 pixels at 30 frames per second in the raw format will need about 255 GB of memory and 566 Mbps channel speed for effective communication. In order to reduce the storage and communication channel bandwidth requirements to manageable levels, data compression techniques are imperative. Data compression in the order of 20 to 40 is normally feasible depending upon the actual picture content and techniques adopted for bringing about the compression.

It is of paramount importance that systems designed for applications communicate with one another effectively and also offer connectivity and compatibility among different services. These requirements are met if these systems are designed to conform to international standards such as JPEG, H.261, HDTV, and MPEG [3]. The development of standards by the ISO, ITU, etc. for audio, image and video, for both transmission and storage, has led to worldwide activity in developing hardware and software systems for a number of diverse applications. Although the standards implicitly address the basic encoding operations, there is enough freedom and flexibility in choosing the algorithms, the actual implementation, and devices. As such, the standards do not stifle the research and development activity, the main objective being maintaining compatibility and interoperability among the systems. The next sub-section describes briefly various standards available currently for compression of still image and motion pictures.

Video Compression Standards

JPEG has been recognized as the most popular and efficient coding scheme for continuous-tone still images. The JPEG compatible fast implementations find applications in color facsimile, high-quality newspaper wire photos, desktop publishing, graphic arts, medical imaging, digital still cameras, imaging scanners, etc. Examples of video sequence (motion picture) standards are H.261 for video telephony and video conferencing, MPEG 1 for digital storage media and MPEG 2 for generic coding of moving video and extending to television broadcasting and communication.

In recent years, additional standards such as JPEG 2000, MPEG 4, and MPEG 7 have also been introduced. The JPEG 2000 standard is intended to complement and not to replace the JPEG standard. Lossless and lossy coding, progressive by resolution and quality, high compression efficiency, error resilience, and lossless color transforms are some of its characteristics. The JPEG 2000 standard is based on discrete wavelet transforms (DWT) and offer higher image quality than is possible with JPEG for the same compression effected. However, all the above mentioned advantages of JPEG 2000 are at the expense of memory size, data access complexity, and processing time when compared to JPEG. MPEG 4 is an international standard that provides core technologies for efficient object-based compression of multimedia content for transmission, storage, and manipulation. MPEG 4, Part 10 (also known as H.264) is based on integer transform, derived from discrete cosine transform used in JPEG, MPEG 1, MPEG 2, H.261 encoders/decoders (codecs). MPEG 7 addresses content description technologies for efficient and effective multimedia retrieval and browsing. Both are essential in developing digital multimedia broadcast and internet multimedia applications. However, these two standards, MPEG 4 and MPEG 7, are more complex and only slower implementations are possible when compared to MPEG 2 and H.264 standards.

The basic operations that bring about image compression, namely, the discrete cosine transform (DCT), quantization (Q) and variable length coding (VLC) are, however, common to all the standards from JPEG to MPEG 2 mentioned earlier. DCT prepares the ground for effective compression. Mapping the image/video signal into the transform domain by itself does not lead to bit-rate reduction. At best, the mapping results in energy compaction in the low frequency range of the transform domain. By cleverly quantizing the coefficients that carry significant information and at the same time coarsely quantizing or dropping the remaining coefficients, the bit rate can be reduced. The resulting quantized DCT coefficients are Huffman coded in the VLC coder, effecting further compression. The resulting compressed, serial bit stream output is then sent out to the channel for onward transmission. Upon receipt of this bit stream, the decoder reconstructs the image by carrying out the inverse operations such as variable length decoding (VLD, inverse quantization (IQ) and inverse discrete cosine transform (IDCT).

Still image processing employs only the DCT, Q, and VLC, exploiting the spatial redundancy. In contrast to this, the motion pictures employ motion estimation and compensation in addition to DCTQ and VLC, effecting more compression owing to the exploitation of temporal redundancy.

Issues Involved in Video Compression

There are two basic issues to be addressed in the design of codecs for video applications: speed of processing and power considerations. With the proliferation of personal computers in multimedia applications and the evolution of digital networks, speed of processing is the most vital need for effective real time communication such as videoconferencing, point to point audio-visual communication, digital cable TV, etc. In spite of galloping technological advances, the internet is pathetically slow to accommodate communication of real time moving pictures, which finds unlimited scope for applications. The same is also true with most of the software and hardware implementations of video compression schemes based on general-purpose or multiprocessor computers and digital signal processors.

Fortunately, implementations based on FPGAs and ASICs hold much promise for high speed processing.

Power consumption is one of the most important criteria in evaluating digital systems, whether portable or not. This stems from a variety of requirements such as prolonging battery life in portable devices, reducing chip packaging, reduction of cooling costs, enhanced reliability of the system, environmental considerations, etc. Increase of clock frequency, operating voltage, or system complexity increases the power consumption drastically. Large power savings are effected by minimizing these parameters as well as through appropriate architectural and algorithmic trade-offs and functional module-level optimizations. The emergence of portable computing and communication devices such as laptop/palmtop computers, cellular phones, videophones, etc. is one of the most important factors driving the need for low power design. For most portable devices, the power consumed in integrated circuits is a significant and increasing portion of the total system power consumption. Thus, the development of low power VLSI design methodologies and tools are inevitable.

In general, the video encoder implementation must have high speed performance, whereas the decoder must have low power characteristics. The emphasis of the present book is on high speed processing rather than on low power so that high resolution pictures may be processed, especially at the encoder end.

The color pictures are represented as a luminance, Y, and two color difference signals, Cb and Cr [3]. The color difference signals are usually sub-sampled with respect to the luminance by 2:1 in both vertical and horizontal directions. This is because the sensitivity of the color component of the human eye is less than that of the luminance component. This sub-sampling of chrominance information leads to further bit-rate reduction in video compression techniques. Other features such as quantization reflecting the human visual system can further contribute to the overall compression. All these features, of course, are accomplished at a price. The result is that the codec complexity increases and the encoding mechanism becomes much more vulnerable to channel noise, requiring sophisticated error detection and correction techniques. The decoder is much less complex than the encoder because most of the decision-making processes are carried out at the encoder. Also, motion estimation and quantization control need to be implemented at the encoder end only. This complex encoder, simple decoder scenario is also appropriate because the decoder can be designed as a mass consumer item.

Evolution of Video Compression Implementations

Various processes such as DCTQ, VLC, motion estimation, etc. will have to keep pace with each other since they are all concurrent and pipelined processes. These demand the development of new and faster algorithms. This book presents the VLSI system design of the core of a video encoder, which involves the development of new and faster algorithm that can be effectively implemented on FPGAs or ASICs. Very few implementations have been reported for a full-fledged video encoder, especially, that which is capable of high speed processing. Although high speed processing is the major thrust of the present book, the design methodology adopted is also in conformity with that for low power design. This scheme can be easily adapted for a low power design by trading-off power with picture size and system clock frequency.

In the next section, we will present the basic steps involved in the design of VLSI systems using FPGAs. We will also present details of some of the leading FPGAs of a couple of vendors.

1.6 Digital System Design Using FPGAs

An FPGA may be viewed as 'sea-of-gates' which can be quickly configured to the desired application right on-the-field. It may also be re-configured at any point of time to another application, provided the external hardware interface circuitry doesn't need any change to suit the new application. Some FPGAs such as Virtex series FPGAs of Xilinx permit even partial reconfiguration, thereby eminently suited for real-time applications needing reconfiguration on-the-fly. Like ASICs, FPGAs can exploit high pipelining and massively parallel circuits, thus outperforming processors, microcontrollers and DSPs, yet remaining competitive. For R&D environment and low volume applications, FPGAs are the right choice. They are far cheaper and development cycle times are lower than that of ASICs. Most companies bring out their systems based on FPGAs first and later on switch to ASICs if the market demands are high. FPGA chip densities range from a few thousand gates (costing under $5 each piece) to over 10 million gates, accommodating right from small designs to very large designs. FPGAs come with different flavors in terms of gate counts, speed grades, input/output pins, packages, operating voltages, etc. However, designers need to watch out for obsolescence. The reader may get specification and application details of FPGAs from the websites of vendors listed in the references.

Development tools are available for FPGA based product designs from the respective vendors: simulator for waveform analysis, synthesis for logic optimization, and mapping the design on an FPGA, place and route for creating bit stream of the design and Programmer for programming the bit stream in an EPROM. The FPGA based digital system design may be realized using the following steps:

1. Formulate the detailed product specification.
2. Develop the detailed hardware architecture.
3. Code the architecture in a hardware design language such as Verilog or VHDL.
4. Compile and simulate the design and verify the functionality.
5. Synthesize to map on to a target FPGA device and optimize the logic.
6. Run the place and route tool for creating bit stream of the design application.
7. Program the bit stream generated in step 6 in a serial EPROM.

8. Design and fabricate the printed circuit board to accommodate the FPGA, the serial EPROM, and other components required for the end application.
9. Solder the components and test the populated FPGA board using the development system, logic analyzer, pattern generator, etc.
10. Download the application bit stream from the development system or the on-board serial EPROM and verify the system functionality.

All these design steps are unfolded gradually, chapter by chapter commencing from Chapter 3. In the next few sub-sections, details of various types of FPGAs from some of the leading vendors are presented, partly in the text and partly in the CD, so that the designer may select the right type of FPGA for his application.

1.6.1 Spartan-3 FPGAs

The Spartan-3 family of field programmable gate arrays of Xilinx is specifically designed to meet the needs of high-volume, cost-sensitive consumer electronic applications. This series offers FPGA densities ranging from 50,000 gates to 5 million gates, as shown in Table 1.1. Spartan-3 FPGAs deliver more functionality and bandwidth than was previously possible. Because of their low cost, Spartan-3 FPGAs are ideally suited to a wide range of consumer electronics applications: broadband access, home networking, display/projection, and digital television equipments, to mention a few. The following are the salient features of Spartan-3 family of FPGAs:

Table 1.1 Summary of Spartan-3 FPGA Attributes (Courtesy of Xilinx Inc.)

Device	System Gates	Equivalent Logic Cells[1]	CLB Array (One CLB = Four Slices)			Distributed RAM Bits (K=1024)	Block RAM Bits (K=1024)	Dedicated Multipliers	DCMs	Maximum User I/O	Maximum Differential I/O Pairs
			Rows	Columns	Total CLBs						
XC3S50[2]	50K	1,728	16	12	192	12K	72K	4	2	124	56
XC3S200[2]	200K	4,320	24	20	480	30K	216K	12	4	173	76
XC3S400[2]	400K	8,064	32	28	896	56K	288K	16	4	264	116
XC3S1000[2,3]	1M	17,280	48	40	1,920	120K	432K	24	4	391	175
XC3S1500[3]	1.5M	29,952	64	52	3,328	208K	576K	32	4	487	221
XC3S2000	2M	46,080	80	64	5,120	320K	720K	40	4	565	270
XC3S4000[3]	4M	62,208	96	72	6,912	432K	1,728K	96	4	712	312
XC3S5000	5M	74,880	104	80	8,320	520K	1,872K	104	4	784	344

Notes:

1. Logic Cell = 4-input Look-Up Table (LUT) plus a "D" flip-flop. "Equivalent Logic Cells" equals "Total CLBs" × 8 Logic Cells/CLB × 1.125 effectiveness.
2. These devices are available in Xilinx automotive versions as described in DS314: Spartan-3 Automotive XA FPGA Family.

3. These devices are available in lower static power versions as described in DS313: Spartan-3L Low Power FPGA Family.

Features

- Low-cost, high-performance logic solution for high-volume, consumer-oriented applications
  - Densities up to 74,880 logic cells (5 million gates)
- Select IO signaling
  - Up to 784 I/O pins
  - 622 Mb/s data transfer rate per I/O
  - 18 single-ended signal standards
  - 8 differential I/O standards
  - Signal swing ranging from 1.14 V to 3.45 V
  - Double Data Rate (DDR) support
  - DDR, SDRAM support up to 333 Mbps
- Logic resources
  - Abundant logic cells with shift register capability
  - Wide, fast multiplexers
  - Fast look-ahead carry logic
  - Dedicated 18 × 18 multipliers
  - JTAG logic compatible with IEEE 1149.1/1532
- Select RAM hierarchical memory
  - Up to 1,872 Kbits of total block RAM
  - Up to 520 Kbits of total distributed RAM
- Digital Clock Manager (up to four DCMs)
  - Clock skew elimination
  - Frequency synthesis
  - High resolution phase shifting
- Eight global clock lines and abundant routing
- Fully supported by Xilinx ISE development system
  - Synthesis, mapping, placement, and routing
- Low-power Spartan-3L Family and Automotive Spartan-3 XA Family variants

Table 1.2 shows the number of RAM blocks, the data storage capacity, and the number of columns for each device.

Table 1.2 Number of Block RAMs by Device (Courtesy of Xilinx Inc.)

Device	Total Number of RAM Blocks	Total Addressable Locations (bits)	Number of Columns
XC3S50	4	73,728	1
XC3S200	12	221,184	2
XC3S400	16	294,912	2
XC3S1000	24	442,368	2
XC3S1500	32	589,824	2
XC3S2000	40	737,280	2
XC3S4000	96	1,769,472	4
XC3S5000	104	1,916,928	4

Spartan-3 FPGA Architectural Overview

The Spartan-3 family architecture consists of fundamental programmable functional elements, CLBs, IOBs, block RAMs. They are as follows:

1. CLBs contain RAM-based LUTs to implement logic and storage elements that can be used as flip-flops or latches. CLBs can be programmed to perform a wide variety of logical functions as well as to store data.
2. IOBs control the flow of data between the I/O pins and the internal logic of the device. Each IOB supports bidirectional data flow and tri-state operation.
3. Block RAM provides data storage in the form of 18-Kbit dual-port blocks. Multiplier blocks accept two 18-bit binary numbers as inputs and calculate the product. Digital Clock Manager (DCM) blocks provide self-calibrating, fully digital solutions for distributing, delaying, multiplying, dividing, and phase shifting clock signals.

These functional elements are organized as shown in Figure 1.1. A ring of IOBs surrounds a regular array of CLBs. The XC3S50 has a single column of block RAM embedded in the array. Those devices ranging from the XC3S200 to the XC3S2000 have two columns and XC3S4000/5000 devices have four RAM columns. Each column is made up of several 18-Kbit RAM blocks. The DCMs are positioned at the ends of the outer block RAM columns. The Spartan-3 family features a rich network of traces and switches that interconnect all the functional elements, transmitting signals among them. Each functional element has an associated switch matrix that permits multiple connections to the routing.

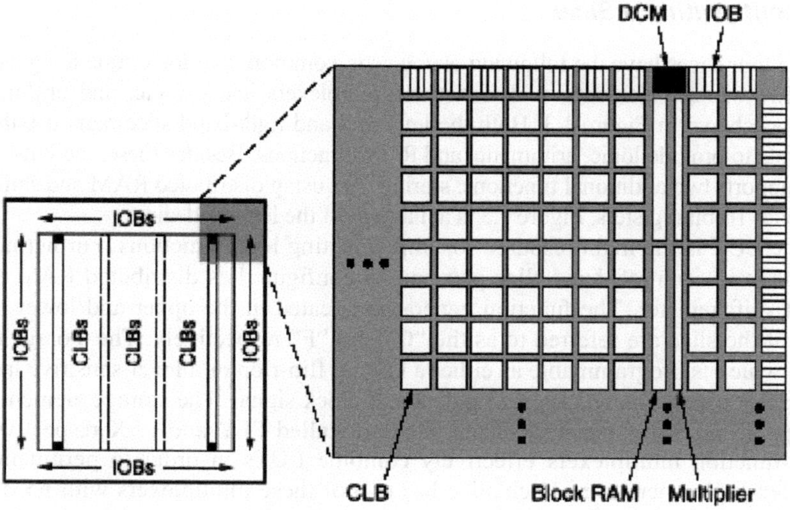

Note:
The two additional block RAM columns of the XC3S4000 and XC3S5000 devices are shown with dashed lines. The XC3S50 has only the block RAM column on the far left.

Fig. 1.1 Spartan-3 family architecture (Courtesy of Xilinx Inc.)

Configuration

Spartan-3 FPGAs are programmed by loading configuration data into static memory cells that collectively control all functional elements and routing resources. Before powering on the FPGA, configuration data is stored externally in a nonvolatile memory such as EPROM and Flash PROM. After applying power, the configuration data is written to the FPGA using any of five different modes: Master Parallel, Slave Parallel, Master Serial, Slave Serial, and Boundary Scan (JTAG).

Overview of Configurable Logic Blocks

The CLB, the main logic resource for implementing digital circuits, comprises four interconnected slices, as shown in Figure 1.2. These slices are grouped in pairs with each pair organized as a column with an independent carry chain. Slices X0Y0 and X0Y1 make up the column-pair on the left, whereas slices X1Y0 and X1Y1 make up the column-pair on the right. For each CLB, the term “left-hand” (or SLICEM) indicates the pair of slices labeled with an even “X” number, such as X0, and the term “right-hand” (or SLICEL) designates the pair of slices with an odd “X” number, e.g., X1.

Elements within a Slice

All the four slices have the following elements in common: two logic function generators, two storage elements, wide-function multiplexers, carry logic, and arithmetic gates, as shown in Figure 1.3. Both the left-hand and right-hand slice pairs use these elements to provide logic, arithmetic, and ROM functions. Besides these, the left-hand pair supports two additional functions: storing data using distributed RAM and shifting data with 16-bit registers. Figure 1.3 is a diagram of the left-hand slice.

The LUT is the main resource for implementing logic functions. Furthermore, the LUTs in each left-hand slice pair can be configured as distributed RAM or a 16-bit shift register. The function generators located in the upper and lower portions of the slice are referred to as the "G" and "F" respectively. The storage element, which is programmable as either a D-type flip-flop or a level-sensitive latch, provides a means for synchronizing data to a clock signal. The storage elements in the upper and lower portions of the slice are called FFY and FFX respectively. Wide-function multiplexers effectively combine LUTs in order to permit more complex logic operations. Each slice has two of these multiplexers with F5MUX in the lower portion of the slice and FiMUX in the upper portion. Depending on the slice, FiMUX takes on the name F6MUX, F7MUX, or F8MUX.

Function Generator

Each of the two LUTs (F and G) in a slice has four logic inputs (A1–A4) and a single output (D). This permits any four-variable Boolean logic operation to be

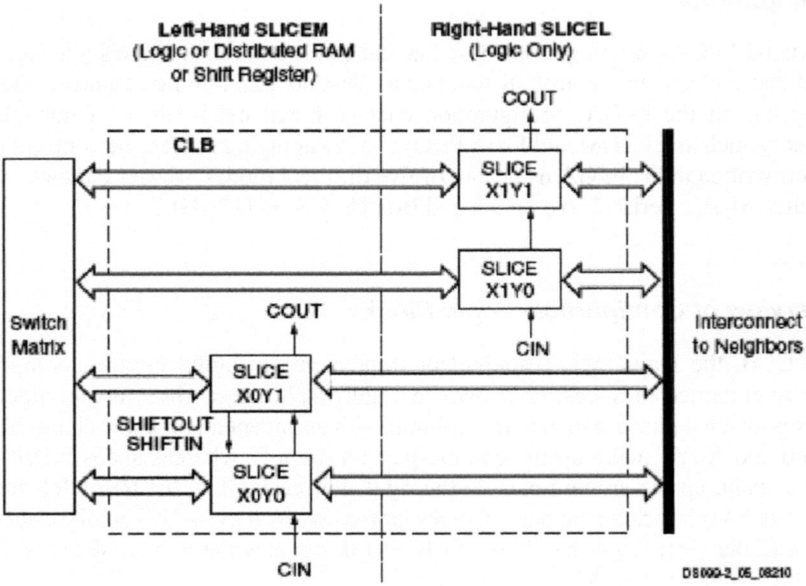

Fig. 1.2 Arrangement of slices within the CLB (Courtesy of Xilinx Inc.)

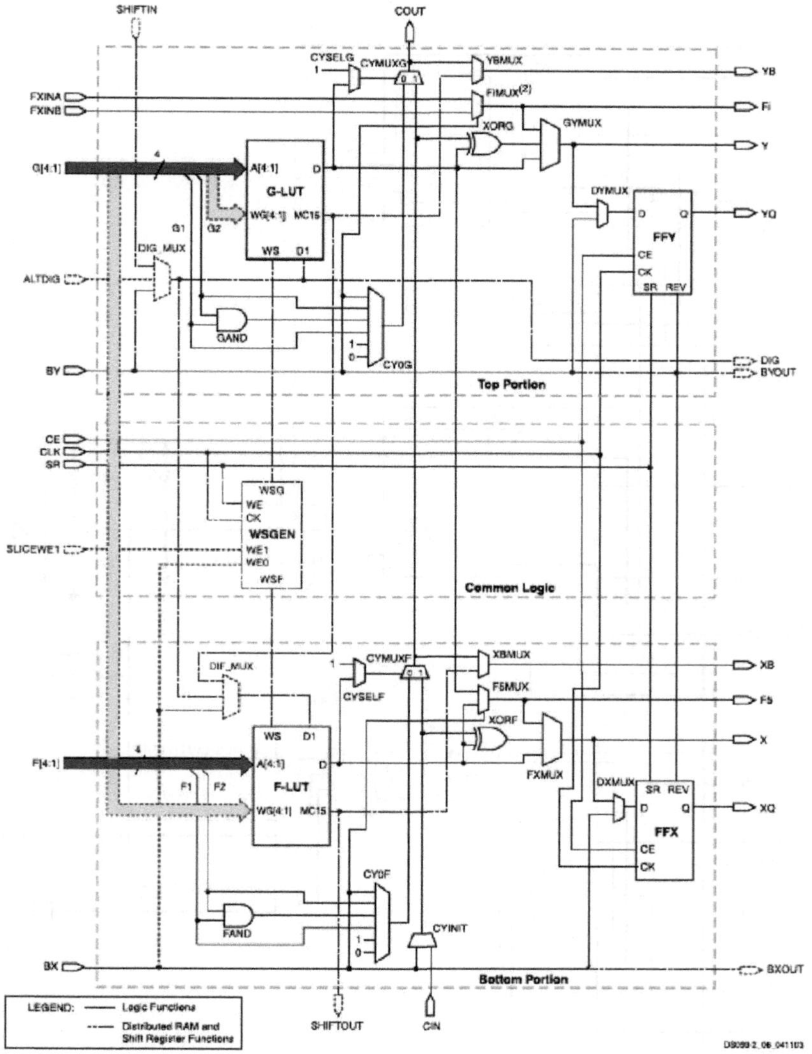

Notes:

1. Options to invert signal polarity as well as other options that enable lines for various functions are not shown.
2. The index *i* can be 6, 7, or 8 depending on the slice. In this position, the upper right-hand slice has an F8MUX, and the upper left-hand slice has an F7MUX. The lower right-hand and left-hand slice has an F6MUX.

Fig. 1.3 Simplified diagram of the left-hand side SLICEM (Courtesy of Xilinx Inc.)

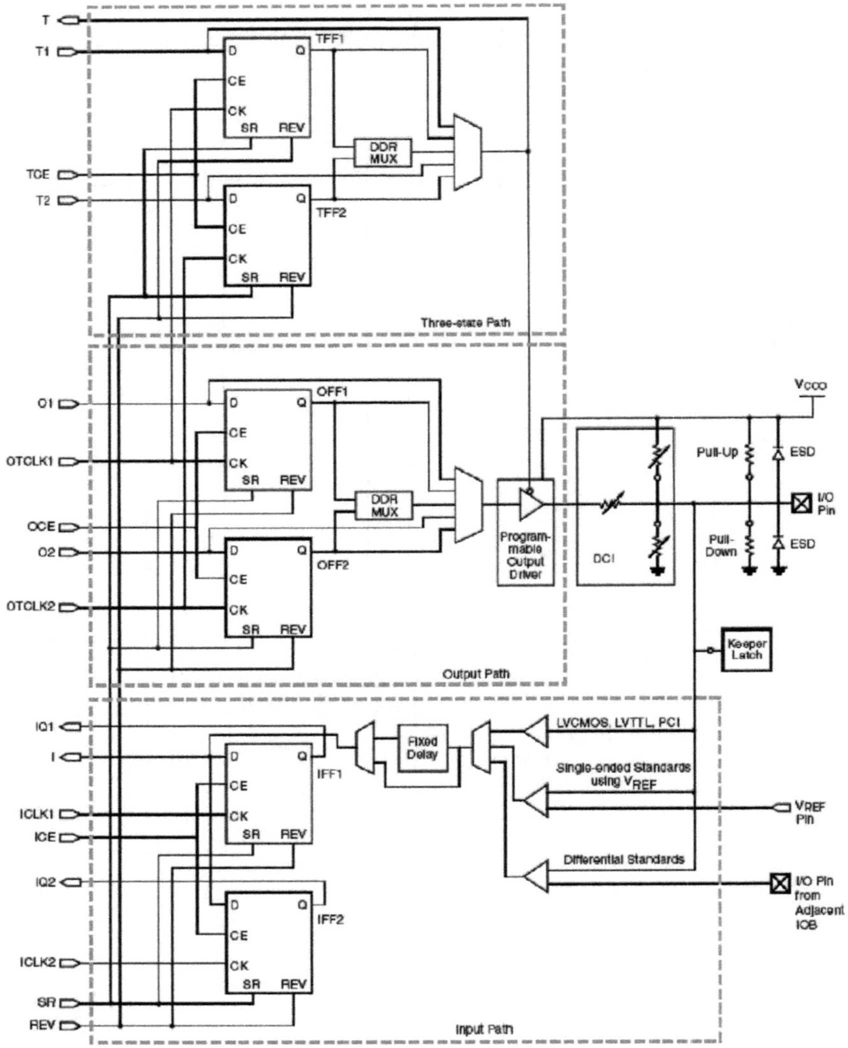

Note:
All IOB signals communicating with the FPGA's internal logic have the option of inverting polarity.

Fig. 1.4 Simplified IOB diagram (Courtesy of Xilinx Inc.)

programmed into them. In addition, wide-function multiplexers can be used to effectively combine LUTs making logic functions with still more input variables. The LUTs also can function as ROM that is initialized with data at the time of configuration. The LUTs can be programmed as distributed RAM, which offers moderate amounts of data buffering anywhere along a data path. One LUT stores 16 bits. A dual-port option combines two LUTs so that memory access is possible from two independent data lines. A distributed ROM option permits pre-loading the memory with data during FPGA configuration. It is possible to program each LUT as a 16-bit shift register, which can be used to delay serial data anywhere from 1 to 16 clock cycles. The four LUTs of a single CLB can be combined to produce delays up to 64 clock cycles. The SHIFTIN and SHIFTOUT lines cascade LUTs to form larger shift registers. It is also possible to combine shift registers across more than one CLB. The resulting programmable delays can be used to balance the timing of data pipelines.

Block RAM Overview

Spartan-3 FPGA devices support block RAM, which is organized as configurable, synchronous 18-Kbit blocks. Block RAM stores relatively large amounts of data more efficiently than the distributed RAM.

IOB Overview

The input/output block provides a programmable, bidirectional interface between an I/O pin and the FPGA's internal logic. A simplified diagram of the IOB's internal structure is shown in Figure 1.4. There are three main signal paths within the IOB: the output path, input path, and tri-state path. Each path has its own pair of storage elements that can act as either registers or latches.

Storage Element Functions

As shown in Figure 1.4, there are three pairs of storage elements in each IOB, one pair for each of the three paths. It is possible to configure each of these storage elements as an edge-triggered D-type flip-flop (FD) or a level-sensitive latch (LD). The storage-element-pair on either the output path or the three-state path can be used together with a special multiplexer to produce Double-Data-Rate (DDR) transmission. This is accomplished by taking data synchronized to the clock signal's rising edge and converting them to bits synchronized on both the rising and the falling edge. The combination of two registers and a multiplexer is referred to as a Double-Data-Rate D-type flip-flop (FDDR).

The clock line OTCLK1 connects the CK inputs of the upper registers on the output and three-state paths. Similarly, OTCLK2 connects the CK inputs for the lower registers on the output and three-state paths. The upper and lower registers on the input path have independent clock lines: ICLK1 and ICLK2. The enable line OCE connects the CE inputs of the upper and lower registers on the output path. Similarly, TCE connects the CE inputs for the register pair on the three-state path and ICE does the same for the register pair on the input path. The Set/Reset (SR) line entering the IOB is common to all six registers, as is the Reverse (REV) line.

Within the Spartan-3 family, all devices are pin-compatible by package and are not pin-compatible with any previous Xilinx FPGA family. When the need for future logic resources outgrows the capacity of the Spartan-3 device in current use, a larger device in the same package can serve as a direct replacement. It is, therefore, important to plan for future upgrades at the time of the board's initial design.

Table 1.3 Xilinx FPGA Product Selector (Continued) (Courtesy of Xilinx Inc.)

Family	Device	Package	I/O	LCs	BRAM	DCM	Mult	eMAC	MGT	PowerPC	EasyPath Option
Virtex-4	LX200	FF1513	960	200,448	6,048	12	96	0	0	0	Yes
Virtex-4	LX160	FF1513	960	152,064	5,184	12	96	0	0	0	Yes
Virtex-4	LX160	FF1148	768	152,064	5,184	12	96	0	0	0	Yes
Virtex-4	FX140	FF1760	896	142,128	9,936	20	192	4	24	2	Yes
Virtex-4	FX140	FF1517	768	142,128	9,936	20	192	4	24	2	Yes
Virtex-4	LX100	FF1513	960	110,592	4,320	12	96	0	0	0	Yes
Virtex-4	LX100	FF1148	768	110,592	4,320	12	96	0	0	0	Yes
Virtex-4	FX100	FF1517	768	94,896	6,768	12	160	4	20	2	Yes
Virtex-4	FX100	FF1152	576	94,896	6,768	12	160	4	20	2	Yes
Virtex-4	LX80	FF1148	768	80,640	3,600	12	80	0	0	0	Yes
Spartan-3	5000	FG900	633	74,880	1,872	4	104	0	0	0	Yes
Spartan-3	4000	FG900	633	62,208	1,728	4	96	0	0	0	Yes
Spartan-3	4000L	FG900	633	62,208	1,728	4	96	0	0	0	
Virtex-4	LX60	FF1148	640	59,904	2,880	8	64	0	0	0	Yes
Virtex-4	LX60	FF668	448	59,904	2,880	8	64	0	0	0	Yes
Virtex-4	FX60	FF1152	576	58,880	4,176	12	128	2	16	2	Yes
Virtex-4	FX60	FF672	352	58,880	4,176	12	128	2	16	2	Yes
Virtex-4	SX55	FF1148	640	55,296	5,760	8	512	0	0	0	Yes
Spartan-3	2000	FG900	565	46,080	720	4	40	0	0	0	Yes
Spartan-3	2000	FG676	487	46,080	720	4	40	0	0	0	Yes
Virtex-4	LX40	FF1148	640	41,472	1,728	8	64	0	0	0	Yes
Virtex-4	LX40	FF668	448	41,472	1,728	8	64	0	0	0	Yes
Virtex-4	FX40	FF1152	448	41,904	2,592	8	48	2	12	2	Yes
Virtex-4	FX40	FF672	352	41,904	2,592	8	48	2	12	2	Yes
Virtex-4	SX35	FF668	448	34,560	3,456	8	192	0	0	0	Yes
Spartan-3E	1600E	FG320	250	33,192	648	8	36	0	0	0	
Spartan-3E	1600E	FG400	304	33,192	648	8	36	0	0	0	
Spartan-3E	1600E	FG484	376	33,192	648	8	36	0	0	0	
Spartan-3	1500	FG676	487	29,952	576	4	32	0	0	0	Yes
Spartan-3	1500	FG456	333	29,952	576	4	32	0	0	0	Yes
Spartan-3	1500	FG320	221	29,952	576	4	32	0	0	0	Yes
Spartan-3	1500L	FG320	221	29,952	576	4	32	0	0	0	
Spartan-3	1500L	FG456	333	29,952	576	4	32	0	0	0	
Spartan-3	1500L	FG676	487	29,952	576	4	32	0	0	0	
Virtex-4	LX25	FF668	448	24,192	1,296	8	48	0	0	0	Yes
Virtex-4	LX25	SF363	240	24,192	1,296	8	48	0	0	0	Yes
Virtex-4	SX25	FF668	320	23,040	2,304	4	128	0	0	0	Yes
Spartan-3E	1200E	FT256	190	19,512	504	8	28	0	0	0	
Spartan-3E	1200E	FG320	250	19,512	504	8	28	0	0	0	
Spartan-3E	1200E	FG400	304	19,512	504	8	28	0	0	0	

Table 1.3 Xilinx FPGA Product Selector (Courtesy of Xilinx Inc.)

Virtex-4	FX20	FF672	320	19,224	1,224	4	32	2	8	1	Yes
Spartan-3	1000	FG676	391	17,280	432	4	24	0	0	0	
Spartan-3	1000	FG456	333	17,280	432	4	24	0	0	0	
Spartan-3	1000	FG320	221	17,280	432	4	24	0	0	0	
Spartan-3	1000	FT256	173	17,280	432	4	24	0	0	0	
Spartan-3	1000L	FT256	173	17,280	432	4	24	0	0	0	
Spartan-3	1000L	FG320	221	17,280	432	4	24	0	0	0	
Spartan-3	1000L	FG456	333	17,280	432	4	24	0	0	0	
Virtex-4	LX15	FF668	320	13,824	864	4	32	0	0	0	
Virtex-4	LX15	SF363	240	13,824	864	4	32	0	0	0	
Virtex-4	FX12	FF668	320	12,312	648	4	32	2	0	1	
Virtex-4	FX12	SF363	240	12,312	648	4	32	2	0	1	
Spartan-3E	500E	PQ208	158	10,476	360	4	20	0	0	0	
Spartan-3E	500E	FT256	190	10,476	360	4	20	0	0	0	
Spartan-3E	500E	FG320	232	10,476	360	4	20	0	0	0	
Spartan-3	400	FG456	264	8,064	288	4	16	0	0	0	
Spartan-3	400	FG320	221	8,064	288	4	16	0	0	0	
Spartan-3	400	FT256	173	8,064	288	4	16	0	0	0	
Spartan-3	400	PQ208	141	8,064	288	4	16	0	0	0	
Spartan-3	400	TQ144	97	8,064	288	4	16	0	0	0	
Spartan-3E	250E	VQ100	66	5,508	216	4	12	0	0	0	
Spartan-3E	250E	TQ144	108	5,508	216	4	12	0	0	0	
Spartan-3E	250E	PQ208	158	5,508	216	4	12	0	0	0	
Spartan-3E	250E	FT256	172	5,508	216	4	12	0	0	0	
Spartan-3	200	FT256	173	4,320	216	4	12	0	0	0	
Spartan-3	200	PQ208	141	4,320	216	4	12	0	0	0	
Spartan-3	200	TQ144	97	4,320	216	4	12	0	0	0	
Spartan-3	200	VQ100	63	4,320	216	4	12	0	0	0	
Spartan-3E	100E	VQ100	66	2,160	72	2	4	0	0	0	
Spartan-3E	100E	TQ144	108	2,160	72	2	4	0	0	0	
Spartan-3	50	PQ208	124	1,728	72	2	4	0	0	0	
Spartan-3	50	TQ144	97	1,728	72	2	4	0	0	0	
Spartan-3	50	VQ100	63	1,728	72	2	4	0	0	0	

Table 1.3 provides the FPGA product selection for the Virtex-4 and Spartan-3 series arranged according to logic cells. In Virtex-4 series, the device number gives the number of logic cells in thousands available in the device and in Spartan-3, the device number gives the number of gate count in millions. For examples, Virtex-4 FX140 device has 142,128 numbers of logic cells and Spartan-3 5000 has 5 M gates available in the device. The package number directly gives the number of pins available in the device. The table also provides other resources such as I/Os, the number of block RAMs, the number of clocks, the number of multipliers and so on. For details of XC4000 series and Virtex II Pro series FPGAs, the reader may refer to Appendix 1 and 2 respectively of the CD.

The Stratix II FPGA family of Altera offers high density and high speed devices. Reader may refer Appendix 3 for a brief write-up of this family of FPGAs. For one of Actel's devices, please refer Appendix 4 of the CD.

1.7 Reconfigurable Systems Using FPGAs

FPGAs may be reconfigured many times during the normal operation of an application, should the need arise. Reconfigurable systems may be classified as having either static or dynamic reconfigurability. A static reconfiguration refers to having the ability to reconfigure a system only once before execution, but once programmed, its configuration remains on the FPGA for the duration of the application. In contrast to this, the dynamic reconfiguration is defined as the selective updating of a sub-section of an FPGA's programmable logic and routing resources, while the remainder of the device's programmable resources continues to function without interruption. There are two basic approaches to implement dynamically reconfigurable applications: full reconfiguration and partial reconfiguration. Systems designed for full reconfiguration are allocated all FPGA resources in each configuration step, application being partitioned into distinct temporal modules of approximately equal size. In other words, for one application, entire execution code will have to be downloaded at one go. This is generally referred to as a coarse grain configuration. In contrast to this, in the partial reconfiguration, only a small amount of code known as fine grain configuration needs to be downloaded.

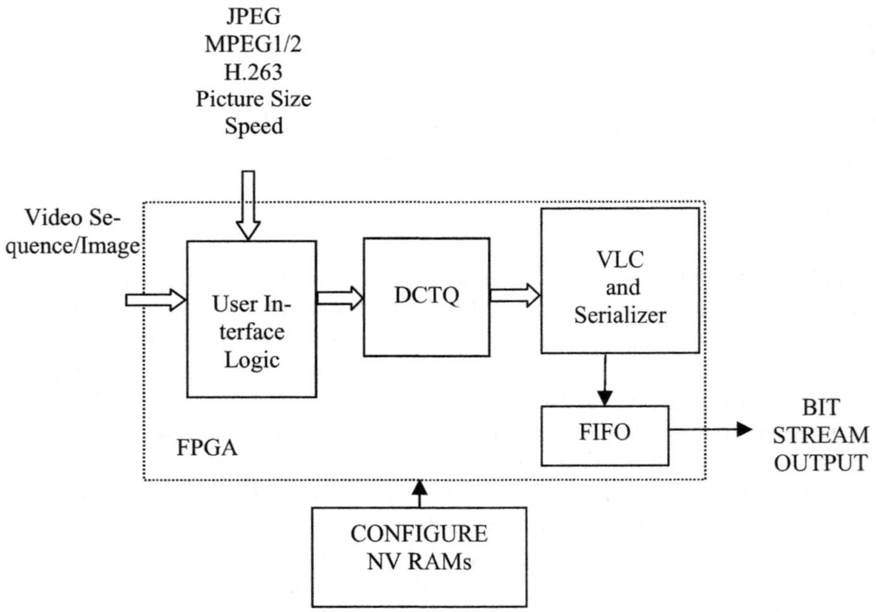

Fig. 1.5 The basic architecture of dynamically reconfigurable video encoder

As an example, a dynamically reconfigurable video encoder may be designed with the following features.

- It can be configured for JPEG, MPEG-1, MPEG-2, or H.263 as per user request.
- The user can select the picture size and processing speed, namely, frame rate.
- It can switch from one application to another dynamically and seamlessly based on user request without missing a single frame.

A scheme of full or partial reconfiguration for different applications like JPEG, MPEG 1 / MPEG 2, and H.263 for an encoder is shown in Figure 1.5. At the encoder, DCTQ and VLC are used, whereas at the decoder end (not shown in figure), VLD and IQIDCT modules are used as described in Section 1.5.1. The reconfiguration can be done by downloading the code stream from nonvolatile RAM at 50 MHz or less using a parallel port at 1 byte per clock cycle. A typical configuration stream size of an FPGA of capacity 300,000 gates is 220,000 bytes. As an example of reconfiguration, let the user who is currently in video conference mode in H.263 format request that a particular MPEG 2 video stream be sent over the serial channel. This can be accomplished dynamically by loading partial MPEG 2 configuration data without disturbing other working circuit and without missing a single frame of video. Once the MPEG 2 stream is complete, one reverts back to video conferencing automatically.

Before we windup this chapter, we will see how the rest of this book is organized.

1.8 Scope of the Book

The main objectives of this book are to present novel algorithms for various projects such as video compression system, etc. and develop new architectures as examples. We will code some of the projects in Verilog and finally simulate, synthesize, place and route using industry-standard tools and implement them on FPGAs to demonstrate the working of the schemes. The algorithms and functional modules are efficiently partitioned such that they are suitable for high speed implementation on FPGAs as well as on ASICs.

1.8.1 Approach

The reader is taken step by step through the complete design of digital VLSI systems using Verilog. The design methodology, however, can be applied to any other hardware design languages such as VHDL. The design starts right from implementing a single digital gate to a massive design consuming well over 100,000 gates. Following a review of basic concepts of digital systems design in the next chapter, a number of design examples are illustrated using conventional digital components such as flip-flops, multiplexers, de-multiplexers, decoders, ROM, programmable array logic, etc. These designs are still being used in small-end applications despite being overshadowed by FPGA based designs. With these foundations, the reader is introduced to Verilog coding of the components mentioned earlier as well as designing systems for small-end applications to start with. These designs are tested using Test Benches, also written in Verilog. All HDL codes the reader wishes to develop for various applications must conform to Register Transfer Level (RTL) Coding Guidelines, without which no chip can work satisfactorily. Separate chapters are dedicated to these aspects.

The sequel to the design is to simulate, synthesize and place and route. Since our sincere interest lies in making the reader a full-fledged engineer, we use the same popular development tools used in the Industries and Research Laboratories, such as Modelsim (a simulation tool of Mentor Graphics), Synplify (a synthesis tool of Synplicity Inc.) and Place and Route and Back Annotation of Xilinx Inc. Equipped with these powerful tools, the elucidation progresses into more and more complex designs such as Memories and Arithmetic Circuits extending into VLSI realms. The concepts, methodologies, etc. delineated in this book may be applied to other tools as well.

Complex Project Designs usually need the development of new Algorithms and Architectures for optimum realization. These issues, which stimulate creative thinking –indispensable for researchers – are thoroughly analyzed and solved in this book. Armed with all the features mentioned earlier, complete Project Designs are presented. This is followed by hands on experience in designing systems using FPGA and input/output boards. Once the path is shown for designing VLSI Systems systematically, numerous Project Designs are suggested for FPGA/ASIC Implementations.

This book gives complete RTL Compliant Verilog codes for several projects which are synthesizable and work on the hardware based on FPGA or ASIC. Most of the Verilog codes developed in this book can be readily used in new projects the reader may undertake in his study or career. The advantages accruing out of this strategy are two-fold: One, the students are trained to suit industries/R&D Laboratories and therefore, industries and other employers need not spend time and money to train them when they are hired. The other advantage is to provide in-plant training in industries, etc. based on this book to retrain old (as well as new) personnel in the new technologies.

Summary

A brief introduction to the evolution of VLSI systems and applications were presented. The merits and limitations of processor based systems and controller based embedded systems were discussed. This was followed by a presentation of FPGA based designs and their advantages over processor and embedded controller based designs. The chapter also discussed in detail various issues involved in one of the latest applications, namely, video compression. An overview of various implementation schemes for video compression was also discussed.

An introduction to digital system design using FPGAs was presented. Following this, detailed descriptions of features and architectures of various types of popular FPGAs manufactured by leading vendors were presented so that the readers may select the right type of FPGAs for their applications. Before commencing the design using Verilog in Chapter 3, the reader is urged to brush-up his digital knowledge, a review of which is presented in the next chapter.

Assignments

1.1 A digital camera can store JPEG images of resolution 1600 × 1200 pixels. Assuming 24 bit true color and an average compression of 10, estimate the memory requirements for storing a maximum of 200 images in the camera.

1.2 An MPEG 2 video encoder is required to be designed for compressing color video sequences in XGA format (resolution: 1024 × 768 pixels). The inputs are applied as luminance (Y) and color (Cb and Cr) components, each of size 8 bits in 4:2:0 format. In this format, for every four blocks of Y, one block of Cb and one block of Cr components are processed, 1 pixel component after another. A block is of size 8 × 8 pixels. One pixel (abbreviation for a picture element) consists of 24 bits in true color. In another format, 4:4:4, components Y, Cb, and Cr are four blocks each. Assuming the processing rate of video as 30 frames per second with a compression of 20, estimate the buffer memory (FIFO) required for storing 1 s of the compressed bit stream in 4:2:0 format, before it is transmitted over a serial channel. The FIFO is of single bit width.

1.3 Estimate the FIFO size required in assignment 1.2 for storing 1 s of the compressed bit stream in 4:4:4 format if compression effected is 10.

1.4 Estimate the FIFO size in assignment 1.2 if the video is changed to SVGA format, whose picture size is 800 × 600 pixels.

1.5 Assuming that each luminance/color component of a pixel is processed every clock cycle, compute the minimum clock frequency required in an FPGA implementation to satisfy 30 frames per second in the assignment 1.2. State your assumptions clearly.

1.6 Repeat assignment 1.5 for the SVGA format.

1.7 A digital cable TV transmitter is required to process up to 50 channels at a picture resolution of 720 × 525 pixels. Each of these channels is time-multiplexed. The picture is non-interlaced and the frame rate is 30 per second. Assuming 4:2:0 format and an average compression of 20, determine the frequency of operation of the video encoder, which transmits compressed video sequence over the cable. Also estimate the transmission rate over a serial channel. Make reasonable assumptions.

1.8 If you were to design the video encoder in assignment 1.7, which of the hardware, processors to ASIC, will you select for implementation? Justify your answer. In case compromises can be made in performance, what other hardware will you choose? Explain.

1.9 A company designed a data acquisition system based on a bought-out FPGA board few years back and marketed the same successfully. All of a sudden, the FPGA board vendor discontinued the product and, instead, offered to supply another type of FPGA board. Discuss how you will assess the suitability of the new board for the application. The company used 64 analog input channels in the data acquisition system spread over a number of boards, which was self fabricated.

1.10 In the assignment 1.9, the FPGA alone changed to a higher capacity with the FPGA board and other resources on the board remaining the same. What will be the repercussions?

1.11 A competitor to the company in the assignment 1.9 made the same product except that they used FPGA from another vendor. All boards including the FPGA board were self-fabricated. In this case, the FPGA became obsolete and another FPGA with a higher chip density and processing speed was available. How will this competitor fare in comparison to the first manufacturer?

1.12 What will be the scenario for both the competitors in assignments 1.9 and 1.11 if they had used an ASIC instead of FPGA?

1.13 An MPEG 1 / MPEG 2 video encoder architecture, which compresses 10-fold an SVGA motion picture in 4:2:0 format is shown in Figure A1. A video sequence is applied to a dual redundant RAM 1 block by block. To start with, one block of Y component of a video frame is completely written into one of the two 64 × 8 bits RAM in 8 clock cycles. When the second block of

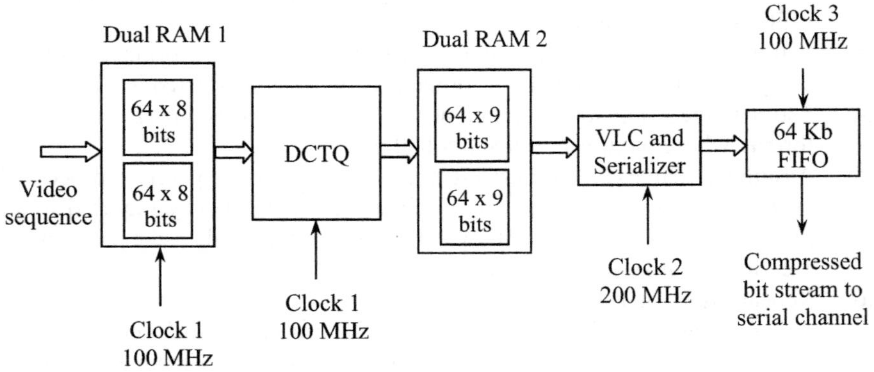

Fig. A1 MPEG video encoder

Y component is written into the other 64 × 8 bits RAM, DCTQ processes simultaneously producing 64 coefficients, storing the same in one of the two 64 × 9 bits RAM in "Dual RAM 2". DCTQ processes one coefficient every clock cycle. When all the 64 coefficients are stored in the first 64 × 9 bits RAM, the next module VLC and serializer takes this stored DCTQ coefficients and generates a compressed single bit stream and stores it in the 64 K bits First-in-First-Out-Memory (FIFO). The compressed bit stream is processed by the serializer, one bit every clock cycle in bursts of 25% duty cycle since VLC processing is usually time consuming. Simultaneously, a second block of DCTQ coefficients are stored in the second RAM in "Dual RAM 2". Thus, input memory, DCTQ, Dual RAM 2, and VLC/serializer can function simultaneously. This is referred to as pipelining. In essence, four blocks of Y, followed by one block of Cb and one block of Cr, are processed sequentially. This pattern is repeated until all the blocks of a frame are processed. Compute the frame rate, i.e., the number of frames per second with the clocks as shown in the figure. State your assumptions clearly.

1.14 Explain how you will provide hand shake signals to co-ordinate various functional modules of the video encoder in assignment 1.13. What is the maximum picture size that can be processed if the frame rate is fixed at 30 per second? Describe a scheme to regulate the bit stream transmitted over the serial channel at 100 mega bits per second.
Hint: Hold processing of individual functional modules or bit stuff 0's in the FIFO after every 16 lines of a frame.

1.15 A dynamically reconfigurable video encoder is required to be designed with the following features.

- It can be configured for JPEG, MPEG 1, MPEG 2, or H.263 as per user request.
- The user can select the picture size and processing speed such as frame rate.
- It can switch from one application to another dynamically and seamlessly based on user request without missing a single frame.

A scheme of full/partial reconfiguration for different applications like JPEG, MPEG 1 / MPEG 2, and H.263 was shown in Figure 1.16 in the text. The full/partial reconfiguration can be done by downloading from nonvolatile RAM at 50 MHz or less using a parallel port at 1 byte per clock cycle. A typical configuration stream size of an FPGA of capacity 300,000 gates is 220,000 bytes as described in Section 1.7. As an example of reconfiguration, let the user who is currently receiving SVGA format MPEG 2 video stream for surveillance application request that the present transmission be switched to higher resolution of 1600 × 1200 pixels in JPEG format in a seamless manner. This can be accomplished dynamically by loading partial JPEG configuration data, which is in the User Interface Logic and the VLC without disturbing other working circuits. The switching shall not loose any data. For JPEG/MPEG 1/MPEG

2 and H. 263, the User Interface Logic are of size 3000 and 2000 gates respectively, while for VLC they are 12,000 gates and 10,000 gates respectively. DCTQ occupies 120,000 gates, while 64 Kbit output FIFO occupies 400,000 gates. Compute the full configuration times for each of the standards as well as time for reconfiguration from SVGA format, MPEG 2 video to 1600 × 1200 pixels, JPEG format. State your assumptions clearly. What are the specific advantages of dynamic reconfiguration when compared to the conventional method of housing all the functions conforming to the above standards? Justify your answer. Name a few more applications where the dynamic reconfiguration can be applied.

Chapter 2

Review of Digital Systems Design

In the first chapter, we presented an introduction to digital VLSI systems design and its applications, which is the main emphasis of this book. We pointed out how FPGA based systems offer better performance than processor based systems, yet remaining competitive. We also presented basic architectures and features of some of the latest FPGAs from a couple of leading vendors so that the reader may select the right type of FPGA for his or her design. Before we go into depths of Verilog based design of digital VLSI systems, we need to brush up our knowledge regarding the digital systems design using the conventional components such as gates, flip-flops, PALs, etc.

In this chapter, we start with the numbering systems followed by twos complement arithmetic and various types of codes that are required in a real system design. We will also be covering in brief Boolean algebra and derivation of functions using minterms and maxterms and Karnaugh map for optimization of logic circuits. This will be followed by the design of combinational and sequential circuits. With these basics, digital system design will be presented using SSI/MSI components. Algorithmic state machine based approach to design is a better alternative to the conventional state graph approach [4, 5]. Designs based on the algorithmic state machine and PAL will also be presented.

2.1 Numbering Systems

Of all the numbering systems, the decimal system is the easiest to comprehend. A decimal number may be expressed as powers of 10. For example, consider a six digit decimal number 987,654, which can be represented as

$$9 \times 100{,}000 + 8 \times 10{,}000 + 7 \times 1{,}000 + 6 \times 100 + 5 \times 10 + 4 \times 1$$

or more concisely as

$$9 \times 10^5 + 8 \times 10^4 + 7 \times 10^3 + 6 \times 10^2 + 5 \times 10^1 + 4 \times 10^0$$

The numbers are 0, 1, 2 up to 9 since in a decimal system, the base is 10. This representation can be easily extended to fractional values as well. For example, the decimal number 99.99 can be represented as

$$9 \times 10^1 + 9 \times 10^0 + 9 \times 10^{-1} + 9 \times 10^{-2}$$

In general, a number may be represented in any numbering system as

$$d_{n-1} b^{n-1} + d_{n-2} b^{n-2} + \cdots + d_1 b^1 + d_0 b^0 + d_{-1} b^{-1} + d_{-2} b^{-2} + \cdots + d_{-n-1} b^{-n-1} + d_{-n-2} b^{-n-2}$$

where d's are digits and b is the number base. As an example, consider the decimal number 98.76, where d_1 and d_0 are 9 and 8 respectively, while d_{-1} and d_{-2} are 7 and 6 respectively. All other digits are zeros. The number base b is 10.

Digital systems rely heavily on binary numbers for their operations. The coefficients of the binary numbers have two values, 0 and 1. Each coefficient is multiplied by powers of 2. As an example, consider the binary number 101010.1010, whose decimal equivalent is

$1 \times 2^5 + 0 \times 2^4 + 1 \times 2^3 + 0 \times 2^2 + 1 \times 2^1 + 0 \times 2^0 + 1 \times 2^{-1} + 0 \times 2^{-2} + 1 \times 2^{-3} + 0 \times 2^{-4} = 42.625.$

Thus, a binary number can be converted to its decimal equivalent by evaluating the sum of the powers of 2 of those coefficients whose value is 1. Similarly, a decimal number can be converted to a binary number by repeated division by 2 for the integer part and by repeated multiplication by 2 for the part after the decimal point. As an example, we will convert the decimal number 42.625 back to binary. To start with, 42 is divided by 2 to get an integer quotient of 21 and a remainder of 0. The quotient is again divided by 2 to get a new quotient 10 and a remainder 1. This process is continued until the quotient is 0. This is followed by multiplying the fractional part 0.625 by 2 to get 1 as the integer and 0.25 as balance. The balance is again multiplied by 2 to get 0 as the integer and 0.5 as balance. This pattern is repeated until the balance is 0. The whole process is as follows:

Quotient	**Remainder**	
42/2		
21/2	0	
10/2	1	
5/2	0	
2/2	1	
1/2	0	
0	1	101010 = Integer answer

	2×0.625
1	2×0.25
0	2×0.5
1	.0

Fraction = .101

The final answer is got by putting the integer and fraction answers together as 101010.101.

Another popular number representation is the octal system. The following is an example of octal to decimal conversion:

$$(765.4)_8 = 7 \times 8^2 + 6 \times 8^1 + 5 \times 8^0 + 4 \times 8^{-1} = (501.5)_{10}$$

Yet another popular numbering system is the hexadecimal system, used especially in microprocessor-based designs. Apart from the decimal numbers 0 to 9, the letters of the alphabet are used to supplement the ten decimal digits since the base of the number is greater than 10. The letters A, B to F are used for digits 10, 11 to 15 respectively. The following is an example of a hexadecimal number converted to a decimal number:

Table 2.1 Conversion of numbers from one system to another

Decimal number (base 10)	Binary number (base 2)	Octal number (base 8)	Hexadecimal number (base 16)
00	0000	00	0
01	0001	01	1
02	0010	02	2
03	0011	03	3
04	0100	04	4
05	0101	05	5
06	0110	06	6
07	0111	07	7
08	1000	10	8
09	1001	11	9
10	1010	12	A
11	1011	13	B
12	1100	14	C
13	1101	15	D
14	1110	16	E
15	1111	17	F
16	10000	20	10

$(FEDCBA)_{16} = 15 \times 16^5 + 14 \times 16^4 + 13 \times 16^3 + 12 \times 16^2 + 11 \times 16 + 10 = (16702650)_{10}$

Table 2.1 presents the conversion of numbers from one system to another system, namely, the decimal, binary, octal, and hexadecimal systems.

2.2 Twos Complement Addition/Subtraction

Twos complement representation is very effective in microprocessor-based designs. This is also true in FPGA and ASIC based designs. Twos complement of a binary number may be evaluated by adding one to the ones complement (which is just performing bit-wise inversion) of the number. For example, the twos complement of the binary number 10011001 may be computed in the following two steps:

Binary number : 10011001
Ones complement : 01100110
Twos complement : 01100111

A faster way to compute the twos complement is to inspect the binary number commencing from the least significant bit (lsb) and retaining all least significant 0's and the first 1 as it is, and by complementing all other higher significant bits.

A number in twos complement notation may be identified as a negative number if the most significant bit (msb) is ‘1’. Otherwise, it is a positive number, whose decimal equivalent is obtained by adding the decimal weights of all the 1’s in the number. The decimal magnitude of a negative twos complement number is obtained by taking the twos complement first and then adding the decimal weights of all the 1’s in the evaluated twos complement. The following examples illustrate the foregoing statements.

Consider the numbers A = 10011001 and B = 01000111 in twos complement notation. The msb of A is ‘1’ and, therefore, it is a negative number. The twos complement is 01100111, as was evaluated earlier. The decimal weights of ‘1’s commencing from lsb are 1, 2, 4, 32, and 64, which sum up to 103 in decimal notation. Therefore, A equals –103 in decimal system. B is a positive number and the decimal weights of 1’s are 1, 2, 4, and 64, which sum up to +71 in decimal notation. In order to become familiar with simple arithmetic using twos complement notation, we will evaluate the following addition/subtraction operations:

(a) A + B,
(b) A – B,
(c) B – A,
(d) –A – B.

(a)

	A	=	10011001	–103
	B	=	+01000111	+71
	A + B = Sum	=	11100000	–32
Twos complement of Sum		=	00100000	–32

The addition of A and B is carried out by the conventional way of adding two unsigned binary numbers. The ‘Sum’ is in twos complement form. ‘1’ in the msb indicates that it is a negative number. Its magnitude can be obtained by evaluating the twos complement and by adding together the weights of all ones. In this case, the only ‘1’ has the weight, 32.

(b) Subtraction can be done by evaluating the twos complement of B first and then adding it to A, ignoring the final carry. Let us make an attempt at evaluating A – B.

	A	=	10011001	–103
Twos complement of	B	=	10111001	–71
	A – B = Sum	=	01010010	–174
	A	=	110011001	–103
Twos complement of	B	=	110111001	–71
	A – B = Sum	=	101010010	–174
Twos complement of Sum		=	010101110	174

The carry generated in Sum is dropped. The msb of ‘Sum’ indicates that the result is a negative number. The twos complement of Sum gives the magnitude of 174. Thus, the result –174 is correct. We can extend the sign bit by one bit if we need to add two numbers. If we wish to add more numbers, we need to extend the sign accordingly.

(c)

	A	=	110011001	–103
Twos complement of	A	=	001100111	+103
	B	=	01000111	+71
	B – A = Sum	=	010101110	+174

(d)

Twos complement of	A	=	001100111	+103
Twos complement of	B	=	110111001	–71
	–A – B = Sum	=	000100000	+32

(c) and (d) are self-explanatory and may be easily verified to be correct.

Just like the twos complement notation, there is another way of representation of signed numbers, the sign-magnitude notation. In this system, the sign is indicated by the msb, while the rest of the bits contain the magnitude of a number. In the sign-magnitude system, addition or subtraction of two numbers follows the rules of ordinary arithmetic. If the signs are the same, then the magnitudes of the two numbers are added and the result gets the same sign. On the other hand, if the signs are different, the smaller magnitude is subtracted from the larger one and the result gets the sign of the larger magnitude. This requires the comparison of the signs and the magnitudes and then performing either addition or subtraction of the numbers. In contrast to this, the addition/subtraction of numbers in the twos complement system requires only addition and does not require either a comparison or a subtraction. Therefore, twos complement system will be extensively used in this book, especially when realizing designs using Verilog.

2.3 Codes

In order to communicate information from one system or sub-system to another, codes are necessary. Some of the most useful codes are as follows.

2.3.1 Binary and BCD Codes

The common type is the binary code represented by ‘n’ bits. The number of codes that can be generated is 2^n. Another popular code is the binary coded decimal (BCD), which requires four bits per decimal digit. The BCD is a straight assignment of the binary equivalent for a single decimal digit. The binary and BCD codes are shown in Table 2.2 for few numbers of decimal numbers. The weights in the binary and BCD codes of size 4 bits are 8, 4, 2, and 1. Consider the decimal number 12. The binary code equivalent can be obtained by selecting the decimal weights, which add up to 12 and by assigning code '1' to the selected decimal weights. '0's are assigned to the non-selected decimal weights. For the decimal number 12, the relevant decimal weights are 8 and 4 and the non-relevant decimal weights are 2 and 1. Therefore, the binary code is 1100.

The decimal to BCD code conversion is similar to the decimal to binary conversion for decimal numbers from 0 to 9. For greater numbers, four more bits per additional decimal digit need to be added as shown in Table 2.2. For 12, we need four additional bits as the msb in order to accommodate '1' in the decimal number 12. Naturally, we assign 0001 for the most significant digit and 0010 for the least significant digit of the BCD number. As another example, we assign the BCD code as 1001 1001 for the decimal number 99. The inverse conversion is obtained by adding up all decimal weights of 1's in the case of the binary number. For the BCD number, all decimal weights of 1's of a BCD digit are independently added up and concatenated. For example, consider the BCD number 1001 1000. All decimal weights of 1's of the most significant BCD digit add up to 9, while all decimal weights of 1's of the least significant BCD digit add up to 8 and, therefore, the decimal number is 98.

Table 2.2 Binary and BCD codes

Decimal number	Binary code 8421	BCD code 8421 8421
0	0000	0000 0000
1	0001	0000 0001
2	0010	0000 0010
3	0011	0000 0011
4	0100	0000 0100
5	0101	0000 0101
6	0110	0000 0110
7	0111	0000 0111
8	1000	0000 1000
9	1001	0000 1001
10	1010	0001 0000
11	1011	0001 0001
12	1100	0001 0010
13	1101	0001 0011
14	1110	0001 0100
15	1111	0001 0101

2.3.2 Gray Code

Gray codes are used in applications such as an optical shaft encoder to track the shaft position digitally. As the shaft rotates, the code changes from one to another in a sequence. If we were to use a binary sequence, there may be changes in more than one bit positions. For example, consider the binary code changing from 0111 to 1000. It may be noted that all the four bits in the binary code change: msb changes from '0' to '1', whereas all other bits change from '1' to '0'. This may produce an error during the transition from one number to the next since one bit may change faster than others. While changing from 0111 to 1000, the intermediate code will be 1111 (erroneous) if the msb changes before other bits. There are similar other possibilities. The advantage of the Gray code is that only one bit in the code changes when going from one sequence to the next. For example, in going from 0111 to 0101, only the second lsb changes from '1' to '0'. Thus, in the Gray code, only one bit changes in value during any transition between two numbers as shown in Table 2.3.

Table 2.3 Four-bit Gray code sequence

Gray code sequence	Decimal equivalent
0000	0
0001	1
0011	3
0010	2
0110	6
0111	7
0101	5
0100	4
1100	12
1101	13
1111	15
1110	14
1010	10
1011	11
1001	9
1000	8

2.3.3 ASCII Code

Usually, we need to send alphanumeric information from one system to another, a keyboard to a computer, for instance. American Standard Code for Information Interchange (ASCII) presents the standard code for the alphanumeric characters. It uses seven bits to code 128 characters as shown in Table 2.4. More characters or symbols up to 128 are also available as extended ASCII codes. This is shown in Table 2.5. The code size that includes the standard ASCII and the extended codes is eight bits.

Table 2.4 ASCII code

Dec	Hex	Char	Dec	Hex	Char	Dec	Hex	Char	Dec	Hex	Char
0	00	Null	32	20	Space	64	40	@	96	60	`
1	01	Start of heading	33	21	!	65	41	A	97	61	a
2	02	Start of text	34	22	"	66	42	B	98	62	b
3	03	End of text	35	23	#	67	43	C	99	63	c
4	04	End of transmit	36	24	$	68	44	D	100	64	d
5	05	Enquiry	37	25	%	69	45	E	101	65	e
6	06	Acknowledge	38	26	&	70	46	F	102	66	f
7	07	Audible bell	39	27	'	71	47	G	103	67	g
8	08	Backspace	40	28	(	72	48	H	104	68	h
9	09	Horizontal tab	41	29	)	73	49	I	105	69	i
10	0A	Line feed	42	2A	*	74	4A	J	106	6A	j
11	0B	Vertical tab	43	2B	+	75	4B	K	107	6B	k
12	0C	Form feed	44	2C	,	76	4C	L	108	6C	l
13	0D	Carriage return	45	2D	-	77	4D	M	109	6D	m
14	0E	Shift out	46	2E	.	78	4E	N	110	6E	n
15	0F	Shift in	47	2F	/	79	4F	O	111	6F	o
16	10	Data link escape	48	30	0	80	50	P	112	70	p
17	11	Device control 1	49	31	1	81	51	Q	113	71	q
18	12	Device control 2	50	32	2	82	52	R	114	72	r
19	13	Device control 3	51	33	3	83	53	S	115	73	s
20	14	Device control 4	52	34	4	84	54	T	116	74	t
21	15	Neg. acknowledge	53	35	5	85	55	U	117	75	u
22	16	Synchronous idle	54	36	6	86	56	V	118	76	v
23	17	End trans. block	55	37	7	87	57	W	119	77	w
24	18	Cancel	56	38	8	88	58	X	120	78	x
25	19	End of medium	57	39	9	89	59	Y	121	79	y
26	1A	Substitution	58	3A	:	90	5A	Z	122	7A	z
27	1B	Escape	59	3B	;	91	5B	[	123	7B	{
28	1C	File separator	60	3C	<	92	5C	\	124	7C	|
29	1D	Group separator	61	3D	=	93	5D	]	125	7D	}
30	1E	Record separator	62	3E	>	94	5E	^	126	7E	~
31	1F	Unit separator	63	3F	?	95	5F	_	127	7F	□

Table 2.5 Extended ASCII code

Dec	Hex	Char	Dec	Hex	Char	Dec	Hex	Char	Dec	Hex	Char
128	80	Ç	160	A0	á	192	C0	└	224	E0	α
129	81	ü	161	A1	í	193	C1	┴	225	E1	ß
130	82	é	162	A2	ó	194	C2	┬	226	E2	Γ
131	83	â	163	A3	ú	195	C3	├	227	E3	π
132	84	ä	164	A4	ñ	196	C4	─	228	E4	Σ
133	85	à	165	A5	Ñ	197	C5	┼	229	E5	σ
134	86	å	166	A6	ª	198	C6	╞	230	E6	µ
135	87	ç	167	A7	º	199	C7	╟	231	E7	τ
136	88	ê	168	A8	¿	200	C8	╚	232	E8	Φ
137	89	ë	169	A9	⌐	201	C9	╔	233	E9	Θ
138	8A	è	170	AA	¬	202	CA	╩	234	EA	Ω
139	8B	ï	171	AB	½	203	CB	╦	235	EB	δ
140	8C	î	172	AC	¼	204	CC	╠	236	EC	∞
141	8D	ì	173	AD	¡	205	CD	═	237	ED	φ
142	8E	Ä	174	AE	«	206	CE	╬	238	EE	ε
143	8F	Å	175	AF	»	207	CF	╧	239	EF	∩
144	90	É	176	B0	░	208	D0	╨	240	F0	≡
145	91	æ	177	B1	▒	209	D1	╤	241	F1	±
146	92	Æ	178	B2	▓	210	D2	╥	242	F2	≥
147	93	ô	179	B3	│	211	D3	╙	243	F3	≤
148	94	ö	180	B4	┤	212	D4	╘	244	F4	⌠
149	95	ò	181	B5	╡	213	D5	╒	245	F5	⌡
150	96	û	182	B6	╢	214	D6	╓	246	F6	÷
151	97	ù	183	B7	╖	215	D7	╫	247	F7	≈
152	98	ÿ	184	B8	╕	216	D8	╪	248	F8	°
153	99	Ö	185	B9	╣	217	D9	┘	249	F9	∙
154	9A	Ü	186	BA	║	218	DA	┌	250	FA	·
155	9B	¢	187	BB	╗	219	DB	█	251	FB	√
156	9C	£	188	BC	╝	220	DC	▄	252	FC	^{n}
157	9D	¥	189	BD	╜	221	DD	▌	253	FD	^{2}
158	9E	₧	190	BE	╛	222	DE	▐	254	FE	■
159	9F	ƒ	191	BF	┐	223	DF	▀	255	FF	□

2.3.4 Error Detection Code

Data communication from one system to another is done using a serial transmission channel. Any external noise picked up en route by the serial link may change some of the bits from 0 to 1 or vice versa. The purpose of an error-detection code is to detect such bit errors. A parity bit is a binary digit that indicates whether the number of '1' bits in the preceding data was even or odd. If a single bit is changed in transmission, the message will change parity and the error can be detected at this point. The generation of parity bit is useful in detecting errors during the transmission. This is done in the following manner. An even parity bit is generated at the transmitter for each message. At the receiver end, the parity of the received data is computed and checked with the received even parity from the transmitter. If the parity of the received information is not even, it means that at least one bit has changed value during the transmission. This method detects any odd combination of errors in each message that is transmitted. The most common occurrence of error is the one bit error. However, an even combination of errors goes undetected. Additional error-detection schemes may be needed to take care of an even combination of errors. Parity is used in many hardware applications where an operation can be repeated in case of an error, or where simply detecting the error is useful. For example, the SCSI bus uses parity to detect transmission errors, and many microprocessor instruction caches include parity protection.

In a serial data transmission, there are two formats: 7 data bits in one format and 8 data bits in the other format. In the first format, there are 7 data bits, a start bit, an even parity bit, and one or two stop bits. Even parity means that the total number of '1' bits is even. This format accommodates all the seven-bit ASCII characters in a byte. In the other format, eight bits of data is used instead of seven bits of data. In a serial communication, parity is generated at the transmitter and checked by the receiver. An UART (universal asynchronous receiver transmitter) is an example. Recovery from the error is usually done by retransmitting the data.

Consider an *even parity scheme* using nine bit codewords. The code comprises 8 data bits followed by a parity bit. The following examples would make the parity scheme clear:

1. The parity of the data 1111 1110 is odd since there are 7 numbers of '1' bits in the data. The parity bit will be 1, giving the codeword 1111 1110 1.
2. The parity of the data 1111 1111 is even as there are 8 numbers of '1' bits. The parity bit is 0, giving the codeword 1111 1111 0.
3. The parity of the data 0000 0000 is even (zero being an even number). The parity bit is 0, giving the codeword 0000 0000 0.
4. A null or non-existent bitstream also has zero '1' bits and, therefore, it would get the parity bit 0 in an even parity scheme.

Parity is a setting used in serial port data transmission. Parity is also used to recover data in redundant array of independent disks. Parity RAM uses parity to detect memory errors. In telecommunication, a Hamming code is used as an error-correcting code. Hamming codes can detect single and double-bit errors, and correct single-bit errors. In contrast to the Hamming codes, the simple parity code cannot detect errors where two bits are transposed; nor can it help correct the errors it can find.

2.4 Boolean Algebra

Boolean algebra is the algebra of two-valued logic, low or '0' and high or '1'. A Boolean algebra is a set with binary operations '+' for addition and '·' for multiplication; a unary operation '–', and elements '0', '1' satisfying the following laws:

1. Commutative laws for addition and multiplication:
   $A + B = B + A$ (for addition)
   $A \cdot B = B \cdot A$ (for multiplication)
   where A and B are two single bit variables.
2. Associative laws for addition and multiplication:
   $(A + B) + C = A + (B + C)$
   $A \cdot (B \cdot C) = (A \cdot B) \cdot C$
   where A, B, and C are single bit variables.
3. Distributive laws both for multiplication over addition and for addition over multiplication:
   $A \cdot (B + C) = (A \cdot B) + (A \cdot C)$
   $A + (B \cdot C) = (A + B) \cdot (A + C)$

The following laws are also satisfied:

4. $(A + B)' = A'B'$
5. $(AB)' = A' + B'$
6. $A + (A \cdot B) = A$
7. $A \cdot (A + B) = A$
8. $A + (A') = 1$
9. $A \cdot (A') = 0$

where A′ (also referred to as $\overline{A}$) is the inverse of A. If A is '0', then A′ is '1' and vice versa. Sl. nos. 4 and 5 are known as DeMorgan's theorems. The above laws

Table 2.6 Verification of commutative, associative, and distributive laws

A	B	C	A+B	B+A	(A + B) + C	A + (B + C)	A + (B·C)	(A+B)· (A + C)	A + (A · B)	A · (A+B)
0	0	0	0	0	0	0	0	0	0	0
0	0	1	0	0	1	1	0	0	0	0
0	1	0	1	1	1	1	0	0	0	0
0	1	1	1	1	1	1	1	1	0	0
1	0	0	1	1	1	1	1	1	1	1
1	0	1	1	1	1	1	1	1	1	1
1	1	0	1	1	1	1	1	1	1	1
1	1	1	1	1	1	1	1	1	1	1

may be easily verified by filling the truth table as illustrated for some of them in Table 2.6. It may be noted that the following rules of additions and multiplications hold good:
$0 + 0 = 0$; $0 + 1 = 1$; $1 + 0 = 1$; $1 + 1 = 1$;
$0 \cdot 0 = 0$; $0 \cdot 1 = 0$; $1 \cdot 0 = 0$; $1 \cdot 1 = 1$

2.5 Boolean Functions Using Minterms and Maxterms

Minterms and maxterms for three binary variables are shown in Table 2.7, as an example. On similar lines, these terms can be found for two, four, or more binary variables. Each variable may appear in either form: A or A′ and so on. There are eight possible combinations of these variables, namely, A′B′C′, A′B′C right up to ABC as shown in the fourth column of the table. Each of these terms is called a 'Minterm' and are designated symbols: m_0, m_1, ···, m_7 respectively. In general, 2^n minterms can be obtained from n variables. The first three columns are arranged as increasing binary numbers of variables. Each minterm is an ANDed term of the variables, wherein each variable is primed if the corresponding bit of the binary number is a '0' and unprimed if the bit is '1'. The minterms are the decimal equivalents of the corresponding binary number of the variables. For example, the minterm 'm_7' corresponds to the binary number ABC = 111. Instead of AND terms, we can also use OR terms as shown in the sixth column. The corresponding symbols are shown in the seventh column. Each of these terms such as A + B + C, A + B + C′, etc. is called a 'Maxterm', with respective symbols: M_0, M_1, etc. The maxterms are the complements of their corresponding minterms. For example, the minterm m_7 is '1' and the maxterm M_7 is '0' (complement of m_7) corresponding to the same binary value ABC = 111.

Any Boolean function can be expressed as a sum of minterms or as a product of maxterms. For example, consider the function F1 formed by the sum of products of variables A, B, and C, Table 2.8:

Table 2.7 Minterms and maxterms for binary variables

			Minterms		Maxterms	
A	B	C	Term	Symbol	Term	Symbol
0	0	0	A′B′C′	m_0	A + B + C	M_0
0	0	1	A′B′C	m_1	A + B + C′	M_1
0	1	0	A′BC′	m_2	A + B′ + C	M_2
0	1	1	A′BC	m_3	A + B′ + C′	M_3
1	0	0	AB′C′	m_4	A′ + B + C	M_4
1	0	1	AB′C	m_5	A′ + B + C′	M_5
1	1	0	ABC′	m_6	A′ + B′ + C	M_6
1	1	1	ABC	m_7	A′ + B′ + C′	M_7

Table 2.8 Truth table of a function to be realized using minterms

A	B	C	F1
0	0	0	1
0	0	1	0
0	1	0	1
0	1	1	0
1	0	0	1
1	0	1	0
1	1	0	1
1	1	1	1

Table 2.9 Truth table of a function to be realized using maxterms

A	B	C	F2
0	0	0	1
0	0	1	0
0	1	0	1
0	1	1	0
1	0	0	1
1	0	1	0
1	1	0	1
1	1	1	1

$F1 = A'B'C' + A'BC' + AB'C' + ABC' + ABC = m_0 + m_2 + m_4 + m_6 + m_7$
F1 may also be expressed in a short form as

$$F1 = \Sigma\ (0, 2, 4, 6, 7)$$

where Σ implies sum (rather OR) of minterms. Let us now consider the function F2′ formed by active low product of sums of variables A, B, and C:
$F2' = (A + B + C')\ (A + B' + C')\ (A' + B + C') = M_1 \cdot M_3 \cdot M_5$
In short, the function may be expressed as follows:
$F2' = \Pi\ (1, 3, 5)$
The product symbol, Π, denotes the ANDing of maxterms. It may be noted that the function F2′ is just the complement of F1. The final result is F2, the complement of F2′ and the same as F1 as shown in Table 2.9. Thus, one may use either the minterms or the maxterms to evaluate a function, whichever is simpler.

2.6 Logic Gates

The symbols and functions of the common types of gates are shown in Figure 2.1. A and B are the inputs and the outputs are F1 to F8. The reader may easily form the truth table for these simple gates using the Boolean expressions presented in the figure.

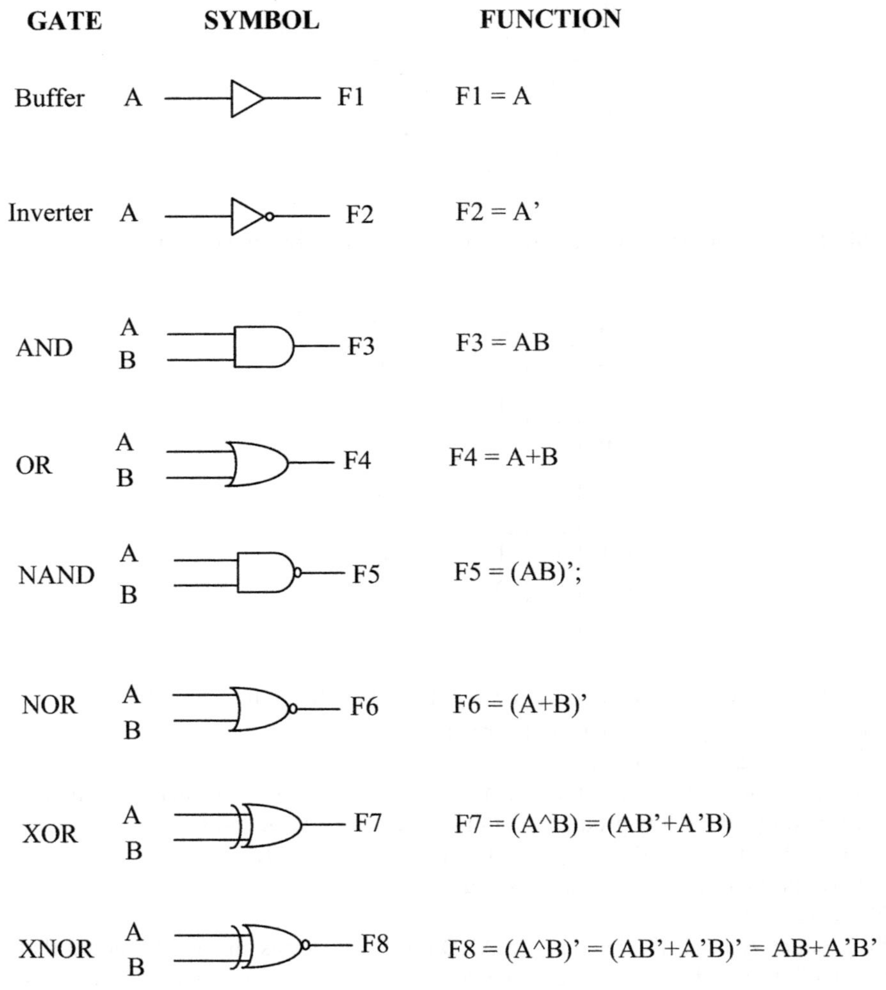

Fig. 2.1 Symbols and functions of the common types of gates

2.7 The Karnaugh MAP Method of Optimization of Logic Circuits

The Karnaugh map or simply K map is made up of squares with each square representing a minterm as shown in Figures 2.2 to 2.4. The signal A is the most significant bit. The K map is useful in optimizing a digital circuit as shown in Figures 2.5 to 2.8. The Boolean function may be simplified by circling 16, 8, 4, 2 or one number of 1's lying as a clutter, single line or multiple lines, be it in a horizontal or a vertical straight line. 1's lying in the first square and the last square in a horizontal or a vertical line can also be circled. However, 1's lying diagonally cannot be circled together. They have to be independently circled as shown in Figure 2.6. The function shown in Figure 2.5 expressed in terms of minterms is F1 = Σ (1, 2, 3). Two pairs of 1's, one each in the horizontal and the vertical directions, offer themselves as good candidates for circling. Inspecting the horizontal 1's, we see that A = 1 and hence the signal 'A' finds a place in the final reduced function F1.

A \ B	0	1
0	m0	m1
1	m2	m3

Fig. 2.2 K map for two signals A and B

C \ AB	00	01	11	10
0	m_0	m_1	m_3	m_2
1	m_4	m_5	m_7	m_6

Fig. 2.3 K map for three signals A, B, and C

CD \ AB	00	01	11	10
00	m_0	m_1	m_3	m_2
01	m_4	m_5	m_7	m_6
11	m_{12}	m_{13}	m_{15}	m_{14}
10	m_8	m_9	m_{11}	m_{10}

Fig. 2.4 K map for four signals A, B, C, and D

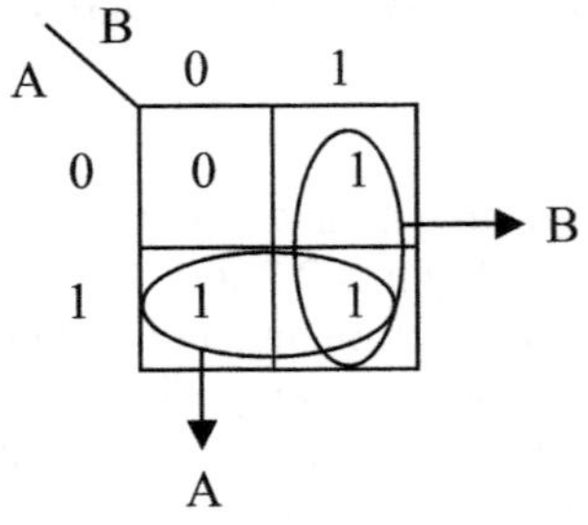

Fig. 2.5 K map reduction for two signals using minterms

Similarly, observing the vertical 1's, we get 'B'. These partial results together form the final reduced result:

$$F1 = A + B$$

Figure 2.6 shows another function F2, which uses three input signals A, B, and C. Looking at the first row, we can circle the last two 1's, and first 1 and the last 1 together. These two combinations produce the partial results, AC′ and B′C′ respectively. The last '1' does not have any neighboring 1's and hence contributes A′BC to the result. The final result is as follows:

$$F2 = \Sigma\,(0, 2, 3, 5)$$

$$F2 = AC' + B'C' + A'BC$$

Finally, we will consider another example of K map reduction for four signals as shown in Figure 2.7. Instantly, we can spot a cluster of eight 1's, which we will circle promptly. This yields 'B' as the partial result. Balance four 1's is at the four corners of the K map. We may regard the corner squares to be neighbors and hence circle all of them to make one integral group. This quad 1's contribute the Boolean expression B′D′. The final result of function

$$F3 = \Sigma\,(0, 1, 2, 3, 5, 7, 8, 9, 10, 11, 13, 15) \text{ is:}$$

$$F3 = B + B'D' = B + D'$$

since B′ is redundant as can be inferred from the K map.

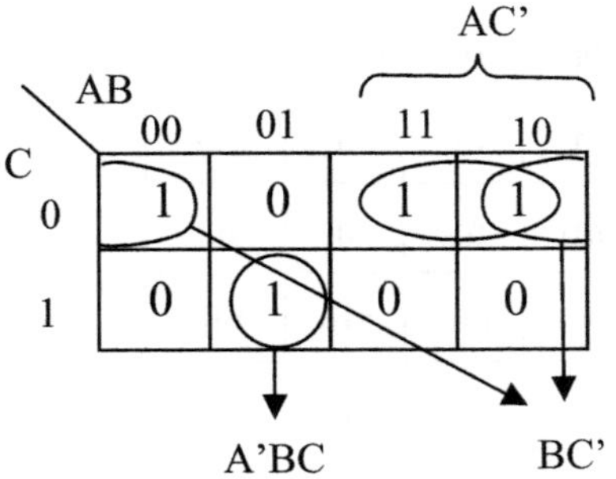

Fig. 2.6 K map reduction for three signals using minterms

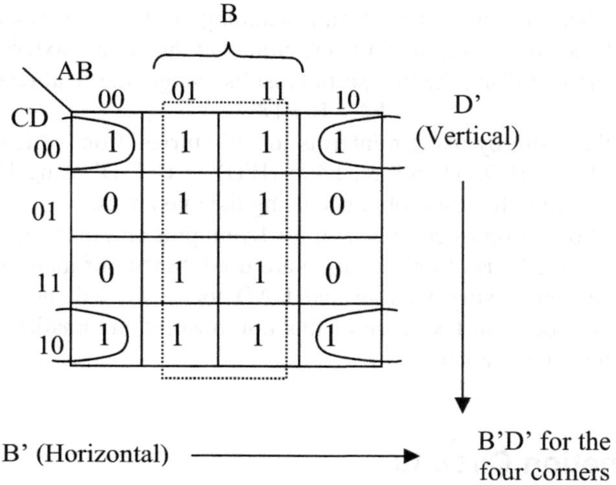

$$F3 = B + B'D' = B + D'$$

Fig. 2.7 K map reduction for four signals using minterms

In the previous examples, the Boolean functions were expressed as the sum of products. Similarly, the product of sums form can also be obtained by circling 0's and combining them as was done before. The combined maxterm inputs must be inverted individually and ORed to get a partial result. Finally, these partial results are ANDed together to form the final result, which is of the 'product of sums' form. To make it clear, we will take the same function F3 evaluated previously using minterms. Consider Figure 2.8, which is the same as Figure 2.7, except that maxterms ('0' entries) are circled. Looking at horizontal '0' entries, we see that 'B' is

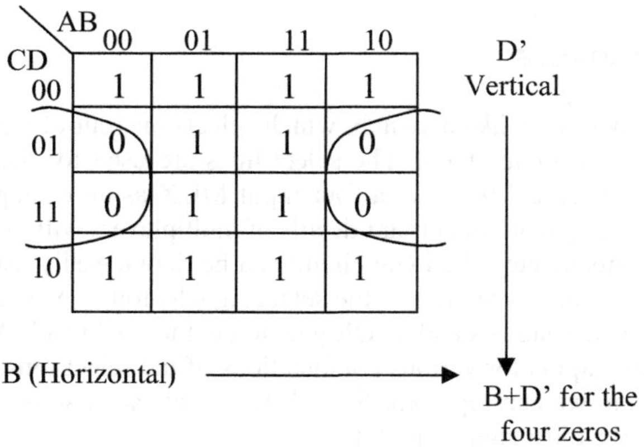

$$F4 = B + D'$$

Fig. 2.8 K map reduction for four signals using maxterms

a constant '0'. Therefore, we get B. Again, scanning circled 0's vertically, we see that D = 1 and hence the partial result is D′. Note that the valid maxterm inputs are inverted individually. ORing the two partial results, we get the final result:

$$F4 = B + D'$$

We can verify the result by using minterms for '0' entries, from which we get the inverted result: F4′ = B′D. Therefore, F4 = (B′D)′ = B + D′ using DeMorgan's theorem, thus verifying the result obtained using the maxterms.

K map reduction becomes complicated for five inputs and more and, therefore, Quine McCluskey methods of optimization are used. Since our main interest is to develop VLSI systems using Verilog and CAD tools, we will not present this method. Synthesis tool will take care of this optimization automatically. We will learn this tool in a later chapter.

2.8 Combination Circuits

Digital systems are of two types: combination and sequential. A combination circuit comprises logic gates, whose outputs depend upon the present inputs without regard to previous inputs. A combination circuit realizes a set of Boolean functions directly. Ideally, the response of a combination circuit is instantaneous. In real practice, however, there is delay owing to propagation of signals through various gates, which combine them logically. There is no memory involved in a combination circuit. In contrast to this, a sequential circuit employs memory elements in addition to logic gates.

The sequential circuits will be covered later on. All gates presented in Section 2.6, multiplexers, demultiplexers, decoders, comparators, half/full adders, etc. are all combination circuits. These circuits from multiplexers onwards will be presented in the following sub-sections.

2.8.1 Multiplexers

A multiplexer (MUX) is like a switch, which selects one out of many inputs, and outputs the selected input signal. The select lines are used for the selection of a particular input. Figure 2.9 shows a four input MUX as an example. On similar lines, the reader may work out circuit details of multiplexers with two inputs, eight inputs, sixteen inputs, etc. The logic circuit can be drawn easily using two inverters to generate active low signals of the select signals, four AND gates combining the inputs with select signals and an OR gate to get the final result, F. The function table shows the outputs for various combinations of the select pins. For brevity in a circuit diagram, we can represent the MUX as a block as shown in Figure 2.9. Note that S1 is the most significant bit.

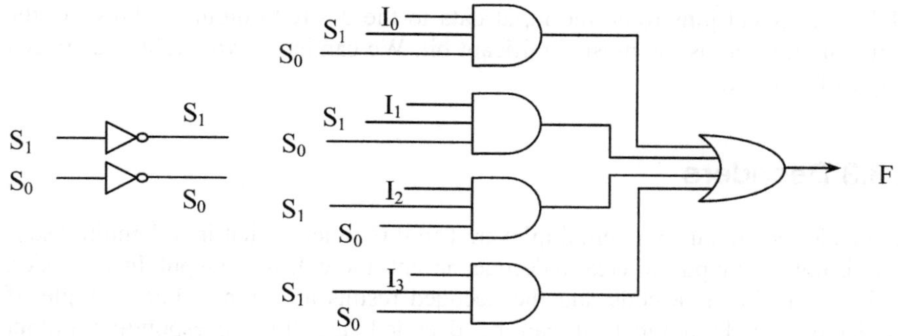

a Logic diagram

S_1	S_0	F
0	0	I_0
0	1	I_1
1	0	I_2
1	1	I_3

b Function table

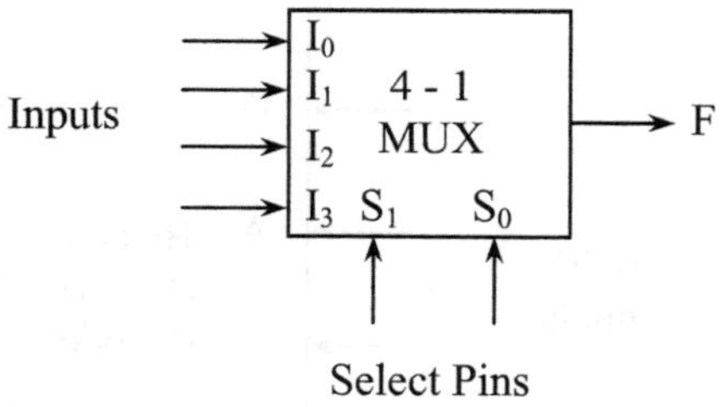

c Block diagram

Fig. 2.9 A four input multiplexer

2.8.2 Demultiplexers

A demultiplexer (DEMUX) is just the inverse of a MUX, which receives a single bit input, and outputs to one of many output lines as shown in Figure 2.10. As in the

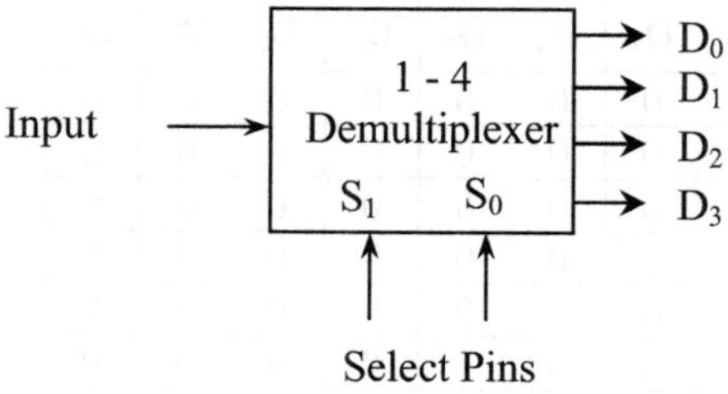

a Block Diagram

S_1	S_0	Input
0	0	D_0
0	1	D_1
1	0	D_2
1	1	D_3

b Function Table

Fig. 2.10 A four output demultiplexer

MUX, the select pins route the input data to the desired output as shown in the function table. S1 is the most significant bit. We can have two, eight and sixteen outputs DEMUX.

2.8.3 Decoders

A decoder is similar to a demultiplexer. The difference is that in a demultiplexer, there is only one input, whereas a decoder accepts more than one input. In a decoder, an input number is a code and the decoded results are output. For example, if we have a BCD number, it can be decoded and the corresponding output

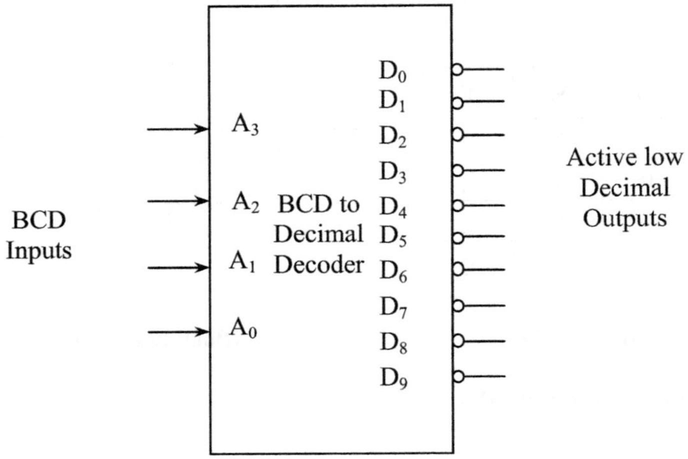

Fig. 2.11 BCD to decimal decoder

Table 2.10 Truth table of BCD to decimal decoder

Inputs				Intermediate outputs									
A_3	A_2	A_1	A_0	D_0	D_1	D_2	D_3	D_4	D_5	D_6	D_7	D_8	D_9
0	0	0	0	1	0	0	0	0	0	0	0	0	0
0	0	0	1	0	1	0	0	0	0	0	0	0	0
0	0	1	0	0	0	1	0	0	0	0	0	0	0
0	0	1	1	0	0	0	1	0	0	0	0	0	0
0	1	0	0	0	0	0	0	1	0	0	0	0	0
0	1	0	1	0	0	0	0	0	1	0	0	0	0
0	1	1	0	0	0	0	0	0	0	1	0	0	0
0	1	1	1	0	0	0	0	0	0	0	1	0	0
1	0	0	0	0	0	0	0	0	0	0	0	1	0
1	0	0	1	0	0	0	0	0	0	0	0	0	1

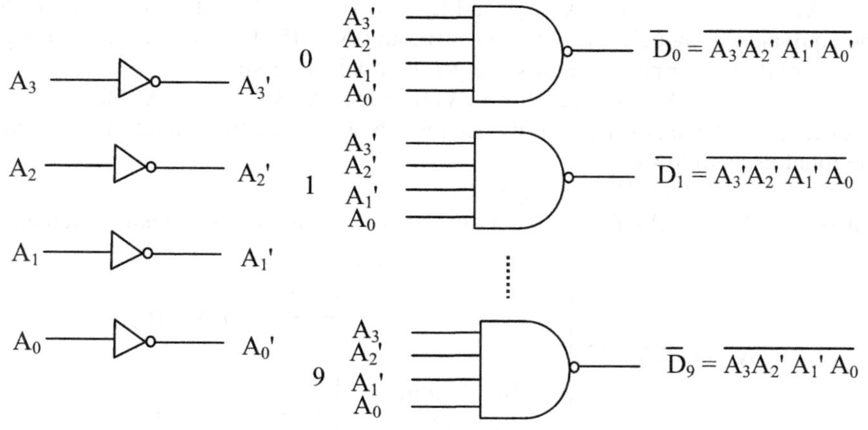

Fig. 2.12 Logic diagram of BCD to decimal decoder

can be activated. Figure 2.11 shows a BCD to Decimal Decoder, wherein D_0 to D_9 are outputs after inverting those signals. For instance, the BCD input 0111 (A_3 is the msb) sets D_7' low, while setting all other outputs high. Table 2.10 shows the truth table for this decoder, where the outputs are shown as D_0 to D_9 instead of their inverses. Note that only one out of these ten outputs is activated for a valid BCD number. The logic for this decoder can be realized using inverters and four input NAND gates as shown in Figure 2.12. For invalid BCD number, no outputs are activated. This can be easily verified from the logic diagram. Similarly, other decoders such as binary to hexadecimal decoder, BCD to seven segment decoder, etc. can be designed.

2.8.4 Magnitude Comparator

A magnitude comparator is a combinational circuit that compares two numbers, A and B. The comparison of two numbers determines if one number is less than or greater than, or equal to the other number. The comparator outputs one of the conditions satisfied $A < B$, $A > B$, or $A = B$, each of which is a single bit. As an example, consider two four-bit numbers A and B represented as $A_3\ A_2\ A_1\ A_0$ and $B_3\ B_2\ B_1\ B_0$, where A_3 and B_3 are the msbs. Since there are only three conditions as mentioned earlier, it is enough if we evaluate any two of the conditions. We will, therefore evaluate $A < B$ and $A = B$. This way, we will reduce the gate count.

We will first evaluate $A < B$. If the msbs are $A3 = 0$ and $B_3 = 1$, it means that $A < B$ and we need not take the trouble of evaluating other bits. This condition may be expressed as $A_3'B_3$. On the other hand, if $A_3 = B_3$, then we repeat the above procedure, in turn, for all the other bits down to the lsbs. These yield expressions $E_3 A_2'B_2$, $E_3 E_2 A_1'B_1$, and $E_3 E_2 E_1 A_0'B_0$ for the subsequent bits, where $E_3 = A_3$

$B_3 + A_3' B_3'$, $E_2 = A_2 B_2 + A_2' B_2'$, and $E_1 = A_1 B_1 + A_1' B_1'$, all of which indicate the equivalence, namely, $A_3 = B_3$, $A_2 = B_2$, and $A_1 = B_1$. Combining (i.e., ORing) together all the four partial results, we get the final expression:

$$(A < B) = A_3'B_3 + E_3A_2'B_2 + E_3E_2A_1'B_1 + E_3E_2E_1A_0'B_0$$

In order to evaluate A = B, we only need to AND together the equality E_3 to E_0 for the four bits as follows:

$$(A = B) = E_3 E_2 E_1 E_0$$

where $E_0 = A_0 B_0 + A_0' B_0' = (A_0 = B_0)$. The last condition can be easily obtained

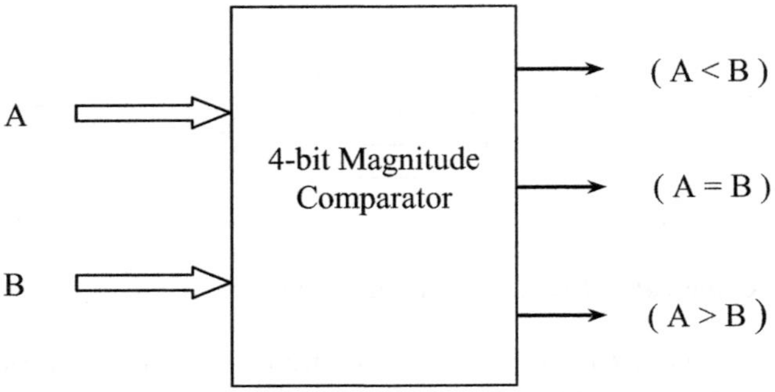

Fig. 2.13 Block diagram of a Four-bit magnitude comparator

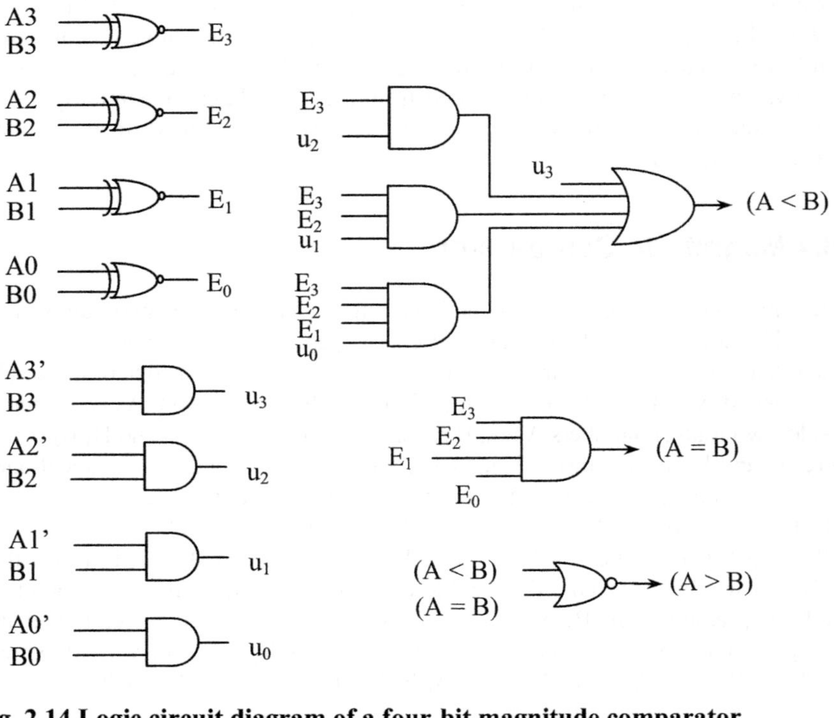

Fig. 2.14 Logic circuit diagram of a four-bit magnitude comparator

by just NORing (A < B) and (A = B) as follows:

$$(A > B) = ((A < B) + (A = B))'$$

This simply means (A > B) is neither (A < B) nor (A = B). The four-bit comparator block diagram and the deduced logic circuit diagram are shown in Figures 2.13 and 2.14 respectively.

2.8.5 Adder/Subtractor Circuits

Arithmetic operations are indispensable in a digital system, be it a processor based system or an FPGA/ASIC based system, especially for data processing applications. The basic arithmetic operations are addition and subtraction of two binary numbers, single or multiple bits. These operations are performed basically using combinational circuits, although they can be sequential circuits as presented in a later chapter on arithmetic circuits.

Half Adder

A half adder performs the addition of two single bits. The truth table of a half adder is shown in Table 2.11. The carry is not shown since we are interested only in a single bit result. Inspecting the '1' output entries of the truth table, we get the Boolean function:

$$\text{Sum\_HA} = A'B + AB' = A \oplus B$$

Table 2.11 Truth table of a half adder

A	B	Sum_HA
0	0	0
0	1	1
1	0	1
1	1	0

The XOR gate realization is shown in Figure 2.15.

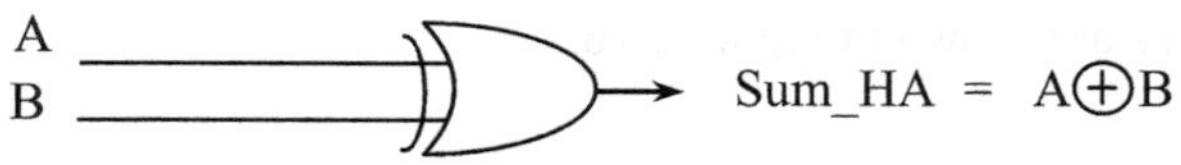

Fig. 2.15 Logic of a half adder

Full Adder

A full adder is a combinational circuit that adds three single bit inputs, A, B, and C. The outputs are carry "C_FA" and the sum "S_FA". The input "C" represents the carry in from the previous lower significant bit position, should they exist. The full adder truth table is shown in Table 2.12.

Table 2.12 Truth table of a full adder

A	B	C	C_FA	S_FA
0	0	0	0	0
0	0	1	0	1
0	1	0	0	1
0	1	1	1	0
1	0	0	0	1
1	0	1	1	0
1	1	0	1	0
1	1	1	1	1

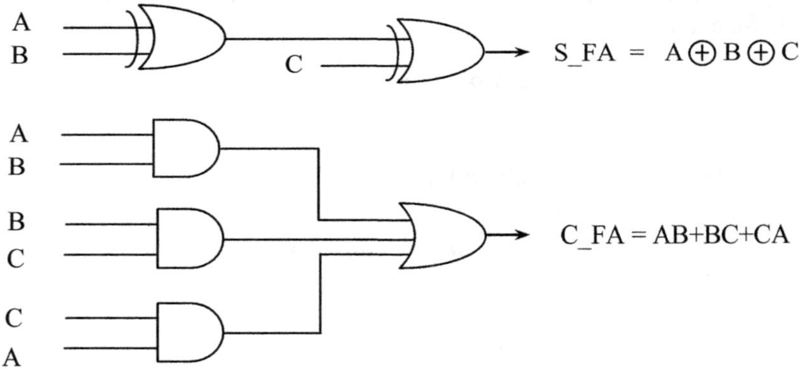

Fig. 2.16 Logic realization of a full adder

The sum and carry outputs are respectively

$$S\_FA = A \oplus B \oplus C$$
$$C\_FA = AB + BC + CA$$

The gate realizations are shown in Figure 2.16.

Half Subtractor

A half subtractor is used for subtracting two single bits, borrowing a '1', if necessary. The truth table of the half subtractor is shown in Table 2.13. The borrow output is not shown in the truth table.

Table 2.13 Truth table of a half subtractor

A	B	S_HS
0	0	0
0	1	1
1	0	1
1	1	0

The Boolean function for the half subtractor output is as follows:

$$S\_HS = A'B + AB' = A \oplus B.$$

It may be noted that the logic for S_HS is the same as that for the output of the half adder as shown in Figure 2.17.

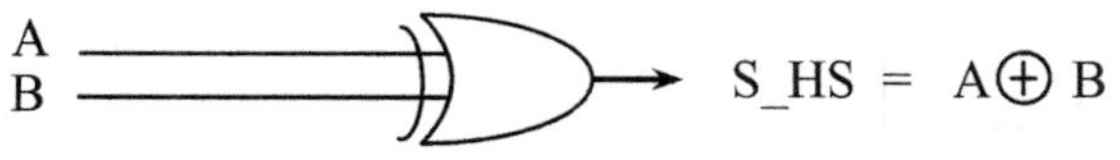

Fig. 2.17 Logic of a half adder

Full Subtractor

A full subtractor, as the name implies, is a combinational circuit that performs a subtraction between two single bits with a borrow-in got from a lower significant stage. In short, the final result is A–B–C. It has three inputs, A, B, and C and two outputs, B_FS (for borrow-out) and S_FS (for subtracted value). The truth table for the full subtractor is shown in Table 2.14. The simplified Boolean output functions for the full subtractor are derived using K map and are as follows:

$$B\_FS = A'B + A'C + BC$$
$$S\_FS = A'B'C + A'BC' + ABC + AB'C'$$

The logic circuit realization is shown (inverters are not shown) in Figure 2.18.

Table 2.14 Truth table of a full subtractor

A	B	C	B_FS	S_FS
0	0	0	0	0
0	0	1	1	1
0	1	0	1	1
0	1	1	1	0
1	0	0	0	1
1	0	1	0	0
1	1	0	0	0
1	1	1	1	1

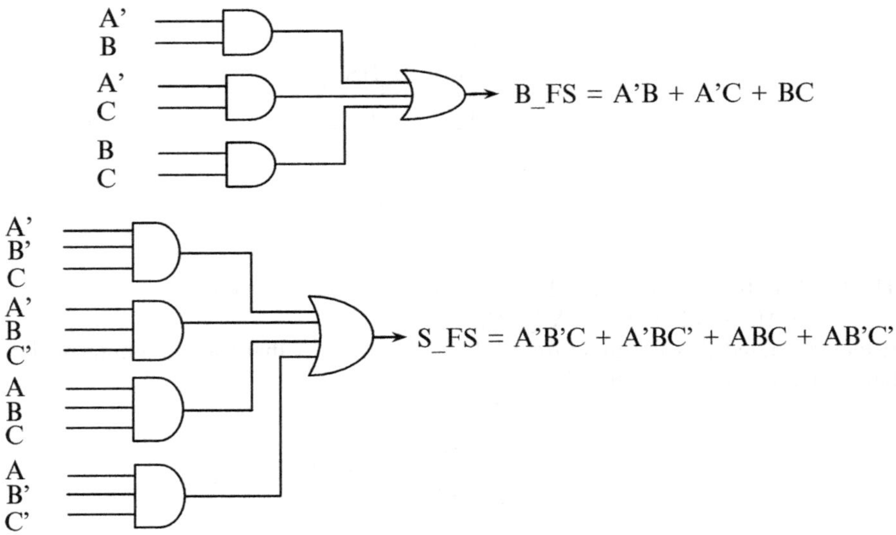

Fig. 2.18 Logic gate realization of a full subtractor

2.8.6 SSI and MSI Components

In the foregoing sections, we have seen the design of a number of gates and circuits. Gates fall under small scale integrated (SSI) circuits category. Most of the circuits considered earlier fall under medium scale integrated (MSI) circuits category. These components are available in IC packages from a number of vendors such as Signetics, Texas instruments, etc. Some of these components are multiplexers (74LS150 to 74LS153), demultiplexers (74LS138, 74LS139, 74LS153, 74LS154), BCD to decimal decoder/driver (74LS145), BCD to seven segment decoders/drivers (7445 to 7448), adder (74LS83), comparator (74LS85), etc. Apart from the combinational circuits, sequential circuits such as flip-flops, counters, etc. are also available. Using these devices and others, we can design cost-effective digital systems for small to medium applications. Some of these components are also used as peripheral devices or interfaces in more complex LSI and VLSI systems. For larger systems, SSI/MSI devices based designs become very complex to produce. Programmable logic devices (PLDs) are more suitable for such applications. We will consider some of these devices in a later section.

2.9 Arithmetic Logic Unit

We have looked at the designs of various gates, adder, and subtractor earlier. Combining all these functions into a single medium scale integrated circuit provides logic

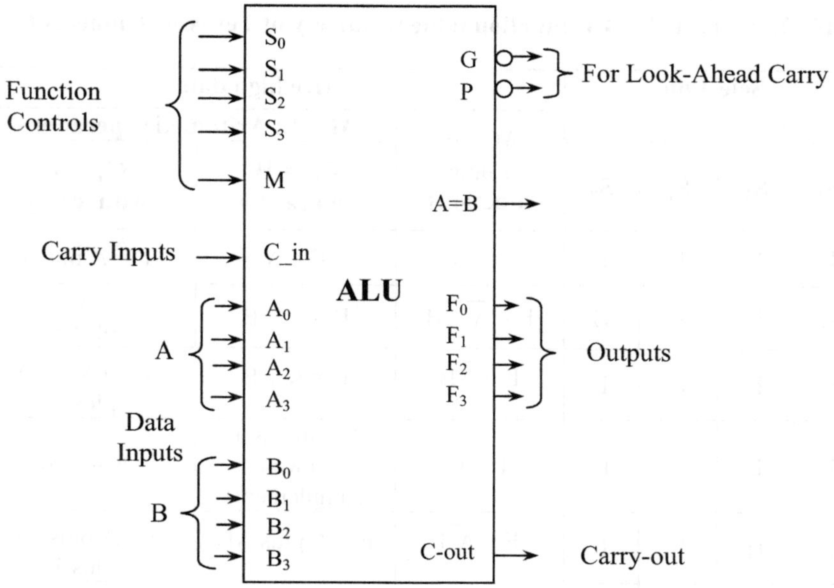

Fig. 2.19 ALU logic symbol (Courtesy of Texas Instruments Inc.)

designers a useful tool. One such example is the 74xx181 of Texas Instruments, whose logic symbol is shown in Figure 2.19. A four-bit select code (S3–S0) and a mode bit (M) are used to decide the operation to be performed on data inputs as shown in Table 2.15. The mode and select code can provide up to 32 different functions as tabulated. As can be seen in the table, the functions are partitioned into two categories: logic and arithmetic. The arithmetic functions are further divided into two groups. Consider the logical operations, for which M = 1: when select = 0000, the four-bit 'A' input is complemented bit-wise and output in F. By changing M to 0, keeping select = 0000, and the carry input high, the output is A plus 1 (which means increment A). Similarly, other functions can be inferred.

2.10 Programmable Logic Devices

A programmable logic device is an integrated circuit with logic gates partitioned into an AND array and an OR array [4–6]. These gates are interconnected to provide sum of products implementation. The un-programmed PLD has all the connections intact. Programming the device breaks the connections to achieve a desired logic function. Figure 2.20 shows three types of PLDs. The programmable read-only memory (PROM/EPROM/Flash ROM) has a fixed AND array and programmable connects for the output OR gates. These devices, hereafter referred to as ROM for short, implement Boolean functions as sum of minterms. The programmable

Table 2.15 74xx181, ALU function table (Courtesy of Texas Instruments Inc.)

Selection				Active high data		
				M = H Logic functions	M = L: Arithmetic operations	
S_3	S_2	S_1	S_0		$C_n' = H$ (no carry)	$C_n' = L$ (with carry)
L	L	L	L	$F = \overline{A}$	F = A	F = A plus 1
L	L	L	H	$F = \overline{A + B}$	F = A + B	F = A + B plus 1
L	L	H	L	$F = \overline{A}B$	$F = \overline{A} + B$	$F = (\overline{A} + B)$ plus 1
L	L	H	H	F = 0	F = minus 1 (Twos complement)	F = zero
L	H	L	L	$F = \overline{AB}$	F = A plus $A\overline{B}$	F = A plus $A\overline{B}$ plus 1
L	H	L	H	$F = \overline{B}$	F = (A + B) plus $A\overline{B}$	F = (A + B) plus $A\overline{B}$ plus 1
L	H	H	L	$F = A \oplus B$	F = A minus B minus 1	F = A minus B
L	H	H	H	$F = A\overline{B}$	$F = A\overline{B}$ minus 1	$F = A\overline{B}$
H	L	L	L	$F = \overline{A} + B$	F = A plus AB	F = A plus AB plus 1
H	L	L	H	$F = \overline{A \oplus B}$	F = A plus B	F = A plus B plus 1
H	L	H	L	F = B	$F = (A + \overline{B})$ plus AB	F = (A + B) plus AB plus 1
H	L	H	H	F = AB	F = AB minus 1	F = AB
H	H	L	L	F = 1	F = A plus A′	F = A plus A plus 1
H	H	L	H	$F = A + \overline{B}$	F = (A + B) plus A	F = (A + B) plus A plus 1
H	H	H	L	F = A + B	F = (A + B) plus A	$F = (A + \overline{B})$ plus A plus 1
H	H	H	H	F = A	F = A minus 1	F = A

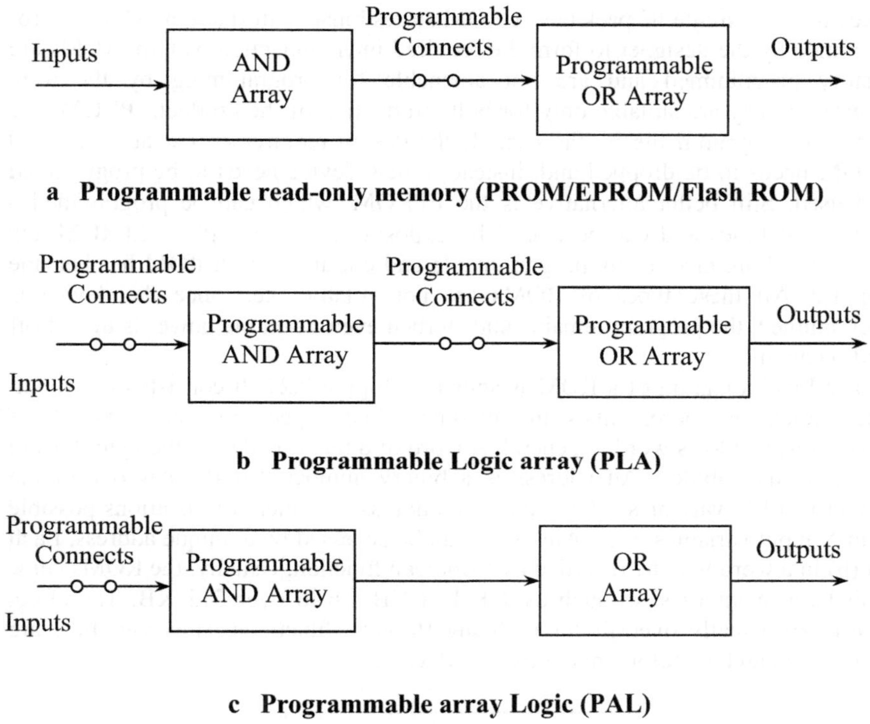

Fig. 2.20 Programmable Logic Devices: ROM, PLA, PAL

logic array (PLA) is an IC, where the AND gates as well as OR gates are programmable. However, programming PLA is complex. The programmable array logic (PAL) has a programmable AND array and a fixed OR array. The product terms in the desired Boolean functions are obtained by programming AND gates. These ANDed terms are summed using an OR gate to realize the Boolean function. The PAL is simple to program and, therefore, more popular, especially in industrial products. In the design using 74 series digital ICs, once the printed circuit is made, we cannot change the logic, if required. This disadvantage is eliminated by using PLDs in the design of digital systems since they can be re-programmed to suit the changed situation, provided the inputs/outputs do not change appreciably. Each of the three types of PLDs is explained in the following successive sub-sections.

2.10.1 Read-Only Memory

A read-only memory is a storage device that fall under any of the categories, MSI and VLSI, depending upon its size. The ROM comprises decoders and the OR gates within a single IC package. A ROM stores binary information, which is programmed by the designer to form the required interconnection pattern. ROMs are factory programmed and are not amenable for programming by the user. Therefore, they are suitable only for bulk production of the product. PROMs are one-time programmable by the user. If the design requirements change, the old PROM needs to be dropped and, instead, a new device needs to be programmed and used. Still better alternative is the EPROM, which can be programmed a number of times and can be erased by exposing it to UV light. EEPROM and Flash ROM are electrically programmable and erasable. Flash ROM has become popular. All these types of ROMs are non-volatile, i.e., once the device is programmed, the program remains undisturbed even when the power is turned off and on again.

The block diagram of a ROM is shown in Figure 2.21. It consists of N input lines called the address lines and W output lines, specifying the width of the device referred to as word in general. A word of 8 bits is called as the byte (B) and 4 bits as the nibble. An address is a binary number that denotes one of the minterms of N variables. The number of addresses or memory locations possible with N input variables is 2^N. Any word can be accessed by a unique address. Each bit (b) in a word may be regarded as a Boolean function. Usually, the ROMs come with large memory sizes such as 4 KB, 8 KB, up to over 512 KB. Therefore, ROMs are usually overkill for realizing Boolean functions. However, they are quite cheap and, therefore, may be cost-effective.

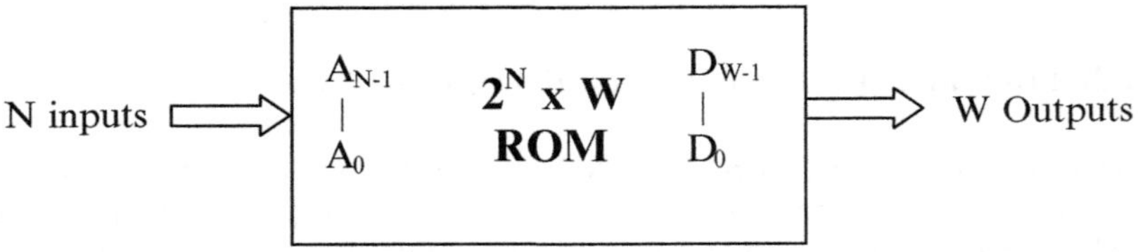

Fig. 2.21 Block diagram of ROM

2.10.2 Programmable Logic Array (PLA)

A PLA has programmable interconnects among inputs and outputs arranged as a matrix as shown in Figure 2.22. These interconnects are in tact initially. The desired Boolean functions are implemented as sum of products by breaking appropriate connections using a programmer. Interconnects are available between all inputs and their complement values to each of the AND gates. They are also available between the outputs of the AND gates and the inputs of the OR gates and also across inverters at the outputs. The size of the PLA is specified by the number of inputs/outputs and the number of product terms. As an example, the figure shows a PLA with 12 inputs and 8 outputs. The designer specifies the Boolean functions and the programmer removes all interconnections, leaving those marked 'X' undisturbed.

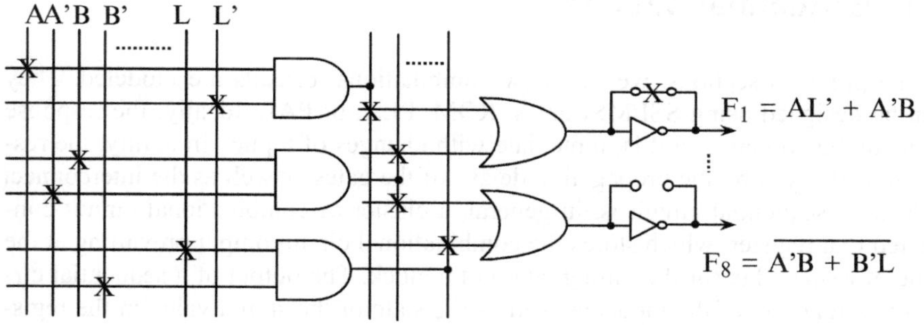

Fig. 2.22 PLA with twelve inputs and eight outputs

2.10.3 Programmable Array Logic (PAL)

The programmable array logic (PAL) is a device with a programmable AND array and a fixed OR array. The PAL is easier to program than the PLA and hence more popular. For simplicity of representation, the un-programmed inputs to each of the AND gates is shown in Figure 2.23. In the actual circuit, however, the vertical lines are independently connected to the inputs of the AND gate. The 12 inputs and 8 outputs PAL, which realizes the same Boolean functions as in Figure 2.22 is shown in Figure 2.24. As in PLA, the designer specifies the Boolean functions and the programmer removes all interconnections, leaving those marked 'X' undisturbed. The inputs of each of the AND and OR gates is limited in PLA/PAL. Later on, we will use a commercially available PAL in an application. The output inverters are different in an actual PAL. PALs are suitable for small and medium sized applications and are more flexible than TTL based ICs. For medium to large designs, FPGAs are far more suitable than PALs, which is the main emphasis of this book.

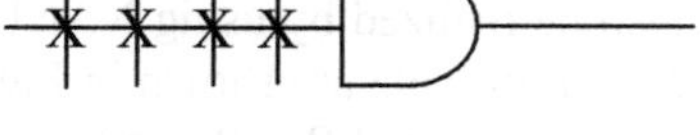

Fig. 2.23 Graphic Symbol of AND Inputs of PAL

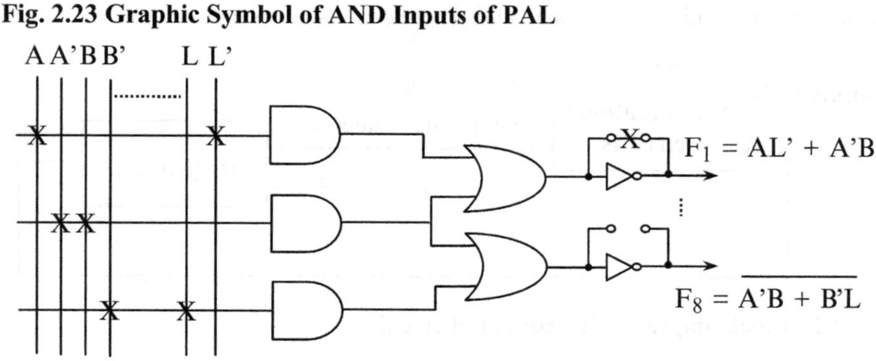

Fig. 2.24 PAL with twelve inputs and eight outputs

2.11 Sequential Circuits

In the previous sections, we saw how combinational circuits are modeled. They can be designed using SSI/MSI gates, ROM, PLA, or PAL. Ideally, the response of a combinational circuit is immediate with changes of inputs. In reality, the response is delayed by the propagation delays of the gates as well as the interconnect delays. A sequential circuit is, in general, a cluster of combinational circuit connected to a register, which stores the combinational circuit output prevailing at the time of rising edge (or the falling edge) of a clock. The output of a sequential circuit is a function of the inputs as well as the state of the stored value in the register. The output of a sequential circuit depends on the present inputs as well as on the past inputs. The register may be a RS flip-flop, a D flip-flop, a JK flip-flop, or a T flip-flop. These flip-flops will be described in later sub-sections. The sequential circuits are referred to as synchronous circuits as they work in tandem with the clock. A block diagram of a sequential circuit is shown in Figure 2.25. It consists of combinational circuits to which registers are connected, with register outputs fed back. The storage elements are flip-flops, capable of storing binary information.

RS Flip-flop

The flip-flops are binary cells, each capable of storing one bit of information. A flip-flop circuit can maintain a binary state '0' or '1' so long as power is applied to the circuit or until directed by an input signal to change state. A common type of flip-flop is called a RS flip-flop or a SR latch. The R and S are the two inputs that are abbreviations for reset and set respectively. Figure 2.26 shows an RS flip-flop circuit using NOR gates. When S = 1 and R = 0, the flip-flop is set, i.e., Q is set to '1' and Q′ is reset. Similarly, when S = 0 and R = 1, the flip-flop is reset. The flip-flop is in store mode for SR = 00 and stores whatever was the previous state as can be seen from the truth table. It may be noted that SR = 11 is an invalid input since the corresponding outputs Q and Q′ must be complement of each other instead of '0' each. The RS flip-flop may also be realized by using NAND gates instead of NOR gates. The logic diagram and its truth table are shown in Figure 2.27. This flip-flop can be set and reset by applying the inputs SR = 01 and 10 respectively. SR = 11 is the store mode, while 00 input is invalid. These circuits are asynchronous.

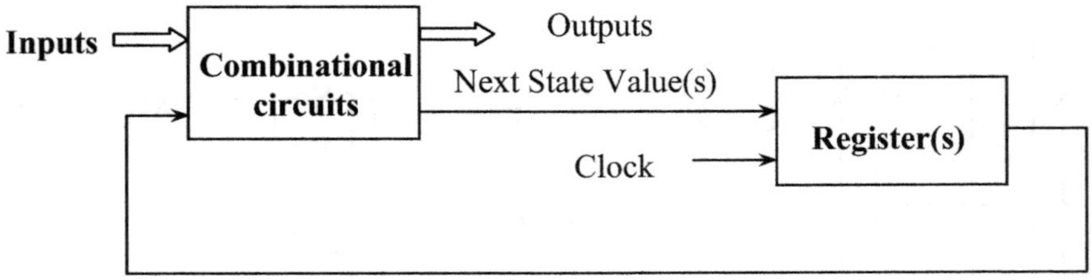

Fig. 2.25 Block diagram of a sequential circuit

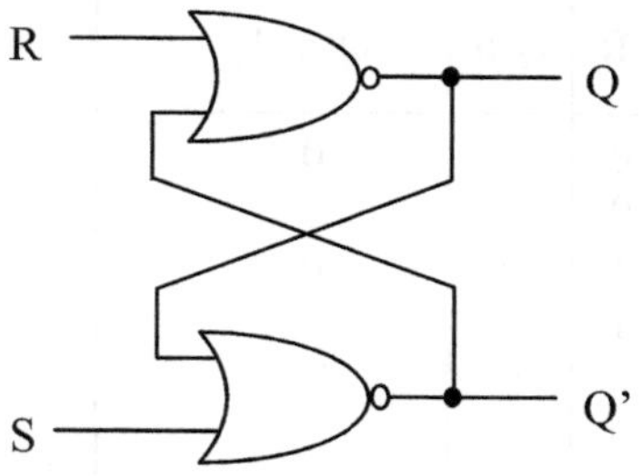

a Logic Circuit

S	R	Q	Q'	
1	0	1	0	(after S =1, R = 0)
0	0	1	0	
0	1	0	1	
0	0	0	1	(after S =0, R = 1)
1	1	0	0	

b Truth table

Fig. 2.26 RS flip-flop circuit using NOR gates

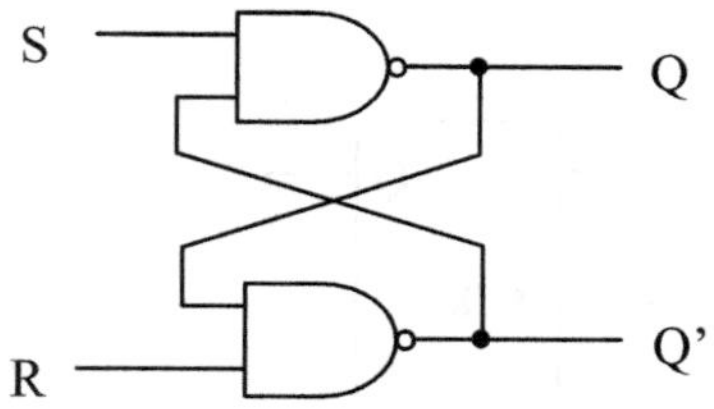

a Logic diagram

S	R	Q	Q'	
1	0	0	1	
1	1	0	1	(after S =1, R = 0)
0	1	1	0	
1	1	1	0	(after S =0, R = 1)
0	0	1	1	

b Truth table

Fig. 2.27 RS flip-flop circuit using NAND gates

Synchronous RS flip-flop can be built by providing a clock input that determines when the state of the circuit is to be changed. The RS flip-flop with a clock input in shown in Figure 2.28a. Q(n) and Q(n + 1) stand for the state of the flip-flop before and after the application of a clock pulse and are referred to as the present and the next state respectively. As shown in the truth table in Figure 2.28b, for the inputs S and R and the present state Q(n), the application of a single pulse causes the flip-flop to go to the next state, Q(n + 1). The last two lines are indeterminate and should not be allowed. This can be met if SR = 0. The characteristic equation of the flip-flop is derived from the K map shown in Figure 2.28c. This specifies the value of the next state as a function of the inputs and the present state. The clocked RS flip-flop may be represented symbolically as shown in Figure 2.28d.

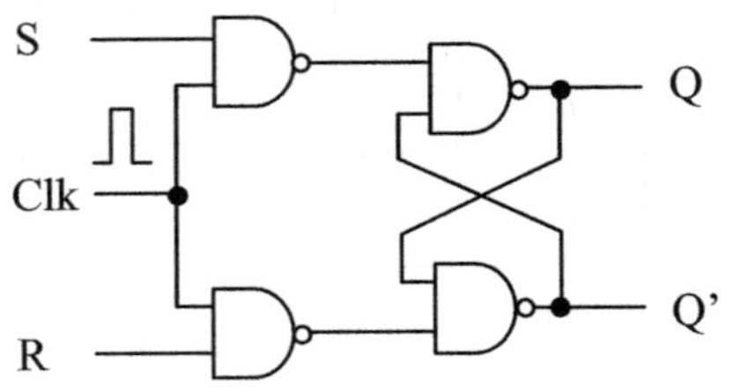

a Logic Diagram

S	R	Q(n)	Q(n+1)
0	0	0	0
0	0	1	1
0	1	0	0
0	1	1	0
1	0	0	1
1	0	1	1
1	1	0	Indeterminate
1	1	1	Indeterminate

b Characteristic Table

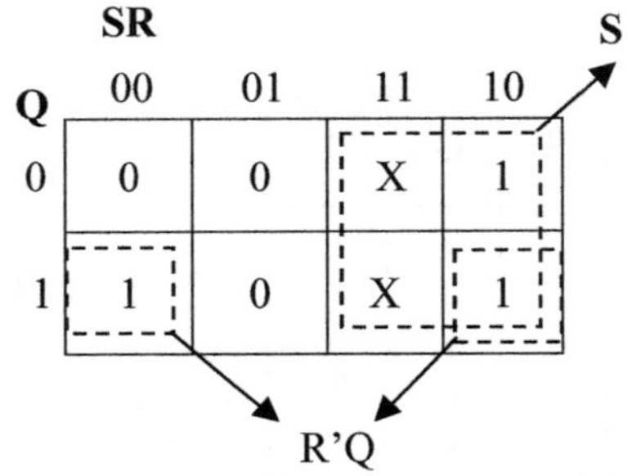

Q(n+1) = S+R'Q
SR = 0

c Characteristic Equation

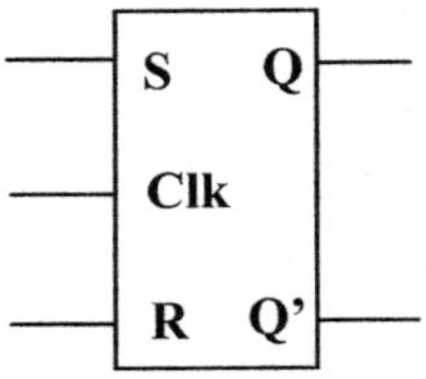

d Symbol

Fig. 2.28 Clocked RS flip-flop

JK Flip-flop

A negative edge triggered JK flip-flop in master–slave configuration is shown in Figure 2.29. The 74LS73 is a commercially available negative triggered dual flip-flop with individual JK inputs. In addition to the J and K inputs, the JK flip-flop has Clock and direct Reset inputs. Positive pulse triggered JK flip-flops such as 7473 and 74H73 are also available commercially. JK information is loaded into the master while the Clock is HIGH and transferred to the slave on the HIGH-to-LOW Clock transition. For these devices the J and K inputs should be stable while the Clock is HIGH for conventional operation. The J and K inputs must be stable one setup time prior to the HIGH-to-LOW Clock transition for predictable operation. The Reset (R_D') is an asynchronous active LOW input. When LOW, it overrides the Clock and the data inputs forcing the Q output LOW and the Q′ output HIGH. The truth table of JK flip-flop is shown in Table 2.16.

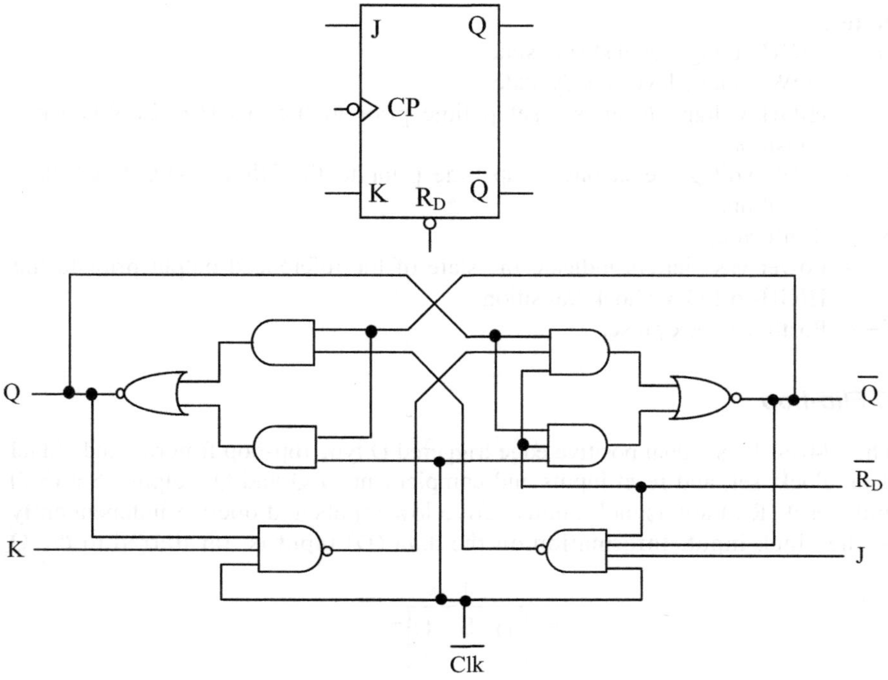

Fig. 2.29 Clocked JK flip-flop

Table 2.16 Truth table of JK flip-flop

Operating Mode	Inputs				Outputs	
	$\overline{R_D}$	$\overline{CP}$	J	K	Q	$\overline{Q}$
Asynchronous Reset (Clear)	L	X	X	X	L	H
Toggle	H	⎍	h	h	$\overline{q}$	q
Load '0' (Reset)	H	⎍	l	h	L	H
Load '1' (Set)	H	⎍	h	l	H	L
Hold 'no change'	H	⎍	l	l	q	$\overline{q}$

Notes:

H = HIGH voltage level steady state.
L = LOW voltage level steady state.
h = HIGH voltage level one setup time prior to the HIGH-to-LOWClock transition.
l = LOW voltage level one setup time prior to the HIGH-to-LOW Clock transition.
X = Don't care.
q = Lower case letters indicate the state of the referenced output prior to the HIGH-to-LOW Clock transition.
⎍ = Positive Clock pulse.

D Flip-flop

The "74LS74" is a dual positive edge triggered D type flip-flop featuring individual data, clock, set, and reset inputs and complementary Q and Q′ outputs. Set (S_D') and Reset (R_D') are asynchronous active low inputs and operate independently of the clock input. Information on the data (D) input is transferred to the Q

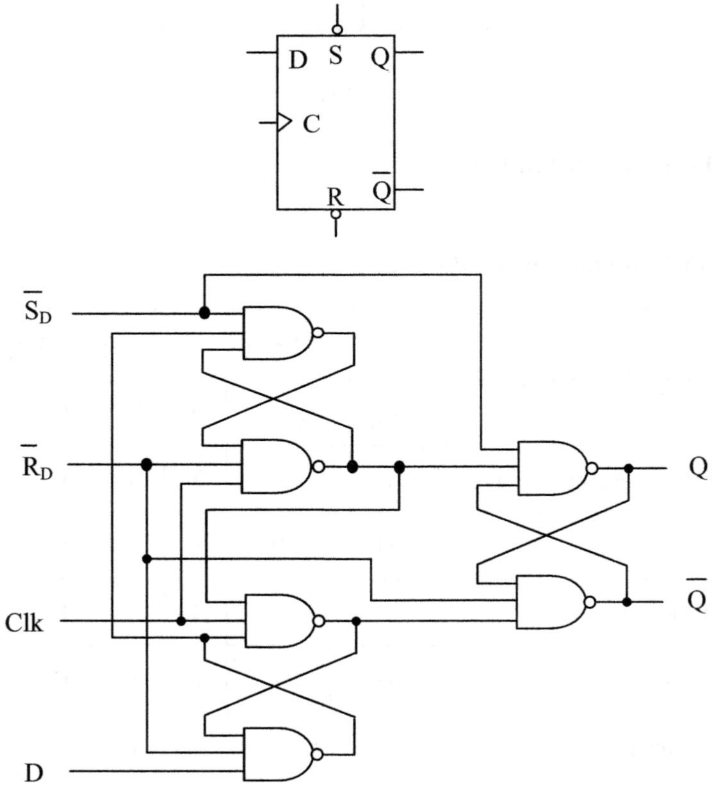

Fig. 2.30 D flip-flop

Table 2.17 Truth table of D flip-flop

Operating Mode	Inputs				Outputs	
	$\overline{S_D}$	$\overline{R_D}$	CP	D	Q	$\overline{Q}$
Asynchronous Set	L	H	X	X	H	L
Asynchronous Reset (Clear)	H	L	X	X	L	H
Undetermined	L	L	X	X	H	H
Load '1' (Set)	H	H	↑	h	H	L
Load '0' (Reset)	H	H	↑	l	L	H

H = HIGH voltage level steady state.
L = LOW voltage level steady state.
h = HIGH voltage level one setup time prior to the LOW-to-HIGH Clock transition.
l = LOW voltage level one setup time prior to the LOW-to-HIGH Clock transition.
X = Don't care.

output on the LOW-to-HIGH (rising edge) transition of the clock pulse. The D inputs must be stable one setup time prior to the LOW-to-HIGH clock transition for predictable operation. Although the clock input is level sensitive, the positive transition of the clock pulse between 0.8 V and 2.0 V levels should be equal to or less than the clock to output delay time for reliable operation. Figure 2.30 shows the logic diagram and the circuit diagram of a D flip-flop. The truth table of D flip-flop is presented in Table 2.17.

T Flip-flop

A toggle or T flip-flop is a single input alternative to a JK flip-flop. The T flip-flop can be realized by using an RS flip-flop and two AND gates to gate the inputs T and the clock as shown in the logic circuit diagram in Figure 2.31a. As the name implies, the output can be toggled by setting the T input as shown in the characteristic table in Figure 2.31b. On the other hand, the next state output Q(n + 1) follows the previous (state) output Q(n) if T is low. The characteristic equation may be obtained from K map as shown in Figure 2.31c.

We have covered four types of flip-flops, RS, JK, D, and T, whose symbols are shown in Figure 2.32. Table 2.18 shows the corresponding characteristics table. These tables may be re-arranged as what are known as excitation tables shown in Table 2.19. From these tables, we can get the desired input(s) corresponding to a set of present state Q(n) and next state Q(n + 1) conditions. This helps in the systematic design of the logic circuit diagram for an application. We will consider a number of applications using these flip-flops in Section 2.15.

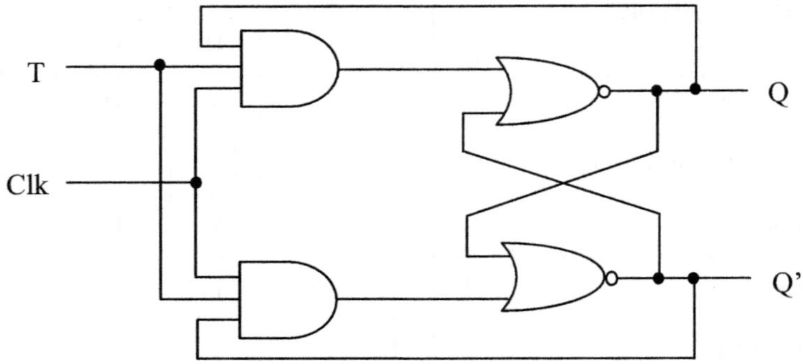

a Logic Diagram

T	Q(n)	Q(n+1)
0	0	0
0	1	1
1	0	1
1	1	0

b Characteristic Table

Q(n) \ T	0	1
0	0	1
1	1	0

$Q(n+1) = TQ(n)' + T'Q(n)$

c Characteristic Equation

Fig. 2.31 T flip-flop

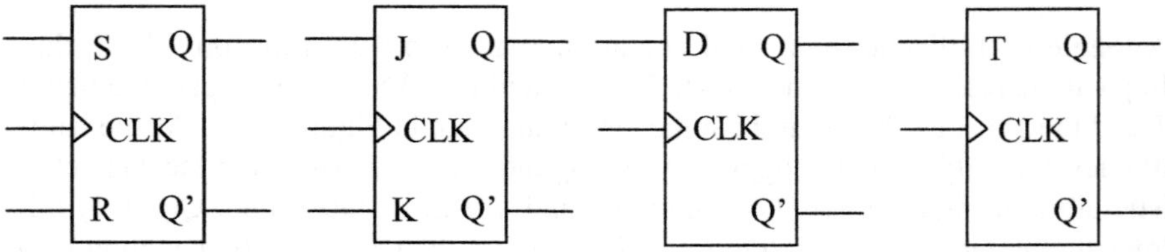

RS FF JK FF D FF T FF

Fig. 2.32 Symbols for flip-flops

Table 2.18 Characteristic tables of flip-flops

RS flip-flop

S	R	Q(n + 1)	Condition
0	0	Q(n)	No change
0	1	0	Reset
1	0	1	Set
1	1	?	Unpredictable

JK flip-flop

J	K	Q(n + 1)	Condition
0	0	Q(n)	No change
0	1	0	Reset
1	0	1	Set
1	1	Q′(n)	Complement

D flip-flop

D	Q(n + 1)	Condition
0	0	Reset
1	1	Set

T flip-flop

T	Q(n + 1)	Condition
0	Q(n)	No change
1	Q′(n)	Complement

Table 2.19 Excitation tables of flip-flops

RS FF

Q(n)	Q(n + 1)	S	R
0	0	0	X
0	1	1	0
1	0	0	1
1	1	X	0

JK FF

Q(n)	Q(n + 1)	J	K
0	0	0	X
0	1	1	X
1	0	X	1
1	1	X	0

D FF

Q(n)	Q(n + 1)	D
0	0	0
0	1	1
1	0	0
1	1	1

T FF

Q(n)	Q(n + 1)	T
0	0	0
0	1	1
1	0	1
1	1	0

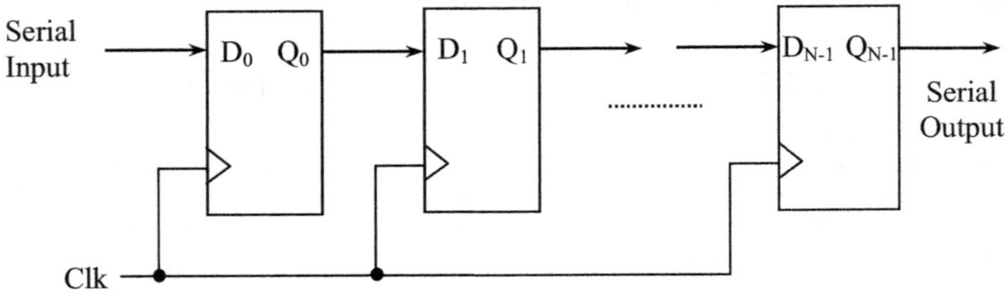

Fig. 2.33 N-bit shift register

Shift Registers

A flip-flop is also referred to as a register. A register can be an array of flip-flops that can store multi-bit binary information. A cascaded register array capable of shifting its stored information is called a shift register. The shift register can be either a right shift or the left shift type. The output of one flip-flop is connected to the input of the next flip-flop. All flip-flops have a common clock input that causes the shift of data from one stage to the next in the chain as shown in Figure 2.33.

2.12 Random Access Memory (RAM)

A RAM is a read-writeable memory and is a collection of storage cells like registers together with address decoding circuits. RAM cells are organized as 8 bits (or a byte), 16 bits (or a word), and so on per location. Each location can be accessed for information transfer to or from at random and hence the name random access memory. The address lines $A_{N-1} - A_0$ select one particular location. The RAM is specified by the number of address lines and the number of bits in each location as shown in Fig. 2.34. Each location is identified by a unique address, from 0 to 2^N-1, where N is the number of address lines. In order to read from or write into the RAM, the chip select CS must be asserted. The RAM may be written into with a data applied at the inputs $D_{W-1} - D_0$ followed by asserting the write pulse at 'W' pin. Similarly, the RAM may be read from a location followed by asserting the read pulse at 'R' pin. The read data manifests at $Q_{W-1} - Q_0$ output pins. The address must be applied at the address bus $A_{N-1} - A_0$ prior to a write or a read. The write/read pulse may be applied only after the address and data inputs are stable. Many vendors are available to supply a wide variety of RAMs. One example is presented in Appendix 5 of CD.

The numbers of locations in a RAM are specified in terms of Kilo (K), Mega (M), or Giga (G) bytes or words as the case may be. K is equal to 1024 or 2^{10}, M is equal to 1048576 or 2^{20}, and G is equal to 2^{30}. The access time of a RAM is the time required to either read or write it.

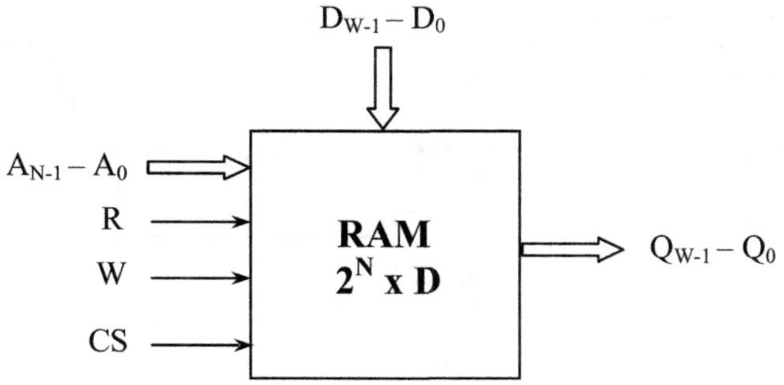

Fig. 2.34 Block diagram of a RAM

2.13 Clock Parameters and Skew

A typical clock waveform used in a digital system is shown in Figure 2.35. Usually, a clock waveform used in a system is a square wave. Occasionally, rectangular waveform may also be used. The time period T is the sum of the on time and off time. Rise time (t_r) and fall time (t_f) are measured respectively by the time taken for the waveform to rise from 10% to 90% and fall from 90% to 10% of the full amplitude as marked on the waveform. The frequency of the waveform is f = 1/T. A relative on and off times is defined by the duty cycle $(T_{ON}/(T_{ON} + T_{OFF})) \times 100\%$. A square wave has a duty cycle of 50%.

Skew refers to the rising edge (or the falling edge) of a clock arriving at different times at register clock inputs in a synchronous sequential circuit that ideally requires the same arrival time at various registers [7, 8]. Skew results because of interconnection delays, whether the design is realized using TTL circuits, processors, FPGAs, or ASICs. This is depicted in Figure 2.36. The figure shows the system clock, CLK, distributed to a number of parts of a digital system with the arrival delayed by small times. All the clock waveforms, CLOCK, CLOCK 1, ..., CLOCK N should be occurring at the same time ideally. Owing to different travel paths, clocks lag behind the original clock. At low clock speeds, skew causes no problem. At high frequency, close to the maximum clock frequency of operation for a circuit, skew causes problems since data to be registered arrives late and hence not likely to be stable. This results in missing the data. This may be minimized if not eliminated by distributing the clock spread in a radial or star like fashion from the clock source rather than connecting all the clock inputs of the registers in a cascade. FPGAs and ASICs have this type of clock distributions, thereby achieving high speeds over 100 MHz. The sequential circuits must meet certain conditions such as hold time and setup times, which is covered in the next section.

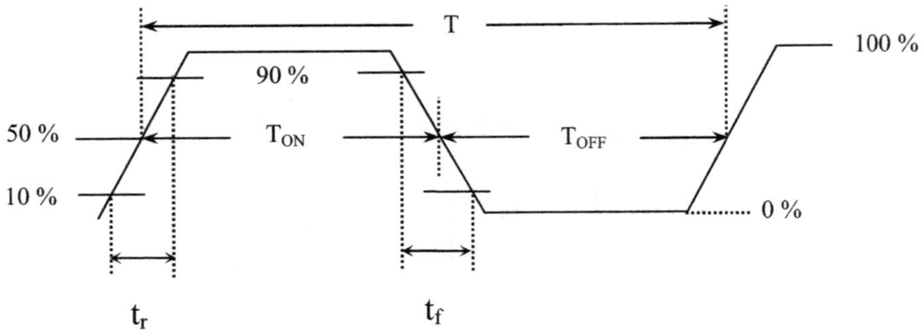

Fig. 2.35 Clock waveform

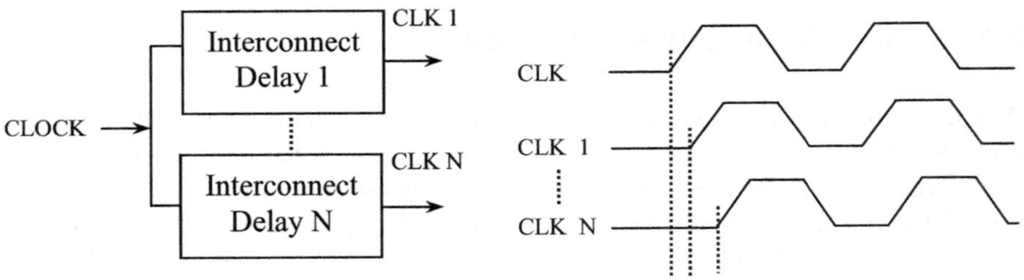

Fig. 2.36 Clock skew

2.14 Setup, Hold, and Propagation Delay Times in a Register

In general, a digital circuit comprises a series of combinational circuit followed by a register such as a D flip-flop. The system clock is directly fed to the clock input of the registers. Setup time, hold time, and propagation delay time are important parameters that need to be taken into account in a flip-flop based design [9]. They are as follows:

Setup time:

This refers to the time between the availability of a stable data input to a flip-flop device and the arrival of clock edge.

Hold time:

This refers to the time that the data input must continue to be stable after the arrival of the clock edge.

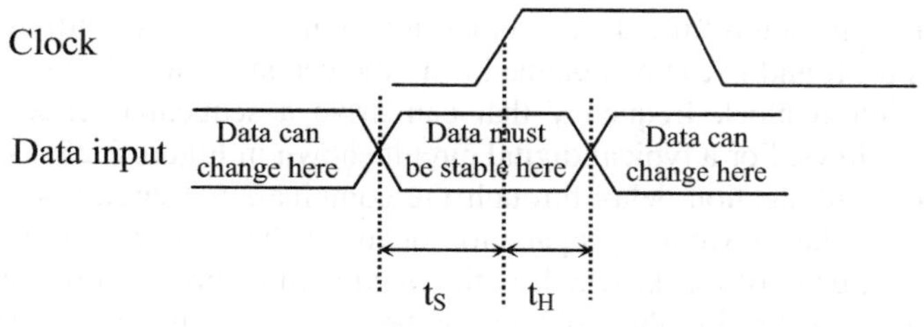

Fig. 2.37 Setup and hold times in a flip-flop

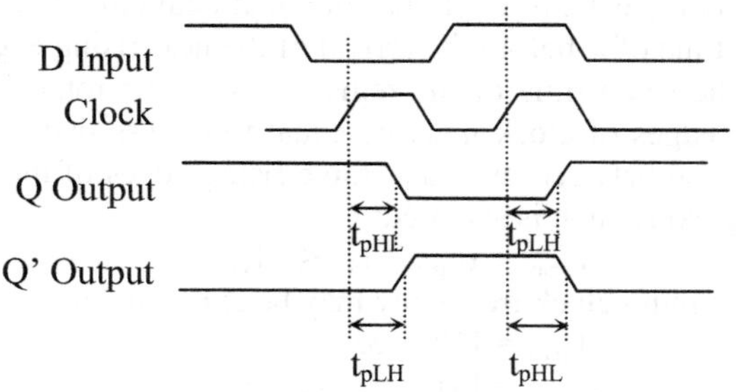

Fig. 2.38 Propagation delay time in a D flip-flop

Figure 2.37 illustrates the setup time (t_S) and hold time (t_H) with reference to the rising edge of clock. Instead of the rising edge, the falling edge may also be used.

Propagation delay:

This is the time between a clock edge (assuming a stable input signal) and the corresponding output across the register. The common propagation delay times are that between the clock edge to the Q and Q′ outputs as shown in Figure 2.38. The time 't_{pLH}' indicates the propagation delay time as the output switches from low to high, while 't_{pHL}' is the delay corresponding to the output switching from high to low. For example, t_{pLH} and t_{pHL} for the 74LS74 are 25 ns and 40 ns respectively, maximum. The longest propagation delay time specified by the manufacturer should be considered while calculating propagation delay paths.

2.14.1 Estimation of Maximum Clock Frequency for a Sequential Circuit

In the last section, we defined the setup (t_S) and hold (t_H) times. The sequential circuits we design need to satisfy these timings if it should work without any problem. 'D' must be stable during the time interval 't_S' before the active clock edge and for the interval 't_H' after the active clock edge, while it can change during other times. If 'D' changes during the forbidden interval, t_S and t_H, the flip-flop may malfunction. Setup time data to clock for high and low are different. Minimum values for t_S and t_H can be obtained from the vendors' data sheets.

The maximum clock frequency that can drive a sequential circuit may be obtained as follows. For a typical digital circuit shown in Figure 2.25, assume that the maximum propagation delay through the combinational circuit is t_{Cmax}. Also assume t_{pmax} as the maximum propagation delay of the register output reckoned from the rising edge of clock, which is the maximum of propagation delays: t_{pLH} and t_{pHL} of the register. Let the clock period be T_{clk}. When the clock (rising edge) arrives, the stable data at register input takes t_{pmax} time to manifest at the register's output, which is fed back to the input of the combinational circuit. This in turn takes t_{Cmax} time to propagate through the combinational circuit. This value must be stable for further time of t_S before the arrival of the next rising edge of the clock in order to satisfy the setup time of the register. Thus, the total time between two successive rising edges of clock must be equal to or less than $t_{pmax} + t_{Cmax} + t_S$. Since T_{clk} is the time between two successive rising edges of the clock, it follows that the following expression is satisfied:

$$t_{pmax} + t_{Cmax} + t_S \leq T_{clk}$$

Therefore the maximum clock frequency may be expressed as

$$F_{max} = 1/\,T_{clk} \text{ or}$$

$$F_{max} = 1/(t_{pmax} + t_{Cmax} + t_S)$$

For example, if $t_{pLH} = 25$ ns and $t_{pHL} = 40$ ns for a D flip-flop, then t_{pmax} is 40 ns. Further, the setup time of the D flip-flop 't_S (H)' is 25 ns. Assuming $t_{Cmax} = 15$ ns for the combinational circuit, the maximum clock frequency is 1/80 ns or 12.5 MHz. In real practice, the maximum frequency will be still lower since we have to include interconnection delay times in t_{pmax} as well as in t_{Cmax}. Thus, in 74LS series based designs, we have to be content with operating frequencies in the order of 10 MHz. Better speeds can be obtained by using 74 S series instead of 74LS series. In FPGAs, however, we can touch high operating frequencies of over 100 MHz.

We also need to pay attention to the parameter, hold time t_H, which is in the order of 5 ns. A hold-time violation and consequent malfunction of the circuit would occur if $t_{pmin} + t_{Cmin}$ is less than the hold time, where t_{pmin} and t_{Cmin} are the minimum of propagation delays of the register and the combination circuit respectively. Therefore, hold time is satisfied if: $t_{pmin} + t_{Cmin} \geq t_H$.

For standard flip-flops, t_{pmin} is greater than t_H and therefore, there is no danger of hold-time violation.

2.14.2 Metastability of Flip-flops

A flip-flop has two states '0' and '1' under normal operating conditions. The Q output can change from a '0' to '1' and vice versa without any problem provided the flip-flop setup and hold times are met. However, if they are not met, a third condition known as the metastable state may exist [7]. It is an undefined state with the voltage level halfway between a logical '0' and a logical '1'. This condition causes problems in the normal operation of a digital system. External inputs to a digital system containing flip-flops often occur asynchronously with respect to signals within the system including clock. As a result, the setup and hold times may be violated. Usual solution is to register the asynchronous inputs and feed the registered inputs to the system flip-flops. Metastable states are to be avoided at any cost for reliable operation of a system.

2.15 Digital System Design Using SSI/MSI Components

A synchronous sequential circuit is made up of combinational circuits and flip-flops or registers as shown in Figure 2.39. The circuit design comprises the selection of any of the flip-flops such as D, JK, or T and then finding a combinational circuit, which together with the flip-flops produces a circuit that meets the desired specifications. The number of states required in the design determines the number of flip-flops in the implementation. The combinational circuit is derived from the state table, which is akin to the excitation tables of flip-flops presented earlier in Table 2.19.

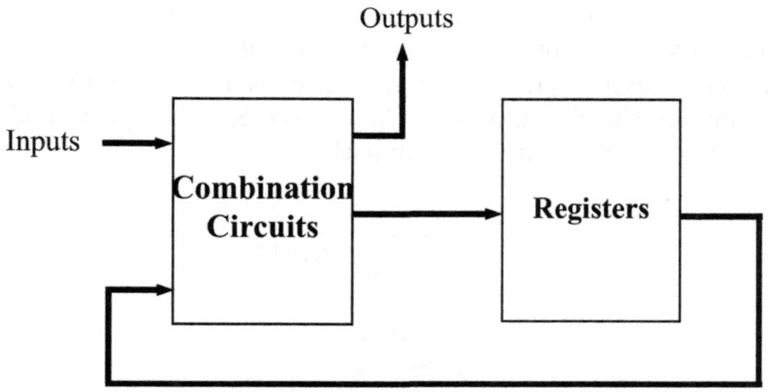

Fig. 2.39 Block diagram of a digital system

In the next few sub-sections, we will present the design of digital systems, which uses one or more of the components: JK, D, T flip-flops, and a ROM.

2.15.1 Two-bit Binary Counter Using JK Flip-flops

To start with, we will design a simple two-bit binary counter using JK flip-flops, whose state diagram is shown in Figure 2.40. As shown therein, the counter starts with '00' value and counts up by one every time the clock strikes so long as the external input 'I' is high. After the count value touches '11', it rolls back to '00' and continues with the same chain of events. At any point of time, when the input goes low, the count value freezes. Once the input is re-applied, the counting continues from where it was held previously. Thus we have a counter that can be controlled by an external input, either to count or to hold. The next step in the design is to use the excitation table of the flip-flop and form what is known as the state table as shown in Table 2.20.

Since the counter size is two bits, we need just two JK flip-flops for realizing the counter. Let us label the flip-flop inputs as J_A, K_A and J_B, K_B corresponding to the two flip-flops A and B. As mentioned before, the design boils down to working out a combinational circuit for the flip-flop inputs. This can be easily arrived at from the state table, wherein the first three columns are formed using the external input 'I' and the present states 'A' and 'B' of the two flip-flops. The last two signals A and B are the non-inverted outputs of the two flip-flops. With the arrival of the clock, these outputs change to 'A^+' and 'B^+' and are referred to as the next state as shown in the next two columns. The desired flip-flop inputs J_A, K_A and J_B, K_B are tabulated thereafter in the state table. These correspond to the outputs of the combinational circuits we are working out.

We will fill the state table now using the excitation table for the JK flip-flop shown in Table 2.19. We shall consider a couple of entries. In the first row entry of Table 2.20 for the flip-flop A, we have a transition from '0' in the present state to '0' in the next state. Referring to Table 2.19, we find that for this transition from 0 to 0, the flip-flop inputs must be J = 0 and K = X. J_A and K_A are, therefore, 0 and X respectively. Similar explanations hold good for 0 and X entries for the first row J_B and K_B. As another example, consider the row I A B = 1 0 1 for the flip-flop B. The present state and the next state for B are 1 and 0 respectively. The corresponding flip-flop inputs are J_B = X and K_B = 1, again obtained from Table 2.19. All other entries are filled similarly.

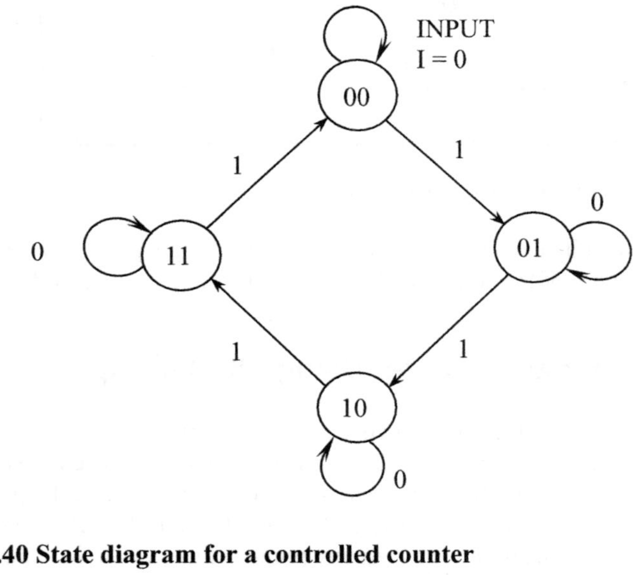

Fig. 2.40 State diagram for a controlled counter

Table 2.20 State table for the controlled two-bits binary counter

External input	Flip-flops, present state		Flip-flops, next state		Flip-flops, inputs			
I	A	B	A^+	B^+	J_A	K_A	J_B	K_B
0	0	0	0	0	0	X	0	X
0	0	1	0	1	0	X	X	0
0	1	0	1	0	X	0	0	X
0	1	1	1	1	X	0	X	0
1	0	0	0	0	0	X	0	X
1	0	1	1	0	1	X	X	1
1	1	0	1	1	X	0	1	X
1	1	1	0	0	X	1	X	1

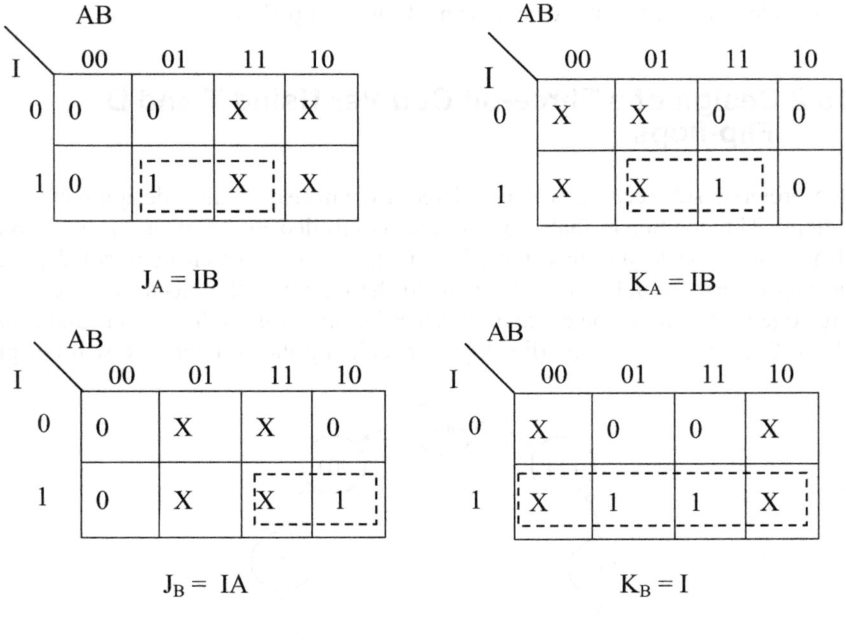

Fig. 2.41 K maps for JK flip-flops' inputs

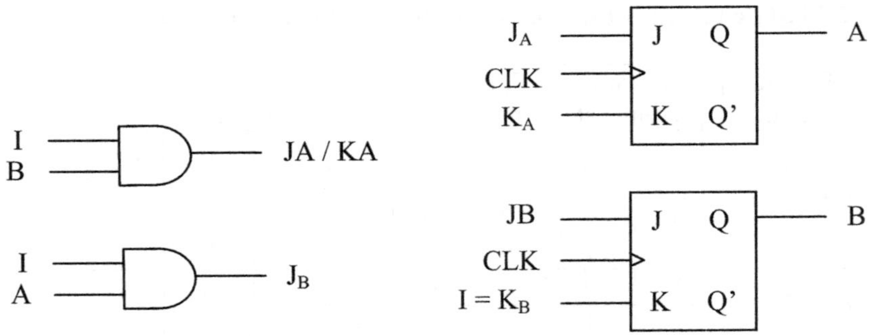

Fig. 2.42 Logic circuit diagram of the controlled 2-bits binary counter

The information from the state table is transferred into the K maps as shown in Figure 2.41. The simplified flip-flop inputs derived are as follows:

$J_A = IB$ $\quad$ $K_A = IB$

$J_B = IA$ $\quad$ $K_B = I$

Using the derived flip-flop inputs, the logic diagram is drawn as shown in Figure 2.42. It consists of two flip-flops and two AND gates. In the next sub-section, we will consider another counter design using T and D flip-flops.

2.15.2 Design of a Three-bit Counter Using T and D Flip-flops

In the previous sub-section, we considered a controlled counter design using JK flip-flops. The counter in that example was controlled by an external input. We will now see how to design a three-bit binary counter, which does not depend upon any external input. The only input to the circuit is the clock. With e very rising edge of the clock, the counter advances by one from '000' to '111' and rolls back to '000'. We need three flip-flops for realizing the counter. We will design

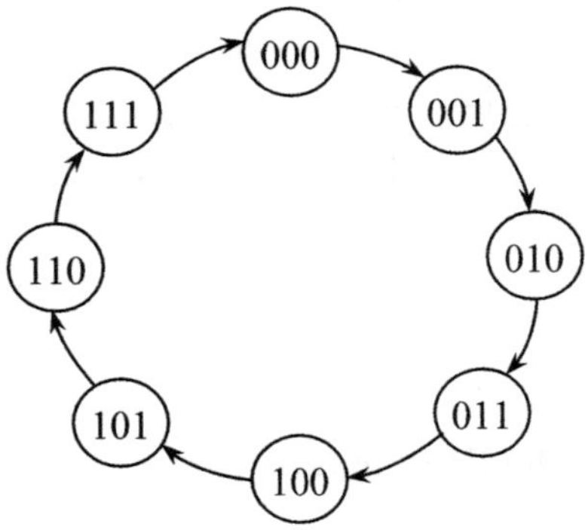

Fig. 2.43 State diagram of a three-bit binary counter

the three-bit counter using T flip-flops first, followed by D flip-flops. The count sequence is shown in the state diagram, Figure 2.43.

The next state of a counter depends entirely on its present state, and the state transition occurs at every rising edge of the clock. The state table for the three-bit counter using T flip-flops as well as D flip-flops is shown in Table 2.21. We use three flip-flops whose outputs are A, B, and C, which also indicate the present states. The present and the next states are filled as in the state table. As shown in the previous design using JK flip-flops, we need to deduce the flip-flop inputs T_A, T_B, T_C, and D_A, D_B, D_C for various possibilities of the present states as shown in the state table. Once again, we use the excitation tables shown in Table 2.19 for the T and D flip-flops. From the state table, for AA^+ = 00 and 11, the corresponding T input is '0'. For AA^+ = 01 and 10 (which indicate the toggling of 0 to 1 and vice versa), T = 1. Using this information, all the T flip-flop inputs in the state table can be filled.

It is much simpler to fill the D inputs since they are exactly the same as the corresponding next states. For example, in the state table, A^+ and D_A columns are identical. Similarly, B^+ and D_B columns and C^+ and D_C columns are identical. K maps may be obtained from the state table. Figure 2.44 shows the K maps and optimized Boolean expressions for the T flip-flop inputs. The logic circuit diagram drawn using these expressions is shown in Figure 2.45. Similarly, the K maps and the logic circuit diagram for D flip-flop realization of the three-bit counter are shown in Figures 2.46 and 2.47 respectively. T flip-flop based design yields much simpler circuit than the D flip-flop based design. However, D flip-flop based designs are more popular in industries than JK and T flip-flop based designs. 74LS74 IC houses two numbers of D flip-flops in a single 14-pin package and may be used together with other gates.

Table 2.21 State table for a three-bit counter using T and D flip-flops

Present state			Next state			T Flip-flops, inputs			D Flip-flops, inputs		
A	B	C	A^+	B^+	C^+	T_A	T_B	T_C	D_A	D_B	D_C
0	0	0	0	0	1	0	0	1	0	0	1
0	0	1	0	1	0	0	1	1	0	1	0
0	1	0	0	1	1	0	0	1	0	1	1
0	1	1	1	0	0	1	1	1	1	0	0
1	0	0	1	0	1	0	0	1	1	0	1
1	0	1	1	1	0	0	1	1	1	1	0
1	1	0	1	1	1	0	0	1	1	1	1
1	1	1	0	0	0	1	1	1	0	0	0

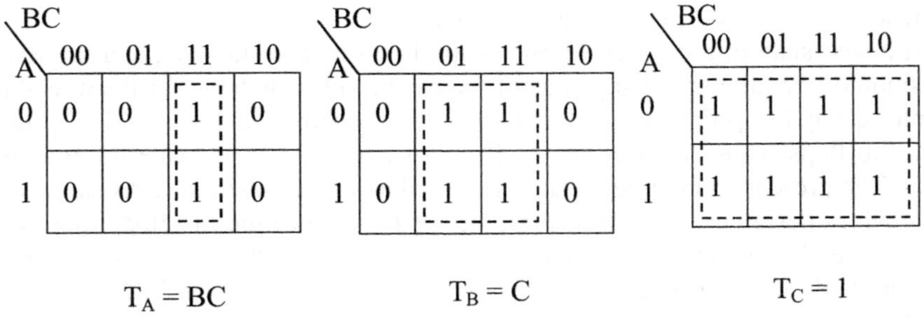

Fig. 2.44 K maps for a three-bit binary counter using T flip-flops

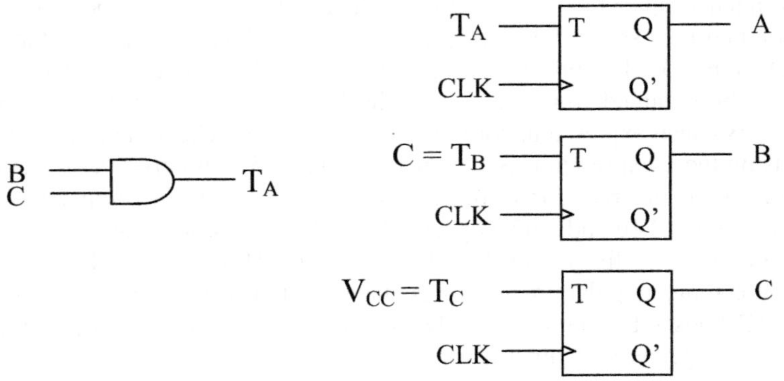

Fig. 2.45 Logic circuit diagram of the three-bit binary counter using Tflip-flops

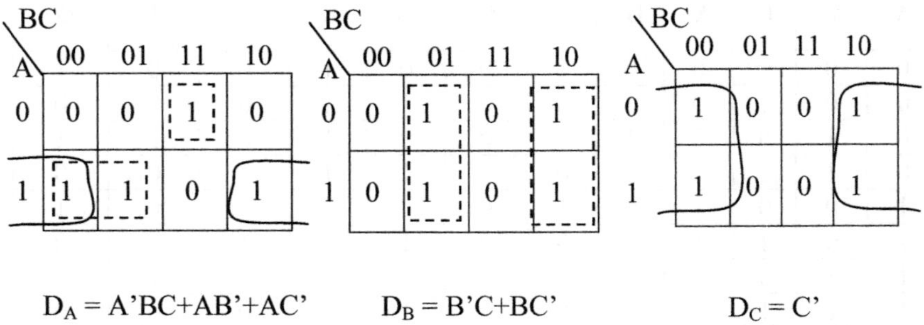

Fig. 2.46 K maps for the three-bit binary counter using D flip-flops

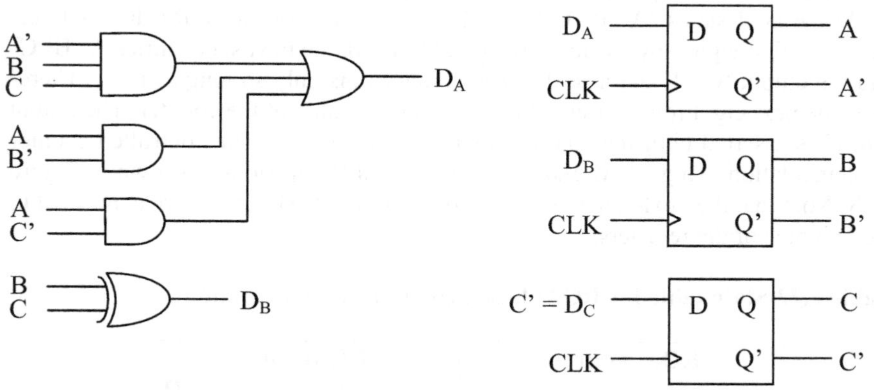

Fig. 2.47 Logic circuit diagram of the three-bit binary counter using D flip-flops

2.15.3 Controlled Three-bit Binary Counter Using ROM and Registers

A digital system comprises a combinational circuit working in tandem with flip-flops or registers. In the previous sections, we used different kinds of registers, namely, JK, T, and D. For the combinational circuit part of the design, we used conventional gates. In lieu of these gates, the ROM can be also be used to implement the combinational circuit part and the flip-flops for the sequential part. The number of inputs to the ROM is equal to the number of flip-flops and the external inputs in the system. The number of outputs of the ROM is equal to the number of flip-flops and the number of external outputs put together. In the previous design, we considered three bit binary counter without any control. In the present design, we will add an external input and thereby run the counter in a controlled manner. If the control input 'I' is active, the counter advances by one at every rising edge of the clock. Otherwise, the flip-flop outputs are cleared and remain in that state until the external input is active. We will also have an output 'OUT' to indicate the active state of the counter.

The number of flip-flops, say D type, required in this design is three with outputs A, B, and C, which also indicate the present state. The next states of these flip-flops are respectively A^+, B^+, and C^+. The state table containing this information is shown in Table 2.22. As mentioned before, a ROM can be used to realize the combinational circuit part of the design. Since we have four inputs I, A, B, and C for the combinational circuit, it follows that we need a ROM with four address inputs A3, A2, A1, and A0. The ROM must have three outputs for the three next states A^+, B^+, and C^+. In addition, it must have one more bit for 'OUT' signal. These ROM data outputs are respectively labeled as D_3 D_2, D_1, and D_0. Therefore, the ROM has four inputs and four outputs and its size is 16×4. In the state table, the input 'I' and the present states 'ABC' together specify the address of ROM while the next states 'A^+ B^+ C^+' and the 'OUT' specify the ROM data outputs. Since the first eight row entries correspond to I = 0, the next state entries A^+ B^+ C^+ and the output 'OUT' are also 0's. The counter runs only so long as I = 1. Therefore, the next eight row entries reflect the running value of the counter. The output 'OUT' is asserted until the control input 'I' is deasserted. The controlled counter implementation using ROM and registers such as D flip-flops is shown in Figure 2.48. Note that the next-states A^+, B^+, and C^+ in the ROM outputs are connected to the D inputs of the registers.

Table 2.22 State table for ROM based counter implementation

ROM address				ROM content			
A_3	A_2	A_1	A_0	D_3	D_2	D_1	D_0
Input	Present state			Next state			
I	A	B	C	A^+	B^+	C^+	OUT
0	0	0	0	0	0	0	0
0	0	0	1	0	0	0	0
0	0	1	0	0	0	0	0
0	0	1	1	0	0	0	0
0	1	0	0	0	0	0	0
0	1	0	1	0	0	0	0
0	1	1	0	0	0	0	0
0	1	1	1	0	0	0	0
1	0	0	0	0	0	1	1
1	0	0	1	0	1	0	1
1	0	1	0	0	1	1	1
1	0	1	1	1	0	0	1
1	1	0	0	1	0	1	1
1	1	0	1	1	1	0	1
1	1	1	0	1	1	1	1
1	1	1	1	0	0	0	1

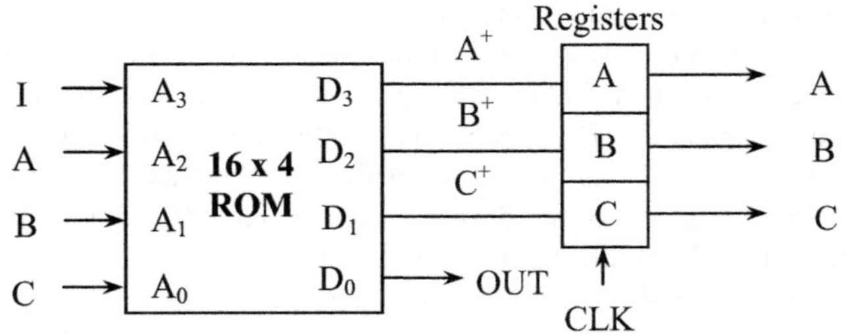

Fig. 2.48 Digital system design using registers and a ROM

Although counters have been used to illustrate the designs in earlier sections, any other sequential system can be implemented. In all the above designs using various flip-flops, ROM, etc., power on reset circuit needs to be added in order to make a practical working circuit. Later on, in the PAL based design, we shall include the reset circuit. Before that, we shall learn what algorithmic state machine is.

2.16 Algorithmic State Machine

In the previous designs, we have used state diagrams to aid our design. An Algorithmic State Machine (ASM) serves the same purpose as the state diagram. In the state diagram approach, usually one input decides the transition from one state to another. If the numbers of inputs increase, the state diagram becomes very complex. In such cases, an ASM chart is a better alternative. An ASM chart resembles a flow chart, but it is quite different functionally. It is a convenient way to specify the hardware sequence of steps and decision paths for an algorithm. The ASM chart is a representation of a state machine, which is another name for a sequential circuit. The finite state machine (FSM) is yet another name for the state machine.

Various building blocks of an ASM chart are shown in Figures 2.49 to 2.51 along with examples. Figure 2.49 shows a rectangular state box, which has the same function as a circle in the state diagram, yet more informative. It provides not only the state code, but also a name for the state which is easy to track and troubleshoot. An example furnished in Figure 2.49b makes this clear. Looking at the code '0000', we can infer that there is a maximum of 16 states and that this state can be referred to by user-friendly name 'INITIALIZE' rather than deal with binary numbers. In addition, we can include one or more outputs and operations such as LOAD and Z = 0 respectively. Any other outputs or register signals of the system are automatically cleared since they do not find a place within the current state box. They shall, however, find place elsewhere.

The next figure shows the condition box and a conditional output. The input to the condition box arrives from a state box, whereas its outputs branch off to a conditional output or a state box. Usually, the conditions are twofold: '0' or '1' or T (for True) or F (for false), although more than one bit condition may be used. More elegant solution for multi-bit condition is to daisy chain the individual bit condition boxes as shown in the example in Figure 2.51. The conditional output or operation derives its input from a decision box and its output gets connected to the next state box as shown in Figure 2.50. Figure 2.51 shows a part of an ASM chart illustrating the use of different blocks. The first state is S0 = 0000, which activates the shift register 'SR' in two ways depending upon the two input signals 'CS' (for chip select) and 'LD' (for parallel load). If CS = 0, SR is cleared and with the arrival of the next clock pulse, the state changes to S1 (0001). Otherwise, if CS = 1 and LD = 1, SR is preset to AA hexadecimal and the state changes to S2 (0010) at the next clock pulse. On the other hand, if CS = 1 and LD = 0, the state S0 does not change. Thus, an FSM can be very efficiently represented by an ASM.

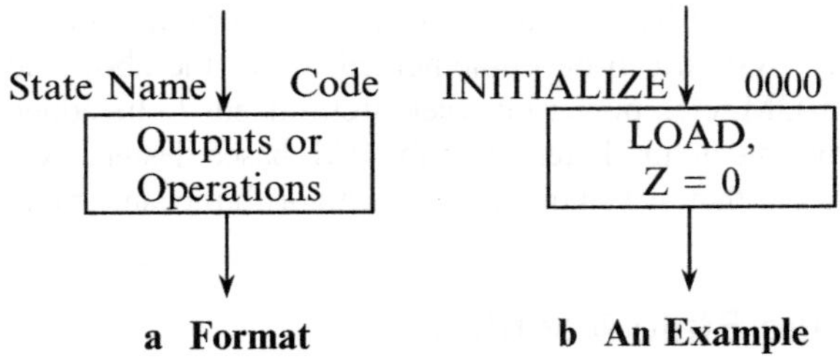

Fig. 2.49 State box of an ASM chart

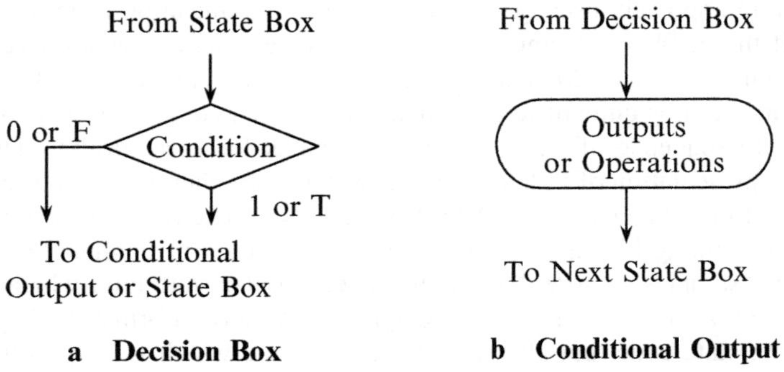

Fig. 2.50 Decision box and conditional output of an ASM chart

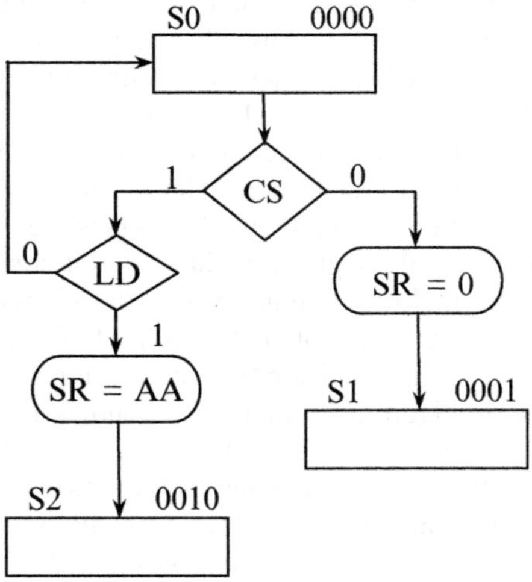

Fig. 2.51 An example of combined decision boxes and conditional outputs

2.17 Digital System Design Using ASM Chart and PAL

We designed a number of small digital systems using the conventional state diagram, flip-flops and ROM in Section 2.15. These designs become cumbersome for complex designs. For such designs, it is far better to use ASM charts and PALs, which we covered in previous sections. A digital system may be viewed as two building blocks: the control logic and the data processor as shown in Figure 2.52. The control logic receives external inputs and generates control signals for the data processor, thus coordinating all the activities in the system. The data processor performs data processing tasks such as add, subtract, multiply, compare, shift, logic, etc. and communicates its status to the control logic, which in turn generates signals for sequencing the operations in the data processor. The data processor receives one or more data inputs and outputs the processed data. All activities in the system are synchronized to the system clock. The control sequence and data processing of a digital system is completely specified by the ASM chart. This will be made clear by a couple of design examples in the next two sub-sections. To start with, we will see how to generate a single clean pulse from a push-button switch using ASM chart. In the next example, we will design a vending machine that caters to soft drinks.

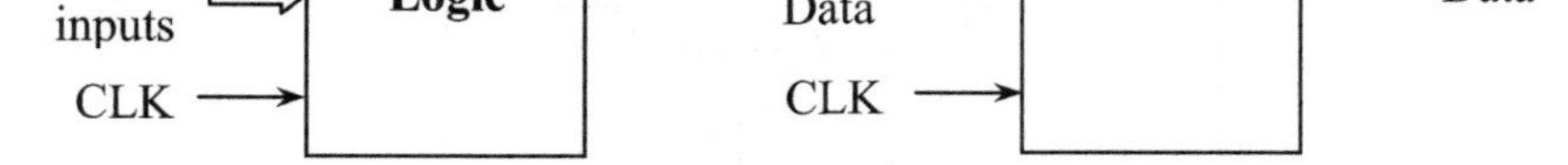

Fig. 2.52 Block diagram of a digital system viewed as control and data processors

2.17.1 Single Pulser Using ASM Chart

A push-button switch can be used to turn on a machine or turn it off. When a push-button switch is pressed, it does not produce a clean contact. Instead, the contact bounces back, makes contact again only to break the contact again. This goes on for a while and finally settles down making a firm contact. If this is directly fed to a digital circuit, multiple pulses will be detected and results in malfunctioning of the system. For instance, if we wish to advance a digital counter by one every time the push button is pressed, we will be annoyed to see multiple counting taking place for a single push of the button. This problem can be eliminated by connecting the push-button switch to RS flip-flops, which debounce the switch and provides a clean single pulse. We still have a problem here. Since a human operator is slow to react and the system clock frequency is high, a single pressing of the switch, even after debouncing, is likely to be recognized as multiple pressings. We must, therefore, develop a scheme such that the digital system we design processes a push-button depression only once. The digital circuit for accomplishing this task may be called a single pulser.

Let us assume that the RS flip-flop debouncer produces logic high when the push button is pressed and logic low in the un-pressed condition. We need to design a circuit to sense the depression of the push button and assert an output signal for one clock pulse duration. The system should not output any other pulse until the operator has released the push button and asserted again. The debounced push-button signal (let us call it 'Deb_PB') is asynchronous since the switch can be pressed at any point of time. We can synchronize this asynchronous signal using a clocked D flip-flop. We shall call this synchronized signal as 'Synch_PB' and the output of the signal pulser circuit as 'Single_Pulse'. The 'Synch_PB' goes high when the push button is pressed.

Now we are ready to draw the ASM chart. To start with, the system needs to wait for 'Synch_PB' to go high. This can be done by starting with a state (code '0') followed by a decision box, which checks whether 'Synch_PB' is high. The state box can be given an appropriate name such as 'DETECT', meaning that it is a state, where the pressing of the push button is detected. If 'Synch_PB' is low, the FSM must remain in the same state DETECT. On the other hand, if 'Synch_PB' goes high, the output "Single_Pulse" must also go high so long as the

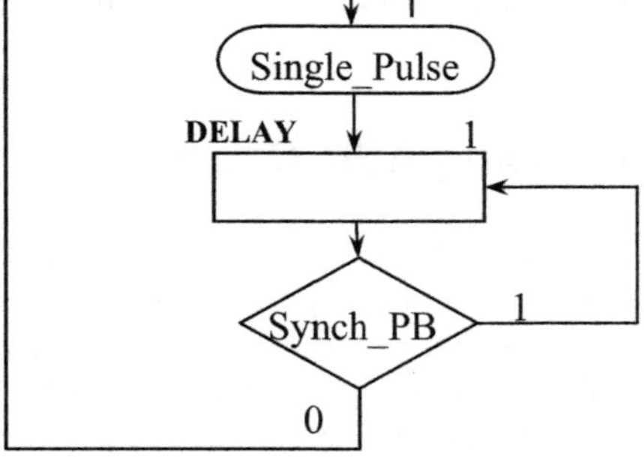

Fig. 2.53 ASM chart of the single pulser

state machine is in DETECT state. This is precisely shown in the ASM chart in Figure 2.53. The 'Single_Pulse' is obviously a conditional output. When the clock strikes, the state changes to 'DELAY', which can be assigned a code '1'. Since the time of transition from '0' state to '1' state is of duration one clock, it follows that the 'Single_Pulse' duration is also one clock period. Once again in the DELAY state, the 'Synch_PB' signal is checked. If 'Synch_PB' is high, the FSM remains in the same state DELAY since this implies that the push button is not yet released. On the other hand, if "Synch_PB" goes low subsequently, the state reverts to DETECT looking for fresh switch depression. Note that the debouncing of the push-button switch cannot be dispensed with since the data synchronizer will output multiple pulses for a single key press if the push-button switch is directly connected.

The next step is to draw a state table, which may be referred to as an ASM table as shown in Table 2.23. The present state, the next state, and the output(s) are exactly similar to that of state table used in earlier designs. The first two columns are unique to the ASM chart. In the first column, we enter the state name for easy identification. The second column known as 'Qualifier' indicates all the Boolean expressions of possible conditions of inputs in a state. The entries of the table may be directly made from the ASM chart. Let us consider a couple of row entries as examples. In the 'DETECT' state, the second row is filled as follows. In this state, Synch_PB = 1, therefore the qualifier is indicated as the active high signal, 'Synch_PB'. The code of the present state is '0' and the next state for this path is '1'. Also, the 'Single_Pulse' output is '1'. All these are entered in the respective columns in the second row. As the next example, we will take the third row. Looking at the ASM chart, the second state name is 'DELAY' and its code is '1'. In this state for one of the conditions, Synch_PB = 0 and, therefore, the qualifier is indicated

Table 2.23 ASM table of single pulser

State name	Qualifier	Present state	Next state	Single_Pulse
Detect	Synch_PB′	0	0	0
	Synch_PB	0	1	1
Delay	Synch_PB′	1	0	0
	Synch_PB	1	1	0

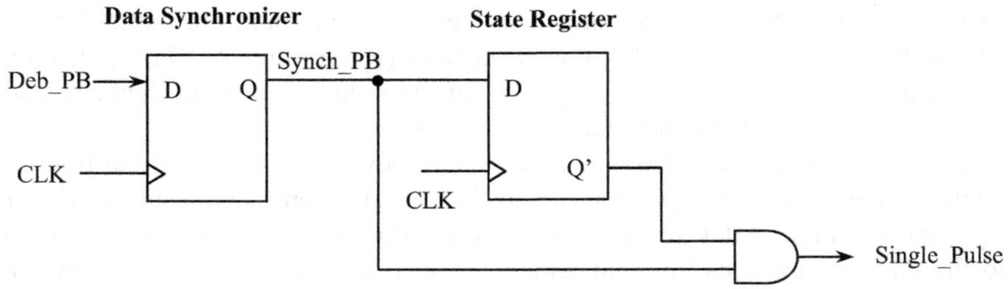

Fig. 2.54 Circuit diagram of the single pulser

as the active low signal: Synch_PB′ in the state table. The output does not get listed and hence it is filled as '0'. On similar lines, other rows of the ASM table can be filled directly. One must not forget to include any of the conditional branches.

The qualifier and the present state(s) form the inputs and the next state(s) and output(s) form the outputs for which we need to deduce the Boolean expressions. Instead of using K maps, mere inspection of the ASM table reveals the Boolean expressions of next state(s) and output(s). Let us assume that we use D flip-flop for the register. Since there are only two states '0' and '1', we need just one D flip-flop, whose input is the next state. Corresponding to the next state = 1, since the present states are different being '0' and '1', we drop the present state. However, the qualifier for these two entries is 'Synch_PB'. Therefore, the next state or the D flip-flop input is Synch_PB. The 'Single_Pulse' output is '1' only for the qualifier, 'Synch_PB' and the present state = 0 (which means the D flip-flop output Q′). Therefore, Single_Pulse = Synch_PB . Q′, which can be realized using an AND gate as shown in Figure 2.54. The circuit diagram of the single pulser shows two D flip-flops, the first one being a data synchronizer, which accepts the debounced push-button switch signal and generates 'Synch_PB' signal with the subsequent rising edge of clock, CLK. The second D flip-flop is the state register.

2.17.2 Design of a Vending Machine Using PAL

We will design a PAL based controller for a vending machine, which caters up to five different types of items such as the canned soft drinks or vegetable drinks. This can be extended to any number or types of items. For this design, we make the following assumptions:

1. A separate mechanism that accepts coin and the type of item from the user is available. This mechanism verifies the correct coin insertion. Otherwise, it ejects the coin.
2. A push-button type BCD switch selects the desired drink. Valid BCD switch setting (for s_2 s_1 s_0 outputs) is 0–4. The settings 5, 6, and 7 roll back to 0, 1, and 2 respectively. This setting selects the desired can.

In Section 2.10.3, we presented the PAL, which can realize any Boolean expression. The commercially available PALs can not only realize combination circuits, but also sequential circuits using D flip-flops. A schematic circuit diagram of the vending system is shown in Figure 2.55. The controller for the vending machine is housed in the PAL. As shown in the circuit diagram, the inputs for the PAL are the system clock (CLK), power-on-reset (RES), three bits (s_2–s_0) from a BCD switch, coin accepted (CA) signal from the coin acceptor mechanism, and outputs RDY′, OS1′–OS5′, and CCA to clear coin accept.

The RES pulse is generated when the system power is applied. At the time of switching power on, the capacitor is not charged and hence the Schmitt trigger output RES is high. After about 1.5 s, the capacitor voltage rises to logical high and the RES output goes low and remains low so long as the power is applied. Thus the RES pulse is generated, which clears D flip-flops inside PAL. Inner circuit details will be presented later on. The clock can be realized by using 555 IC. The clock frequency used is low since we need to activate a solenoid for outputting a can. A single digit BCD switch is used to select the type of drink by setting a code using the push buttons + and –. Since the design allows only 5 different types of can drinks, 0 to 4 setting will be enough. However, we will allow 5–7 setting for rolling back to 0–2 and 8, 9 setting for 0, 1 respectively. The coin acceptor accepts coin only if RDY lamp is on. When it accepts the coin, it asserts the coin accept 'CA' signal, which is fed to the PAL controller. The controller activates one of the five output solenoids OS1 to OS5 dispensing the user desired can set in the BCD switch. Thereafter PAL asserts the CCA signal for the coin acceptor to clear the CA signal, without which the can dispenser would output multiple cans instead of a single can. Figure 2.55 also depicts the sequence of getting the desired drink. All the outputs are active low to suit the PAL outputs.

Figure 2.56 shows the state diagram (also called as the state graph) for the vending machine. The FSM has one initial state S_0 and five different output states S_1 to S_5. In S_0 state, RDY lamp is activated to indicate that the controller is ready to dispense the cans. The state machine remains in this state so long as the coin is not accepted (CA = 0). Once the coin is accepted, the machine checks the user selected can type (s = 0–9) and the control branches to the corresponding state, where the relevant solenoid is activated as shown in the state graph. The states are assigned one-hot codes, which have only one '1' entry with the rest being 0's, in order

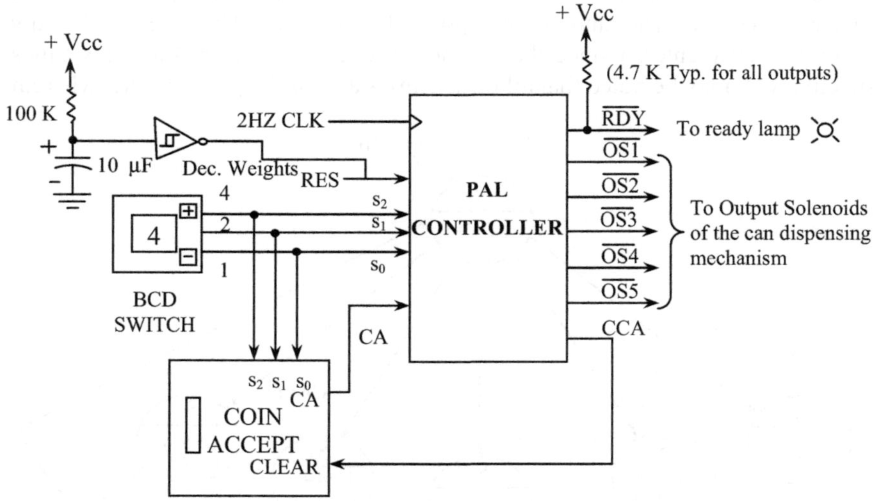

Usage:

1. Wait for RDY lamp to switch ON.
2. Set BCD switch to the desired value.
3. Insert the correct coin and collect the desired can.

Fig. 2.55 Circuit diagram of vending machine using PAL

to optimize the circuit as will be seen in the chapter on Synthesis. These assignments along with basic control signal definitions are as follows:

State	One-hot Assignment	
S_0	1 0 0 0 0 0	$s = s_2\ s_1\ s_0$ => BCD switch to select the desired item
S_1	0 1 0 0 0 0	CA = Coin Accepted
S_2	0 0 1 0 0 0	RDY = System is Ready to Accept Coin
S_3	0 0 0 1 0 0	RES = Power On Reset Signal
S_4	0 0 0 0 1 0	CCA = Clear Coin Accept
S_5	0 0 0 0 0 1	

The next step is to draw a state table from the state diagram presented earlier. Since we used one-hot assignment for the six states, we need six registers for the implementation. The present states and the next states are respectively labeled as ABCDEF and $A^+B^+C^+D^+E^+F^+$. We infer the following from the state diagram. The inputs are CA and $s_2\ s_1\ s_0$ and the outputs are RDY, CCA, OS1–OS5. Table 2.24 shows the state table incorporating the above details. We have two blocks of entries; one for the initial state S_0 and the other for each of the states from S_1 to S_5. For the input CA = 0, without regard to the value set in the BCD switch, the next state continues to be S_0. The RDY output is '1', while other outputs are '0'. This information obtained from the state diagram is entered in the first row of the table. The next eight rows are filled as follows. Inspecting the state diagram, we need to cover for the case, CA = 1. There are five possibilities of next states: S_1 to S_5 corresponding to the BCD switch settings $s_2\ s_1\ s_0$. The settings $s_2\ s_1\ s_0 =$ 5–7 are fold backs as explained earlier and are incorporated in the table. The cases s = 8 and 9 are not explicitly entered since these conditions are inherent in 0 and 1 settings respectively. It may be noted that all these entries are for the present state, wherein

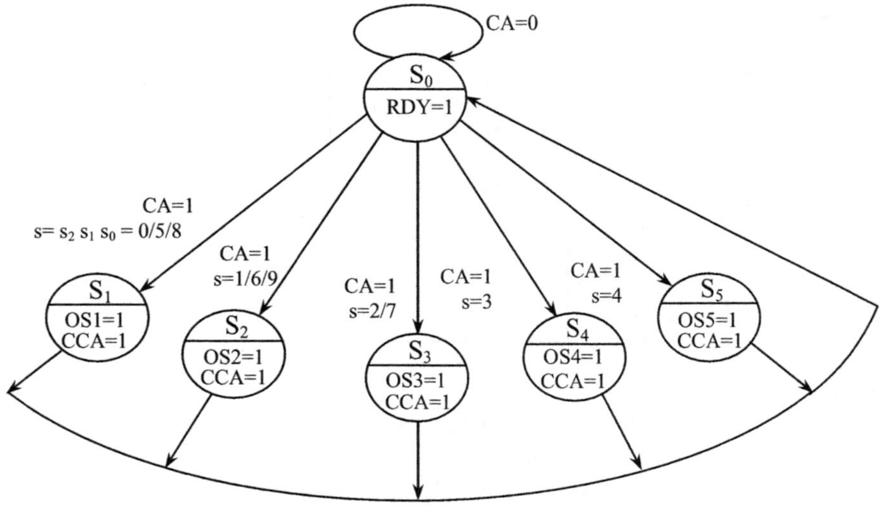

Fig. 2.56 State graph for vending machine

RDY output alone is high, while all other outputs are 0's. These details are entered in the table accordingly.

Coming to the entries for the present state S_1, the next state is S_0 and the two outputs to be turned on are CCA and OS1, regardless of CA and 's' settings. Accordingly, we make these entries for the tenth row. It may be noted that all other outputs OS2–OS5 are not listed in the state diagram and hence we fill 0's for these outputs. Similarly all other present states, S_2 to S_5 are filled.

Boolean expressions for the next states and outputs may be deduced by mere inspection of the state table. We shall illustrate a couple of examples. Let us consider the expression for the next state A^+. The output is '1' for the condition ABCDEF = 100000 and CA = 0, from which we get the first term: A B′ C′ D′ E′ F′ CA′ by inverting the corresponding signal for '0' value. For '1' value we retain the signal as it is. The output is also '1' for each of the present states S_1 to S_5. Therefore, we get a second term (B + C + D + E + F). RES′ = CCA. The signal RES′ is included in order to activate this term only for normal working when the reset signal is not present. We need one more term RES to force A^+ value to '1' when reset is applied. Combining all the three terms, we get the expression for A^+.

Table 2.24 State table for the vending machine

Present state	Inputs	Next state	Outputs						
ABCDEF	CA s_2 s_1 s_0	A^+ B^+ C^+ D^+ E^+ F^+	RDY	CCA	OS1	OS2	OS3	OS4	OS5
S_0 10 00 00	0 x x x	S_0 1 0 0 0 0 0	1	0	0	0	0	0	0
	1 0 0 0	S_1 0 1 0 0 0 0	1	0	0	0	0	0	0
	1 0 0 1	S_2 0 0 1 0 0 0	1	0	0	0	0	0	0
	1 0 1 0	S_3 0 0 0 1 0 0	1	0	0	0	0	0	0
	1 0 1 1	S_4 0 0 0 0 1 0	1	0	0	0	0	0	0
	1 1 0 0	S_5 0 0 0 0 0 1	1	0	0	0	0	0	0
	1 1 0 1	S_1 0 1 0 0 0 0	1	0	0	0	0	0	0
	1 1 1 0	S_2 0 0 1 0 0 0	1	0	0	0	0	0	0
	1 1 1 1	S_3 0 0 0 1 0 0	1	0	0	0	0	0	0
S_1 01 00 00	x x x x	S_0 1 0 0 0 0 0	0	1	1	0	0	0	0
S_2 00 10 00	x x x x	1 0 0 0 0 0	0	1	0	1	0	0	0
S_3 00 01 00	x x x x	1 0 0 0 0 0	0	1	0	0	1	0	0
S_4 00 00 10	x x x x	1 0 0 0 0 0	0	1	0	0	0	1	0
S_5 00 00 01	x x x x	1 0 0 0 0 0	0	1	0	0	0	0	1

As one more example, let us evaluate B^+. This output is '1' for $s_2\ s_1\ s_0 = 000$ or $s_2\ s_1\ s_0 = 101$ and for G = A B′ C′ D′ E′ F′ . CA . RES′. Combining these, we get the expression for B^+. In a similar manner, we can deduce other next states and outputs. All the Boolean expressions for the next states and the outputs are listed in the following:

$A^+ =$ A B′ C′ D′ E′ F′. CA′ + CCA + RES,

$B^+ = G\,.\,(s_2'\, s_1'\, s_0' + s_2\, s_1'\, s_0)$, where G = A B′ C′ D′ E′ F′ . C A RES′

$C^+ = G\,.\,(s_2'\, s_1'\, s_0 + s_2\, s_1\, s_0')$

$D^+ = G\,.\,(s_2'\, s_1\, s_0' + s_2\, s_1\, s_0)$

$E^+ = G\,.\,s_2'\, s_1\, s_0$

$F^+ = G\,.\,s_2\, s_1'\, s_0'$

RDY = A . RES′,

OS1 = B . RES′, OS2 = C . RES′, OS3 = D . RES′, OS4 = E . RES′, OS5 = F . RES′,

CCA = (B + C + D + E + F) . RES′

PAL Selection

We need to take stock of the I/O requirements for the design in order to arrive at the right device of PAL from a manufacturer. In Figure 2.55, we presented the overall circuit diagram for the vending machine using PAL. We need the following I/Os for our PAL implementation:

6 Nos. of Inputs: CLK, RES, CA, s_2, s_1, s_0

6 Nos. of Registered Outputs: RDY, OS1, OS2, OS3, OS4, and OS5

1 No. of Combination Output: CCA

Scanning through the PAL catalog, such as that of Monolithic Memories [10], we can eliminate all PALs not having register outputs and those that have more or less I/Os than we require. We then see the following types of PAL suiting our I/O requirements with certain compromises. They are 16R8, 16R6, 16R4, 16RP8A, 16RP6A, 16RP4A. Of these, 16R6 and 16RP6A are just adequate. Let us select one of the two, namely, 16R6. It has 10 inputs including a clock input, 6 D flip-flops (registers) and 2 combination outputs. A blank PAL diagram is presented in Appendix 6 on CD, which may be used for assignments or mini projects. Studying the circuit, we see that all outputs are inverted and therefore, we need to generate active low outputs instead of active high outputs we deduced as Boolean expressions earlier.

Figure 2.57 presents the 16R6 programmed for our application with the active low outputs. They are pulled high by resistor arrays so that during power on reset condition, they are in the inactive high state. The CLK input is connected to pin 1 of the IC. Internally, it is connected to the clock inputs of the D flip-flops via a buffer. The horizontal lines are connected to the AND gates (not shown in the circuit of the PAL), whereas the vertical lines are connected to I/Os and their inverted signals. Thus the horizontal/vertical lines are arranged as a matrix, whose inter junctions may be programmed. 'X' indicates a programmed connection. It may be noted that all the outputs are connected via inverting tri-state buffers. Their tri-state control lines are connected to RES′ signal so that during power on reset condition, these buffers are tri-stated. The reader may verify that the programmed connections conform to the next states and the outputs derived earlier.

Medium 20 Series, 16R6* PAL Logic Diagram (Courtesy: Monolithic Memories)

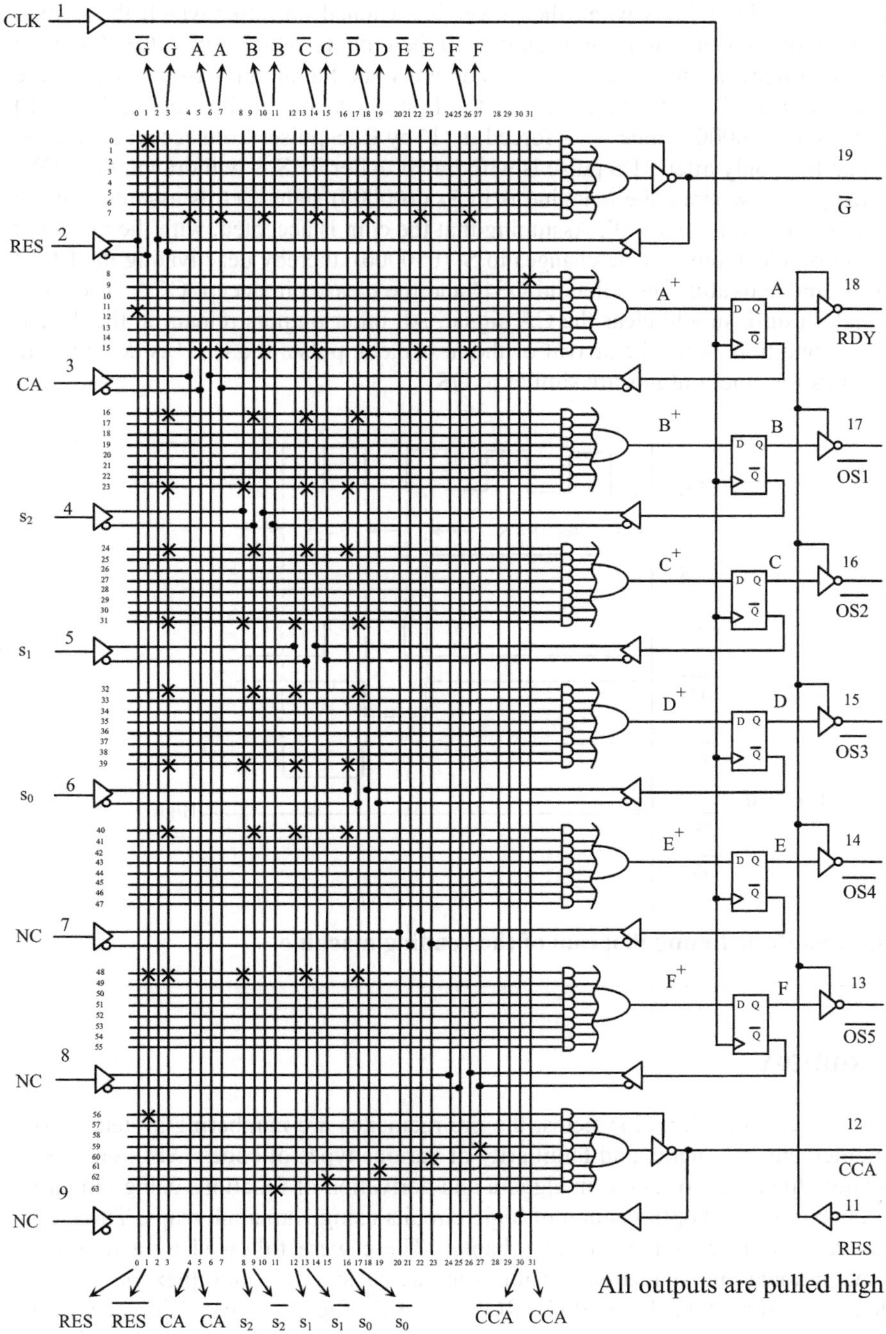

Fig. 2.57 Realization of the vending machine using PAL

Figure 2.58 presents the timing diagram of the vending machine. RES pulse is applied for about 1.4 s when the power is switched on, after which the normal working of the vending machine starts. All the outputs, RDY′, OS1′–OS5′, remain inactive (high) during the time RES is active since the output tri-state buffers are disabled by RES. With the arrival of the first clock, the D flip-flops (ABCDEF) are preset to 100000 since A^+ is forced to '1' by RES pulse. Coin accept signal CA can go high only after RES pulse is withdrawn. After RES is withdrawn, the RDY′ signal goes low since the machine is in S_0 (100000) state. Let us assume that the BCD switch 's' is set to '0'. Assuming that the coin is accepted, with the arrival of the fourth clock pulse, state changes to S_1 (010000), thereby deactivating the RDY′ signal and activating the solenoid OS1′ and the clear coin accept CCA. The CCA signal, in turn, should clear the CA signal. All other outputs remain in the deactivated condition. With the arrival of the next clock pulse, the OS1′ is deactivated. Timings for other outputs are similar to OS1′.

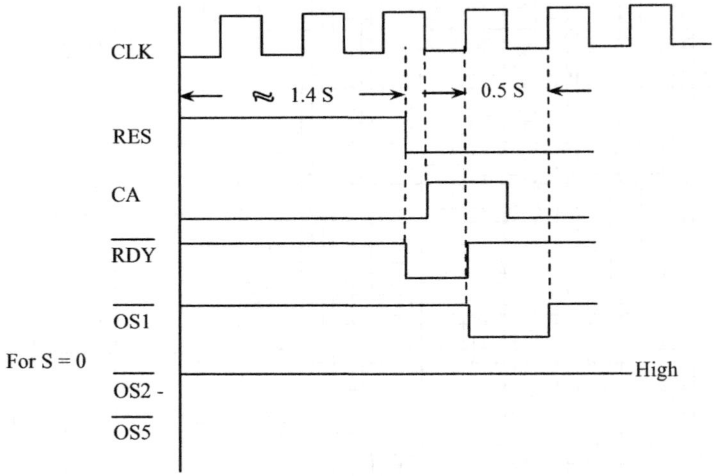

Fig. 2.58 Basic timing diagram of the vending machine

Summary

For the design of digital systems, the pre-requisites are numbering systems, twos complement arithmetic, and familiarity of various types of codes. Also, one needs to know the basics of Boolean algebra and derivation of functions using minterms and maxterms and optimization of logic circuits using Karnaugh maps. These fundamentals were presented in this chapter. These were followed by a review of combinational and sequential circuits such as basic gates, multiplexers, demultiplexers, comparators, PLA, PAL, ROM, D/JK/T flip-flops, etc. With the aid of these basics, designs of small digital systems were presented using SSI/MSI components. The algorithmic state machine based approach to design was shown to be a better alternative to the conventional state graph approach. Designs based on the algorithmic state machine and PAL were also presented. Equipped with these fundamentals, the reader should not have any difficulty in understanding the rest of this book. We commence Verilog designs in the next chapter.

Assignments

2.1 Realize a four-bit adder using half adders as the building block.

2.2 There are two ways of converting a two input NAND gate to an inverter. What are they?

2.3 The output of the circuit shown in Figure A2.1 is equal to

a) 1
b) 0
c) A'B + AB'
d) (A ⊕ B)' ⊕ (A ⊕ B)'
e) None of these

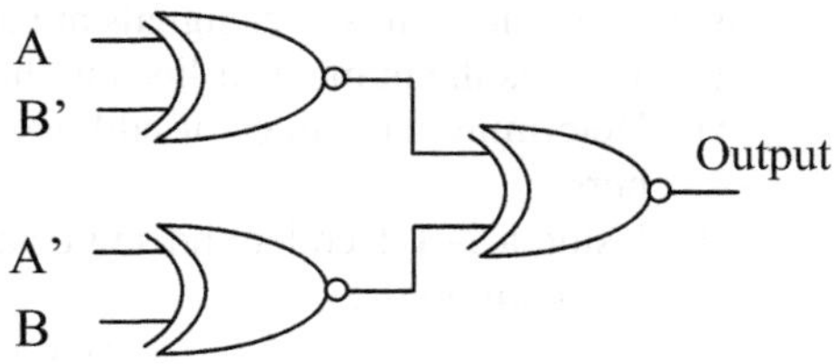

Fig. A2.1

2.4 There are two switches to control the light in a long corridor. You must be able to turn the light on while entering the corridor from any direction and turn it off at the other end when you leave. Draw a circuit which satisfies these conditions. If there are three switches that can turn on and off a light in a room, how will you wire them?

2.5 Show how NAND gates can be used to build the logic circuit for $Y = A + BC'$.

2.6 Realize $Y = AB + CD$ using only NAND gates.

2.7 Prove the following Boolean expressions without using truth tables:
  i) $A' + AB = A' + B$
  ii) $A + A'B = A + B$

2.8 Assume that signals A, B, C, D, and D′ are available. Using a single 8 to 1 multiplexer and no other gate, implement the following Boolean function:

$$f(A, B, C, D) = B\,C + A\,B\,D' + A'\,C'\,D$$

2.9 A Boolean function is given as sum of products: $F = \Sigma m\,(3, 4, 5, 6)$, where A, B, and C are inputs.
  a) Implement this function using an 8:1 MUX.
  b) What will be the minimized sum of products expression for F?

2.10 A water tank has a float and two electrically operated outlet valves. The first valve is to be opened if the float reaches a first level and both the valves are to be opened if the float reaches a higher level. Both the valves are to be closed and the water pump activated if the water level goes below the lower level. The pump is switched off when the float touches the higher level. Assume that sensors can sense the two levels of the float only if the float is close to the respective levels. Design a control circuit to operate the two valves and the pump. State your assumptions clearly.

2.11 A ROM is required to be used to implement the Boolean functions given below:

$F_1 = ABCD + A'B'C'D'$

$F_2 = (A + B)(A' + B' + C)$

$F_3 = \Sigma\ 13, 15 + \Sigma_d\ 3, 5$

a) What is the minimum size of the ROM required?

b) Determine the data in each location of the ROM.

2.12 A CPU has parallel address and data bus, RD′ and WR′ signals. Two ROMs of size 4K words each and two RAMs of sizes 16K and 8K words respectively are required to be connected to the CPU. The memories are to be connected such that they are memory mapped as shown in Figure A2.2. Assume that the chip select signals are active low.

a) What is the number of lines in the address bus of the CPU?

b) Determine the values of addresses, X, Y, Z, and W as decimal numbers.

c) Using a 2–4 decoder and some additional gates, draw a circuit for the decoding logic.

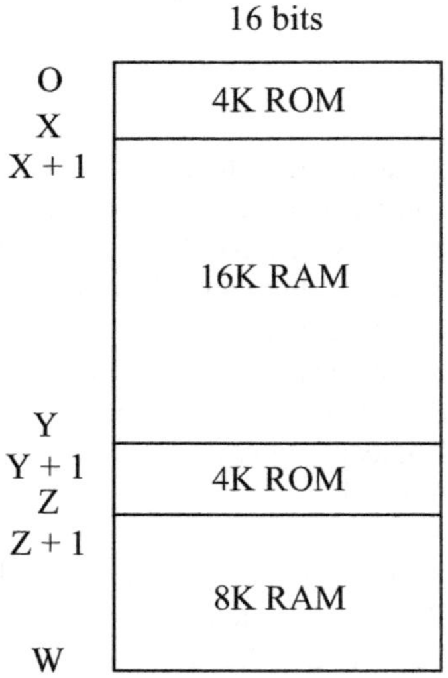

Fig. A2.2 Memory map addressing

2.13 A code converter is required to be designed to convert a 5421 BCD code to the normal 8421 BCD code. The input BCD combinations for each digit are given in Figure A2.3. A block diagram of the converter is also shown alongside.

a) Draw K maps for outputs, D_3, D_2, D_1, and D_0.

b) Obtain minimized expressions for the outputs.

Decimal	5421 BCD Code			
	A	B	C	D
0	0	0	0	0
1	0	0	0	1
2	0	0	1	0
3	0	0	1	1
4	0	1	0	0
5	1	0	0	0
6	1	0	0	1
7	1	0	1	0
8	1	0	1	1
9	1	1	0	0

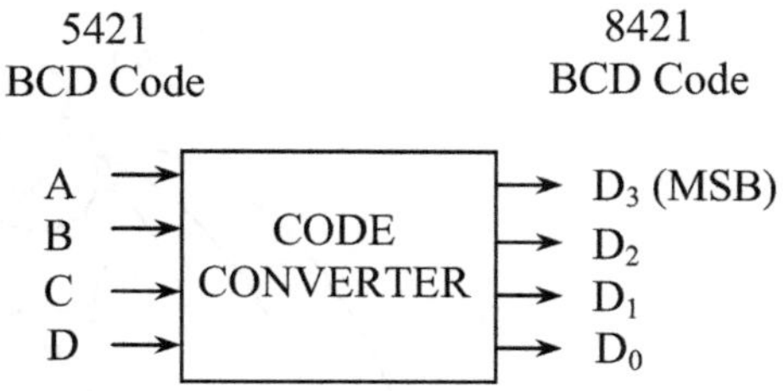

Fig. A2.3 Code converter

2.14 What is the difference between a latch and a register? For the same input, how would the output look for a latch and for a register?

2.15 Design a divide-by-three sequential circuit with 50% duty cycle. (Hint: Double the clock frequency).

2.16 Suppose you have a combinational circuit between two registers driven by a clock. What will you do if the delay of the combinational circuit is greater than the clock period?

2.17 A circular wheel with half painted white and the other half painted black on the same side and mounted on the rotating shaft of a motor can be used to find the direction of rotation of the motor. There are two sensors located slightly apart facing the surface of the wheel and are asserted for white and deasserted for black passing before them. Design a circuit to detect the direction of wheel rotation.

2.18 Design a divide-by-five circuit using a state machine. The clock has 50% duty cycle and the output waveform need not be symmetrical.

2.19 A synchronous up/down decade counter is required to be designed. The counter must count up or down in binary, depending on the value of a control input signal. The counter shall count up for control input = 0. Otherwise, the counter shall count down. The counter shall wrap around from 9 to 0 while counting up and from 0 to 9 while counting down. During counting up, if the terminal count is reached, an output shall be set. Similarly another output shall be set if the terminal count is reached while counting down. Draw the state diagram and the state table for this counter. What is the easiest way to realize the circuit of this counter? Why?

2.20 An FSM is shown in Figure A2.4. How many flip-flops are required if we are to realize the FSM? What is the purpose of this FSM? Formulate a state table assuming implementation using D flip-flops.

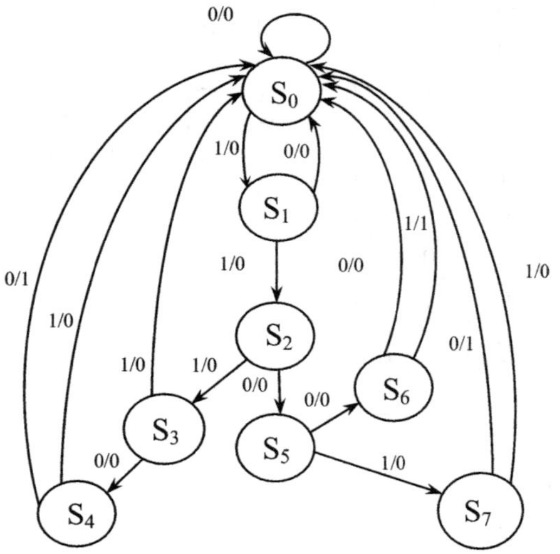

Fig. A2.4 FSM graph

2.21 Figure A2.5 shows a state machine to detect even or odd numbers of 1's or 0's in a three-bit incoming data. Explain how it detects even or odd sized three-bit patterns. Draw a state table for the implementation using D flip-flops.

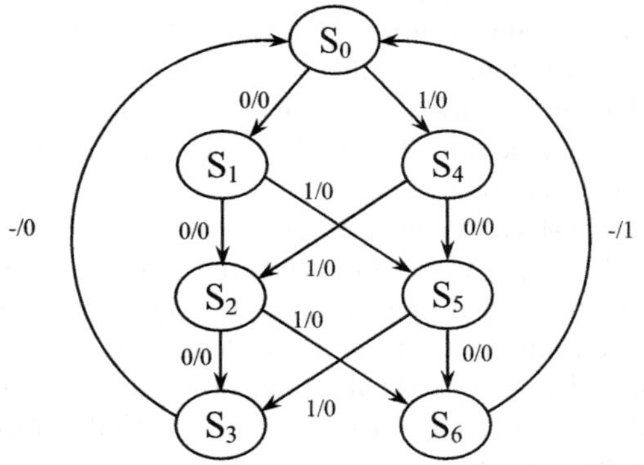

Fig. A2.5 State diagram of an even/odd pattern detector

2.22 Implement the following flip-flops using RS flip-flop:
- (i) D flip-flop
- (ii) T flip-flop
- (iii) JK flip-flop

2.23 Implement the following flip-flops using JK flip-flop:
   (i) D flip-flop
   (ii) T flip-flop

2.24 Modulo 5 counter may be implemented using any flip-flop. Formulate a truth table for such a counter using JK, T, and D flip-flops.

2.25 Realize the circuit diagrams for the assignment 2.24.

2.26 The state diagram and the state table for an FSM are shown in Figure A2.6. Realize the circuit diagram. What is the function of the circuit?

Present State			Next State		J_A	K_A	J_B	K_B
A	B	U	A	B				
0	0	0	1	1	1	ϕ	1	ϕ
0	1	0	0	0	0	ϕ	ϕ	1
1	0	0	0	1	ϕ	1	1	ϕ
1	1	0	1	0	ϕ	0	ϕ	1
0	0	1	0	1	0	ϕ	1	ϕ
0	1	1	1	0	1	ϕ	ϕ	1
1	0	1	1	1	ϕ	0	1	ϕ
1	1	1	0	0	ϕ	1	ϕ	1

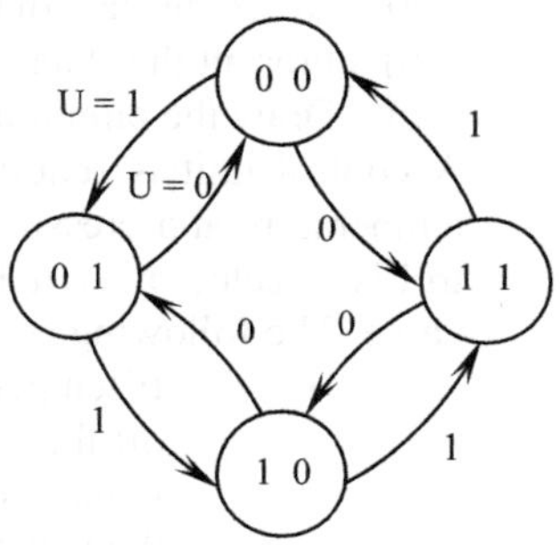

Fig. A2.6 State table and FSM

2.27 Modulo 10 Gray code, decade counter sequence shown in Figure A2.7 can be implemented using EPROM. Draw the state table showing the EPROM contents clearly. Include a power on reset in the complete circuit diagram of your design. Realize the design using D flip-flops.

0 → 1 → 3 → 2 → 6 → 14 → 10 → 11 → 9 → 8

Fig. A2.7 Modulo 10 Gray code decade counter

2.28 Realize the Modulo 10 Gray code decade counter in the assignment 2.27 using a PAL, not necessarily a commercially available one. List the Boolean expressions deduced and show PAL programming clearly.

2.29 A new clocked XY flip-flop is defined with two inputs, X and Y in addition to the clock input. The flip-flop functions as follows:
If XY = 00, the flip-flop changes state with each clock pulse.
If XY = 01, the flip-flop state Q becomes 1 with the next clock pulse.
If XY = 10, the flip-flop state Q becomes 0 with the next clock pulse.
If XY = 11, no change of state occurs.
   a) Write the truth table for the XY flip-flop.
   b) Write the excitation table for the XY flip-flop.

c) It is desired to convert a JK flip-flop into the XY flip-flop. Draw a circuit diagram to show how you will implement the XY flip-flop.

2.30 A clocked sequential circuit has three states A, B, and C and one input X. As long as the input X is 0, the circuit alternates between the states A and B. If the input X becomes 1 (either in state A or in state B), the circuit goes to state C and remains in state C as long as X continues to be 1. The circuit returns to state A if the input becomes 0 once again and from then onwards repeats the sequence. Assume that the state assignments are A = 00, B = 01, and C = 10.

a) Draw the state diagram of the circuit.
b) Present the state table for the circuit.
c) Draw the circuit using D-flip-flops.

2.31 A control unit is required to be designed for a chemical process, where temperature and pressure are the two parameters to be controlled. The control is effected by switching on or off a heater and by opening or closing a valve. The following control rules apply:

i) If temperature and pressure are in the normal range, switch off the heater and close the valve.
ii) If the temperature is normal and pressure is too high, open the valve and close it if the pressure is low.
iii) If the temperature is high and the pressure is low, turn off the heater and close the valve.
iv) If the pressure is normal and the temperature is low, turn on the heater and, turn it off if the temperature is too high.
v) If the pressure is high and the temperature is low, open the valve and turn on the heater.
vi) If both temperature and pressure are low or too high, let an alarm ring.

Design the system to the above specifications.

2.32 A state graph was used for the design of a vending machine controller in the text. Instead of the state graph, draw an ASM chart. How will the state table be affected?

2.33 Black jack dealer is a game played by a dealer and one or more players using cards [8]. In this game, cards are assigned values of 1 for Ace, 2 to 10 for said numbers, and 10 for face cards. An Ace may have the value of 1 or 11 during a play, whichever is advantageous. The dealer picks cards one at a time, counting Ace as 11, until his score is greater than 16. If the dealer's score doesn't exceed 21, he 'stands' (i.e., he wins the round) and his play of the hand is finished. On the other hand, if the dealer's score is greater than 21, he is 'broke' (i.e., he looses the game). However, the dealer must revalue an Ace from 11 to 1 to avoid going broke and must continue picking cards (called 'hits') until the count exceeds 16. The goal of this assignment is to design hardware (H/W) that acts as a 'dealer' following the above-mentioned rules of the game. The H/W must have a provision for inputting the (Play) card value for which a four-bit binary input is required. One can either use four independent toggle switches or a thumb wheel switch or a push-button operated switch or a DIP switch. We further need one push-button switch to signal the H/W that the card value set can be accepted (Card Ready) for further processing. We need two digits, seven segment displays for displaying the score, as well as LEDs for indicating the status: HIT, READY (to play), STAND, and BROKE. This is depicted in Figure A2.8.

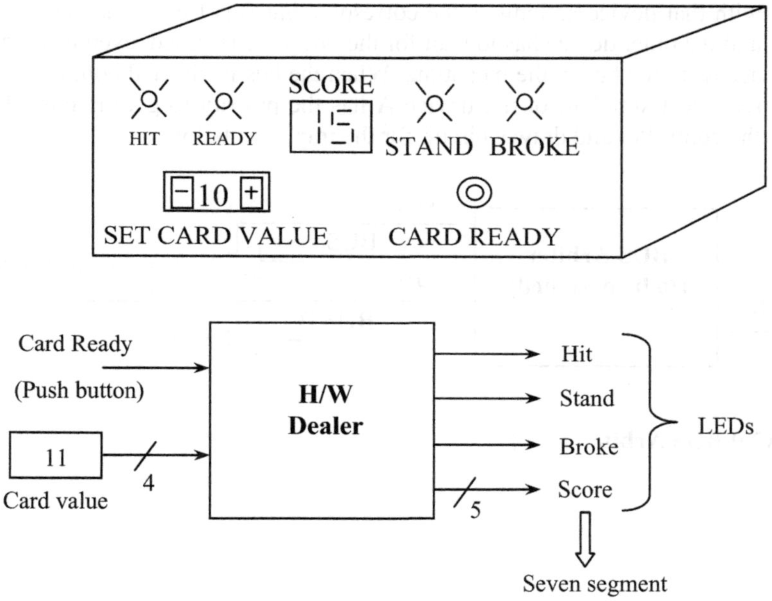

Fig. A2.8 Block diagram of the black jack dealer

Draw a detailed hardware architecture, an ASM chart, a state table and deduce the Boolean expressions of outputs and realize the black jack dealer using 16R4 or 16R6 PAL. You may use any extra gates or D flip-flops if the PAL resources are not adequate. Draw a detailed circuit diagram. You may use a block diagram for adder/subtractor.

2.34 A train station has three platforms marked 1 to 3. A train approaching the station in any of the two directions is to be routed to one of the three platforms. If all the three platforms are empty, then the approaching train is to be routed to platform 1. On the other hand if it is occupied, the train is to be routed to platform 2. Only if both platforms 1 and 2 are occupied, the train is routed to platform 3. A switching control system is required to be designed which will set the appropriate rail track points and turn on signal lights. Each platform has a sensor which is turned on if a train is in that platform. If a train approaches a light signal, the corresponding sensor is activated. The controls required for departing trains from the platforms may be ignored. What is the easiest way to design? Design the control system accordingly.

2.35 A bus arbiter is to be designed to control the access to a common bus by two devices A and B as shown in Figure A2.9. RA and RB are Bus Request signals from devices A and B respectively and GA and GB are the corresponding Bus Grant signals. When the bus is idle, the arbiter grants it to the device which has requested it. Once the bus is granted to a device, it stays with that device as long as the corresponding Bus Request signal is active and the other device has to wait for the bus to be released, even if it requires access to the bus in the meantime. When the bus is idle and both the devices request it simultaneously, device A has the priority to get the bus. Design the controller and draw a circuit for the implementation.

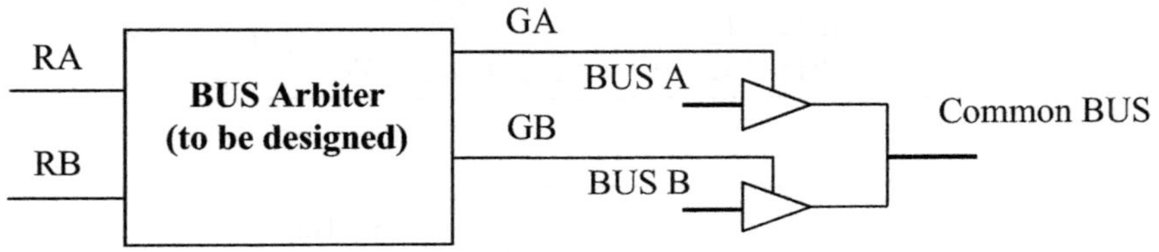

Fig. A2.9 BUS Arbiter

Chapter 3

Design of Combinational and Sequential Circuits Using Verilog

3.1 Introduction to Hardware Design Language

Having had a good review of digital circuit design in the previous chapter, it is time for us to pick up hardware design language (HDL) coding [11–17]. A digital system is primarily a combination of combinational and sequential circuits put together in any mix. To start with, we will be learning the design of simple combinational circuits using Verilog followed by more complex circuits. As we progress further, we will be designing sequential circuits. In Chapter 4, we will see how to write effective test benches so that we may test the functionality of our design. This will be followed by RTL coding guidelines, a pre-requisite for successful working of the hardware that we wish to design. Subsequent three chapters will give you hands-on experience on various industry standard CAD tools, namely, the simulation tool, the synthesis tool, the place and route tool, and the back annotation. Later on, you will be learning the design of memories and arithmetic circuits, development of algorithms and architectures, etc. Thereafter, we will proceed to the design of digital VLSI systems. The penultimate chapter will give an insight into the working of the actual hardware using FPGAs. In the final chapter, a number of projects are suggested for implementation.

Before we go into the coding of combinational circuits, we will review the evolution of the HDLs. Primarily, HDLs were used in order to speed up design cycle times. Hitherto, schematic circuit diagrams were used, and they were very handy for designers who were accustomed to designing systems using discrete digital ICs such as 74 series TTLs. This approach offers cost effective systems so long as the design size is small in the order of few thousand gates. However, when it comes to the design of VLSI circuits involving well over 10,000 gates, the designer will have to struggle with a number of drawing sheets of the size of A1. The readability suffers while wading through hundreds of drawing sheets. The usage of Karnaugh map and Quine McCluskey methods of circuit optimization also prove to be cumbersome for circuits of such complexity. There are other disadvantages such as very high circuit entry time, time consuming preparation of documents, etc. These disadvantages manifest in spite of using CAD based schematic design entry.

The disadvantages in schematic design approach opened up the need for introducing HDLs in circuit design. The HDL has brought about a revolution in the realms of digital design. The HDL based design reduces the cycle time dramatically. From experience, we can say that there is a speeding up of design cycle time for VLSI systems by at least five times. HDL provides very concise representation of circuits in contrast to the schematic logic circuit diagrams. This is made possible by using what are called behavioral, RTL, and data flow statements. All these lead to a very concise description of the digital hardware we design. In contrast to this, in schematic approach, you have to build circuits gate by gate, and it will take quite an effort to achieve the same end result. Further, there is no need for Karnaugh map and Quine McCluskey methods of optimization in HDL based designs because synthesis tool is supposed to take up this role automatically.

HDL designs are portable from one vendor platform to another. For example, let us say that you have developed your HDL design on Altera platform to start with. Later on, when the occasion demands, you can always migrate to another platform, say, Xilinx without the need to redesign your codes. HDL based designs are also technology independent. Few years back, 0.65 μm technology was in vogue; thereafter 0.5 μm and 0.35 μm came into existence, followed by 0.105 μm technology. Presently, the fabricators are embarking on 0.0309 μm technology. All these technological changes will have no effect on these HDL designs so long as the design is free from technology dependence. That means, whatever you have designed earlier when the technology was less advanced, the same design will work precisely in the present day technology, and most assuredly on new technologies that are yet to emerge in the future. However, you have to be very cautious in wielding the HDL tool in the design, taking care to avoid technology-dependent delay circuits, and designing only RTL compliant circuits. We will discuss these issues in depth when we deal with RTL coding guidelines.

Two most popular HDLs used currently are Verilog and VHDL. Recently, Cadence has come up with a HDL using mixed analog and digital design. They call it Verilog AMS. You can implement A to D converters, D to A converters, PLLs and the rest of analog circuits, and you can mix them with digital circuits. It is needless to emphasize that Verilog and VHDL can be used in circuit design ranging from SSI to VLSI. You may regard 50 transistors or less to be falling in the SSI category; below 500 as MSIs; below 5000 as LSIs; beyond that as VLSIs/ULSIs. Roughly, four or five transistors may be regarded as representing a (two input NAND) gate.

We will be using Verilog in this book for digital VLSI systems design. Verilog is a hardware description language developed originally by the Gateway Design Automation in 1984. Cadence popularized it later on. In 1987, synthesis was introduced by Synopsys, one of the leading vendors in ASIC platform. Verilog has become an industry standard because of its simplicity: you can quickly learn Verilog; in fact it is easier than learning VHDL in our experience. It has 'C' like structure and very fast design cycle times. What is meant by C like structure is that you can use 'if, else' statements, 'case' statements, etc. The input/output structures are also more or less similar. Of course, there are small differences, which you will be learning gradually.

Verilog has been very popular in hi-tech areas of USA, in the west coast, whereas VHDL is popular in the eastern coast. The reason behind this may be that industries prefer Verilog as a means for faster implementation, whereas institutions prefer VHDL. It is only a general observation, and there are always exceptions to this view. You as a designer can start with Verilog first, and having mastered it, switch over to VHDL later on when the occasion demands. The basic design concepts and methodology dealt in this book are, however, equally applicable to both Verilog as well as VHDL. Both Verilog and VHDL conform to IEEE standards.

We have already seen that Verilog is a hardware design language, and is very much akin to C. However, you should bear in mind that Verilog is a hardware design tool and not a software design tool, whereas C is clearly a programing language, which basically runs sequentially unless you veto it by 'call', 'jump', and similar instructions. So also are assembly languages for microprocessors and DSP processors. Even though Verilog code resembles C program, you have to remember that they are all coded as concurrent statements. It is exactly the same as a digital circuit design that makes use of the conventional TTL ICs, etc. working concurrently. As a matter of fact, the hardware design works several times faster than the same design implemented in software processed by a computer such as the Pentium Processor or the Digital Signal Processor.

Verilog allows different levels of abstraction. One is 'behavioral', which we have mentioned earlier, using 'if ', 'else' structures; 'for', 'while' loops, etc. There is another structure called 'data flow' structure, which is basically concerned with the flow of data from one register to another. We will see quite many examples as we progress. For those who are used to gate level implementation earlier, they can continue to use gate level primitives in a limited way. For faster implementations, you often need extra timing closures, which can be achieved by using gate level primitives. However, for bigger designs, Register Transfer Level (RTL) coding practice will have to be adopted, which is the main emphasis of this design book. RTL conformance is the core of digital design as such. If you violate RTL coding guidelines, the synthesis tool will promptly reject or report errors or warnings. This is only to ensure that the final product that you are going to tape out or deliver will be really working on the hardware it is meant for.

Verilog also features switches such as NMOS or PMOS switches [11] in case you need them for specific application. The problems with the switches are that they are technology dependent and, therefore, these are not covered in this book. What will be adopted in this book is the RTL type and occasionally going for data flow and behavioral types, and on very rare occasions we will use the gate level primitives. All Verilog design codes in this book, unless otherwise specified, will be fully RTL compliant so that they may readily work on the hardware when implemented.

3.2 Design of Combinational Circuits

Combination circuits can be designed by using either 'assign' statements or 'always' block statements as described in the following sub-sections.

3.2.1 Realization of Basic Gates

We will start with the design of combinational circuits using simple 'assign' statements. It is as simple as the statement 'assign out = A + B ;' for evaluating the sum of A and B and transferring the result to 'out'. Some of the examples that we will go forth now are for very primitive gates. In order to speed up your learning, we will first consider only the core or the main Verilog statements in order to realize the hardware. Later on, when you have mastered these cores, we will garnish them with other Verilog statements to make a full-fledged code. To start with, we would like to make a simple buffer. What all you need to do is assign the input, say, A, to an output, F1. The development tools take care to translate this simple statement into a real hardware, the buffer. As shown in Figure 3.1, the corresponding Verilog statement for this is:

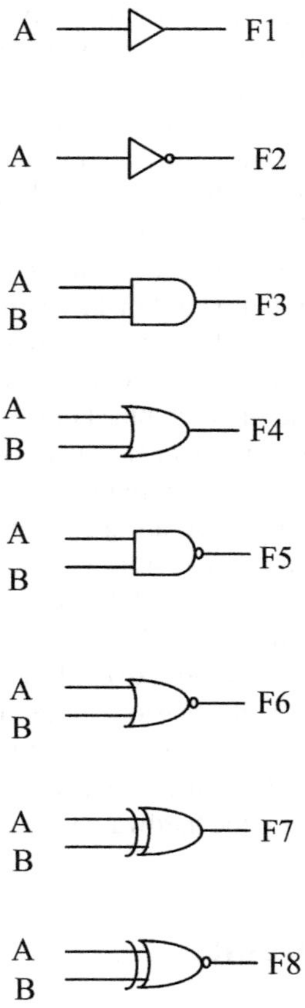

GATE	SYMBOL	VERILOG STATEMENT
Buffer	A → F1	assign F1 = A ;
Inverter	A → F2	assign F2 = !A ;
AND	A, B → F3	assign F3 = A&B ;
OR	A, B → F4	assign F4 = A|B ;
NAND	A, B → F5	assign F5 = !(A&B) ;
NOR	A, B → F6	assign F6 = !(A|B) ;
XOR	A, B → F7	assign F7 = (A^B) ;
XNOR	A, B → F8	assign F8 = !(A^B) ;

Fig. 3.1 Basic gates realization in Verilog

assign F1 = A ;

The 'assign' word is mandatory in order to assign any statement. The output is on the left hand side of the '=' symbol. It is almost like a C statement with the exception that 'assign' is a new statement implying a hardware field. Similarly, if you want an inverter what all you have to do is just replace 'A' with '!A'. This exclamation mark is an inversion (Not) signal, and this is basically a logical inversion. You can also use '~' in lieu of '!' even though '~' is primarily used for multi-bit precision, implying bit wise negation. If you wish to do AND or OR with two inputs, what you require is '&' for AND operation and '|' for OR operation. Similarly, whatever gates we have already considered, we can get their inversions as well. For example, a NAND gate can be derived by complementing (!) the result after AND operation. For exclusive or (XOR), the symbol is '^'. XNOR realization is just the inverse of XOR. All the gate realizations we have covered so far are summarized in Figure 3.1.

3.2.2 Realization of Majority Logic and Concatenation

Now that we have seen how to use assign statements for simple combinational circuits, we can also implement the same using 'always' statement. The always statement is a block of instructions which you will see later on. First, consider a simple circuit such as a majority logic shown in Figure 3.2. This is nothing but the realization of logic: F9 = AB + BC + CA. In order to realize this, you need three numbers of two input AND gates and one number of three input OR gate. The next example we are going to see is how to concatenate different signals. For

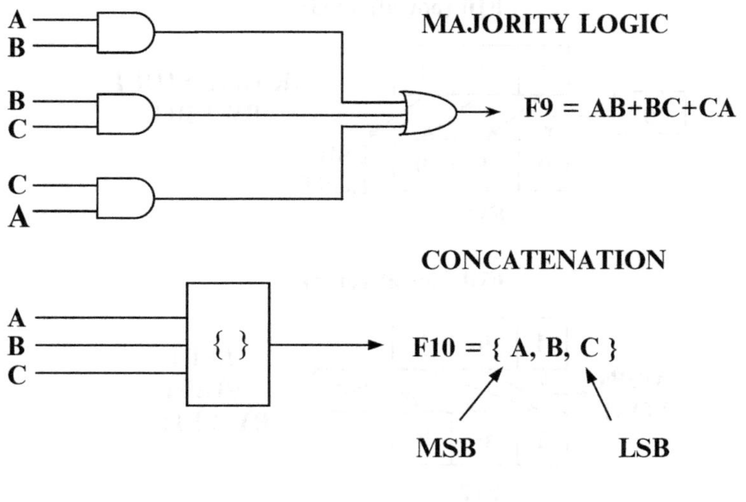

Fig. 3.2 Circuit diagram of majority logic and concatenation

example, let A, B, and C signals be one bit each. Concatenation is just putting the signals together as a single multi-bit signal in the order you want it. The concatenation symbol is the flower brackets '{ }', and you need to separate the signals requiring concatenation with a comma.

The concatenated result of the three signals A, B, and C is expressed as follows:

F10 = {A, B, C} ;

where A is the MSB. Always, the signal listed left most will be the MSB and, naturally, it follows suit that the right most signal bit is the LSB. For example, if we assume A, B, and C as 1, 0, and 1 respectively, then F10 = 101 after concatenation. One may be tempted to call 'signals' such as A, B, C, etc. as variables. That is because of our past habit with the C language. Here, in hardware design, you don't speak in terms of variables but in terms of signals. For examples, any digital node is a signal and whatever input/output (I/O) you have in your design is also a signal.

3.2.3 Shift Operations

The next two examples we will consider are 'right shift' and 'left shift'. For instances, we will shift by just one bit for 'right' and two bits for 'left' shift. Let us say that we have a signal which is three bits wide, namely, F10 we have seen before. What happens after right shift is shown in Figure 3.3. After the shift, the LSB is lost and the vacated MSB is forced to a zero. In the case of left shift by two bits, the two MSB bits are lost, and zeros will occupy the bits that are vacated. The Verilog code for the four examples covered so far may be combined into one compact 'always' block as shown in the following:

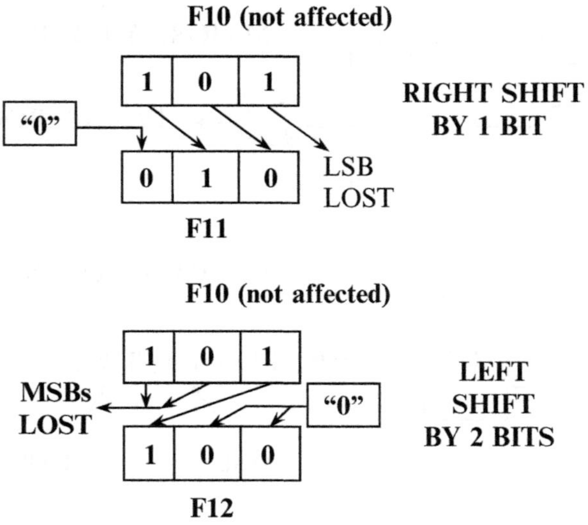

Fig. 3.3 Basics of shift operations

Verilog_code 3.1

```
always @ (A or B or C or F10)
        begin
                F9  = (A&B)|(B&C)|(C&A) ;        // Realize AB+BC+CA.
                F10 = {A, B, C} ;                // Concatenate A, B and
                                                 // C to get 3 bit result.
                F11 = F10 >> 1 ;                 // Right shift by one bit.
                F12 = F10 << 2 ;                 // Left shift by two bits.
        end
```

The first code realizes the majority logic, which is F9 = AB + BC + CA. Note that '&' stands for AND, '|' for OR, and that is how we get AB + BC + CA. A good practice is to use brackets '( )' as shown for the signal F9. They enhance not only the readability, but also guide the synthesis tool, which aspect we will learn later on. You would notice that there is an 'always @' statement in the code. What all it means is, whenever the inputs A or B or C or F10 changes, only then the results are evaluated, and not otherwise. If you have multiple statements, as in the present case, you need to put a 'begin' statement and an 'end' statement indicating that the whole set of statements is a block. In one block, you can put as many statements as you want. If a statement is complete, you end it up with a semicolon. If you forget this, the compiler will promptly report error. Although the multiple statements are written sequentially (one followed by another), in real hardware, they are put as actual gates as the case may be and, therefore, these statements in a block may be regarded as 'concurrent statements'. The same is also true for 'assign' statements and also other statements in a complete design.

The next statement is for concatenation, which we have already seen. This is exactly the same as the one we saw pictorially earlier. Suppose you want signal C to be the MSB, and A as LSB, instead of the present order, then you need to put C first, followed by B and A in that order. In addition to this, let us say that you want to add three zeros to the LSB; all we need to do is to put a comma after A and add three zeros as follows:

F10 = {C, B, A, 0, 0, 0 } ;   or more concisely as   F10 = {C, B, A, 3{0}} ;

Note that in the case of shift operations, F11 = F10 >> 1 ; for example, the source F10 itself is not affected. Also note that we can use any number of bits to shift and it does the job at one stroke. '//' can be used for writing line comments.

3.2.4 Realization of Multiplexers

Now, let us go into the coding of a slightly more complex component such as a multiplexer. A two input MUX, with I0 and I1 as the data inputs and A as the select input is shown in Figure 3.4. If A = 0, the MUX automatically selects I0;

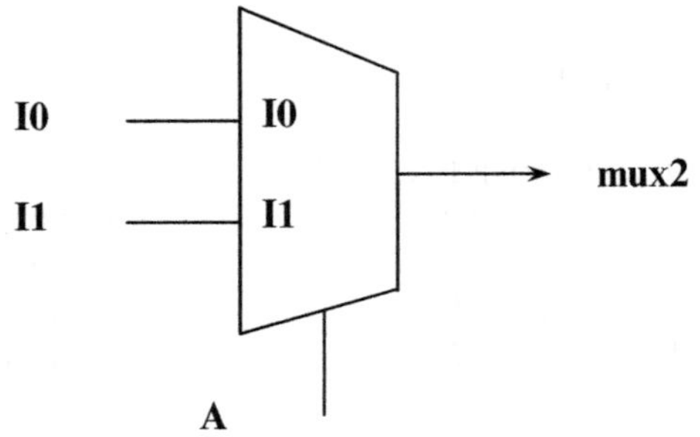

assign mux2 = (A = = 1) ? I1 : I0 ; // mux2 = I1 if A = 1, otherwise mux2 = I0.

Fig. 3.4 Two input MUX using 'assign' statement

otherwise it selects I1, and outputs it on to the signal 'mux2'. In order to write the code, we need only one assign statement as shown in the figure. 'A' is the select pin of the MUX, which may also be written as (A = = 1) or (A = = 1'b1). In the last two cases, a comparator is actually put by a synthesis tool, which tool we will learn in a later chapter. Most designers normally prefer the last option, in general, since it throws light on the number of bits present in the select signal 'A'. '1' before 'b indicates the precision, b stands for binary, and '1' after b represents the logic high state of the signal 'A'. Instead of 'b', 'h' or 'o' can also be used for hexadecimal and octal numbers respectively. '?' is mandatory for representing the MUX structure. After '?', the higher order input (I1) separated by ':', and followed by the lower order input (I0) are written.

We will now see how a four input MUX is coded using nested assign statement. I0, I1, I2, and I3 are the four inputs as shown in Figure 3.5. In order to access these inputs one at a time, two selector pins B and C are required. 'B' is the

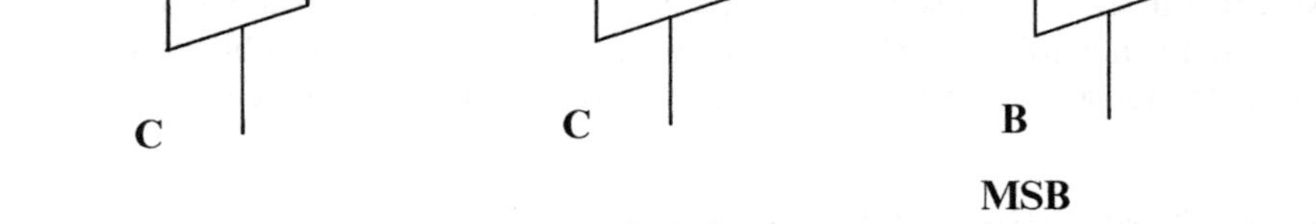

assign mux4 = B ? (C ? I3 : I2) : (C ? I1 : I0);
// Note: Avoid using nesting. Instead, use 'case' for more than 2 inputs.

Fig. 3.5 Four input MUX using nested assign statements

MSB. Take due care of the MSB and the LSB. Otherwise, the wrong input will be accessed. This case is precisely the same as two input MUX which we have used earlier, except that the inputs and outputs are different. I0, I1 inputs of the last MUX are derived from the first two MUX outputs. The signal C selects either the I0 input or the I1 input and outputs to the signal 'mux4', provided the signal B is low. On the other hand, if B is high, mux4 is assigned to the input I2 or I3. It is better to avoid nesting like this since there are two cascaded MUX delays resulting in lower speed of operation. A better speed performance can be obtained by using 'case' statements. Therefore, we will restrict MUX assign statement only for two input processing.

We will consider the realization of an eight input MUX in order to illustrate the use of 'case' statement. I0–I7 are the inputs, A, B, and C are the select pins, where A is the MSB, and mux8 is the final output as shown in Figure 3.6. We will be using the 'case' statement within 'always' block in the following code. Always block will be executed whenever any of the inputs, A–C, I0–I7 change. In the code, all these inputs listed are separated by 'or'. Note that this is not a logical statement but plain English (caution: don't use '|' or anything other than 'or'). Case is checked based on the concatenated value of A, B, and C. {A, B, C} = 000 binary value corresponds to the address of the input, I0, and so on. Each address is three bits wide and is represented as 3'b inside the case statement. Depending upon the dynamic value of this address, the signal mux8 is assigned one of the input values from I0 to I7. Since A, B, and C signals may also be don't cares (x) or high impedance (z), a 'default' statement must also be included as shown. The statement, 'case' must be terminated by a corresponding 'endcase' statement. Within the always block, a 'begin' and 'end' must be added if multi-statements are used within the block. The code for the eight input MUX is as follows:

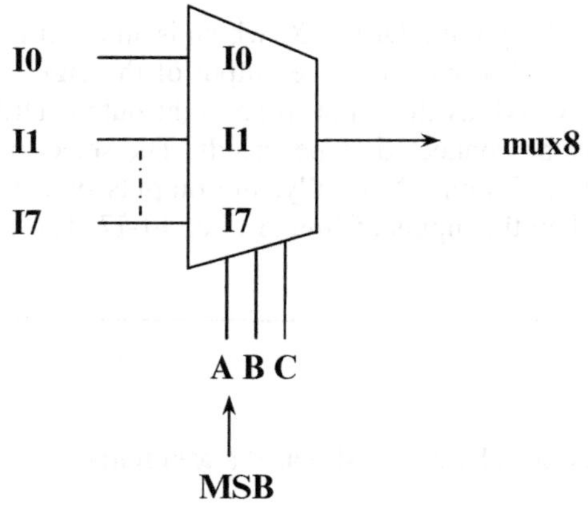

Fig. 3.6 Eight input MUX

Verilog_code 3.2

```
//        Eight input MUX code using 'always' and 'case' statements.
//        Fastest possible hardware implementation.

always @ (A or B or C or I0 or I1 or I2 or I3 or I4 or I5 or I6 or I7)
        begin
                case ({A, B, C})
                 3'b000: mux8 = I0 ;        // Read the input addressed by ABC
                 3'b001: mux8 = I1 ;        // and output the same to mux8.
                 3'b010: mux8 = I2 ;
                 3'b011: mux8 = I3 ;
                 3'b100: mux8 = I4 ;
                 3'b101: mux8 = I5 ;
                 3'b110: mux8 = I6 ;
                 3'b111: mux8 = I7 ;
                 default: mux8 = 0 ;        // The value can be I0 or any other.
                endcase
        end
```

The case statement described here is very much similar to 'C code' except that this is a hardware design, where the compiled code translates as an eight input MUX.

3.2.5 Realization of a Demultiplexer

Next, we will see the design of a DEMUX, which is an exact counter part of MUX as shown in Figure 3.7. We will feed the output of the eight input MUX (mux8) we have just now covered, as the input to an eight output DEMUX even though any other signal may be connected as per needs. The select pins are the same as that used in the MUX design. Naturally, the outputs of the DEMUX, D0–D7, must be none other than the inputs of 'mux8', i.e., I0–I7. The code for DEMUX is as follows:

Verilog_code 3.3

```
//        DEMUX using 'always' and 'case' statements.

 always @ (A or B or C or mux8)
   begin
        case ({A, B, C})
                                // Read the input into D0, etc., and clear other outputs.
        3'b000: begin           D0 = mux8 ; D1 = 0; D2 = 0; D3 = 0;
                                D4 = 0; D5 = 0; D6 = 0; D7 = 0; end
```

```
        3'b001: begin   D1 = mux8 ; D0 = 0; D2 = 0; D3 = 0;
                        D4 = 0; D5 = 0; D6 = 0; D7 = 0; end
        3'b010: begin   D2 = mux8 ; D0 = 0; D1 = 0; D3 = 0;
                        D4 = 0; D5 = 0; D6 = 0; D7 = 0; end
        3'b011: begin   D3 = mux8 ; D0 = 0; D1 = 0; D2 = 0;
                        D4 = 0; D5 = 0; D6 = 0; D7 = 0; end
        3'b100: begin   D4 = mux8 ; D0 = 0; D1 = 0; D2 = 0;
                        D3 = 0; D5 = 0; D6 = 0; D7 = 0; end
        3'b101: begin   D5 = mux8 ; D0 = 0; D1 = 0; D2 = 0;
                        D3 = 0; D4 = 0; D6 = 0; D7 = 0; end
        3'b110: begin   D6 = mux8 ; D0 = 0; D1 = 0; D2 = 0;
                        D3 = 0; D4 = 0; D5 = 0; D7 = 0; end
        3'b111: begin   D7 = mux8 ; D0 = 0; D1 = 0; D2 = 0;
                        D3 = 0; D4 = 0; D5 = 0; D6 = 0; end
        default: begin  D0 = 0; D1 = 0; D2 = 0; D3 = 0; D4 = 0;
                        D5 = 0; D6 = 0; D7 = 0; end
endcase
end
```

This appears to be more complicated than the MUX design since we have multiple statements within each case. The always block and case statements are used here as we had done in the MUX design before. Note that the 'mux8' is an input. 'mux8' input is assigned to the particular DEMUX output depending upon the ABC select pins. For example, if ABC = 111, mux8 is output to D7. It may be noted here that the other outputs, viz., D0 through D6 are each assigned low. Although this is done deliberately to every case value in order to enhance the readability and reliability of the code, this need not be done this way alone. A simpler

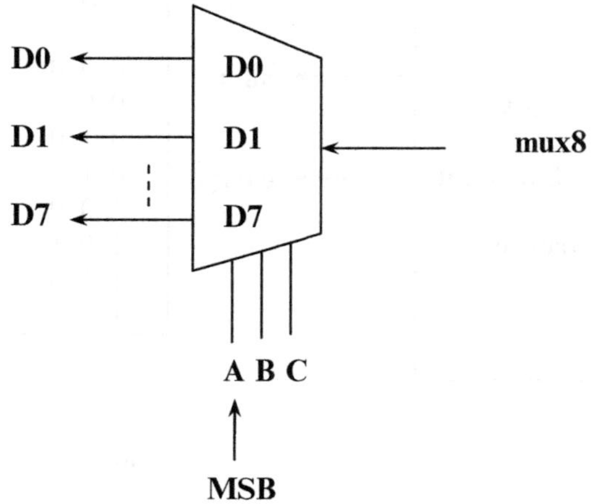

Fig. 3.7 Eight output DEMUX

way of writing is to insert the following 'D' statements between 'always @ (A or B or C or mux8) begin' and 'case ({A, B, C})'.

D0 = 0 ; D1 = 0; D2 = 0; D3 = 0;
D4 = 0; D5 = 0; D6 = 0; D7 = 0;

// Common statement for clearing all outputs.
You may then remove all 'D' outputs assigned to '0' in every case statement.

3.2.6 Verilog Modeling of a Full Adder

We will consider a full adder and code it in different ways in order to understand behavioral, data flow and structural realizations of the same. As shown in Figure 3.8, we have three inputs, A, B, and C, where C may be regarded as the carry in. When the three bits are added, we get a 'carry out' and a 'sum'. The truth table can be easily filled as follows: In the traditional way, inputs are entered as a binary progression starting with '000' entry. A faster way to fill inputs A B C are to fill vertically starting with four '0's followed by four '1's for A; alternately two '0's and two '1's for B followed by alternately one '0' and one '1' for C. The outputs can be filled by treating them as two-bit result, adding two numbers first [(A+B)], followed by adding the result to the last number [(A + B) + C]. As an example, let us see the addition of the last row, which may be expressed as: (1 + 1) + 1 = (10) + 1 = 11, all in binary notation. Similarly, all rows of the truth table can be filled fast. By inspecting the truth table, we recognize the outputs, 'sum' and 'carryo' as XOR and majority logic of the three inputs respectively. A majority logic is recognized if there are two or more '1' (logic high) input entries for the corresponding 'high' output.

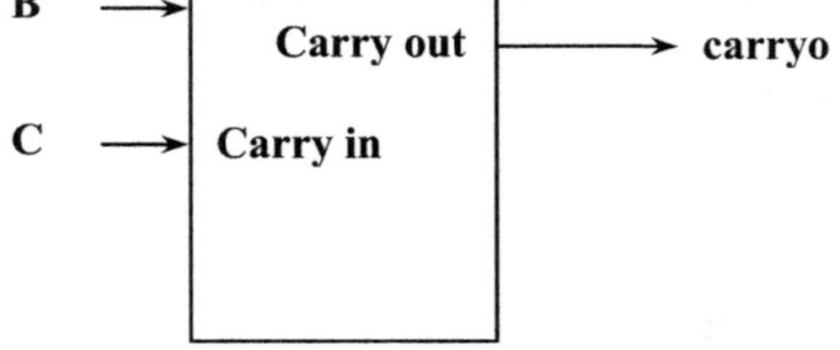

TRUTH TABLE

A B C	carryo	sum
0 0 0	0	0
0 0 1	0	1
0 1 0	0	1
0 1 1	1	0
1 0 0	0	1
1 0 1	1	0
1 1 0	1	0
1 1 1	1	1

sum = (A^B)^C

carryo = AB+BC+CA

Fig. 3.8 Realization of a full adder

There are three ways of implementations for the full adder. The first one is the behavioral level of realization expressed by a single 'assign' statement:

Verilog_code 3.4

```
assign sum_total  =  (A + B) + C;        // Realize sum, carry being inherent.
```

The parenthesis is generally put in order to make the synthesis tool more efficient in optimization, which aspect will be discussed in RTL coding guidelines. This type offers the most concise representation of addition, and holds good for multi-bit precision as well, unlike other two types that will be discussed shortly. Another type of realizing the full adder is the data flow structure as shown in the following:

Verilog_code 3.5

```
assign sum    = (A^B) ^C ;                // Realize sum.
assign carryo = (A&B)|(B&C)|(C&A) ;       // Realize carry out, AB + BC + CA.
```

This type was pictorially depicted earlier in Figure 3.8. The above two types will be very handy while pipelining a design, which we will learn later on. The third type is the structural realization using primitive gates as shown in Figure 3.9. This is very close to the schematic circuit diagram representation. Verilog code for the same is as follows:

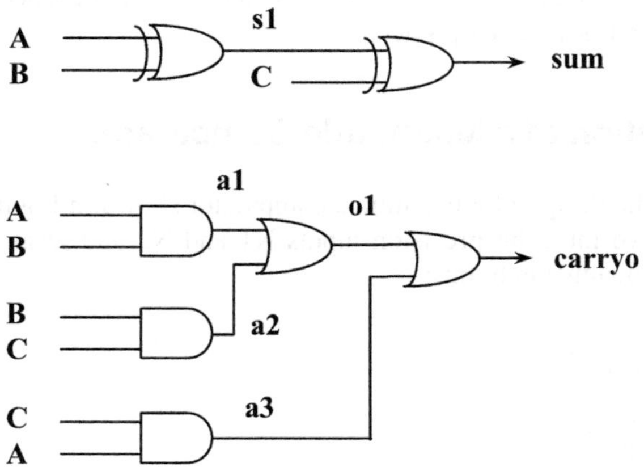

Fig. 3.9 Structural level realization of full adder using primitive gates

Verilog_code 3.6

```
xor (s1, A, B) ;            // Realize sum = (A^B)^C using gate
xor (sum, s1, C) ;          // primitives.
// The first entry 'sum' is for output and, the other two are for inputs for all
// primitive gates.
and (a1, A, B) ;            // Also compute AB,
and (a2, B, C) ;            // BC,
and (a3, C, A) ;            // CA
or (o1, a1, a2) ;           // and OR them together
or (carryo, o1, a3) ;       // to realize carryo = AB + BC + CA.
```

Note that s1, a1, a2, a3, and o1 are all intermediate output signals. Other gates that can be used in a structural design are 'nand', 'nor', 'xor', 'xnor', 'buf', and 'not'. As far as possible, usage of this approach must be minimized since it leads to the writing of long codes, and are unwieldy for large designs. However, this structure can be effectively used for the implementation of tri-state buffers/inverters such as the following:

```
bufif0 u1 (out, in, sel) ;   // out = in if sel = 0, otherwise out is tri-stated
bufif1 u2 (out, in, sel) ;   // out = in if sel = 1, otherwise out is tri-stated
notif0 u3 (out, in, sel) ;   // out = ! in if sel = 0, otherwise out is tri-stated
notif1 u4 (out, in, sel) ;   // out = ! in if sel = 1, otherwise out is tri-stated
```

where 'u1', 'u2', etc. stand for the instantiations similar to the convention adopted in schematic circuit diagrams, 'out' for the tri-stated (high impedance, z) output, 'in' for the input, and 'sel' for chip select.

3.2.7 Realization of a Magnitude Comparator

We will examine the design of a magnitude comparator shown in Figure 3.10. Let us say, we have two multi-bit precision inputs N1 and N2 and wish to generate outputs for the following conditions:

(i) N1 greater than N2,
(ii) N1 less than N2,
(iii) N1 equal to N2,
(iv) N1 not equal to N2,
(v) N1 less or equal to N2, and
(vi) N1 greater or equal to N2.

Verilog code for this example is as follows:

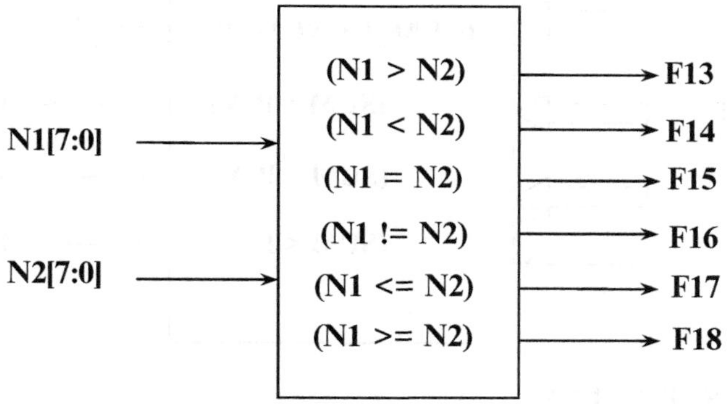

Fig. 3.10 Block diagram of an eight bit magnitude comparator

Verilog_code 3.7

// Realization of a magnitude comparator

```
always @ (N1 or N2)
        begin
          F13 = (N1 > N2) ;          // Set output if N1 is greater than N2.
          F14 = (N1 < N2) ;          // Set output if N1 is less than N2.
          F15 = (N1 = = N2) ;        // Set output if N1 is equal to N2.
          F16 = (N1 != N2) ;         // Set output if N1 is not equal to N2.
          F17 = (N1 <= N2) ;         // Set output if N1 is less than or equal to N2.
          F18 = (N1 >= N2) ; // Set output if N1 is greater than or equal to N2.
        end
```

The codes are straightforward and self-explanatory. Logical equivalence uses two equal to symbols as shown. The compiler will report error if only one equal to symbol is used. When the prescribed condition [(N1 = = N2), for example] is satisfied, a logical high is assigned to the specified output [namely, F15].

3.2.8 A Design Example Using an Adder and a Magnitude Comparator

Before we wind up the combinational circuit design, let us consider one more design example using some of the examples we have already covered. As shown in Figure 3.11, let us say that we wish to compute the sum (SUM[8:0]) of two multi-bit precision numbers, NUM_1 and NUM_2, and compare with a preset value,

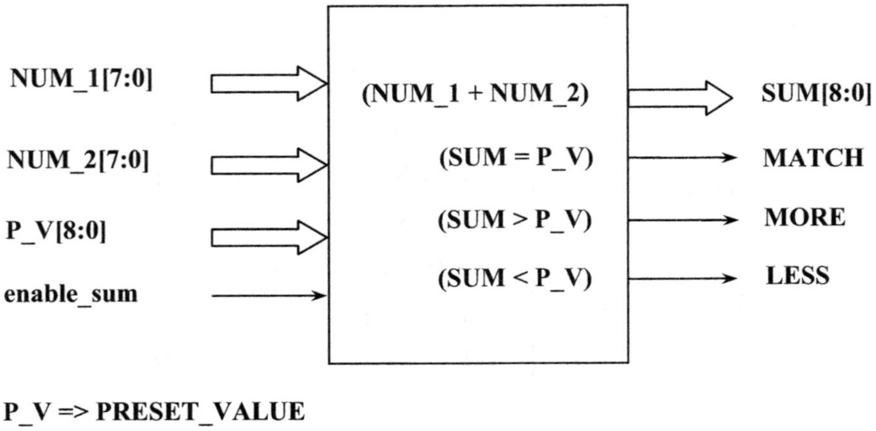

Fig. 3.11 Block diagram of a design example

P_V[8:0]. If SUM equals the preset value, the output MATCH is generated; else if SUM is greater than the preset value, MORE is generated. Otherwise, the signal LESS is issued. Outputs must be valid only if 'enable_sum' is active, otherwise clear all the outputs.

We will now see how to write the code for this application. Even though we could have used 'assign' statements for this problem, we will use 'always' block since we need output only when there is a change in any of the inputs. The always block contains the list of all the input signals, namely, enable_sum, NUM_1, NUM_2, and PRESET_VALUE. In order to compute the sum, we will use the behavioral model of a full adder we have seen earlier. Similarly, we will use a MUX model to generate the final outputs, taking the signal, enable_sum, into account. Note that the MUX model used here does not use 'assign' statement as done before. This approach facilitates the writing of a very compact code, yet readable as well as synthesizable. Verilog code is as follows:

Verilog_code 3.8

// Verilog code for design example shown in Figure 3.11.

```
always @ (enable_sum or NUM_1 or NUM_2 or PRESET_VALUE)
begin
        SUM[8:0] = enable_sum ? (NUM_1[7:0] + NUM_2[7:0]) : 9'd0 ;
                        // Compute sum if enabled, otherwise output '0'.
        MATCH   = enable_sum ? (SUM = = PRESET_VALUE) : 1'b0 ;
                        // Set output if SUM is equal to
                        // PRESET_VALUE, only if enabled.
        MORE    = enable_sum ? (SUM > PRESET_VALUE) : 1'b0 ;
                        // Set output if SUM is greater than
```

```
                    // PRESET_VALUE, only if enabled.
     LESS      = enable_sum ? (SUM < PRESET_VALUE) : 1'b0 ;
                    // Set output if SUM is  less than
                    // PRESET_VALUE, only if enabled.
end
```

The first statement for SUM contains the signal widths as well, so that the readability is enhanced, even though the code will work if precision is not mentioned explicitly. Statement like (SUM > PRESET_VALUE) returns a '1' if the mentioned condition is satisfied, otherwise not. All the four statements are processed concurrently.

3.3 Verilog Modeling of Sequential Circuits

3.3.1 Realization of a D Flip-flop

We will now see how to model sequential circuits. The simplest of flip-flops is the D flip-flop shown in Figure 3.12. When a reset signal is applied to the flip-flop, the output Q is cleared. At the rising edge of the clock, whatever value is present at the D input is stored in Q. We will use asynchronous, active low for the reset signal and positive edge for the clock for registering D input of the flip-flop throughout the book since these are popular practices in industries. Coding this is quite easy. We will use the 'always' block as in the combinational circuits realization, but with this difference. As this is a sequential circuit, we specify positive edge for the clock as 'posedge clk' in always statement along with active low reset signal as 'negedge reset_n' since in this type of 'always' statement list, only the edge transitions are allowed. The Verilog code is as follows:

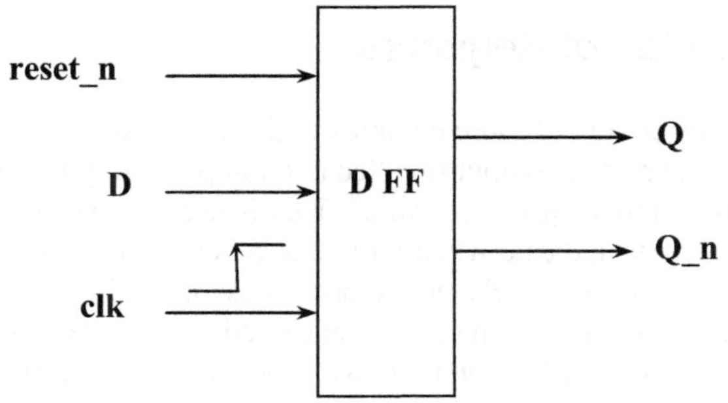

Fig. 3.12 D flip-flop with reset

Verilog_code 3.9

// Realization of a D flip-flop with a reset control using 'always' block.

```
always @ (posedge clk or negedge reset_n)
begin
        if (reset_n = = 1'b0)
                begin
                        Q       <= 0 ;   // Clear register when system is reset.
                        Q_n     <= 1 ;
                end
        else
                begin
                        Q       <= D ;   // Store or register D input
                        Q_n     <= !D ;  // and its complement.
                end
end
```

In big systems, the signal 'reset_n' may be a master reset, mounted on a control panel, which will be usually connected via long cable. In the event when the cable snaps, the reset pin must be pulled to a safe value. This can be easily done if a pull-up resistor is installed at the reset pin in the PCB housing the sequential circuits. This requires active low signal for the reset signal. Usually, the term 'register' is used instead of a flip-flop. In the code for assigning a signal, the symbol '<=' is used. Don't mistake it for less than or equal to assignment. It is known as a non-blocking statement, whereas in 'assign' statements discussed earlier, we had used '=' symbol called the blocking statement. In an 'always' block, if we use '=' instead of '<=', the compiler tool will complain.

3.3.2 Realization of Registers

We will now see the coding of more complex registers. What you see in Figure 3.13 are two registers: one register output is called pixeloutp_valid: p for previous and the other is the desired output, 'pixelout_valid'. This is used in one of the applications such as video scaling to indicate when an image pixel is valid. As per the application, 'pixelout_valid' must be delayed by one clock pulse. In order to accomplish this, we have 'pixeloutp_valid', which gets registered in advance by one clock pulse. The Verilog code for this application is shown in Verilog_code 3.10.

The pixelout data is valid only within a window. The start and end of the window is determined by the conditions:

set_pixout = (!A)(B)(!C) and reset_pixout = ABC. These conditions are realized using assign statements as shown in the code. There are two 'always' blocks: the first one is the 'pixeloutp_valid' register and the second is the 'pixelout_valid'

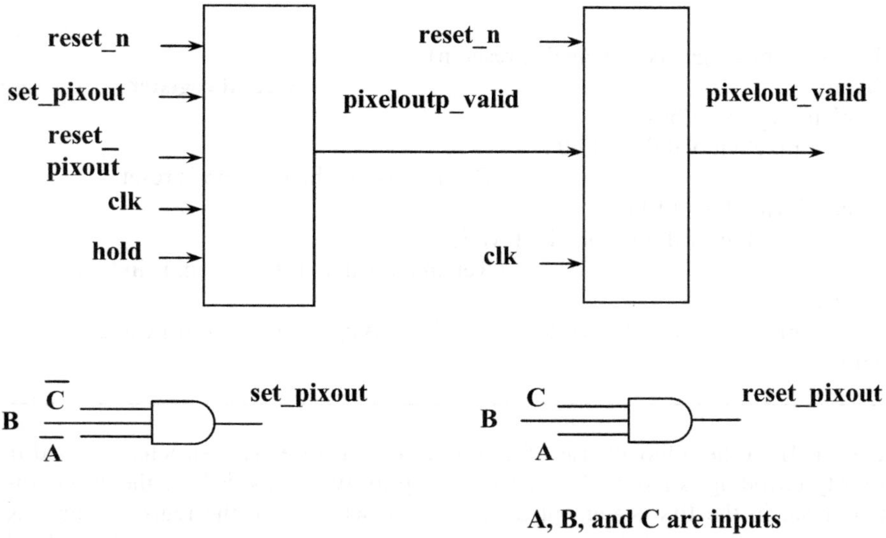

Fig. 3.13 Realization of registers

Verilog_code 3.10

// Realization of registers using 'always' block

```
assign set_pixout   = (A = = 1'b0) && (B = = 1'b1) && (C = = 1'b0) ;
                                        // Pre-compute (not A) and (B) and (not C)
                                        // which is a condition for setting pixout_valid.
assign reset_pixout  = (A = = 1'b1) && (B = = 1'b1) && (C = = 1'b1) ;
                              // Pre-check ABC status.
                              // ABC = 1 is a condition for resetting pixout_valid.
always @ (posedge clk or negedge reset_n)
begin                                                    // First register
  if (reset_n == 1'b0)
        pixeloutp_valid <= 1'b0 ;       // Clear register when system is reset.
  else if (hold == 1'b1)
        pixeloutp_valid <= pixeloutp_valid ;
                                        // Retain the value if the system is in hold.
  else if (set_pixout == 1'b1)
        pixeloutp_valid <= 1'b1 ;       // Set or
  else if (reset_pixout == 1'b1)
         pixeloutp_valid <= 1'b0;       // reset when the conditions are satisfied.

  else
        pixeloutp_valid <= pixeloutp_valid ;          // Otherwise, don't disturb.
end
```

```
always @ (posedge clk or negedge reset_n)
 begin                                          // Second register
   if (reset_n == 1'b0)
         pixelout_valid <= 1'b0 ;
                                   // Clear register when system is reset.
   else if (hold == 1'b1)
         pixelout_valid <= pixelout_valid ;
                                   // Retain the value if the system is in hold.
   else
         pixelout_valid <= pixeloutp_valid ; // Assign previous (clk) value.
 end
```

register. Both these blocks use 'if–else if–else' structure. This structure is used if priority encoding is required. The top most priority is provided for the asynchronous reset in the first statement using 'if ', in which case the register output is cleared (may be regarded as system initialization). The next in priority is the hold signal. This is similar to the hold signal in most microprocessors. Whenever a 'hold' signal is applied, the previous register content is frozen, so that the process may resume from where it was suspended when the hold signal was withdrawn. This is taken care of in the first 'else if ' statement. In the next two 'else if ' statements, the register is set or reset in that order of priority. This holds good only for the 'pixeloutp_valid' register and not for 'pixelout_valid'. The last statement 'else' is the lowest priority. In this else case, 'pixeloutp_valid' is not disturbed, whereas 'pixelout_valid' register receives the 'pixeloutp_valid' contents. The latter statement is responsible for bringing about one clock delay for 'pixelout_valid' with respect to the 'pixeloutp_valid' register.

It is a good practice to have meaningful comments throughout your code: what is obvious in a statement, don't repeat it verbatim. Instead, your comment must be as far as possible simple English statement narrating a story. Also, give apt names to signals, modules, and files in your design. In case, you want to give a description running to many lines, you can use /* */ as in 'C'. Note that each of the 'if' or 'else if' statements would create a nested two input multiplexer and, thereby, slowing the system operation. From experience, we suggest that you don't exceed four or five such nesting. Smaller the number of nesting, higher will be the speed of operation. If you have many signal outputs (registers) in one sequential 'always' block, debugging the code will be a frustrating experience. It is, therefore, highly recommended that one should have only one register in one 'always' block as shown in the code. One register output feeding into another register is known as the 'pipelining'. In the present case, pipelining is not the aim. It is only to delay a signal, say, 'pixelout_valid' so as to keep pace with another data signal, namely, pixel_data (not shown in this design). Pipelining will be covered in depth in the later chapters.

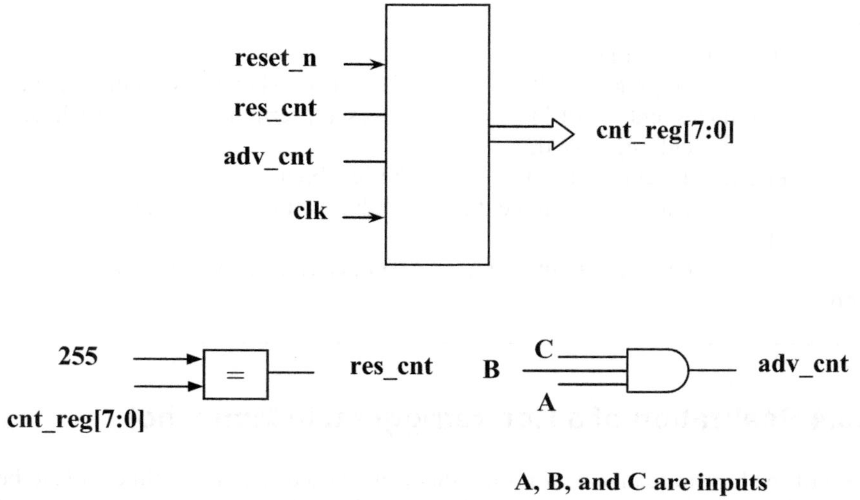

Fig. 3.14 Realization of a Counter

3.3.3 Realization of a Counter

We will go on to the next example of realizing a counter which is also made up of flip-flops. It has an asynchronous reset input, reset_n, and an 8-bit counter of width [7:0] as shown in Figure 3.14. The output is named as 'cnt_reg' so that it may be readily identified as a register. The counter can advance by one at the rising edge of 'clk' so long as the counter is enabled by the signal 'adv_cnt' for the condition: ABC = 1. When the running 'cnt_reg' equals 255, the counter is reset since the signal, 'res_cnt', is asserted.

The Verilog code for the counter is as follows: Signals, res_cnt, adv_cnt, and cnt_next are realized using 'assign' statements as shown in the Verilog_code 3.11. Since cnt_reg is sequential in nature, it is realized in an 'always @(posedge clk)' block. The functioning of each statement will be clear by reading the line comments. In order to speed up the operation, the counter is pre-incremented using the 'assign' statement, instead of advancing within the always block.

Verilog_code 3.11

```
// Verilog code to realize a counter
assign res_cnt = (cnt_reg  = = 255) ;          // Condition for resetting the counter.
assign adv_cnt = (A = = 1'b1)&(B = = 1)&(C = = 1) ;
                                     // Condition for Pre-incrementing the counter.
assign cnt_next = cnt_reg + 1 ;      // Pre-increment the counter.

always @ (posedge clk or negedge reset_n)
```

```
begin
        if (reset_n == 1'b0)
                cnt_reg <= 8'd0 ;            // Initialize when the system is reset.
        else if (res_cnt == 1'b1)            // Reset if terminal count is reached.
                cnt_reg <= 8'd0 ;
        else if (adv_cnt == 1'b1)            // If enabled,
                cnt_reg <= cnt_next ;        // advance the counter once.
        else
                cnt_reg <= cnt_reg  ;        // Otherwise, do not disturb.
end
```

3.3.4 Realization of a Non-retriggerable Monoshot

We will see how to realize a monoshot that can be used as a timer. Since it is to be designed as a non-retriggerable monoshot, once it is triggered, future triggers shall have no effect so long as it is running. In other words, you can trigger it only when it is not running, i.e., when the output, 'delay_out', is not high. The monoshot, as shown in Fig. 3.15, can be triggered by applying a rising edge signal at the 'trigger' input. 'cntd_reg' is an 8-bit counter that increments by one at the rising edge of 'clk', provided the timer is running. This counter is reset and the 'delay_out' signal goes low once the counter touches the value 255. The monoshot produces 255 clock cycles delay for the preset value, cntd_reg = 255, and for this duration it turns on the 'delay_out' signal. If we want longer delays, we can either cascade more number of counters or increase the bit precision of the counter. The signal, reset_n, is the asynchronous reset input, which clears the counter, 'cntd_reg', and the 'delay_out' signal when the system is powered on.

Verilog code for this design is as follows:

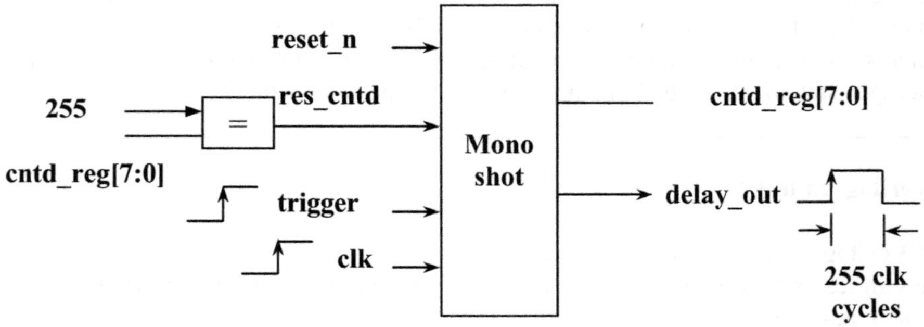

Fig. 3.15 Block diagram of a non-retriggerable monoshot

Verilog_code 3.12

// Verilog code for the non-retriggerable monoshot.

```
assign res_cntd  = (reset_n = = 1'b0) || (cntd_reg  = = 255) ;
                                              // Condition for resetting the counter.
                          // Change preset value 255 if you need another delay.
assign run_delay = (triggerp = = 0) && (trigger = = 1) ;
                                              // Detect the positive edge of trigger.
assign cntd_next = cntd_reg + 1 ;             // Pre-increment the  counter.

always @ (posedge clk or posedge res_cntd)
begin
   if (res_cntd == 1)                         // Initialize when the system is reset
                                              // or if the terminal count is reached.
                                              // This has the top most priority.
       begin
              cntd_reg <= 8'd0 ;
             delay_out <= 0 ;
               triggerp <= 0 ;
       end
   else if (delay_out = = 1)                  // This implies the timer is running.
                                              // This has the second priority.
       begin
              cntd_reg <= cntd_next ;         // Advance the count by one if
                                              // the timer is still running.
              triggerp <= trigger ;           // Preserve the current state of trigger.
       end
  else if (run_delay = = 1)                   // This is the rising edge of trigger.
                                              // This has the lowest priority.
       begin
            delay_out <= 1 ;                  // Start the delay if the positive
                                              // edge of trigger is detected.
              triggerp <= trigger ;           // Preserve the current state of trigger.
       end
  else
       begin
             cntd_reg <= cntd_reg ;           // Otherwise, don't disturb.
            delay_out <= delay_out ;
              triggerp <= trigger ;           // Preserve the current state of trigger.
       end
end
```

The first assign statement combines 'reset_n' signal and the counter value advancing to 255 to generate the reset signal for the counter. The second statement detects the positive edge of the trigger input if the present value is high and the previous value (register 'triggerp') is low. Since we have only three registers in this design, namely, cntd_reg, delay_out, and triggerp, only a single 'always' block is used, instead of providing a separate 'always' block for each of the three registers as suggested before. Note that we have used 'posedge res_cntd' in the always block, instead of the usual 'negedge reset_n'. Follow this argument very closely. When the running counter, cntd_reg, value is 254 at the rising edge of the 'clk', then the statement 'else if (delay_out = = 1)' alone is satisfied since the timer is already running and, therefore, the statement 'cntd_reg <= cntd_next;' advances the cntd_reg to 255. Immediately, the signal 'res_cntd' goes high, satisfying the condition: 'posedge res_cntd' in the always block, thus resetting the cntd_reg as well as the timer output. The above process happens in a flash; as a result you can notice only a very sharp pulse for the res_cntd signal. In other words, the count 254 and 255 take place practically at the same time. Thus, the timer (rather the counter, cntd_reg) starts counting from 0 through 254 only, which explains why we get a time delay of 255 clock cycles. Later on, when we examine the waveforms in Chapter 6, we will see that the above discussion is indeed true.

The statement 'else if (run_delay = = 1)' detects the rising edge of the trigger only if the timer is not running, and starts the timer by asserting its output, 'delay_out'. Note that this statement will not be processed if 'delay_out' is already high since the statement 'else if (delay_out = = 1)' has a higher priority. One must not forget to push the current value of the 'trigger' to the previous value of trigger (triggerp) at every 'else if' or 'else' statement block as shown in the code. Otherwise, the rising edge of trigger may be detected at every rising edge of 'clk' and hamper with the satisfactory working of the timer.

3.3.5 Verilog Coding of a Shift Register

We saw earlier, how to model shift registers using combinational circuit. In the present treatment, we will use sequential circuit to effect the actual shifting operation. As shown in Figure 3.16, **a** right shift register consists of a register of specific width, say, 16 bits represented by [15:0], which can be preset with a value, 1010101010101010, for instance, and a shift control to bring about right shifting of the register at the positive edge of 'clk'. The register can be cleared asynchronously by the input, 'reset_n'.

There are two ways of realizing the shift register as shown in the following codes, Verilog_code 3.13 and 3.14. Both these methods use an 'assign' statement and an 'always' block. The two methods differ only in the 'assign' statements used to perform right shift operation in advance. In the first method, 'data_out1 >> 1' effects the shifting, which approach is the same as the one we have used in combinational circuits earlier. In the alternative approach, we use concatenation:
{1'b0, data_out1[15:1]}.

It may be noted that after right shift, the MSB of the result is forced to a zero. This can be done by the first entry, 1'b0. After right shift, data_out1[0] is discarded, and data_out1[15:1] becomes the new content for the bits [14:0] of the end result.

Therefore, the final result is:
dataout2_next[15:0] = {1'b0,data_out1[15:1]}. This is a straightforward assignment without really involving any shift.

In the always block, any data that is desired is preset whenever the system is reset by asserting the signal, reset_n. During the normal course of operation, at the rising edge of 'clk', the pre-shifted value is transferred to the final output, data_out1[15:0] or data_out2[15:0] if 'shift' signal is asserted. The shift signal is asserted synchronously with the clock only when the occasion demands it. Otherwise, it is de-asserted.

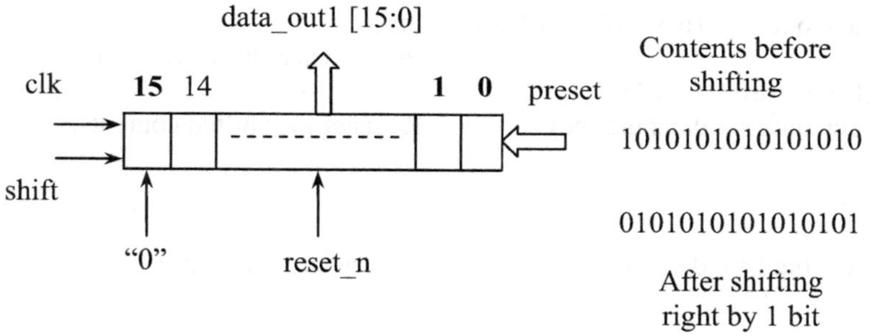

Fig. 3.16 Block diagram of a right shift register

Verilog_code 3.13

// Realization of a right shift register using 'assign' and 'always' block

```
assign dataout1_next = (data_out1 >> 1) ;
                                    // Pre-shift right the contents of data_out1
                                    // register by one bit.
always @ (posedge clk or negedge reset_n)
  begin
        if (reset_n == 1'b0)
        data_out1 <= 16'b1010_1010_1010_1010 ;
                                            // Preset when the system is reset.
/*      Underscore inserted in between the binary values are to improve the
        readability – Compiler accepts the same without complaining.        */
        else if (shift == 1'b1)
        data_out1 <= dataout1_next ;        // Register the shifted contents.
        else
        data_out1 <= data_out1 ; // Otherwise, don't shift.
  end
```

Verilog_code 3.14

// Alternate realization of the right shift register

```
assign dataout2_next[15:0] = ({1'b0, data_out1[15:1]}) ;
                                        // Pre-shift right the contents of data_out2
                                        // register by one bit.

always @ (posedge clk or negedge reset_n)
 begin
        if (reset_n == 1'b0)
        data_out2 <= 16'b1010_1010_1010_1010 ;
                                        // Preset when the system is reset.
        else if (shift == 1'b1)
        data_out2 <= dataout2_next ;    // Register the shifted contents.

        else

        data_out2 <= data_out2 ;        // Otherwise, don't shift.
 end
```

3.3.6 Realization of a Parallel to Serial Converter

Often, we need to convert parallel information into serial bits, and vice versa. For instance, in the computer keyboard, we type a character, and the controller recognizes it as a (parallel) byte of information, converts it into serial bits, and sends it to the host computer. Let us see how to realize such a parallel to serial converter. We can use a shift register to accomplish this. We can load a multi-bit precision data, say, set_data [15:0] into the shift register 'sr' at the positive edge of 'clk' with 'load' control asserted as shown in Figure 3.17. Once the shift register is loaded, the load input must be withdrawn in order to start the conversion. Depending upon which bit (LSB or MSB) we need to transmit first, signal 'rl_n' must be held high or low. For example, rl_n is kept low (meaning left shift of register) so that MSB is sent out first.

The conversion starts when the input 'shift' is asserted. Shift and rl_n signals must be maintained till the conversion is complete. If rl_n is '0', then the MSB is sent out to 'data_out' pin for serial transmission. Otherwise (meaning right shift), the LSB of the data word is sent out. At what point of time the 'data_out' is valid is indicated by 'data_valid' signal. When all the 16 bits are transmitted, one bit every 'clk' cycle, 'data_valid' signal is de-asserted and the output 'eoc' (abbreviation for end of conversion) is asserted. Whenever 'reset_n' is applied, the module is initialized. Verilog code for this module is as follows:

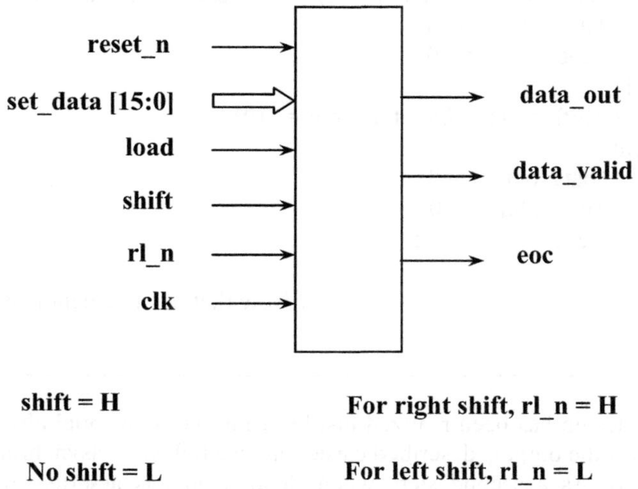

Fig. 3.17 Block diagram of a parallel to serial converter

Verilog_code 3.15

// Code for parallel to serial converter

```
always @ (posedge clk or negedge reset_n)
begin
        if (reset_n == 0)
          begin
               data_out        <= 0 ;    // Clear when the system is reset.
               data_valid      <= 0 ;
               eoc             <= 0 ;
          end
        else if (load == 1)
          begin
               sr              <= set_data ;       // Preset or clear registers.
               cnt_ps_reg      <= 16 ;
               data_out        <= 0 ;
               data_valid      <= 0 ;
               eoc             <= 0 ;
          end
        else if ((shift == 1) && (cnt_ps_reg != 0))
          begin
               sr          <=  rl_n ? (sr >> 1) : (sr << 1) ;
                                               // Register the shifted contents.
               data_out    <= rl_n ?  sr[0] : sr[15]; // Select LSB or MSB.
```

```
                cnt_ps_reg  <= cnt_ps_reg - 1; // Keep track of the bits to be sent.
               data_valid   <= 1 ;
                eoc         <= 0 ;
          end
        else if ((shift == 1) && (cnt_ps_reg == 0))
          begin
                data_out    <= 0 ;
               data_valid   <= 0 ;
                eoc         <= 1 ;
          end
        else                    ;              // Note that no statement is written.
end
```

The entire design has been realized just by using one sequential always block. To start with, all the outputs described earlier are cleared when asynchronous reset input is applied. As a second step, when 'load' signal is asserted, input data, 'set_data [15:0]', is transferred to the shift register, 'sr[15:0]', and the outputs are cleared. A counter, 'cnt_ps_reg', is preset to 16, which keeps track of the number of bits to be transmitted. In the next step, 'load' is withdrawn, and 'shift is asserted. The control branches off to the statement block: else if ((shift = = 1) && (cnt_ps_reg != 0)) begin. Inside this statement block, the shift register data is shifted by one bit right or left, and the 'data_out' is transmitted with 'data_valid' asserted.

The logic must normally be put outside the always block using 'assign' statements for improving the speed of operation, even though it is put inside 'else if ' statement in the code. The running counter is decremented by one. It is this branch which is processed repeatedly at every rising edge of the 'clk' till all the bits are sent out. At this point of time, 'cnt_ps_reg' becomes zero. With the arrival of the next 'clk', the last 'else if ' statement is satisfied, thus signaling the end of conversion and de-asserting 'data_valid' signal. The penultimate statement is 'else;', rather a blank statement since we have nothing more to process. The hardware realized in this design is basically registers, a counter and multiplexers.

3.3.7 Realization of a Model State Machine

A finite state machine (FSM) is one of the most important components in the design of a sequential circuit. We will consider a model of a typical state machine. The state diagram of such a model is presented in Figure 3.18. It consists of four states: S0 through S3. The corresponding binary states are 00 to 11, marked within each state. Each of these states is identified by a lamp: Z0, Z1, Z2, or Z3. A logical high ('1') turns on a lamp. The state of the machine depends upon two inputs: In1 and In2. To start with, let us assume that the machine is in state S0. This state continues so long as In1 = 0. If In1 changes, state changes to S2. From this state, the machine may switch to either S1 state or S3 state depending upon In2 value.

In S1 state, there are three possibilities of state change: S1 state is continued for In1 = 0, In2 = 1, whereas for In1 = 1, In2= 1, the state changes to S3. Instead, if In2 becomes '0', then it goes back to the S0 state, from where we started. A similar explanation holds good for the state S3. The Verilog code for the FSM is as follows:

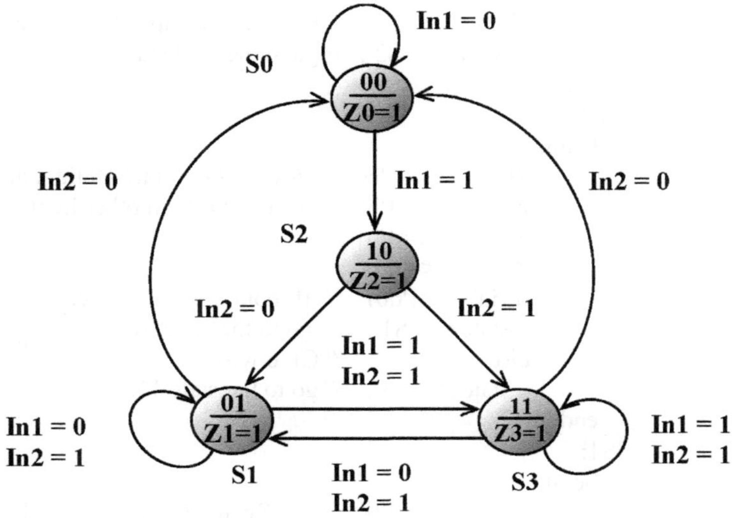

Fig. 3.18 Model state machine

Verilog_code 3.16

// Model for Sequential Machines

```
always @ (posedge clk or negedge reset_n)
begin
  if (reset_n == 1'b0)
        begin
                        Z0       <= 1'b0 ;           // Switch off all the
                                                     // lights to start with.
                        Z1       <= 1'b0 ;
                        Z2       <= 1'b0 ;
                        Z3       <= 1'b0 ;
                        state    <= `S0 ;            // Initialize the state when the
                                                     // system is reset.
        end
  else
        case(state)
              `S0:
                 begin
```

```
                Z0      <= 1'b1 ;         // Switch ON state 00 light and
                Z1      <= 1'b0 ;         // switch OFF all other lights.
                Z2      <= 1'b0 ;
                Z3      <= 1'b0 ;
                if (in1 == 1'b0)   // If input 1 is not active, continue
                   state <= `S0 ;  // to remain in the state 00.
                else               // However, if input 1 is active,
                   state <= `S2 ;  // go to the next state.
     end
   `S 2:
     begin
                Z0      <= 1'b0 ; // Switch ON  state 10 light and
                Z1      <= 1'b0 ; // switch OFF all other lights.
                Z2      <= 1'b1 ;
                Z3      <= 1'b0 ;
                if (in2 == 1'b0)    // If input 2 is not active,
                  state <= `S1 ;    // go to the state 01.
                else                // Otherwise,
                  state <= `S3 ;    // go to the state 11.
     end
   `S1:
     begin
                Z0      <= 1'b0 ;       // Switch ON state 01 light and
                Z1      <= 1'b1 ;
                Z2      <= 1'b0 ;       // switch OFF all other lights.
                Z3      <= 1'b0 ;
                if (in2 == 1'b0)               // If input 2 is not active,
                  state <= `S0 ;               // go to the state 00.
                else if (in1 == 1'b1)          // If input 1 is active,
                  state <= `S3 ;               // go to the state 11.
                else
                  state <= `S1 ;      // Otherwise, remain in the state 01.
      end
   `S3:
      begin
                Z0      <= 1'b0 ;     // Switch ON state 11 light and
                Z1      <= 1'b0 ;
                Z2      <= 1'b0 ;     // switch OFF all other lights.
                Z3      <= 1'b1 ;
                if (in2 == 1'b0)               // If input 2 is not active,
                 state <= `S0 ;                // go to the state 00.
                else if (in1 == 1'b0)          // If input 1 is not active,
                  state <= `S1 ;               // go to the state 01.
                else
                 state <= `S3 ;       // Otherwise, remain in the state 11.
      end
```

```
                default:  state <= `S0 ;  // Otherwise, remain in the  state 00.
        endcase
end
```

The FSM can be conveniently realized by using the 'always' block and 'case' statements. As in other applications we have considered earlier, system reset clears all outputs Z0–Z3, and initializes state to S0. Once the reset is withdrawn, the FSM becomes active going from one state to another as we have seen before. These states are all covered in the 'case' statements. With the arrival of 'clk' after reset is withdrawn, the machine enters 'S0' state switching on Z0 lamp. If in1 = 0, the 'state' continues to be in S0, otherwise, it changes to S2 state. These controls are brought about by 'if ', 'else' statements. Similar explanations hold good for all other states: S1 to S3. Towards the end, we also put a default state to take care of tri-state or don't care conditions that may occur for the signal, 'state'.

3.3.8 Pattern Sequence Detector

Next, we will consider the design of a pattern sequence detector. Let us say, we wish to detect the occurrence of '0110' sequence in a serial bit stream input, 'in'. The desired output is as follows for the corresponding input pattern listed as an example:

```
Input pattern applied:     11110100110110001101....................
Output desired:            00000000001001000010....................
```

Immediately after every occurrence of the pattern 0110, the output must be one. This being a sequential circuit, we use an always block active at the positive edge of clock and as usual, we have a system reset as well. Being a serial sequence detector, it has only one input and one output. This design is basically an FSM realization. Therefore, we use similar structure, namely, always block and case statements we used in the case of the model state machine. The Verilog code for this application is straightforward and is as follows:

Verilog_code 3.17

// Pattern sequence detector

```
always @ (posedge clk or negedge reset_n)

begin
  if (reset_n == 1'b0)
                begin
```

```
                    out              <= 0 ;
                                          // Switch OFF output to start with.
                    psd_state   <= 0 ;          // Initialize the state when the
                                                // system is reset.
              end
else
          case (psd_state)
              0:  begin
                    out          <= 0 ;         // Switch OFF output.
                    psd_state <=  in ? 0 : 1 ;  // Change the state to '1' if
                                                // the input is '0'.
                                        // Remain in the state '0' otherwise.
                  end
              1:  begin                 // Enter for first occurrence of '0'.
                    out          <= 0 ;         // Switch OFF output.
                    psd_state <=  in ? 2 : 1 ;  // Change the state to '2' if
                                                // the input is '1'.
                                        // Remain in the state '1' otherwise.
                  end
              2:  begin                 // Enter for first occurrence of '01'.
                    out          <= 0 ;         // Switch OFF output.
                    psd_state <= in ? 3 : 1 ;   // Change the state to '3' if
                                                // the input is '1',
                                          // otherwise change the state to '1'.
                  end
              3:  begin           // Enter for first occurrence of '011'.
                    out          <= in ? 0 : 1 ;  // Switch ON the output if
                                                  // the input is '0110',
                                                  // otherwise switch it OFF.
                    psd_state <= in ? 0 : 1 ;   // Change the state to '1' if
                                                // the input is '0'.
                                        // Otherwise change the state to '0'.
                  end
          default:      psd_state          <= 0 ;
                                   // Remain in the state 0 for invalid states.
          endcase
end
endmodule
```

When the system is reset, the output, 'out' and state, 'psd_state', are initialized. In the previous example, we used actual names such as S0, S1, etc. for the states. We can use even decimal numbers for states, which will be easier to deal with. In the normal mode of operation when reset is not present, the machine remains in '0' state as long as the input is '1', realized by a MUX statement. If the input changes to ‘0’, then the state changes to ‘1’. As long as the input continues to be ‘0’, it remains in the state ‘1’, which is the state to detect the first occurrence of ‘0’ in the input bit stream. When input is ‘1’, the state changes to ‘2’. Similarly, state ‘2’ entry means first occurrence of ‘01’ in the bit stream. Pattern ‘010’ will take it back to state ‘1’. On the other hand, input pattern sequence ‘011’ changes the state to ‘3’. In this state, if input ‘0’ is detected, it means that the desired sequence, ‘0110’ has occurred. Therefore, the ‘out’, which was ‘0’ in all other states, is turned on to ‘1’. Since the last input is ‘0’, state changes to ‘1’ looking for ‘01’ pattern again. On the other hand, if the pattern detected was ‘0111’, the state changes to ‘0’, a reset condition, which we started to begin with. A default condition is also accounted for, which takes the machine to a safe ‘0’ state.

3.4 Coding Organization

In the foregoing treatments, we discussed only the core of specific designs for the combinational and sequential circuits. In order to convert them into working codes, we need to add some more statements and conform to certain specifications. Any design, be it combinational or sequential, may be regarded as a block as shown in Figure 3.19. The design will naturally have a name, say, ‘module_name’, inputs and outputs (I/Os) as shown. This is similar to viewing an IC with I/Os as pins. In Verilog, you can do the same precisely by using statements as follows. Design is referred to as a ‘module’. This shall be followed by identifying the design, say, ‘module_name’. Thereafter, list all the I/Os within brackets as shown. Next, we identify the actual inputs and outputs. For examples:

input reset_n ;
output out_1 ;

Declare the inputs and outputs in your design. In the next step, nets or wires (any signal used in ‘assign’ statements) and registers (any signal used in ‘always’ blocks) are declared. Examples are:

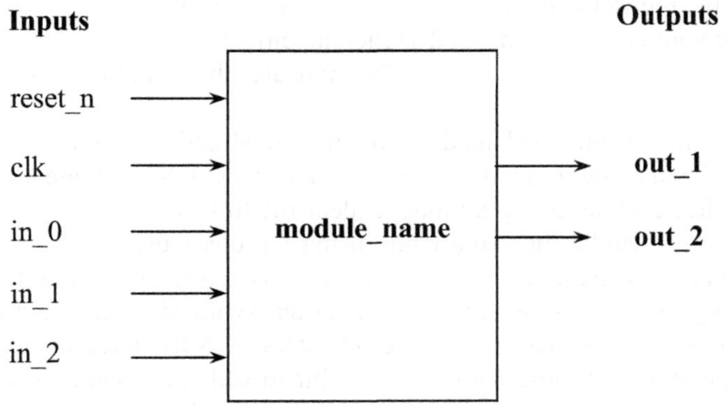

Fig. 3.19 Structure of a design module

```
wire    F1 ;           // in Figure 3.1,
reg  mux8 ;            // in Figure 3.6 and
reg  delay_out ;       // in Figure 3.15.
```

This is followed by Verilog description of your combinational and sequential circuits design which we covered in detail in Sections 3.2 and 3.3. Finally, you must signal the end of design by the key word: 'endmodule'. These are summarized as follows:

```
// Coding Organization

module module_name (// List your inputs and outputs here.
                       reset_n,
                       clk ,
                       in_0 ,
                       in_1 ,
                       in_2 ,
                       out_1 ,
                       out_2
                   );
// Note: Module statement above has one to one correspondence with
// Figure 3.19.

// Declare the inputs and outputs here as follows.
input reset_n ;
input clk ;
input in_0 ;
input in_1 ;
input in_2 ;
output out_1 ;
output out_2 ;

// Declare the register and wire (or net) as per your design.
// Describe your combinational and sequential circuits.
endmodule                  // This indicates the end of design.
```

Each of the designs explained in combinational and sequential circuits earlier could have been made into independent modules (files). Since there are very many such modules, each requiring an independent file to house the design, it would be cumbersome to handle the same while using various tools. Further, you need to write as many test benches, as there are designs. Each of these will have to be taken through various tools such as simulation, synthesis, place and route, back annotation, etc., as we shall see in later chapters. A better alternative to this is to create just two design files, one for the combinational circuits and the other for the sequential circuits, thereby lightening our burden. Incidentally, this approach gives a feel of handling larger and real design applications, which we will cover in depth in the last few chapters. You may use any of the standard editors such as WordPad, Vim, Vi, etc. for your Verilog design entry and, being commonplace, will not be described in this book.

3.4.1 Combinational Circuit Design

All the combinational circuits we have seen before have been put together in a single file and the same is as follows. All the codes are adequately commented and self-explanatory. Give proper spacing, indentation, and meaningful names for the files and signals to enhance the readability of your codes. Note that all I/O signals are declared as such after declaring the module. All the output signals in 'assign' statements are declared as 'wire' and the output signals in 'always' statements are declared as 'reg'. The data width is also declared as appropriate. For examples, N1 input is declared as width, [7:0], and width of 'reg' is [2:0] for F10. No width need be mentioned for single bit signals.

Verilog_code 3.18

```
/*          Combinational circuit realization using Verilog codes

The file name of the following code is:       comb_ckts.v
'.v' is the extension to indicate that the design file is in Verilog.

Note:    Verilog is case sensitive. A signal 'A' is different from signal 'a'.

To start with, declare the module you wish to design.
Note that the design file name is the same as the module name, 'comb_ckts', even
though different names may be used. This style reduces confusion later on.
*/
/*          means multiple line comments like C.        */
// means line comment.

module comb_ckts (                    // Declare the design module and list the
                                      // inputs and outputs (I/O).
                          A ,         // I/O s can be written in any order.
                          B ,
                          C ,
                          I0 ,
                          I1 ,
                          I2 ,
                          I3 ,
                          I4 ,
                          I5 ,
                          I6 ,
```

```
            I7 ,
            N1 ,
            N2 ,
            enable_sum ,
            NUM_1 ,
            NUM_2 ,
            PRESET_VALUE ,
            F1 ,
            F2 ,
            F3 ,
            F4 ,
            F5 ,
            F6 ,
            F7 ,
            F8 ,
            F9 ,
            F10 ,
            F11 ,    // Note that these I/O s are separated by
            F12 ,    // commas except the last.
            mux2 ,
            mux4 ,
            mux8 ,
            D0 ,
            D1 ,
            D2 ,
            D3 ,
            D4 ,
            D5 ,
            D6 ,
            D7 ,
            sum_total ,
            sum_df ,
            carryo_df ,
            sum ,
            carryo ,
            F13 ,
            F14 ,
            F15 ,
            F16 ,
            F17 ,
            F18 ,
            SUM ,
            MATCH ,
            MORE ,
            LESS
    );
```

```
input     A ;     // Declare the Inputs and Outputs of the module.
input     B ;
input     C ;
input     I0 ;
input     I1 ;
input     I2 ;
input     I3 ;
input     I4 ;
input     I5 ;
input     I6 ;
input     I7 ;
input     [7:0]       N1 ;     // Mention the size of the input -
                               // [7] is the MSB
input     [7:0]       N2 ;     // and [0] is the LSB.
input     enable_sum ;
input     [7:0]       NUM_1 ;
input     [7:0]       NUM_2 ;
input     [8:0]       PRESET_VALUE ;

output    F1 ;
output    F2 ;
output    F3 ;
output    F4 ;
output    F5 ;
output    F6 ;
output    F7 ;
output    F8 ;
output    F9 ;
output    [2:0]       F10 ;
output    [2:0]       F11 ;
output    [2:0]       F12 ;
output    mux2 ;
output    mux4 ;
output    mux8 ;
output    D0 ;
output    D1 ;
output    D2 ;
output    D3 ;
output    D4 ;
output    D5 ;
output    D6 ;
output    D7 ;
output    [1:0]                sum_total ;
output    sum_df ;
output    carryo_df ;
output    sum ;
```

```
output          carryo ;
output          F13 ;
output          F14 ;
output          F15 ;
output          F16 ;
output          F17 ;
output          F18 ;
output          [8:0]                   SUM ;
output          MATCH ;
output          MORE ;
output          LESS ;

wire            F1 ;            // Declare nets (combinational circuit outputs).
wire            F2 ;            // F1 through F9 are all single bit outputs.
wire            F3 ;
wire            F4 ;
wire            F5 ;
wire            F6 ;
wire            F7 ;
wire            F8 ;

reg             F9 ;                    // Declare registers.
reg             [2:0]           F10 ;   // F10 through F12 are all three bit
                                        // outputs.
reg             [2:0]           F11 ;
reg             [2:0]           F12 ;
reg             D0 ;
reg             D1 ;
reg             D2 ;
reg             D3 ;
reg             D4 ;
reg             D5 ;
reg             D6 ;
reg             D7 ;

wire            mux2 ;
wire            mux4 ;

reg             mux8 ;

wire            [1:0]                   sum_total ;
wire            sum_df ;
wire            carryo_df ;
wire            sum ;
wire            carryo ;
wire            s1 ;
```

```
wire            a1 ;
wire            a2 ;
wire            a3 ;
wire            o1 ;

reg             F13 ;
reg             F14 ;
reg             F15 ;
reg             F16 ;
reg             F17 ;
reg             F18 ;
reg                     [8:0]                SUM ;
reg                     MATCH ;
reg                     MORE ;
reg                     LESS ;

// Combinational circuits using 'assign' statements.
assign F1 = A ;                        // Verilog code for buffer,
assign F2 = !A ;                       // inverter,
assign F3 = A&B ;                      // AND,
assign F4 = A|B ;                      // OR,
assign F5 = !(A&B) ;                   // NAND,
assign F6 = !(A|B) ;                   // NOR,
assign F7 = (A^B) ;                    // XOR, and
assign F8 = !(A^B) ;                   // XNOR.

// Combinational circuits using 'always' statements.
always @ (A or B or C )
// F9 through F12 are computed only if there is change in any of the three
// inputs, A, B, C or F10.
        begin
                F9  = (A&B)|(B&C)|(C&A) ;       // Realize AB + BC + CA.
                F10 = {A, B, C} ; // Concatenate A, B and C to get 3 bit result.
                F11 = F10 >> 1 ;                // Right shift by one bit.
                F12 = F10 << 2 ;                // Left shift by two bits.
        end

// Verilog models for Multiplexers
// Two input MUX using 'assign' statement.
assign mux2 = (A == 1) ? I1 : I0 ; // mux2 = I1 if A = 1, otherwise mux2 = I0.

// Four input MUX using 'assign' statement, nested.
assign mux4 = B ? (C ? I3 : I2) : (C ? I1 : I0) ;

// Avoid using nesting. Instead, use case for more than 2 inputs.
```

```
// Eight input MUX using 'always' and 'case' statements.
// Fastest possible hardware implementation.

always @ (A or B or C or I0 or I1 or I2 or I3 or I4 or I5 or I6 or I7)
begin
        case ({A, B, C})
                3'b000:  mux8 = I0 ;  // Read the input addressed by ABC.
                3'b001:  mux8 = I1 ;
                3'b010:  mux8 = I2 ;
                3'b011:  mux8 = I3 ;
                3'b100:  mux8 = I4 ;
                3'b101:  mux8 = I5 ;
                3'b110:  mux8 = I6 ;
                3'b111:  mux8 = I7 ;
                default: mux8 = 0 ;
        endcase
end

// Eight input DEMUX using 'always' and 'case' statements.
// This is a simple inverse of a MUX.
always @ (A or B or C or mux8)
begin
        case ({A, B, C})
        3'b000: begin    D0 = mux8 ; D1 = 0; D2 = 0; D3 = 0; D4 = 0; D5 = 0;
                         D6 = 0; D7 = 0; end
                // Read the input into D1, etc., and clear all other outputs.
        3'b001: begin D1 = mux8 ; D0 = 0; D2 = 0; D3 = 0; D4 = 0; D5 = 0;
                      D6 = 0; D7 = 0; end
        3'b010: begin D2 = mux8 ; D0 = 0; D1 = 0; D3 = 0; D4 = 0; D5 = 0;
                      D6 = 0; D7 = 0; end
        3'b011: begin D3 = mux8 ; D0 = 0; D1 = 0; D2 = 0; D4 = 0; D5 = 0;
                      D6 = 0; D7 = 0; end
        3'b100: begin D4 = mux8 ; D0 = 0; D1 = 0; D2 = 0; D3 = 0; D5 = 0;
                      D6 = 0; D7 = 0; end
        3'b101: begin D5 = mux8 ; D0 = 0; D1 = 0; D2 = 0; D3 = 0; D4 = 0;
                      D6 = 0; D7 = 0; end
        3'b110: begin D6 = mux8 ; D0 = 0; D1 = 0; D2 = 0; D3 = 0; D4 = 0;
                      D5 = 0; D7 = 0; end
        3'b111: begin D7 = mux8 ; D0 = 0; D1 = 0; D2 = 0; D3 = 0; D4 = 0;
                      D5 = 0; D6 = 0; end
        default: begin D0 = 0; D1 = 0; D2 = 0; D3 = 0; D4 = 0; D5 = 0;
                       D6 = 0; D7 = 0; end
        endcase
end
```

```
// Realization of a full adder
// Behavioral level realization
assign sum_total    =    (A + B) + C ;          // Realize sum.

// Data flow level realization
assign sum_df    = (A^B) ^C ;                   // Realize sum.
assign carryo_df = (A&B)|(B&C)|(C&A) ; // Realize carry out, AB + BC + CA.

// Structural level realization
xor (s1, A, B) ;                                // Realize (A^B) ^C using gate
xor (sum, s1, C) ;                              //  primitives.
and (a1, A, B) ;   and (a2, B, C) ;   and (a3, C, A) ;
                                                // Compute AB, BC, and CA
or (o1, a1, a2) ;
or (carryo, o1, a3) ;                           //Realize carry out.

// Realization of a magnitude comparator
always @ (N1 or N2)
        begin
        F13 = (N1 > N2);// Set output if N1 is greater than N2.
        F14 = (N1 < N2);// Set output if N1 is less than N2.
        F15 = (N1 == N2);      // Set output if N1 is equal to N2.
        F16 = (N1 != N2);      // Set output if N1 is not equal to N2.
        F17 = (N1 <= N2);      // Set output if N1 is less than or equal to N2.
        F18 = (N1 >= N2);   // Set output if N1 is greater than or equal to N2.
        end

always @ (enable_sum or NUM_1 or NUM_2 or PRESET_VALUE)
        begin
        SUM[8:0] = enable_sum ? (NUM_1[7:0] + NUM_2[7:0]) : 9'd0 ;
                               // Compute sum if enabled, otherwise output 0.
        MATCH    = enable_sum ? (SUM == PRESET_VALUE) : 1'b0 ;
        // Set output if SUM is equal to PRESET_VALUE, only if enabled.
        MORE     = enable_sum ? (SUM > PRESET_VALUE) : 1'b0 ;
        // Set output if SUM is greater than PRESET_VALUE, only if enabled.
        LESS     = enable_sum ? (SUM < PRESET_VALUE) : 1'b0 ;
        // Set output if SUM is  less than PRESET_VALUE, only if enabled.
        end
endmodule                              // This indicates the end of design.
```

3.4.2 Sequential Circuit Design

As was done in combinational circuits, all the sequential circuits considered earlier may be put in a single file as presented in the following. Output signals in 'always' statements are declared as ‘reg’. All other declarations are similar to that described in combinational circuits.

Verilog_code 3.19

```
/*                    Sequential circuit realization using Verilog codes
 Note that the design file name is 'seq_ckts.v'.
*/
`define  S0              3'd0    // Define the state of the controller
`define  S1              3'd1    // for the model state machine.
`define  S2              3'd2
`define  S3              3'd3

module seq_ckts (                          // Declare the design module.
                    D ,                    // I/O s can be written in any order.
                    Q ,
                    Q_n ,
                    A ,
                    B ,
                    C ,
                    clk ,
                    reset_n,
                    hold,
                    shift ,
                    in1 ,
                    in2 ,
                    trigger ,              // Monoshot trigger input.
                    pixelout_valid ,       // Register output.
                    cnt_reg ,
                    delay_out ,            // Monoshot trigger output.
                    data_out1 ,            // Shift registers'  outputs.
                    data_out2 ,
                    set_data ,             // Inputs and
                    load ,
                    shift_ps ,
                    rl_n ,
                    data_out ,             // outputs of Parallel to
                    data_valid ,
                    eoc ,                  // serial converter.
                    Z0 ,     // Model state M/C state indicator outputs.
                    Z1 ,
                    Z2 ,
                    Z3 ,
                    in ,                   // Pattern detector I/O,
                    out                    // 1 bit each.
```

```
                );

input           D ;         // Declare the Inputs and Outputs of the module.
input           A ;
input           B ;
input           C ;
input           clk ;
input           reset_n ;
input           hold ;
input           shift ;
input           in1 ;
input           in2 ;
input           trigger ;
input           in ;

output          Q ;
output          Q_n ;
output          pixelout_valid ;
output          [7:0]           cnt_reg ;
output          delay_out ;
output          [15:0]          data_out1 ;
output          [15:0]          data_out2 ;
output          Z0 ;
output          Z1 ;
output          Z2 ;
output          Z3 ;
output          out ;

input           [15:0]          set_data ;          // Inputs and
input           load ;
input           shift ;
input           rl_n ;

output          data_out ;                          // outputs of Parallel to
output          data_valid ;
output          eoc ;                               // serial converter.

reg             Q ;             // Declare registers
reg             Q_n ;

wire            set_pixout ;    // Declare nets (combinational circuit outputs).
wire            reset_pixout ;

reg             pixeloutp_valid ;
reg             pixelout_valid ;
```

```
wire        res_cnt ;
wire        adv_cnt ;
wire        [7:0]       cnt_next ;
wire        res_cntd ;
wire        run_delay ;
wire        [7:0]       cntd_next ;

reg         [7:0]       cnt_reg ;
reg         [7:0]       cntd_reg ;
reg         delay_out ;
reg         triggerp ;

wire        [15:0]      dataout1_next ;
wire        [15:0]      dataout2_next ;

reg         [1:0]       state ;
reg         [15:0]      data_out1 ;
reg         [15:0]      data_out2 ;
reg         Z0 ;
reg         Z1 ;
reg         Z2 ;
reg         Z3 ;
reg         out ;
reg         [1:0]       psd_state ;
reg         data_out ;
reg         data_valid ;
reg         eoc ;
reg         [15:0]      sr ;
reg         [4:0]       cnt_ps_reg ;
```

// Realization of a D flip-flop with a reset control using 'always' block.

```
always @ (posedge clk or negedge reset_n)
begin
        if (reset_n == 1'b0)
                begin
                  Q      <= 0 ;           // Clear register when system is reset.
                  Q_n    <= 1 ;
                end
        else
                begin
                  Q      <=  D ;          // Store or register D input.
                  Q_n    <= ~D ;
                end
end
```

```
// Realization of registers using 'always' block.
assign set_pixout   = (A == 1'b0) && (B == 1'b1) && (C == 1'b0);
                                    // Pre-compute (not A) and (not B) and (C).
assign reset_pixout = (A == 1'b1) && (B == 1'b1) && (C == 1'b1) ;
                                    // Pre-compute ABC.
always @ (posedge clk or negedge reset_n)
begin
  if (reset_n == 1'b0)
         pixeloutp_valid <= 1'b0 ;          // Clear register when system is reset.
  else if (hold == 1'b1)
         pixeloutp_valid <= pixeloutp_valid ;        // Retain the value if the
                                                     // system is in hold.
  else if (set_pixout == 1'b1)
          pixeloutp_valid <= 1'b1 ;         // Set or reset when the
  else if (reset_pixout == 1'b1)
         pixeloutp_valid <= 1'b0 ;                   // conditions are satisfied.
  else
         pixeloutp_valid <= pixeloutp_valid ;        // Otherwise, don't disturb.
end

always @ (posedge clk or negedge reset_n)
begin
  if (reset_n == 1'b0)
         pixelout_valid <= 1'b0 ;                    // Clear register when
                                                     // system is reset.
  else if (hold == 1'b1)
         pixelout_valid <= pixelout_valid ;  // Retain the value if the system is
                                             // in hold.
  else
         pixelout_valid <= pixeloutp_valid ; // Assign previous (clk) value.
end

// Realization of a counter using 'always' block.
assign res_cnt = (cnt_reg == 255) ;         // Condition for resetting the
                                            // counter.
assign adv_cnt = (A == 1'b1)&(B == 1)&(C == 1) ;
                                  // Condition for Pre-incrementing the counter.
assign cnt_next = cnt_reg + 1 ;             // Pre-increment the counter.

always @ (posedge clk or negedge reset_n)
begin
  if (reset_n == 1'b0)
         cnt_reg <= 8'd0 ;                  // Initialize when the system is reset.
  else if (res_cnt == 1'b1)         // Reset if terminal count is reached.
         cnt_reg <= 8'd0 ;
  else if (adv_cnt == 1'b1)
```

```
          cnt_reg <= cnt_next ;      // Advance the count by one if the timer is
                                     // still running.
  else
          cnt_reg <= cnt_reg  ;      // Otherwise, don't disturb.
end
```

// Realization of a non-retriggerable Monoshot delay using a counter.
// This produces 255 clock cycles delay for the preset value, cntd_reg = 255.
// For longer delays, change the cntd_reg width, and its preset value.

```
assign res_cntd = (reset_n == 1'b0) || (cntd_reg  == 255) ;
                                          // Condition for resetting the counter.
assign run_delay = (triggerp == 0)&&(trigger == 1) ;
                                            // Detect the positive edge of trigger.
assign cntd_next = cntd_reg + 1 ;           // Pre-increment the counter.

always @ (posedge clk or posedge res_cntd)
begin
  if (res_cntd == 1)                        // Initialize when the system is reset
                                            // or if the terminal count is reached.
        begin
              cntd_reg        <= 8'd0 ;
              delay_out       <= 0 ;
              triggerp        <= 0 ;
        end
  else if (delay_out == 1)
        begin
              cntd_reg        <= cntd_next ; // Advance the count by one if
                                              // the timer is still running
              triggerp        <= trigger ;
        end
  else if (run_delay== 1'b1)
        begin
              delay_out       <= 1 ;    // Start the delay if the positive
                                        // edge of trigger is detected.
              triggerp        <= trigger ;  // Preserve the current state of
        end                                 // trigger.
  else
        begin
              cntd_reg        <= cntd_reg ;      // Otherwise, don't disturb.
              delay_out       <= delay_out ;
              triggerp        <= trigger ;
        end
end
```

```
// Realization of a shift register using 'assign' statement and 'always'
// block.
assign dataout1_next = (data_out1 >> 1) ;   // Pre-shift right the contents of
                                             // data_out1 register by one bit.
always @ (posedge clk or negedge reset_n)
begin
  if (reset_n == 1'b0)
        data_out1 <= 16'b1010_1010_1010_1010 ;  // Initialize when the
                                                 // system is reset.
  else if (shift == 1'b1)
        data_out1 <= dataout1_next ;            // Register the shifted
                                                // contents.
  else
        data_out1 <= data_out1 ;                // Otherwise, don't shift.
end

assign dataout2_next[15:0] = ({1'b0, data_out1[15:1]}) ;
                           // Pre-shift right the contents of data_out1
                           // register by one bit.
always @ (posedge clk or negedge reset_n)
begin
  if (reset_n == 1'b0)
        data_out2 <= 16'b1010101010101010 ;     // Initialize when the
                                                // system is reset.
  else if (shift == 1'b1)
        data_out2 <= dataout2_next ;      // Register the shifted contents.
  else
        data_out2 <= data_out2 ;          // Otherwise, don't shift.
end

// Code for Parallel to serial converter
always @ (posedge clk or negedge reset_n)
begin
  if (reset_n == 0)
        begin
            data_out      <= 0 ;   // Clear when the system is reset.
            data_valid    <= 0 ;
            eoc           <= 0 ;
        end
  else if (load == 1)
        begin
            sr[15:0] <= set_data[15:0] ; // Preset or clear registers.
            cnt_ps_reg    <= 16 ;
            data_out      <= 0 ;
            data_valid    <= 0 ;
            eoc           <= 0 ;
```

```
        end
  else if ((shift == 1) && (cnt_ps_reg != 0))
        begin
            sr[15:0] <= rl_n ? (sr[15:0] >> 1) : (sr[15:0] << 1) ;
                                                    // Register the shifted contents.
            data_out <= rl_n ?  sr[0] : sr[15]; // Select LSB or MSB.
            cnt_ps_reg[4:0] <= cnt_ps_reg[4:0] - 1 ;
                                                    // Keep track of the bits to be sent.
            data_valid <= 1 ; eoc <= 0 ;
          end
  else if ((shift == 1) && (cnt_ps_reg == 0))
        begin
                data_out        <= 0 ;
                data_valid      <= 0 ;
                eoc             <= 1 ;      // End of conversion.
        end
  else                  ;
 end

// Model for sequential machines
always @ (posedge clk or negedge reset_n)
begin
  if (reset_n == 1'b0)
        begin
                        Z0      <= 1'b0 ;           // Switch off all the
                                                    // lights to start with.
                        Z1      <= 1'b0 ;
                        Z2      <= 1'b0 ;
                        Z3      <= 1'b0 ;
                        state   <= `S0 ;            // Initialize the state when the
                                                    // system is reset.
        end
  else
        case(state)
                `S0:
                  begin
                        Z0      <= 1'b1 ;  // Switch ON state 00 light and
                        Z1      <= 1'b0 ;  // switch OFF all other lights.
                        Z2      <= 1'b0 ;
                        Z3      <= 1'b0 ;
                        if (in1 == 1'b0)   // If input 1 is not active, continue
                          state <= `S0 ;   // to remain in the state 00.
                        else               // However, if input 1 is active,
                          state <= `S2 ;   // go to the next state.
                  end
```

```
        `S2:
          begin
                Z0      <= 1'b0 ; // Switch ON  state 10 light and
                Z1      <= 1'b0 ; // switch OFF all other lights.
                Z2      <= 1'b1 ;
                Z3      <= 1'b0 ;
                if (in2 == 1'b0)           // If input 2 is not active,
                  state <= `S1 ;           // go to the state 01.
                else                       // Otherwise,
                  state <= `S3 ;           // go to the state 11.
          end
        `S1:
          begin
                Z0      <= 1'b0 ;        // Switch ON state 01 light and
                Z1      <= 1'b1 ;
                Z2      <= 1'b0 ;        // switch OFF all other lights.
                Z3      <= 1'b0 ;
                if (in2 == 1'b0)           // If input 2 is not active,
                  state <= `S0 ;           // go to the state 00.
                else if (in1 == 1'b1)      // If input 1 is active,
                  state <= `S3 ;           // go to the state 11.
                else
                  state <= `S1 ;       // Otherwise, remain in the state 01.
           end
         `S3:
           begin
                Z0      <= 1'b0 ;     // Switch ON state 11 light and
                Z1      <= 1'b0 ;
                Z2      <= 1'b0 ;     // switch OFF all other lights.
                Z3      <= 1'b1 ;
                if (in2 == 1'b0)           // If input 2 is not active,
                 state <= `S0 ;            // go to the state 00.
                else if (in1 == 1'b0)      // If input 1 is not active,
                  state <= `S1 ;           // go to the state 01.
                else
                 state <= `S3 ;    // Otherwise, remain in the state 11.
           end
        default: state <= `S0 ;    // Otherwise, remain in the  state 00.
    endcase
end
```

// Another design example – Pattern sequence detector

```
// Pattern to be detected:  0110
// Input pattern applied:   11110100110110001101 ....................
// Output desired:          00000000001001000010 ....................

always @ (posedge clk or negedge reset_n)
```

```
begin
  if (reset_n == 1'b0)
                  begin
                           out         <= 0 ;    // Switch OFF output to start with.
                           psd_state  <= 0 ;   // Initialize the state when the
                                                // system is reset.
                  end
  else

             case(psd_state)
                  0:  begin
                           out         <= 0 ;            // Switch OFF output.
                           psd_state  <=  in ? 0 : 1 ;  // Change the state to '1' if
                                                         // the input is '0'.
                                                    // Remain in the state '0' otherwise.
                      end
                  1:  begin                        // Enter for first occurrence of '0'.
                           out         <= 0 ;           // Switch OFF output.
                           psd_state <=  in ? 2 : 1 ;  // Change the state to '2' if
                                                        // the input is '1'.
                                                   // Remain in the state '1' otherwise.
                      end
                  2:  begin                        // Enter for first occurrence of '01'.
                           out         <= 0 ;            // Switch OFF output.
                        psd_state <= in ? 3 : 1 ;        // Change the state to '3' if
                                                         // the input is '1',
                                                   // otherwise change the state to '1'.
                      end
                  3:  begin                        // Enter for first occurrence of '011'.
                           out         <= in ? 0 : 1 ;  // Switch ON the output if
                                                        // the input is '0110',
                                                        // otherwise switch it OFF.
                           psd_state <= in ? 0 : 1 ;    // Change the state to '1' if
                                                        // the input is '0'.
                                                   // Otherwise, change the state to '0'.
                      end
             default:      psd_state <= 0 ;  // Otherwise, remain in the state 0.
             endcase
end
endmodule
```

In order to enable the reader to design Verilog based systems faster, a Verilog quick reference design card is presented in Appendix 7 on the CD.

Summary

A brief introduction of the evolution of hardware design language was presented. Verilog was introduced as a tool for realizing digital systems design. The advantages of Verilog coding over the traditional schematic circuit diagram approach were established, especially when the design of VLSI circuits crossing over 50,000 transistors mark is encountered. A number of design examples were illustrated for both combinational and sequential circuits. These examples cater to the frequently used digital circuits in a system design, especially in industries. Only the cores of the designs were initially presented to expedite the learning process. These modules were later on integrated into full-fledged codes, ready for testing. These designs are tested using test benches, which will be covered in depth in the next chapter. Register Transfer Level coding, vital for designing chips that work successfully, is the main emphasis of this design book, and is discussed in a later chapter.

Assignments

Note: Verilog codes for each of the assignments may be either separately written or all of them combined into a single design file as had been presented in the book as per your convenience.

3.1 Can you realize Schmitt trigger and open collector buffers/inverters using Verilog? If so, explain how. If not, what do you suggest for their implementation? What are the possible applications for them?

3.2 Tristate buffers can be implemented in Verilog using primitive gates. Write Verilog codes for circuits in Figure A3.1.

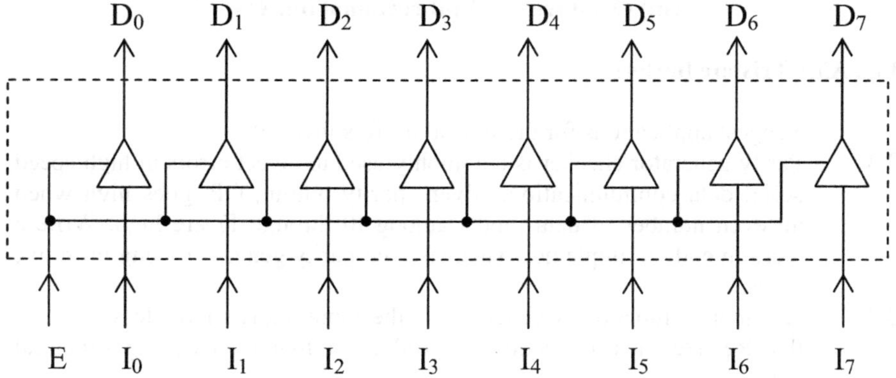

(i) Octal tristate buffers

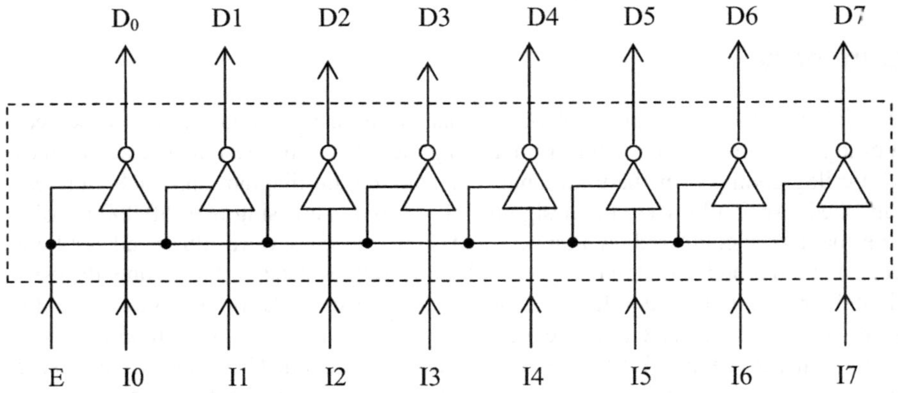

(ii) Octal tristate inverters

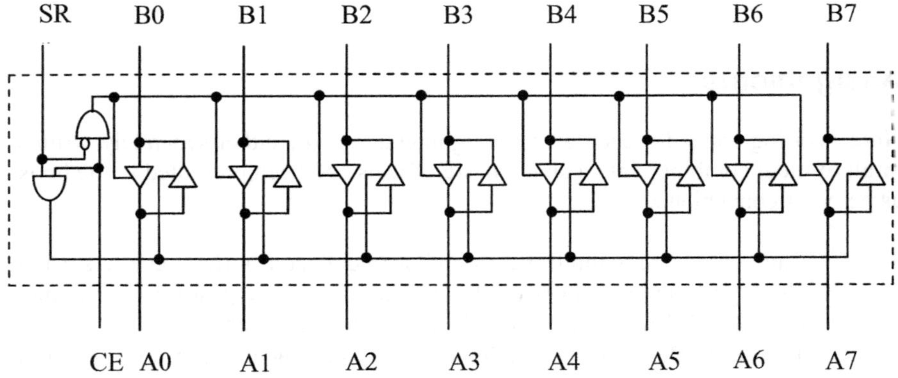

(iii) Octal tristate bi-directional buffers

Fig. A3.1 Tristate buffers

Suggest applications for these octal buffers/inverters.

3.3 Parity generator/checker is commonly used to detect errors in high-speed serial data communication. 'Even' parity output, OP, goes high when an even number of data inputs among I0 through I7 are high. Write a Verilog code to implement such an even parity generator using primitive gates.

3.4 A minority function is generated if the input signals have less 1's than 0's. Realize such a function in Verilog for four inputs using structural gates.

3.5 Write Verilog codes to realize the following functions:

(i) $\sum(0, 2, 4, 6, 9, 10, 13, 15) + \sum(3, 5, 7, 11)$

don't
cares

(ii) Max terms: $\prod(1, 3, 5, 7, 9, 11, 13, 14)$

(iii) Min terms: $\sum(0, 10, 40, 70, 100, 127)$

(iv) Min terms or Max terms, whichever require less number of codes:
$\sum(0, 1, 2, 4, 6, 7, 9, 11, 13, 15)$

3.6 Write a Verilog code to implement circuit in Figure A3.2 without using primitive gates.

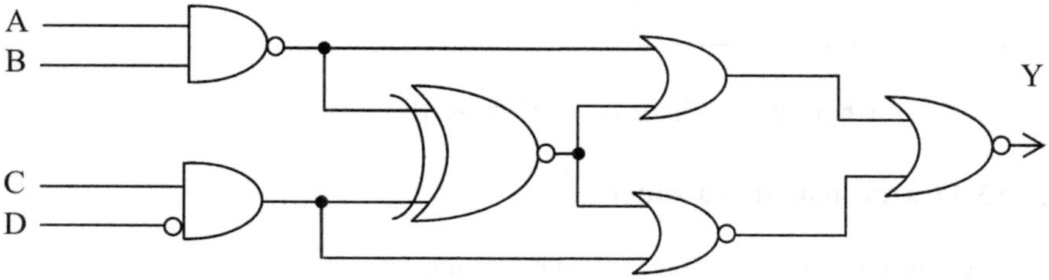

Fig. A3.2 Combination circuit

3.7 Truth table for an 8-input priority encoder is furnished. Realize the priority encoder in Verilog. All signals except EO are active high. X is don't care (H or L).

Truth table for an 8-input priority encoder

Inputs									**Outputs**				
EI	**I0**	**I1**	**I2**	**I3**	**I4**	**I5**	**I6**	**I7**	**A2**	**A1**	**A0**	**GS**	**E0**
L	X	X	X	X	X	X	X	X	L	L	L	L	L
H	L	L	L	L	L	L	L	L	L	L	L	L	H
H	X	X	X	X	X	X	X	H	H	H	H	H	L
H	X	X	X	X	X	X	H	L	H	H	L	H	L
H	X	X	X	X	X	H	L	L	H	L	H	H	L
H	X	X	X	X	H	L	L	L	H	L	L	H	L
H	X	X	X	H	L	L	L	L	L	H	H	H	L
H	X	X	H	L	L	L	L	L	L	H	L	H	L
H	X	H	L	L	L	L	L	L	L	L	H	H	L
H	H	L	L	L	L	L	L	L	L	L	L	H	L

3.8 A full adder circuit may be realized using two half adders as shown in the Figure A3.8.

For multi-bit precision, this circuit and the expressions in Figure A3.3 may be used.

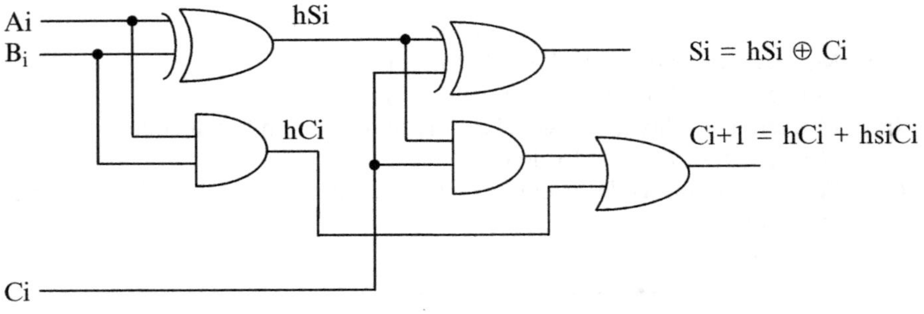

where $hS_i = A_i \oplus B_i$, $hC_i = A_i B_i$

Fig. A3.3 Carry look-ahead adder

Carry for 2-bit precision: $C_2 = hC_1 + hS_1 C_1$
Carry for 3-bit precision: $C_3 = hC_2 + hS_2 hC_1 + hS_2 hS_1 C_1$
Carry for 4-bit precision: $C_4 = hC_3 + hS_3 hC_2 + hS_3 hS_2 hC_1 + hS_3 hS_2 hS_1 hC_1$

and so on.

These are referred to as a look-ahead carry generator. C_4 does not have to wait for C_3 and C_2 carries to propagate. In fact, C_4, C_3, and C_2 propagate simultaneously. Implement Verilog code for adding two 4-bit numbers using the look-ahead carry generator, and compare its performance with the behavioral implementation illustrated in the text.

3.9 If A = '10101010', what is A&(A >> 2) & (A << 2)? Write a Verilog code.

3.10 Write Verilog codes to rotate an input, IN[15:0],

a. left by *n* bits
b. right by *n* bits.

'*n*' can be either 2 or 4 selected by the user.

3.11 Realize the debouncing circuit shown in Figure A3.4 using Verilog. Explain how the circuit works.

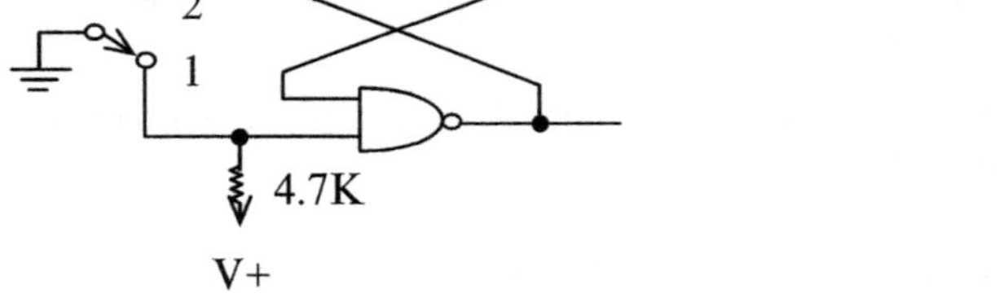

Fig. A3.4 Push button debouncing circuit

3.12 Design a circuit to generate a single 'clk' pulse every time the push button switch shown in Figure A3.5 is pressed. The debounce time of the switch may be assumed to be 2.5 ms. The 'clk' frequency is 50 MHz. Include an asynchronous, active low signal for resetting the circuit. Write a Verilog code for the design.

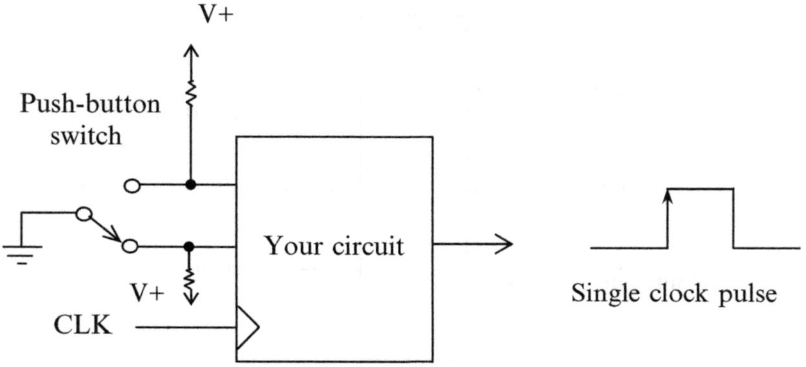

Fig. A3.5 Single clock pulse generation

3.13 A 16-bit re-triggerable monoshot is depicted in Figure A3.6. Realize the complete design in Verilog.

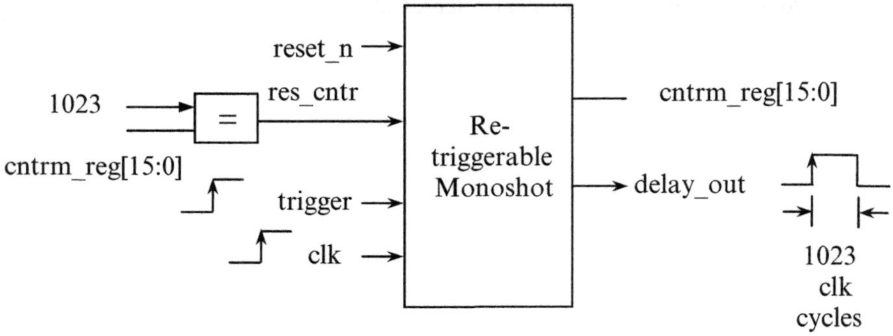

Fig. A3.6 Re-triggerable monoshot

3.14 An asynchronous serial input to parallel converter is shown in Figure A3.7. The incoming serial data arrives at the input pin, 'input_data', D0 bit first and D7 bit last, whereas the converted parallel data appears at the 'out_data' pins. The outputs are synchronized to the rising edge of the system clock, 'clk'. The validity of the parallel data is signaled by a data valid signal, 'out_data_valid' for a 'clk' period duration. Serial to parallel conversion can be commenced if mark state (active high) goes low at the 'input_data' pin. Use one start bit (active low) and one stop bit (active high) in the incoming serial data. Each bit is of three 'clk' periods duration. Include an 'error' flag of single 'clk' period duration if the stop bit is not received. The serial bit stream reception must correct all by itself with the arrival of mark state. Draw an ASM chart and realize the complete design in Verilog.

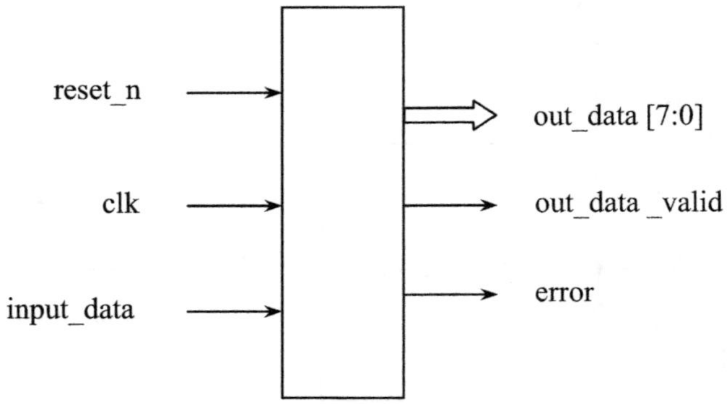

Fig. A3.7 Block diagram of an asynchronous serial to parallel converter

3.15 Baud rate (bits/second) is a measure of the communication speed in a serial channel. What is the baud rate for the design of assignment 3.14? Explain how standard baud rates such as 300, 600, 1200, 2400, 4800, 9600, etc. up to 614400 can be achieved.

Chapter 4

Writing a Test Bench for the Design

In the previous chapter, we have seen how to design combinational and sequential circuits. The question is, how do we test them? We need to apply appropriate stimulus to the design in order to test it. This can be done by writing another Verilog code called the 'Test Bench'. This is written as a separate file, different from the design file(s). Though not necessary, it is easier for identification if we give the same name as the top design file, of course, with an extension, '_test' or 'tb'; readily revealing itself as a test bench. For example, the test bench name for the design 'comb_ckts.v' can be 'comb_ckts_test.v'. Note that this is a '.v' file, indicating that it is a Verilog file. Stimulus is nothing but the application of various permutations and combinations of inputs at various points of time and, looking for correct results produced by the design. As you are the designer, you know precisely how a particular circuit functions. Therefore, you can cross check whether it is functioning as per your design specification or not. The functionality of the design can be easily tested if we can view waveforms. We will see how to generate waveforms using simulation in a later chapter. In the present chapter, we will concentrate on how to write a test bench [15]. Verification engineers need to develop expertise in writing effective test benches for designs, even more than the design engineers.

4.1 Modeling a Test Bench

A model of a test bench is shown in Figure 4.1. The design may comprise of just one module or several modules depending upon the complexity of the application. As mentioned earlier, the test bench applies appropriate test pattern to the inputs of the design under test and checks the outputs of the design so as to verify its functionality. In order to get a quick grasp of what a test bench is, we will first consider a very simple design which has one number of two-input AND gate as shown in Figure 4.2. The truth table is also given alongside. The functionality of the design can be easily verified using the timing diagram shown in Figure 4.3. The Verilog code of the AND gate design presented is self-explanatory. Figure 4.4 shows the connection between the test bench and the design. 'A' and 'in' are the stimuli generated by the test bench in accordance with the timing diagram shown in Figure 4.3. Note that the test bench stimuli need not bear the same names as the corresponding signals of the design. The test bench for the AND gate design is given in Verilog_code 4.2. It is adequately commented. 'and_2in.v' is the design

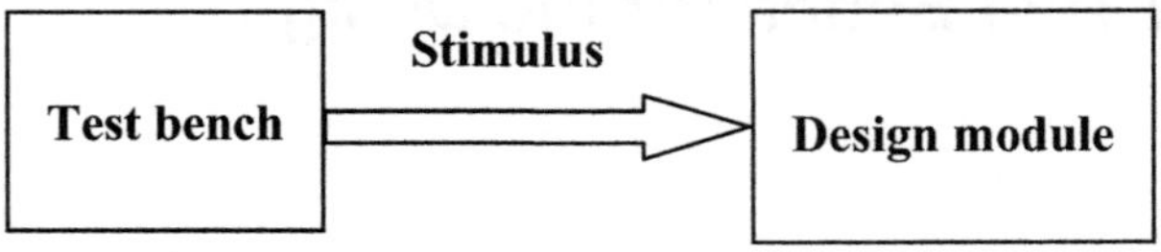

Fig. 4.1 Model of a test bench

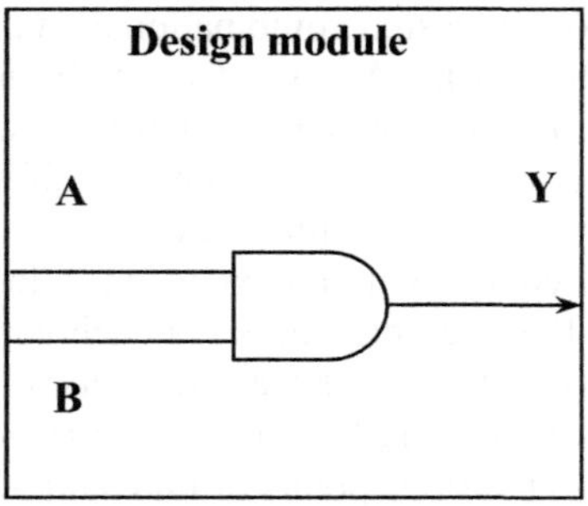

TRUTH TABLE

A B	Y
0 0	0
0 1	0
1 0	0
1 1	1

Fig. 4.2 Two-input AND gate with its truth table

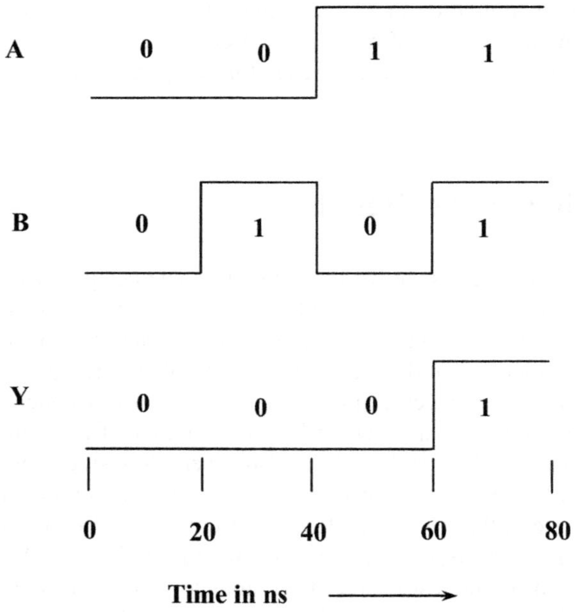

Fig. 4.3 Timing diagram of AND gate

file, which must be included in the test bench, 'and_2in_test.v'. '`include' is a reserved word used for the purpose of identifying one of the files included in the test bench or a design as the case may be. It would be a good practice to just include the top design in the test bench, while including all other submodules in the top design. This way, we need only compile the test bench in the simulation tool and the top design in the synthesis tool, instead of compiling all the modules. We will learn these tools in Chapter 6 and Chapter 7 respectively. The test bench follows the timing diagram shown in Figure 4.3 closely. As shown therein, the various combination of inputs are applied at 20 ns interval. `timescale specifies the time base along with its resolution. 'Module' and 'endmodule' have the same meanings as in the design. In the declaration, module and_2in_test, no inputs/outputs (I/Os) are listed since the test bench is the top most module. 'A' and 'in' are the stimuli we need to apply. Since we need to hold their values for specific time, say, 20 ns, they are declared as 'reg'. The design output, 'Y' which appears as 'out' in the test

---

Verilog_code 4.1

// A sample Verilog design module to explain the use of a test bench

```
module and_2in (A, B, Y);                // Declare the design module.

    input           A, B ;               // Declare the design inputs
    output          Y ;                  // and output.
    wire            Y ;                  // Declare the output as net.
    assign Y = A & B ;                   // Realize the 2 input AND gate.

endmodule                                // The end of design.
```

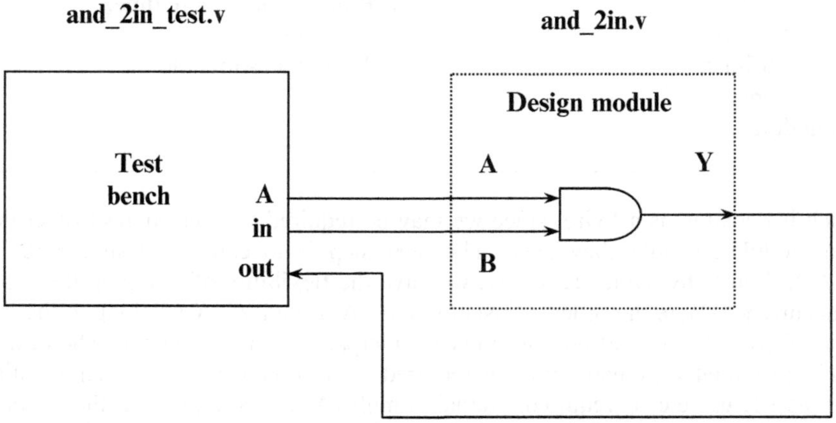

Fig. 4.4 Interconnection of the test bench and the design module

Verilog_code 4.2

```
// Test bench for functional checking of two input AND gate design
// This is put in a separate file, 'and_2in_test.v'

`include 'and_2in.v'              // This is the design file.
`timescale 1ns/100ps              // Time base is in nano seconds and its
                                  // resolution is in pico seconds.
module and_2in_test ;             // Declare the test module.

reg                 A, in ;
           // Declare inputs as registers since we need to hold the values.
wire                out ;
           // Declare output(s) as wire (meaning net) since we need to
           // inter-connect other submodules, should they exist.

and_2in u1(        // Call the design module.
                    // u1 stands for the first instantiation.

                    .A(A) ,          // Connect ports by name.
                    .B(in) ,
                    .Y(out)
                    ) ;
initial
        begin
                A = 0; in = 0;       // Apply stimulus at time 0.
        #20     A = 0; in = 1;       // Change inputs at time 20 ns,
        #20     A = 1; in = 0;       // 40 ns, and
        #20     A = 1; in = 1;       // 60 ns.
        #40                          // Run for some more time,
        $stop ;                      // and stop.
        $finish ;                    // Terminate simulation.
        end
endmodule
```

bench is declared as a 'wire' since we may be required to inter-connect other design modules, should they exist. The next step is to call the design module, 'and_2in' ports by name. This way, we have the flexibility of changing the order of occurrence. Note the order of instantiation. 'A, B and Y' I/Os belong to the design, whereas 'A, in and out' are their counterparts in the test bench. The design can be instantiated as many times as required, every time using a different identity such as u1, u2, etc. if required. Actual stimuli (A, in) is applied in the 'initial' block as shown. The block is identified by a 'begin' followed by an 'end'.

To start with at '0' time, 'A' and 'in' are both forced to logical '0' as per the timing diagram. Every 20 ns, a new test data is applied by using #20 before the stimulus. The time unit is not mentioned explicitly since it has been already declared using the `timescale. 100 ps, i.e., 0.1ns is the resolution set and, therefore, we can set a higher precision for time such as #20.5, if required. The time mentioned using #20 etc. are cumulative since blocking statements (#20 A = 0, for example) are used. At zero time, the design inputs 'A' and 'in' are 00, at 20 ns they are 01, at 40 ns they are 10 and at 60 ns they are 11. Before the simulation is stopped using $stop and $finish, we need some more time, say, 20 or 40 ns after the input data 11 is applied so as to hold the last input values till it is processed. There are two ways to verify the correctness of the design output. One way is to use the simulation tool and display the waveforms and compare it with Figure 4.3 and, the other method is to add more codes such as 'display' and 'monitor' in the test bench to check the output, 'out' every time the stimulus is applied. Since the second method is tedious for big designs, we will use the simulation method to view the outputs at every step. These will be covered in depth in a later chapter on simulation.

4.2 Test Bench for Combinational Circuits

The test bench that checks the functionality of the combinational circuits is presented in Verilog_code 4.3. The design file is declared by `include statement. Note the presence of the single reverse quote. The design file will have to be specified within double quotes. Otherwise, the Verilog compiler will report errors and not permit us to proceed with the simulation. You have to specify the file name completely. If it is in some other directory or folder, you have to give the entire path. However, it is a good practice to keep all the files of a project in the same directory and operate from that directory. After declaring the module name 'comb_ckts_test', all design inputs are declared as 'reg' and outputs as 'wire' for the reasons mentioned earlier. Size of each of the signals used is also reported. For examples, N1 width is 8 bits [7:0] with bit [7] as the MSB and F10 width is 3 bits [2:0] with bit [2] as the MSB.

At the next level, the design module, 'comb_ckts' is called. Suppose we have a circuit diagram for an application, making use of several TTL gates such as 74LS00, 74LS245, etc. and, any other ICs. They are identified as U1, U2, and so on. The same nomenclature may be applied in Verilog code while calling different modules, such as 'comb_ckts', 'seq_ckts', etc. Each of these designs or submodules may be called any number of times as per the needs of the application. They are referred to as instantiation. In the Verilog_code 4.3, u1 is the instantiation of the design. In this test, we need to call the design only once to check its functionality. Multiple instantiations are required only in bigger designs, which will be covered in the chapter on project design. All the I/Os in the design are listed by calling the ports by name as we had explained in the model test bench before. Note that the I/Os are separated by commas except the last. The stimulus to the design is applied in the 'initial' block. Any number of 'initial' blocks may be present in a test bench as per needs. In general, they work concurrently. A, B, and C are inputs used for various gate realizations using 'assign' and 'always' statements. They are also used as the select pins for the MUX/DEMUX realization. It would, therefore, be convenient if we change their values in steps of one starting from '000'. At zero time, we apply ABC = 000, at 20 ns 001 and so on up to 111 at 140 ns. I0 through I7 are used as inputs to the three types of MUX and are cleared to start with. At 10 ns, 30 ns, etc. apply a pulse of duration 10 ns to I0, I1, etc. in that order.

Verilog_code 4.3

```
// Test bench for checking combinational circuit realization
// This test bench may be housed in 'comb_ckts_test.v' file

`include 'comb_ckts.v'              // This is the design file.

module comb_ckts_test ;             // Declare the test module.

reg                  A ;            // Declare all design inputs as 'reg'
reg                  B ;            // so that they may hold the values
reg                  C ;            // till they are changed again.
reg                  I0 ;
reg                  I1 ;
reg                  I2 ;
reg                  I3 ;
reg                  I4 ;
reg                  I5 ;
reg                  I6 ;
reg                  I7 ;
reg     [7:0]                       N1 ;
reg     [7:0]                       N2 ;
reg                  enable_sum ;
reg     [7:0]                       NUM_1 ;
reg     [7:0]                       NUM_2 ;
reg     [8:0]                       PRESET_VALUE ;

wire                 F1 ;    // Declare nets (combinational circuit outputs).
wire                 F2 ;    // F1 through F9 are all single bit outputs.
wire                 F3 ;
wire                 F4 ;
wire                 F5 ;
wire                 F6 ;
wire                 F7 ;
wire                 F8 ;
wire                 F9 ;           // Declare as wire.
```

```
wire    [2:0]       F10 ;         // F10 through F12 are all three bit
                                  // outputs.
wire    [2:0]       F11 ;
wire    [2:0]       F12 ;
wire    [1:0]       sum_total ;
wire                sum_df ;
wire                carryo_df ;
wire                sum ;
wire                carryo ;
wire                F13 ;
wire                F14 ;
wire                F15 ;
wire                F16 ;
wire                F17 ;
wire                F18 ;
wire    [8:0]       SUM ;
wire                MATCH ;
wire                MORE ;
wire                LESS ;

comb_ckts      u1(                // Instantiate the design module.
                    .A(A) ,       // u1 stands for the first instantiation.
                    .B(B) ,       // Only one instantiation is used in this
                    .C(C) ,       // test bench.
                    .I0(I0) ,     // I/Os are called by name.
                    .I1(I1) ,
                    .I2(I2) ,
                    .I3(I3) ,
                    .I4(I4) ,
                    .I5(I5) ,
                    .I6(I6) ,
                    .I7(I7) ,
                    .N1(N1) ,
                    .N2(N2) ,
                    .enable_sum(enable_sum) ,
                    .NUM_1(NUM_1) ,
                    .NUM_2(NUM_2) ,
                    .PRESET_VALUE(PRESET_VALUE) ,
                    .F1(F1) ,
                    .F2(F2) ,
                    .F3(F3) ,
                    .F4(F4) ,
                    .F5(F5) ,
                    .F6(F6) ,
                    .F7(F7) ,
                    .F8(F8) ,
```

```
                    .F9(F9) ,
                    .F10(F10) ,
                    .F11(F11) ,
                    .F12(F12) ,
                    .mux2(mux2) ,
                    .mux4(mux4) ,
                    .mux8(mux8) ,
                    .D0(D0) ,
                    .D1(D1) ,
                    .D2(D2) ,
                    .D3(D3) ,
                    .D4(D4) ,
                    .D5(D5) ,
                    .D6(D6) ,
                    .D7(D7) ,
                    .sum_total(sum_total) ,
                    .sum_df(sum_df) ,
                    .carryo_df(carryo_df) ,
                    .sum(sum) ,
                    .carryo(carryo) ,
                    .F13(F13) ,
                    .F14(F14) ,
                    .F15(F15) ,      // Note that these I/Os are
                                     // separated by commas except the last.
                    .F16(F16) ,
                    .F17(F17) ,
                    .F18(F18) ,
                    .SUM(SUM) ,
                    .MATCH(MATCH) ,
                    .MORE(MORE) ,
                    .LESS(LESS)
            );

initial
        begin
                A = 1'b0 ;           // At time zero, let the inputs be 000 binary.
                B = 1'b0 ;
                C = 1'b0 ;
                I0 = 0 ; I1 = 0 ; I2 = 0 ; I3 = 0 ; I4 = 0 ; I5 = 0 ; I6 = 0 ; I7 = 0 ;
                N1 = 200 ;           // Test for the condition, N1 > N2.
                N2 = 199 ;
                NUM_1 = 0 ;
                NUM_2 = 0 ;
                enable_sum = 0 ;
                PRESET_VALUE = 0 ;
```

```
#10   I0 = 1 ;           // At 10 ns, apply a pulse of width 10 ns.

      NUM_1 = 250 ;  // Apply inputs to test equality,
      NUM_2 = 250 ;  // NUM_1 + NUM_2 = PRESET_VALUE
      enable_sum = 1 ;
      PRESET_VALUE = 500 ;
#10   A = 1'b0 ;                 // At time 20 ns, let the inputs be 001.
      B = 1'b0 ;
      C = 1'b1 ;
      I0 = 0 ;
      N1 = 100 ;                 // Test for the condition, N1 < N2.
      N2 = 199 ;
      NUM_1 = 250 ;
      NUM_2 = 251 ;  // NUM_1 + NUM_2 > PRESET_VALUE
      enable_sum = 1 ;
      PRESET_VALUE = 500 ;
#10   I1 = 1 ;           // At 30 ns, apply a pulse of width 10 ns.
      NUM_1 = 255 ;
      NUM_2 = 250 ;
      enable_sum = 0 ;
      PRESET_VALUE = 500 ;
#10   A = 1'b0 ;                 // At time 40 ns, let the inputs be 010.
      B = 1'b1 ;
      C = 1'b0 ;
      I1 = 0 ;
      N1 = 255 ;                 // Test for the condition, N1 = N2.
      N2 = 255 ;
      NUM_1 = 100 ;  // NUM_1 + NUM_2 < PRESET_VALUE
      NUM_2 = 255 ;
      enable_sum = 1 ;
      PRESET_VALUE = 500 ;
#10   I2 = 1 ;           // At 50 ns, apply a pulse of width 10 ns.
#10   A = 1'b0 ;         // At time 60 ns, let the inputs be 011.
      B = 1'b1 ;
      C = 1'b1 ;
      I2 = 0 ;
#10   I3 = 1 ;           // At 70 ns, apply a pulse of width 10 ns.
#10   A = 1'b1 ;         // At time 80 ns, let the inputs be 100.
      B = 1'b0 ;
      C = 1'b0 ;
      I3 = 0 ;
#10   I4 = 1 ;           // At 90 ns, apply a pulse of width 10 ns.
#10   A = 1'b1 ;         // At time 100 ns, let the inputs be 101.
      B = 1'b0 ;
      C = 1'b1 ;
      I4 = 0 ;
```

```
        #10    I5 = 1 ; I1 = 1 ;  // At 110 ns, apply a pulse of width 10 ns.
        #10    A = 1'b1 ;         // At time 120 ns, let the inputs be 110.
               B = 1'b1 ;
               C = 1'b0 ;
               I5 = 0 ; I1 = 0 ;
        #10    I6 = 1 ;           // At 130 ns, apply a pulse of width 10 ns.
        #10    A = 1'b1 ;         // At time 140 ns, let the inputs be 111.
               B = 1'b1 ;
               C = 1'b1 ;
               I6 = 0 ;
        #10    I7 = 1 ;           // At 150 ns, apply a pulse of width 10 ns.
        #10    I7 = 0 ;
        #50                       // Run for some more time
        $stop ;                   // and stop.
        end
endmodule
```

N1 and N2 are two numbers we wish to compare. We apply different inputs at 0, 20, and 40 ns so that we may verify the circuit functionality for the conditions: (i) N1 > N2, (ii) N1 < N2, and (iii) N1 = N2. In the next application, two numbers NUM_1 and NUM_2 are summed up and compared with a PRESET_VALUE. At zero time, enable_sum input is forced to '0'. Therefore, all the relevant outputs will be cleared. At 10, 20, and 40 ns, the enable_sum is made high so that the summing/comparing process may take place. At 30 ns, enable_sum input is forced to '0' deliberately to check whether the outputs are cleared again. Towards the end of the test bench, we add #50 to allow the simulation run for some more time. $finish has not been used in this example.

4.3 Test Bench for Sequential Circuits

Test bench for checking sequential circuit is similar to that for combinational circuits discussed in the last section except that stimuli are different. Test bench is presented in Verilog_code 4.4. As shown in the test bench, the first statement defines a variable called 'clkperiodby2' (meaning half time period) so that we may produce a clock operating at 50 MHz. This is followed by the inclusion of the design file and declaration of the module. As done before, all design inputs are declared as 'reg' to hold the value till it is changed to another value, and all design outputs are declared as 'wire'. The design module is invoked thereafter, calling ports by name. The design is instantiated (u1 refers to the instantiation) only once as had been done in the case of combinational circuits testing. As done before, we apply various stimuli at different points of time in 'initial' block so that all possible combinations are checked.

'reset_n' is an active low system reset that clears all the flip-flops in the design. This signal is applied for 20 ns commencing at 20 ns. Normal circuit operation, therefore, commences at 40 ns. D flip-flop input is forced to '0' and '1' at 40 and 60 ns respectively so that the flip-flop may be checked for correct functioning. A, B, and C inputs are used in the register, pixeloutp_valid, and in the counter, cnt_reg, realization. The register is set for ABC = 010 at 40 ns and reset for ABC = 111 at 140 ns, whereas the counter is incremented by one for the latter condition. ABC is advanced by one every 20 ns commencing from 000.

Verilog_code 4.4

// Test bench for checking sequential circuit realization

```
`define clkperiodby2 10   // 10 ns is the half time period (50 MHz clock).
`include 'seq_ckts.v'                           // This is the design file.

module seq_ckts_test ;                          // Declare the test module.

reg                 D ;                         // Declare all inputs as reg.
reg                 A ;
reg                 B ;
reg                 C ;
reg                 clk ;
reg                 reset_n ;
reg                 hold ;
reg                 shift ;
reg                 in1 ;
reg                 in2 ;
reg                 trigger ;
reg                 in ;
reg      [15:0]     set_data ;
reg                 load ;
reg                 shift_ps ;
reg                 rl_n ;

wire                pixelout_valid ;            // Declare outputs as nets.
wire     [7:0]      cnt_reg ;
wire                delay_out ;
wire     [15:0]     data_out1 ;
wire     [15:0]     data_out2 ;
wire                Z0 ;
wire                Z1 ;
wire                Z2 ;
wire                Z3 ;
wire                out ;
wire                data_out ;
wire                data_valid ;
```

```
wire                    eoc ;

seq_ckts          u1(                      // Instantiate the design module.
                        .D(D) ,            // u1 stands for the first instantiation.
                        .Q(Q) ,
                        .Q_n(Q_n) ,
                        .A(A) ,
                        .B(B) ,
                        .C(C) ,
                        .clk(clk) ,
                        .reset_n(reset_n) ,
                        .hold(hold) ,
                        .shift(shift) ,
                        .in1(in1) ,
                        .in2(in2) ,
                        .trigger(trigger) ,
                        .pixelout_valid(pixelout_valid) ,
                        .cnt_reg(cnt_reg) ,
                        .delay_out(delay_out) ,
                        .data_out1(data_out1) ,
                        .data_out2(data_out2) ,
                        .set_data(set_data) ,        // Inputs and
                        .load(load) ,
                        .shift_ps(shift_ps) ,
                        .rl_n(rl_n) ,
                        .data_out(data_out) ,        // outputs of parallel to
                        .data_valid(data_valid) ,
                        .eoc(eoc) ,                  // serial converter.
                        .Z0(Z0) ,
                        .Z1(Z1) ,
                        .Z2(Z2) ,
                        .Z3(Z3) ,
                        .in(in) ,
                        .out(out)
        );
initial
        begin
                D = 1 ;
                A = 1'b0 ;          // At time zero, let the inputs be 000 binary.
                B = 1'b0 ;
                C = 1'b0 ;
                clk = 0 ;                  // Initialize clk, reset_n and hold.
                reset_n = 1 ;
                trigger = 0 ;              // Monoshot input – not applied.
                hold = 0 ;
                shift = 1'b0 ;             // Don't start the shift operation.
```

```
        set_data = 16'hAAAA ;
        load = 1 ;               // Load the above data.
        shift_ps = 0 ;           // Don't start the shifting as yet.
        rl_n = 0 ;               // Means left shift.
        in1 = 0 ;                // Remain in state 00 (S0).
        in2 = 0 ;
        in  = 1 ;                // Input of Pattern sequence detector.
#20     D = 0 ;
        A = 1'b0 ;               // At time 20 ns, let the inputs be 001.
        B = 1'b0 ;
        C = 1'b1 ;
        reset_n = 0 ;            //Apply Reset.
#20     D = 0 ;
        A = 1'b0 ;               // At time 40 ns, let the inputs be 010.
        B = 1'b1 ;
        C = 1'b0 ;
        reset_n = 1 ;
#20     D = 1 ;
        A = 1'b0 ;               // At time 60 ns, let the inputs be 011.
        B = 1'b1 ;
        C = 1'b1 ;
        hold = 1 ;               // Suspend the Register processing.
        trigger = 1 ;     // Apply trigger to monoshot, i.e., start the timer.
        load = 0 ;
        shift_ps = 1 ;
        rl_n = 1 ;               // Means right shift by one bit.
#20     A = 1'b1 ;               // At time 80 ns, let the inputs be 100.
        B = 1'b0 ;
        C = 1'b0 ;
        hold = 0 ;               // Withdraw the hold, i.e.,
                                 // allow the process to run.
#20     A = 1'b1 ;               // At time 100 ns, let the inputs be 101.
        B = 1'b0 ;
        C = 1'b1 ;
        trigger = 0 ;
#20     A = 1'b1 ;               // At time 120 ns, let the inputs be 110.
        B = 1'b1 ;
        C = 1'b0 ;
#20     A = 1'b1 ;               // At time 140 ns, let the inputs be 111.
        B = 1'b1 ;
        C = 1'b1 ;
        shift = 1'b1 ;     // Start the shift operation for the shift register.
        in1 = 1 ;          // Test sequential machine – Go to state 10
                           // (S2) from 00 (S0).
#40     in1 = 0 ;          // Go to state 01 (S1).
        in2 = 1 ;
```

```
#40   in1 = 1 ;           // Go to state 11 (S3) from state 01 (S1).
      in2 = 1 ;
#40   in1 = 0 ;           // Go back to state 01 (S1).
      in2 = 1 ;
#40   in1 = 1 ;           // Go to state 11 (S3) from state 01 (S1).
      in2 = 1 ;
      trigger = 1 ;             // Apply trigger again at 300 ns.
#40   in1 = 0 ;                       // Go to state 00 (S0).
      in2 = 0 ;
#40   in1 = 1 ;                       // Go to state 10 (S2).
      in2 = 1 ;
#40   in1 = 1 ;                       // Go to state 11 (S3).
      in2 = 1 ;
#40   in1 = 0 ;                       // Go to state 01 (S1).
      in2 = 1 ;
#40   in1 = 0 ;                       // Go to state 00 (S0).
      in2 = 0 ;
      trigger = 0 ;                   // Withdraw trigger at 500 ns.

// Another design example – Pattern sequence detector
// Pattern to be detected:  0110
// Input pattern applied:   1111010011011000l101 .....................
// Output desired:          0000000000l001000010 .....................

#25   in  = 1 ;               // Apply first input pattern (for 'psd')
                              // before every positive edge of the clk.
      set_data = 16'h5555 ;   // Change data for 'ps' at 525 ns.
      load = 1 ;
      shift_ps = 0 ;
      rl_n = 1 ;
#20   in  = 1 ;               // Apply input pattern (for 'psd')
      load = 0 ;
      shift_ps = 1 ;          // and start left shifting at 545 ns.
      rl_n = 0 ;
#20   in  = 1 ;          // Continue applying input pattern (for 'psd').
#20   in  = 1 ;
#20   in  = 0 ;               // 605 ns
#20   in  = 1 ;
#20   in  = 0 ;
#20   in  = 0 ;               // 665 ns
#20   in  = 1 ;
#20   in  = 1 ;
#20   in  = 0 ;               // 725 ns
#20   in  = 1 ;
#20   in  = 1 ;
```

```
        #20     in = 0 ;                    // 785 ns
        #20     in = 0 ;
        #20     in = 0 ;                    // 825 ns
        #20     in = 1 ;
        #20     in = 1 ;
        #20     in = 0 ;                    // 885 ns
        #20     in = 1 ;                    // 905 ns
        #200                                // Run for some more time
        $stop ;                             // and stop.
        end

always
  #`clkperiodby2 clk <= !clk ;              // Toggle to get a free running clk.

endmodule
```

The 'trigger' signal for the non-retriggerable monoshot is applied at 60 ns and withdrawn at 100 ns. The 'trigger' signal is reapplied at 300 ns while the timer is still running to check whether the non-retriggerable feature is working. The shift operation of the shift register commences at 140 ns. The data to be shifted is inherent in the design and, therefore, the test bench need not apply it. The inputs to parallel to serial converter are set_data, load, shift_ps and rl_n. They are initialized right at the beginning as mentioned in the comments. However, the shifting operation commences only at 60 ns. Change data at 525 ns and start left shifting at 545 ns onwards at every rising edge of the 'clk'. 'in1' and 'in2' are inputs to influence the model sequential machine. They are initialized to '0' at the start remaining in the 'state' at 'S0'. The sequential machine starts going from one state to another from 140 ns onwards, depending upon the inputs. Functional testing can be made comprehensive by applying all possible combinations of stimuli.

Pattern sequence detector is the last application covered in this test bench. The signal 'in' is the only (serial) input to this part of the design. Input pattern is applied at every positive edge of the 'clk' commencing from 525 ns. 200 ns after the last input is applied, the simulation is stopped. There is no hard and fast rule for giving this allowance. Anything more than 20 ns will do. The statement #`clkperiodby2 clk <= !clk; within always block means toggle the 'clk' every clkperiodby2 ns, so that we may generate a free running clock running at 50 MHz. By changing 'clkperiodby2' value, we can get any other frequency of operation.

In order to help the reader to design Verilog based systems faster, a Verilog quick reference design card is presented in Appendix 7 of CD. A quick reference for the test bench is also presented in Appendix 8. In addition, a more detailed HDL coding guidelines used in industries is presented in Appendix 9.

Summary

Test benches are forte of verification engineers. This chapter showed how to write an effective test bench. The basic concept of a test bench was shown by presenting a simple design and a model test bench for testing the design exhaustively. For bigger designs, an elaborate test may prove to be difficult. In such cases, the test may be carried out for a range of inputs covering minimum, maximum, center, and few other input values applied judiciously. In the previous chapter, designs for combinational and sequential circuits were dealt. Test benches for the same were presented in this chapter. Usually, the test bench size will be smaller than that of the design. For successful working of a system, RTL coding techniques are inevitable. This is covered in the next chapter.

Assignments

4.1 In the assignments of Chapter 3, it was suggested that Verilog codes for each of the assignments might be either separately written or all of them combined into a single design file. Depending upon the scheme adopted by you, write independent test benches or an integrated test bench to conduct elaborate test on the designs presented in those assignments.

4.2 The following circuits are known as 'ladder diagrams' used in Relay based logic control panel or programmable logic controller (PLC) based open loop industrial controls. The AC or DC field supply is applied across L and N lines. In the first circuit, a motor M may be started or stopped by pressing the appropriate push button switches. Note that an auxiliary contact of the motor is fed back in parallel to the START button. When the START button is pressed, the circuit is complete, thus energizing the motor and subsequently closing the contact M. The circuit is, therefore, completed even after the START push button is released. You need to press STOP push button if you want to stop the motor. In addition to the push button contacts shown, other types of switches are represented as ---| |--- for normally open and ---|/|--- for normally closed contacts. LS represent limit switches, IN stands for field inputs, CR for Control relays or Contactors, and SOL for solenoids. You may regard contacts in series as logical 'AND' and parallel contacts as logical 'OR'. Write Verilog codes for these circuits. Also write a test bench, which checks all the possible combinations for the ladder diagrams (see Figure A4.1).

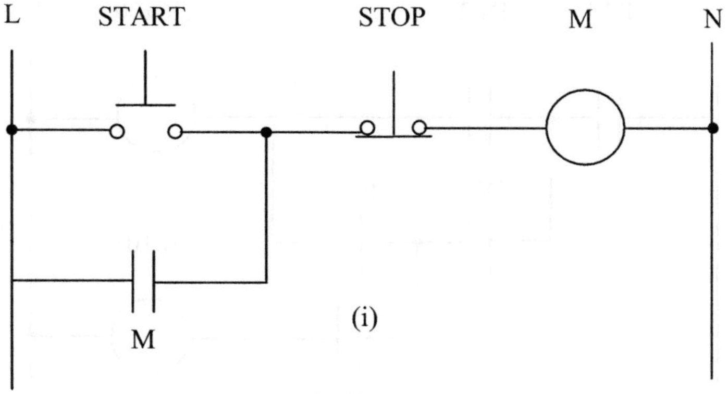
L
START
STOP
M
N
M

(i)

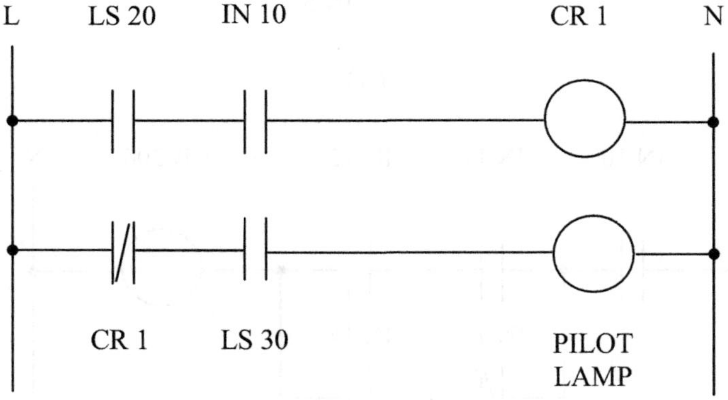
L
LS 20
IN 10
CR 1
N
CR 1
LS 30
PILOT LAMP

(ii)

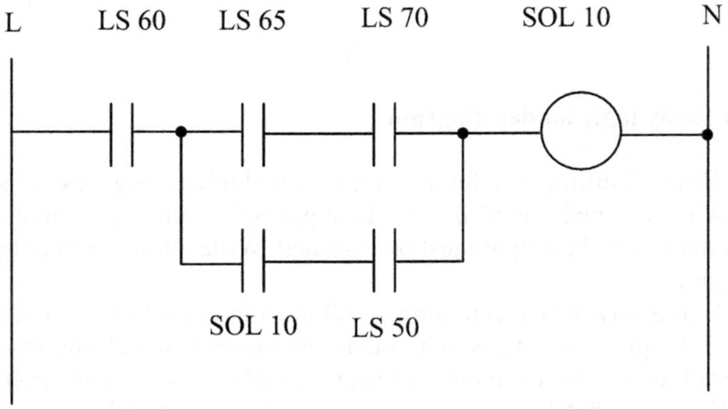
L
LS 60
LS 65
LS 70
SOL 10
N
SOL 10
LS 50

(iii)

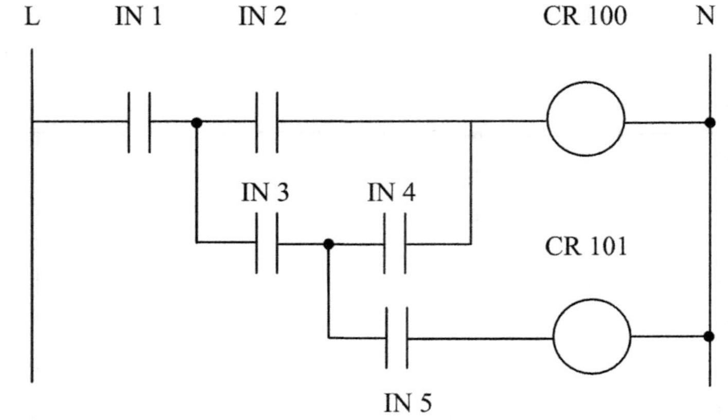

(iv)

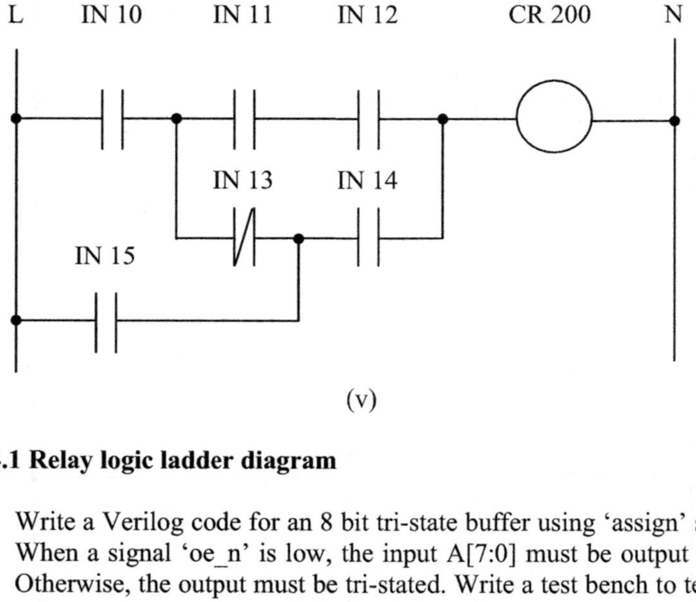

(v)

Fig. A4.1 Relay logic ladder diagram

4.3 Write a Verilog code for an 8 bit tri-state buffer using 'assign' statement. When a signal 'oe_n' is low, the input A[7:0] must be output to Y[7:0]. Otherwise, the output must be tri-stated. Write a test bench to test the circuit.

4.4 A company, which is putting a stall in an industrial exhibition, wants you to design a system, which counts the number of persons who visit the stall as well as the number of people inside the stall at any point of time. You may install lights/photo-electric cells or any other sources/sensors for detecting people entering and leaving the stall. Realize the design using Verilog. Also write a test bench to test all possible combinations.

4.5 Often in certain applications, we need to apply specified set of values. The best way to apply these is to use 'initial' block of statements as shown in the following example:

```
initial
        begin
                        test_pattern = 1;
                #20     test_pattern = 0;
                #30     test_pattern = 1;
                #40     test_pattern = 1;
                #50     test_pattern = 0;
                #60     test_pattern = 1;
        end
```

Draw a timing diagram for the 'test_pattern' signal.

4.6 Another way to apply specified set of values is shown in the following example:

```
initial
        begin
                test_pattern =          1;
                test_pattern = #20      0;
                test_pattern = #30      1;
                test_pattern = #40      1;
                test_pattern = #50      0;
                test_pattern = #60      1;

        end
```

Draw a timing diagram for the 'test_pattern' signal in this case.

4.7 Yet another way to apply specified set of values is shown in the following example:

```
initial
        begin
                test_pattern <=   1;
                test_pattern <= #20     0;
                test_pattern <= #50     1;
                test_pattern <= #90     1;
                test_pattern <= #140    0;
                test_pattern <= #200    1;

        end
```

Draw a timing diagram for the 'test_pattern' signal in this case.

4.8 'initial' block of statements processes only once, whereas 'always' block of statements processes in a repeated manner. The 'always' way to apply specified set of values is shown in the following example:

```
parameter idle = 300;

initial
        begin

        #1000 $stop;
        end

always
        begin
                        test_pattern <=   10;
                        test_pattern <= #20        20;
                        test_pattern <= #50        30;
                        test_pattern <= #90        40;
                        test_pattern <= #140       50;
                        test_pattern <= #200       60;
                        # idle;
        end
```

Draw the waveform for this example.

4.9 Generate a clock signal with different on–off timings, say, TON = 10 ns and TOFF = 20 ns. Use parameter for on–off timings and 'always' block to realize the same. Can you get a 50% duty cycle for this clock? If so, how?

4.10 Using a 'repeat' loop, a fixed number of clock pulses can be generated. Code such a clock. Parameterize the ON, OFF times and the fixed number of clock pulses. Draw the waveform for a 50% duty cycle clock for ten cycles.

4.11 Many applications demand two or more clocks with phase delays among them. Generate two clock signals with a phase delay of 5 ns and different on–off timings, say, TON = 10 ns and TOFF = 20 ns. Use parameter for phase delay, on–off timings and 'always' block to realize the same.

4.12 Write Verilog codes to realize a pattern sequence detector which detects any of the three sequences 0101, 1010, and 1100. Also write a test bench for the same.

Chapter 5

RTL Coding Guidelines

We have so far seen how to model combinational and sequential circuits in Verilog, which are vital ingredients in any digital VLSI system design. The ultimate aim of the designer is to finally map the design on an FPGA device or implement as an ASIC, and this is possible only if you follow certain guidelines. A popular guideline is known as the RTL Coding Guideline, where RTL stands for Register Transfer Level, signifying that data transfers in a system take place via registers [17]. It is basically adhering to synchronous design practices, and it signifies the regulation of data flow, and how the data is processed. Since we deal with a synchronous design, it should run smoothly through Simulation, Synthesis, and finally on place and route tools, which we will learn in subsequent chapters. In order to do this, we have to isolate the asynchronous and sequential circuits. The combinational circuits fall under the category of asynchronous circuits. We have actually followed the RTL coding style in our designs dealt in an earlier chapter. Therefore, the codes developed there will run smoothly in all the tools mentioned above.

5.1 Separation of Combinational and Sequential Circuits

Basically, RTL coding style is describing the circuits in terms of its registers (REG) and the combinational logic (COMB) between them as shown in Figure 5.1a. Any complex combinational circuit, which usually slows down the system speed, can be further broken down into simpler circuits (COMB_1, COMB_2, etc.) of approximately same propagation delay times and sandwiching registers (REG_1, REG_2, etc.) in between two adjoining combinational circuits to improve the overall processing speed. This is shown in Figure 5.1b. The same pattern is followed successively to build a system. The interposing registers are referred to as pipeline registers, which we will discuss in depth later on when we deal with arithmetic circuits.

5.2 Synchronous Logic

The RTL coding guidelines primarily consist of 'DOs' and 'DONTs' for building digital circuits. For instance, the common mistake we make in a design is to take a

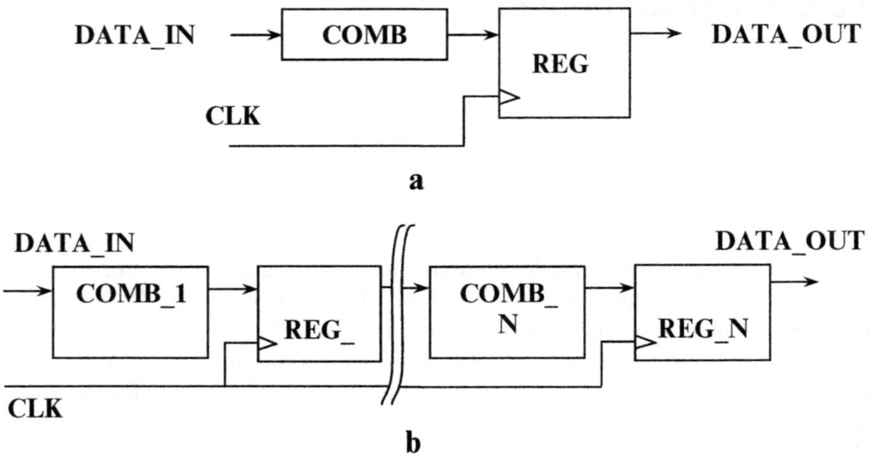

Fig. 5.1 RTL coding – separation of combinational and sequential circuits and pipelining

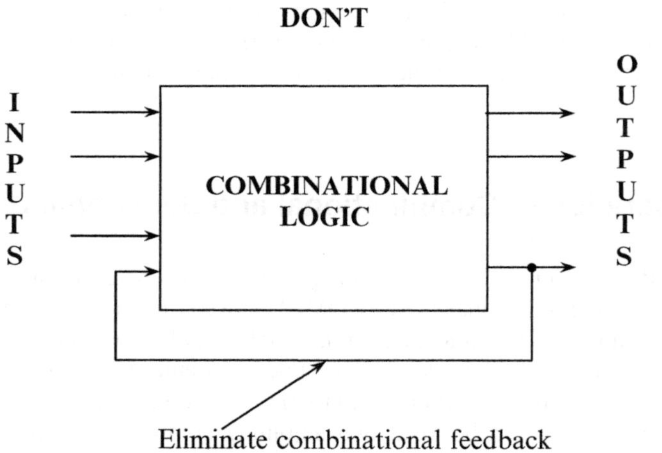

Fig. 5.2 Asynchronous logic – combinational feedback

combinational output and feed it back to one of its inputs as shown in Figure 5.2. This is detrimental in making a working chip since asynchronous feedbacks lead to racing problems and result in unpredictable functioning of the circuit. The remedy for this problem is by breaking the feedback and passing the signal through a D flip-flop as shown in Figure 5.3. The system clock is connected directly to the CLK input of the flip-flop. Note that, by doing so, no functionality is changed except for a clock cycle delay introduced, which we can always afford to spend.

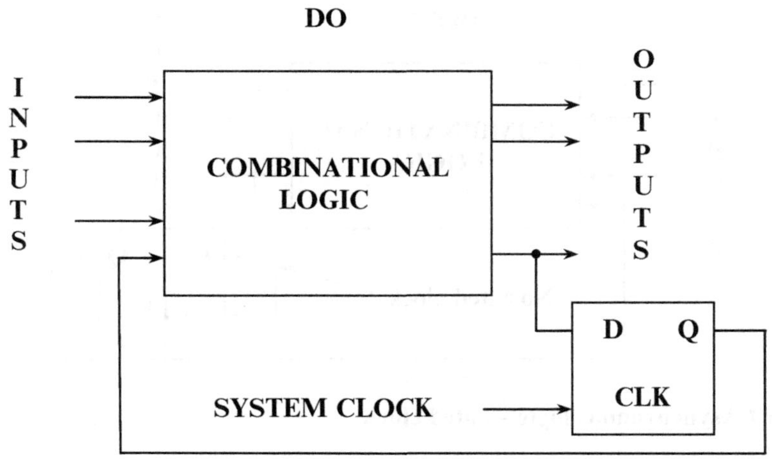

Fig. 5.3 Synchronous logic – eliminates racing and glitches

Further, it eliminates racing and glitches, which is normally present in asynchronous circuits. This will result in a system that will work perfectly when mapped onto a device, whereas the previous circuit will not work at all.

5.3 Synchronous Flip-flop

Another practice a novice designer adopts is to gate the system clock in the manner shown in Figure 5.4. This introduces skew in the clock. A VLSI system usually has innumerable numbers of registers that are connected to the system clock. The clock, owing to gate delays and interconnection path delays in the chip, aggravated by gating the clock, arrives at different points of time to each of these registers resulting in the violation of setup and hold times. The solution is once again by breaking the gating of the clock and, instead, by introducing a MUX in the data path as shown in Figure 5.5. Logic must be incorporated in data input instead of gating the clock. When the signal SELECT is asserted, the DATA fed to the higher order input pin of the MUX is selected and directed to the D input of the flip-flop, and is registered as DATA OUT at the following system clock edge. This continues so long as SELECT is active. Since the flip-flop output is also connected to I0 input of the MUX, the last DATA that was registered in the flip-flop remains stored when SELECT goes low. Thus, the circuit function is precisely the same as that shown for Figure 5.4, while eliminating the clock skew, since the system clock is directly fed without any gate delays. Price we have paid is just a two-input MUX for getting a reliable operation.

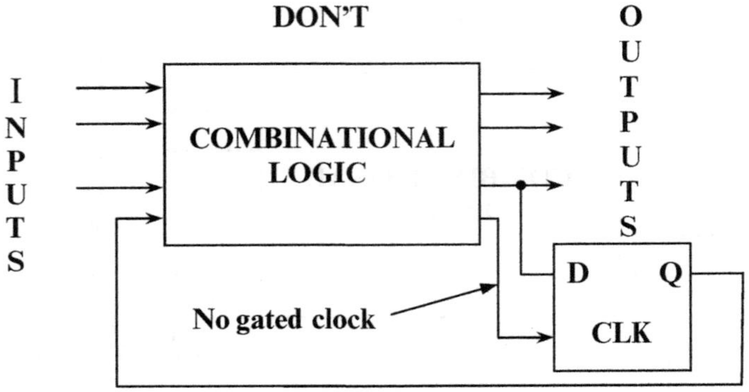

Fig. 5.4 Asynchronous logic – gated clock

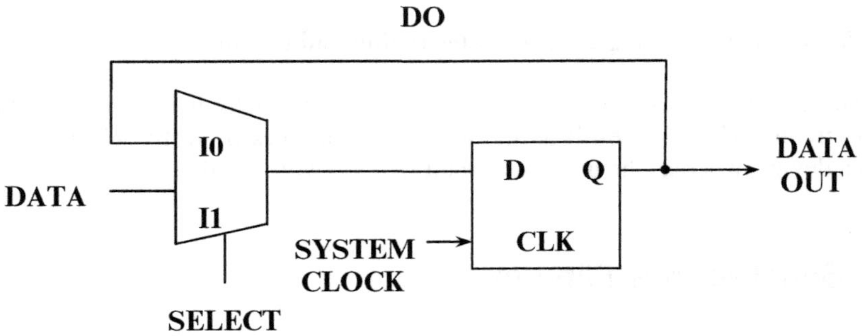

Fig. 5.5 Synchronous flip-flop – logic incorporated in data input

5.4 Realization of Time Delays

Another design mistake commonly committed is to generate a pulse using gate delays as shown in Figure 5.6. Frequently, we need to create a single pulse when a push button is pressed or a long durational pulse for a typical application such as a photographic timer or an industrial timer. The traditional way the designers adopt is to put N numbers of buffers (or inverters), 1 through N, in order to achieve a delay of N times the propagation delay of each buffer. After buffering, the signal 'InD' is inverted and fed to one of the inputs of the AND gate, while the signal 'In' is connected to the other input. This produces the desired output, 'Out', whose pulse width is $N \times t_p$, where t_p is the propagation delay of a buffer. Of course, if the input 'In' is from a push button switch, we need to debounce it before the same is used. Unfortunately, the buffer gate delays in this circuit are technology dependent. For example, if you had used earlier 0.65 μm technology and produced a delay of 100 nanosecond, then with the new technology, say, 0.09 μm technology,

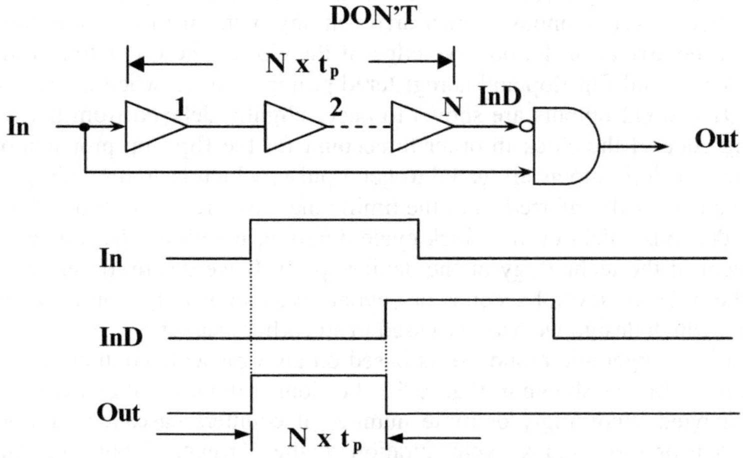

Fig. 5.6 Delay realized using gates – technology dependent

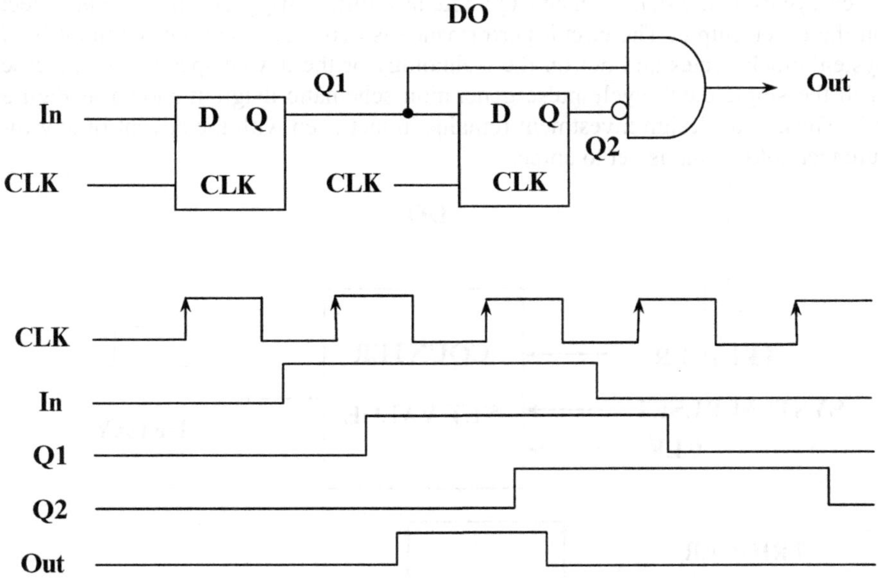

Fig. 5.7 Synchronous: technology independent single clock delay

the delay might crash to 8 to 10 times lower, which may not be sufficient for the application for which it was originally intended. The same is also true if you changed to a faster device without changing the design accordingly. A better way to produce a single pulse is by using two flip-flops as shown in Figure 5.7.

As shown in Figure 5.7, the input 'In' is applied to the D input of the first flip-flop and, being asynchronous, it may arrive at any point of time. Q1 follows the input with the arrival of the positive edge of the clock. The Q1 output in turn is applied to a second flip-flop and is registered promptly at Q2 when the next clock arrives. Q1 and Q2 outputs are shown to occur slightly delayed from the respective rising edge of the clock in order to account for the flip-flop propagation delays. The flip-flop outputs are gated to get a pulse, which lasts for a single clock cycle as can be easily inferred from the timing diagram presented. Since the delay obtained depends solely on the clock cycle duration, it follows that this circuit is independent of the technology or the device speed. If we desire to get large timings of the order of several seconds or beyond, we may use the non-retriggerable monoshot, which design we had discussed in an earlier chapter.

The non-retriggerable monoshot is based on an 8-bit wide counter, which has inbuilt time setting as shown in Figure 5.8. For longer duration of delay, the width can be changed accordingly or more number of counters cascaded. Applying a pulse of one or more clock cycle duration to the 'Trigger' input can start the monoshot or the timer. Once started, the timer output 'DELAY' goes high and remains as such until the lapse of the set delay, after which the output goes low. It has system reset and clock as inputs. For a setting of 255, you will get exactly 255 clock cycles as the delay. Once triggered, any further triggers will have no effect on the timer output. The circuit performance is dependent only on the number of system clock cycles and not on the technology or the device speed as is the case with the single clock cycle pulse generation schematic diagram shown in Figure 5.7. Thus, your design investment remains in tact even with the advent of any future technology that is yet to come.

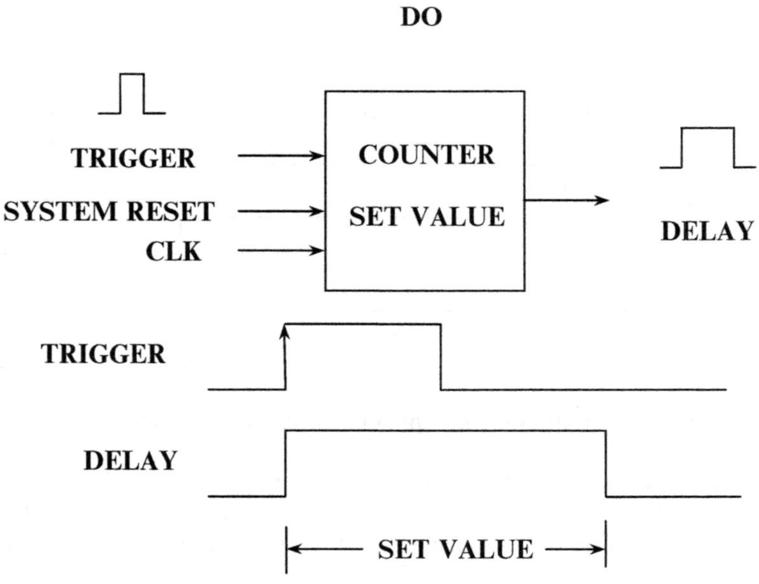

Fig. 5.8 Non-retriggerable monoshot – technology independent

5.5 Elimination of Glitches Using Synchronous Circuits

We discussed earlier the occurrence of glitches in digital circuits. A glitch, which is a narrow pulse, is an uninvited guest, which we would like to eliminate. For example, in the circuit shown in Figure 5.9, a glitch is produced at the node D1 when inputs In0 and In1 are both high. Consider the case when SEL goes from high to low. SEL* responds by going from low to high after the propagation delay through the inverter. It is this delay, which is responsible for producing the unwanted glitch at D1. If we are to use this signal for further processing (such as a clock input for a register), it may so happen that the circuit samples at the undesired region of the glitch, thus causing erroneous functioning of the circuit (such as registering a data or a signal when we do not want it to register). This is true even if the SEL signal is synchronous to the CLK. The circuit will not malfunction only if we can eliminate the glitch. This can be easily accomplished by using a D flip-flop as shown. The flip-flop ensures that the glitch is avoided at the rising edge of the next CLK. Thus, the flip-flop output 'Out' is free from the glitch. SEL signal must be synchronous to the CLK.

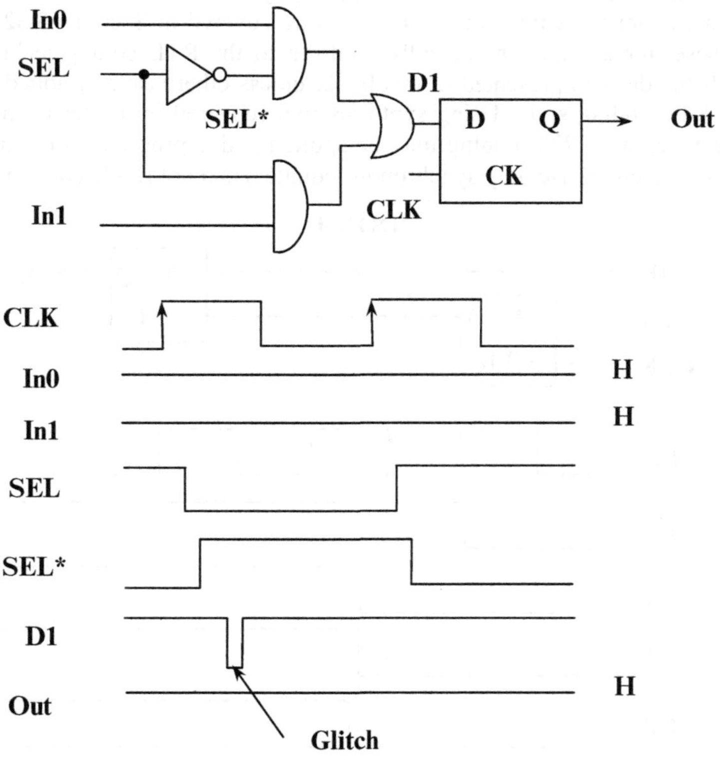

Fig. 5.9 Synchronous circuit – no glitch

5.6 Hold Time Violation in Asynchronous Circuits

Another mistake a designer usually commits is to use a ripple counter as shown in Figure 5.10. Q1 output of the first flip-flop is fed to the clock input of the next flip-flop. This may result in the violation of hold time if D2 is an asynchronous input. To start with, let us say that D1 and D2 inputs are low. With the arrival of the positive edge of clock, the flip-flop output Q1 is cleared. Let us further assume that D1 goes high thereafter. Q1 registers this D1 value (i.e., high) with the arrival of the subsequent positive edge of clock after a propagation delay 't_p' between CLK and Q1. Since D2 is asynchronous, D2 and Q1 may change simultaneously as shown in the waveform. When the second flip-flop encounters the rising edge at its clock input, the data applied at D2 input must be stable at least for the hold time requirement of the flip-flop, otherwise the hold time is violated.

Looking at the waveform, we observe that before the rising edge of Q1, D2 is low and stable, thus satisfying the setup time requirements. However, it changes from low to high exactly when the clock Q1 also changes, thus violating the hold time. There would have been no violation if D2 had remained low for some more time greater than the hold time. After all, what you desire to implement is a register or a counter basically, which designs were covered in Sections 3.32 and 3.33 and, these implementations actually conform to the RTL coding techniques. In fact, all the designs presented in this book, unless otherwise mentioned, conform to the RTL coding style. Using synthesis tool (covered in a later chapter), non-conformance with RTL coding may be spotted and appropriate corrective action applied. Therefore, use only synchronous counters and not ripple counters.

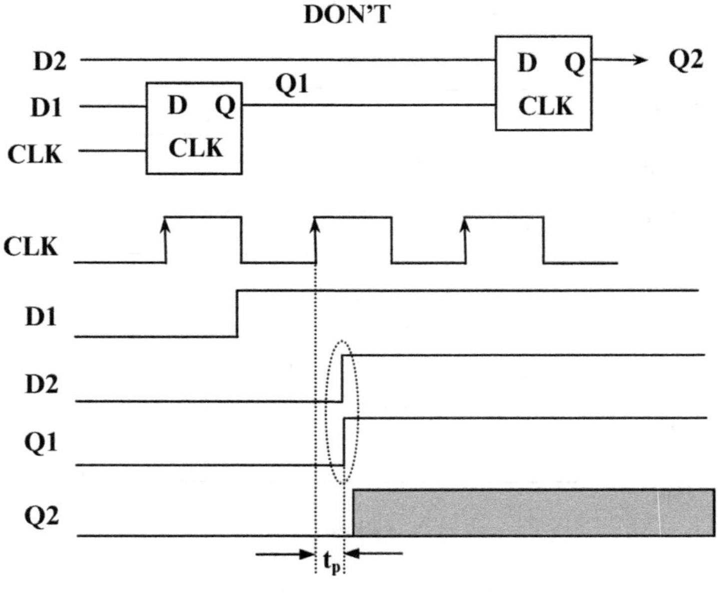

Fig. 5.10 Asynchronous circuits – hold time violation

5.7 RTL Coding Style

In a large design, there may be several modules or sub-modules. It is illegal to nest modules, i.e., write module within a module. But you may call a module within a module. For example, the following Verilog code is prohibited, and the compiler will report error for putting another module such as module_2 within the module_1:

```
module        module_1       (// List the first module I/Os here)
                    // Declare the first module I/Os, wires, and registers.
                    // Write the required combination and sequential
                    // logic for the first module.
     // Calling the module_2 as follows is illegal.
     module         module_2       (// List second module I/Os here)
                    // Declare the second module I/Os, wires, and registers.
                    // Write the required combination and sequential
                    // logic for the second module.
     endmodule

endmodule
```

Instead, the following method of calling the module_2, module_3, etc. within module_1 is perfectly legal:

```
module        module_1       (// List the first module I/Os here) ;

                    // Declare the first module I/Os, wires, and registers.
                    // Write the required combination and sequential
                    // logic for the first module.
module_2      U1 (// List second module I/Os here, calling ports by name) ;
module_3      U2 (// List third module I/Os here, calling ports by name) ;
// Call other modules, if any. U1, U2, etc. are instantiations.
// Note the presence of ';' at the end of each of the statements.

endmodule     // This signifies the end of module_1. Note that there is no space
              // between 'end' and 'module'.
```

A model code which follows the RTL coding guidelines, some of which we have already discussed earlier, is shown in Verilog_code 5.1. We shall call this module as 'rtl_coding' to signify that it describes the RTL coding style. Once you identify the actual module, then list the inputs and outputs in any order you like, separated by commas as shown in Verilog_code 5.1. All the inputs and outputs used in the module are declared as inputs or outputs as the case may be. Outputs are declared as registers or wires if they occur in 'always' blocks and 'assign' statements respectively. In Section 5.1, we have seen that the basic RTL coding style is to break any complex circuit into a series of combination circuits and registers so that data may flow like a perennial river, thus improving the processing speed of an implementation. The first circuit described in the code is nothing but the realization of a D flip-flop, which design we have already seen in Chapter 3. The purpose here is to show the need to separate out combinational and sequential circuits. The first part is an 'always' block, which is just a combinational circuit realization you are already familiar with. Whenever the input d1 or d2 changes, then this combinational block will be processed. After 'begin:' statement, we may declare a 'COMBINATIONAL_CIRCUIT' for enhancing the readability. An exclusive or (XOR) of the two inputs, d1 and d2, is assigned to 'd' which forms the D input of the flip-flop.

Another 'always' block named 'SEQUENTIAL_CIRCUIT', which functions only at positive edge of the clock, outputs the data 'd' to 'Q'. Also, its complemented value is output to 'Q_n'. When we look at the schematic circuit diagram generated by the synthesis tool later on, we will actually find two flip-flops, 'Q' and 'Q_n', and not just one. The flip-flop 'Q_n' can be eliminated by removing the statement 'Q_n <= !d ;' in the sequential circuit if one wishes to reduce the flip-flop count. In lieu of the flip-flop, we need to use an inverter by using an 'assign' statement outside the 'always' block:

```
assign Q_n = !d ;
```

Thus, as a designer, you need to experiment and know the limitations of a tool and then, find out ways and means to circumvent those limitations. Whatever signals are used in an 'always' block, they are declared as registers. 'Q' and 'Q_n' are flip-flops, generally called as registers. Segregating the combinational and sequential circuits thus will facilitate easy reading of the codes and thereby minimize comment writing. This self-documenting feature is the essence of RTL coding style.

We have considered 'if' statements earlier. We will see some more aspects of the same next. We are all familiar with a two way switch as shown in Figure 5.11. The switch is of the type known as the single pole double throw (SPDT). Using this switch, you can select one of the two signals, A or B, depending upon the switch position, and deliver it as the 'OUT' signal. The digital analogy for this switch is the MUX. Here, the A and B inputs are precisely the same as in the mechanical switch, so also the output. The mechanical control of switching from one position to another is simulated by a select control, 'SEL'. A logic '0' at this pin selects the A input designated as '0' input and directs the same to 'OUT'. Similarly, SEL = 1 routes the '1' (B) input to the output. In Verilog coding, 'if' statement infers the multiplexer. The MUX is a combinational circuit, which can be realized by using an always block as shown in Verilog_code 5.1. If SEL is high, then B input is sent to the output, 'out1'. Otherwise, A is selected. All the inputs (SEL, A and B) that influence the output in an always block need to be listed as shown in the code.

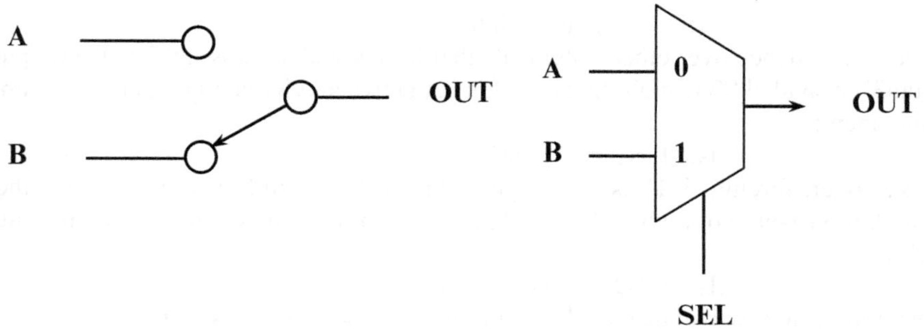

a Two-way Switch b MUX

Fig. 5.11 Switch/MUX analogies

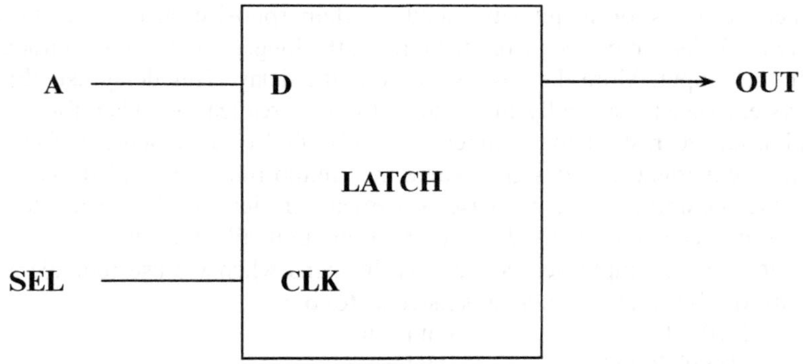

Fig. 5.12 Latch is inferred unless all signals are assigned in all branches of an 'if–elseif' statement block

In this always block, we have considered all the possibilities, i.e., SEL = 0 as well as SEL = 1. In the next block, whose output is 'out2', let us see what happens if we omit the 'else' statement. When 'SEL' goes high, the output 'out2' registers the value 'A' and remains latched on to this value even if 'SEL' goes low subsequently. Latches are, therefore, inferred unless all signals are assigned in all branches. Figure 5.12 shows the standard latch, where the output, 'OUT' follows the 'D' input (A) so long as the clock, 'SEL' is high. The moment the 'SEL' signal goes low, the output freezes at the value of 'A' prevailing at that moment. Avoid all latches in your design since they pass on glitches in the circuit.

Next, we will see how to realize a priority encoder using 'if–elseif' statements. Consider four inputs 'in0' through 'in3', which is required as the desired output, 'out3', based on priority determined by three control signals S2 to S0. S2 signifies the top most priority and S0 the least priority. If more than one such signal is asserted, the topmost priority prevailing at the time will determine the final output. This can be easily realized as shown in the Verilog_code 5.1. The very first statement in the always block, namely, if (S2 == 1) out3 = in0 ;

takes precedence over other statements that follow and assigns 'in0' as the output, 'out3', provided S2 is high. In case S2 is not active and S1 is high, then the second statement:

else if (S1 == 1) out3 = in1 ;

takes over, forcing 'in1' as the output. This is the second priority. On the other hand, if S1 were not high and S0 high, then the third priority statement comes into play:

else if (S0 == 1) out3 = in2 ;

Naturally, 'in2' will appear as the output in this case. The last statement:

else out3 = in3 ;

is processed outputting 'in3' only if none of the signals S2–S0 are not asserted. In summary, whatever you put as the first statement in an always block is automatically assigned the top priority. Subsequent statements will have lower and lower priority, with the last statement being assigned the least priority. After synthesis, the Verilog codes for the priority encoder metamorphosises as the circuit shown in Figure 5.13. As can be seen from the circuit, the longest delay to the output, 'out3' is for 'in3' input. More the nesting, more is the delay. This delay usually affects the system speed, especially in sequential circuit realization. Therefore, it would be advisable to restrict the number of 'if–elseif–else' statements to four or five based on experience. In designs, where this thumb-rule is exceeded, one may explore the possibility of using 'case' statements in lieu of 'if–else if' statements. 'case' statements must be used if conditions are mutually exclusive.

In the next example, let us see what happens when we use redundant conditions in 'if–else if–else' statements listed as follows.

```
if (P < Q)          out4 = R ;
else if (P > Q)     out4 = S ;
else if (P == Q)  out4 = T ;
```

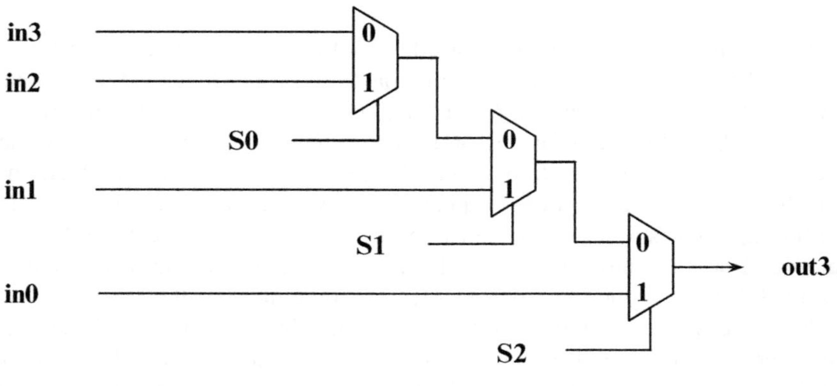

Longest delay to 'out3' is for 'in3' input

Fig. 5.13 Priority encoder

Here, the last statement which checks whether P is equal to Q or not is a redundant statement since the first two statements together have already checked that P is equal to Q. Therefore, this statement may be replaced by the following statement:

```
else     out4 = T ;
```

In the Verilog code, it is shown as a separate circuit whose output is 'out5'. Both the circuits use three numbers of two inputs MUX. However, the 'out5' circuit uses only two comparators instead of three used in the case of 'out4'. This will become clear when we run the synthesis tool later on.

Before we wind up RTL coding guidelines, let us consider the usage of a couple of Verilog Directives for 'case' statements:

- Synopsys full_case
- Synopsys parallel_case

Full_case indicates that all cases are specified even if they don't consider all possibilities. It may be noted that the full_case infers a multiplexer and does not infer a latch as shown in the Verilog_code 5.1. Do not use default for the Synopsys full_case as you do in the normal 'case' statements without Verilog Directives. In the code, one-hot assignment (i.e., only one '1' entry in the signal such as SELECT = 010) is shown first, followed by regular assignment which is not one-hot. The following codes infer latches for the outputs since all possible cases are not specified:

```
case (SELECT)

        3'b001:  LATCH = A ;
        3'b010:  LATCH = B ;
        3'b100:  LATCH = C ;
endcase
```

Note that in this case, neither Verilog Directive is present nor default specified.

The last but one block, 'case', in Verilog_code 5.1 shows the Synopsys parallel_case. It indicates that all cases listed are mutually exclusive to prevent priority-encoded logic. When SELECT equals AA, the output is assigned '001' value. Similarly, the output is assigned either '010' value for SELECT = BB or '100' for SELECT = CC. Three latches are created by the synthesis tool for this block of codes. These latches may be eliminated and a MUX created instead if Synopsys parallel_case and full_case directives are combined as shown in the last always block. The chip area is considerably lower in this case. Therefore, this combination is a better choice than the parallel_case. The consolidated Verilog code is as follows.

Verilog_code 5.1

```
/*                          RTL coding style
 To start with, declare the module you wish to design.
 Note that the design file name is the same as the module name, 'rtl_coding'.
 .v is the extension to indicate that the design file is in Verilog.
```

```
*/
module rtl_coding (                              // Declare the design module.
                   d1 ,
                   d2 ,
                   clk ,
                   Q ,
                   Q_n ,
                   SEL ,
                   A ,
                   B ,
                   C ,
                   out1 ,
                   out2 ,
                   S2 ,
                   S1 ,
                   S0 ,
                   in0 ,
                   in1 ,
                   in2 ,
                   in3 ,
                   out3 ,
                   P ,
                   QQ ,
                   R ,
                   S ,
                   T ,
                   PP ,
                   QQQ ,
                   RR ,
                   SS ,
                   TT ,
                   out4 ,
                   out5 ,
                   SELECT ,
                   SELECTN ,
                   OUT_F_OH ,
                   OUT_F ,
                   LATCH ,
                   AA ,
                   BB ,
                   CC ,
                   OUT_P ,
                   OUT_PF       // Note the absence of ',' for the last I/O.

          );
```

```
input                  d1 ;      // Declare the Inputs and Outputs of
input                  d2 ;      // the module.
input                  clk ;
input                  SEL ;
input                  A ;
input                  B ;
input                  C ;
input                  S2 ;
input                  S1 ;
input                  S0 ;
input                  in0 ;
input                  in1 ;
input                  in2 ;
input                  in3 ;
input       [7:0]      P ;
input       [7:0]      QQ ;
input       [7:0]      R ;
input       [7:0]      S ;
input       [7:0]      T ;
input       [7:0]      PP ;
input       [7:0]      QQQ ;
input       [7:0]      RR ;
input       [7:0]      SS ;
input       [7:0]      TT ;
input       [2:0]      SELECT ;
input       [2:0]      SELECTN ;
input       [2:0]      AA ;
input       [2:0]      BB ;
input       [2:0]      CC ;

output                 Q ;
output                 Q_n ;
output                 out1 ;
output                 out2 ;
output                 out3 ;
output      [7:0]      out4 ;
output      [7:0]      out5 ;
output                 OUT_F_OH ;
output                 OUT_F ;
output                 LATCH ;
output      [2:0]      OUT_P ;
output      [2:0]      OUT_PF ;

reg                    d ;
reg                    Q ;
reg                    Q_n ;
```

```
reg                 out1 ;
reg                 out2 ;
reg                 out3 ;
reg      [7:0]      out4 ;
reg      [7:0]      out5 ;
reg                 OUT_F_OH ;
reg                 OUT_F ;
reg                 LATCH ;
reg      [2:0]      OUT_P ;
reg      [2:0]      OUT_PF ;

// Separate Combinational and Sequential Circuits.
// Easy to read and self-documenting.

always @ (d1 or d2)
    begin: COMBINATIONAL_CIRCUIT     // Realize XOR of the two
        d = d1^d2 ;                  // inputs whenever they change state.
    end

always @ (posedge clk )
    begin: SEQUENTIAL_CIRCUIT
        Q       <=    d ;     // Q & Q_n are two different
        Q_n     <=    !d ;    // flip-flops.
    end

// IF statement infers multiplexer
always @ (SEL or A or B)
        begin
            if (SEL)
                out1 = B ;        // Both possibilities are
            else
                out1 = A ;        // taken into account.
        end

// Latches are inferred unless all signals are assigned in all branches
    always @ (SEL or A or B)
        if (SEL)
            out2 = A ;        // Second possibility (SEL = 0)
                              // is ignored, resulting in the
                              // creation of undesirable latch.

// Priority encoders are inferred by IF–ELSE–IF statements
always @ (S2 or S1 or S0 or in0 or in1 or in2 or in3)
    begin
        if (S2 == 1)
```

```
                out3 = in0;          // Top priority.
            else if (S1 == 1)
                out3 = in1;          // Second priority.
            else if (S0 == 1)
                out3 = in2;          // Third priority.
            else
                out3 = in3;          // Lowest priority.
    end

// Redundant conditions must be removed.
// Don't code in the following manner.
always @ (P or QQ or R or S or T)
begin
    if (P < QQ)
        out4 = R;
    else if (P > QQ)
        out4 = S;
    else if (P == QQ)                // This is a redundant condition
        out4 = T;                    // which may be removed.
    else
        out4 =  out4 ;
end

// Instead, do code it this way.
always @ (PP or QQQ or RR or SS or TT)
begin
    if (PP < QQQ)
        out5 = RR ;
    else if (PP > QQQ)
        out5 = SS ;
    else                         // The redundant statement is removed
        out5 = TT ;              // resulting in the reduction of hardware.
end

// CASE statements must be used if conditions are mutually exclusive.
// Verilog Directives
// 'full_case' indicates that all cases are specified.
// Infers a multiplexer – does not infer a latch.
    always @ (SELECT or A or B or C)
      begin
        case (SELECT)                        // synopsys full_case
            3'b001:  OUT_F_OH = A ;          // One hot assignment –
            3'b010:  OUT_F_OH = B ;          // do not use default.
            3'b100:  OUT_F_OH = C ;
        endcase
      end
```

```
        always @ (SELECTN or A or B or C)
          begin
                case (SELECTN)                     // synopsys full_case
                     3'b001:  OUT_F = A ;          // Not one hot.
                     3'b011:  OUT_F = B ;
                     3'b110:  OUT_F = C ;
                endcase
          end

// The following codes infer latches for the outputs since all possible cases are
// not specified. Also, 'default' is not specified.
        always @ (SELECT or A or B or C)
          begin
                case (SELECT)
                          3'b001:  LATCH = A ;
                          3'b010:  LATCH = B ;
                          3'b100:  LATCH = C ;
                endcase
          end

// parallel_case prevents priority-encoded logic but infers a latch.
        always @ (SELECT or AA or BB or CC)
          begin
                case (SELECT)                      // synopsys parallel_case
                          AA :  OUT_P = 3'b001 ;
                          BB :  OUT_P = 3'b010 ;
                          CC :  OUT_P = 3'b100 ;
                endcase
         end

// Combining parallel_case and full_case directives.
// parallel_case directive prevents priority-encoded logic.
// Infers a multiplexer and not a latch due to full_case directive.

        always @ (SELECT or AA or BB or CC)
          begin
                case (SELECT)              // synopsys parallel_case full_case
                          AA :  OUT_PF = 3'b001 ;
                          BB :  OUT_PF = 3'b010 ;
                          CC :  OUT_PF = 3'b100 ;
                endcase
         end
endmodule                                          // End of design.
```

The test bench for Verilog_code 5.1 is straightforward and self-explanatory, and is as follows. The test bench may be put in a separate file named 'rtl_coding_test.v'. The stimulants are applied using the 'initial' block as we have seen before. A block of codes can be repeated a number of times by using the code 'repeat'. For example, the following codes using 'repeat(3)' repeats the same set of codes embedded within 'begin' and 'end' three times. The objective here is to apply three positive going pulses at the input 'in1'.

```
repeat(3)
begin
#5 in1 = 0 ;            // Apply a positive going input.
#5 in1 = 1 ;
#5 in1 = 0 ;
end
```

The test bench for testing the rtl_coding design is given in Verilog_code 5.2.

Verilog_code 5.2

```
/*                          RTL coding style test bench
This test bench is put into a file called 'rtl_coding_test .v'.
Note that the design file name is 'rtl_coding.v'.
*/

`define clkperiodby2 10
`include "rtl_coding.v"                // This is the design file.

module  rtl_coding_test  ;

reg                 clk ;
reg                 d1 ;
reg                 d2 ;
reg                 SEL ;
reg                 A ;
reg                 B ;
reg                 C ;
reg                 S2 ;
reg                 S1 ;
reg                 S0 ;
reg                 in0 ;
reg                 in1 ;
reg                 in2 ;
reg                 in3 ;
reg     [7:0]       P ;
reg     [7:0]       QQ ;
```

```
reg     [7:0]       R ;
reg     [7:0]       S ;
reg     [7:0]       T ;
reg     [7:0]       PP ;
reg     [7:0]       QQQ ;
reg     [7:0]       RR ;
reg     [7:0]       SS ;
reg     [7:0]       TT ;
reg     [2:0]       SELECT ;
reg     [2:0]       SELECTN ;
reg     [2:0]       AA ;
reg     [2:0]       BB ;
reg     [2:0]       CC ;

wire                Q ;
wire                Q_n ;
wire                out1 ;
wire                out2 ;
wire                out3 ;
wire    [7:0]       out4 ;
wire    [7:0]       out5 ;
wire                OUT_F_OH ;
wire                OUT_F ;
wire                LATCH ;
wire    [2:0]       OUT_P ;
wire    [2:0]       OUT_PF ;

// Instantiate the rtl_coding design module.

rtl_coding      U1(                     // Declare the design module.
                    .d1(d1) ,           // Call ports by name so that
                    .d2(d2) ,           // I/Os can be written in any order.
                    .clk(clk) ,
                    .Q(Q) ,
                    .Q_n(Q_n) ,
                    .SEL(SEL) ,
                    .A(A) ,
                    .B(B) ,
                    .C(C) ,
                    .out1(out1) ,
                    .out2(out2) ,
                    .S2(S2) ,
                    .S1(S1) ,
                    .S0(S0) ,
                    .in0(in0) ,
                    .in1(in1) ,
```

```
                    .in2(in2) ,
                    .in3(in3) ,
                    .out3(out3) ,
                    .P(P) ,
                    .QQ(QQ) ,
                    .R(R) ,
                    .S(S) ,
                    .T(T) ,
                    .PP(PP) ,
                    .QQQ(QQQ) ,
                    .RR(RR) ,
                    .SS(SS) ,
                    .TT(TT) ,
                    .out4(out4) ,
                    .out5(out5) ,
                    .SELECT(SELECT) ,
                    .SELECTN(SELECTN) ,
                    .AA(AA) ,
                    .BB(BB) ,
                    .CC(CC) ,
                    .OUT_F_OH(OUT_F_OH) ,
                    .OUT_F(OUT_F) ,
                    .LATCH(LATCH) ,
                    .OUT_P(OUT_P) ,
                    .OUT_PF(OUT_PF)
          );

initial
begin
          clk   <=    0 ;
#50       d1    <=    0 ;       // Apply all possible combination of inputs to
          d2    <=    0 ;  // d1 & d2 to test the realization of d = d1 XOR d2
                      // and SEQUENTIAL_CIRCUIT, Q = d ; Q_n = NOT d.
#20       d1    <=    0 ;
          d2    <=    1 ;
#20       d1    <=    1 ;
          d2    <=    0 ;
#20       d1    <=    1 ;
          d2    <=    1 ;

// Check whether IF statement infers multiplexer, and
// whether latches are inferred unless all variables are assigned in all branches.
// Also, check CASE statements with Verilog Directives such as
// full_case and parallel_case, independently as well as together.

#20       SEL   <=    0 ;       // Don't start the tests right now.
```

```
#20   SEL      <=   1 ;        // Start the tests by applying various
      A        <=   0 ;        // combinations of inputs.
      B        <=   0 ;
      C        <=   0 ;
#20   A        <=   0 ;
      B        <=   0 ;
      C        <=   1 ;
      SELECTN  <=   3'b001 ;
      SELECT   <=   3'b001 ;
      AA       <=   3'b001 ;
      BB       <=   3'b010 ;
      CC       <=   3'b100 ;
#20   A        <=   0 ;
      B        <=   1 ;
      C        <=   0 ;
      SELECT   <=   3'b010 ;
      AA       <=   3'b001 ;
      BB       <=   3'b010 ;
      CC       <=   3'b100 ;
#20   A        <=   0 ;
      B        <=   1 ;
      C        <=   1 ;
#20   A        <=   1 ;
      B        <=   0 ;
      C        <=   0 ;
      SELECT   <=   3'b100 ;
      AA       <=   3'b001 ;
      BB       <=   3'b010 ;
      CC       <=   3'b100 ;
#20   A        <=   1 ;
      B        <=   0 ;
      C        <=   1 ;
      SELECTN  <=   3'b011 ;
#20   A        <=   1 ;
      B        <=   1 ;
      C        <=   0 ;

#20   A        <=   1 ;
      B        <=   1 ;
      C        <=   1 ;
      SELECTN  <=   3'b110 ;
#20   C        <=   0 ;

// Check whether Priority encoders are inferred by IF–ELSE–IF statements.
      in0      =    0 ;              // Initialize inputs.
      in1      =    0 ;
```

```
        in2    =    0 ;
        in3    =    0 ;
#20     SEL    <=   0 ;           // Disable the previous tests.
        A      <=   0 ;
        B      <=   0 ;
        C      <=   0 ;
        SELECT <=   3'b010 ;
        AA     <=   3'b001 ;
        BB     <=   3'b010 ;
        CC     <=   3'b100 ;
        S2     <=   1 ;           // Priority encoder test starts here.
        S1     <=   1 ;
        S0     <=   1 ;

    repeat(3)
    begin
        #5 in0  =   0 ;           // Apply three pulses to input.
        #5 in0  =   1 ;
        #5 in0  =   0 ;
    end

#20     S2     <=   0 ;
        S1     <=   1 ;
        S0     <=   1 ;
        in0    <=   0 ;           // Withdraw previous input and

    repeat(3)
    begin
        #5 in1  =   0 ;           // apply another input.
        #5 in1  =   1 ;
        #5 in1  =   0 ;
    end

#20     S2     <=   0 ;
        S1     <=   0 ;
        S0     <=   1 ;
        in1    <=   0 ;           // Withdraw previous input and

    repeat(3)
    begin

        #5 in2  =   0 ;           // apply another input.
        #5 in2  =   1 ;
        #5 in2  =   0 ;
    end
```

```
#20     S2      <=      0 ;
        S1      <=      0 ;
        S0      <=      0 ;
        in2     <=      0 ;             // Withdraw previous input and

    repeat(3)
    begin
        #5 in3  =       0 ;             // apply another input.
        #5 in3  =       1 ;
        #5 in3  =       0 ;
    end

#20     S2      <=      1 ;
        S1      <=      0 ;
        S0      <=      0 ;
        in3     <=      0 ;             // Withdraw previous input and

    repeat(3)
    begin
        #5 in0  =       0 ;             // apply another input.
        #5 in0  =       1 ;
        #5 in0  =       0 ;
    end

#20     S2      <=      0 ;
        S1      <=      1 ;
        S0      <=      0 ;
        in0     <=      0 ;             // Withdraw previous input and

    repeat(3)
    begin
        #5 in1  =       0 ;             // apply another input.
        #5 in1  =       1 ;
        #5 in1  =       0 ;
    end

#20     S2      <=      1 ;
        S1      <=      1 ;
        S0      <=      0 ;
        in1     <=      0 ;             // Withdraw previous input and

    repeat(3)
    begin
        #5 in0  =       0 ;             // apply another input.
        #5 in0  =       1 ;
        #5 in0  =       0 ;
```

```
    end

#20     S2      <=      1 ;
        S1      <=      0 ;
        S0      <=      1 ;

    repeat(3)
    begin
        #5 in0  =       0 ;                 // Apply the same input.
        #5 in0  =       1 ;
        #5 in0  =       0 ;
    end

// Check whether a redundant condition increases the gate count.
#20     P       <=      249 ;
        QQ      <=      250 ;
        R       <=      100 ;
        S       <=      150 ;
        T       <=      200 ;
        PP      <=      249 ;
        QQQ     <=      250 ;
        RR      <=      100 ;
        SS      <=      150 ;
        TT      <=      200 ;
#20     P       <=      255 ;
        QQ      <=      250 ;
        PP      <=      255 ;
        QQQ     <=      250 ;
#20     P       <=      250 ;
        QQ      <=      250 ;
        PP      <=      250 ;
        QQQ     <=      250 ;

#1000                   $stop ;
end

always
  #`clkperiodby2 clk <= ~clk ;             // Toggle to get a free running clk.

endmodule
```

Summary

Every designer is vitally interested in making his or her design work when implemented as a hardware, which uses FPGA or ASIC. This requires a high degree of discipline or care while designing such systems. The design can work on the chip only if it conforms to RTL coding guidelines. The present chapter discussed RTL coding techniques in depth, which is basically adhering to synchronous design practices. RTL approach deals with the regulation of data flow, and how the data is processed using register transfer level as the primary means. Since we deal with a synchronous design, it should naturally run smoothly through various tools such as simulation, synthesis and place and route, which tools are described at length in succeeding chapters.

Assignments

5.1 Code the following Boolean expressions in terms of registers and combinational logic between them such that the code results in the best possible processing time. Your codes must be RTL compliant.

a. $X = (A \odot B) + (A \oplus B) + (C \odot D) + (C \oplus D)$

b. $Y = (A \odot B)\ (A \oplus B)\ (C \odot D)\ (C \oplus D)$

c. $Z = \sum (0, 1, 5, 9, 10, 15, 19, 21, 25, 29, 30, 35, 37, 39, 40, 43, 48, 49, 50, 55, 58, 60, 61, 63)$

5.2 Write a test bench for the assignment 5.1.

5.3 Will direct implementation of the following expressions yield RTL compliant code? If yes, explain how and write the code. Otherwise, code them for RTL compliance.

U = (CNT_1 >= 128) (CNT_5 = 100) + W

V = (CNT_1 >= 128) (CNT_2 = 100) (CNT_3 = 200) (CNT_4 = 400) (CNT_5 = 100)

W = (CNT_1 = 500) U

5.4 Write a test bench for the RTL compliant design you have written for assignment 5.3.

5.5 Write Verilog code for the circuit given in Figure 5.5 in the text and a test bench for the circuit.

5.6 Code for the circuit given in Figure 5.9 in the text and a test bench for verifying the circuit functionality.

5.7 A keyboard is used in a computer or in a musical instrument. A key depressed can be recognized if an input port, key_in [3:0], and an output port, key_out [3:0], are connected in a matrix with a normally open key connected across the intersection of an input and an output line as shown in the Figure A5.1. The computer adopts what is known as the 'two-key lock out' system, and the musical instrument adopts the N key roll over system. In the former system, when more than one key is pressed simultaneously, only the last key released will be recognized, whereas in the latter, all the keys pressed are recognized so long as they are not released. The keys require 2.5 ms debouncing time. You may adapt the software debouncing approach used in microprocessor-based designs. Explain clearly your design methodology using ASM chart and realize the RTL Verilog code to implement both the systems.

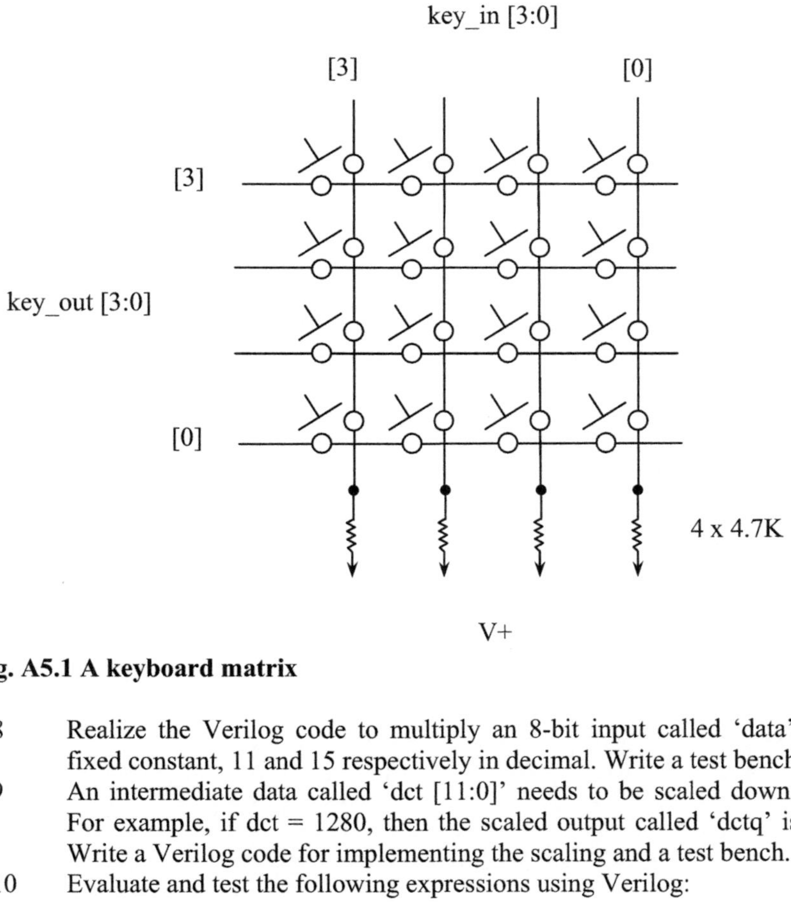

Fig. A5.1 A keyboard matrix

5.8 Realize the Verilog code to multiply an 8-bit input called 'data' by a fixed constant, 11 and 15 respectively in decimal. Write a test bench.

5.9 An intermediate data called 'dct [11:0]' needs to be scaled down by 8. For example, if dct = 1280, then the scaled output called 'dctq' is 160. Write a Verilog code for implementing the scaling and a test bench.

5.10 Evaluate and test the following expressions using Verilog:

a. i1 – i2 + 2 i3 – 2 i4

b. 100 i1,

where i1 to i4 are 8 bit unsigned numbers.

Chapter 6

Simulation of Designs – Modelsim Tool

In Chapter 4, we showed how to write test benches so that we may simulate our designs presented in Chapter 3 as well as any other designs. Prior to starting the simulation, we will see how the design flows for VLSI circuits. This will be followed by a discussion on design methodology that may be adopted for solving problems effectively. Finally, we will learn the simulation tool and apply it to verify our designs covered earlier as well as those that will be covered in later chapters.

6.1 VLSI Design Flow

Before taking up the design of a product, we need to assess the demand for the product based on market research. Having thus identified the product, we have to formulate what is called preliminary specifications. We need to discuss with pros-

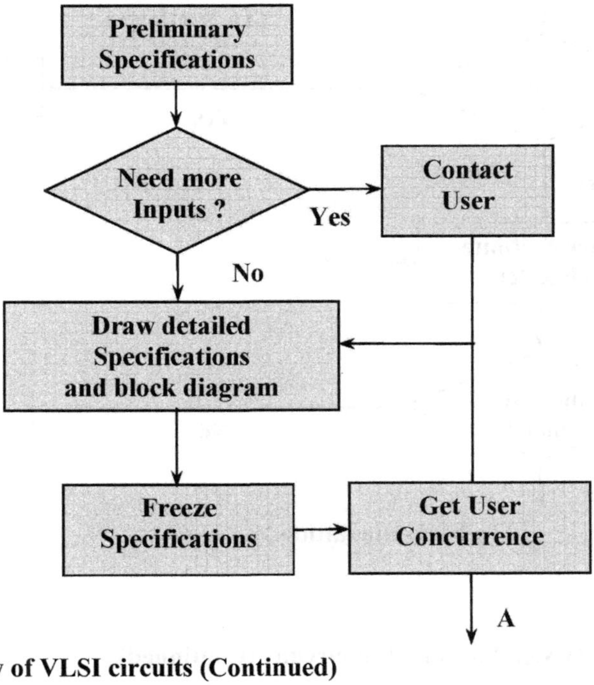

Fig. 6.1 Design flow of VLSI circuits (Continued)

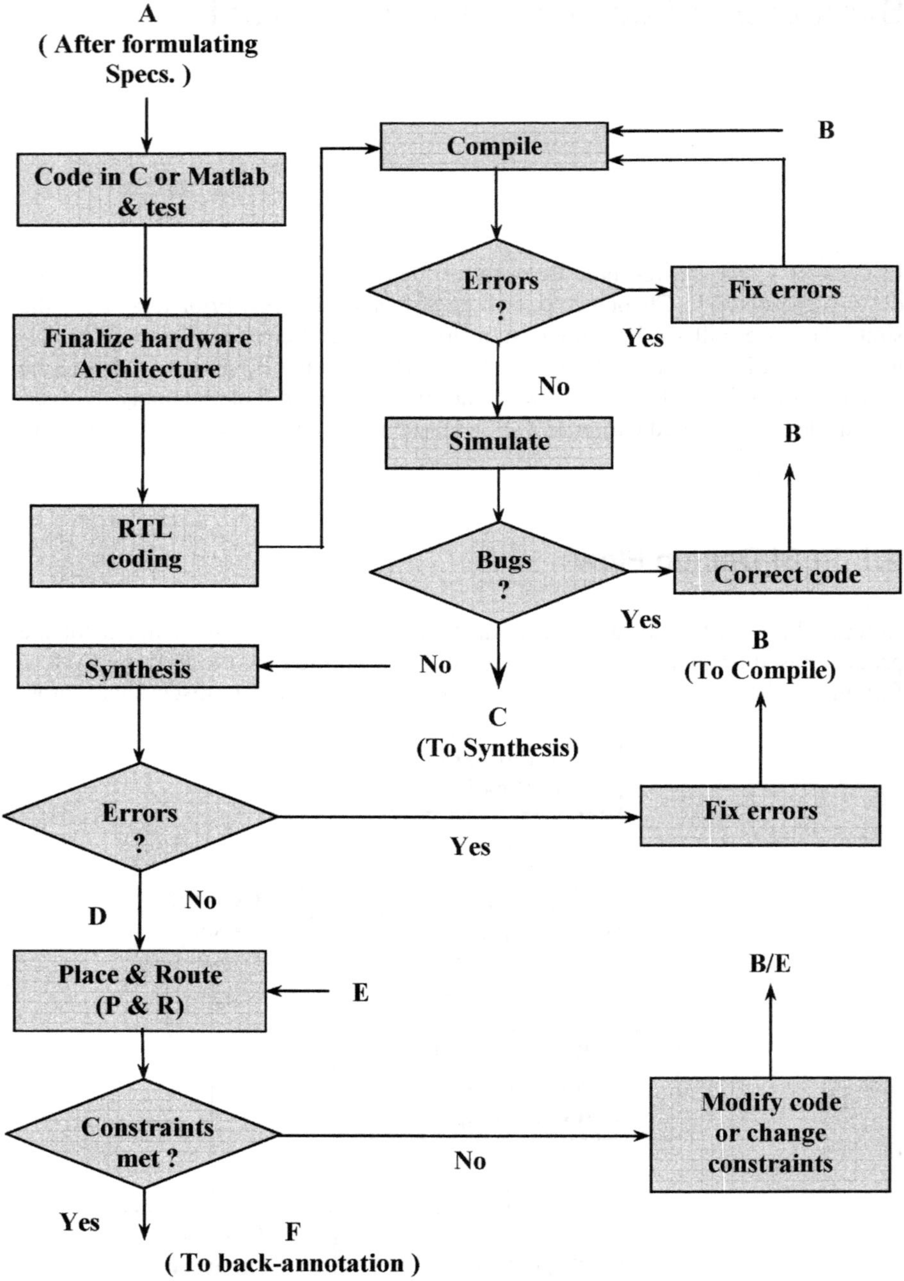

Fig. 6.1 Design flow of VLSI circuits (Continued)

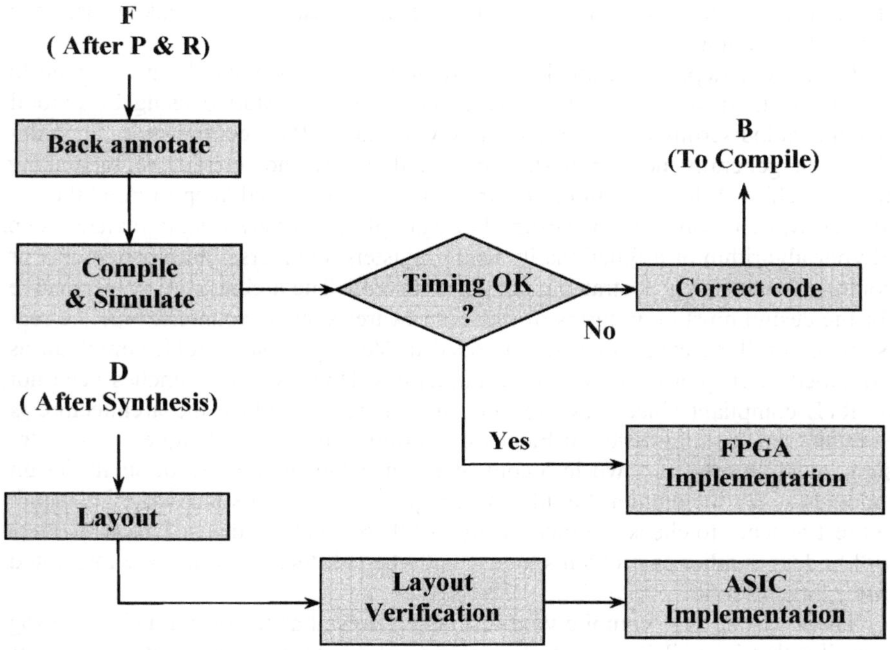

Fig. 6.1 Design flow of VLSI circuits

pective users, and gather more inputs as we progress with drawing a detailed specification as shown in Figure 6.1. We also need to draw a block diagram of the system we wish to design at this stage. This is an iterative process that needs to be done with care before we can freeze the specifications. Finalizing the specifications at this stage is of paramount importance since the design involves the development of hardware such as printed circuit board (PCB), integration, testing, etc.; rework of which is time consuming and costly. Naturally, this calls for timely user approval of the finalized specifications. The above hardware development cycle involving PCB, etc. is not shown in Figure 6.1 since our main interest lies in the development of Verilog codes for FPGA or ASIC implementation.

Once you are done with detailed specification of the system, you may have to prove new concepts that you have proposed for implementation. For example, you might have developed a new algorithm or a new architecture, which you are attempting to implement. These are still unproven and, in order to prove their suitability for the design, you need to use higher-level language such as C or Matlab and test the codes for all possible combinations that you are likely to encounter later on in the actual hardware when you implement the design. Even standards can't be taken for granted and needs to be tested before undertaking time consuming Verilog codes. When you are sure that your concept is working, then it is right time for you to start the hardware design. However, for simple designs, the above step of proving the concept may be bypassed since it can be easily taken care of in the Verilog coding.

In the next step, the detailed hardware architecture is worked out. We tend to design the hardware the way we have coded in C or Matlab using behavioral statements indiscriminately, which is likely to violate RTL coding style. Such designs are generally not synthesizable and, therefore, not conducive to making working chips. While designing the architecture, you should keep in mind the actual hardware involved in the design. For examples, you may wish to realize a step of your algorithm or a functionality using registers, counters, and pipeline data or control flow registers at strategic points, and so on. The architecture you conceive for the design must be in terms of these hardwares such as counters, pipeline registers, controllers, etc., which when coded in Verilog must be RTL compliant as explained in chapter on RTL coding guidelines. However, test benches need not be RTL compliant since they are not parts of designs. Once the architecture is through, start coding different blocks conforming to RTL techniques. Large designs must be broken down into convenient, manageable chunks of small design modules as per the dictum, 'divide and conquer'. Where possible, write independent test benches to check the functionality of these small modules. This way, there will be less number of problems to sort out when all these modules are integrated into a large design.

The next step is to compile your test bench as well as the design using Verilog compiler that is available in simulation or synthesis tool or independent compiler like NC Verilog of Cadence. The compiler reports syntax errors, miss-spelling, etc. Fix errors, if any, and recompile them till they are error free. After that, you can go on to the next stage, simulation, wherein you use the built-in waveform viewer to analyze the functionality of your design. If you encounter bugs in your design, you have to fix them; go back to compiler to repeat the previous step. These steps are iterative in nature. Beginners, especially, have frustrating experience fixing errors at the incipient stage. Perhaps, the best way to overcome this hurdle will be to take a known good working code (such as any of the codes developed in this book), create deliberate errors, study the error reported and thereby learn to fix them. Once you are through with this exercise, you will be at home later on. If there are no bugs, you go to the next step known as the synthesis. Modelsim, signal scan are some of the tools used for simulation.

Using a synthesis tool, you can map on to a particular target device if the device is an FPGA or a vendor technology library if the design is for an ASIC implementation. Synplify and Leonardo Spectrum are some of the synthesis tools for FPGA platform and DC compiler of Synopsys is for synthesis of ASIC design. Synopsys tool also has built-in simulation. The main purpose of synthesis is to perform logic optimization on your design. Once again, you may encounter errors, which will have to be fixed. In addition to the syntax errors, the tool will report RTL non-compliance, frequency of operation specified is too high, setup time violation, etc. If there are errors, you need to compile, simulate, and synthesize again after applying appropriate corrective action.

In the case of ASIC design flow, you need to branch off to layout from here. In synthesis, you will get gate level net list. For example, in FPGA realms, the tool will be mapping your design in terms of look up tables (LUTs), MUX, primitive gates pertaining to a particular type of FPGA, and so on. For FPGA design, an electronic data information format (EDF) file will be created finally by the synthesis tool, which you can use as input for the next tool called the place and route (P & R). Here too, if there are errors, you have to go back for compilation, simulation, and synthesis once again till your design is totally free from errors. At this stage, the types of errors normally encountered are constraints such as speed of operation, power, and area are not met. Depending upon the errors reported, you need to correct your code and go back for compilation, etc. or change constraints and continue the exercise starting from P & R. Place and route is vendor specific. Xilinx P & R tool is an example. EDF file generated during synthesis will be input for place and route tool. In addition to clock constraints, desired input/output (I/O) pins can also be specified by the designer.

The next step is back annotation if all constraints are met. The real gate delays come into play only after back annotation. Therefore, the maximum safe frequency of operation for a design can be determined only after the back-annotated code is tested in the simulation producing the correct results. There is a tendency among the beginners to do simulation without taking the design through synthesis, place and route, and back annotation and, jumping to conclusion that the design works at a very high speed of several hundred Mega Hertz. This is a sure way to indulge in self-deception since one can simulate the design at any desired speed in the order of GHz and, therefore, is required to be curbed right in the incipient stage. What are practical on FPGAs are 50 to 100 MHz operation speeds depending upon the gate delays in the design.

On ASIC platform, depending upon the technology used, one may achieve two to four times the operating frequency that can be achieved on FPGAs for the same design. This aspect will be made clear in the later chapters when we actually see the back-annotated results for some of the designs. If the desired timing is not achieved after simulation of back annotated design, then one must explore the possibility of adding more pipelining stages, modify or correct code accordingly and repeat the tool iteration starting from compilation till the desired timing is achieved. One may also switch over to faster devices if available. If the desired timing is not achieved even after a number of iterations, then one must settle down for the maximum possible frequency of operation and compromise on the specifications accordingly. When the technology improves, one may hope to get better performance. If the timing constraint is finally met, FPGA implementation may be taken up.

The FPGA implementation step requires the populated PCB with the target FPGA mounted and duly tested to ensure the healthiness of the entire hardware. Trouble shooting of the hardware may be undertaken using logic analyzer, pattern generator, and a development system. Usually, when the specifications of the system are frozen, the above-mentioned hardware is also developed simultaneously as far as feasible with the development of Verilog code. This, naturally, requires an assessment of the right package, size, and speed of the target FPGA beforehand. For the same type and package of FPGA, it is possible to migrate from smaller capacity to higher capacity in a limited way without the need to rework the PCB. In short, the whole design process is an iterative process; you have to do it again and again till you get a totally bug free code that will work on your circuit board finally in accordance with the specifications formulated. After place and route, a bit stream is generated. This bit stream is meaningful only if timing constraint is met and needs to be downloaded into the FPGA either from an on-board EPROM or from the development system using parallel or serial port to configure the FPGA to your application. If you are an intellectual property (IP) core designer, you need to supply only the bit stream along with proper documents, and there will be no need to deliver the actual hardware.

In the case of ASIC design, after running through synthesis using Synopsys, layout and its verification are carried out using Cadence back-end tool and taping out the design to the fabricator. As in the case of FPGA design, the PCB is developed using the fabricated IC and tested, thus completing the ASIC implementation. In ASIC platform, in addition to clock constraints, you can specify power as well as chip area as constraints. As mentioned before, the hardware and Verilog coding may be developed simultaneously. Normally, a team of engineers will be developing the hardware and another team developing Verilog codes, if the project size is large. Both the teams need to interact with each other effectively in order to deliver the final product to the satisfaction of the user.

In Altera FPGA development platform, there is what is called localized parametric modules (LPM), whereas in Xilinx they have Logiblox and Coregen modules. They are tailor made design modules for different applications such as multipliers, filters, etc. Although they are menu driven and speed up the development cycle, they are all vendor-specific and, therefore, cannot be used if you intend migrating from one FPGA vendor to another or proceed to ASIC implementation. Therefore, this design book does not use any of the vendor-specific modules. The approach adopted in this book enables the Verilog codes developed to work on any FPGA platform as well as on the ASIC platform without the need to change the codes.

6.2 Design Methodology

Before we start on a particular design, we need a strategy or methodology for designing systems efficiently. For instance, we may be designing a very complex system requiring the development of long codes. As pointed out before, we need to break up large designs into smaller modules so that debugging the codes is made easy. Any design has a top level module, which may call lower level modules, which in turn may call still lower level modules, and so on. In the examples we covered earlier, namely, combinational and sequential circuits design, we had only top level and no lower level modules. However, if we have bigger designs such as those we will be dealing in the later chapters, it is a good practice to down size those designs to smaller submodules and further submodules, and so on as per our system requirements. In this fashion, you can go to any number down below. Here, the idea is that we need to manage only small modules at a time, while developing the design. There are two approaches basically: Bottom-up design methodology and top-down design methodology.

Once you have coded and tested the modules of lowest level, you can put them together to form a higher level and test that higher module. This way, you can move your design from bottom till you eventually reach the top as shown in Figure 6.2. For example, let us say that we have a large design called 'dctq' which requires a maximum of two levels below the top design, dctq, as shown in Figure 6.3. Lowest level module in this design hierarchy is 'ram_rc', which is being called twice by the module 'dualram'. In addition to this module, the top design instantiates multipliers such as 'mult8ux8s', adders such as 'adder12s' and registers, 'dctreg2x8xn', ROMs such as 'romc' and a controller named 'dctq_controller'. If none of these modules are already available at the time of commencing our design, in this methodology, we need to start the design from any of the lowest level module, ram_rc for instance. Instead of ram_rc, we can start from mult8ux8s or adder12s or any other module at the same level since these modules do not depend upon each other. It may be noted that dualram module can be designed only after ram_rc module is developed and satisfactorily tested and not before. Similarly, the top design, dctq, can be developed and tested only after all other modules are completed.

Another approach to the design is the top-down methodology as shown in Figure 6.4. Of all the modules, the top design is closest to the product specification. Therefore, in this method, we start the design with the top level module and move down. This approach is handy if submodules, which you or other design team members have developed earlier for some other project, are already available to you. Often, it is necessary that some of these modules be modified to suit the present application. At every stage, naturally, you need to develop a test bench to check the functionality pertaining to that stage of development. This way, when

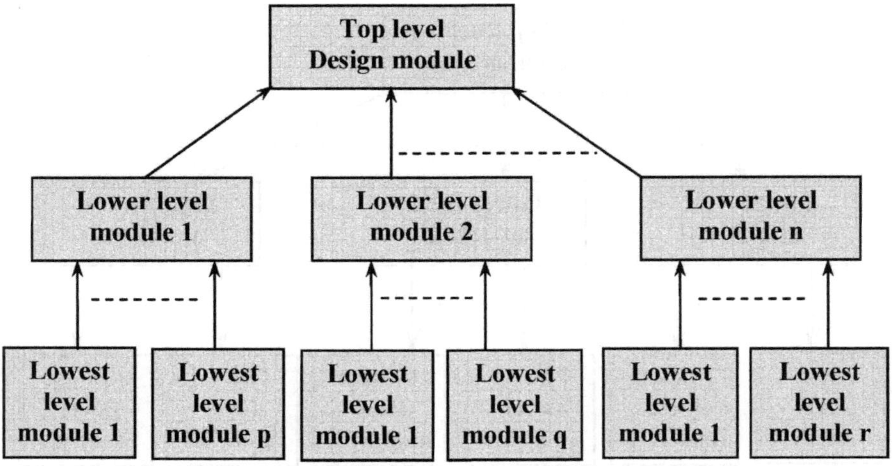

Fig. 6.2 Bottom-up design methodology

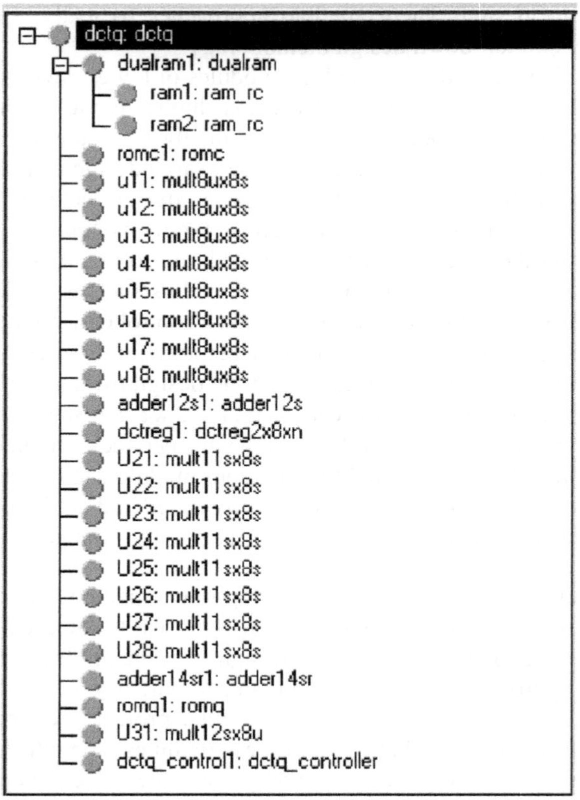

Fig. 6.3 An example of a design hierarchy

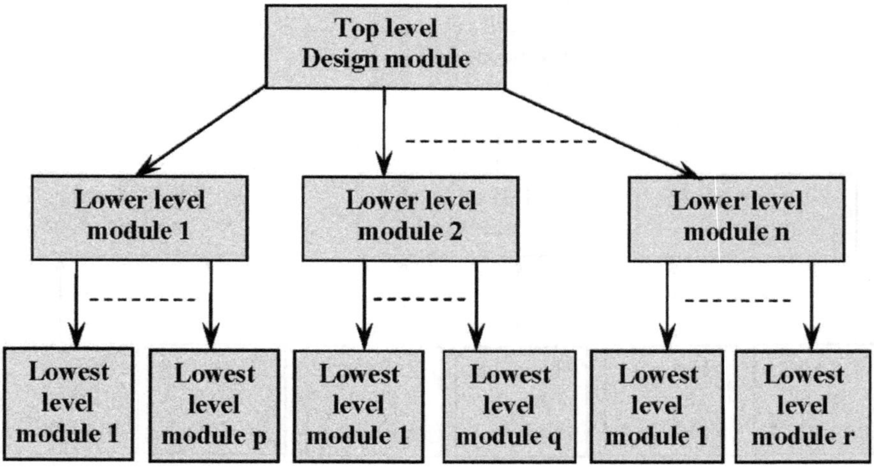

Fig. 6.4 Top-down design methodology

you integrate all the modules finally, it is not likely to present problems. We can't possibly say which of the two methodologies is the best approach for a design. In real practice, a mix of both these methodologies is required making use of the available designs on hand and proceeding on either direction to achieve the desired goal in minimum possible time.

6.3 Simulation Using Modelsim

Let us now gain hands on experience with simulation. We will be using ModelSim [18] for simulating our designs. Update latest versions periodically since there may be useful new features. A command summary of this tool is presented as a ready reckoner towards the end of the chapter. To start with, we will take for simulation the simple design, two-input AND gate, whose test bench was developed in an earlier chapter.

Creating a Project: Double click on the ModelSim icon (Figure 6.5) on your desktop to open the simulator. Two windows open as shown in Figures 6.6 and 6.7. To begin with, we need to create a new project, so click on "Jumpstart" followed by "Create a Project" in the Welcome window. Another window shown in Figure 6.8 also opens. You can also create the project by clicking-on "File => New => Project" and keying-in 'and_2input' in "Project Name" field. Also use "Browse"

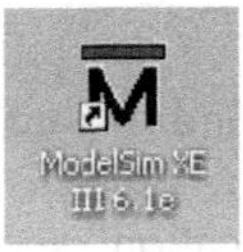

Fig. 6.5 Icon of ModelSim simulator

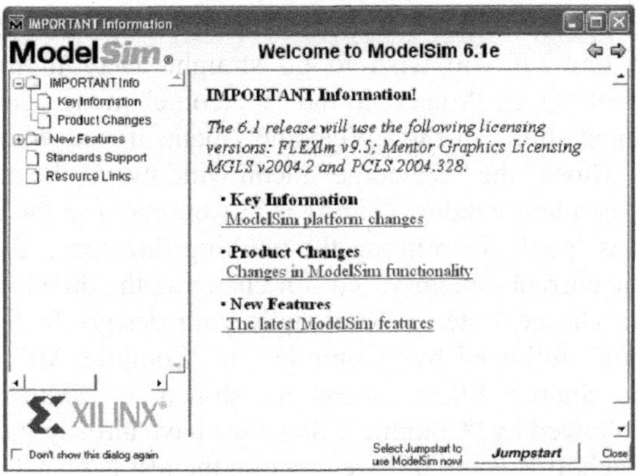

Fig. 6.6 Welcome message of ModelSim

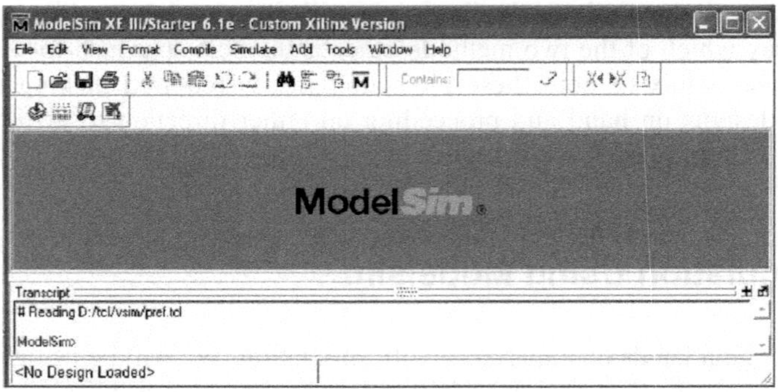

Fig. 6.7 Opening menu of ModelSim main window

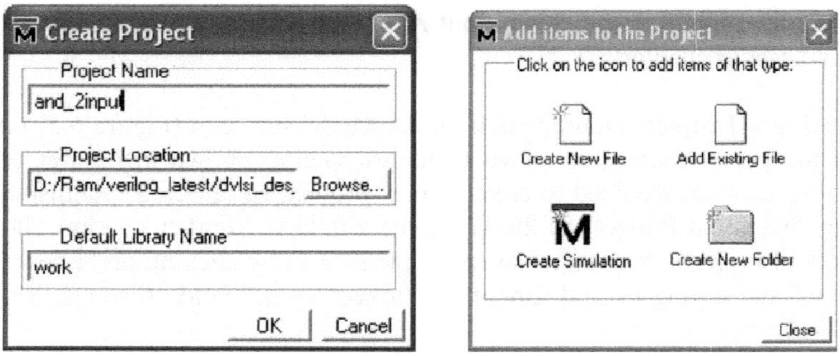

Fig. 6.8 Create (new) project window and add existing file

to select the desired "Project Location" such as 'D:\ram\verilog_latest\dvlsi_des_ Verilog', where your design is residing. Making sure that "work" is specified in "Default Library Name" field, click on "OK". Another window called "Add items to the Project" opens. Click on "Add Existing File". In a new window "Add file to a Project" that opens, select the desired test bench 'and_2in_test.v' using "Browse". Next time, if you wish to go straight away to the same project, 'and_2input', select "Open Project" in the "Welcome" menu. Click on OK. Click on "Open Documentation" in the "Welcome" menu if you wish to use the tool documentation. "Close" the "Welcome" menu window. At the command prompt, "modelsim>" in the main window (Figure 6.7), you may use the operating system commands such as 'pwd' for printing the working directory, 'dir' for displaying the contents of the current directory, 'cd' for changing the directory, etc.

Compilation: The next step is to compile your design. In the main window, click on "Compile" followed by "Compile" or "Compile All". A new window called "Compile Source Files" opens as shown in Figure 6.9. Click on "and_2in_test" followed by "Compile". Since we have already included the design file in the test bench, it is enough if we compile the test bench. The tool will compile the design as well as the test bench in that order automatically. You can ensure this by clicking on "Compile" => "Compile Summary". Errors, if any, will

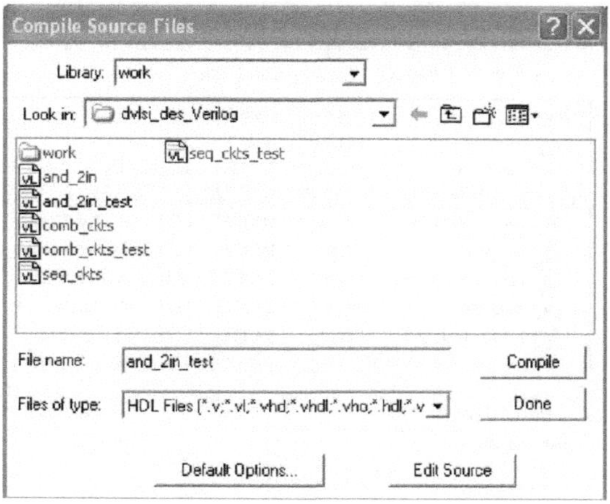

Fig. 6.9 Compile window

be reported in the main window. If there are errors, you will have to correct the errors in your code using "Edit Source" in "Compile Source Files" window and re-run the compiler. Click on "Done" to dismiss the compilation window. In the next chapter, we will see how to tackle these errors. Finally, when there are no errors encountered, you can do the simulation.

Simulation: In order to start the simulation of the design, click on "Simulate => Start Simulation". The window named "Start Simulation" shown in Figure 6.10 pops-up. Click on "+" on the left of 'work'. Click on the desired test bench followed by "OK" to load the test bench and the design. While loading the design, errors or warnings may be encountered. If your system time is different from that of the server, which has the license (for ModelSim SE/PE versions), error will be reported. The remedy is to set the system time correctly. Warnings such as Module 'and_2in' do not have a timescale directive in effect, but previous modules (and_2in_test) do may be ignored. The timescale is mentioned only in the test bench and not in the design, since synthesis tool will ignore all timescale settings in the design.

Waveform Analysis: The main purpose of simulation is to get the timing diagram so that our design may be tested functionally. In the main menu, click on "View => Workspace" to view the Workspace in the main window. Make sure that a tick mark appears to the left of "Workspace". In this window, click on "View => Debug Window => Objects" followed by "View => Debug Window => Wave" to open the Objects (Signals) and Waveform windows showing the timing diagram (see Figures 6.11 and 6.12). In "Objects" pane, click on "Add => Wave => Signals in Design" to display all the signals in the test bench as well as in the design in the waveform window. 'A', 'in', and 'out' are the signals in the test bench, whereas their counterparts in the design are 'A', 'B', and 'Y'. 'u1' is the instance we have used in the test bench to call the design. In order to get the wave forms, click on the icon marked "Run-all" in the wave window. The source file

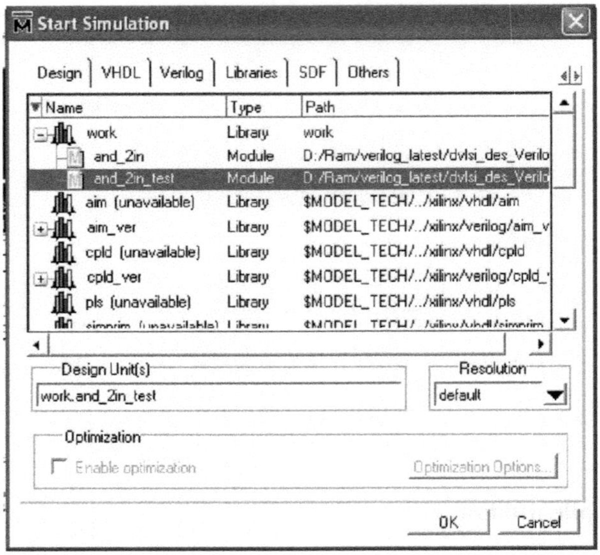

Fig. 6.10 Simulation window

(test bench) is also opened, which you can close if you do not need it. Click on "Zoom Full" icon in the "Wave" window to view the entire timing diagram. You can use "Zoom in" and "Zoom out" to get the display size convenient for analysis. The simulation result for the two-input AND design is shown in Figure 6.13. Upon inspection of the displayed waveforms, you will see that the result tallies with the timing diagram and truth table shown for the design, and_2in, in the chapter on test bench. It may be noted that design signals A, B, and Y are exactly the same as the corresponding signals A, in and out of the test bench. You can use "Restart" to clear the display and "Run-all" (in the main or wave window) to run and capture the waveform once again, if you wish. You can also use "Run" icon to advance the waveform in small steps, which you can key-in in the "Run Length" field in

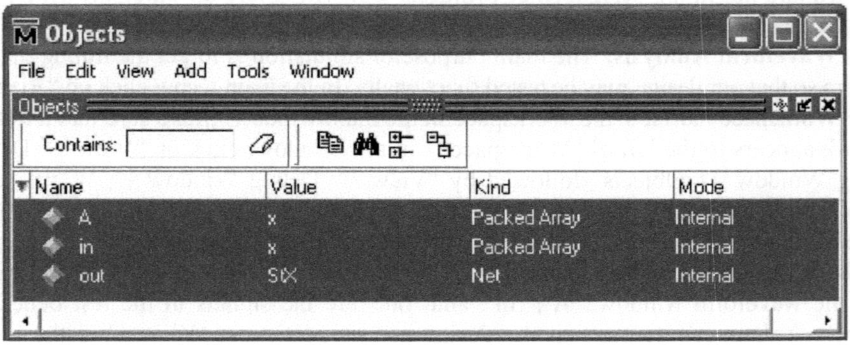

Fig. 6.11 Signals window

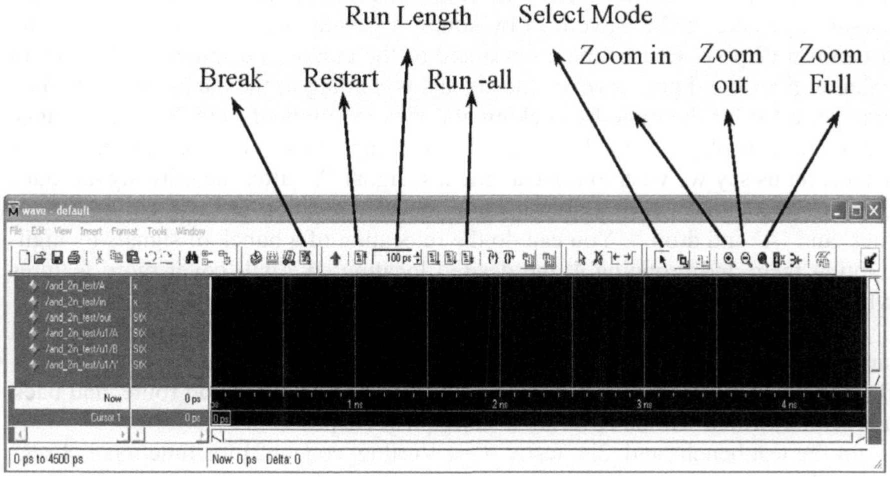

Fig. 6.12 Wave window

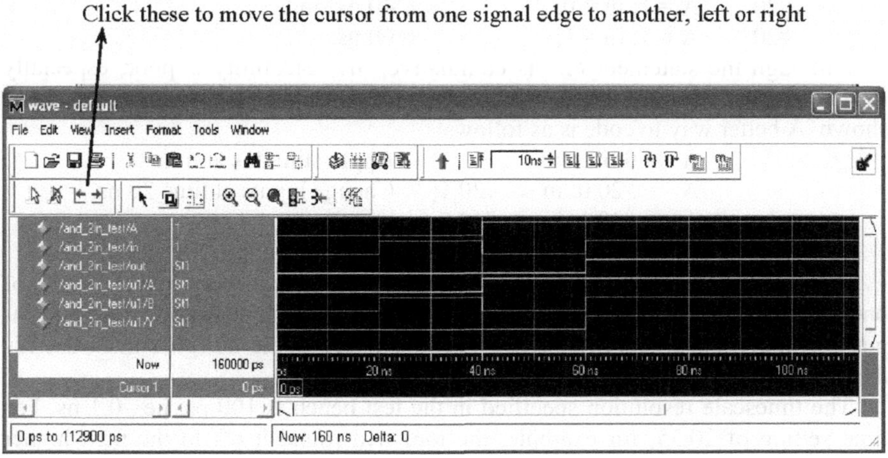

Fig. 6.13 Simulation result of 'and_2in' design

the Wave window. Click on "Select Mode" if "Run" keys are not energized in the wave window.

Let us investigate some more features of the simulation tool. If the waveforms are displayed in the fashion shown, it will be too crowded for you to view large designs. Fortunately, we have a way out of this and that is by using the format menu. In the wave window, click on the first column, signal 'A' and, holding the shift key, click again on the last signal 'Y' to highlight all the signals. Click on "Format" and "Height" to open a window marked "Wave Height". Enter, say, 55 pixels in "Height" field followed by "OK" and note that the signals are separated making the display uncluttered, thus enhancing the readability. We can put the cursor anywhere in the waveform by simply clicking at the desired place on the waveform display. Exact time co-ordinate of the cursor is displayed. The second column gives the digital level of the signals prevailing at the cursor position. You may get a feel of the same by clicking at different points of time. You can change the order of display of signals if you wish to compare two or more signals. For instance, let us say we wish to relocate the last signal 'Y' after the third signal 'out'. Click on the signal 'Y' and drag it up to the new position in between the signals 'out' and 'A' and drop it. You can do the relocation of a bunch of signals by highlighting them and dropping at the desired location. We will cover more features when we simulate other designs.

It may be noted that there are no gate delays involved in the present level of simulation since no vendor specific device is mapped. Gate delays will get reflected only after running other tools such as synthesis, place and route, and back annotation, which will be covered in subsequent chapters.

In the test bench 'and_2in_test.v' (see Verilog_code 4.2) for functional checking of the design 'and_2in.v', we applied stimulus as follows:

```
      A = 0; in = 0;        // Apply stimulus at time 0.
#20   A = 0; in = 1;        // Change inputs at time 20 ns,
#20   A = 1; in = 0;        // 40 ns, and
#20   A = 1; in = 1;        // 60 ns.
```

Although the statement #20 is cumulative, the readability is poor, especially for long codes. We have to keep track of the actual time by proper commenting as shown. A better way to code is as follows:

```
A <= 0; in <= 0;            // Apply stimulus at time 0.
A <= #20 0; in <= #20 1;    // Change inputs at time 20 ns,
A <= #40 1; in <= #40 0;    //40 ns, and
A <= #60 1; in <= #60 1;    //60 ns.
```

Note that "=" sign is replaced by "<=" sign needing two key strokes instead of one. You have to pay a further price by adding the timing statement for every input. For long codes, this is not convenient. Therefore, the designer must use his discretion to use one of the two methods effectively.

The timescale resolution specified in the test bench is 100 ps, i.e., 0.1 ns. For time setting of 20.35, for example, the tool will round it off to the next higher value, 20.4.

6.3.1 Simulation Results of Combinational Circuits

We have seen the design of combinational circuits in Chapter 3 and its test bench in Chapter 4. We will now see the simulation results for the design, 'comb_ckts.v'. Just as we had created a project for two-input AND gate earlier, we need to create a new project for combination circuits as explained before. This is followed by compilation and loading of the test bench 'comb_ckts_test.v'. Invoke the waveform and run the simulation. Results are shown in figures starting from Figure 6.14. From now onwards, the waveforms will be shown with white background instead of black shown earlier. This is only to enhance the readability. Rearrange the waveforms to facilitate easy analysis. The realization of basic gates is shown in Figure 3.1, where signals A and B are the inputs and F1 through F8 are the outputs. The output F1 is a buffered output of A and, therefore, the waveforms for these signals are the same as seen in the figure. Similarly, F2 is the inverse of A for all input data. F3 being the AND operation of the two inputs A and B, it is high only for A = B = 1. F4 is A OR B and hence it is '0' only for A = B = 0. F5 and F6 are just the inverses of F3 and F4 respectively being 'Not AND' and 'Not OR'. F7 is exclusive OR of A and B. Therefore, it is high only for A not equal to B. F8 is XNOR (denoting equivalence) and hence it is the inverse of F7.

F9 depicts the majority logic among the three inputs A, B, and C. This means that the output is logic '1' if two or more ones are present among the inputs. This may be easily verified by inspecting the waveforms shown in Figure 6.15. F10 is concatenation of the three inputs A, B, and C in that order progressing in binary starting from 000 and changing every 20 ns. F11 and F12 are respectively F10 right shifted by one bit and left shifted by two bits as can be easily verified from the figure.

A, B, and C are the select pins used for MUX. Figure 3.4 in Chapter 3 showed a two-input MUX with A used as the select signal. I0 appears at the output mux2 if A = 0. Otherwise, mux2 output is the same as I1. This can be verified by inspecting Figure 6.16. In Figure 3.5, a four-input MUX was shown. I0 through I3 are routed to its output, mux4, depending upon the select signals B and C. For BC =00,

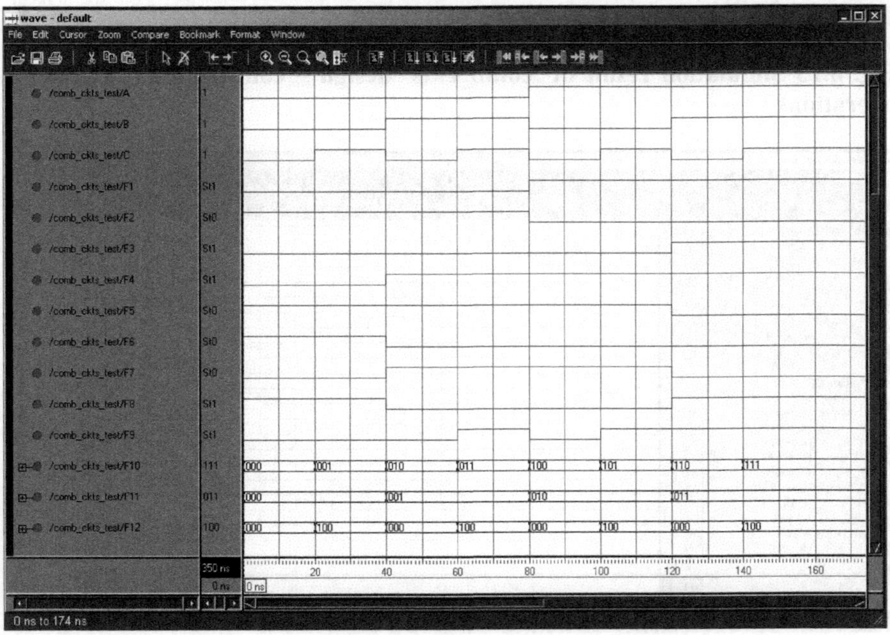

Fig. 6.14 Simulation result of 'comb_ckts' design – basic logic gates

I0 is selected and so on as is seen in Figure 6.16. Eight-input MUX shown in Figure 3.6 can also be verified in a similar fashion. An easier way to verify it is by combining it with the eight-output DEMUX. Thereby, the DEMUX is also functionally verified. It may be noted that the select pins for the MUX and DEMUX are the same. Since the mux8 output of the eight-input MUX is connected to the input of the DEMUX, it follows that its outputs D0 to D7 are precisely the same as I0 to I7 input signals of the MUX. This is indeed true as is revealed from Figure 6.16.

Next in line in the combinational circuits design is the full adder. In Section 3.26, we considered three different ways of implementation for the same. In all these cases, the inputs are A, B, and C. C may be regarded as the carry in and the others as the two input bits needing to be added. The first one is the behavioral level of realization, wherein the output is sum_total [1:0], higher order bit [1] being the carry out and bit [0], the sum. The result tallies with the truth table given in Figure 3.8. The second realization is the data flow structure whose outputs are 'sum_df' and 'carryo_df' ('df' standing for data flow) and, the last is the

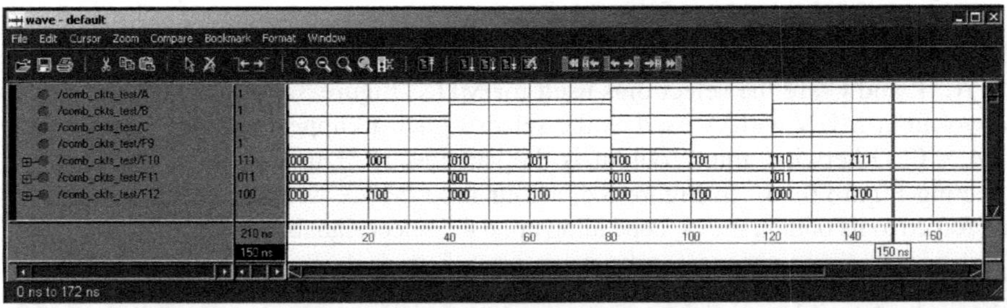

Fig. 6.15 Simulation result of 'comb_ckts' design – concatenation and shift operations

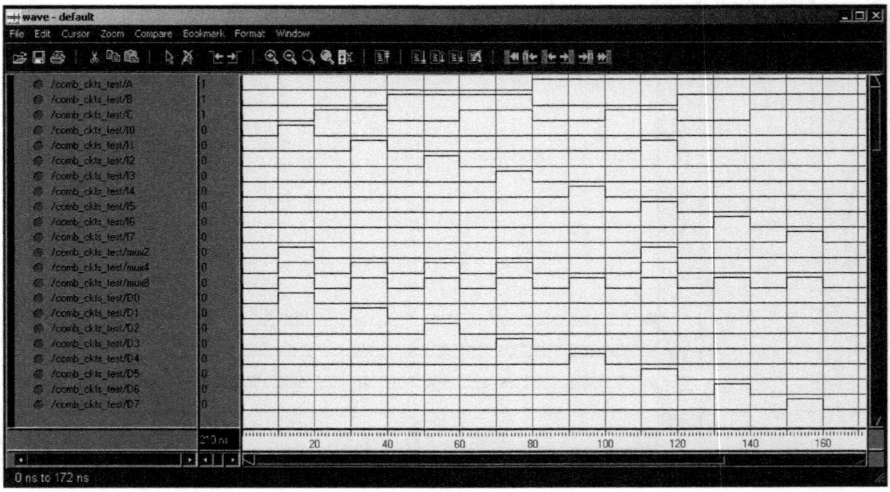

Fig. 6.16 Simulation result of 'comb_ckts' design – MUX and DEMUX

structural realization using primitive gates with the corresponding outputs 'sum' and 'carryo'. The sum and carry outputs of the three types must be the same and can be easily seen from the waveforms shown in Figure 6.17. Visually comparing the waveforms sum_total [1], carryo_df, and carryo reveals that they are exactly the same. So is the case with signals sum_total [0], sum_df, and sum. These signals may be relocated one below the other to make the comparison still easier. You can move from one transition to another, be it positive or negative transition. For instance, select or highlight signal 'C' by clicking on the signal on the simulator waveform and clicking on "Find previous transition" or "Find next transition" as per your need. This will be very handy while analyzing sequential circuits.

The next part of the design is a magnitude comparator shown in Figure 3.10. The simulation results for the same are shown in Figure 6.18. Two numbers, N1 and N2, are compared, and outputs F13 through F18 are set accordingly as shown in Figure 3.10. In the waveform, select N1 and N2 and click on "Format => Radix => Unsigned" to view them as unsigned decimal numbers. This way, it will be convenient to make the comparison faster. F13 to F15 outputs are straightforward. They reflect the three conditions N1 > N2, N1 < N2, and N1 = N2 respectively. Note that F15 and F16 waveforms are inverses of each other since the first one

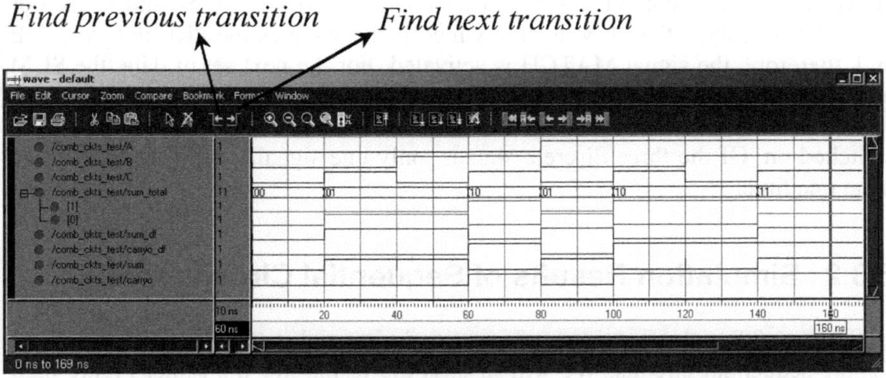

Fig. 6.17 Simulation result of 'comb_ckts' design – full adder

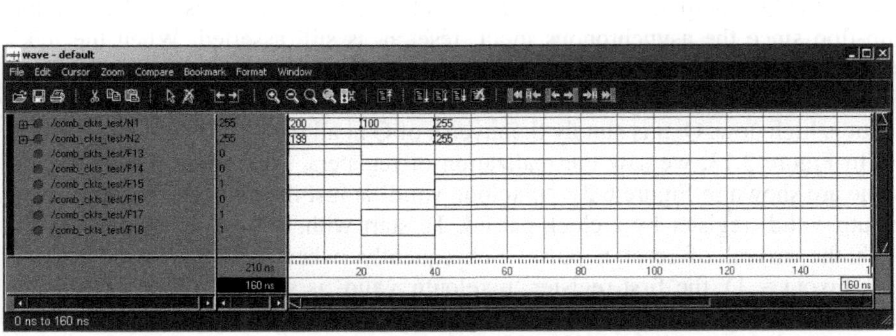

Fig. 6.18 Simulation result of 'comb_ckts' design – magnitude comparator

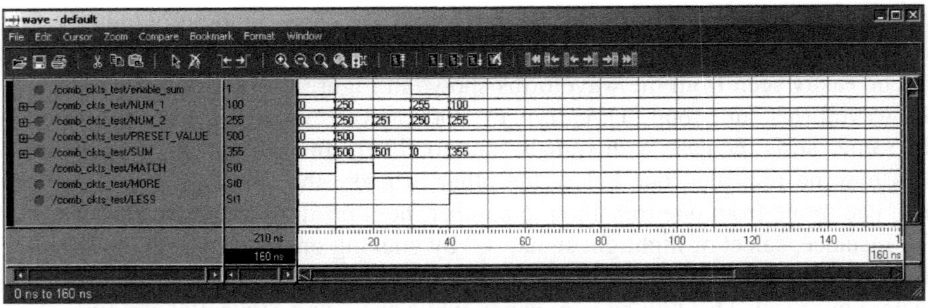

Fig. 6.19 Simulation result of 'comb_ckts' design – an example

looks for equality and the other for inequality. Similarly, F14 (N1 < N2) and F18 (N1 >= N2) are inverses of each other. So is the case with F13 (N1 > N2) and F17 (N1 <= N2). Looking this way, analysis will be easier and faster.

A simple design example using a magnitude comparator was shown in Figure 3.11. The simulation results for that application are shown in Figure 6.19. As shown therein, all the outputs are cleared so long as the signal enable_sum is not active. The SUM is computed only for data applied at 10 ns, 20 ns, and 40 ns. For the first set of data, the SUM is 500, which is the same as the PRESET_VALUE and, therefore, the signal MATCH is activated. For the next set of data, the SUM is 501 and, being greater than the PRESET_VALUE, MORE is turned on this time. Similarly, for the last set of data, the SUM is 355 and the signal LESS is switched on. Of the three discrete signals, only one signal among them is turned on at one time.

6.3.2 Simulation Results of Sequential Circuits

In Sections 3.3 and 4.3, we considered the design and its test bench respectively for the sequential circuits. We will discuss the results for the same. Figure 6.20 shows the simulation result of a D flip-flop with reset, a block diagram of which was shown in Figure 3.12. A low pulse at the reset_n pin clears the Q output and presets the Q_n output. A positive edge of 'clk' signal at 30 ns has no effect on the flip-flop since the asynchronous input, reset_n, is still asserted. When the 'clk' strikes again at 50 ns, the 'D' input is (presently logic '0') assigned to Q output. Similarly, with the rising edge of 'clk' the next time, the Q output is set since D input is high now. Q_n is simply the inverse of Q at all times.

In Figure 3.13, we saw the realization of registers. Simulation results for the same are shown in Figure 6.21. 'pixelout_valid' is just the delayed output of 'pixeloutp_valid' register by a clock period. To start with, both registers are cleared with the arrival of reset_n pulse. Subsequently, when ABC is forced to 010 (set_pixout = 1), the first register 'pixeloutp_valid' is set at the following rising edge of 'clk' signal. Since the 'hold' signal is applied for 20 ns commencing from 60 ns, the register 'pixelout_valid' is not set at 70 ns with the rising edge of 'clk', but only with the subsequent clock when 'hold' is withdrawn. This 'hold effect'

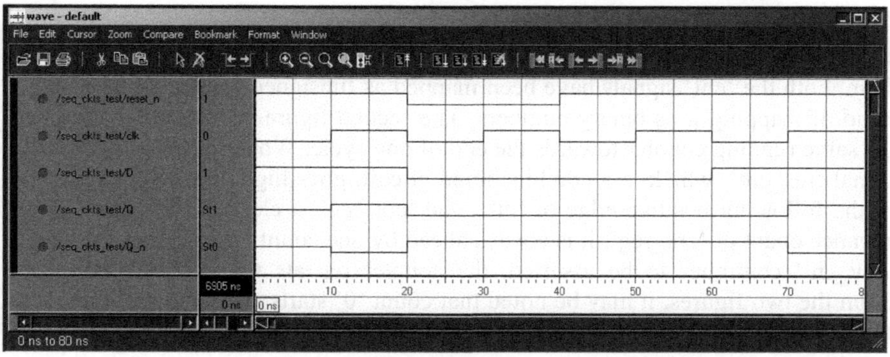

Fig. 6.20 Simulation result of sequential circuits – D flip-flop with reset

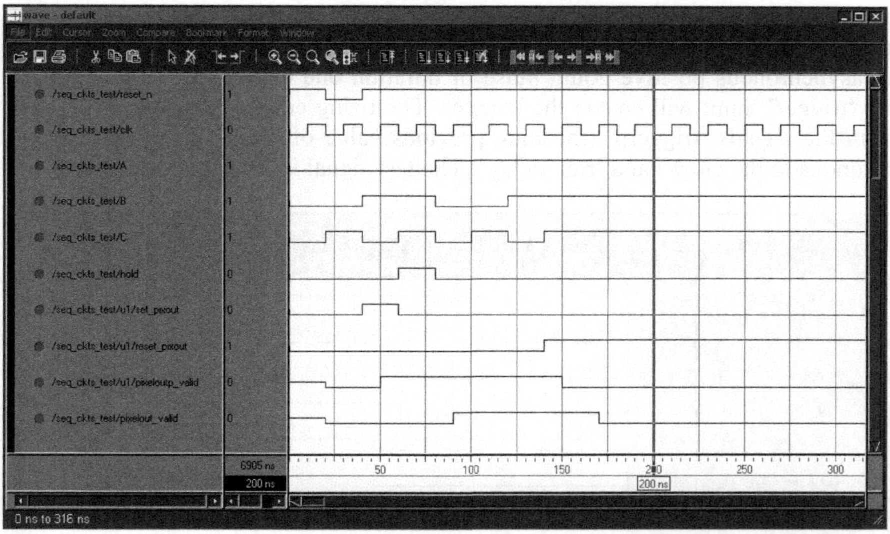

Fig. 6.21 Simulation result of sequential circuits – Realization of registers

can be easily inferred by observing the delay of one clock period between the negative edges of the two registers, ‘pixeloutp_valid’ and ‘pixelout_valid’, when ‘hold’ is not asserted. When hold was asserted, the delay between the positive edges of the two registers was two clock periods. ‘pixeloutp_valid’ goes low with the following rising edge of ‘clk’ (at 150 ns) after ‘reset_pixout’ goes high (ABC = 111).

Figure 3.14 depicts the realization of a counter and Figure 6.22 its simulation results. A low pulse at ‘reset_n’ pin clears the counter, ‘cnt_reg’. The counting starts only at the positive edge of ‘clk’ (@ 150 ns) following ABC = 111, i.e., when ‘adv_cnt’ goes high. Of course, the signal ‘res_cnt’ must remain low for counting to take place. It may be noted that ‘cnt_next’ is an advance increment of 'cnt_reg' since that signal is realized as a combinational circuit (using 'assign' statement). The counting takes place at every rising edge of the 'clk'. In the waveform, both the 'cnt' signals have been mapped as 'unsigned' decimal numbers instead of mapping it as binary numbers. The second figure gives the waveform of the same running counter towards the end of one cycle. When 'cnt_reg' is 255, the signal 'res_cnt', which is a combinational circuit, goes high and, with the arrival of the following positive edge of 'clk', the 'cnt_reg' is cleared. 'cnt_next' being advance count of 'cnt_reg', it is always ahead by one count value. Since the signal 'adv_cnt' continues to be asserted, the counter repeats the counting non-stop. From the two figures, it may be noted that count '0' starts at 130 ns and the count '255' ends at 5250 ns, thus taking 256 clock cycles to complete one round of counting.

We will now consider the simulation of non-retriggerable monoshot we designed in Section 3.3.4. The test bench for the same was discussed in Section 4.3. Figure 6.23 shows the simulation results of the non-retriggerable monoshot. The monoshot is based on the operation of an 8-bit counter, 'cntd_reg'. To start with, applying a negative going pulse at the pin 'reset_n' clears this counter. Applying an asynchronous positive going pulse of duration one clock period minimum to the 'trigger' input will enable the counter. The rising edge of 'trigger' is derived from the signals 'triggerp' (meaning previous value of trigger input and is synchronous to the clock) and 'run_delay'. The last signal is detected as high only at

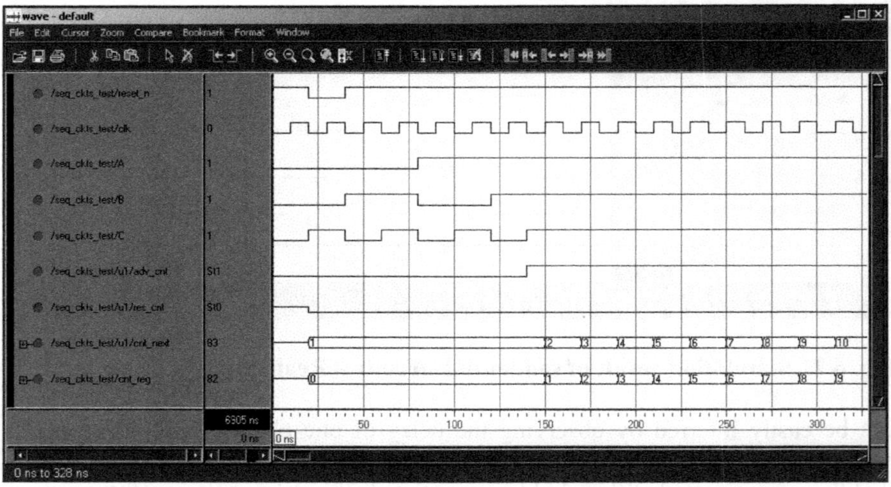

Fig. 6.22 Simulation result of sequential circuits – Realization of a counter (Continued)

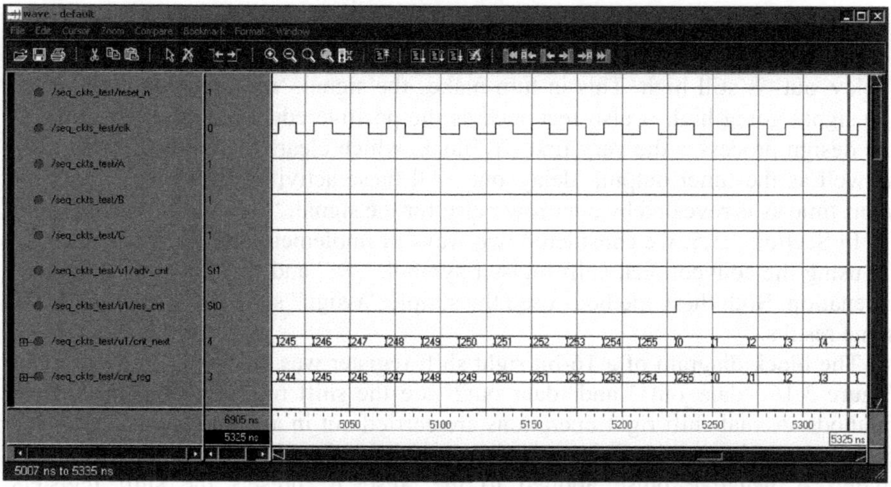

Fig. 6.22 Simulation result of sequential circuits – Realization of a counter

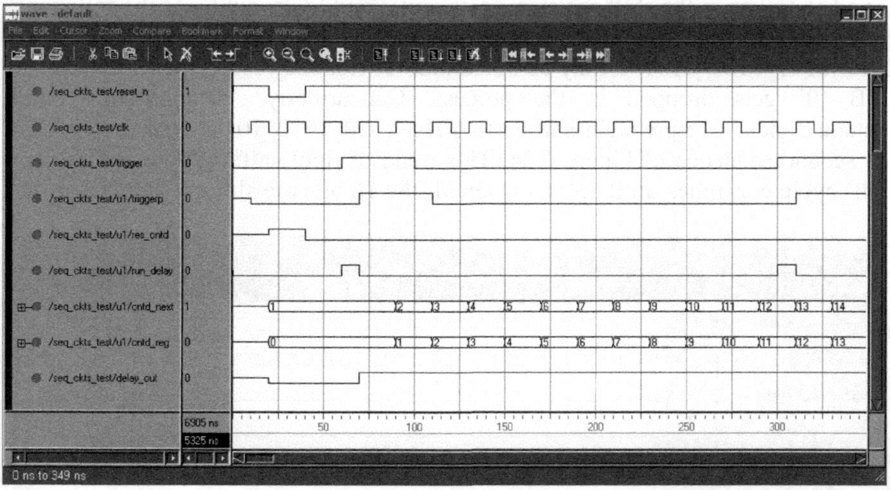

Fig. 6.23 Simulation result of sequential circuits – realization of a non-retriggerable monoshot (Continued)

(70 ns) the rising edge of the 'clk'. Immediately, the counter is enabled and the timer output, 'delay_out', goes high. 'cntd_next' is the advance increment counter of 'cntd_reg'. The counter updates its value every rising edge of the 'clk' until the end of set delay, namely, 255 cycles. The 'trigger' is applied again at 300 ns to see whether the timer is not triggered. Inspection of the first waveform reveals that this is true since 'cntd_reg' continues to run and is not cleared at 310 ns. Also, there is no change in the output, 'delay_out' at this point of time. Towards the end

of timing (at 5170 ns), the signal 'cntd_next' is 255 and register 'cntd_reg' is 254. At this point of time, 'clk' signal rises high, advancing 'cntd_reg' to 255 since 'delay_out' is still high. This in turn makes the signal, 'res_cntd' go high. Since the signal going high is also reckoned as the positive edge, the 'always' block in the design processes the very first (if) block, which clears the counter, 'cntd_reg' as well as the timer output, 'delay_out'. All these activities take place in a very short time as is revealed by a narrow pulse for the signal, 'res_cntd' at 5170 ns.

In Section 3.3.5, we considered two ways of implementing a shift register, one by using the conventional shift register symbol, ">>" and the other by using concatenation. Both these methods used the simple "assign" statements to produce the same results.

The block diagram of a 16-bit right shift register was shown as an example in Figure 3.16. 'data_out1' and 'data_out2' are the shift registers used in the two methods. Actual shift right operations are carried out in advance by using combinational circuits, whose outputs are 'dataout1_next' and 'dataout2_next' respectively. A negative pulse applied to the 'reset_n' presets the shift registers 'data_out1' and 'data_out2' to 1010_1010_1010_1010 as shown in Figure 6.24. 'dataout1_next' and 'dataout2_next' exhibit the single bit right shift effected in anticipation. However, no shift operations take place until the signal 'shift' is asserted at 140 ns in the shift registers 'data_out1' and 'data_out2'. At the following rising edge of 'clk' (@ 150 ns), the contents of shift register is right shifted to 0101_0101_0101_0101. It may be noted that '0' fills the vacated MSB bit and the LSB '0' gets dropped in the process. Concurrently, 'dataout1_next' and 'dataout2_next' are pre-right shifted to 0010_1010_1010_1010. This is shown in the second sequence of Figure 6.24. This trend of right shifting by one bit every 'clk' cycle continues until 450 ns when all the 16 bits are shifted. Hereafter, the

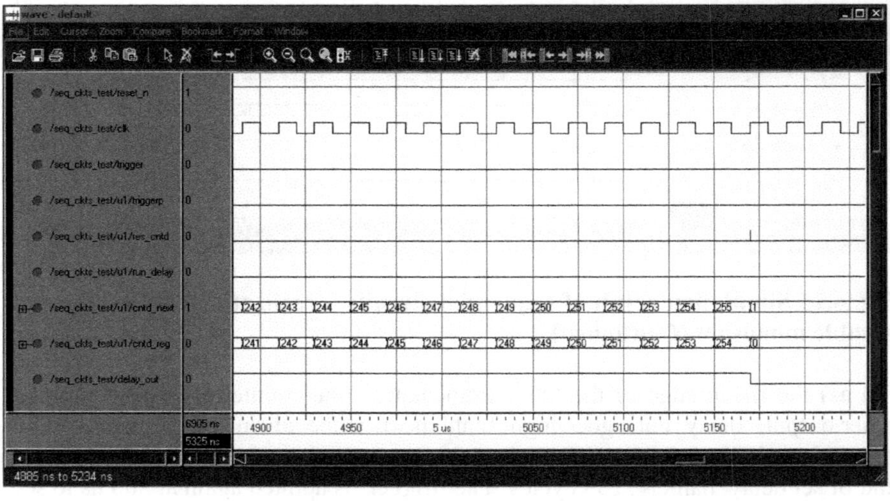

Fig. 6.23 Simulation result of sequential circuits – realization of a non-retriggerable monoshot

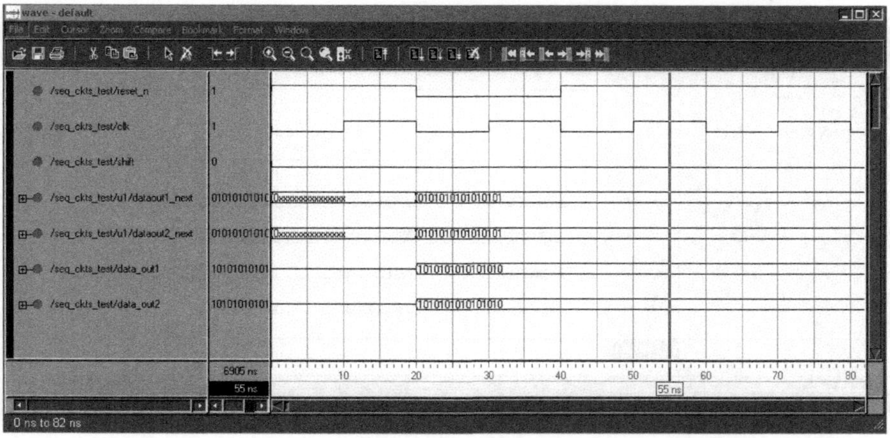

Fig. 6.24 Simulation result of sequential circuits – realization of shift register (Continued)

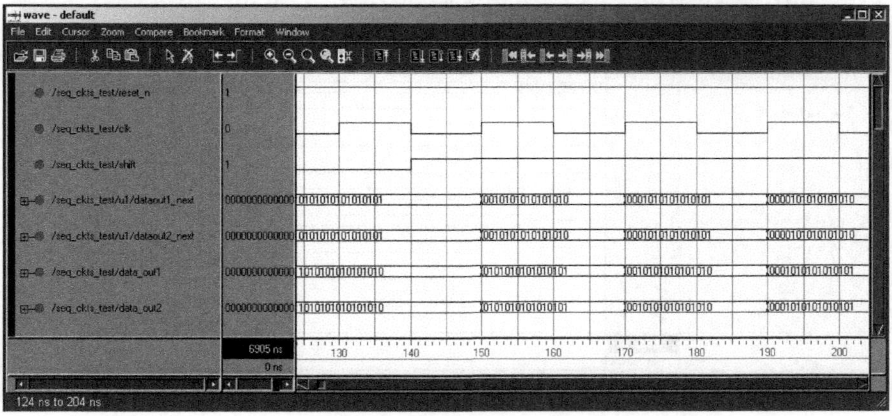

Fig. 6.24 Simulation result of sequential circuits – realization of shift register (Continued)

shift registers will be stuck at 0000_0000_0000_0000 since the new vacated bits continue to be zeros.

The next design we saw in Section 3.36 was a parallel to serial converter and whose block diagram was shown in Figure 3.17. Figure 6.25 shows the simulation results for the same. With the application of the reset pulse, all the outputs, data_out, data_valid, and eoc are switched off. The shift register, sr, is pre-loaded with a pattern, say, 1010_1010_1010_1010 when the 'load' signal is applied. The data is actually input from the signal, set_data [15:0]. This takes effect at 50 ns with the rising edge of 'clk' after the reset pulse is withdrawn. We don't need to mind that 'load' has already taken effect at 10 ns before the reset pulse is applied.

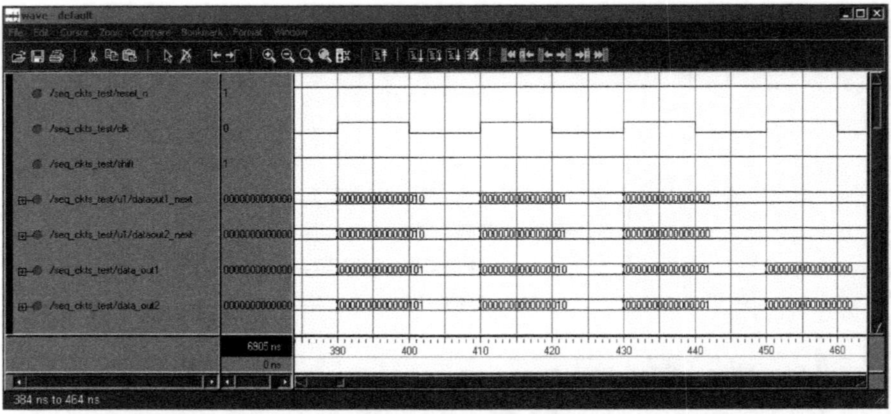

Fig. 6.24 Simulation result of sequential circuits – realization of shift register

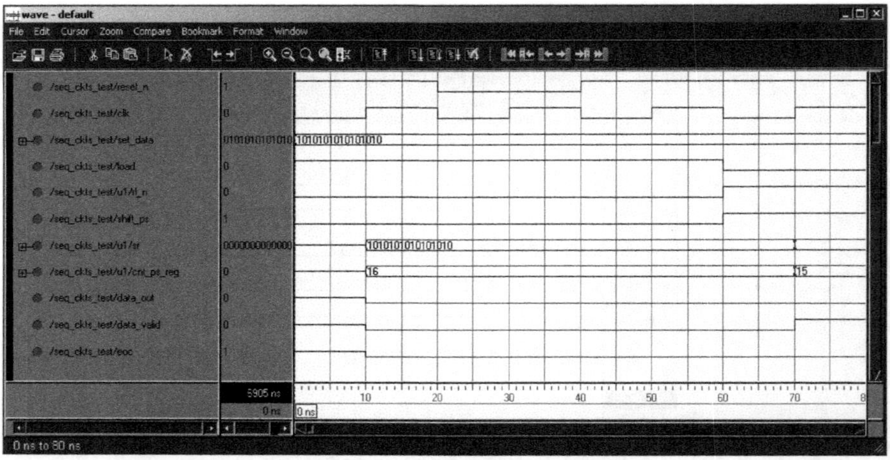

Fig. 6.25 Simulation result of sequential circuits – realization of parallel to serial converter (Continued)

An internal counter, ‘cnt_ps_reg’, that keeps track of the number of bits remaining to be sent over a serial channel output (data_out) is also preset with the total number of bits, say, 16, to be transmitted. The aim of this design is to transmit a bit stream out from a parallel word acquired from the input, set_data. The first bit in the stream will be either the MSB or the LSB of the parallel information, depending upon the type of shift implemented. For the transmission of MSB first, left shift needs to be done, whereas for the LSB first, right shift will have to be done. This is taken care of by the signal, rl_n. For a right shift, it is high. Otherwise, it is low. To start with, let us say that we want to transmit the LSB first in order to effect the parallel to serial conversion. This requires that we set ‘rl_n’ signal high.

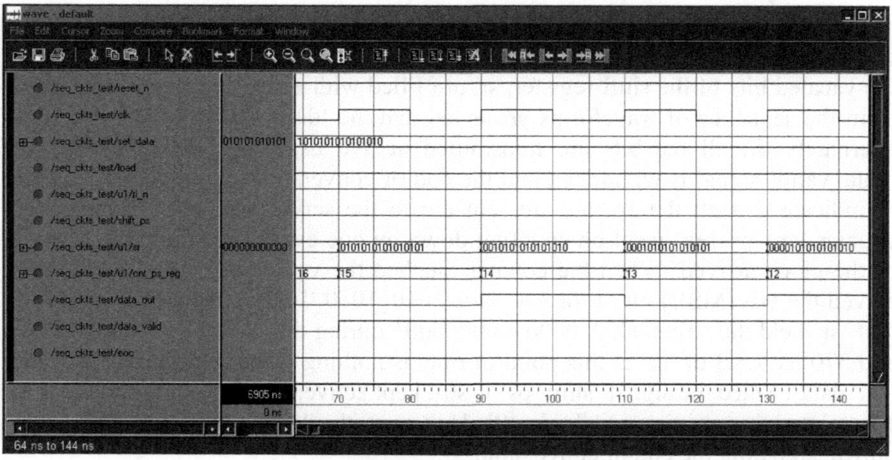

Fig. 6.25 Simulation result of sequential circuits – realization of parallel to serial converter (Continued)

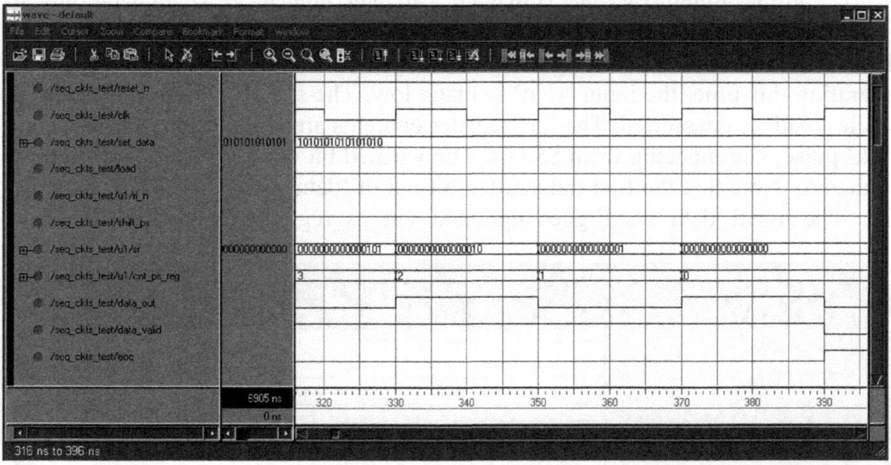

Fig. 6.25 Simulation result of sequential circuits – realization of parallel to serial converter (Continued)

Actual shift operation starts only if the signal 'shift_ps' is set high and 'load' low at 60 ns. With the arrival of the 'clk' at 70 ns, 'sr' is right shifted to 0101_0101_0101_0101, transmitting the LSB (logic '0') onto the output pin, 'data_out' with 'data_valid' signal asserted. Since the first bit is transmitted now, the counter, 'cnt_ps_reg' is decremented by one to a value 15, which implies that we need to transmit 15 more bits.

These operations of right shifting, outputting, and updating the counter take place every 20 ns till all the 16 bits of the parallel word are transmitted. Note that the vacated bits of the shift register, sr, are filled with zeros. It may also be noted from the sequence of waveforms presented that the 'data_out' goes low and high alternately till all the bits are transmitted at 390 ns. When this happens, the 'data_valid' signal is withdrawn and the end of conversion signal 'eoc' goes high to indicate that all the 16 bits are sent out to the serial channel. All along, the counter 'cnt_ps_reg' went on counting down by one every time the positive edge of the clock arrived. When the counter touched the value "0", the 'data_out' received the last (MSB) bit of the set_data, 1010_1010_1010_1010 at 370 ns. Note that 'sr' held the value 0000_0000_0000_0001 during the interval between 350 ns and 370 ns; LSB of 'sr' at this point of time is nothing other than the MSB of the 'set_data'. Since the signal 'shift_ps' is still kept active, the right shift continues to take place, transmitting out logic '0'. However, the 'data_valid' signal is withdrawn, signaling the end of conversion.

So far, we have seen the simulation results for right shift operation. We will now see the same for left shift operation. This time, a different data, 0101_0101_0101_0101 is loaded in to the shift register, sr, through the input 'set_data' and by asserting the 'load' signal at 525 ns. Also, the signal 'shift_ps' is deasserted for obvious reasons. However, 'set_data' appears at 'sr' register only at the following 'clk' edge at 530 ns. The counter 'cnt_ps_reg' is also initialized to 16 as was done before. The signal 'eoc' is also cleared. Since we need left shift operation this time, the input 'rl_n' is made low. The signal 'shift_ps' is asserted while 'load' is deasserted. The 'sr' register contents are shifted left; one bit every 'clk' pulse, commencing from 550 ns. The vacated bit on the LSB of 'sr' is filled with zero. Note that the first transmitted bit out of 'data_out' is the MSB, namely, '0'. The signal 'data_valid' goes high, and 'cnt_ps_reg' counts down every clock

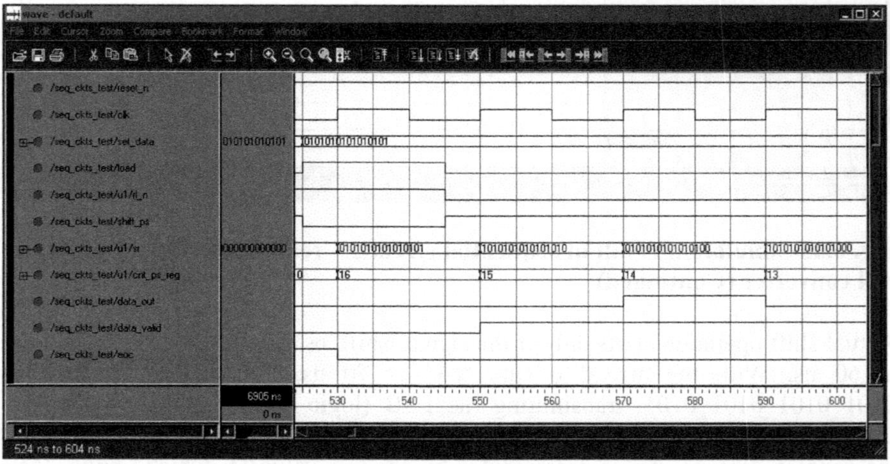

Fig. 6.25 Simulation result of sequential circuits – realization of parallel to serial converter (Continued)

cycle. Towards the end, the 'data_out' receives the LSB of the 'set_data', namely, '1' at 850 ns. With the arrival of the next clock pulse at 870 ns, the 'data_valid' goes low and 'eoc' goes high since all the 16 bits of parallel data are transmitted with the MSB as the first outgoing bit.

A model state machine was pictorially depicted in Figure 3.18. The Verilog design and its test bench were presented in Verilog_code 3.16 and Verilog_code 4.4 respectively. At 0 ns, the input 'in1' was forced to '0' in the test bench. Following this, the 'clk' arrives at 10 ns. Towards the end of the design, the statement "default: state <= `S0 ;" takes effect and hence the 'state' is initialized to S0 state. The circuit starts functioning only after the application of the 'reset' pulse when all the 'Z' outputs are cleared and the 'state' remains in 'S0' as shown in the simulation waveform, Figure 6.26. So long as reset pulse continues to be active, that long the Z0 output remains cleared although the 'state' continues to be in 'S0' as can be inferred from the design. The output Z0 goes high only with the arrival of the rising edge of 'clk' at 50 ns since reset pulse is with drawn only after 40 ns. At 140 ns, the input 'in1' is applied and is sensed with the arrival of the positive edge of 'clk' when the 'state' changes to S2 (10). Since this state is recognized only at the next rising edge of clock at 170 ns, the Z0 output continues to be high till the next clock strikes.

In S2 state, the input 'in2' is '0' and hence the output Z2 is turned on, while the output Z0 is turned off. The next state is S1 as can be seen in the waveform and crosschecked from the state graph presented in Chapter 3. When the next clock arrives at 190 ns, Z1 is turned on and Z2 is turned off since the input 'in2' is asserted in the meanwhile, thus forcing the machine to remain in the S1 state. This continues until in1, in2 are made 1,1 and subsequently recognized at 230 ns. The next state is the S3 state. With the arrival of the next clock, the output Z3 is turned

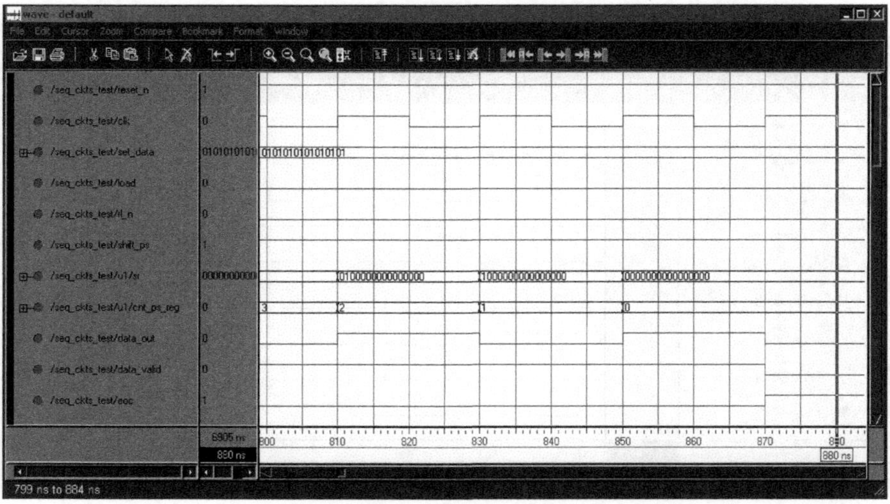

Fig. 6.25 Simulation result of sequential circuits – realization of parallel to serial converter

on and all other outputs are cleared since the state machine is now in S3 state. At 260 ns, prior to the arrival of the positive clock edge, in1, in2 are set to 0,1. When the clock arrives, the machine state changes to S1 as can be seen both from the state graph and the timing diagram. Similarly, in S1 (01) state, which is recognized at 290 ns, the output Z1 is switched on while switching off all other outputs. The input is again changed to in1 = 1 (in2 continues to remain at '1') at 300 ns. The rising edge of clock at next transition is S0 since 'in2' is forced to '0'.

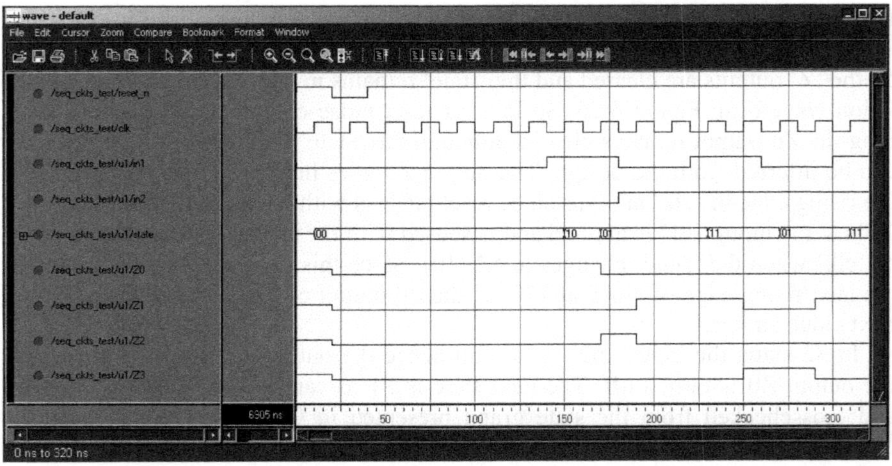

Fig. 6.26 Simulation result of sequential circuits – realization of a model state machine (Continued)

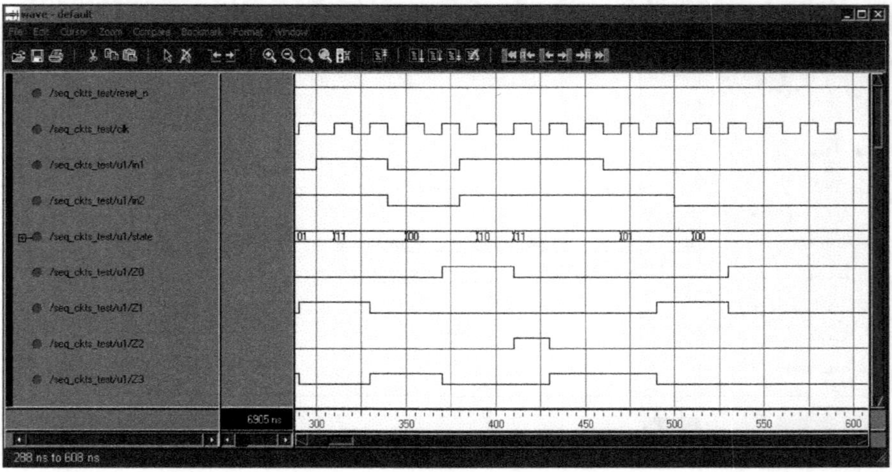

Fig. 6.26 Simulation result of sequential circuits – realization of a model state machine

We have so far checked all transition paths except for the paths S2 to S3 and S1 to S0. S0 to S2 transition is effected by making in1 = 1. Similarly, forcing in2 to '1', the transition from S2 to S3 takes place. The inputs in1 in2 = 0 1 causes the transition from S3 to S1. In S1 state, when in2 = 0 and the clock arrives, the machine state changes to S0. In each of the states, one of the appropriate Z outputs is turned on.

Figure 6.27 presents the simulation results of a pattern sequence (0110) detector, whose design and test bench were presented in Verilog_code 3.17 and Verilog_code 4.4 respectively. An active low pulse applied at 'reset_n' pin clears the output register 'out' as well as the 'psd_state', where psd stands for pattern sequence detector. Initially, starting from 525 ns, the input 'in' is held high since the first serial pattern we need to detect is '0' (MSB) in the sequence 0110. This corresponds to the application of '1111' pattern using the test bench. By inspecting the test bench and the waveforms, we observe the following:
Input pattern applied: 11110100110110001101....................
Output: 00000000001001000010....................
It may be noted that input pattern is applied regularly every 20 ns, commencing from 525 ns, with the last data '1' applied at 905 ns. The first '0110' pattern occurs in the time range: 665 ns–725 ns, while the second and third patterns occur in the ranges: 725 ns–785 ns and 825 ns–885 ns respectively. A mere glance at the input patterns in the waveform will be sufficient to spot out the desired pattern sequence, 0110, easily. The output 'out' goes high at the following rising edge of 'clk' for a clock cycle duration after the desired pattern sequence is encountered. A close examination reveals that the machine states change in accordance with the design presented in Verilog_code 3.17.

The simulation tool may be quickly learnt from a summary of the commands listed in the following.

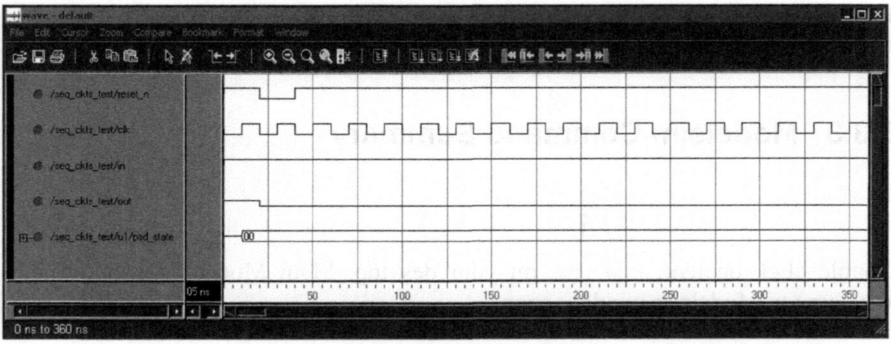

Fig. 6.27 Simulation result of sequential circuits – realization of a pattern sequence (0110) detector (Continued)

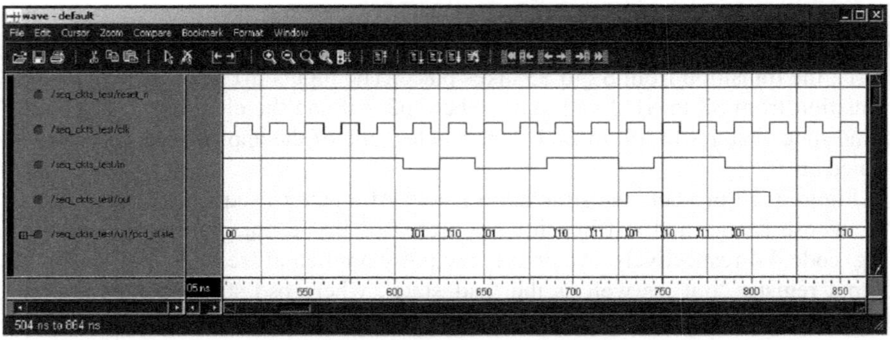

Fig. 6.27 Simulation result of sequential circuits – realization of a pattern sequence (0110) detector (Continued)

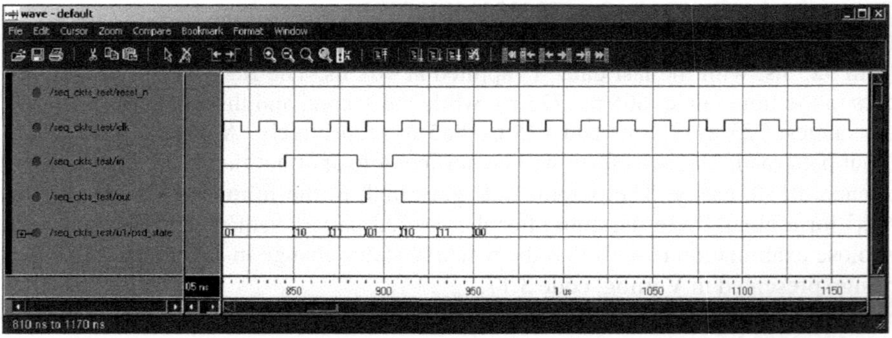

Fig. 6.27 Simulation result of sequential circuits – realization of a pattern sequence (0110) detector

6.3.3 Modelsim Command Summary

Double click on icon  on your desktop. Main Modelsim window and Welcome to Modelsim window open.

1. Click on "Jumpstart" followed by "Create a Project" in the Welcome window. Another window also opens. You can also create the project by clicking-on "File => New => Project" and keying-in 'and_2input' (as an example) in "Project Name" field. Also use "Browse" to select the desired "Project Location" such as 'D:\ram\verilog_latest\dvlsi_des_Verilog', where your design is residing. Making sure that "work" is specified in "Default Library Name" field, click on "OK". Another window called "Add items to the Project"

opens. Click on "Add Existing File". In a new window "Add file to a Project" that opens, select the desired test bench such as 'and_2in_test.v' using "Browse".

Note: Next time, if you wish to go straight away to the same project, 'and_2input', select "Open Project" in the "Welcome" menu. Click on OK. Click on "Open Documentation" in the "Welcome" menu if you wish to use the tool documentation. "Close" the "Welcome" menu window.

2. At the command prompt, "modelsim>" in the main window, you may use the operating system commands such as 'pwd' for printing the working directory, 'dir' for displaying the contents of the current directory, 'cd' for changing the directory, etc.
3. **Compilation:** In the main window, click on "Compile" followed by "Compile". A new window called "Compile Source Files" opens. Click on the design file (for e.g., and_2in_test) followed by "Compile". If there are errors, you will have to correct the errors in your code using "Edit Source" in "Compile Source Files" window and rerun the compiler. Click on "Done" to dismiss the compilation window.
4. **Simulation:** Click on "Simulate => Start Simulation". The window named "Start Simulation" pops-up. Click on "+" on the left of 'work'. Click on the desired test bench followed by "OK" to load the test bench and the design. Fix errors, if any.
5. **Waveform Analysis:** In the main menu, click on "View => Workspace" to view the Workspace in the main window. Make sure that a tick mark appears to the left of "Workspace". In this window, click on "View => Debug Window => Objects" followed by "View => Debug Window => Wave" to open the Objects (Signals) and Waveform windows showing the timing diagram. In "Objects" pane, click on "Add => Wave => Signals in Design" to display all the signals in the test bench as well as in the design in the waveform window. In order to get the waveforms, click on the icon marked "Run-all" in the wave window. The source file (test bench) is also opened, which you can close if you do not need it.
6. Click on "Zoom Full" icon in the "Wave" window to view the entire timing diagram. You can use "Zoom in" and "Zoom out" to get the display size convenient for analysis. You can use "Restart" to clear the display and "Run-all" (in the main or wave window) to run and capture the waveform once again, if you wish. You can also use "Run" icon to advance the waveform in small steps, which you can key-in in the "Run Length" field in the Wave window. Click on "Select Mode" if "Run" keys are not energized in the wave window.
7. We can put the cursor anywhere in the waveform by simply clicking at the desired place on the waveform display. Exact time co-ordinate of the cursor is displayed. The second column gives the digital level of the signals prevailing at the cursor position. You may get a feel of the same by clicking at different points of time. You can change the order of display of signals if you wish to compare two or more signals. For instance, let us say we wish to relocate the last signal 'Y' after the third signal 'out'. Click on the signal 'Y' and drag it up to the new position in between the signals 'out' and 'A' and drop it.

8. You can also use “Cut” or “Copy” (one or more signals) and “Paste” it (all in “Edit” menu) at the desired place. You can also use “Edit => Find” for locating the desired signal.
9. Place signals, for example, reset_n, clk, D, Q, and Q_n one below another. Highlight all these signals by clicking on first signal followed by “Shift + Clicking” on the last signal. Click on “Format => Height”. Wave height window opens. Key in 50 followed by clicking on “Apply” and “OK”. All the selected signals are spaced by 50 pixels. Height of the waveforms, however, remains the same.
10. Click on “Format => Radix => Binary or decimal or hexadecimal or octal as per your requirement. All selected signals are set to the desired number system.
11. Use ◁, ▷ arrows and ■ to move the waveform along the time axis. Analyze the waveforms (timing diagram) to check the functionality. Click the two blue arrows on the wave window to move the cursor from one signal edge to another, left or right after highlighting the signal and clicking inside the waveform to get the cursor.
12. To save the waveform format, click on Floppy disc symbol or “File => Save Format”, and in the opened window, type “ur_design.do” and click “Save”.
13. This completes one session. To exit the ModelSim, click on “File => Quit” or X and “Yes”.
14. If you want to use Modelsim again to resume the same project, double click on icon  on your desktop and “Open a Project” as explained in Sl. no. 1. NOTE. Load the test bench as described in 4.
15. In the prompt field, type “do ur_design.do” and enter, the waveform we saved in 13. The desired waveform with the previously set format is retrieved. Click on “Run-all” and continue functional testing of other parts.

Summary

VLSI Design flow was presented along with the design methodology so that one may become conversant with various steps involved in designing a product. Several designs were considered in Chapter 3 and their test benches in the following chapter. This was followed by RTL coding guidelines in Chapter 5. The next logical step in the design flow is the simulation, vital for testing one’s design. All the Verilog designs presented in the third chapter were analyzed using waveforms in the present chapter. Industry standard Modelsim tool of Mentor Graphics was employed for the simulation. A command summary of the Modelsim tool, which serves as a quick reference while using the tool, was furnished. Although the functionality of a design is checked using simulation, it does not test time critical paths or furnish insights into gate delays, unless back annotated, since simulation does not map a target chip such as an FPGA, where the design will have to reside ultimately. These features are possible with synthesis tool, which is presented in the next chapter.

Assignments

6.1 For the assignments 4.1 to 4.12 of Chapter 4 and 5.1 to 5.10 of Chapter 5, run the simulation tool to establish the working of the designs and the test benches considered.

6.2 Write RTL compliant Verilog code to generate random numbers in the following ranges:
(i) 0 to 511
(ii) 0 to 1 with a resolution of 0.001
(iii) –100 to +100.
Test your designs.

6.3 Realize the circuit shown in Figure A6.1 using Verilog. Include a signal for clearing the flip-flop. What type of circuit is this? Test it.

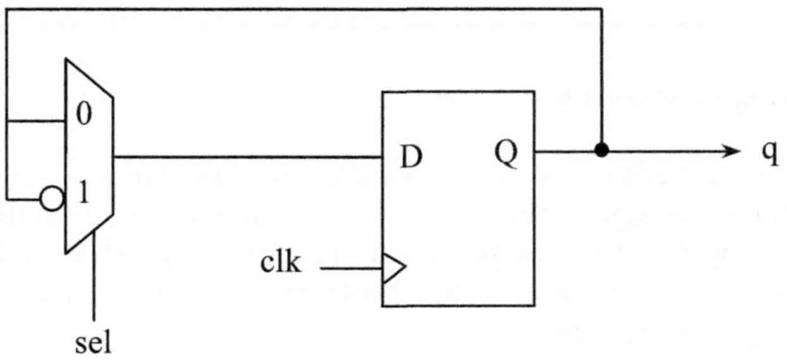

Fig. A.6.1 Sequential circuit

6.4 A presettable BCD/Decade, Up/Down counter is shown in Figure A6.2. For up counting, up_dn = 1, otherwise, it is down counting. It can be changed only if chip enable, ce, is low or 'clk' is not rising. 'ce' is active high. LOAD (asynchronous, active high) presets the input D3–D0 into the Q3–Q0 flip-flops. Counting takes place at the rising edge of clk if 'ce' is high. Terminal count (TC) is normally low and goes high when the counter reaches '0' in the count down mode or reaches '9' in the count up mode. Realize the presettable decade counter using Verilog and test it.

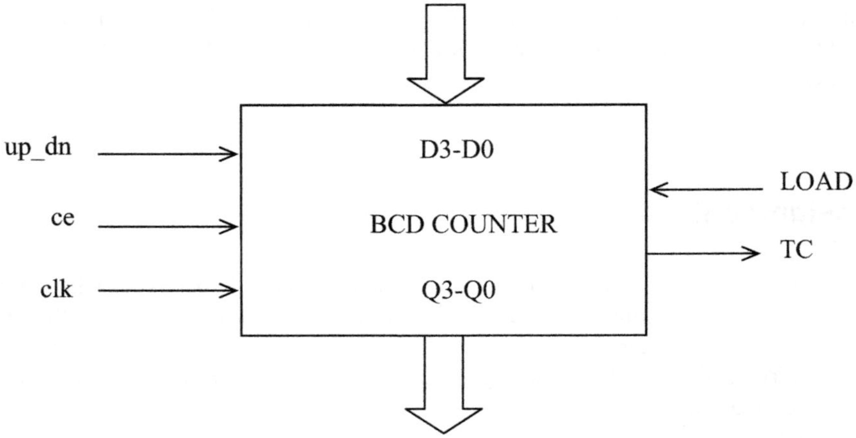

Fig. A6.2 Presettable BCD counter

6.5 A modulo 10 Gray code decade counter sequence is as follows:

0 => 1 => 3 => 2 => 6 => 14 => 10 => 11 => 9 => 8

Fig. A6.3 Gray code decade counter

Realize the Verilog code for the counter shown in Figure A6.3 and test it.

6.6 The following signals (Figure A6.4) are required to be generated for operating a stepper motor. Design a circuit using Verilog, which fulfils the above requirements. Also write a test bench and simulate to ensure the correct working of the design.

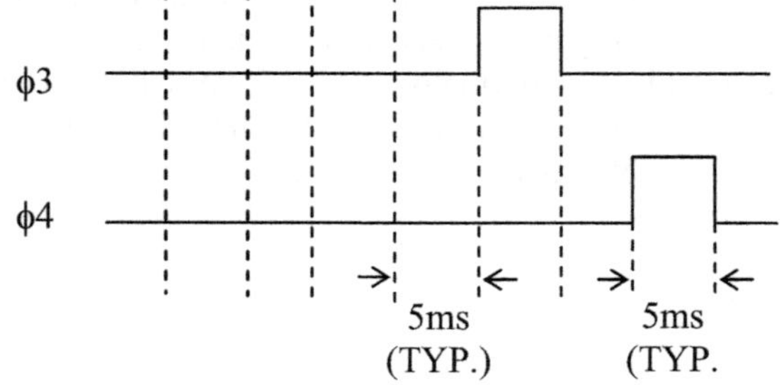

Fig. A6.4 Stepper motor timing

6.7 Timers are used in various applications such as industrial process timing, machine tool controls, heating controls (ovens, refrigerators, etc.), photography, injection molding, motor/press controls, etc. Timers are primarily of three types:

(i) **On delay timers:** Time delay is initiated by switching on a contact as shown in Figure A6.5(i). The output is energized at the end of the set time delay. The timer resets if the contact is switched off prematurely.

(ii) **Off delay timers:** Closing of the control contact energizes the output. Opening of the contact initiates the time delay, at the end of which the output is de-energized as shown in Figure A6.5(ii). If the control contact is re-closed during the time delay, the timer resets but the output remains energized.

(iii) **Interval delay timers:** Closing of the control contact energizes the output and initiates the time delay, at the end of which the output is de-energized as shown in Figure A6.5(iii). The contact may be open if required soon after the output goes high. One can use this type of timer for photography by providing a normally open push button switch for the control contact.

Design the three types of timers using Verilog for timing in the range: 0 to 999.9 s. Note that all switches must be debounced. Write a test bench and test them by simulation.

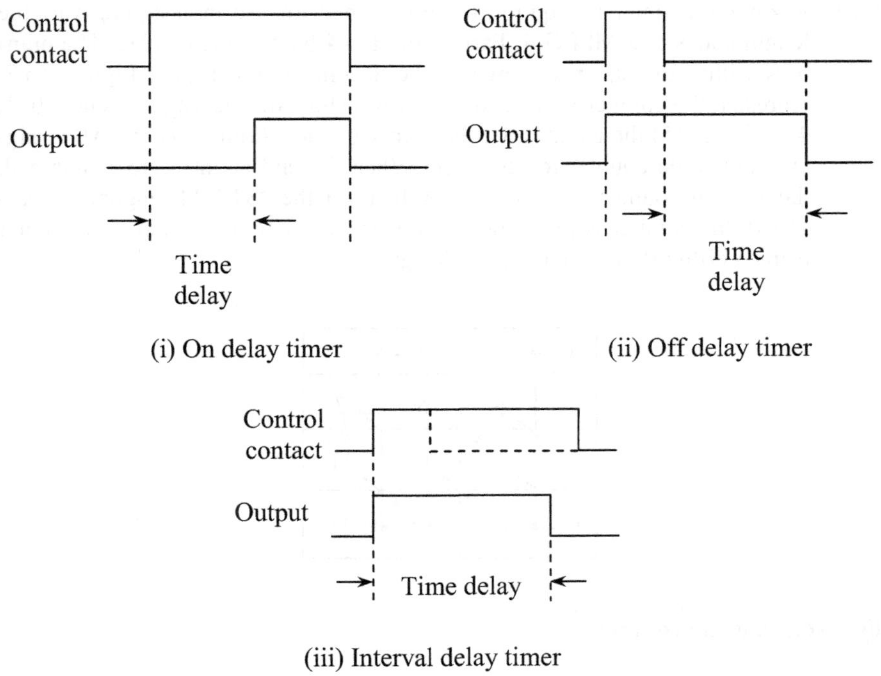

Fig. A6.5 Waveforms of timers

6.8 A beeping sound alarm is required to be generated as shown in Figure A6.6 for one of the applications called 'Alarm Annunciator'. Assuming 50% duty cycle, write Verilog code for the same and simulate to show the working of the same.

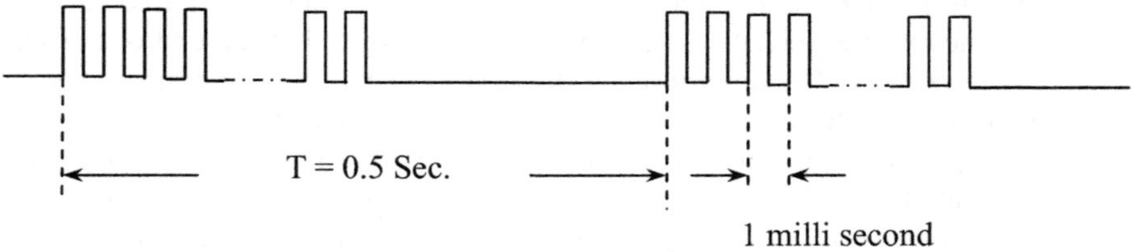

Fig. A6.6 Beeping sound alarm

6.9 A boiler plant for power generation has several processes for monitoring the temperature and pressure using appropriate sensors. Whenever the temperature and pressure are both high or both low for each of the processes for 10 seconds or more, an output is switched on and a common alarm is also sounded. The audio alarm may be silenced if acknowledged. The outputs and the alarm are cleared automatically if normal operation is restored. Assuming four such processes, write a Verilog code for realizing an integrated controller for this application. Simulate to prove the soundness of the design.

6.10 A zig-zag counter is required to be designed for use in assigning variable length codes in a MPEG-4, Part 10 or H.264 based video codec. The numbers within the squares shown in the accompanying figure, Figure A6.7, represent the counter states and the arrows show the zig-zag sequence (0, 1, 4, 8, 5, ... 15) the counter will have to sequence through. A "START" signal allows the counter to sequence. When the end count of 15 is attained, the zig-zag counter freezes at 15. Whenever the "START" signal is deasserted, the counter is reset and remains in "0" state. Design such a counter using Verilog RTL and test your design.

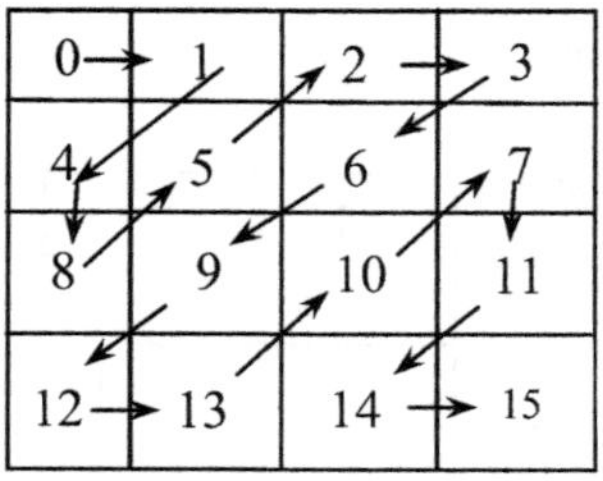

Fig. A6.7 Zig-zag counter

Chapter 7

Synthesis of Designs – Synplify Tool

In the previous chapter, we learnt the design flow of VLSI circuits. This was followed by a discussion on design methodology for solving problems effectively. We gained hands on experience on the simulation tool and verified our designs presented earlier. Designs are meaningful only if they are mapped on to a target chip such as an FPGA. The present chapter deals with a synthesis tool, which not only maps the HDL design on an FPGA but also brings about an efficient optimization of logic, thus conserving a substantial number of gates.

7.1 Synthesis

Synplify Pro 8.5, 7.1 and 7.7.1 tools of Synplicity Inc. [19] are used for synthesis in this book since it is a popular tool, especially in industries. Synplify supports FPGA devices of all vendors. The next sub-section highlights the features of this tool.

7.1.1 Features of Synthesis Tool

The Synplify tool accomplishes the following tasks in a nut-shell:

- Mapping of device
    i. Vendor
    ii. Type of device
    iii. Actual device – required capacity and package
    iv. Speed
- Compilation of the HDL design
- Logic optimization
- View of schematic circuit diagram
    i. RTL (Register Transfer Level) view
    ii. Technology view
- Creates optimized Verilog file for simulation
- Generates EDIF (Electronic Data Information Format) file ready for vendor specific place and route

Xilinx, Altera, and Actel are some of the leading FPGA vendors. Each of these vendors has a range of devices in terms of FPGA type, chip capacity, type of package, and speed grade. The Synplify tool is completely menu driven and the above selections may be made using drop-down menu. The drudgery involved in Karnaugh map or Quine McCluskey types of reductions are conspicuous by their absence while using this tool. The tool brings about the logic optimization automatically when the designer runs his design, thus relieving a heavy burden from the designer's shoulders. Designers, therefore, need not stretch too much on manual optimization of the HDL codes. Designers, on the other hand, need to optimize complex algorithms before developing the code. As an example of this kind, the reader may refer to algorithmic reduction of discrete cosine transform and quantization presented in Chapters 11 and 12.

Often, designers would like to see how their HDL designs are shaping in terms of circuitry. The synthesis tool provides the RTL view of schematic circuit diagrams just by a mouse click. Clicking on a part of the circuit diagram, the corresponding HDL code is displayed. If the designer wants to view more detailed circuit diagram displays involving primitives used in the targeted FPGA, one need only click on the technology view. Propagation delays may also be viewed. Once the synthesis tool is run, an ".edf" file is created for the design, which is used as the input in the place and route tool for creating the ".bit" file. We will be learning the place and route tool in the next chapter. In the synthesis tool, optimized Verilog file is also generated if that option is chosen. This file can be run on the simulation tool to verify that the functionality of the design is still preserved after optimization. It must be remembered that synthesis can be run only on the design and not on the test bench. In the next section, we will learn synthesis of designs using the Synplify tool.

7.2 Analysis of Design Examples Using Synplify Tool

The synthesis tool may be opened by double clicking on the Synplify icon on the desktop:

The opened main window is shown in Figure 7.1. To start with, a new project must be opened. In order to open a new project, click on "File => New". A new window opens as shown in Figure 7.2. Click on the project file. As an example, we will synthesize the combination circuits we developed in Chapter 3. Type "comb_ckts" in the "file name" field and also type the desired "file location", where you wish the new project to reside. Click OK and the new project window opens. Click on "New Implementation" button. "Options for implementation window" opens as shown in Figure 7.3. In that window, click on the "Device" button to select the desired FPGA device. Select "Xilinx Virtex" in "technology" field, "XCV800" in "part" field, "-4" in "speed" field and "HQ240" in "package" field. This selects the Xilinx Virtex series FPGA, XCV800-4 as the target device. Click

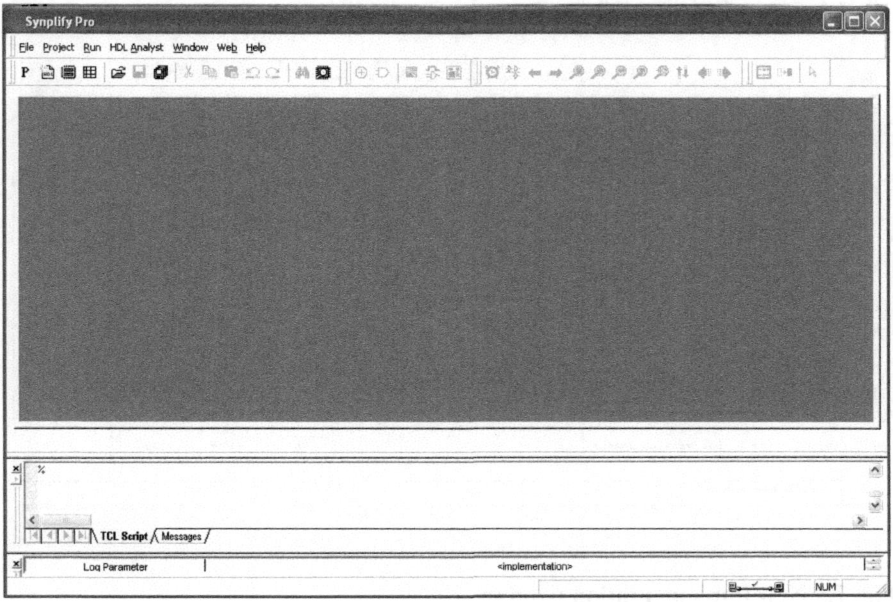

Fig. 7.1 Synplify main window

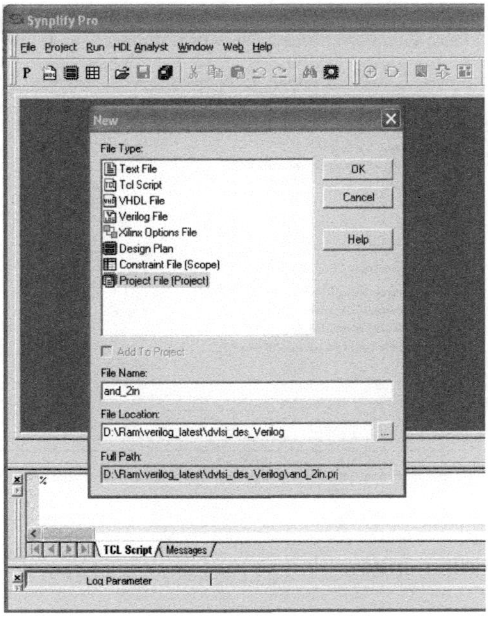

Fig. 7.2 Create new project file

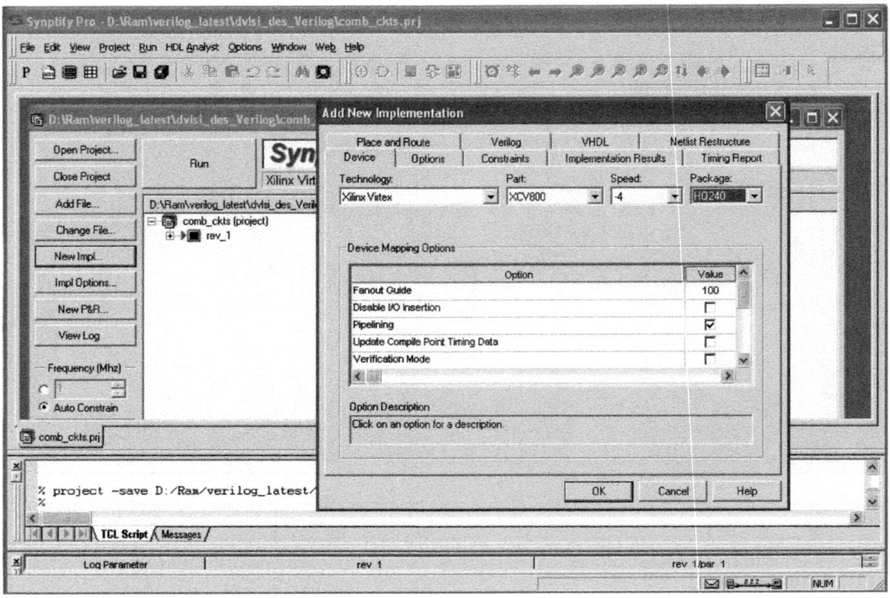

Fig. 7.3 New implementation window

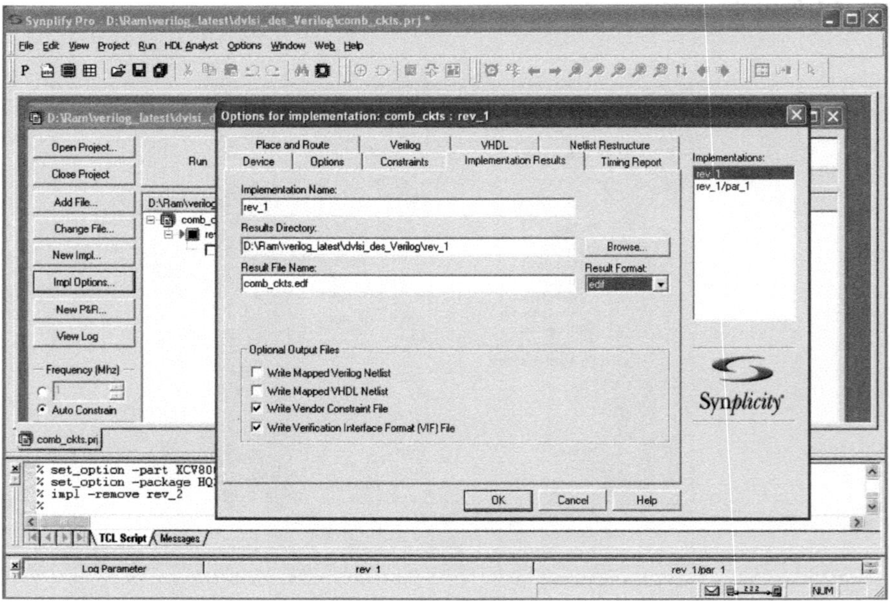

Fig. 7.4 Implementation option

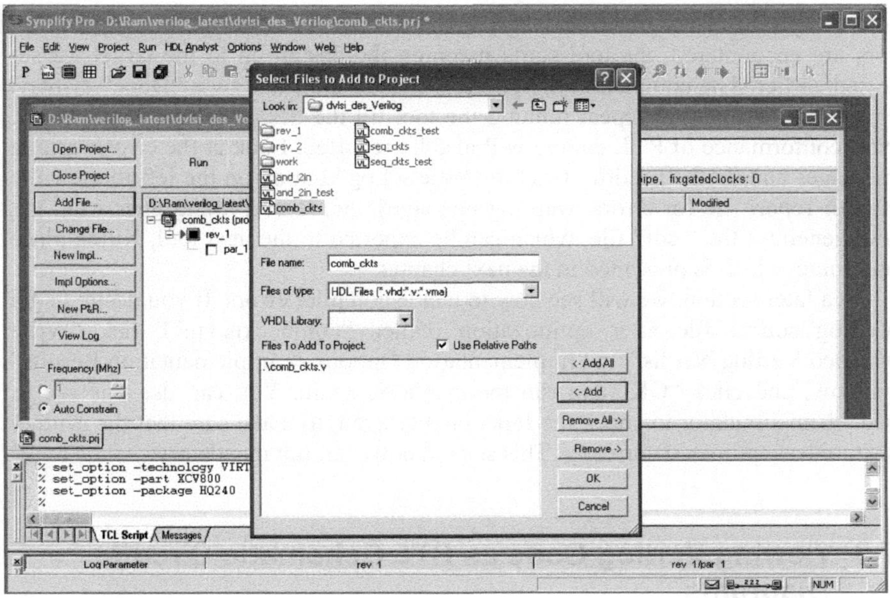

Fig. 7.5 Add design files

"OK". You can change to any other vendor device according to your needs such as the capacity and the speed subsequently by clicking on "Implementation Options". Implementation options can be chosen from the drop-down menu. The options are to be found in the "technology field". In the Options for implementation window, click on "Implementation Results" and select "EDIF" in "Results Format" field as shown in Figure 7.4. This outputs the synthesized codes in EDIF format. If the design is "comb_ckts.v", then the synthesized output will appear as "comb_ckts.edf". In "Optional Output Files", tick "Write Vendor Constraint File" if it is not already ticked. Click on "OK".

The next step is to add all the files in a design. To add a file, click on the "Add File" button in the main window. A window named "Select Files to Add to Project" opens as shown in Figure 7.5, displaying all the available files, and we need to select the desired file from the list. The file may be with the ".v" or ".vhd" extension depending upon the HDL we use in our design. After selecting the desired file, they will be displayed in the window along with the entire path, where it resides. Click on "Add" and "OK". A design usually consists of several files in a hierarchy. In that case, we have to "Add" all of them one after another. In the current design of the "comb_ckts" or the "seq_ckts" that we have designed, we have only one file for the design. We can also select all the files and click "Add all" and click on "OK". A better option is to make sure that all the sub-modules of your design are included in the Top Design and add only the top design.

The maximum frequency of operation desired for our design may be specified in the field marked "Frequency (MHz)" in the main window. Click on the "Run" button on the top of the window to start the synthesis. To start with, the synthesis tool compiles the design and displays syntax and semantic errors, if any. If no errors are encountered, the tool starts mapping the design on to the target we selected earlier. The completion of synthesis will be indicated by a “Done” display. Fix errors, if any, and repeat running the tool till the design is free of all errors. Non-conformance of RTL coding will also be reported. Look at the error/warning messages and do the needful. Click on “View Log” button on the left to view the Log or report file for errors, warning messages, etc. If everything goes well, the tool generates the “.edf” file, which can be exported to the next tool, Xilinx place and route, which is presented in the next chapter.

In a later section, we will see how to tackle compiler errors. If you desire to get Verilog source file after optimization (called “comb_ckts.vm”), tick “Write Mapped Verilog Net list” in “Implementation Options -> Implementation Results” window, and click “OK” and run the synthesis again. You can use this file in Modelsim simulator to check the functionality again to make sure that the synthesis tool has optimized correctly. This step is, however, not mandatory.

7.3 Viewing Verilog Code as RTL Schematic Circuit Diagrams

Click on “⊕” button on the top to view the RTL schematic circuit diagram of the design. Use zoom features “1”, “+”, “–“, “F” to zoom in or out. For instance, for 100% zoom in, click on 1 and click again on the schematic to zoom. Use ⇨,⇦ arrows to advance from one drawing sheet to another if more than one sheet is present. Study the circuit diagram(s) to make sure that all the functionalities you have designed are in tact. If you wish to see the Verilog code corresponding to a part of the circuit diagram, double click on the circuit. Click on  button on the top to view the Technology schematic circuit diagram of the design. This shows all the primitive cells used in the target FPGA. Use features such as zoom and advance from one sheet to another. Use ⇧ or ⇩ arrows to push or pop hierarchy. Study the circuit diagram to make sure that all the functionalities we have designed are in order. Click on  to see cumulative critical paths and slack time respectively on the circuit. A command summary of the Synplify tool is presented towards the end of this chapter. We will now analyze the designs we presented in previous chapters.

In Chapter 4, we presented a very simple design, a two-input AND gate. We will run the Synplify tool for this design first. Subsequently, we will see the synthesis results for the other designs, “comb_ckts”, “seq_ckts”, and “rtl_coding”. Synplify Pro 7.7.1 or 8.5 has been used for synthesis of these designs with XC3S200 FT256-4 as the target FPGA. The log report generated for the AND design by running the Synplify tool is as follows:

@I: “D:\ram\book\dvlsi_sys_verilog\and_2in.v”
Verilog syntax check successful!
Selecting top level module and_2in

Synthesizing module and_2in
Writing Analyst data base D:\ram\book\dvlsi_sys_verilog\rev_3\and_2in.srm
Writing EDIF Netlist and constraint files
START OF TIMING REPORT
Top view: and_2in
Requested Frequency: 100.0 MHz
Wire load mode: top
Paths requested: 5
Constraint File(s):
Performance Summary :
Resource Usage Report for and_2in
Mapping to part: xc3s200ft256-4
I/O primitives: 3
IBUF 2 uses
OBUF 1 use
I/O Register bits: 0
Register bits not including I/Os: 0 (0%)
Mapping Summary:
Total LUTs: 1 (0%)
Mapper successful!

The RTL view and the technology view of the AND gate are shown in Figures 7.6 and 7.7 respectively. EDIF file created by the tool is "and_2in.edf" as can be seen

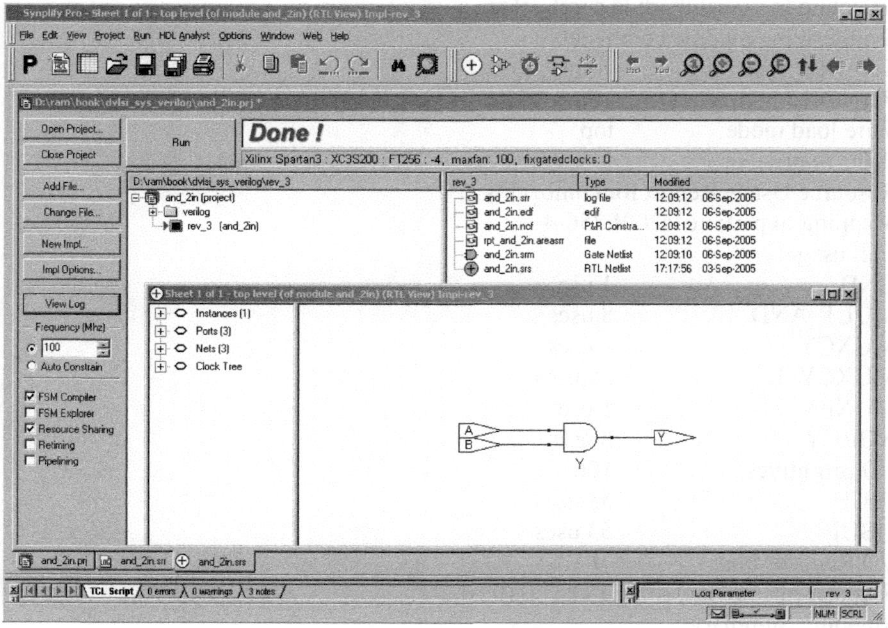

Fig. 7.6 RTL view of two-input "AND" gate

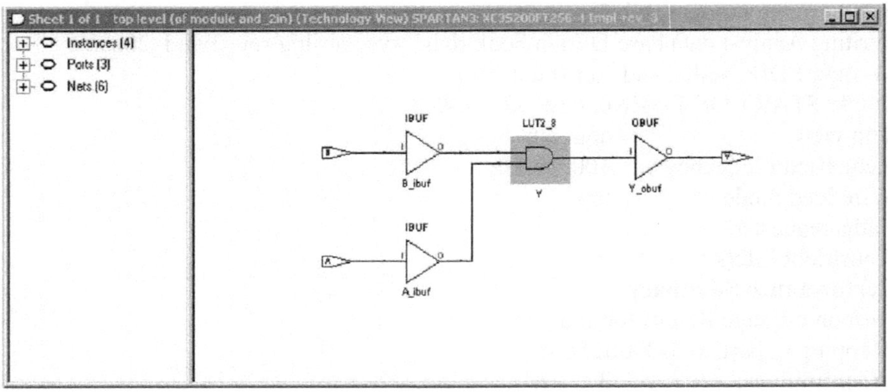

Fig. 7.7 Technology view of two-input "AND" gate

in Figure 7.6. The number of input buffers and output buffer used in the design are respectively 2 and 1 as revealed in the technology view. This is tallying with the log report presented earlier. The number of look-up table (LUT) consumed by this design is one.

We will now consider the next design, namely, "comb_ckts". Synplify log report for the same is as follows:

@I:: "D:\RAM\book\Verilog_Ch_3\comb_ckts.v"

Verilog syntax check successful!

Selecting top level module comb_ckts

Synthesizing module comb_ckts

Top view:	comb_ckts
Requested Frequency:	100.0 MHz
Wire load mode:	top
Paths requested:	5

Resource Usage Report for comb_ckts

Mapping to part: xc3s200ft256-4

Cell usage:

GND	1 use
MULT_AND	8 uses
MUXCY	4 uses
MUXCY_L	38 uses
MUXF5	1 use
XORCY	8 uses
I/O primitives:	106
IBUF	53 uses
OBUF	53 uses
I/O Register bits:	0

Register bits not including I/Os: 0 (0%)

Mapping Summary:

Total LUTs: 80 (2%)

The report gives primitive cells used in the design. All the combinational circuits we designed in Chapter 3 consume only 80 LUTs. Figure 7.8 shows the RTL view of the design. As seen in the RTL view, function “F1” is the buffered signal of “A”. Similarly, RTL views of all other gate realizations F2 to F8 may be verified with our design. F9 is the result of behavioral type full adder output, and the signal sum_total [1], which is none other than the carry. F10 is the concatenated signals {A, B, C}. F11 and F12 are right shift of F10 by 1 and left shift by 2 respectively. F13 to F18 are the magnitude comparator outputs, whose inputs are N1 and N2. The multiplexer outputs for 2 inputs, 4 inputs, and 8 inputs are “mux2”, “mux4”, and “mux8” derived from select pins A, B C and A B C respectively. Inputs are respectively I0, I1, I0–I3, and I0–I7. The “mux8” is fed as the input of the 8 output DEMUX, whose outputs are D0-D7, the same as the inputs I0–I7. Finally, a design example using full adder and a magnitude comparator may be verified. In this example, the sum (SUM [8:0]) of two multi-bit precision numbers, NUM_1 and NUM_2, is compared with a preset value, P_V [8:0]. If SUM equals the preset value, the output MATCH is generated; else if SUM is greater than the preset value, MORE is generated. Otherwise, the signal LESS is generated. Outputs are valid only if “enable_sum” is active, otherwise all the outputs are cleared. The three signals: MATCH, MORE, and LESS are not presented. Reader may run the tool and browse through the RTL view in order to find these signals.

Details for Synplify log report for “seq_ckts” are as follows. The tool issues warning signals. The reader should cross-check with the design to heed the warning and take corrective action or to ignore the warning. For example, the synthesis

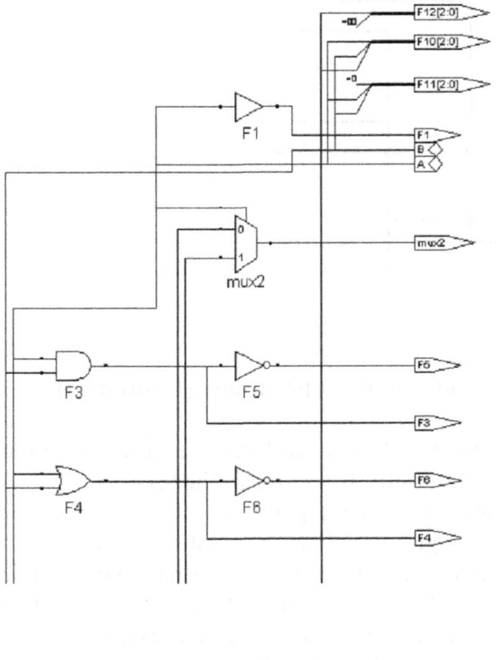

Fig. 7.8 RTL view of “comb_ckts” design (Continued)

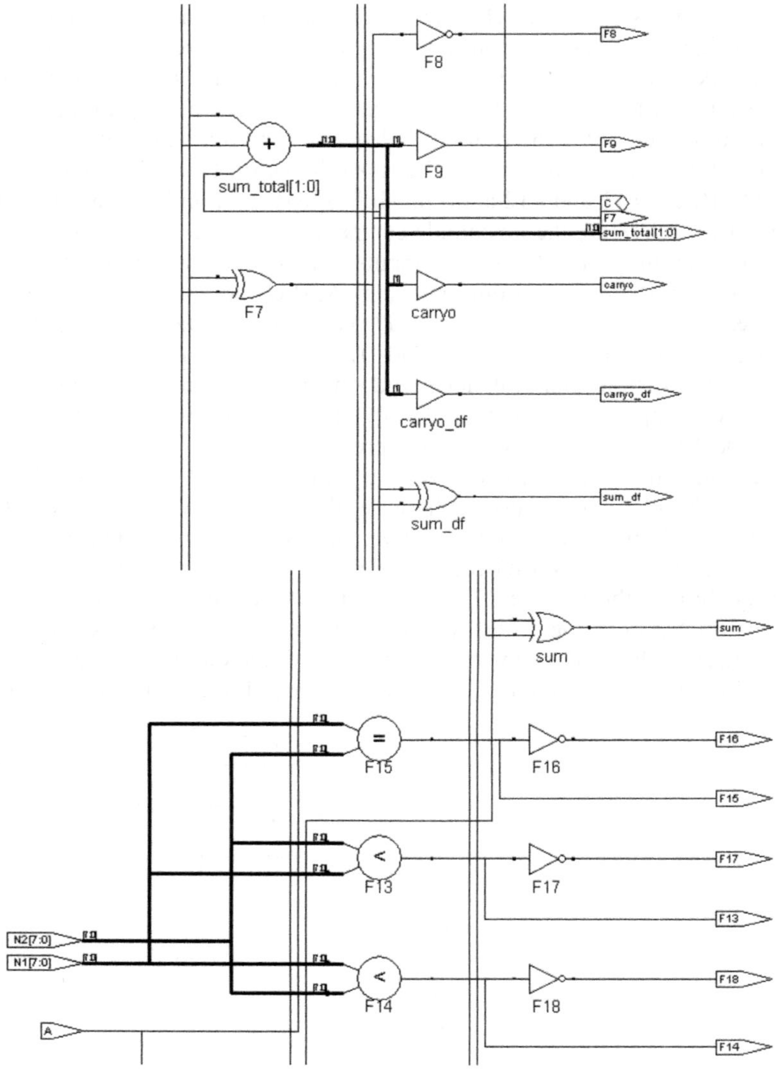

Fig. 7.8 RTL view of "comb_ckts" design (Continued)

tool warns us to check whether we forgot set/reset assignment for the signals, sr [15:0] and cnt_ps_reg [4:0]. If you recall the "seq_ckts" design we learnt in chapter 3, we designed a circuit for parallel to serial conversion, in which we have used a 16-bit shift register "sr" and a counter "cnt_ps_reg". In the Verilog code, for the reset condition, i.e., reset_n = 0, we have not initialized these two signals since we are interested only in presetting the desired values to "set_data" and 16 respectively, which is being done if "load" signal is asserted. Therefore, we can afford to

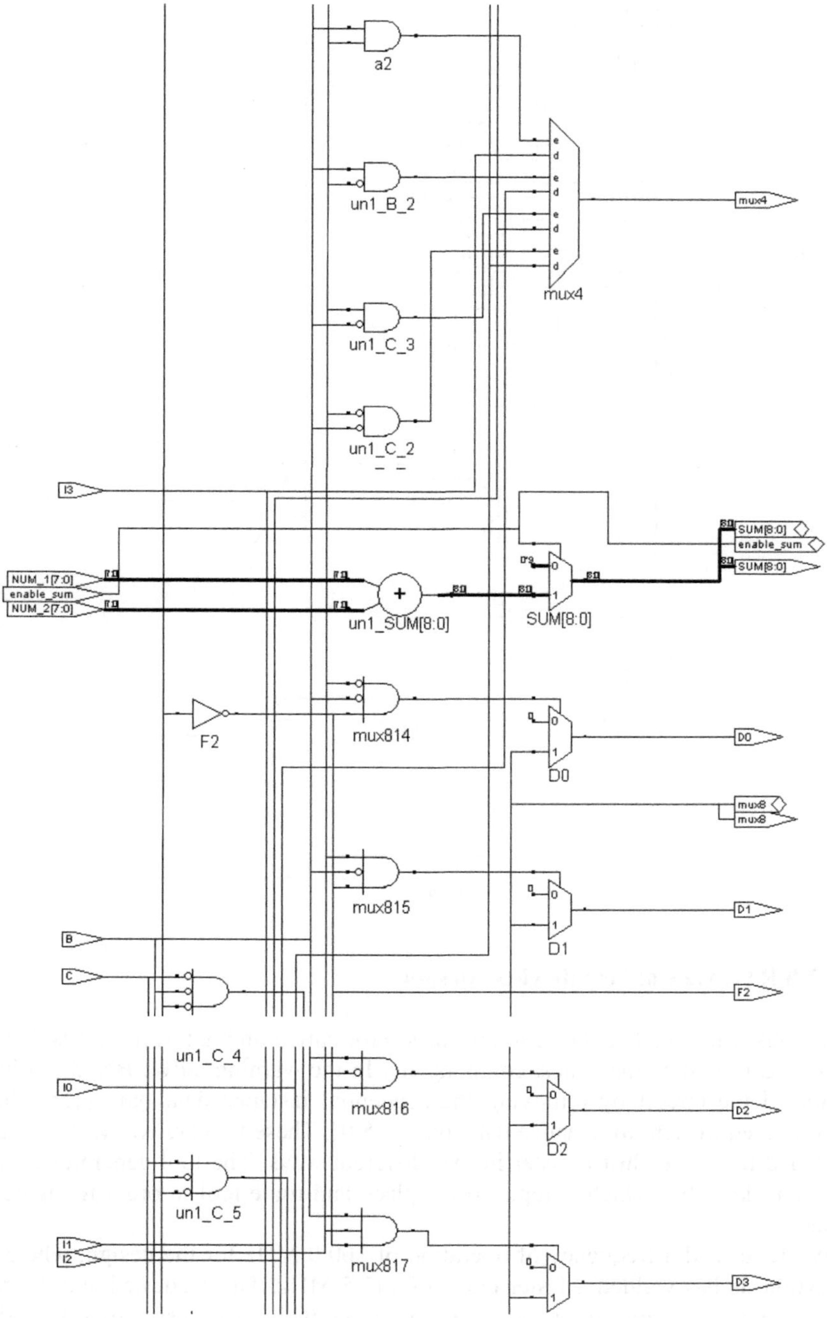

Fig. 7.8 RTL view of "comb_ckts" design (Continued)

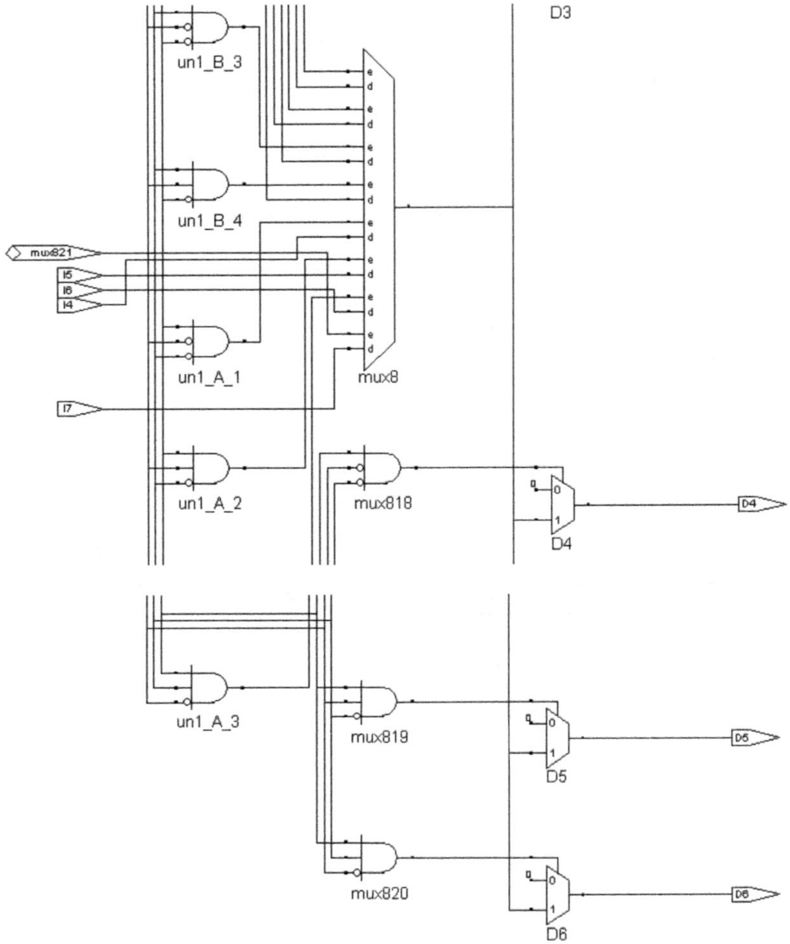

Fig. 7.8 RTL view of "comb_ckts" design

ignore this warning. The tool also provides information and notes. It reports that it has recognized state machines, counters, etc. In the warning, @W: BN132, it has optimized the circuit by removing the sequential instance data_out1 [15:0] because it is equivalent to instance data_out2 [15:0]. These two are equivalent since we coded the same shift register in two different ways. The tool generates EDIF file "seq_ckts.edf", which is input to the place and route tool in order to create a bit stream.

We requested a frequency of operation of 100.0 MHz for the design, whereas the synthesis has yielded a faster clock of 147.5 MHz. The requested and the reported frequencies are 10 ns and 6.781 ns respectively in terms of time periods. The difference known as the slack time is 3.219 ns. The slack time must be positive. Otherwise, the device cannot meet the requested frequency of operation. The tool provides elaborate timing reports such as slack time for every signal in the design.

Studying these, we can identify the long path delays and introduce pipeline registers in-between and, hence, speed up the system if we wish to do so. Thereafter, the tool reports various primitive cells used in the FPGA. The number of LUTs consumed by the "seq_ckts" design is just 78 (approximately 2% of 200,000 gates). A sample RTL view (Figure 7.9) only is presented since the whole RTL view of the design is unreadable (owing to low resolution provided by the tool) and the zoomed-in version exceeds the page width of this book. The reader may, therefore, run his or her code and browse the RTL view in order to check the design thoroughly. One can see the state graphs of a design using the "FSM Viewer" by right clicking on state machine of RTL view and clicking on "View FSM". These are shown for "state" and psd_state" of the "seq_ckts" design in Figures 7.10 and 7.11. The log report for "seq_ckts" is as follows:

Log report of "seq_ckts.v"

```
$ Start of Compile
@I::"D:\RAM\book\dvlsi_sys_verilog\seq_ckts.v'"
Verilog syntax check successful!
Selecting top level module seq_ckts
Synthesizing module seq_ckts
@W: CL112
"D:\RAM\book\dvlsi_sys_verilog\seq_ckts.v":382:0:382:5|Feedback mux created
for signal sr[15:0]. Did you forget the set/reset assignment for this signal?
@W: CL112
"D:\RAM\book\dvlsi_sys_verilog\seq_ckts.v":382:0:382:5|Feedback mux created
for signal cnt_ps_reg[4:0]. Did you forget the set/reset assignment for this signal?
@N: CL201 :"D:\RAM\book\dvlsi_sys_verilog\seq_ckts.v":543:0:543:5|Trying to
extract state machine for register psd_state
Extracted state machine for register psd_state
State machine has 4 reachable states with original encodings of:
  00
  01
  10
  11
@N: CL201 :"D:\RAM\book\dvlsi_sys_verilog\seq_ckts.v":434:0:434:5|Trying to
extract state machine for register state
Extracted state machine for register state
State machine has 4 reachable states with original encodings of:
  00
  01
  10
  11
@W: BN132 :"d:\ram\book\dvlsi_sys_verilog\seq_ckts.v":354:0:354:5|Removing
sequential instance data_out1[15:0], because it is equivalent to instance
data_out2[15:0]
Encoding state machine work.seq_ckts(verilog)-psd_state[3:0]
original code -> new code
  00 -> 00
```

```
  01 -> 01
  10 -> 10
  11 -> 11
Encoding state machine work.seq_ckts(verilog)-state[3:0]
original code -> new code
  00 -> 00
  01 -> 01
  10 -> 10
  11 -> 11
@N:"d:\ram\book\dvlsi_sys_verilog\seq_ckts.v":382:0:382:5|Found counter in
view:work.seq_ckts(verilog) inst cnt_ps_reg[4:0]
@N: "d:\ram\book\dvlsi_sys_verilog\seq_ckts.v":282:0:282:5|Found counter in
view:work.seq_ckts(verilog) inst cnt_reg[7:0]
@N: "d:\ram\book\dvlsi_sys_verilog\seq_ckts.v":309:0:309:5|Found counter in
view:work.seq_ckts(verilog) inst cntd_reg[7:0]
Writing EDIF Netlist and constraint files
Found clock seq_ckts|clk with period 10.00ns

##### START OF TIMING REPORT #####
Top view:                    seq_ckts
Requested Frequency:         100.0 MHz
Wire load mode:              top
Paths requested:             5
Performance Summary
*******************
Worst slack in design: 3.219
               Requested   Estimated   Requested  Estimated  Clock
Starting Clock Frequency   Frequency   Period     Period     Slack
-------------------------------------------------------------------------
seq_ckts|clk   100.0 MHz   147.5 MHz   10.000     6.781      3.219
-------------------------------------------------------------------------
Detailed Report for Clock: seq_ckts|clk
Starting Points with Worst Slack
*************************
                Starting                                      Arrival
                Reference
Instance        clock          Type   Pin   Net               Time     Slack
----------------------------------------------------------------------------------
cnt_ps_reg[2]   seq_ckts|clk   FDRE   Q     cnt_ps_reg[2]     0.720    3.219
cnt_ps_reg[3]   seq_ckts|clk   FDRE   Q     cnt_ps_reg[3]     0.720    3.219
cnt_reg[0]      seq_ckts|clk   FDCE   Q     cnt_reg_c[0]      0.720    3.546
cnt_reg[1]      seq_ckts|clk   FDCE   Q     cnt_reg_c[1]      0.720    3.546
cnt_reg[2]      seq_ckts|clk   FDCE   Q     cnt_reg_c[2]      0.720    3.546
cnt_reg[3]      seq_ckts|clk   FDCE   Q     cnt_reg_c[3]      0.720    3.546
cnt_reg[6]      seq_ckts|clk   FDCE   Q     cnt_reg_c[6]      0.720    3.546
cnt_reg[7]      seq_ckts|clk   FDCE   Q     cnt_reg_c[7]      0.720    3.546
```

cnt_ps_reg[0]	seq_ckts|clk	FDRE	Q	cnt_ps_reg[0]	0.720	3.938
cnt_ps_reg[1]	seq_ckts|clk	FDRE	Q	cnt_ps_reg[1]	0.720	3.938

Ending Points with Worst Slack
****

Instance	Starting Reference clock	Type	Pin	Net	Required Time	Slack
sr[0]	seq_ckts|clk	FDE	CE	N_106_i	9.398	3.219
sr[1]	seq_ckts|clk	FDE	CE	N_106_i	9.398	3.219
sr[2]	seq_ckts|clk	FDE	CE	N_106_i	9.398	3.219
sr[3]	seq_ckts|clk	FDE	CE	N_106_i	9.398	3.219
sr[4]	seq_ckts|clk	FDE	CE	N_106_i	9.398	3.219
sr[5]	seq_ckts|clk	FDE	CE	N_106_i	9.398	3.219
sr[6]	seq_ckts|clk	FDE	CE	N_106_i	9.398	3.219
sr[7]	seq_ckts|clk	FDE	CE	N_106_i	9.398	3.219
sr[8]	seq_ckts|clk	FDE	CE	N_106_i	9.398	3.219
sr[9]	seq_ckts|clk	FDE	CE	N_106_i	9.398	3.219

Worst Path Information
****

Path information for path number 1:

Requested Period:	10.000
- Setup time:	0.602
= Required time:	9.398
- Propagation time:	6.179
= Slack (critical) :	3.219
Number of logic level(s):	4

Resource Usage Report for seq_ckts

Mapping to part: xc3s200ft256-4

Cell usage:

FDC	11 uses
FDCE	30 uses
FDE	16 uses
FDP	1 use
FDPE	8 uses
FDRE	4 uses
FDSE	1 use
GND	1 use
MUXCY_L	18 uses
VCC	1 use
XORCY	21 uses
I/O primitives:	82
IBUF	30 uses

OBUF 52 uses
BUFGP 1 use
I/O Register bits: 2
Register bits not including I/Os: 69 (1%)
Global Clock Buffers: 1 of 8 (12%)
Mapping Summary:
Total LUTs: 78 (2%)

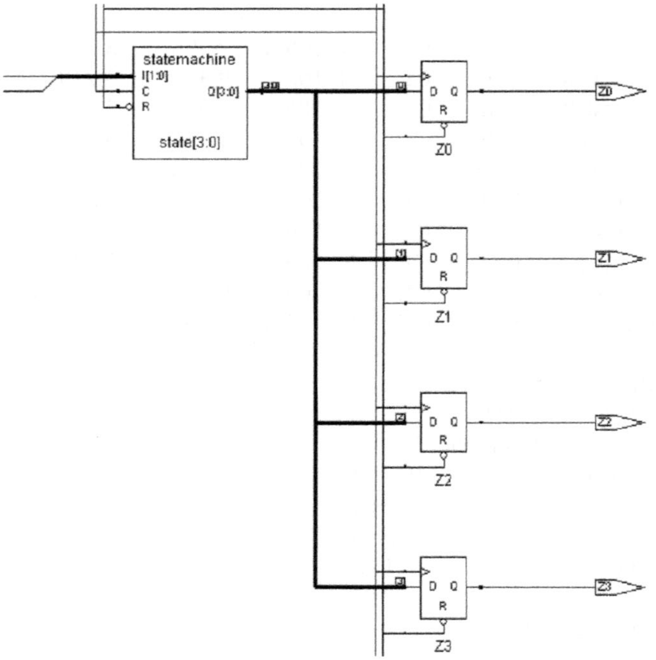

Fig. 7.9 A sample RTL view of the "seq_ckts.v" design

Synplify log report for "rtl_coding" design presented in Chapter 5 is as follows: The tool issues warning signals such as "Latch generated from always block for signal OUT_P[2:0], probably caused by a missing assignment in an if or case stmt". This is true since we deliberately created a latch as we mentioned in the comment of the code:

// parallel_case prevents priority-encoded logic but infers a latch.

This does not conform to RTL coding style and should not be used. The way out of this problem is by combining parallel_case and full_case directives as shown in the "OUT_PF" signal towards the end of "rtl_coding" design. More details of these "synopsys cases" will be presented in Section 7.3. Similarly, the reader will have to ponder over other warnings to look for potential problems and fix problems, if any. The EDIF file generated for this design is "rtl_coding.edf". The

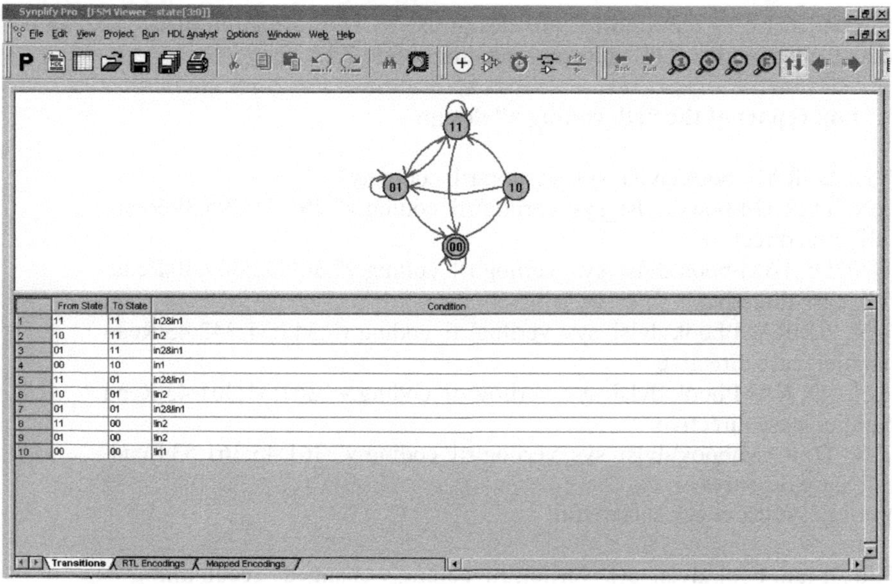

Fig. 7.10 FSM view of "state" used in the "seq_ckts.v" design

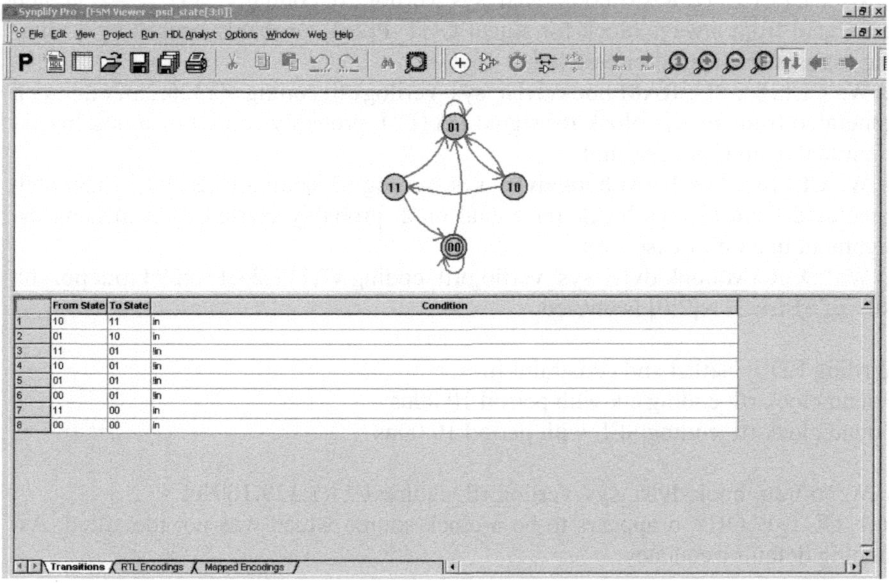

Fig. 7.11 FSM view of "psd_state" used in the "seq_ckts.v" design

number of LUTs reported for this design is 79. The formation of latches can be visually seen in the RTL view presented in Figure 7.12.

Log report of the "rtl_coding.v" design

@I::"D:\RAM\book\dvlsi_sys_verilog\rtl_coding.v"
@N:"D:\RAM\book\dvlsi_sys_verilog\rtl_coding.v":295:31:295:39|Read full_case directive
@N:"D:\RAM\book\dvlsi_sys_verilog\rtl_coding.v":309:32:309:40|Read full_case directive
@N: "D:\RAM\book\dvlsi_sys_verilog\rtl_coding.v":342:31:342:43|Read parallel_case directive
@N: "D:\RAM\book\dvlsi_sys_verilog\rtl_coding.v":361:31:361:43|Read parallel_case directive
@N:"D:\RAM\book\dvlsi_sys_verilog\rtl_coding.v":361:45:361:53|Read full_case directive
Verilog syntax check successful!

File D:\ram\book\dvlsi_sys_verilog\rtl_coding.v changed - recompiling
Selecting top level module rtl_coding
Synthesizing module rtl_coding

@W: CL118 : "D:\RAM\book\dvlsi_sys_verilog\rtl_coding.v":342:2:342:5|Latch generated from always block for signal OUT_P[2:0], probably caused by a missing assignment in an if or case stmt
@W: CL118 : "D:\RAM\book\dvlsi_sys_verilog\rtl_coding.v":326:2:326:5|Latch generated from always block for signal LATCH, probably caused by a missing assignment in an if or case stmt
@W: CL118 : "D:\RAM\book\dvlsi_sys_verilog\rtl_coding.v":220:2:220:3|Latch generated from always block for signal out2, probably caused by a missing assignment in an if or case stmt
@W: "D:\RAM\book\dvlsi_sys_verilog\rtl_coding.v":112:23:112:29|Input port bit <0> of SELECTN[2:0] is unused

Writing EDIF Netlist and constraint files
Found clock rtl_coding|clk with period 10.00ns
Found clock rtl_coding|SEL with period 10.00ns

@W: "d:\ram\book\dvlsi_sys_verilog\rtl_coding.v":1:1:329:16|Net un1_OUT_F_OH7_n appears to be a clock source which was not identified. Assuming default frequency.

START OF TIMING REPORT #####[

Top view:	rtl_coding
Requested Frequency:	100.0 MHz
Wire load mode:	top

Paths requested: 5
Resource Usage Report for rtl_coding

Mapping to part: xc3s200ft256-4
Cell usage:

FD	2 uses
GND	1 use
LD	1 use
LDCP	3 uses
LD_1	1 use
MUXCY	2 uses
MUXCY_L	14 uses
MUXF5	1 use

I/O primitives:	137
IBUF	106 uses
IBUFG	1 use
OBUF	30 uses
BUFG	1 use
BUFGP	1 use

I/O Register bits:	2
Register bits not including I/Os:	0 (0%)
Global Clock Buffers:	2 of 8 (25%)

Mapping Summary:
Total LUTs: 79 (2%)

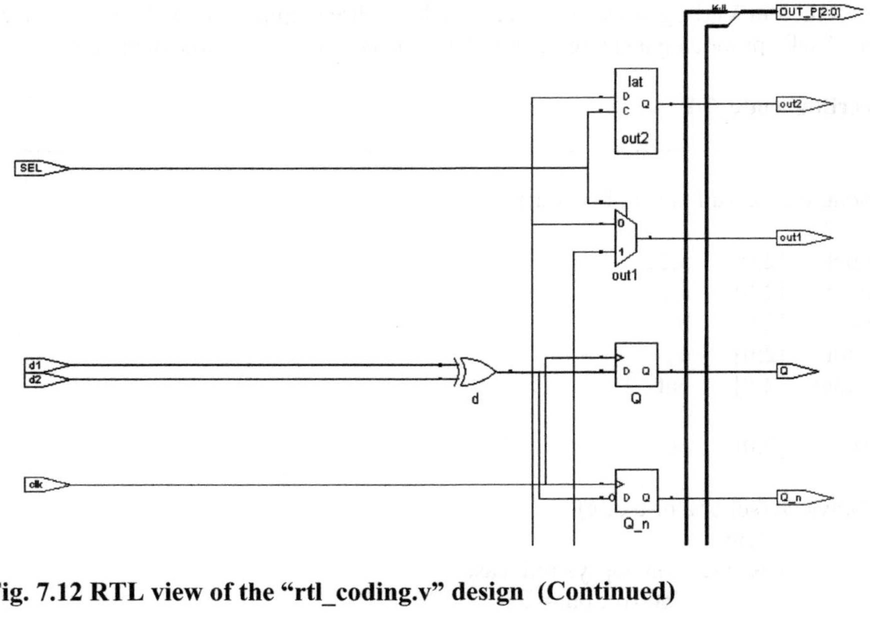

Fig. 7.12 RTL view of the "rtl_coding.v" design (Continued)

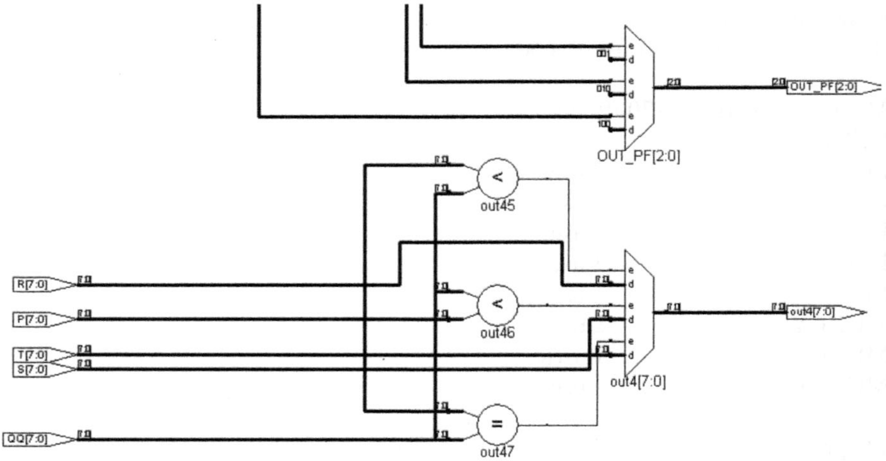

Fig. 7.12 RTL view of the "rtl_coding.v" design

7.4 Optimization Effected in Synopsys Full and Parallel Cases

So far, we have been discussing RTL coding guidelines for Synopsys full case, Synopsys parallel case, etc. Next we will consider each of these files independently and see how optimization really works for the Synopsys full and parallel cases. First consider the source code for the full case. The coding is very simple as presented in Verilog_code_7.1. Here, we have three inputs a, b, and c and a select pin, "sel", producing a single output, "out". This is a "case" implementation.

Verilog_code_7.1

```
module case_full (sel, a, b, c, out) ;

input    [2:0]   sel ;
input    [2:0]   a ;
input    [2:0]   b ;
input    [2:0]   c ;
output   [2:0]   out ;

reg      [2:0]   out ;

always @ (sel or a or b or c)
         begin
         case (sel) //synopsys full_case
                  3'b001  : out = a ;
```

```
                3'b010  : out = b ;
                3'b100  : out = c ;           // Infers a MUX
//              default   : out = 0 ;
        endcase
        end
endmodule
```

Verilog_code_7.2

```
module case_nofull (sel, a, b, c, out) ;

input   [2:0]   sel ;
input   [2:0]   a ;
input   [2:0]   b ;
input   [2:0]   c ;
output  [2:0]   out ;

reg     [2:0]   out ;

always @ (sel or a or b or c)
        begin
        case (sel)
                3'b001  : out = a ;
                3'b010  : out = b ;
                3'b100  : out = c ;           // Infers a latch
//              default   : out = 0 ;
        endcase
        end
endmodule
```

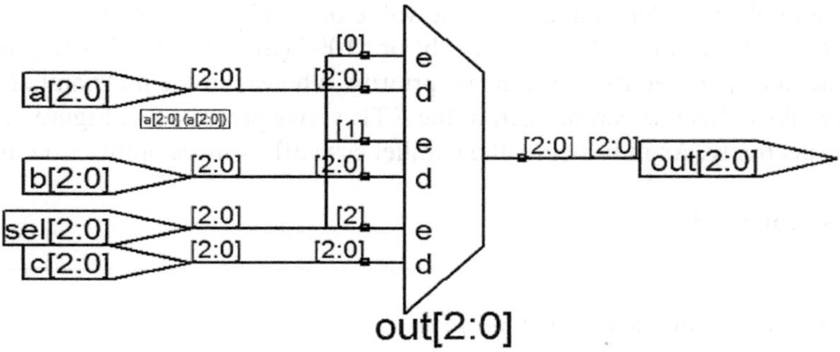

Fig. 7.13 RTL view of the "synopsys full_case"

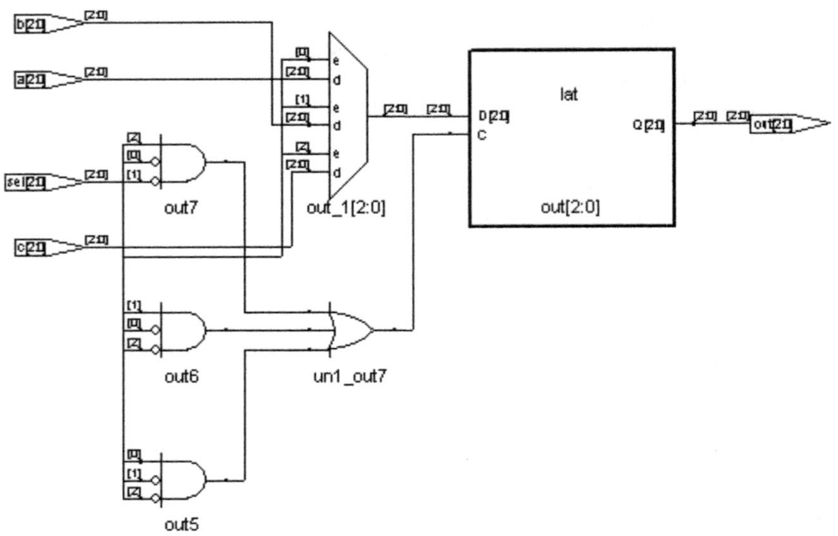

Fig. 7.14 RTL view of the "synopsys nofull_case"

It should be noted here that "//synopsis full_case" is a reserved phrase. Don't mistake "//" for a regular comment. This code simply outputs the input values a, b, and c depending upon the select pin, which has one hot assignment. In order to study the effect of the above key phrase, remove the phrase as shown in Verilog_code_7.2. Run the two codes in Synplify to get the RTL views as shown in Figures 7.13 and 7.14. Comparing the two figures, we clearly see that we can reduce chip complexity substantially to the tune of four gates, and a latch which is taboo during chip implementation, if we use the special phrase, //synopsys full_case.

Similarly, we can have a parallel case and a no parallel case as shown in Verilog_code_7.3 and 7.4 respectively. In the parallel case, the key phrase is //synopsys parallel_case as shown in Verilog_code_7.3. Note that this is dropped in no parallel case. Depending upon the value of "sel" signal, the output, "out" is "001" (for sel = a) or "010" (for sel = b) or "100" (for sel = c), all being one-hot assignments. "parallel_case" means no priority, whereas in the no "parallel_case", priority takes effect as can be seen in the RTL views presented in Figure 7.15 and 7.16 respectively. As in full case, the parallel case offers more optimized circuit.

Verilog_code_7.3

```
module case_parallel (sel, a, b, c, out) ;

input   [2:0]   sel ;
input   [2:0]   a ;
input   [2:0]   b ;
```

```
input   [2:0]   c ;
output  [2:0]   out ;

reg     [2:0]   out ;

always @ (sel or a or b or c)
        begin
        case (sel) //synopsys parallel_case
                                        // parallel_case means no priority.
                a : out = 3'b001 ;
                b : out = 3'b010 ;
                c : out = 3'b100 ;
//              default   : out = 3'b000 ;
        endcase
        end
endmodule
```

Verilog_code_7.4

```
module case_noparallel (sel, a, b, c, out) ;

input   [2:0]   sel ;
input   [2:0]   a ;
input   [2:0]   b ;
input   [2:0]   c ;
output  [2:0]   out ;

reg     [2:0]   out ;

always @ (sel or a or b or c)
        begin
        case (sel)
                        // No parallel_case means priority is present.
                a : out = 3'b001 ;
                b : out = 3'b010 ;
                c : out = 3'b100 ;
//              default   : out = 3'b000 ;
        endcase
        end
endmodule
```

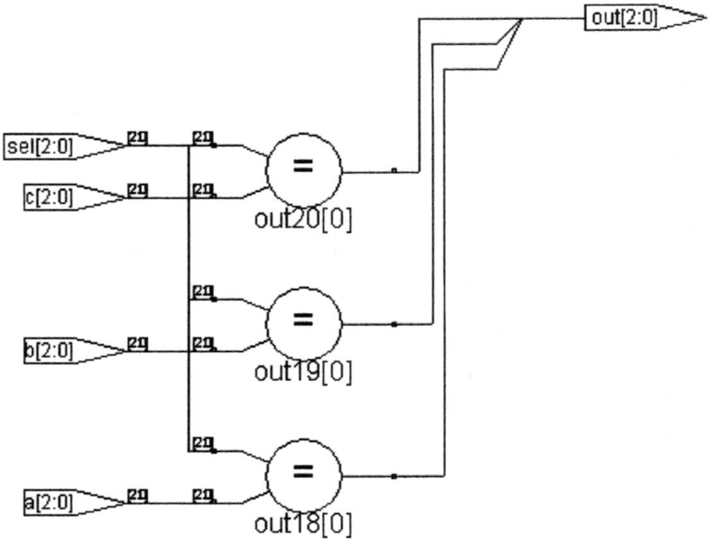

Fig. 7.15 RTL view of the "synopsys parallel_case"

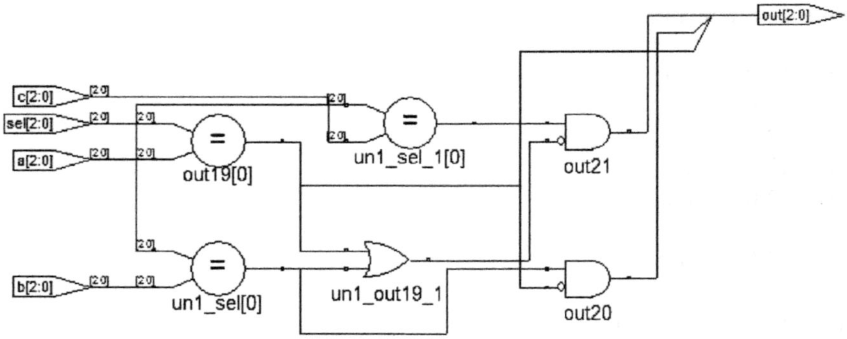

Fig. 7.16 RTL view of the "synopsys noparallel_case"

7.5 Performance Comparison of FPGAs of Two Vendors for a Design

By changing the target FPGAs and running the Synplify tool, we can assess the suitability of an FPGA for our application. We will run the tool for our design, "seq_ckts", using XCV100EPQ240-8 FPGA of Xilinx and EPF10K100ARC240-1 EPLD of Altera, both of the same capacity and speed. Performance summary of these devices are as follows. The Xilinx device is faster (100 MHz) than the

Altera's (76 MHz) device, while the design size is more or less the same. However, we need to complete the entire design of a project before we can make a comparison in order to arrive at the suitability of a device for our project.

Performance Summary for Xilinx' FPGA, XCV100EPQ240-8

Starting Clock	Requested Frequency	Estimated Frequency	Requested Period	Estimated Period	Slack
clk	100.0 MHz	100.1 MHz	10.000	9.985	0.015
System	100.0 MHz	105.6 MHz	10.000	9.467	0.533

Resource Usage Report for seq_ckts
Mapping to part: xcv100epq240-8
Cell usage:

FDC	11	uses
MUXCY_L	17	uses
XORCY	20	uses
FDCE	30	uses
FDE	21	uses
FDP	1	use
FDPE	8	uses
GND	1	use
VCC	1	use

I/O primitives:

IBUF	30	uses
OBUF	52	uses
BUFGP	1	use

I/O Register bits: 12
Register bits not including I/Os: 59 (2%)
Global Clock Buffers: 1 of 4 (25%)

Mapping Summary:
Total LUTs: 77 (3%)

Performance Summary of Altera's FPGA, EPF10K100ARC240-1

Design view:work.seq_ckts (Verilog)
Selecting part epf10k100arc240-1

Total LUTs: 116 of 4992 (2%)

Logic resources:	116 LCs of 4992 (2%)
Number of Nets:	262
Number of Inputs:	771
Register bits:	70 (59 using enable)
EABs:	0 (0% of 12)
I/O cells:	83

Starting Clock	Requested Frequency	Estimated Frequency	Requested Period	Estimated Period	Slack
clk	100.0 MHz	75.9 MHz	10.000	13.167	–3.167*
System	100.0 MHz	128.7 MHz	10.000	7.767	2.233

* Negative slack time implies, the device cannot meet 100 MHz operation

Details:

Cells in logic mode:	69
Cells in arith mode:	8
Cells in cascade mode:	6
Cells in counter mode:	13
DFFs with no input combinational logic:	20 (uses cell for routing)
LUTs driving both DFF and logic:	17

7.6 Fixing Compilation Errors in Modelsim and Synplify Tools

Beginners are usually confronted with compilation errors while running the simulation tool or the synthesis tool. Even proficient designers often get perplexed while fixing the problems, especially when the tools do not throw light on the problem. Much of these irritants may be mitigated if we systematically learn how to fix these problems step by step. This can be done by deliberately injecting errors in an already working source file such as any of the Verilog code files presented in this book. In this section, we will precisely do this to expose the types of common errors we commit. This exercise would by no means be an exhaustive coverage of errors. Based on this experience, the reader can devise his or her own ingenious ways of creating errors and thereby learn to fix them.

All the Verilog codes developed in this book are available in a CD. These files may be copied into a convenient folder for normal working with the tools. In order to learn fixing compilation errors, make a copy of the desired files in a different working directory, say, "Compiler_errors". This will avoid the original disk files from getting corrupted. We will first make one error, study the compiler report and record the error message and restore the original file. We can then experiment with the next error. While experimenting with compiler errors, we will use Synplify for synthesis and Modelsim for simulator. Each of these errors, enclosed within double quotes: " ", is dealt point-wise as follows:

1. As an example, we will take the design file, "seq_ckts.v", presented in Chapter 3 and create deliberate errors. One of the most common errors is using " ' " instead of " ` ". Let us change the very first character of the first statement accordingly, leaving other parts untouched:
   `define S0 3'd0 // Define the state of the controller
   as
   'define S0 3'd0 // Define the state of the controller
   Save the design file and run the test bench in the Modelsim. As the design file is included in the test bench, the tool will compile the design file also. The following is the compiling error report by Modelsim:

   # ERROR: seq_ckts.v(11): near "'d": expecting: MACROMODULE MODULE PRIMITIVE (*
   # ERROR: seq_ckts.v(10): Illegal digit for specified base in numeric constant
   # WARNING[10]: seq_ckts.v(445): Macro `S0 is undefined
   # ERROR: seq_ckts.v(445): near ";": expecting: IDENT
   # WARNING[10]: seq_ckts.v(452): Macro `S0 is undefined
   # ERROR: seq_ckts.v(452): near ":": expecting: ";"
   # WARNING[10]: seq_ckts.v(462): Macro `S0 is undefined
   # WARNING[10]: seq_ckts.v(498): Macro `S0 is undefined
   # WARNING[10]: seq_ckts.v(518): Macro `S0 is undefined
   # WARNING[10]: seq_ckts.v(530): Macro `S0 is undefined

   The error occurs in line 11 as reported and pin-points to: near "'d". Double clicking on the error report opens the source file, highlighting the error. This naturally results in further errors wherever "S0" is used. There is a difference in running the Modelsim and Synplify tools. In Synplify, the test bench has no role to play and hence, we will run only the design file, "seq_ckts.v". In the case of Synplify, the errors are recorded in the log file. We observe that the same syntax error is reported by Synplify tool. Before we experiment with the next error, we will restore the working file at every step.

2. Miss-spelling of the file name: In the module declaration, the "seq_ckts" may be miss-spelt as "seq_cktss":
   module seq_cktss(// Declare the design module.
   When we compile with this error in Modelsim, we do not get any error. This error may be caught at the time of loading the design. Modelsim reports: Instantiation of the given file name failed. This is because we have changed the

file name by miss-spelling. In later versions of Modelsim, even this is not reported. The Synplify Tool accepts the miss-spelling.

3. We will next make the mistake of putting a "comma" at the end of last port in the listing within the module declaration:
out,

Modelsim reports: Too few port connections.
Loading work.seq_ckts_test
Loading work.seq_ckts
** Warning: (vsim-3017) D:/ram/book/Verilog_Ch_3/seq_ckts_test.v(109): [TFMPC] - Too few port connections. Expected 32, found 31.
Region: /seq_ckts_test/u1

The tool was obviously expecting more ports to be listed. The Synplify tool accepts the comma. Restore the original file.

4. Change the name of the signal, say "clk" to "clock" as follows:
input clock ;

Modelsim reports "clk" is undefined as follows:
-- Compiling module seq_ckts
ERROR: seq_ckts.v(214): Undefined variable: clk
ERROR: seq_ckts.v(601): Identifier must be declared with a port mode: clk
-- Compiling module seq_ckts_test

In Synplify, it reports: signal clock is missing from port list. Double clicking on the report opens the same line, input clock;

5. Remove a semicolon in the next input declaration as follows:
input reset_n

Modelsim report is as follows:
-- Compiling module seq_ckts
ERROR: seq_ckts.v(69): near "input": expecting: ";"

Line 69 is the next statement, input hold; which is not recognized because of the missing semicolon.
Synplify tool reports: expecting delimiter or semicolon. Double clicking on the report opens the same line 69.

6. So far, we have been dealing with only single bit precision signals. We will now try multi-bit signals. Let us change the output specification of "cnt_reg [7:0]" as in the following two statements:
output cnt_reg ;

reg [6:0] cnt_reg ;

The Modelsim and Synplify compilers are silent. They don't report anything wrong. But Modelsim tracks the error while loading (or simulating) the design:
Loading work.seq_ckts_test
Loading work.seq_ckts
** Warning: (vsim-3015) D:/ram/book/Verilog_Ch_3/seq_ckts_test.v(109): [PCDPC] - Port size does not match connection size (port 'cnt_reg').

All the errors we created so far were in the design file. Now let us create a couple of errors in the test bench to get a different feel.

1. Let us change the declaration of the input stimulant from "reg" to "wire" as follows:
   wire D ; // Declare all inputs as reg.

   Modelsim reports of illegal reference to net (meaning wire).
   ERROR: D:/ram/book/Verilog_Ch_3/seq_ckts_test.v(116): Illegal reference to net: D
   # ERROR: D:/ram/book/Verilog_Ch_3/seq_ckts_test.v(138): Illegal reference to net: D
   # ERROR: D:/ram/book/Verilog_Ch_3/seq_ckts_test.v(146): Illegal reference to net: D
   # ERROR: D:/ram/book/Verilog_Ch_3/seq_ckts_test.v(153): Illegal reference to net: D
   However, in the case of Synplify, the test bench is not relevant.

2. One of the most common errors is forgetting to key-in "endmodule". Modelsim reports:
   # ERROR: D:/ram/book/Verilog_Ch_3/seq_ckts_test.v(1): near "EOF":syntax error
   Double clicking on the report, it points to the top of the module rather than the end.

On similar lines, the reader may create various types of errors, one at a time, and gradually learn to fix each of these problems.

7.7 Synplify Command Summary

A summary of the Synplify tool commands is presented in the following to be used as a ready reckoner.

1. Double click on icon 

on your desktop. Synplify window opens.

2. Click on "File => New" to open a new project. A "New" window opens. Click on Project File. Type comb_ckts in "File Name" field and also type the desired "File Location" or browse and select, where you wish the new project to reside. Click on "OK". The project window opens.
3. Click on "New Implementation". "Add New Implementation" window opens. In that window, click on "Device". Select "Xilinx Virtex" in "Technology" field, "XCV800" in "Part" field, "-4" in "Speed" field, and HQ240 in "Package" field. You can change to any other vendor device according to your specific needs using "Impl Options". Click on "OK".
4. Click on "Add File" menu on the left. A "Select Files to Add to Project" window opens. Click on the desired design; say "comb_ckts" listed in the window. Also, click on "Add" button on the right to include the design file. The selected file is displayed. Click "OK". In this manner, add all the files in your design. While adding files, add starting from the lowest level of files going up to the top design file. Alternatively, make sure that all the sub-modules of your design are included in the Top Design and add only the top design. This is a better option.

   CAUTION:

   Never use the test bench for synthesis. Only your design is valid.
5. Set the desired operating frequency using the "Frequency (MHz)" menu on the left. Click on "Auto Constrain" to maximize speed.
6. Click on the "Run" button on the top of the window to start the synthesis. "Compiling" followed by the "Mapping" (of your design) will be displayed on the window. The completion of synthesis will be indicated by a "Done" display. Fix errors, if any, and repeat 6. Also, look at the warning messages and do the needful.

Note:

Tick "Write Mapped Verilog Net list" in "Implementation Options => Implementation Results" window, and click "OK" before running the Synthesis, if you desire to get Verilog source file after optimization. You can use this file in Modelsim simulator to check the functionality again to make sure that the Synthesis tool has optimized correctly. When the synthesis is run, an ".edf" file is automatically created, which must be used in the place and route tool such as the Xilinx P&R tool in order to create a ".bit" file.

7. Click on "View Log" button on the left to view the Log or report file for errors, warning messages, etc.
8. Click on "⊕" button on the top to view the RTL schematic circuit diagram of the design. Use zoom features "1", "+", "–", "F" to zoom in or out. For instance, for 100% zoom in, click on 1 and click again on the schematic to zoom. Use ⇨ , ⇦ arrows to advance from one sheet to another. Study the circuit diagram to make sure that all the functionalities you have designed are in tact.
9. Click on ⊐▷ button on the top to view the Technology schematic circuit diagram of the design. Use features such as zoom and advance from one sheet to another as in 8. Use ⇧ or ⇩ to push or pop hierarchy. Study the circuit dia-

gram to make sure that all the functionalities you have designed are in order. Click on  to see cumulative critical paths and slack time (in ns) respectively on the circuit.

10. Click on Options => Xilinx => Start ISE Project Navigator to open the Xilinx Place and Route Tool.

Summary

This chapter covered the synthesis of designs using Synplify tool, widely used in industries. The salient features of synthesis are mapping of an FPGA device, logic optimization, and viewing schematic circuit diagrams of the Verilog code. The tool creates optimized Verilog file and Electronic Data Information Format (EDIF) file, which may be used for simulation using Modelsim and to run vendor specific place and route tool such as Xilinx P&R respectively. Synplify tool supports all types of FPGAs. Errors were created deliberately and correction applied in order to learn and fix compiling and simulation errors in Modelsim and Synplify tools. A command summary of the Synplify tool was furnished for quick reference while using the tool. EDIF file is exported to the next tool, the place and route tool, for creating a bit stream of the design. Bit stream is downloaded into FPGAs for configuring it for the desired application. These features are shown in later chapters. The place and route tool is presented in the next chapter.

Assignments

7.1 For the design assignments 3.2 to 3.14 in Chapter 3, run the synthesis tool to establish the RTL conformance of the designs. Present the RTL views and important synthesis results.

7.2 For the design assignments 4.2, 4.4, and 4.12 in Chapter 4, run the synthesis tool to establish the RTL conformance of the designs and present the important synthesis results.

7.3 For the assignments 4.9 to 4.11 in Chapter 4, run the synthesis tool and discuss your observation.

7.4 For the test bench in assignment 4.12 in Chapter 4, run the synthesis tool and record your observation. What is your inference?

7.5 In Synopsys full and parallel cases in the text (Verilog_code_7.1 to Verilog_code_7.4), "default" statements were commented. Uncomment these and run the Synplify tool. Discuss your observation with the aid of RTL views.

7.6 In the text, it was shown that fixing of compiler errors can be made easy if deliberate errors were created and the corresponding tool reports were analyzed. Create some more errors and run Modelsim and Synplify tools and discuss your observations.

7.7 Design a controller for an elevator in a two-story building using the ASM chart shown in Figure A7.1. The controller must respond to call push-button switches on each floor (FS1 and FS2) and floor-select push-button switches (LFS1 and LFS2) within the car. When the car lands on the floors of the building, signals are generated from the limit switches LS1 and LS2. The door should open when the car lands on the floors and close after a delay. The controller should also generate control signals to move the car "Up" and "Down" as shown in Figure A7.1. An output signal "Door" opens the door when the lift arrives at one of the floors. A limit switch "Door_closed" is closed when the door is completely shut. The lift is in the ground floor with the door closed to start with. All switches are debounced. State your assumptions clearly. Code the design in Verilog and run the synthesis, and discuss the results.

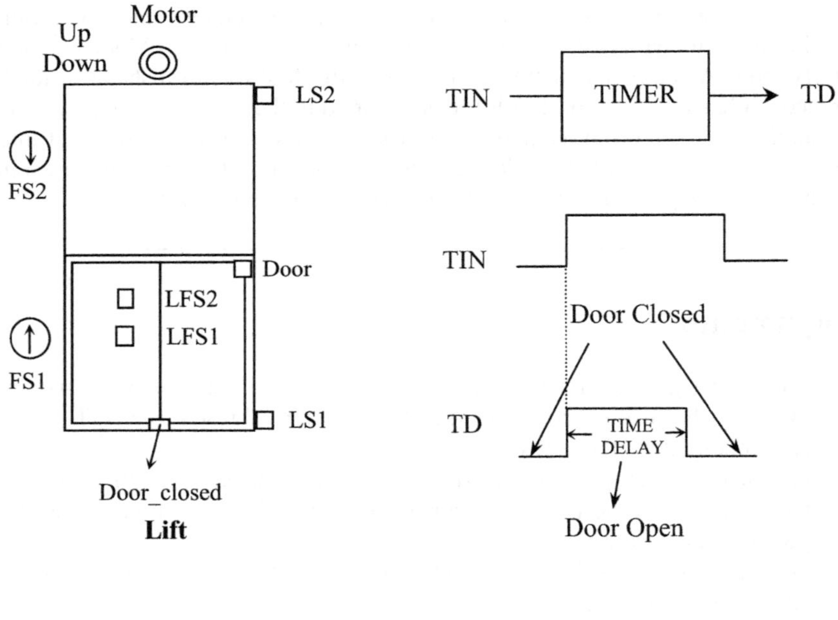

Fig. A7.1a Controller specifications of an elevator (Continued)

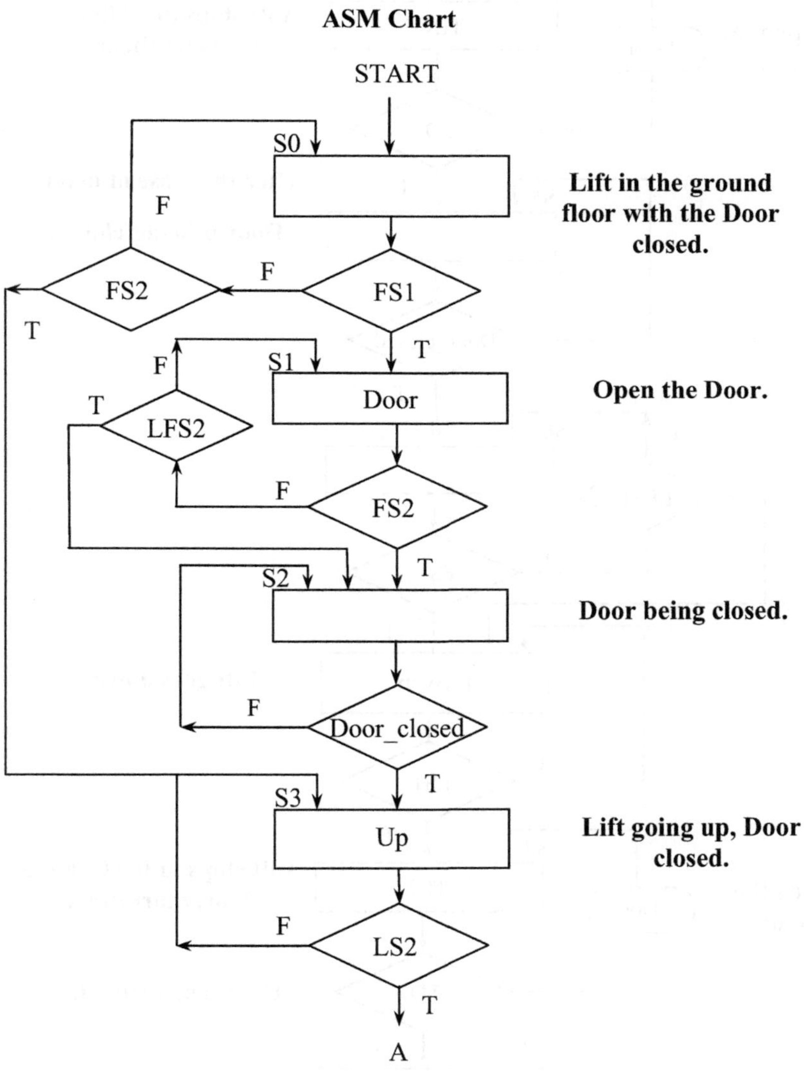

Fig. A7.1b ASM chart of the elevator controller (Continued)

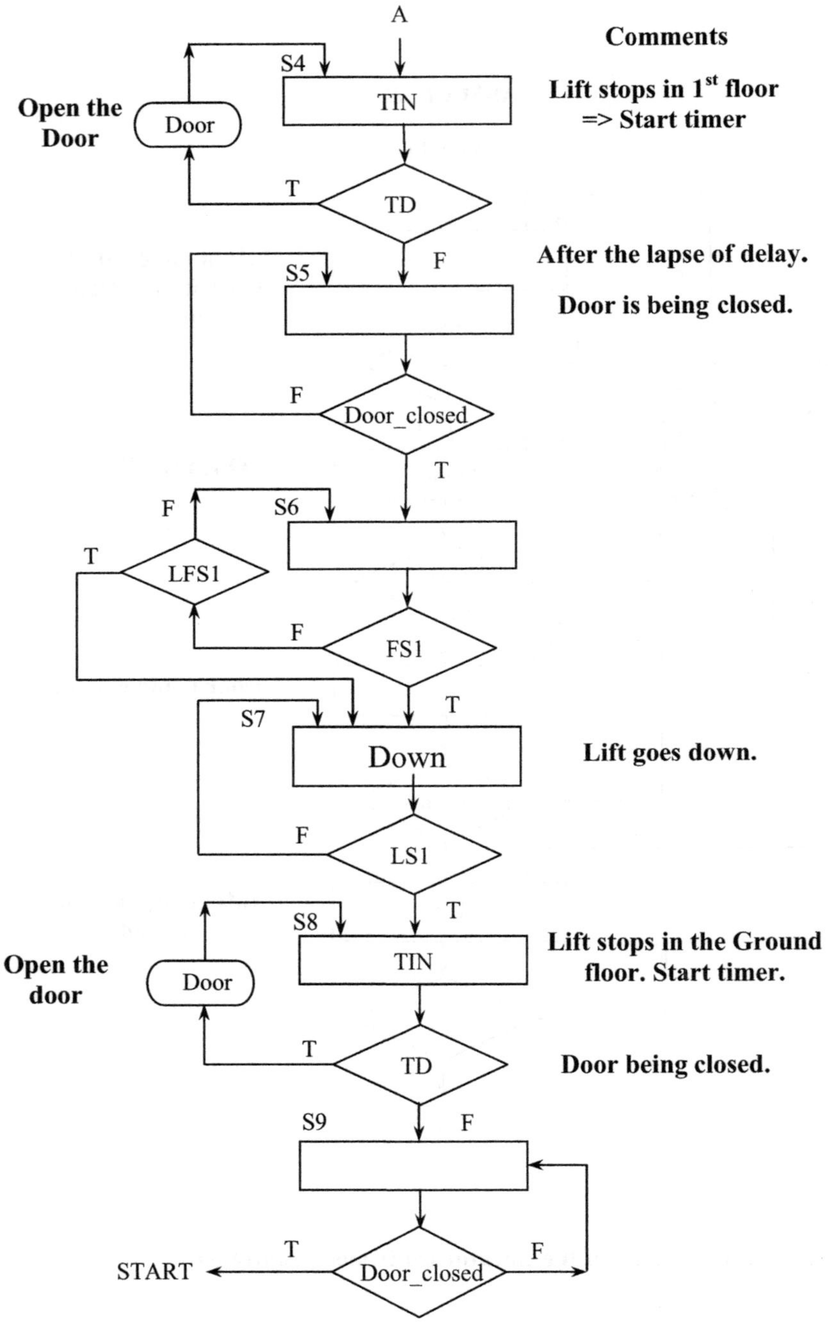

Fig. A7.1b ASM chart of the elevator controller

7.8 The state graph of a pattern sequence detector is shown in Figure A7.2. In this detector, the output becomes "1" and remains as "1" thereafter when at least two 0's and at least two 1's have occurred as inputs regardless of the order of occurrence. Code it in Verilog and present the RTL view and important parameters of the synthesis results.

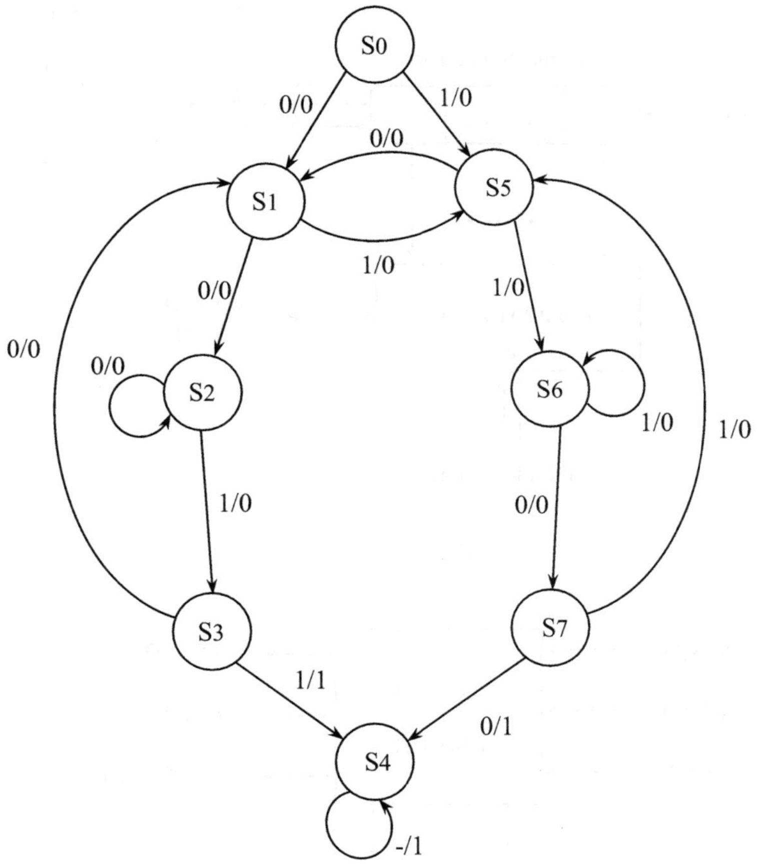

Fig. A7.2 Pattern sequence detector

7.9 Design a combination lock using an ASM chart. The following sequence is used for the operation of the lock:
A binary code of fixed length is used to open the lock. Each bit of the binary code is set by the use of a switch and entered serially by pressing a READ switch. After entering the required bits, an OPEN switch is pressed. The lock opens if the code matches with the predetermined code. An ERROR indication is on when the code entered is wrong or the user tries to enter data other than the fixed length of bits. The lock can be realized using an "M" bit shift register and a comparator as shown. Combination lock size

is "M" bits fixed length. Basic architecture of the combination lock is shown in Figure A7.3. Write Verilog code and present the RTL view and important parameters of the synthesis results.

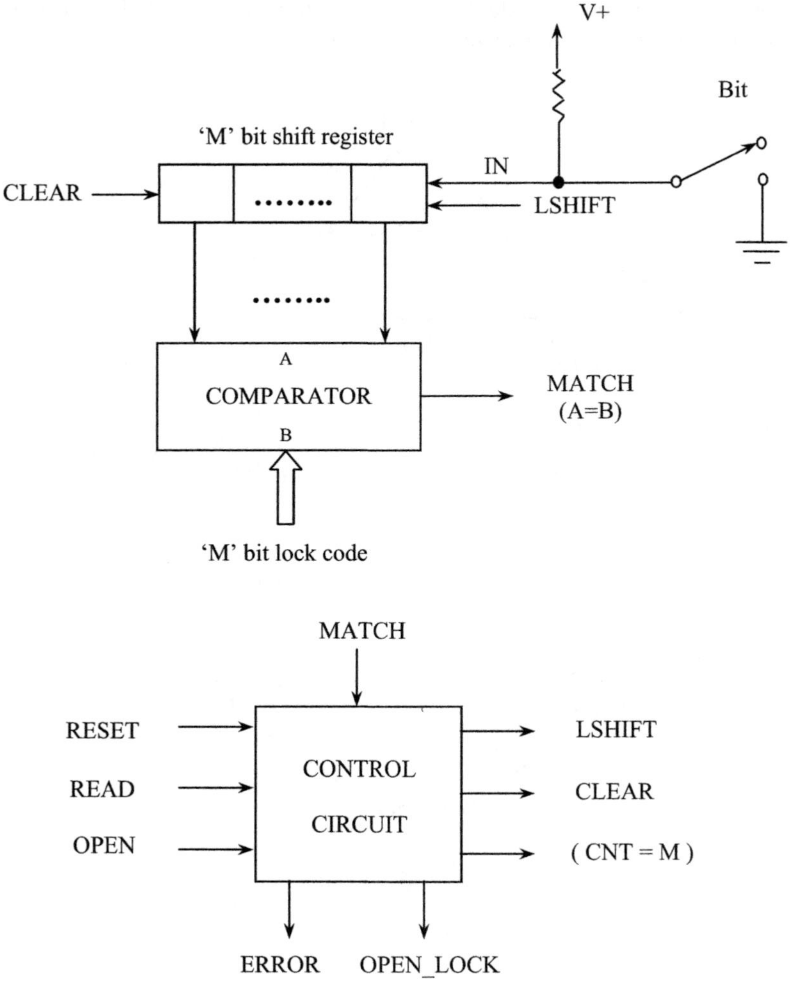

Fig. A7.3 Serial combination lock

7.10 Design a controller for a vending machine, which dispenses up to ten different types of items such as canned sweet drinks, fruit/vegetable drinks, biscuits, chocolates, chips, peanuts, etc. A separate mechanism that accepts coins from the user may be assumed. It verifies the insertion of correct coin. Otherwise, it ejects the coin. A 4-bit push-button BCD switch serves as the input selection for the type of drinks/snacks the user desires.

Sequence of operation:

1. Wait for RDY lamp to switch ON.
2. Set BCD switch to the desired item.
3. Insert the correct coin and collect the desired can/snack.

A schematic circuit diagram of the vending system is shown in Figure A7.4. Develop an ASM chart and realize the vending machine controller using Verilog. Your code must be RTL compliant. Present the RTL view and important parameters of the synthesis results.

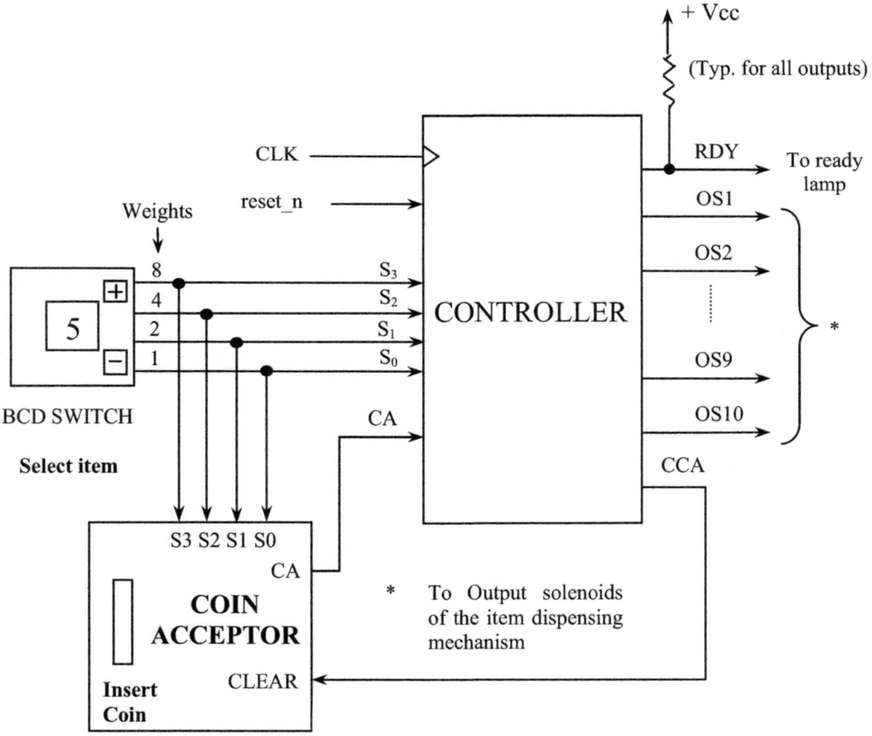

Fig. A7.4 Vending machine controller

Chapter 8

Place and Route

In the previous chapters, we learnt how to simulate and synthesize our designs using Modelsim and Synplify tools respectively. Synplify produces an ".edf" output file of our design, which is input to the next tool called the place and route. This tool generates ".bit" file that may be downloaded into FPGA mounted on a printed circuit board designed and fabricated as per the system requirements of our design. We will be using Xilinx place and route tool [20] in order to realize our designs on FPGAs.

8.1 Xilinx Place and Route

There are two versions in the Xilinx place and route tool: one is the design manager, which is an older version the user may have and the other is the navigator. Both these tools may be invoked from the Synplify tool or by clicking project navigator icon on the desktop. In the recent versions, we do not get the design manager but only the project navigator. We will, therefore, learn both the tools, which serve the same purpose of generating ".bit" files. We will first learn the design manager for clear understanding of the tool. For the present treatment, we will assume that our design is "seq_ckts.v", from which the Synplify tool created the corresponding EDIF file, "seq_ckts.edf". In the Synplify "options" menu, go to Xilinx, where we have three options:

Design manager
Project navigator
Floor planner

Click on the "Design Manager" to open the Xilinx window as shown in Figure 8.1. Another window called new version also opens along with the main window. The input for the Xilinx P&R tool is the ".edf" file of our design. Since we have opened it from the Synplify window, it has automatically taken the relevant file, "seq_ckts.edf ". The tool has created a separate version and revision for the design in order to avoid overwriting inadvertently. Instead of this, we can also create a new version/revision or new design, should we so desire. The FPGA device we selected in Synplify tool is automatically mapped by the place and route tool. We can change to any other device if we wish to do so before running the place and route tool.

Choose "Custom" in "constraints file" column using the drop down menu. Another window called "constraints file" opens as shown in Figure 8.2. Using browse

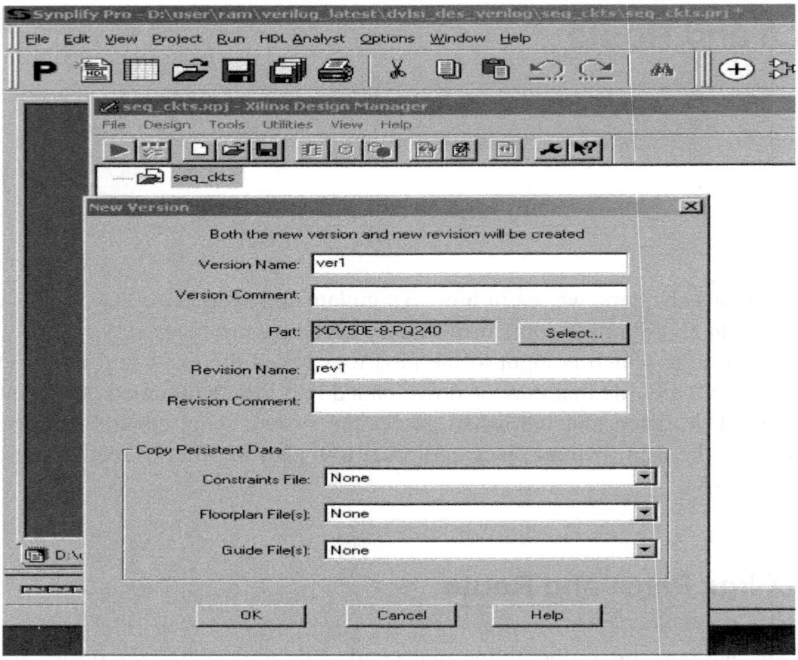

Fig. 8.1 Xilinx design manager window

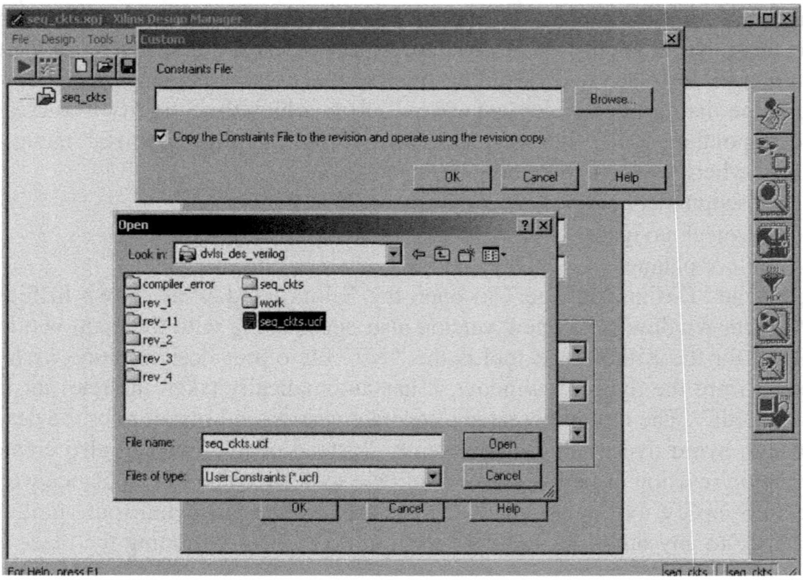

Fig. 8.2 Constraints file window

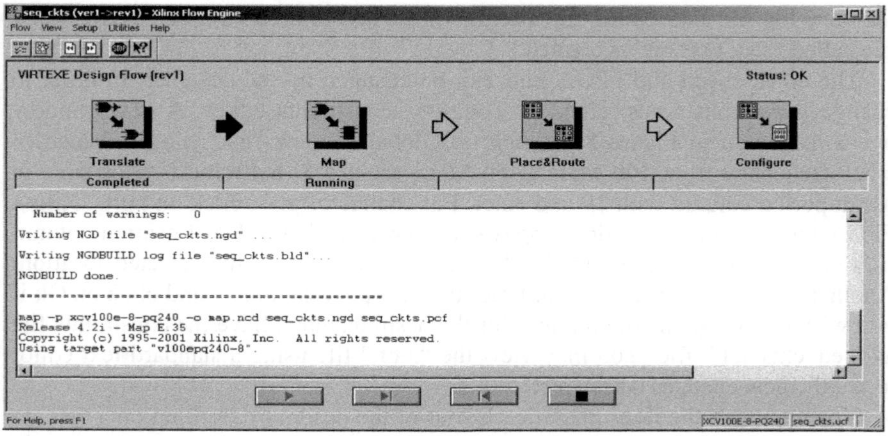

Fig. 8. 3 Xilinx flow engine

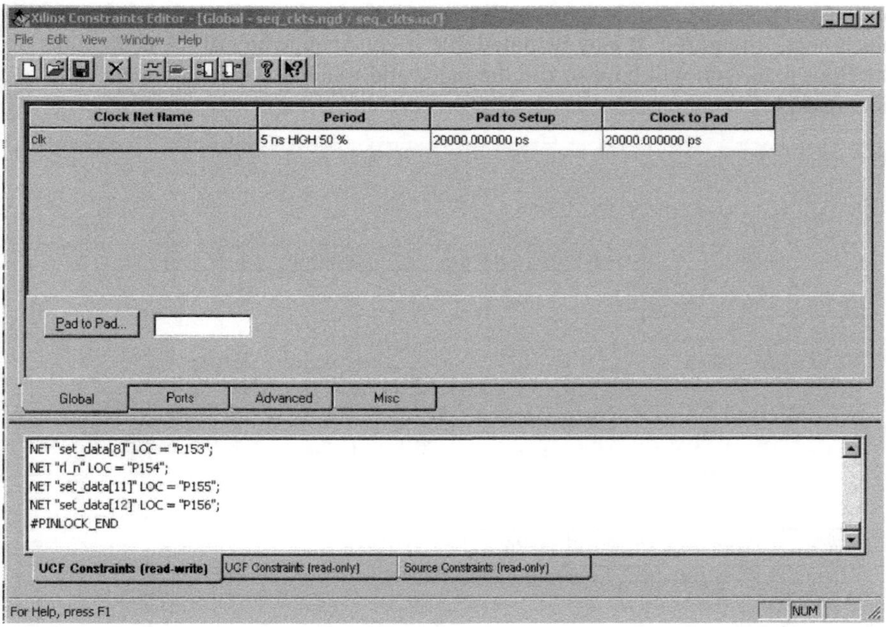

Fig. 8.4 Clock constraints window

button, select the relevant "seq_ckts.ucf" file. Click on open and "OK" twice. In order to start the implementation of the place and route, click on the dark arrow on the top left. A window called Xilinx flow engine opens as shown in Figure 8.3. The progress of translation, mapping, place and routing, and configuration are displayed in this window. The implementation details may be viewed by clicking on the view log file and reports file after the place and route is completed. An output file called "seq_ckts.bit" is created.

The clock speed and FPGA pins can be changed by the designer. In order to change these constraints, click on Tools => constraints editor. A new window opens as shown in Figure 8.4. Click on Global to view "clk". For a change of clock frequency from 100 MHz to 50 MHz, replace 5 (half clock period time in ns) in period column with 10 and enter. For change of pins, click on ports. Xilinx FPGA pin constraints window appears as shown in Figure 8.5. Location column lists all the pins of signals in the design. Double click on the selected location. Another window opens as shown. Enter the new pin assignment and click on Ok if you wish to change the pin number for the desired signal. Save the constraints file as "seq_ckts.ucf " file. You may view the ".ucf " file using a standard text editor to see the new changes take effect.

You can open the floor plan, i.e., layout of the design by clicking on Tools => Floorplanner in the main menu shown in Figure 8.1. The floor plan is shown in Figure 8.6. It shows the placement of various components (primitive cells) of our design and the pin connections. If we wish to change the component locations and pin connections, we can do so towards the end of a project. Use help to change the placement, if required. It may be noted that if you change any of the constraints or the floor plan, you will have to run the place and route as well as the back annotation again.

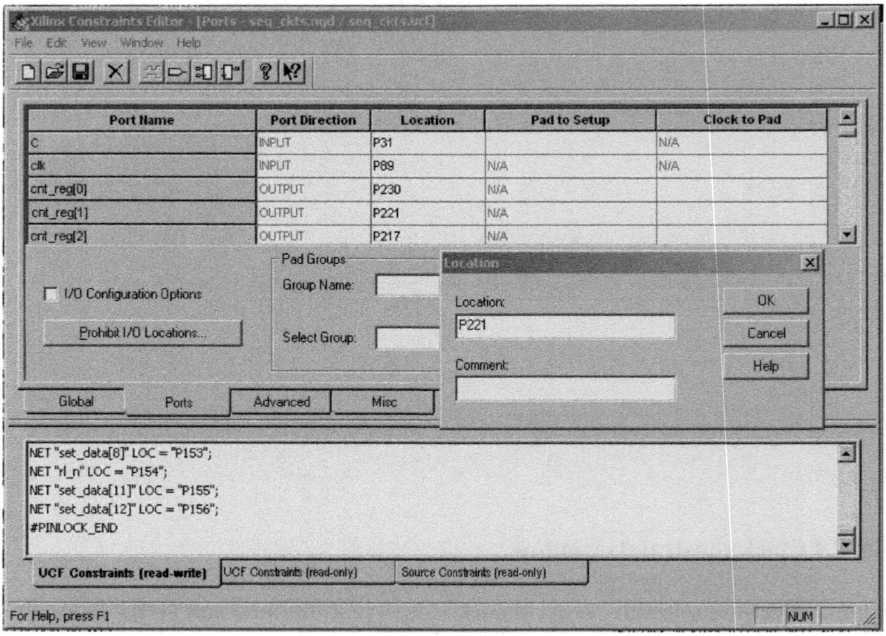

Fig. 8.5 Xilinx FPGA pin constraints window

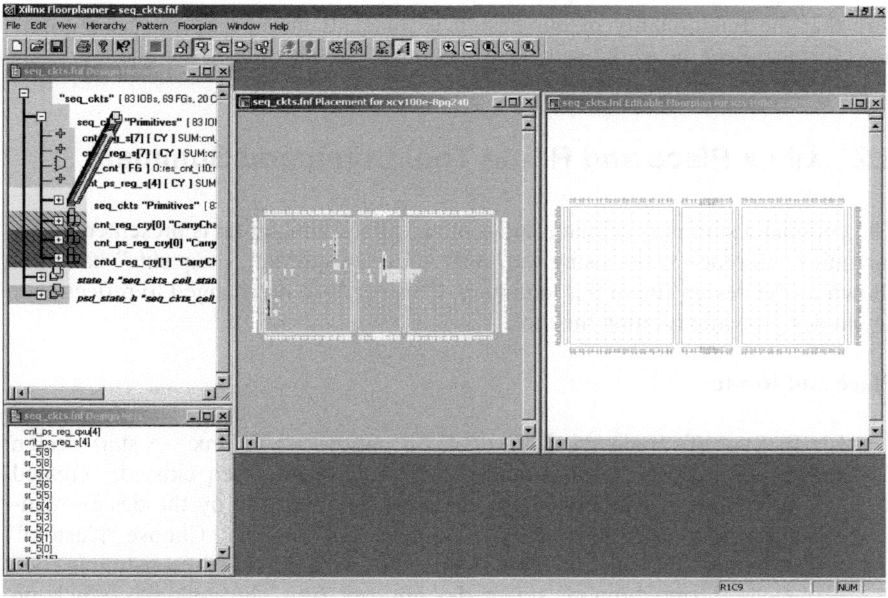

Fig. 8. 6 Xilinx floor plan

We can view the log files by clicking on Utilities => report browser menu. Towards the end of the log file, we can see that the tool has created "seq_ckts.ncd" file for use in the back annotation process. It may be noted that the timing report of Xilinx P&R tool gives more accurate results than the synthesis timing report, which we got earlier using Synplify tool. The P&R tool also reports errors, if any, timing details, the maximum frequency of operation, the gate count, etc.

The above treatment of using the tool is summarized in the next section. These are covered in steps starting from 1 to 4. The next two steps, 5 and 6, are for generating back annotated file, "seq_ckts_banno.v, which are self-explanatory. The back annotated file must be simulated using Modelsim in order to make sure that our design is still working at the maximum frequency of operation reported by the P&R tool. These are covered in steps 7 to 9. It may be noted that after back annotation, the actual gate delays take effect. "Vlog" is a command in the ModelSim tool in order to compile the test bench, which includes the back annotated design file, along with the primitive cell library (such as LUTs, MUX, etc.) used in the FPGA. Make sure that the three files mentioned in step 7 are present in the current working directory in ModelSim. Check it by using the "dir" command. When you execute the "Vlog" command, it compiles the test bench, the back annotated design file and the relevant library modules.

The next step 10 is to create a new work directory and load the back annotated design. The waveforms of the back annotated design can be viewed by following step 11. Analyze these waveforms as we had done before in Chapter 6 and make sure that the functionality of our design is intact. It may be noted that the new waveforms reflect the actual gate delays, being the result of back annotation.

8.2 Xilinx Place and Route Tool Command Summary

The command summary of place and route and back annotation is presented below for quick reference while using the tool. As an illustration, "seq_ckts" has been shown as the design file in this summary. This needs to be changed to reflect your actual design while running the tool.

Place and Route

1. In Synplify main window, click on options => Xilinx => start design manager. Xilinx design manager window opens. "seq_ckts.edf" created by synthesis is automatically taken as the input file by the design manager. Also another window called new version opens. Choose "Custom" in "constraints file" column and another window called "constraints file" opens. Using browse, select the relevant (seq_ckts.ucf) file. Click on open and "Ok" twice. If everything is ok in the Xilinx design manager, then the revision number, say, rev1 will be displayed.
2. Click on the dark arrow on the top left to start the implementation of the place and route. After it is completed, click on the view log file and reports file to get the implementation details. If you are done, click OK to dismiss the implement status window. An output file called "seq_ckts.bit" is created. This file is downloaded into the FPGA housed on the target circuit board while checking your design on the hardware later on. This file is what is supplied by the IP core developer along with the documentation.
3. To change the constraints such as clock speed and pins, click on Tools => constraints editor. A new window opens. Click on Global to view "clk". For change of clock from 100 MHz to 50 MHz, replace 5 in period column with 10 and enter. For change of pins, click on ports and in location column corresponding to the desired signal in the design. Double click on the selected location. Another window opens. Enter the new pin assignment and click on OK. Save the constraints file with an extension, ".ucf". You may view the ".ucf" file to see the new changes take effect.
4. Click on tools => Floorplanner to open the floor plan or layout of the design. Use help to change the placement, if required.

Note: If you change any of the constraints or the floor plan, you will have to run the place and route as well as the back annotation again.

Back Annotation

5. Open a DOS command window and move to the directory where your Xilinx P&R files are located. In the DOS prompt, key in the following command and execute to convert ".ngd" file to ".nga" file. Make sure that the ".ncd" and the ".pcf " files of the design are present in the current working directory. The command is:
ngdanno –o seq_ckts.nga –p seq_ckts.pcf seq_ckts.ncd
6. Convert the ".nga" file into back annotated ".v" file by executing the command:
ngd2ver seq_ckts.nga seq_ckts_banno.v
"seq_ckts.sdf " file is also created. "seq_ckts_banno.v" is the back annotated file of our original design, "seq_ckts.v".
7. Preferably, make a new directory and copy all the annotated files, "seq_ckts_banno.v", "seq_ckts.sdf" and the test bench, "seq_ckts_test.v", into the same.
8. Edit "seq_ckts_banno.v" file and comment out two statements as follows and save it:
Wire GSR ; // mglbl. GSR;
Wire GTS ; // mglbl. GTS;
Also in the test bench, change
` include seq_ckts.v
to
` include seq_ckts_banno.v
9. In ModelSim, compile the Xilinx library and back annotated files by the command:
vlog –y C:/Xilinx/Verilog/src/simprims+libext+. v seq_ckts_test.v
10. Create a new work directory and load the work file using the command:
vsim work.seq_ckts_test
When we execute this command, we can see the list of all the modules included in the test bench.
11. Use View => wave to get the waveform for analysis. Use View => Signals to open the signals window, in which click on "Add" and "Wave" and "Signals in Design" and, click on "Run-All" to study the displayed waveforms. Note that the new waveforms reflect the actual gate delays, being the result of back annotation. Make sure that all circuit functionalities of our design are preserved.

8.3 Place and Route and Back Annotation Using Xilinx Project Navigator

The project navigator tool of Xilinx ISE 6.1i/7.1i/8.2i serves the same purpose as the design manager. The project navigator can be opened either from Synplify tool as described earlier or by double clicking on the icon:

on the desktop. The navigator main window is shown in Figure 8.7. Within the main window, two more windows, "Sources in Project" and "Processes for Source", also open. In case they don't open, click on "View => Project Workspace" in the main menu. Various choices such as synthesis, implementation, programming (bit) file, etc. are shown in project work space in the two figures of 8.7. If these choices are not shown, click on "Project => Toggle Paths" to make other options visible. A step by step procedure for using this tool is presented as command summary.

The first step is for invoking the tool. The design can be implemented by following the second step. Look into errors and warnings, if any, and take the remedial action as you had done while using the Xilinx design manager earlier. Step 3 describes the method of generating the bit stream for configuring an FPGA. Floor planning is a process of choosing the best grouping and connectivity of logic in a

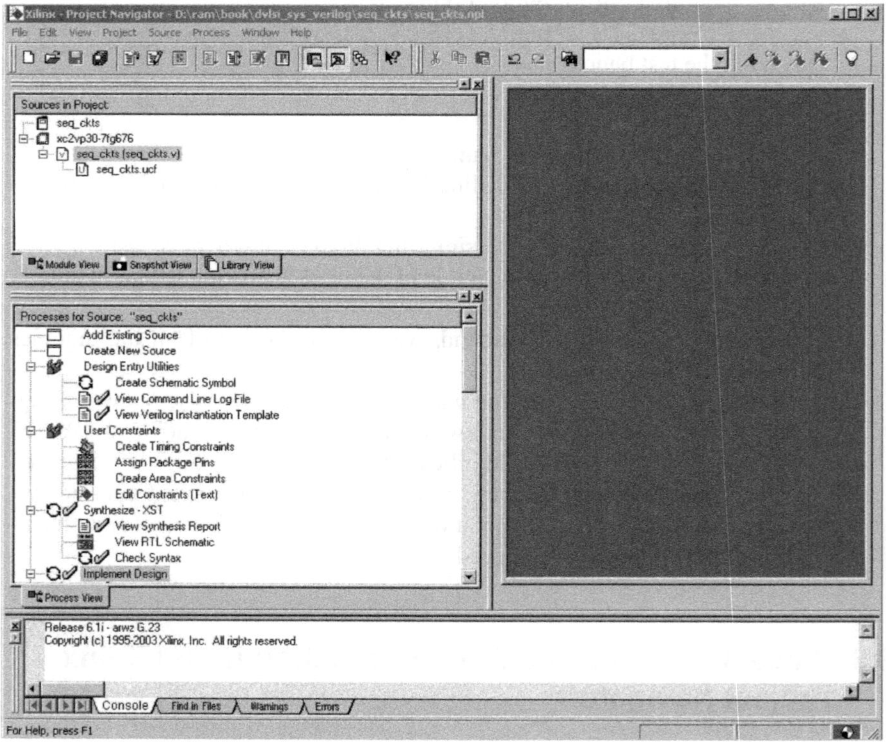

Fig. 8.7 Xilinx project navigator (Continued)

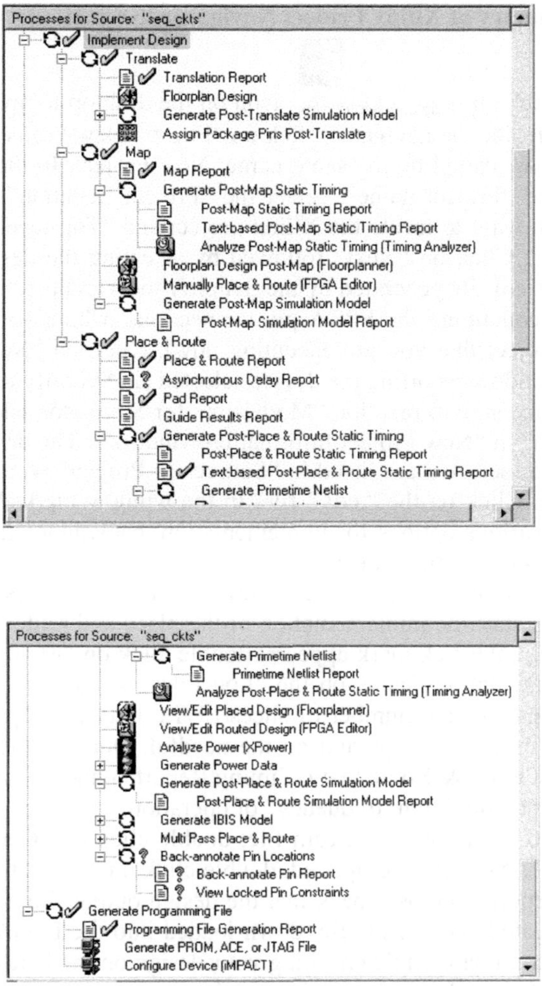

Fig. 8.7 Xilinx project navigator

design. It is also the process of manually placing blocks of logic in an FPGA, where the goal is to increase density, routability, or performance. Refer help facility for using this feature. Steps 4 to 9 cover the back annotation of "yourdesign". Substitute "yourdesign" with the actual design file name, "seq_ckts", for example. While executing step 9, select the frequency of operation so that it is less than or equal to the maximum frequency of operation reported in step 2. Otherwise, the results will not be correct and, in Modelsim, the tell-tale signals are usually "Hold time" or "Setup time" violation. The command summary of project navigator follows.

Command Summary of Xilinx Project Navigator ISE 6.1i/7.1i

1. Double click on Project Navigator.lnk icon on the desktop to open the navigator window. Inside the navigator, click on "File => New Project". This opens a new window called by the same name, New Project. In that window, type the desired "Project name", select the "Project Location", i.e., the folder where you want to locate the Xilinx project and "Top-level Module Type" as "EDIF". Click on "Next" followed by selecting the desired "Input Design" (the .edf file generated by the Synplify tool) and the "Constraints file" (.ucf file containing the FPGA pin connections, which you have keyed in for the project that you are executing now). Click on "Next". This opens another window reporting the device selected (in Synplify tool) and "EDIF" opted. Make sure to mention "Modelsim" for Simulator selection and click on "Finish" in "New Project Information window". The device, ".edf " and ".ucf " files selected are listed in "Sources in Project" window in the main Navigator. Click on the ".edf " file. In the window marked "Processes for Source", various options for implementation, back annotation and creation of bit stream (.bit) files appear.
2. Double click on the "Implement Design" on the "Processes for Source" window to start the implementation of the place and route. After it is completed, i.e., after tick mark appears, double click on "Map Report" to open the implementation details on the "Log Files" window on the right. It reports the gate count, number of slices/LUTs used in the project, etc. Also look into errors/warnings and do the needful. Double click on the "Text-based Post Place & Route Static Timing Report" under "Place & Route" for viewing the "Maximum frequency" of operation.
3. In order to create the bit stream file, double click on "Generate Programming File". After it is completed, double click on "Programming File Generation Report" to cross-check that the design output file with ".bit" extension is created. This is the file down loaded into the FPGA housed on the target circuit board while checking your design on the hardware later on. If security of your design is required, right click on "Generate Programming File => Properties", which opens the window named "Process Properties". Click on "Read back Options". In the drop down menu of "Security – Value", select "Disable Read back and Reconfiguration".

Back annotation

4. In the navigator, right click on "Post Place & Route Simulation Model" followed by "Properties", which opens a window called "Process Properties". Under "General Simulation Model Properties", select "Modelsim_Verilog" in "Value" field opposite "Simulation Model Target". Click on "OK". Double click on "Post Place & Route Simulation Model" to generate your design_timesim.v, yourdesign_timesim.sdf files. These are the back anno-

tated files, which you need to simulate again in order to check whether your design is working at the maximum frequency reported in Sl. No. 2 or not.

In case you have ISE6.1 version or ISE7.1 version of the navigator, right click on "Generate Post Place & Route Simulation Model" to open the window "Process Properties/Simulation Model Properties". Select "ModelSim SE (Verilog)" in "Value" field opposite "Simulation Model Target". Click on "OK". Double click on "Generate Post Place & Route Simulation Model" to generate yourdesign_timesim.v and yourdesign_timesim.sdf back annotated files. Watch out for changes from version to version and apply accordingly.

5. Copy yourdesign_timesim.v, yourdesign_timesim.sdf and c:/Xilinx/verilog/src/glbl.v into the folder where your design files are residing. In the test bench, change `include "yourdesign.v" to `include "yourdesign_timesim.v" and `define clkperiodby2 10 to change the operating frequency equal to or less than the maximum frequency reported in Sl. No. 2. The operating frequency may be obtained by computing: (500/ clkperiodby2) MHz. Change 10 accordingly in the statement:
`define clkperiodby2 10.
6. Open Modelsim, File => New Project. Enter the Project Name and Location of Project.
7. Click on Design => Compile. Select the files yourdesign_test.v and glbl.v. Click on "Default Options" in the "Compile HDL Source Files" dialog box followed by "Verilog" option. Also, click on the button "Library Search..." and specify the library directory as
C:/Xilinx/verilog/src/simprims. Click on "Open".
Then click on the button "Extension..." and select ".v". Click on "OK" and then Compile. Click on "Done".
8. Click on Design => Load Design or Simulate as the case may be. Click on "glbl" and then click on "Add". Do the same for "yourdesign_test". Then click on "Load".
9. Open the waveform, run simulation and analyze the timing diagrams as you have done before to ensure that your design is working perfectly.

Notes:

1. If you change any of the constraints or the floor plan, you will have to run the place and route, back annotation, and bit stream generation again. Whenever you change the design file or the constraint file, run the Synplify and then the navigator tools again. In order to run "Implement Design" etc. again, right click on the same and click "Rerun" or "Rerun All".
2. Double click on "Floor Plan Design" in Implement Design/Translate to relocate pins, components, etc., if you wish to improve the timing. Seek help if you wish to learn more about these features.
3. Synthesis, place and route results and back annotated waveforms are presented for various designs from Chapter 9 onwards.

If you desire to start with source file, instead of EDF file, the first two steps will be as follows:

1. Creation of new project is the same as done before. After the project is created, click on "Project => Add Source". Another window named "Add Existing Sources" opens. In that window, choose the desired HDL source file, say, "traffic_controller.v". In order to include user constraints file, click on "Project => Add Source" again. In the window "Add Existing Sources" that opens, select the desired UCF file, say, "tc_12seq_rt.ucf". UCF may be created by using any standard text editor such as the word pad, "Vi" etc. The above features are shown in Figure 8.8. The ".ucf" can be displayed as shown in the figure by double clicking on "Edit Constraints (Text)" in the "Processes for Source" window.

   Note: If the same .v or .ucf file is selected again, the tool reports error. To come out of this error, click on "Project => Cleanup Projects File". Also, if pins specified in ".ucf" file does not agree with the device selected, then the tool reports error. Therefore, select the device/pins correctly.
2. The design can be synthesized by double clicking on "Synthesize – XST". After the synthesis is complete, it is ticked along with "View Synthesis Report" and "Check Syntax". Double click on "View Synthesis Report", which opens the said report on the right. Browsing it, we get the following details:
   1. Synthesis options summary
   2. HDL compilation
   3. HDL analysis
   4. HDL synthesis
      4.1 HDL synthesis report
   5. Advanced HDL synthesis
   6. Low level synthesis
   7. Final report
      7.1 Device utilization summary
      7.2 Timing report

   In synthesis options summary, the target device used, say, xcv800-4-hq240, and whether the design conforms to RTL (RTL Output : Yes) are reported. The device utilization summary reports as follows:

Selected device :	v800hq240-4			
Number of slices:	139	out of	9408	1%
Number of slice Flip Flops:	81	out of	18816	0%
Number of 4 input LUTs:	252	out of	18816	1%
Number of bonded IOBs:	18	out of	170	10%
Number of GCLKs:	1	out of	4	25%

Under timing report, a timing summary presents the maximum frequency of operation possible with the selected FPGA device. For example, for the design "traffic_controller", which will be presented in the penultimate chapter, the following is the timing summary reported by the tool.

Timing Summary :

Speed grade: –4 minimum period: 15.629 ns (maximum frequency: 63.984 MHz)

Minimum input arrival time before clock: 12.196 ns
Maximum output required time after clock: 8.938 ns
Maximum combinational path delay: No path found

Double clicking on "View RTL Schematic" under the menu "Synthesize – XST", a block diagram of the design is displayed in a new window called "Xilinx ECS". The RTL circuit diagram of the design is displayed by double clicking on the block diagram.

Note: Other steps from 2 to 9 are the same as that presented in Command summary of Xilinx project navigator.

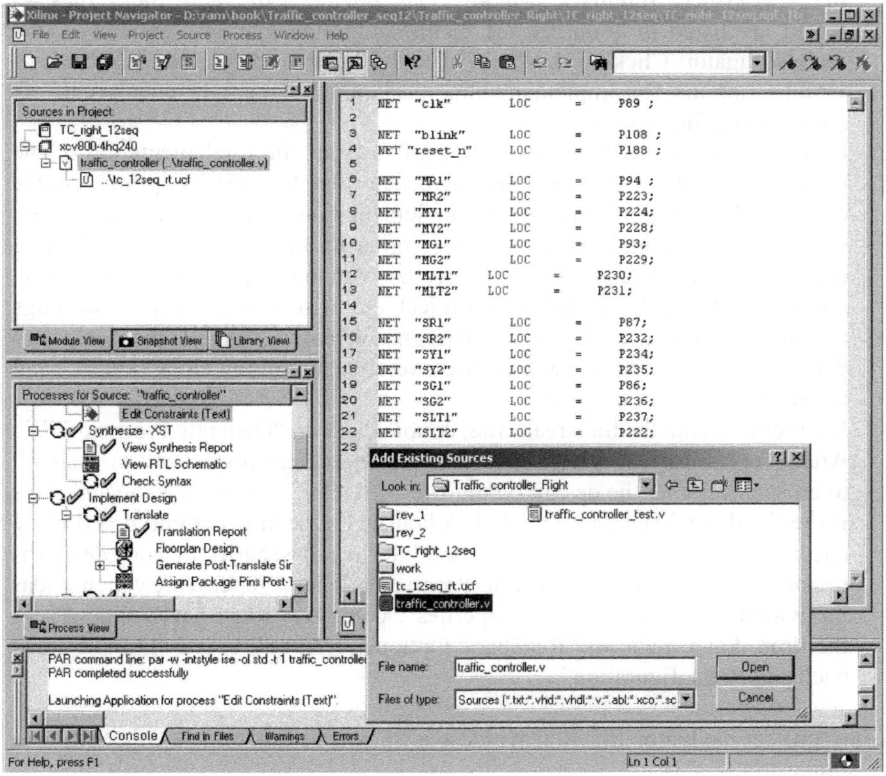

Fig. 8.8 Xilinx project navigator – Addition of ".v" and ".ucf" files

Command Summary of Xilinx Project Navigator ISE 8.2i

1. Double click on 

icon on the desk top to open the navigator window. Inside the navigator, click on "File => New Project". This opens a new window called by the name, "New Project Wizard". In that window, type the desired "Project name", select the "Project Location", i.e., the folder where you want to locate the Xilinx project and "Top-level Source Type" as "EDIF". Click on "Next" followed by selecting the desired "Input Design" (the .edf file generated by the Synplify tool) and the "Constraints file" (.ucf file containing the FPGA pin connections, which you have keyed in for the project that you are executing now). Click on "Next". This opens another window reporting the device (properties) selected (in Synplify tool) and "EDIF" opted. You may change the device if you wish. Make sure to mention "Modelsim XE or SE" as the case may be for "Simulator" selection and click on "Finish" in "New Project Summary" window. The device ".edf" and ".ucf" files selected are listed in "Sources for Synthesis/Implementation" window in the main navigator. Click on the ".edf" file. In the window marked "Processes", various options for implementation, back annotation, and creation of bit stream (.bit) files appear.
2. Double click on the "Implement Design" to start the implementation of the place and route. After it is completed, i.e., after tick marks appear, double click on "Map Report" to open the implementation details on the "Log Files" window on the right. You can also get similar information by double-clicking on "View Design Summary" in "Processes" window. It reports the gate count, number of slices/LUTs used in the project, etc. Also look into errors/warnings and do the needful. Click on the "Design Overview/Timing Constraints" under "FPGA Design Summary" for viewing the "Maximum clock period" (and hence frequency) of operation.
3. In order to create the bit stream file, double click on "Generate Program File". After it is completed, double click on "Programming File Generation Report" to cross-check that the design output file with ".bit" extension is created. This is the file down loaded into the FPGA housed on the target circuit board while checking your design on the hardware later on. If security of your design is required, right click on "Generate Program File => Properties", which opens the window named "Process Properties". Click on "Read back Options". In the drop down menu of "Readback Options – Value", select "Disable Read back and Reconfiguration".

Xilinx project navigator ISE 8.2i window is shown in Figure 8.9. Other features are similar to ISE 7.1i.

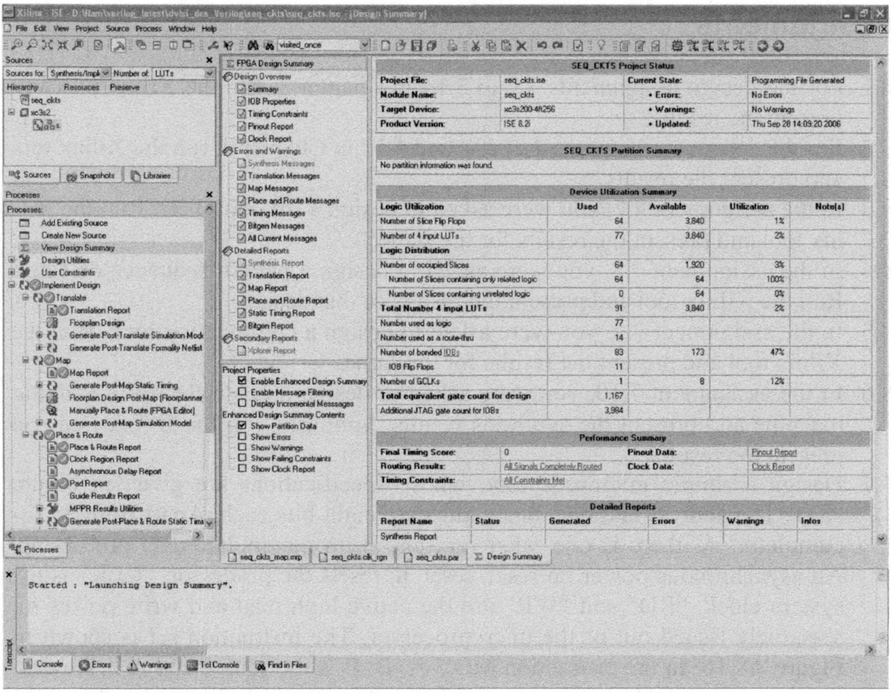

Fig. 8.9 Xilinx project navigator – ISE 8.2i

Summary

This chapter covered the place and route tool of Xilinx, widely used in industries. The salient features of Xilinx P&R tool are creation of ".bit" file from EDIF file created by the Synplify tool, change of specification of user constraints such as clock speed and FPGA pins, remapping of the target FPGA device if desired, back annotation and floor planning. The back annotated file is simulated again in Modelsim to ensure that the design is working correctly at the maximum frequency reported by the place and route tool. Report file generated by the P&R tool gives the maximum frequency of operation possible as well as the gate count (chip complexity) for the design. Command summaries of the Xilinx P&R design manager tool and navigator tools were furnished as a ready reckoner. With the three industry standard tools we have learnt so far, namely, the simulation using Modelsim, synthesis using Synplify, and place and route using Xilinx, we are well equipped to undertake Digital VLSI Systems design, be it based on FPGA or ASIC. To start with, we will learn how to design memories in the next chapter.

Assignments

8.1 For the design assignments 3.2 to 3.14 in Chapter 3, run the Xilinx tool and present the results.

8.2 For the design assignments 4.2, 4.4, and 4.12 in Chapter 4, run the Xilinx tool and present the results.

8.3 In the assignment 7.7, you were asked to design a lift controller. Run the Xilinx tool and present the results for the same.

8.4 In the assignment 7.8, you were asked to design a pattern sequence detector. Run the Xilinx tool and present the results of that design.

8.5 In the assignment 7.9, you were asked to design a combination lock. Run the Xilinx tool and present the results for the same.

8.6 In the assignment 7.10, you were asked to design a controller for a vending machine and present the synthesis results. Similarly, run the Xilinx tool and present the results.

8.7 Design a simple microprocessor, whose specifications are given in Figure A8.1. The address and data bus widths are eight bits each. Apart from an accumulator, A, there is one other register, B, to manipulate data processing. An asynchronous power on reset, reset_n, resets the processor. "CLK" is the system clock, "RD" and "WR" are the active high read and write pulses respectively issued out of the microprocessor. The instruction set is shown in Figure A8.1b. In the instruction MOV A, B; B is the source and A is the destination as per INTEL format. The instruction set is followed by the ASM chart in A8.1c. The basic read/write timing diagram is shown in Figure A8.1d. Implement the design in Verilog and present the Synplify and Xilinx results.

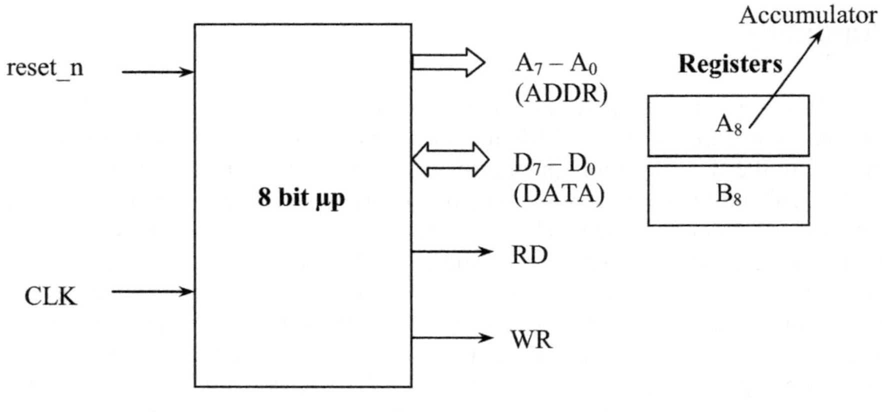

a A simple microprocessor architecture

Fig. A8.1 A simple microprocessor (Continued)

	Instruction	Op Code (1st Byte)	2nd Byte
	MOV A, B	0	
	MOV B, A	1	
	MVI A, #data	2	data
Load A from memory	LDA addr	3	addr
Store A into memory	STA addr	4	addr
Bit-wise logic	AND A, B	5	
Bit-wise logic	OR A, B	6	
Bit-wise logic	NOT A	7	
Bit-wise logic	XOR A, B	8	
A<= A+B	ADD A, B	9	
A<=A-B	SUB A, B	10	
	JMP addr	11	addr
	JZ addr	12	addr

b Instruction set

Fig. A8.1 A simple microprocessor (Continued)

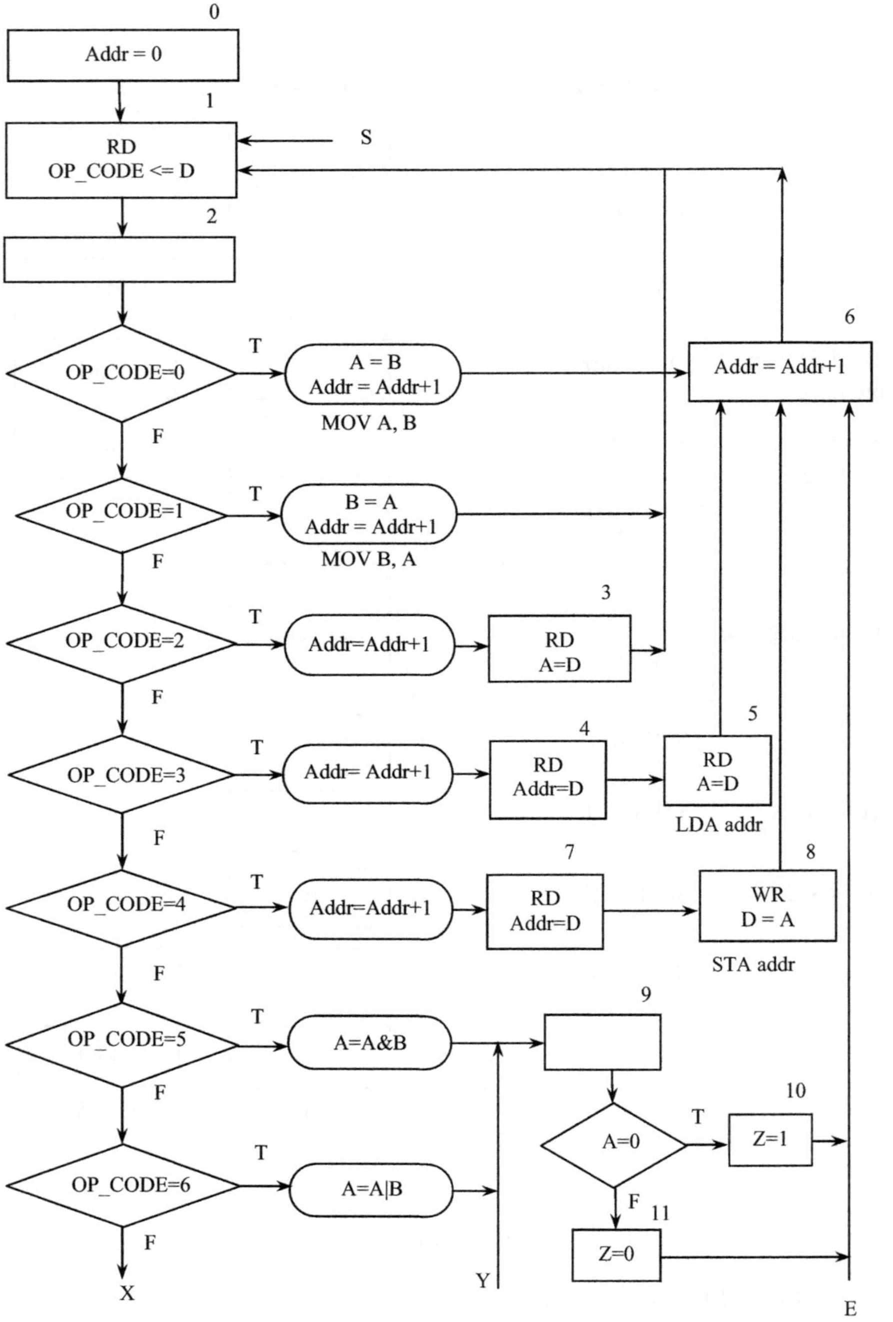

c ASM Chart

Fig. A8.1 A simple microprocessor (Continued)

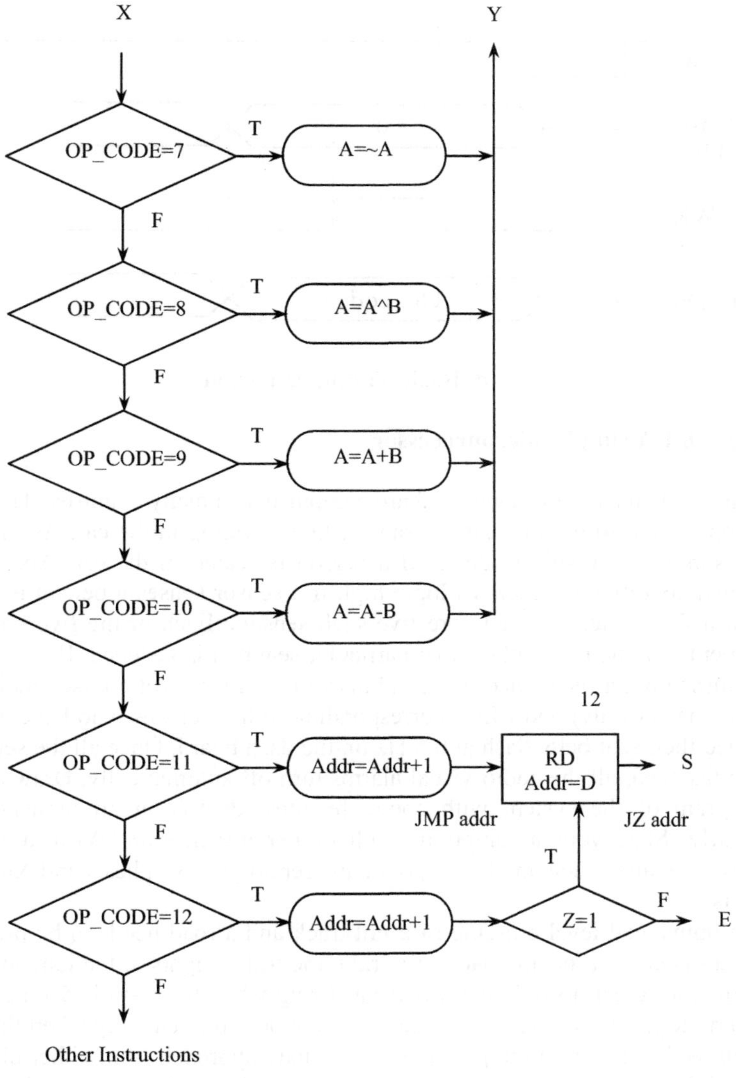

c ASM Chart

Fig. A8.1 A simple microprocessor (Continued)

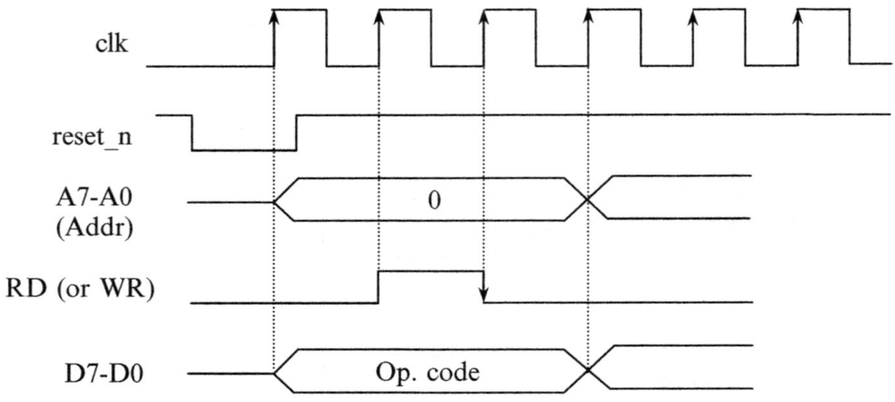

d Basic Timing Diagram

Fig. A8.1 A simple microprocessor

8.8 Car seat belts are life savers and are compulsory in many countries. However, people often forget to wear the same while traveling in the car. Assume that sensors are available to detect if a person is seated in the car. Appropriate signal conditioner generates logic high if a sensor senses a person is seated. For a five seater car, there are five such sensors. Each of the five seat belts generates logic high only if that particular seat belt is fastened. If the ignition is turned on, a piezo-electric sound alarm is triggered (of course, audible inside the car only) and LEDs corresponding to the persons who have not fastened their seat belts flash at 0.5 Hz on the dash board. Once all the seat belts are fastened, all the audio-visual alarms turn off automatically. Draw a block diagram of the system with above features identifying all inputs/outputs clearly. State your assumptions with proper justification. Write a Verilog code to realize such an alarm system and report your synthesis and Xilinx results.

8.9 An unmanned level crossing of a rail track and a road needs to be protected by a system. At the approach of a train, the traffic lights at the junction must turn from green to red, and a bell must ring when the train is 5 miles away from the junction. The railway gates must close automatically when the same train is 4 miles from the junction. Trains may approach from either direction on different tracks. Red traffic lights are different for the two directions, whereas the green light and the bell are common. Inductive proximity switches buried under each of the rail tracks at 5 miles and 4 miles from either side of the junction sense the train above and send logic high signal to the controller at the junction. The red lights change back to green, bell switches off and gates open when the end of the train is 5 miles or more away from the junction. Clearly state your assumptions and describe your design. Realize the controller using Verilog and present your synthesis and Xilinx results. Also present the RTL view.

8.10 Draw an ASM chart for the traffic light controller shown in Figure A8.2. The timing starts if a signal, STRT_TMR, is asserted for a clock period. When the timing is complete, an output, TIME_OVER, is set. This is timed for 25 s or 5 s depending upon another input, T25_5N. If it is high, it is 25 s; otherwise 5 s. The 5 s timing is for delaying the switching on of green lights. Assume that the timer is available. MG and MR are main road green and red lights, whereas SG and SR are side road green and red lights respectively. There are inductive sensors buried under the side roads to sense traffic in the side roads. SENSOR is high if there is any vehicle on the side road. Only straight traffic and right turns are allowed. No free right. Once the main traffic is allowed, it must persist at least for 25 s. Realize the traffic light controller using Verilog and present your synthesis and Xilinx results.

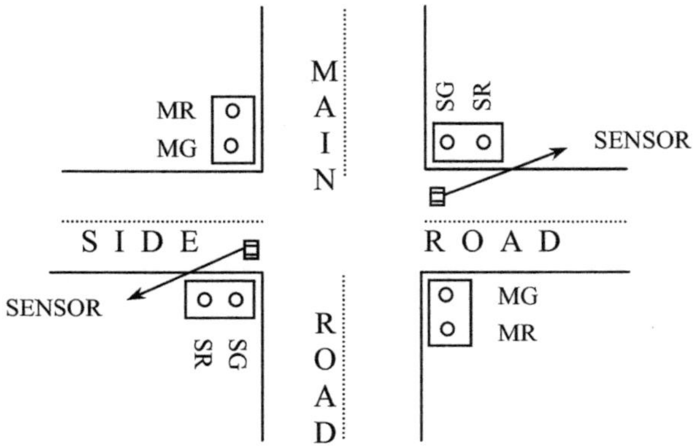

Fig. A8.2 Traffic light controller with sensors

Chapter 9

Design of Memories

Memory design is one of the challenging areas, where we need to take extreme care while designing systems for ASIC and FPGA implementations. We know that there are basically two types of memories: ROM and RAM. The organization of the memory would depend upon a particular application. The conventional memories that we use either have single address and single data output for ROM or separate read and write addresses and data bus for a dual port RAM. In real time systems such as video processing systems, these conventional memories may not be of help since the typical applications often demand access to two memory locations simultaneously, write word-wise and read column-wise, etc. These requirements call for tailor made memory design solutions for these applications. These application-specific requirements, arising mainly due to the need for efficient implementation of computationally intensive algorithms, are addressed in this chapter.

9.1 On-chip Dual Address ROM Design

Consider an application, where the system ROM is required to supply two different data concurrently. This need can be fulfilled if the ROM is designed to have dual address and dual data output. The application further needs synchronous and pipelined operation in order to achieve fast processing speeds. Therefore, we need a clock input. Two addresses are provided so that we may fetch two locations concurrently. The number of bits in the address bus will decide the number of locations in the ROM. Let us say that we provide 3 bits for the address and, therefore, the ROM contains eight locations. Further, let us take data width as 64 bits. The data that will be read from the ROM table are 'dout1' and 'dout2' corresponding to the addresses, 'addr1' and 'addr2' respectively as shown in Figure 9.1. It should be noted that the ROM content is only a single block of 8 × 64 bits even though two addresses are involved in the design. This requirement arises in one of the design applications, Discrete Cosine Transform and Quantization (DCTQ) processor used in JPEG, MPEG 1, MPEG 2, H.263 based still image/video compression codecs. We will cover this design application in depth in a later chapter. Other designs we will consider subsequently are single address ROM, dual redundant RAM, etc. These designs will also be used in the same application mentioned earlier. Let us see how the present dual address ROM design flows.

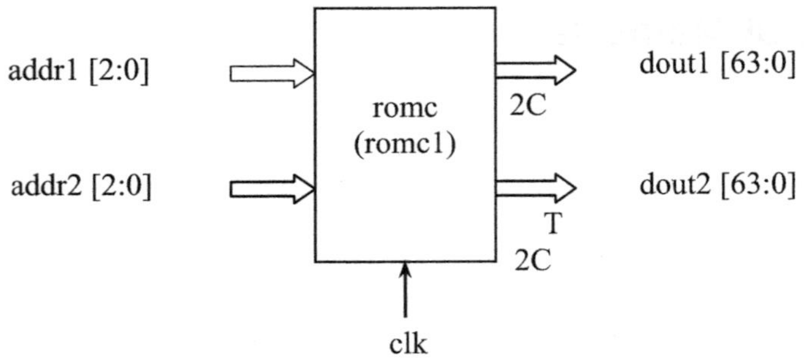

Fig. 9.1 On-chip dual address ROM design

In DCTQ application mentioned earlier, we need to compute cosine transform coefficients from a block of data and cosine terms (C) and its transpose (C^T). This can be realized by using two separate, single addressed ROMs for storing cosine values and the transpose of the cosine values. A better way in terms of chip area will be a single ROM with dual address since C and C^T contents are precisely the same. The transpose can be accessed by using the address 'addr2', while C can be accessed by using 'addr1'. Since the DCTQ computation requires both C and C^T simultaneously for processing different steps of the DCT algorithm, we need the dual address. It may be noted that this is not a conventional approach of realizing a ROM, which uses single address. The ROM requirements may be summarized as follows with minor modification:

- The ROM stores the cosine terms $2 \times C$ instead of C in order to improve accuracy.
- Two-stage pipelining for $2 \times C$ matrix of cosine terms to keep pace with dual RAM used in the DCTQ design.
- ROM size is 8×64 bits. Two locations, each of size 64 bits, can be accessed and output to the data bus, 'dout1' and 'dout2' simultaneously using the two addresses, 'addr1' and 'addr2' respectively.

Since we store twice the value of the required cosine terms in the ROM with a view to improve the accuracy of computation, we should not forget to divide the final DCTQ coefficients by two. In order to match the speed of dual RAM, also used in the DCTQ design, which outputs the block of data delayed by two clock cycles, we introduce two-stage pipelining in the dual ROM design.

9.1.1 Verilog Code for Dual Address ROM Design

The Verilog code can be realized either by using the register array, 'reg mem' or by the 'case' statements. We will use the 'case' statements in 'always' block as shown in the Verilog code_9.1 since double address is easy to handle in the case statement approach. To begin with, adequate explanation of how the ROM is organized is furnished as comments. This is followed by declaring the module, namely 'romc', listing and identifying the inputs and outputs of the design. 'dout1_next' and 'dout2_next' are declared as 'reg' since they appear as outputs in 'always' blocks. 'dout1_reg1' and 'dout2_reg1' are outputs in the first stage of pipeline. Similarly, 'dout1' and 'dout2' are outputs in the second stage of pipeline. 'loc0' through 'loc7' ROM data are realized using 'assign' statements and hence declared as 'wire'. The maximum value of signed data in an 8-bit fixed point representation is E7 H close to the maximum of FF H, which explains why we have multiplied C by two instead of any other integer. The next two are 'always' blocks using 'case' statements with 'addr1' and 'addr2' respectively to read one out of the eight locations. 'dout1_next' and 'dout2_next' are the outputs accessed using the dual address. These outputs are delayed by two clock cycles after the addresses are applied by the two pipeline stages. It may appear that 'assign' statements are redundant since the ROM table could be directly put into the case blocks. But that would have meant duplicating the same ROM data needing more data entry. Of course, we pay a small price in terms of additional combinational circuits using 'assign' statements. More details will be provided in the DCTQ design in the chapter on design applications.

Verilog_code_9.1

```
/*        ROMC Design.
This code is put into a file named "romc.v".
This is the code for a ROM that stores the cosine, 2 × C or 2 × C^T matrix.

Design incorporates two pipelining stages in order to keep pace with Dual RAM
used in the DCTQ design. ROM size is 8 × 64 bits. Two locations can be accessed
and output to the data bus, 'dout1' and 'dout2' simultaneously using the two ad-
dresses, 'addr1' and 'addr2' respectively. Each location stores eight numbers of
cosine terms. Thus, in eight such locations, a total of 64 terms are stored.
*/

module romc     (         clk,
                          addr1,
                          addr2,
                          dout1,
                          dout2
                );

        input                   clk ;          // Declare I/Os.
        input       [2:0]       addr1 ;
        input       [2:0]       addr2 ;
        output      [63:0]      dout1;
        output      [63:0]      dout2;

        reg         [63:0]      dout1_next ;// This is the next value that will
                                            // be loaded into dout1_reg1.
```

```
        reg        [63:0]      dout2_next ;
        reg        [63:0]      dout1_reg1 ;   // First pipeline registers.
        reg        [63:0]      dout2_reg1 ;
        reg        [63:0]      dout1;         // Second pipeline registers,
        reg        [63:0]      dout2;         // i.e., final outputs.

        wire       [63:0]      loc0 ;         // ROM data declared as nets.
        wire       [63:0]      loc1 ;
        wire       [63:0]      loc2 ;
        wire       [63:0]      loc3 ;
        wire       [63:0]      loc4 ;
        wire       [63:0]      loc5 ;
        wire       [63:0]      loc6 ;
        wire       [63:0]      loc7 ;

//2 × C or 2 × C^T ROM table organized as 8 × 64 bits.
    assign    loc0    =    64'h5B5B5B5B5B5B5B5B ; // ROM data -
    assign    loc1    =    64'h7E6A4719E7B99682 ;  // eight numbers
    assign    loc2    =    64'h7631CF8A8ACF3176 ;  // of 8 bits data
    assign    loc3    =    64'h6AE782B9477E1996 ;  // per location.
    assign    loc4    =    64'h5BA5A55B5BA5A55B ;
    assign    loc5    =    64'h4782196A96E77EB9 ;
    assign    loc6    =    64'h318A76CFCF768A31 ;
    assign    loc7    =    64'h19B96A827E9647E7 ;

always @ (loc0 or loc1 or loc2 or loc3 or loc4 or loc5 or loc6 or loc7 or
          addr1 or addr2)
begin
 case (addr1)          // Addressed data is accessed whenever there is a change
                       // in any of the inputs in the always statement.
                       // addr1 serves as the address to read C matrix data.
        3'b000 :        dout1_next     =     loc0 ;
        3'b001 :        dout1_next     =     loc1 ;
        3'b010 :        dout1_next     =     loc2 ;
        3'b011 :        dout1_next     =     loc3 ;
        3'b100 :        dout1_next     =     loc4 ;
        3'b101 :        dout1_next     =     loc5 ;
        3'b110 :        dout1_next     =     loc6 ;
        3'b111 :        dout1_next     =     loc7 ;
        default :       dout1_next     =     loc0 ;
 endcase
 case(addr2)            // addr1 serves as the address to read C matrix data.
        3'b000 :        dout2_next     =     loc0 ;
        3'b001 :        dout2_next     =     loc1 ;
        3'b010 :        dout2_next     =     loc2 ;
        3'b011 :        dout2_next     =     loc3 ;
```

```
        3'b100 :      dout2_next    =      loc4 ;
        3'b101 :      dout2_next    =      loc5 ;
        3'b110 :      dout2_next    =      loc6 ;
        3'b111 :      dout2_next    =      loc7 ;
        default:      dout2_next    =      loc0 ;
endcase
end

always @ (posedge clk)      // First pipeline stage
begin
        dout1_reg1     <=      dout1_next ;     // Pipeline registers.
        dout2_reg1     <=      dout2_next ;
end

always @ (posedge clk)      // Second pipeline
begin
        dout1  <=      dout1_reg1 ;     // Data outputs read using addr1
        dout2  <=      dout2_reg1 ;     // and addr2 respectively.
end
endmodule
```

9.1.2 Test Bench for Dual Address ROM Design

The ROM we have designed can be tested by writing a test bench as follows. Since the ROM permits fast access, we shall have a clock running at 100 MHz frequency indicated by 5 (ns) for clock period by two in the Verilog_code 9.2. We will use the back annotated '.v' file obtained using Xilinx Place and Route tool to get a feel of the access time of the ROM. In this test bench, we change the two addresses 'addr1' and 'addr2' every 10 ns. since the frequency of operation is 100 MHz. The module is declared as 'romc_test'. The ROM outputs are declared as 'wire', while the clock and address inputs as 'reg' in the test bench. This is followed by calling the design 'romc'. Stimulants are applied in the 'initial' block. The clock goes high after 5 ns. Therefore, we apply the first set of address inputs at 7 ns in order to avoid the rising edge of the clock. Subsequently, we apply the address inputs every 10 ns. The ROM contents are read for all combinations of addresses in the range 0 to 7. Different addresses applied to 'addr1' and 'addr2' are deliberate. Towards the end, we toggle the 'clk' to obtain 100 MHz operation.

Verilog_code_9.2

```
/*        Test bench for ROMC Design. Put this in a file named "romc_test.v".

`define clkperiodby2 5     // Required to  generate 100 MHz clock.
```

```
`include 'romc_banno.v'  // Design file is romc.v and back annotated file is
                         // romc_banno.v

module romc_test (        dout1,
                          dout2
                   );

        output     [63:0]      dout1;
        output     [63:0]      dout2;

        reg                    clk ;
        reg        [2:0]       addr1 ;
        reg        [2:0]       addr2 ;

romc   romc1(
                         .clk(clk),
                         .addr1(addr1),
                         .addr2(addr2),
                         .dout1(dout1),
                         .dout2(dout2)
                         );
initial
begin
                 clk              =        1'b0 ;
                 // Read the ROM contents for all combinations of addresses.
        #7        addr1  =      3'b000 ;
                  addr2  =      3'b111 ;
        #10       addr1  =      3'b001 ;
                  addr2  =      3'b110 ;
        #10       addr1  =      3'b010 ;
                  addr2  =      3'b101 ;
        #10       addr1  =      3'b011 ;
                  addr2  =      3'b100 ;
        #10       addr1  =      3'b100 ;
                  addr2  =      3'b011 ;
        #10       addr1  =      3'b101 ;
                  addr2  =      3'b010 ;
        #10       addr1  =      3'b110 ;
                  addr2  =      3'b001 ;
        #10       addr1  =      3'b111 ;
                  addr2  =      3'b000 ;
        #40                               // Run for some more time
 $stop ;                                  // before stopping the simulation.
end
```

```
always
  #`clkperiodby2 clk <= ~clk ;                    // Generate 100 MHz clock.
endmodule
```

9.1.3 Simulation Results of Dual Address ROM Design

The simulation results are shown in Figure 9.2 to Figure 9.4. Note that we have used back annotated file obtained after running the synthesis and place and route tools, and hence the gate delays must be revealed in the waveforms. Close examination of the timing diagram in Figure 9.2 reveals that the address changes occur after a delay of about 2 ns, reckoned from the rising edge of 'clk' signal. The data outputs appear faster than the corresponding addresses as can be seen in Figures 9.3 and 9.4, where the waveforms are zoomed closer than that in Figure 9.2. Let us examine the data output by ROM. For instance, in Figure 9.2, the data read out from ROM at 37 ns is 5b5b5b5b5b5b5b5b as shown in the figure. This is in conformity with the design given in Verilog_code_9.1, where the ROM content is 'loc0' in 'assign' statement for the address, 'addr1'. The simulation tool reports the data in lower case although we have used upper case in our design file. The first address, addr1 = 0, is applied at 7 ns in the test bench, Verilog_code_9.2. This just misses the rising edge of the 'clk' signal occurring at 5 ns. This address is, therefore, recognized at the following positive edge of 'clk' at 15 ns. Since the ROM incorporates two stages of pipelining, which implies two clock delays, the

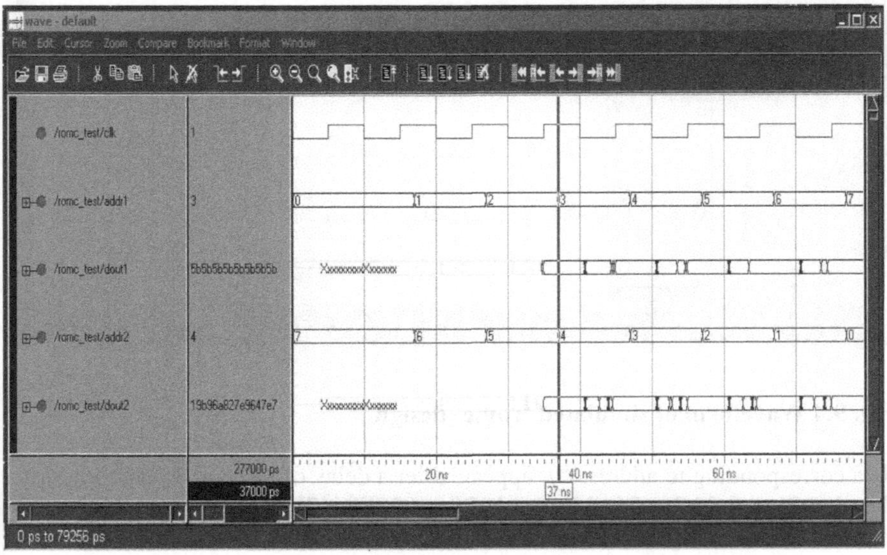

Fig. 9.2 Waveform of simulated 'romc' design

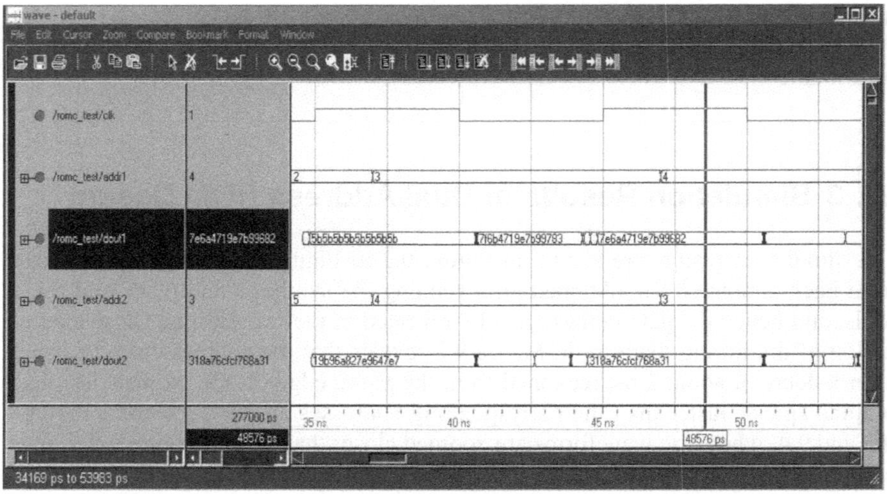

Fig. 9.3 Waveform of simulated 'romc' design

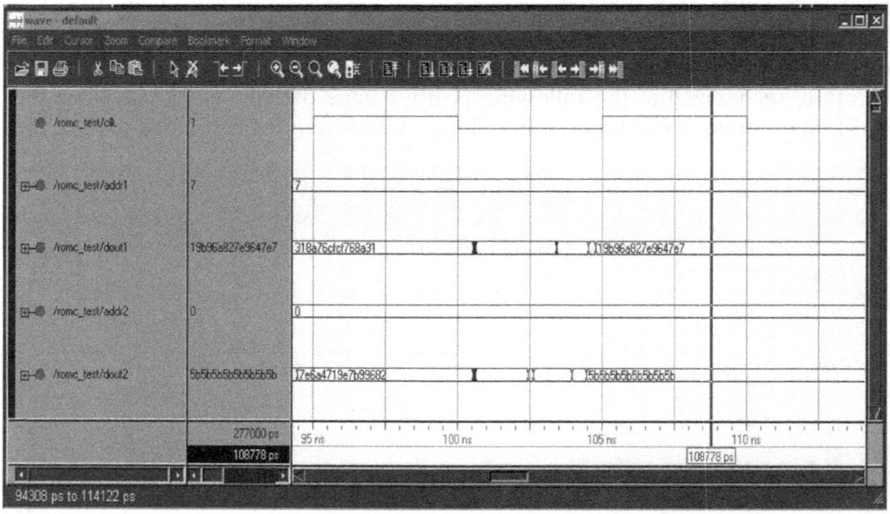

Fig. 9.4 Waveform of simulated 'romc' design

data corresponding to address '0' appears after a delay of 20 ns at 35 ns. Similarly, the 'dout2' for addr2 = 7 at 35 ns is 19B96A827E9647E7. The cursor, however, is shown at 37 ns in Figure 9.2. Other results may be similarly verified.

9.1.4 Synthesis Results for Dual Address ROM Design

The synthesis results of Synplify tool is as follows. We have mapped onto a device, which we will probably use for the design of DCTQ later. From now onwards, for all the designs pertaining to the DCTQ such as the dual RAM, arithmetic circuit designs, etc., we will use the very same device with the highest speed available presently. The device type is XCV600 EHQ 240-8. It is a 240-pin package and Virtex-E series FPGA of Xilinx. The maximum estimated frequency of operation reported by the tool is very high, about 160 MHz. However, this is only an approximate estimate and we need to take only the place and route tool report to be close to the actual. The synthesis report gives the flip-flop usage, the number of input and output buffers used and the total number of LUTs for the ROM design. If mapping is successful, the report asserts to that effect. The reader may view the RTL and technology view to see how the circuit looks. Double clicking on the circuit opens the corresponding Verilog code of our design.

Performance summary
Worst slack in design: 13.771

Starting Clock	Requested Frequency	Estimated Frequency
clk	50.0 MHz	160.5 MHz

Requested Period	Estimated Period	Slack	Clock Type
20.000	6.229	13.771	Inferred

Resource usage report for romc
Mapping to part: xcv600ehq240-8
Cell usage:

FDS	38	uses
FDR	8	uses
FDRS	26	uses
FD	142	uses
GND	1	use
VCC	1	use

I/O primitives:

IBUF	6	uses
OBUF	128	uses

BUFGP	1	use

I/O register bits: 110
Register bits not including I/Os: 104 (0%)
Global clock buffers: 1 of 4 (25%)
Mapping summary:
Total LUTs: 68 (0%)

9.1.5 Xilinx P&R Results for Dual Address ROM Design

The Xilinx place and route tool reports the number of slices and 4 input LUTs. The total equivalent gate count for the design is about 2100. The frequency of operation has come down after running the place and route to 138 MHz. However, the frequency report for overall project design only counts finally. The tool also creates a bit stream output (romc.bit) required to configure the FPGA mounted on a functional circuit board.

Xilinx P&R report:
map -p xcv600e-8-hq240 -o map.ncd romc.ngd romc.pcf
Using target part "v600ehq240-8".
Removing unused or disabled logic. . .
Running cover. . .
Writing file map.ngm. . .
Running directed packing. . .
Running delay-based packing. . .
Running related packing...
Writing design file 'map.ncd'. . .
Design summary:

Number of errors:	0			
Number of warnings:	0			
Number of slices:	86	out of	6,912	1%
Number of slices containing unrelated logic:	0	out of	86	0%
Number of slice flip-flops:	104	out of	13,824	1%
Number of 4 input LUTs:	68	out of	13,824	1%
Number of bonded IOBs:	134	out of	158	84%
IOB flip-flops:	110			
Number of GCLKs:	1	out of	4	25%
Number of GCLKIOBs:	1	out of	4	25%
Total equivalent gate count for design:	2,120			
Additional JTAG gate count for IOBs:	6,480			

Timing summary:
Minimum period: 7.226 ns
(maximum frequency: 138.389 MHz)

Minimum input arrival time before clock: 8.726 ns
Minimum output required time after clock: 9.873 ns
Saving bit stream in "romc.bit".

9.2 Single Address ROM Design

We will cover the design of another type of ROM used in DCTQ application as shown in Figure 9.5. This ROM, with single address and single data output, is used to store inverse quantization values. In quantization process, we need to divide the DCT coefficients by the corresponding quantization values. However, division can also be implemented as multiplication if we take the inverse of the quantization values. We therefore, store the inverse of the quantization values in the ROM. The data table is arranged as eight locations of size, 64 bits. While reading the ROM, only one byte is retrieved at a time. Thus, this implementation is different from the conventional design storing byte-wise. Since it is read as 64 × 8 bits, i.e., $2^6 \times 8$ bits, we need 6 bits of address. The ROM specification may be summarized as follows:

- The ROM stores the inverse of the quantization values (8 bits, unsigned).
- It is organized as 8 × 64 bits, and can be read byte-wise.

9.2.1 Verilog Code for Single Address ROM Design

'romq' is declared as the module at the beginning and the inputs/outputs listed. This is followed by declaring the I/Os, registers, and wires as shown in the code. Locations 'loc0' to 'loc7' are assigned the ROM data, each of size 64 bits. In the 'always' block that follows, these locations are assigned to the memory, 'mem [0]' to 'mem [7]'. The next 'always' block separates the 64 bits data in the 'mem' as bytes. MSB of 'mem [a]', i.e., 'mem_data [63:56]', is assigned as the LSB byte, 'byte_data [0]'. Similar explanation holds good for other bytes. Finally, the 'data_byte' is registered as 'd'.

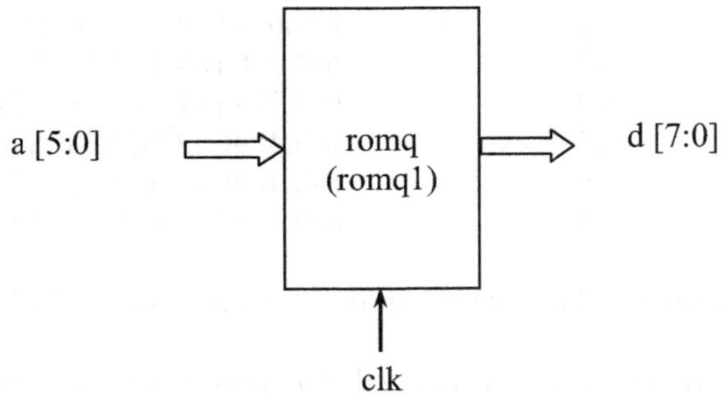

Fig. 9.5 Single address ROM design

Verilog_code_9.3

```
/*      ROMQ Design.
        This code can be put in a file named 'romq.v'.
        This ROM stores the inverse of quantization values (8 bits, unsigned).
        Although organized as 8 × 64 bits, it is byte-addressed (64 × 8 bits) while
        reading.
*/
module romq (clk, a, d);

        input                   clk;        // Declare I/Os.
        input       [5:0]       a;          // This is the 6-bit address and
        output      [7:0]       d;          // data output of ROM.

        reg         [7:0]       d;          // Declare as the register.
        wire        [7:0]       d_next ; // Declare as the wire.

        reg         [63:0]      mem      [7:0] ;   // ROM organized as 8x64
        reg         [7:0]       byte_data  [7:0] ; // bits, but read byte-by-byte.

        wire        [63:0]      mem_data ; // Declare 'assign' outputs as wire.
        wire        [63:0]      loc0 ;
        wire        [63:0]      loc1 ;
        wire        [63:0]      loc2 ;
        wire        [63:0]      loc3 ;
        wire        [63:0]      loc4 ;
        wire        [63:0]      loc5 ;
        wire        [63:0]      loc6 ;
        wire        [63:0]      loc7 ;

        assign      loc0    =   64'hFF806C5D4F4C473C ;
        assign      loc1    =   64'h80805D554C473C37 ;
        assign      loc2    =   64'h6C5D4F4C473C3C36 ;
        assign      loc3    =   64'h5D5D4F4C473C3733 ;
        assign      loc4    =   64'h5D4F4C47403B332B ;
        assign      loc5    =   64'h4F4C47403B332B23 ;
        assign      loc6    =   64'h4F4C473C362D251E ;
        assign      loc7    =   64'h4C473B362D251E19 ;

always @ (loc0 or loc1 or loc2 or loc3 or loc4 or loc5 or loc6 or loc7)
begin
        // Bytes from each row is accessed in a raster scan order (MSB first, etc).
          mem       [0]     =   loc0 ;
          mem       [1]     =   loc1 ;
```

```
        mem     [2]     =     loc2 ;
        mem     [3]     =     loc3 ;
        mem     [4]     =     loc4 ;
        mem     [5]     =     loc5 ;
        mem     [6]     =     loc6 ;
        mem     [7]     =     loc7 ;
end

always @ (mem_data)
begin
byte_data [0]     =     mem_data [63:56] ; // MSB is assigned as
byte_data [1]     =     mem_data [55:48] ;  // LSB.
byte_data [2]     =     mem_data [47:40] ;
byte_data [3]     =     mem_data [39:32] ;
byte_data [4]     =     mem_data [31:24] ;
byte_data [5]     =     mem_data [23:16] ;
byte_data [6]     =     mem_data [15:8] ;
byte_data [7]     =     mem_data [7:0] ;  // LSB is assigned as MSB.
end
assign mem_data   =     mem [a[5:3]] ;          // Get 64 bits data.
assign d_next     =     byte_data [a[2:0]] ;    // Get byte data.

always @ (posedge clk)
        d     <=     d_next ; // Register byte data.
endmodule
```

9.2.2 Test Bench for Single Address ROM Design

The test bench for single address ROM design is housed in a file named 'romq_test.v' and is described in Verilog_code_9.4. The first statement on definition specifies the half period of clock in ns, which corresponds to an operational frequency of 100 MHz. File 'include' specifies the back annotated source file so that we may incorporate the actual gate delays in simulation. 'romq_test' is declared as the module with 'd' as the 8-bit output. Inputs, 'clk' and 'a' are declared as 'reg' in the test bench. Address is of size 6 bits for accessing 64 locations, each of width 8 bits. The next statement invokes the design, 'romq' calling I/Os by name. 'count' stands for a counter and is of type integer, which is used in 'for' loop subsequently. The next group of statements is placed in the 'initial' block. 'clk' is cleared at 0 ns and the initial address 'a' is applied at 7 ns. The statement 'a = count' is the only statement in the 'for' loop, which changes the address 'a' every 10 ns. The address advances in steps of 1 starting from 0 right up to 63. The statement in 'always' block, clk <= ~clk, is for generating a 100 MHz clock.

Verilog_code_9.4

```
// Test Bench for ROMQ Design. Place this in a file named 'romq_test.v'.

`define clkperiodby2 5          // Specify the frequency of operation as 100 MHz.
`include "romq_banno.v"              // Use the back annotated version of
                                     // romq.v for testing the design.
module romq_test  (d);               // The test module is declared.

output   [7:0]     d ;               // So also the output.

reg                clk ;             // The inputs are declared as
reg      [5:0]     a ;               // 'reg' in a test bench.

romq   romq1(    .clk(clk),                    // romq1 is an instantiation of
                 .a(a),                        // the design, romq.
                 .d(d)      /                  / The ports are called by name.
                 ) ;
integer count ;                                // count is an integer variable.

initial
begin
           clk           =       1'b0 ;    // Initialize the clock and
#7         a             =       0 ;       // address.
for (count = 0; count < 64; count = count+1)          // count = 0–63.

#10         a            =       count ;          // Apply new address
                                                  // every 10 ns a = 0–63.
#200                                              // Stop after some time.
$stop ;
end

always
 #`clkperiodby2  clk <= ~clk ;     // Generate 100 MHz clock.

endmodule
```

9.2.3 Simulation Results of Single Address ROM Design

The simulation results of single address ROM design are shown in Figures 9.6 and 9.7. As in the previous design, we have used back annotated file. From the design presented in Verilog_code_9.3, we see that the first seven bytes of the ROM are FF, 80, 6C, 5D, 4F, 4C, 47. This is easily verified by inspecting 'd' waveform in

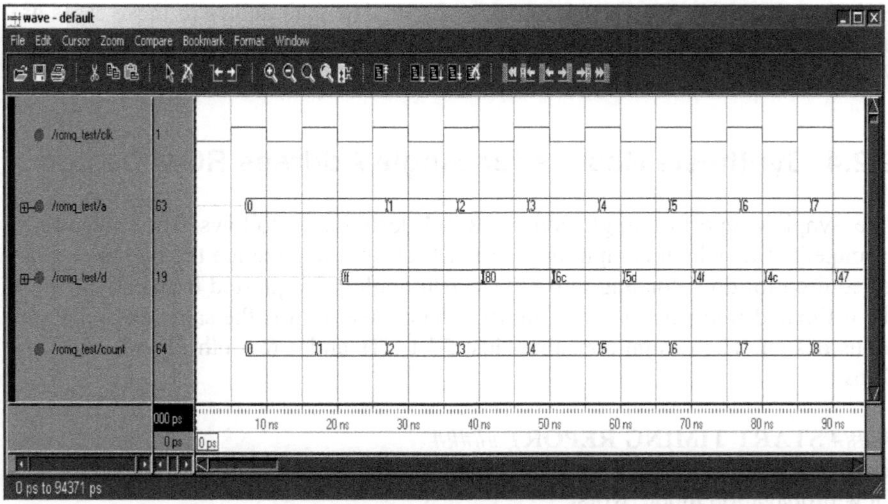

Fig. 9.6 Simulation results of back annotated 'ROMQ' Design at the beginning

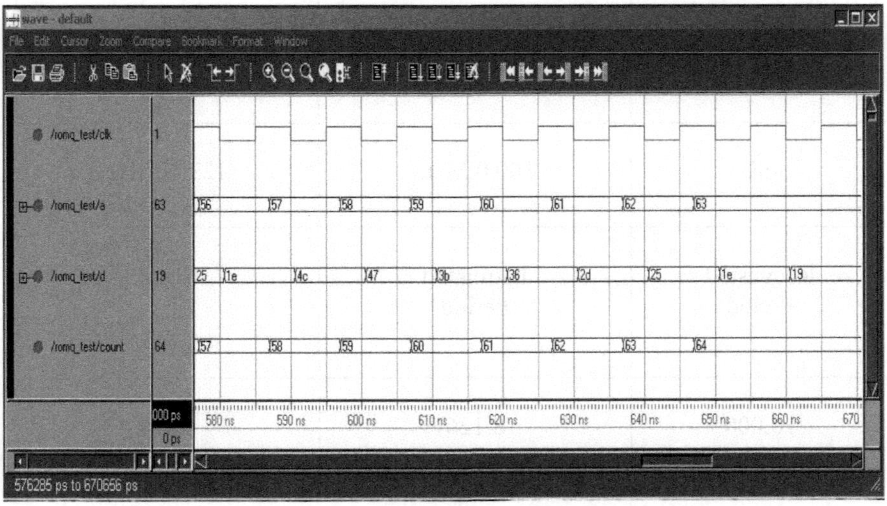

Fig. 9.7 Simulation results of back annotated 'ROMQ' towards the end

Figure 9.6. Since we have used the back annotated design file, the address 'a' is delayed by 2 ns, while data 'd' is delayed by 11 ns with reference to the rising edge of the 'clk'. It may be noted that 'a' is delayed by a 'clk' cycle with reference to 'count', since we have used #10 in the statement: #10 a = count in the test bench, Verilog_code_9.4. Similarly, Figure 9.7 presents the ROM contents of last eight locations: 4C, 47, 3B, 36, 2D, 25, 1E, 19. The gate delays in these cases are nearly the same as in the first seven data. The reader may simulate the code and verify the correctness of other data, not shown.

9.2.4 Synthesis Results for Single Address ROM Design

The Synplify report of single address ROM design is as follows. The tool creates 'romq.edf' file, which is input into Xilinx P&R tool to generate the bit stream that is required for downloading into FPGA. Although the requested frequency is high, the estimated frequency is even higher. The device used is the same as in the previous design. The design consumes just 37 LUTs and a few other gates and flip-flops.

START TIMING REPORT
Top view: romq
Slew propagation mode: worst
Paths requested: 5
Worst slack in design: 1.874

Starting Clock	Requested Frequency	Estimated Frequency
clk	100.0 MHz	123.1 MHz
Requested Period	Estimated Period	Slack
10.000	8.126	1.874

Mapping to part: xcv600ehq240-8
Cell usage:

FDR	1	use
MUXF5	11	uses
MUXF6	4	uses
FD	7	uses

I/O primitives:

IBUF	6	uses
OBUF	8	uses

BUFGP	1	use

I/O register bits: 8
Global clock buffers: 1 of 4 (25%)
Total LUTs: 37 (0%)

9.2.5 Xilinx P&R Results for Single Address ROM Design

The place and route for the design is presented as follows. Number of 4 input LUTs reported by this tool is 35, and is more accurate than the Synplify tool. The tool also reports the gate count for the design as 319, which information cannot be had in the Synplify tool. Also, the frequency reported is much higher than that reported by the Synplify tool. The P&R tool creates 'romq.bit' file.

Design summary:

Using target part 'v600ehq240-8'.

Number of slices:	18	out of	6,912	1%
Number of slices containing unrelated logic:	0	out of	18	
Number of 4 input LUTs:	35	out of	13,824	1%
Number of bonded IOBs:	14	out of	158	8%
IOB flip-flops:	8			
Number of GCLKs:	1	out of	4	25%
Number of GCLKIOBs:	1	out of	4	25%

Total equivalent gate count for design: 319
Additional JTAG gate count for IOBs: 720

Timing summary:

Minimum input arrival time before clock: 6.568 ns (152 MHz)
Minimum output required time after clock: 5.633 ns

Saving bit stream in 'romq.bit'.

9.3 On-chip Dual RAM Design

We were looking into the design of ROMs earlier. Now, we will see the design of RAMs, which falls in the VLSI category, as the chip area in the design will be much more than the equivalent of 50,000 transistors. The requirement of dual RAM with a particular memory organization as presented in this section arises from the needs of the design application, DCTQ, which we mentioned in the ROM designs earlier. The dual RAM consists of two RAMs, each of which stores the image information. This information will be written from a host computer such as a PC into one of the RAMs through peripheral connect interface (PCI) bus. Initially, one of the double memory buffers, RAM 1, is filled and once it is full, the image information is written to the second RAM. While the second memory, RAM 2, is being written into, the RAM 1 will be read concurrently to process the

DCTQ coefficients. The design requirement arises from these concurrent operations of acquiring image data and DCTQ computations. An important feature in this design is that the data is written at the rate of 64 bits per clock cycle. The data transfer is through the PCI interface with 64 bits data bus. In order to process DCTQ, we need to write a block of image data consisting of 64 pixels. We will, therefore, need eight clock cycles to write a block of information since we can write eight bytes per cycle. One pixel data size is one byte for monochrome and three bytes for color image or picture. This design is for processing monochrome picture or color motion picture.

The pin diagram of the dual RAM is shown in Figure 9.8. The validity of the input data signals, 'di [63:0]' is signaled by 'din_valid'. Also, the validity of each of the 8 bits in the 64 bit data bus is indicated by the 'be [7:0]' pin referred to as the byte enable signal. In order to write a pixel block, we need only 3 bits address, wa [2:0], corresponding to eight locations, each location being 64 bits in width. Data is written at the positive edge of 'pci_clk' signal.

The signal, 'rnw', meaning read negative (low) and write positive (high), is used to configure one of the RAMs in write mode while the other RAM block is configured in the read mode. That is how we achieve concurrent processing of both writing and reading of the double buffer. It should be noted that RAM 1 and RAM 2 are dual RAMs of size 8 × 64 bits each. If RAM 2 is in read only mode, then RAM 2 is automatically configured to the write only mode and vice versa. The RAM is written row-wise and read column-wise. This is due to the complexity of the DCT algorithm, which is discussed at length in later chapters on the development of algorithms and design of architectures. 'ra [2:0]' is the read address to process DCTQ reckoned at the rising edge of 'clk' signal. The column-wise data read appears at the output, 'do [63:0]'. It may be noted that the design does not include a data out valid signal since this is kept track in a controller design which will be discussed in a later chapter on design applications.

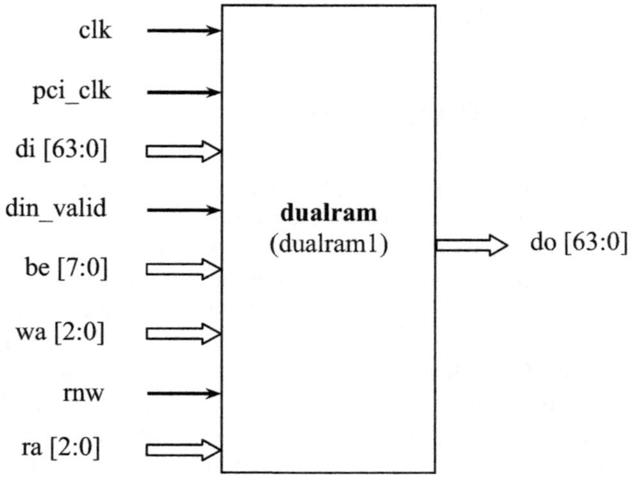

Fig. 9.8 On-chip dual RAM design

9.3.1 Verilog Code for Dual RAM Design

The RTL code of dual RAM design is presented in Verilog_code_9.5. Two RAMs, ram1 and ram2, comprise the dual RAM. The control input signal, rnw, configures the two RAMs in write only mode and read only mode alternately. For instance, a 'high' configures ram1 in write mode and ram2 in read mode and vice versa. Once configured, the RAM in write mode gets all the eight locations written. Each location is of size 64 bits. The data is written row-wise. While reading a ROM, it is done column-wise. This arises owing to the order of computation of DCT mentioned earlier. The include statement identifies the single RAM design. This is followed by the declaration of the dualram module and the listing of I/Os along with their identification. The 'wire' and 'reg' signals are also identified. The dual RAMs, ram1 and ram2, are instantiated by invoking the 'ram_rc' module twice. The 'switch_bank' signal, which is the inverted signal of 'rnw', configures ram2 in read mode if rnw is high and in write mode if it is low. Naturally, rnw configures ram1 in write mode for rnw = 1 and in read mode for rnw = 0. The RAM data output 'do2 or 'do1' is registered after a clock cycle delay at the positive edge of 'clk_sys'. This delay is purposely introduced to keep pace with 'romc' design discussed in an earlier section. It may be noted that both 'dualram' and 'ramc' designs are used in the DCTQ design.

Verilog_code_9.5

```
/*
This is the RTL code for Dual RAM Design.
Place this in a file named "dualram.v".
ram1 and ram2 are the dual RAMs, 8 × 64 bits each.
If ram1 is in read only mode, then ram2 is automatically configured to the write
only mode, and vice versa.
The RAM is written row-wise and read column-wise.
*/
`include "ram_rc.v"                                  // This is the individual RAM
                                                     // code file.
module dualram (                                     // Declare the module and
                        clk,                         // I/O ports.
                        pci_clk,
                        rnw,
                        be,
                        ra,
                        wa,
                        di,
                        din_valid,
                        do
            );
```

```
    input               clk ;           // System clock.
    input               pci_clk ;       // PCI clock for inputting data, di
                                        // synchronously.
    input               rnw ;           // Sets one RAM in write only mode
                                      // and the other RAM in read only mode.
    input               din_valid ;     // Data in (di) valid.
    input     [7:0]     be ;            // Byte enable.
    input     [2:0]     ra, wa ;        // Read/write address.
    input     [63:0]    di ;            // Data input and
    output    [63:0]    do ;            // Data output of dual RAM.

    wire                switch_bank;    // Declare net outputs.
    wire      [63:0]    do1 ;
    wire      [63:0]    do2 ;
    wire      [63:0]    do_next ;

    reg       [63:0]    do;             // Declare registered outputs.
    reg                 rnw_delay ;

assign switch_bank = ~rnw;            // Configure ram1/ram2 for read and write
                                      // mode respectively to start with.

ram_rc  ram1 (          .clk(clk),        // Instantiate the first RAM.
                                          // 'rc' stands for write row-wise and
                        .pci_clk(pci_clk),// read column-wise.
                        .rnw(rnw),        // If rnw =1,  ram1 is configured for
                        .be(be),          // write mode. Otherwise, read mode.
                        .ra(ra),
                        .wa(wa),
                        .di(di),
                        .din_valid(din_valid),
                        .do(do1)
             ) ;

ram_rc  ram2(           .clk(clk),        // Instantiate the second RAM.
                        .pci_clk(pci_clk),
                        .rnw(switch_bank), // If rnw =1, ram2 is configured for
                        .be(be),           // read mode. Otherwise, write mode.
                        .ra(ra),
                        .wa(wa),
                        .di(di),
                        .din_valid(din_valid),
                        .do(do2)
              ) ;
assign do_next    =     (rnw_delay) ? do2 : do1 ; // Read ram2 or ram1.
```

```
always @ (posedge clk)
begin
        rnw_delay        <= rnw ;        // Delay the rnw signal by one clock.
        do               <= do_next ;    // Register the selected RAM output.
end
endmodule
```

The dual RAM design we covered earlier included the individual RAM design file. This design is presented in Verilog_code_9.6. The design module is called 'ram_rc'. All I/Os are declared as inputs or outputs as the case may be. This is followed by identification of various signals used in the design as 'wire' or 'reg'. The statement 'reg [63:0] mem [7:0] ;' refers to a register array arranged as 8 × 64 bits, which serves as a ROM of width 64 bits and eight locations. The signal 'addr' in the first 'assign' statement uses the read or write address of the RAM. It is interpreted in accordance with the signal 'rnw'. If it is high, it is write address. Otherwise, it is read address. In the second 'assign' statement, the content of location addressed by 'addr' is temporarily stored in 'mem_data [63:0]', and is used in the last 'always' block. The next eight 'assign' statements store each of the eight locations so that individual byte data may be extracted and read column-wise as shown in the 'always' statement using 'case' statement. For instance, for addr = 0, the 'column' extracts the MSBs [63:56] of each of the eight locations 'loc0' to 'loc7' and concatenates, i.e., arranges them in that order.

The 'always @ (posedge pci_clk)' block is primarily used to write data into the RAM only if be7 = 1, and so on. This condition is satisfied only if be[7] = 0 and rnw = 1 and din_valid =1 using eight 'assign' statements before the always block mentioned earlier. The MSB data of 'mem [addr]' is assigned either 'di[63:56]' or 'mem_data[63:56]' for 'be7' value equal to '1' or '0'. Similar explanation holds good for all other bytes. The 64 bit value is obtained by putting all the eight bytes together by concatenation. The statement 'assign do_next = (rnw) ? do : column ;' reads column-wise data from RAM if rnw = 0. Otherwise, the original content 'do' is not disturbed. The last 'always @ (posedge clk)' block registers the output 'do' using 'do_next' as its input.

Verilog_code_9.6

```
// This is the individual RAM design.
// Place this in a file named 'ram_rc.v'.
// This is a single block RAM, called twice by 'dualram.v'
// RAM size: eight locations of width 64 bits.
// Writing is done by row addressing, and reading by column addressing.

module ram_rc (            clk,        // Declare the module and list
                       pci_clk,        // I/Os.
                           rnw,
```

```
                    be,
                    ra,
                    wa,
                    di,
                    din_valid,
                    do
                );

    input               clk ;               // Declare I/Os.
    input               pci_clk ;
    input               rnw ;
    input               din_valid ;
    input      [7:0]    be ;
    input      [2:0]    ra ;
    input      [2:0]    wa ;
    input      [63:0]   di ;
    output     [63:0]   do ;

    reg        [63:0]   do ;                // Declare registered outputs.

    wire       [63:0]   mem_data ;          // Declare all wire signals.
    wire       [63:0]   do_next ;
    wire       [2:0]    addr ;
    wire       63:0]    loc0 ;
    wire       [63:0]   loc1 ;
    wire       [63:0]   loc2 ;
    wire       [63:0]   loc3 ;
    wire       [63:0]   loc4 ;
    wire       [63:0]   loc5 ;
    wire       [63:0]   loc6 ;
    wire       [63:0]   loc7 ;
    wire                be0 ;
    wire                be1 ;
    wire                be2 ;
    wire                be3 ;
    wire                be4 ;
    wire                be5 ;
    wire                be6 ;
    wire                be7 ;

    reg        [63:0]   column ;
    reg        [63:0]   mem [7:0] ;// Declare register array, 8×64 bits.

assign addr = (rnw) ? wa : ra ;      // Get write address (rnw = 1) or read address.
assign mem_data = mem [addr] ;  // Fetch the  memory content.
```

```
        assign    loc0        = mem [0] ;  // Intermediate store for memory.
        assign    loc1        = mem [1] ;
        assign    loc2        = mem [2] ;
        assign    loc3        = mem [3] ;
        assign    loc4        = mem [4] ;
        assign    loc5        = mem [5] ;
        assign    loc6        = mem [6] ;
        assign    loc7        = mem [7] ;

always @ (addr or loc0 or loc1 or loc2 or loc3 or loc4 or loc5 or loc6 or loc7)
begin
case (addr)                       // Read the RAM column-wise.
3'b000:
column = {loc0[63:56], loc1[63:56], loc2[63:56], loc3[63:56], loc4[63:56],
                                  loc5[63:56], loc6[63:56], loc7[63:56]} ;
3'b001:
column = {loc0[55:48], loc1[55:48], loc2[55:48], loc3[55:48], loc4[55:48],
                                  loc5[55:48], loc6[55:48], loc7[55:48]} ;
3'b010:
column = {loc0[47:40], loc1[47:40], loc2[47:40], loc3[47:40], loc4[47:40],
                                  loc5[47:40], loc6[47:40], loc7[47:40]} ;
3'b011:
column = {loc0[39:32], loc1[39:32], loc2[39:32], loc3[39:32], loc4[39:32],
                                  loc5[39:32], loc6[39:32], loc7[39:32]} ;
3'b100:
column = {loc0[31:24], loc1[31:24], loc2[31:24], loc3[31:24], loc4[31:24],
                                  loc5[31:24], loc6[31:24], loc7[31:24]} ;
3'b101:
column = {loc0[23:16], loc1[23:16], loc2[23:16], loc3[23:16], loc4[23:16],
                                  loc5[23:16], loc6[23:16], loc7[23:16]} ;
3'b110:
column = {loc0[15:8],  loc1[15:8], loc2[15:8],  loc3[15:8],  loc4[15:8],
                                    loc5[15:8],  loc6[15:8],  loc7[15:8]} ;
3'b111:
column = {loc0[7:0], loc1[7:0], loc2[7:0], loc3[7:0], loc4[7:0], loc5[7:0],
                                             loc6[7:0],   loc7[7:0]} ;
default :
column = {loc0[7:0], loc1[7:0], loc2[7:0], loc3[7:0], loc4[7:0], loc5[7:0],
                                             loc6[7:0],   loc7[7:0]} ;
endcase
end

assign   be7    =       (!be[7]) & rnw & din_valid ;
                                  // Enable write only if be7 = 1, and so on.
assign   be6    =       (!be[6]) & rnw & din_valid ;
assign   be5    =       (!be[5]) & rnw & din_valid ;
```

```
assign  be4    =      (!be[4]) & rnw & din_valid ;
assign  be3    =      (!be[3]) & rnw & din_valid ;
assign  be2    =      (!be[2]) & rnw & din_valid ;
assign  be1    =      (!be[1]) & rnw & din_valid ;
assign  be0    =      (!be[0]) & rnw & din_valid ;

always @ (posedge pci_clk)
begin                      // Write into RAM only if be7 = 1, and so on.
                           // Otherwise, don't disturb the RAM contents.
 mem [addr] <=     { ( (be7) ? di[63:56] : mem_data[63:56] ),
                     ( (be6) ? di[55:48] : mem_data[55:48] ),
                     ( (be5) ? di[47:40] : mem_data[47:40] ),
                     ( (be4) ? di[39:32] : mem_data[39:32] ),
                     ( (be3) ? di[31:24] : mem_data[31:24] ),
                     ( (be2) ? di[23:16] : mem_data[23:16] ),
                     ( (be1) ? di[15:8]  : mem_data[15:8] ),
                     ( (be0) ? di[7:0]   : mem_data[7:0] )
               } ;
end

assign do_next = (rnw) ? do : column ;
                                 // Read column-wise from RAM only if rnw = 0.
                                 // Otherwise, don't disturb.
always @ (posedge clk)
         do <=  do_next ;                          // Register the output.
endmodule
```

9.3.2 Test Bench for the Dual RAM Design

The following is the test bench to verify the functionality of the dual RAM. We include the back annotated file and examine only the output 'do'. As usual, we will operate at 50 MHz. We have two clock signals: one is the 'pci_clk' for writing image data into the RAM and the other one is the system clock, 'clk', for reading the data from RAM. As shown in the Verilog_code_9.7, the two define statements give the half period for the two clocks. All the inputs are declared as registers and the dual RAM design is instantiated, calling ports by name. The signal 'rnw' decides whether a RAM is in read mode or in write mode. The stimulants are applied in the 'initial' block. To start with, various signals are initialized at zero time. The first block of data is written into RAM 1, 64 bits at a time corresponding to one row of a block of image. Each row of data is applied with different 'wa' and is written every 20 ns since the frequency of operation is 50 MHz. The data for the write address 'wa = 1' is applied at 17 ns in order to avoid changing data at the rising edge of 'pci_clk'. Thereafter, the signal 'rnw' is toggled and the above process of writing is repeated for RAM 2. Simultaneously, RAM 1 which was written earlier is read back by applying progressively increasing read addresses 'ra' every 20 ns. Similarly, two more blocks are written followed by reading it back to verify whether the dual RAMs are working properly. While reading the fourth block, no further block is written. The last two 'always' blocks are used to run the two clocks in the design. The test may be repeated for 'din_valid = 0' and 'be [7] = 1' and so on up to 'be [0] = 1'. For these conditions, the concerned RAM must be inhibited from writing.

Verilog_code_9.7

```
// This is the test bench for dual RAM design.
// Place it in a file named 'dualram_test.v'.

`define clkperiodby2 10
`define clkby2 10
`include "dualram.v"                              // This is the design file.

module dualram_test (    do                       // Declare the module.
                         );

         output  [63:0]    do ;                   // Declare the output.

         reg               clk ;                  // Declare the input signals.
         reg               pci_clk ;
         reg               rnw ;
         reg               din_valid ;
         reg     [7:0]     be ;
         reg     [2:0]     ra, wa ;
         reg     [63:0]    di ;

dualram    u1 (                                   // Invoke the design
                        .clk(clk),                // calling ports by name.
                        .pci_clk(pci_clk),
                        .rnw(rnw),
                        .be(be),
                        .ra(ra),
                        .wa(wa),
                        .di(di),
                        .din_valid(din_valid),
                        .do(do)
               );
initial                                           // Apply stimulants.
begin
 clk              =        1'b0 ;
 pci_clk          =        1'b0 ;
```

```
rnw             =       1'b0 ;
din_valid       =       1'b1 ;    // Change to 1'b0 if  write is to be inhibited.
be              =       8'h00;    // Change "0" to "1" if byte write is to
                                  // be inhibited.
      wa   =  3'd0;   di  =   64'h0 ;
#17   wa   =  3'd1;   di  =   64'h123456789abcdef0 ;
                                  // Write first block of data into ram1.
#20   wa   =  3'd2;   di  =   64'h7E6A4719E7B99682 ;
#20   wa   =  3'd3;   di  =   64'h7631CF8A8ACF3176 ;
#20   wa   =  3'd4;   di  =   64'h6AE782B9477E1996 ;
#20   wa   =  3'd5;   di  =   64'h5BA5A55B5BA5A55B ;
#20   wa   =  3'd6;   di  =   64'h4782196A96E77EB9 ;
#20   wa   =  3'd7;   di  =   64'h318A76CFCF768A31 ;
#80   rnw  =  1'b1 ;              // Switch the roles of the RAM banks.
      ra = 3'd0;   wa = 3'd0;   di = 64'h5BA5A55B5BA5A55B ;
                                  // Write second block of data into ram2.
                                  // Simultaneously read from ram1.
#20  ra = 3'd1;   wa = 3'd1;   di = 64'h4782196A96E77EB9 ;
#20  ra = 3'd2;   wa = 3'd2;   di = 64'h318A76CFCF768A31 ;
#20  ra = 3'd3;   wa = 3'd3;   di = 64'h19B96A827E9647E7 ;
#20  ra = 3'd4;   wa = 3'd4;   di = 64'h7E6A4719E7B99682 ;
#20  ra = 3'd5;   wa = 3'd5;   di = 64'h7631CF8A8ACF3176 ;
#20  ra = 3'd6;   wa = 3'd6;   di = 64'h6AE782B9477E1996 ;
#20  ra = 3'd7;   wa = 3'd7;   di = 64'h5BA5A55B5BA5A55B ;
#80  rnw = 1'b0 ;   ra = 3'd0;   wa = 3'd0;  di = 64'haa5500ff0055aaff ;
                                  // Write third block of data into ram1
                                  // and simultaneously read from ram2.
#20  ra = 3'd1;   wa = 3'd1;   di = 64'h4782196A96E77EB9 ;
#20  ra = 3'd2;   wa = 3'd2;   di = 64'h318A76CFCF768A31 ;
#20  ra = 3'd3;   wa = 3'd3;   di = 64'h19B96A827E9647E7 ;
#20  ra = 3'd4;   wa = 3'd4;   di = 64'h7E6A4719E7B99682 ;
#20  ra = 3'd5;   wa = 3'd5;   di = 64'h7631CF8A8ACF3176 ;
#20  ra = 3'd6;   wa = 3'd6;   di = 64'h6AE782B9477E1996 ;
#20  ra = 3'd7;   wa = 3'd7;   di = 4'h5BA5A55B5BA5A55B ;
#80  rnw = 1'b1 ;   ra = 'd0;     wa = 3'd0;   di = 64'h0;
                                  // Switch the roles of the RAM banks again.
                                  // Write fourth block of data into ram2 and
                                  // simultaneously read from ram1.
#20    ra = 3'd1;     wa = 3'd1;     di = 64'h123456789abcdef0;
#20    ra = 3'd2;     wa = 3'd2;     di = 64'h7E6A4719E7B99682 ;
#20    ra = 3'd3;     wa = 3'd3;     di = 64'h7631CF8A8ACF3176 ;
#20    ra = 3'd4;     wa = 3'd4;     di = 64'h6AE782B9477E1996 ;
#20    ra = 3'd5;     wa = 3'd5;     di = 64'h5BA5A55B5BA5A55B ;
#20    ra = 3'd6;     wa = 3'd6;     di = 64'h4782196A96E77EB9 ;
#20    ra = 3'd7;     wa = 3'd7;     di = 64'h318A76CFCF768A31 ;
#80    rnw = 0 ;      ra = 3'd0;     // Read fourth block of data from ram2.
```

```
#20     ra = 3'd1;
#20     ra = 3'd2;
#20     ra = 3'd3;
#20     ra = 3'd4;
#20     ra = 3'd5;
#20     ra = 3'd6;
#20     ra = 3'd7;
#100
$stop ;                         // Stop testing after a while.
end

// Run the clocks.
always
 #`clkperiodby2 clk       <= ~clk ;

always
 #`clkby2 pci_clk         <= ~pci_clk ;

endmodule
// din_valid and "be [7]" to "be [0]" checked for all combinations – OK.
```

9.3.3 Simulation Results of Dual RAM Design

The dual RAM design and its test bench were presented in Verilog_code_ 9.5/9.6 and Verilog_code_9.7 respectively. The simulation results of the design are shown in Figure 9.9 to Figure 9.14. In the test bench, we applied the write address at 0 ns, 17 ns, 37 ns, etc. The corresponding 64-bit data written was 64'h0000000000000000, 64'h123456789abcdef0, 64'h7E6A4719E7B99682, and 64'h7631CF8A8ACF 3176. Inspecting the row marked 'di' in Figure 9.9, we find that the data are tallying with those mentioned above. These correspond to the write addresses 0 through 3. It may be noted that the byte enable, be = 00 H and di_valid is active. The signal 'rnw' is low configuring RAM 1 in write only mode. Since the read address 'ra' was not applied for the first block of data, the output 'do' is in don't care state. It may be recalled from the design that writing into the selected RAM takes place at the positive edge of 'pci_clk' and reading at the positive edge of 'clk'. Similarly, the reader may verify that data writes for wa = 4 to 7 shown in Figure 9.10 are precisely the same as that in the test bench.

The reading back of the RAM is column-wise as shown in the row marked 'do' in Figures 9.11 and 9.12. The second block of data written is 64'h5BA5A55B5BA5A55B at location 0, etc. commencing from 217 ns. The signal 'rnw' is changed to 'high' at 217 ns, thus configuring the first RAM already written to read mode and the second RAM to write mode. The read address, ra = 0, is applied at the same time. It may be noted that the first location data is 00127E766A5B4731 H since it is designed to be read column-wise. This data is the same as the first column of the first block of data written in the test bench. The data appears at 250 ns although the address was established at 217 ns. This is owing to two 'clk' delays, one each in the individual RAM design 'ram_rc' and the other in the top design 'dualram'. The corresponding two rising edges of the 'clk' occur at 230 ns and 250 ns. The first data appears after a latency of 33 ns after the read address is applied. However, subsequent data appear at the 'clk' rate of 20 ns. This advantage accrues from pipelining the design. Similarly, the writing of the

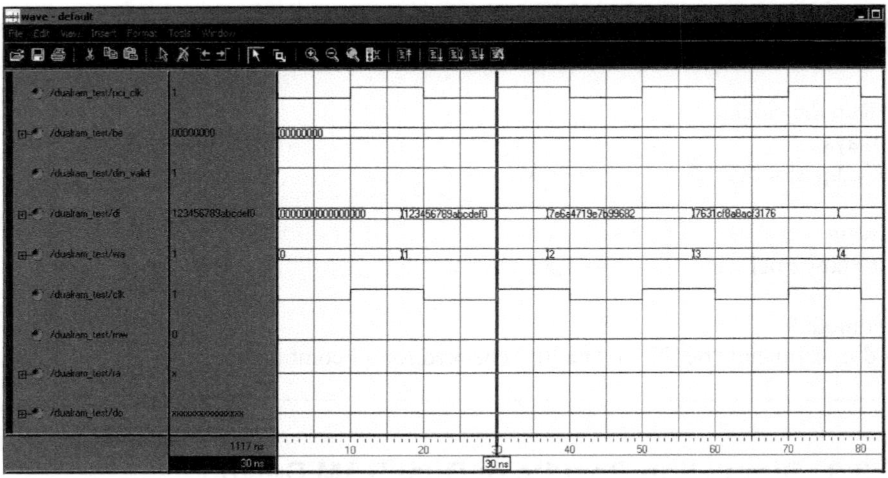

Fig. 9.9 Simulation results of dual RAM design – writing of first block

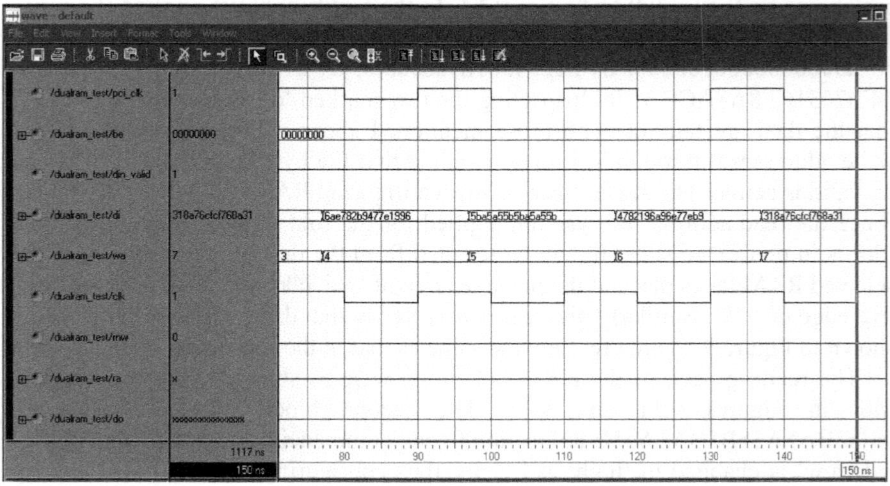

Fig. 9.10 Simulation results of dual RAM design – writing of first block

the second and the third blocks and concurrent reading of first and second blocks respectively may be verified by comparing waveforms shown in Figures 9.13 and 9.14 and the corresponding data listed in the test bench.

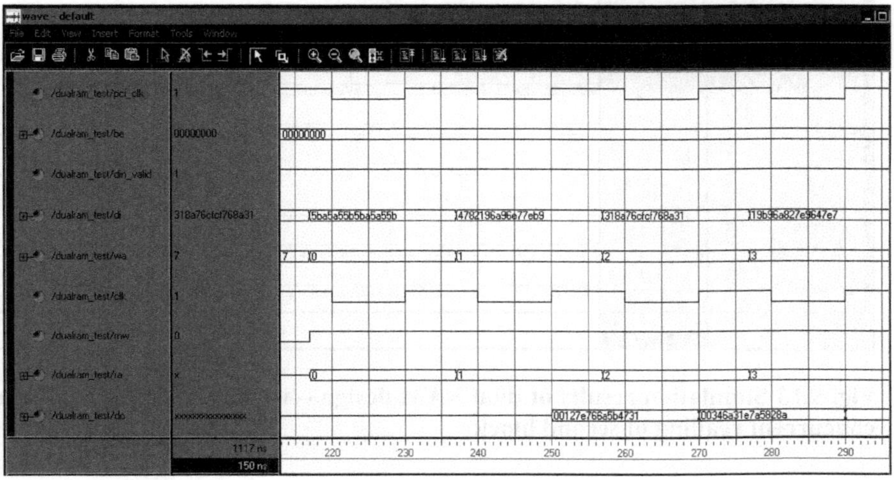

Fig. 9.11 Simulation results of dual RAM design – writing of second block and concurrent reading of first block

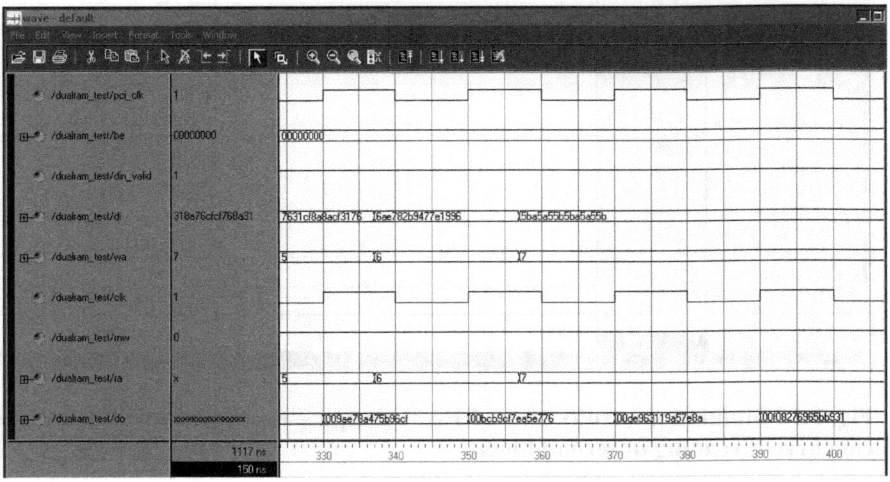

Fig. 9.12 Simulation results of dual RAM design – writing of second block and concurrent reading of first block

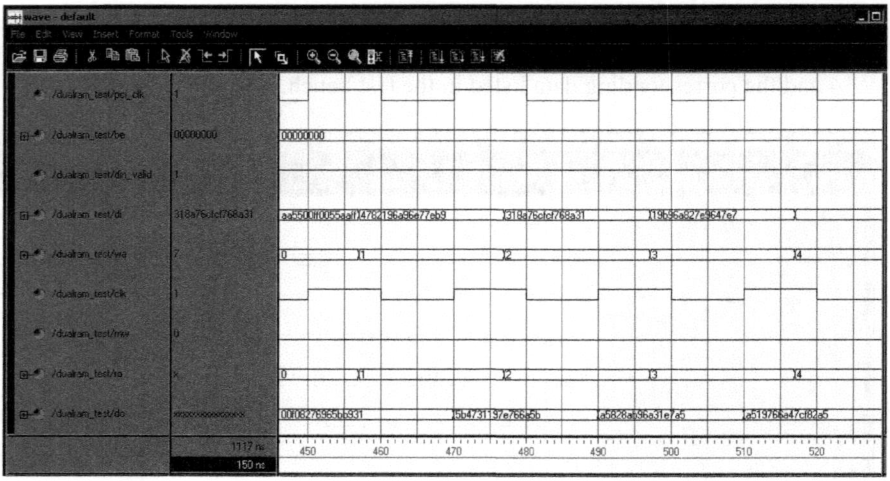

Fig. 9.13 Simulation results of dual RAM design – writing of third block and concurrent reading of second block

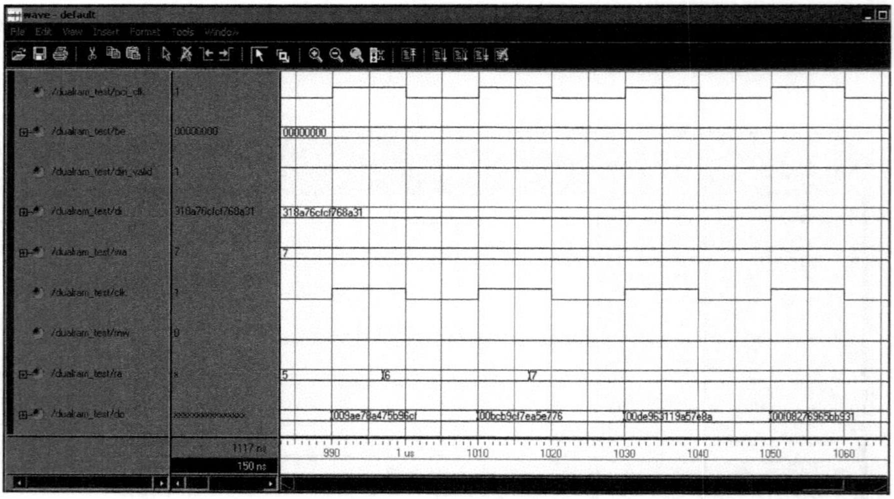

Fig. 9.14 Simulation results of dual RAM design – writing of third block and concurrent reading of second block

9.3.4 Synthesis Results for the Dual RAM Design

The Synplify results of dual RAM design are as follows. From the Synplify results tabulated, the estimated clock frequencies for writing and reading are 124 MHz and 160 MHz respectively. However, the frequency of operation reported by Xilinx place and route is more accurate and is presented in the next section. One hundred and twenty eight numbers of single port RAMs (RAM16X1S) have been used for the design.

Synthesis results:

@I::“D:\user\ram\verilog_latest\dvlsi_des_verilog\dualram.v”

@I:“D:\user\ram\verilog_latest\dvlsi_des_verilog\dualram.v”:“D:\user\ram\verilo g_latest\dvlsi_des_verilog\ram_rc.v”

Verilog syntax check successful!

START TIMING REPORT

Top view: dualram

Slew propagation mode: worst

Paths requested: 5

Constraint file(s):

@N| This timing report estimates place and route data. Please look at the place and route timing report for final timing.

@N| Clock constraints cover all FF-to-FF, FF-to-output, input-to-FF, and input-to-output paths associated with a particular clock.

Performance summary

Worst slack in design: 1.185

Starting Clock	Requested Frequency	Estimated Frequency
clk pci_clk	100.0 MHz 100.0 MHz	159.8 MHz 123.9 MHz
Requested Period	Estimated Period	Slack
clk : 10.000 pci_clk : 10.000	6.260 8.074	3.740 1.926

Mapping to part: xcv600ehq240-8

Cell usage:

VCC	3	uses
GND	3	uses
MUXF5	256	uses
MUXF6	128	uses
FDE	1152	uses
FD	65	uses
BUF	12	uses

I/O primitives:

IBUF	80	uses
OBUF_F_24	64	uses
BUFGP	2	uses

I/O register bits: 64

Register bits not including I/Os: 1152 (8%)

RAM/ROM usage summary

Single port rams (RAM16×1S): 128

Global clock buffers: 2 of 4 (50%)

Total LUTs: 925 (6%)

The RTL view of the design reported by the Synplify tool is shown in Figure 9.15.

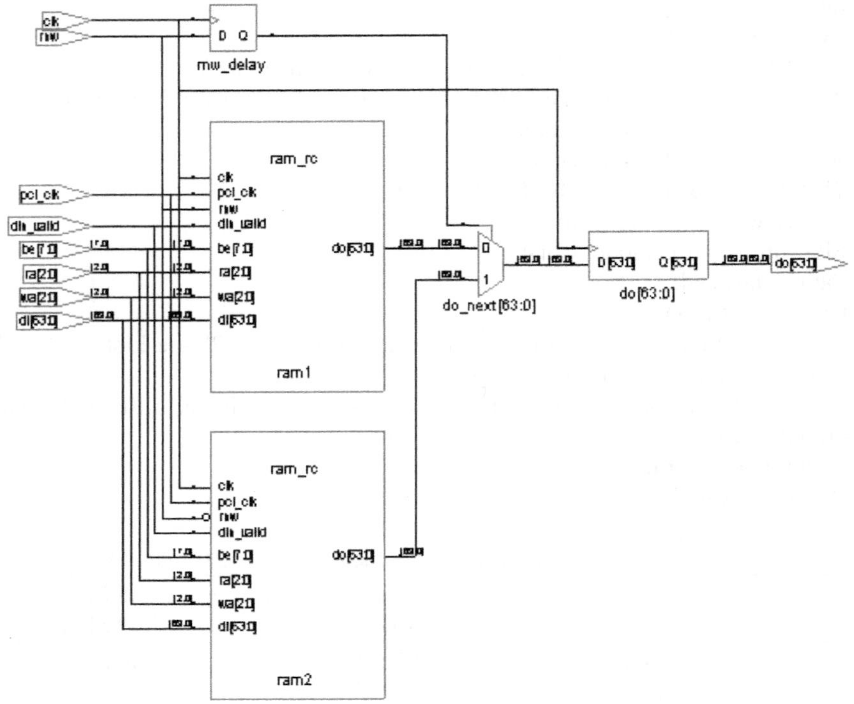

Fig. 9.15 RTL view of dual RAM design reported by Synplify tool

9.3.5 Xilinx P&R Results for the Dual RAM Design

The place and route tool report presented in this section reports the same numbers (128) of 16 × 1 RAMs as was reported in the report of synthesis tool earlier. In earlier designs such as ROMs, we have seen gate count equivalents of up to about 2000. In the present design of dual RAM, we can see a big leap in the total count of (32,114) gates. This clearly falls under the VLSI category. Maximum frequency reported by this tool is 99 MHz. However, what counts is the frequency report by place and route tool for the overall design such as the DCTQ (and not a part of design), which will be covered in chapter on design applications.

Design summary:

Number of slices:	928	out of	6,912	13%
Number of slice flip-flops:	1,152	out of	13,824	8%
Total number of 4 input LUTs:	935	out of	13,824	6%
Number used as 16 × 1 RAMs:	128			
Number of bonded IOBs:	144	out of	158	91%
IOB flip-flops:	65			

Number of GCLKs:	2	out of	4	50%
Number of GCLKIOBs:	2	out of	4	50%

Total equivalent gate count for design: 32,114
Additional JTAG gate count for IOBs: 7,008
Minimum period: 10.095 ns
(Maximum frequency: 99.059 MHz)

9.4 External Memory Controller Design

We have so far seen the designs of on-chip memories, ROMs and RAMs. If an application such as video scaling, demands a large memory, it is better to locate these memories external to an FPGA since on-chip memories in the order of 32 KB and above slows down the system speed considerably. The FPGA size can also be scaled down accordingly if the memory is located external to FPGA. This scheme is viable since fast RAMs of the order of 10 ns access times are available commercially. Refer Appendix 5 on CD for one of the suitable RAMs. Depending upon the actual application requirement, the designer can add more number of these devices. Now we will see how to interface FPGA/ASIC with the external RAMs.

9.4.1 Design of an External RAM Controller for Video Scalar Application

The block diagram of a controller required for an application, video scalar, which interfaces with an external RAM is shown in Figure 9.16. We wish to have a synchronous design; therefore a system clock 'clk_out' is used. The controller design has reset (active low) and hold pins as in a microprocessor based design. In addition to these signals, we have an enable pin, 'endram', for the external RAM. We also have a chip enable (ce_n), a write enable (wr_n), and a read enable (oe_n), all of which are active low signals. There is also a signal 'rwn' to configure the

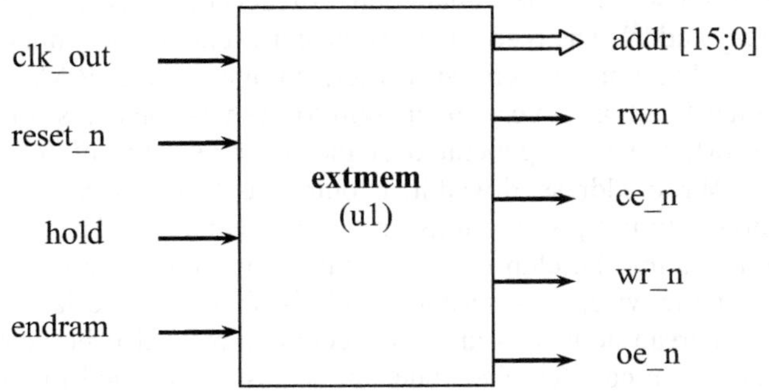

Fig. 9.16 External RAM controller

external RAM in write mode or in read mode. An address bus of 16 bits is provided so that we may access an external RAM of size 64 KB whether we write or read, one location at a time or a block of locations as a burst. A low at the pin 'rwn' sets the RAM in write mode, while a high sets it in read mode.

9.4.2 Verilog Code for External RAM Controller Design

Verilog_code_9.8 presents the RTL code for External RAM Controller used in Video Scalar application. First, we define the maximum address, 'drpixaddr', in the external RAM and then declare the module name and list all the inputs/outputs that we saw in the block diagram earlier. This is followed by the declaration of all the I/Os. All the signals used in the design are declared as 'wire' or 'reg' as the case may be. Also, we need the read address (drraddr) and write address (drwaddr) separately for the controller although externally one 'addr' bus is adequate. A high on 'rwn' implies that the 'addr' is the read address. Otherwise, it is 'drwaddr'.

The first assign statement 'res_addr' resets the address bus, 'addr' when maximum address is encountered. Using external RAM, we will not be able to achieve higher speed than the on-chip RAM. We need to scale it down by a factor of two as explained later. For example, if we have a system clock of 100 MHz, we can access the external RAM only at 50 MHz rate. In order to do this, we have counters separately for write (wr_cnt) as well as read (rd_cnt). These counters are enabled by signals 'enwr_cnt' and 'enrd_cnt' respectively. The write/read counters are only toggling flip-flops. They are configured as counters so that by increasing their widths, slower external RAM may be used as per the application requirement. The first sequential always block realizes the 'rwn' signal. This signal is initialized at the start and can be held if a 'hold' signal is asserted. When all the RAM locations are accessed in a burst, 'rwn' signal is toggled so that the RAM may be configured to read mode from write mode, for instance. The 1 bit counter 'wr_cnt' is changed every falling edge of 'clk_out' signal. The write pulse 'wr_n' for RAM is generated only if the write counter is enabled, i.e., enwr_cnt = 1. Otherwise, the write pulse is disabled.

The RAM write address 'drwaddr' is advanced, if enabled by the signal 'enwaddr'. It is reset when the terminal count is reached, indicated by the setting of the signal 'res_waddr'. The read counter, 'rd_cnt', function is similar to that of the write counter, 'wr_cnt', described earlier. Similarly, the RAM read address (drraddr) signal function is very much akin to the write address 'drwaddr'. Read signal for RAM, 'oe_n' is generated at the negative edge of 'clk_out' only if enrd_cnt = 1. Write address, drwaddr, is output as 'addr' if rwn = 0. Otherwise, the read address, drraddr, is output using 'assign' statement. The external RAM is selected by activating the chip enable signal, ce_n (active low), if the address is within range. Otherwise, it is disabled. This is shown as the last 'assign' statement. The write/read addresses and their counters may also be combined into a single address and a counter. The code can be easily modified for accessing dual or multiple external RAMs.

Verilog_code_9.8

```
// This is the design of an external RAM controller for a video scalar. Put this
// code in a file named 'extmem.v'.
`define max_drpixaddr 65535              // This is the last address of RAM.

module extmem (              clk_out,      // Clock signal.
                             reset_n,      // Reset input.
                             hold,         // Signal to hold processing.
                             addr,         // Address input.
                             endram,       // Enable RAM signal.
                             rwn,          // Read/write select signal.
                             ce_n,         // Chip enable,
                             wr_n,         // write and
                             oe_n          // output enable or read signal.
                           );
      input                    clk_out ;           // Declare inputs and
      input                    reset_n ;
      input                    hold ;
      input                    endram ;
      output      [15:0]       addr ;              // outputs of the design.
      output                   rwn ;
      output                   ce_n ;
      output                   wr_n ;
      output                   oe_n ;

      wire        [15:0]    addr ;                 // Declare signals used in
      wire                     ce_n ;             // 'assign' statements as 'wire'.

      reg                      rwn ;               // Declare signals in 'always'
      reg                      wr_n ;              // block as 'reg'.
      reg                      oe_n ;
      reg         [15:0]       drwaddr ;           // Write and
      reg         [15:0]       drraddr ;           // read address.

      wire        [15:0]       drwaddr_next ;      // Pre-advance write and
      wire        [15:0]       drraddr_next ;      // read address.
      wire                     enwaddr ;
      wire                     enwr_cnt ;
      wire                     enraddr ;
      wire                     wr_cnt_next ;
      wire                     res_addr ;
      wire                     res_waddr ;

      reg                      wr_cnt ;
```

```
        wire                    enrd_cnt ;
        wire                    rd_cnt_next ;

        reg                     rd_cnt ;

        wire                    res_raddr ;

assign res_addr = (addr == `max_drpixaddr)&
                  ( (wr_cnt == 1'b1)|(rd_cnt == 1'b1) )& (ce_n == 1'b0) ;
                  // Reset address when maximum address is encountered.

always @ (posedge clk_out or negedge reset_n)
begin
        if      (reset_n == 1'b0)
                rwn      <=      1'b0 ;          // rwn = 1 for read.
                                                 // Otherwise, write.
        else if (hold == 1'b1)
                rwn      <=      rwn ;
        else if (res_addr == 1'b1)
                rwn      <=      !rwn ;          // Change write to read
                                                 // mode or vice versa.
        else
                        rwn      <=      rwn ;
end

assign enwr_cnt = (endram == 1'b1)&(rwn == 1'b0)&(ce_n == 1'b0) ;
                                         // Condition for write counter.
always @ (negedge clk_out or negedge reset_n)
begin                                    // Write signal for RAM,
                                         // effective at the falling edge.
        if (reset_n == 1'b0)
                wr_n     <=      1'b1 ;  // Initialize.
        else if (hold == 1'b1)
                wr_n     <=      wr_n ;  // Hold the write pulse.
        else if (enwr_cnt == 1'b1)
                wr_n     <=      !wr_n ; // Generate RAM write pulse,
                                         // if enabled.
        else
                wr_n     <=      1'b1 ;  // Otherwise, disable write pulse.
end

assign wr_cnt_next = wr_cnt + 1 ; // Pre-advance the write counter.

always @ (posedge clk_out or negedge reset_n)
```

```
begin
        if (reset_n == 1'b0)
                wr_cnt  <=      1'b0 ;
                                // Counter to slow down RAM write by a factor of two.
        else if (hold == 1'b1)
                wr_cnt  <=      wr_cnt ;
        else if (wr_cnt == 1'b1)
                wr_cnt  <=      1'b0 ;          // Reset for maximum count.
        else if (enwr_cnt == 1'b1)
                wr_cnt  <=      wr_cnt_next;    // Advance the write counter.
        else
                wr_cnt  <=      wr_cnt ;        // Otherwise, don't disturb.
end

assign drwaddr_next = drwaddr + 1 ;             // Pre-advance the write address.
assign enwaddr = (endram == 1'b1)& (rwn == 1'b0)&(wr_cnt == 1'b1)
                                                &(ce_n == 1'b0) ;
assign res_waddr = (drwaddr == max_drpixaddr)&(wr_cnt == 1'b1)
                                                &(ce_n == 1'b0) ;
                                // Conditions for enabling and resetting write address.
always @ (posedge clk_out or negedge reset_n)
begin                           // Write address, drwaddr, for RAM.
        if (reset_n == 1'b0)
                drwaddr         <=      16'd0 ;
        else if (hold == 1'b1)
                drwaddr         <=      drwaddr ;
        else if (res_waddr == 1'b1)  // Reset when the terminal count is reached.
                drwaddr         <=      16'd0 ;
        else if (enwaddr == 1'b1)
                drwaddr         <=      drwaddr_next ;
                                // RAM write address is incremented, if enabled.
        else
                drwaddr         <=      drwaddr ; // Otherwise don't disturb.
end

assign rd_cnt_next = rd_cnt + 1 ;               // Pre-advance read counter.
assign enrd_cnt = (endram == 1'b1)&(rwn == 1'b1)&(ce_n == 1'b0) ;
                                // Condition for enabling the read counter.
always @ (posedge clk_out or negedge reset_n)
begin
        if (reset_n == 1'b0)
                rd_cnt          <=      1'b0 ;
                                // Counter to slow down RAM read by a factor of two.
        else if (hold == 1'b1)
                rd_cnt          <=      rd_cnt ;
        else if (rd_cnt == 1'b1)
```

```
                rd_cnt  <=      1'b0 ;    // Reset read counter.
        else if (enrd_cnt == 1'b1)
                rd_cnt  <=      rd_cnt_next ; // Advance the read counter.
        else
                rd_cnt  <=      rd_cnt ;
end

assign drraddr_next = drraddr + 1 ;          // Pre-advance the read address.
assign enraddr = (endram == 1'b1)&(rwn == 1'b1)&(rd_cnt == 1'b1)&
                                                    (ce_n == 1'b0) ;
assign res_raddr = (drwaddr ==`max_drpixaddr)&(rd_cnt == 1'b1)&
                                                   (ce_n == 1'b0) ;
                        // Conditions for enabling and resetting read address.
always @ (posedge clk_out or negedge reset_n)
begin                                        // Read address for RAM.
        if (reset_n == 1'b0)
                drraddr <=      16'd0 ;
        else if (hold == 1'b1)
                drraddr <=      drraddr ;
        else if (res_raddr == 1'b1)
                drraddr <=      16'd0 ;
                                // Reset when the terminal count is reached.
        else if (enraddr == 1'b1)
                drraddr <=      drraddr_next ;  // Increment the RAM address.
        else
                drraddr <=      drraddr ;
end

always @ (negedge clk_out or negedge reset_n)
begin                                               // Read signal for RAM.
        if (reset_n == 1'b0)
                oe_n            <=      1'b1 ;
        else if (hold == 1'b1)
                oe_n            <=      oe_n ;
        else if (enrd_cnt == 1'b1)
                oe_n            <=      !oe_n  ; // RAM read signal.
        else
                oe_n            <=      1'b1 ; // Otherwise, disable read pulse.
end

assign addr = (rwn == 1'b0)  ?  drwaddr : drraddr ;
        // Write address is output if rwn = 0. Otherwise, read address is output.
assign ce_n = (addr <= `max_drpixaddr)  ? 1'b0 : 1'b1 ;
                // Select RAM if the address is within range. Otherwise, disable.
endmodule
```

9.4.3 Test Bench for External RAM Controller Design

Now, we will look into the coding of the test bench for the external RAM controller. As usual, we will have a 100 MHz clock operation, and so we define the clock period by two as 5 ns. We have included the design file, 'extmem.v' in its back annotated form. The test bench simulates the external RAM of size 64 K × 24 bits. Data width is 24 bits in order to accommodate three color (RGB) pixel information required for video scalar application. We have 65,536 locations in the RAM, and each location takes two-clock cycles for either a read or a write operation. The test is conceived as a Go–No Go test, that is, whether the test has passed or failed since a large number of locations are involved in the test.

The design is called by addressing ports by name. The test inputs are applied in the 'initial' block. After the specified time, the test bench will automatically stop processing. The signal 'rwn' is 0 for the write mode. Otherwise, it is in read mode. Initially, we write the same data in all the locations. Thereafter, change the 'rwn' signal and start reading in order to check whether the data is in tact. While checking the contents of memory locations, it may so happen that some of the data are corrupted. These are indicated by setting an error flag. If no error is encountered during the test, we display the message 'External RAM Test PASS'. On the other hand, if any error is encountered, we display a failure message. If error is encountered in any location except the last, the error flag is set. The 'data' read from the external RAM is obtained by the last 'assign' statement. The test bench is presented in Verilog_code_9.9.

Verilog_code_9.9

```
// The test bench for external RAM controller is named 'extmem_test.v'

`define clkperiodby2 5           // Program the clock to run at 100 MHz.
`define max_drpixaddr 65535      // This is the last location of the external RAM.
`define test_data 24'h555555
                 // Change this data to 'aaaaaa', '000000' and 'ffffff' in turn and
                 // run the test again.
`include "extmem_banno.v"         // This is the external memory controller
                                  // design after back annotation.
module extmem_test (                      // Declare the test module and list the
                              addr,       // the outputs.
                              ce_n,
                              wr_n,
                              oe_n
                        );
      output      [15:0]      addr ;      // Declare the outputs.
      output                  ce_n ;
      output                  wr_n ;
      output                  oe_n ;
```

```
        reg                      clk_out ;// Declare the inputs as registers

        wire        [15:0]       addr ;    // and combinational circuit signals
        wire                     ce_n ;    // as wires.
        wire                     wr_n ;
        wire                     oe_n ;

        reg                      reset_n ;
        reg                      hold ;
        reg                      endram ;
        reg                      error ;

        wire        [23:0]       data ;
    // Data width is 24 bits to accommodate three color (RGB) pixel information.

        reg         [23:0]    mem [`max_drpixaddr:0] ;
                                 // Simulate external RAM, 64 K × 24 bits.

        extmem u1 (              // Invoke the design and call  ports by name.
                    .clk_out(clk_out),
                    .reset_n(reset_n),
                    .hold(hold),
                    .addr(addr),
                    .endram(endram),
                    .rwn(rwn) ,
                    .ce_n(ce_n),
                    .wr_n(wr_n),
                    .oe_n(oe_n)
                  );
initial
begin
         clk_out        =        1'b0 ;    // Initialize all input signals.
         reset_n        =        1'b1 ;
         hold           =        1'b0 ;    // Repeat the test by asserting hold.
         endram         =        1'b0 ;
        #10  reset_n    =        1'b0 ;
        #30  reset_n    =        1'b1 ;
        #16  endram     =        1'b1 ;
        #2800000              // Run long enough to test all the 64 K locations.
        $stop ;
end

always @ (rwn or addr)
begin
        if (rwn == 0)
```

```
                mem [addr]      <=      `test_data ;
                // Write the same data into all locations of the external RAM.
                // Note that the design automatically takes care to advance the
                // address.
        else    ;       // Otherwise, don't disturb.
end

always @ (addr)
// Read data from the external RAM and check with the written data for all
// locations.
        case ({(rwn == 1), (mem [addr] == `test_data ),
                        (addr == `max_drpixaddr )})
                3'b111: // This state corresponds to checking the last location.
                        begin
                                if (error == 0)
                        // If no error is encountered, the design passes the test.
                                $display ("External RAM test: PASS") ;
                                else
                                        // If error is encountered, then the test fails.
                                begin
                                  $display ("External RAM test: FAIL") ;
                                  error <= 0 ;
                                end
                        end
                3'b101:
      $display ("External RAM test: FAIL  => Last address @   %d", addr) ;
                3'b100: error <= 1 ;
                        // If error is encountered in any location except the last,
                        // the error flag is set.
                default: ;              // Take care of other possibilities.
        endcase
assign data     =       mem [addr] ;            // Get the data.

always
        #`clkperiodby2 clk_out <= ~clk_out ;    // Run the clock continuously.

endmodule
```

9.4.4 Simulation Results for External RAM Controller Design

The simulation results for the external RAM controller design are shown in Figures 9.17 to 9.20. The first figure shows the back annotated design loaded (identified by FPGA primitive components such as LUT4, INV, BUF, etc.) and a Go–No Go result. The display, External RAM test: PASS, indicates that the external RAM has passed the test. This implies that all the 65,536 locations, each of size 24 bits, are successfully written and read back.

We are justified in running the simulation at 100 MHz as can be seen from the waveforms displayed in Figure 9.18. The memory access is at 50 MHz. This means that we can use a standard RAM of access time of about 10 ns available in the market. The rwn signal is low at the time of writing as can be seen in the figure. The data to be written, 555555 H, appears in signal 'data' as a result of the two statements, mem [addr] <= `test_data and assign data = mem [addr], present in the test bench. It should be noted that the address is changing once every two clock cycles at the rising edge. The waveform shows the writing through the address range 0 to 5. Also, the active high enable external RAM (endram) and the active low chip select (ce_n) are activated and the write pulse is applied commencing from the zero address after the address stabilizes.

The write signal, wr_n, goes low only after about 7 ns from falling edge of 'clk_out' signal although as per the design it should have occurred at the negative edge of 'clk_out'. This is clearly due to the actual gate delays in the FPGA since we have used the back annotated design in the simulation. This fact can be verified by running the simulation by including the design source file, 'extmem.v', in the test bench in lieu of the back annotated file, 'extmem_banno.v'. The reader may also attempt running the simulation with back annotated design at a much higher frequency than 100 MHz to see whether the design works or not. The write

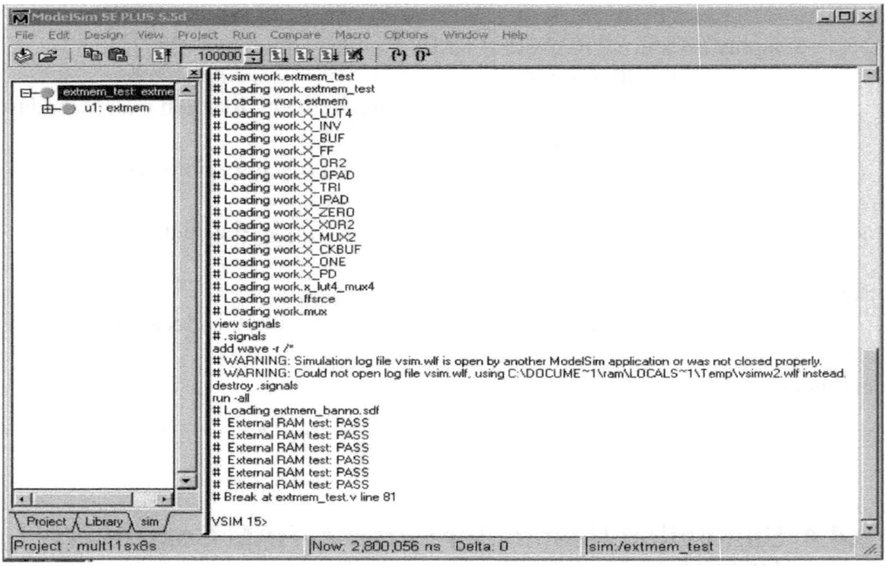

Fig. 9.17 Simulation results for external RAM controller design – Go–No Go test

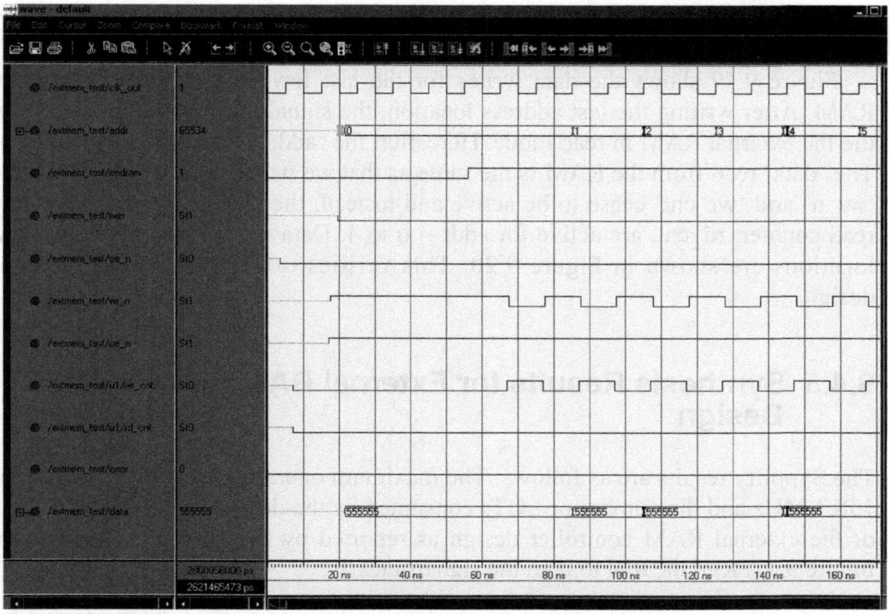

Fig. 9.18 Simulation results of external RAM controller design – commencement of data writes

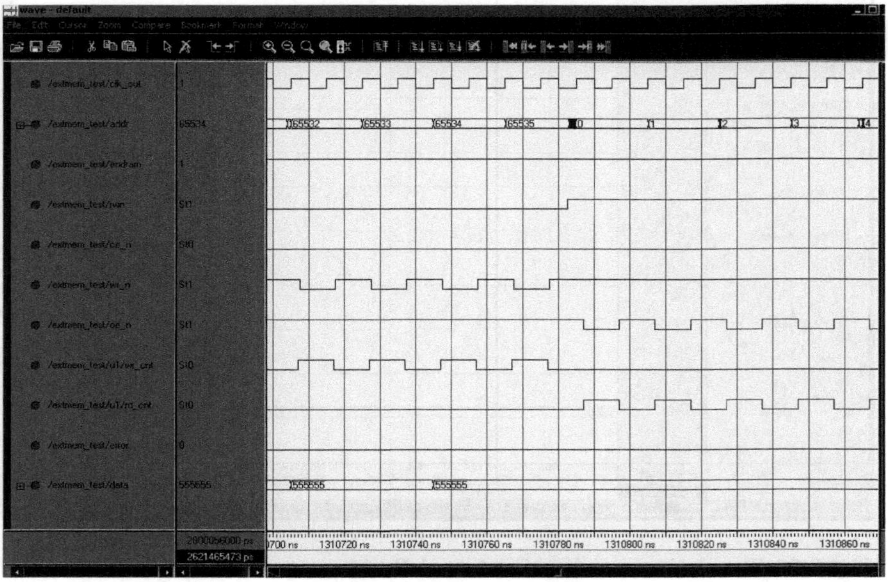

Fig. 9.19 Simulation results of external RAM controller design – data writes of last few locations and commencement of read back

counter, wr_cnt, appears as an inverted signal of the write signal, wr_n, since only 1-bit is used for wr_cnt.

Figure 9.19 shows the data writes for the last few locations of the external RAM. After writing the last address location, the signal rwn is toggled to configure the external RAM in read mode. Hereafter, the 'addr' reflects the read address. The 'data' read from the RAM is the same as that we wrote earlier. Also, note that 'wr_n' and 'wr_cnt' cease to be active and instead, the read pulse, oe_n, and the read counter, rd_cnt, are active for addr = 0 to 4. Data reads towards the last few locations are shown in Figure 9.20. This verifies the correct working of our design.

9.4.5 Synthesis Results for External RAM Controller Design

The Synplify results are as follows. The maximum operating frequency reported is 138.3 MHz and the number of LUTs consumed by the design is just 55. RTL view of the external RAM controller design as reported by Synplify tool is shown in Figure 9.21. Run the tool to read the signals clearly.

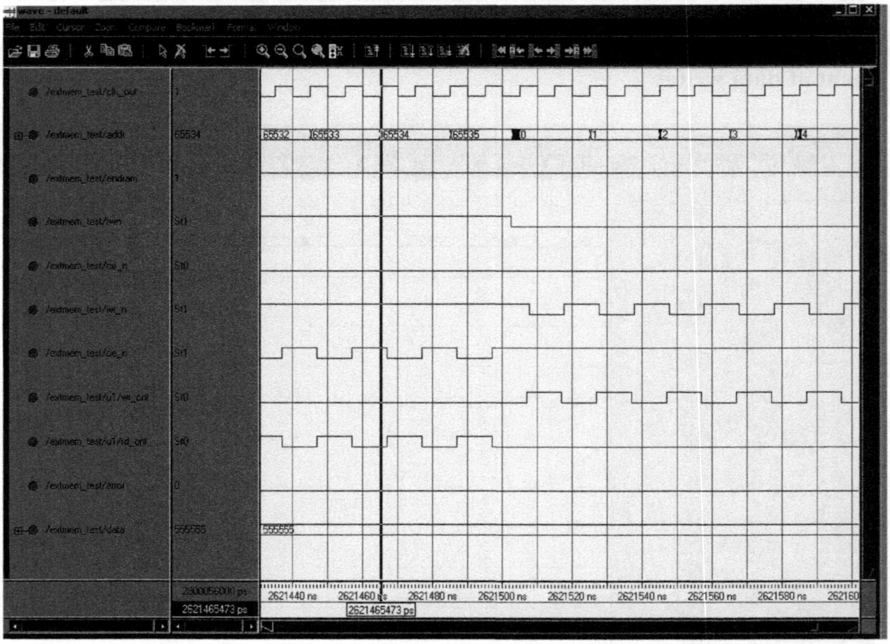

Fig. 9.20 Simulation results of external RAM controller design – data reads towards the last few locations

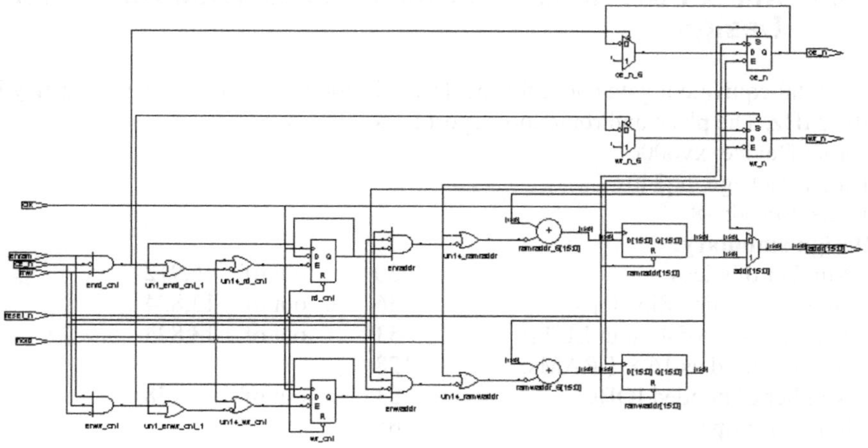

Fig. 9.21 RTL view of external RAM controller design

Performance summary:
Worst slack in design: 1.315

Starting Clock	Requested Frequency	Estimated Frequency
clk	100.0 MHz	138.3 MHz
Requested Period	Estimated Period	Slack
10.000	7.233	2.767

Resource usage report for extmem
Mapping to part: xcv600ehq240-8
Cell usage:

MUXCY_L	28 uses
XORCY	30 uses
FDCE	32 uses
FDPE_1	2 uses
FDC	2 uses
GND	1 use

I/O primitives:

IBUF	5 uses
OBUF	18 uses
BUFGP	1 use

I/O register bits: 0
Register bits not including I/Os: 36 (0%)
Global clock buffers: 1 of 4 (25%)
Total LUTs: 55 (0%)

9.4.6 Xilinx P&R Results for the External RAM Controller Design

The total equivalent gate count for the design is 600 and the operating frequency is 100 MHz. The place and route tool report is as follows:
Target Device: xv600e
Target Package: hq240
Target Speed: –8

Design summary:

Number of slices:	30	out of	6,912
Number of slice flip-flops:	36	out of	13,824
Total number of 4 input LUTs:	53	out of	13,824
Number used as 16 × 1 RAMs:	128		
Number of bonded IOBs:	23	out of	158
IOB flip-flops:	65		
Number of GCLKs:	1	out of	4
Number of GCLKIOBs:	1	out of	4
Total equivalent gate count for design:	600		
Additional JTAG gate count for IOBs:	1152		

Summary

Memory design is one of the most important aspects of a VLSI system design. This chapter showed the way to design various types of on-chip ROMs and RAMs, some of them unconventional, in order to meet the special requirements of a particular application. The size of memory that could be incorporated on-chip is usually limited by the order of a few tens of Kilo Bytes with the currently available FPGAs. This limitation is fast changing with the advances in the technology. In applications, where large memories are called for, external memories such as the commercially available RAMs, ROMs, Flash RAMs, etc. may be used. Towards this end, a controller design that interfaces with an external memory was presented in the text. However, the access speed of external memory falls by a factor of two when compared to on-chip memory. On the other hand, on-chip memory increases the chip area consumed. Therefore, the designer must consider carefully the pros and cons before making the choice for on-chip or external memory in a system design. Another important design in a system is the arithmetic circuit design, which is presented in the next chapter.

Assignments

9.1 A ROM can be used to multiply two binary numbers by splitting the address lines to accommodate the two numbers. Implement using Verilog such a multiplier for multiplying two signed numbers, each of size 4 bits. Verify your results by simulation. Will this be an efficient implementation if used for two 8-bit, unsigned numbers? Discuss.

9.2 A simple squaring circuit may be designed using ROM. Implement such a circuit using Verilog for squaring numbers up to 15 if unsigned, and –8 to +7 if signed. Use a single ROM of minimum possible size. Verify your results by simulation.

9.3 Implement in Verilog a ROM based square root circuit for unsigned numbers in the range 0 to 15. Provide three digits after decimal point. Write a test bench and verify your results.

9.4 Realize a circuit using Verilog to compute the cube of a BCD number. Provide an error flag and clear the result if the input number exceeds its range. Write a test bench and present your simulation results.

9.5 Single address ROM shown in the text is organized as 8 × 64 bits but accessed byte-wise. Redesign the same with 64 locations, each of size 8 bits. The ROM is read as a byte at one time. Verify your design.

9.6 ROM can be tested by finding out the 'check sum' or 'signature' for the specified address range. The signature can be either 1 byte or two bytes in width. It is obtained by adding the succeeding locations (bytes) at a time, ignoring the carry every time. In this manner all the bytes are added up (accumulated) to get a single byte check sum. In the case of double byte check sum, addition is carried out on a word by word basis, ignoring the carry generated at every addition. A word consists of two bytes, with lower order address as the MSB. The final check sum in this case will naturally be a double byte. The check sum, expressed in hexadecimal, provides a very convenient, 'Go–No Go' test. This also serves the purpose of identifying the 'Program' or 'Data' stored in a ROM. Write a Verilog code to implement the computation of the single byte as well as the double byte check sums for a ROM of size 64 KB organized as bytes. The address range must be user specified.

Example:

Address range: 2000 H to 2007 H.

Address	ROM data
2000 H	11110000
2001 H	01111000
2002 H	00111100
2003 H	00011110
2004 H	00001111
2005 H	10000111
2006 H	11000011
2007 H	11100001

Single byte check sum for the above example is 'FC H', and the double-byte signature is 'FFFE H'.

9.7 A test pattern generator is required to be designed with the following specifications:

a. Accept specified number of 16 bit data inputs by the user. A maximum of 64 such numbers can be specified.
b. Output 16 bit pattern starting from the first user specified data.
c. Delay by user specified time, say, 10 ms or 100 ms.
d. Output the next 16 bit pattern specified by the user.
e. If all the user specified patterns are output start a fresh cycle from Step a. Otherwise, repeat from Step c.

Draw an elaborate specification and realize the Verilog design of the pattern generator and test it.

9.8 RAMs can be tested for soft errors in the following way called the Ramp Test. This test is conducted at one location. To start with, '0's are written into all the data bits in a byte and read back to make sure that the read data is in tact. The data is incremented by '1' and the above process is repeated until the maximum possible data (255 decimal for a byte based test) is encountered. The user specifies the address location for which the ramp test is conducted. Write and test the Verilog code to accomplish this test. The result must be indicated by pass or fail.

9.9 RAMs can also be tested using what is called the Walk Test. This test is also conducted at one byte location. The user specifies the address location as well as one byte data to start the test. Write the user specified data in to the address location of the RAM under test, and check the RAM back to ensure that the written data is correctly read back. At the second step, rotate the data in the RAM right by 1 bit and check the written data is correct. Continue this test till all the data bits are covered. Realize this design and write a test bench in Verilog to test the same for a RAM organized as bytes. Assuming the specified data is 11110000 at RAM location 1000 H, for example, the test has to walk through the following data pattern:

RAM data @ 1000 H

11110000
01111000
00111100
00011110
00001111
10000111
11000011
11100001

9.10 A sequential system can be implemented using a ROM and D flip-flops. The ROM can be used to implement the combinational circuit part of the system, while the flip-flops serve as the registers. The number of inputs to the ROM is equal to the sum of the number of flip-flops and the number of external inputs. The number of outputs of the ROM is equal to the sum of the number of flip-flops and the number of external outputs. Such a ROM based system is shown in Figure A9.1a. The ROM truth table is identical to the state table with 'present state' and 'inputs' specifying the address of ROM and 'next state' and 'outputs' specifying the ROM outputs. The next-state values must be connected from the ROM outputs to the register inputs. Realize the Verilog RTL for a programable counter, whose state diagram is shown in Figure A9.1b. With power on reset, the counter must be cleared. For in1 = 0 and in2 = 0, the counter functions as an Up counter, while it functions as a Down counter for in1 = 1 and in2 = 0. It can also count in steps of two when 'in2' is high. An output is set when the counter touches the last count in every mode. Write a test bench and test your design.

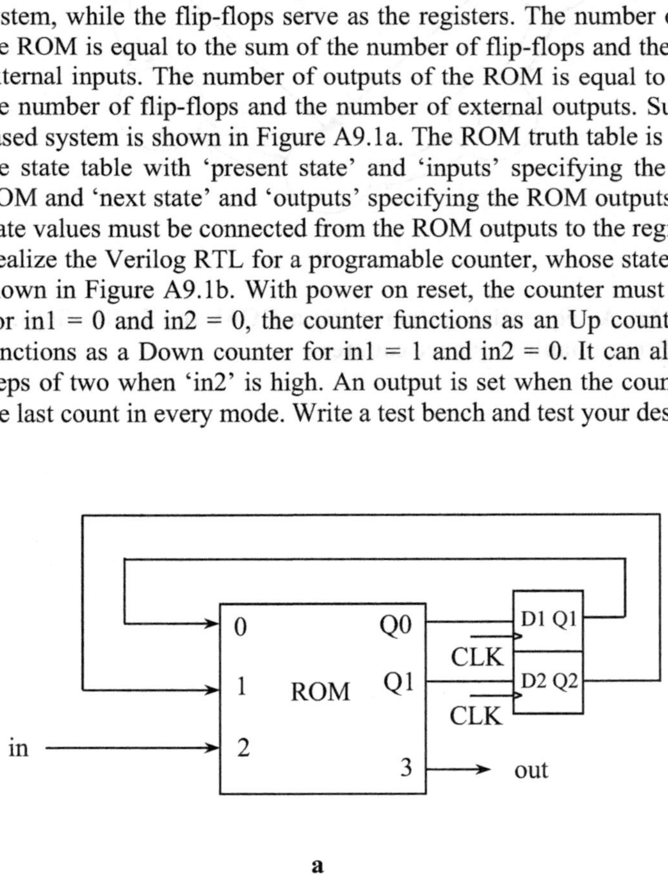

a

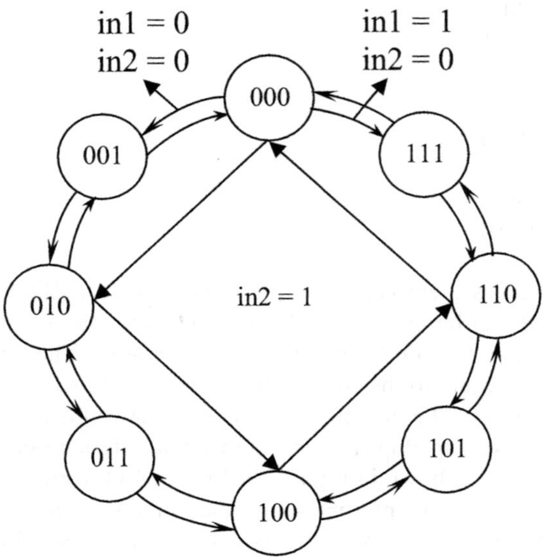

b

Fig. A9.1 (a) A sequential system using a ROM and D flip-flops. (b) State graph of Programable sequence counter

Chapter 10

Arithmetic Circuit Designs

One of the most important categories of circuits that we need to design for FPGA or ASIC implementation is the arithmetic circuits. Basic arithmetic circuits are add, subtract, multiply, divide; unsigned or signed. All these circuits are computationally intensive and, therefore, conventional methods are not sufficient. We will have to base our designs on what is popularly known as pipelining in order to speed up the processing. The throughput is substantially improved by building a high degree of parallelism in our designs, of course at the cost of additional chip area. The arithmetic circuits presented in this chapter basically stems from specific applications such as DCTQ, which design will be presented in a later chapter. We will cover fixed-point arithmetic and not floating point arithmetic in our designs. This is because fixed-point arithmetic is simple and takes minimum chip area.

In the traditional approach, processes such as add, subtract, multiply, etc., are treated as a single process, which may take considerable amount of time for processing when configured on an FPGA or an ASIC. Suppose that one of the processes such as a multiplier takes 100 ns for processing as shown in Figure 10.1. If we are to adopt the traditional method of processing for computationally intensive applications like video compression, video scaling, etc., we will have to make large compromise on the specifications such as picture resolution, frame rate, etc. In most cases, real time operations are severely hampered if implemented in the traditional way. For example, a high resolution color picture of size 1600 × 1200 pixels in the traditional way can be processed only at about 3 frames per second, thus falling far short of 30 frames per second prescribed in the standards such as MPEG 2 for motion picture compression. This means that we need to speed up processing ten-fold, unfortunately using the same speed grade of FPGA for want of another. Fortunately, adopting high pipeline and massively parallel circuit approach in our design makes this feasible in seemingly impossible task. This approach has more advantages than disadvantages. The method of pipelining is explained in the next section.

10.1 Digital Pipelining

Consider a pipe carrying oil or water from one place to another. In order to bring about this, we need a motor to pump the liquid. This process will naturally have some delay before the liquid is available for use at the end of the pipe. This delay may be referred to as latency. Once the pipeline is full, the vital liquid is available to the consumer continuously like a perennial river. This analogy of pipelining may be effectively applied to flow of data in a digital system. In this digital pipelining, we have data or control signals, etc., flowing through registers that may be regarded as pipes and the system clock as the driving motor. Thus, the data, etc., are carried from one part of a circuit to another via a series of registers which are clocked. Data flows from one register into another whenever the clock strikes. Enroute, the data may undergo any type of process such as add, subtract, multiply, compare, etc. By this means, any complex algorithm can be solved, often with spectacular speed-up of processing time.

Pipelined approach is basically dividing an entire process into small and roughly equal time consuming sub-processes such that the total processing time of these sub-processes equals the total processing time of the entire process. For example, Figure 10.1a shows the traditional approach of processing an operation such as a multiplier in about 100 ns. In the pipelined approach, we divide this process into ten sub-processes, each of approximately 10 ns processing time. After each sub-process, we add a register with a clock signal. As shown in the figure, the input data is applied to Proc. 1, which process is completed, say in 10 ns. The result of this sub-process is registered in Reg. 1 at the positive edge of the clock. This is subjected to a sequel Proc. 2 followed by registering in Reg. 2. This is repeated up to Proc. 10, registering the desired final result in Reg. 10. Thus, the data flows in a digital pipeline from input to the final output, traveling from Reg. 1 to Reg. 10 successively, and undergoing various processes on the way. Since the data will have to travel through ten registers, we will have to wait for ten clock pulses

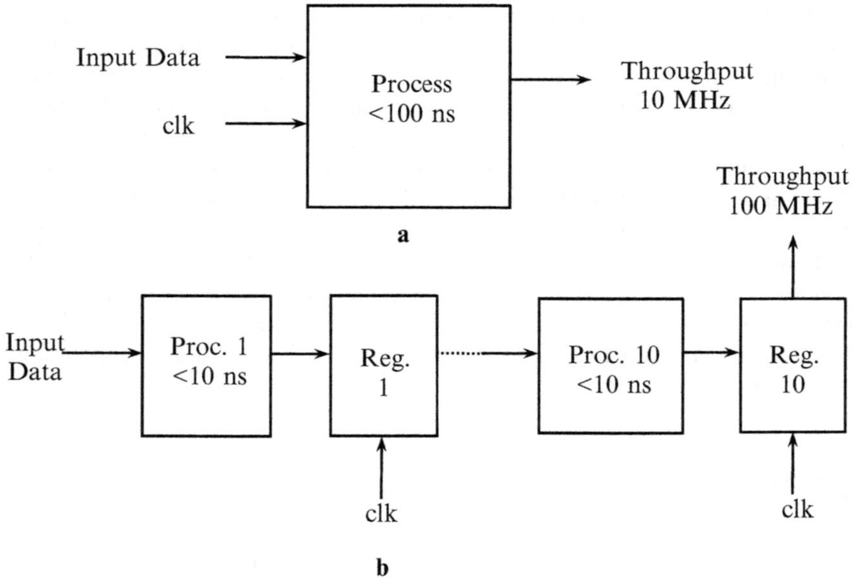

Fig. 10.1 (a) Traditional approach. (b) Pipelined approach

Time (ns)	Input	Reg. 1	Reg. 2	………	Reg. 10
0	Data1				
10	Data2	Proc.1_1			
20	Data3	Proc.1_2	Proc.2_1		
⋮	⋮	⋮	⋮		
100	Data11	Proc.1_10	Proc.2_9	………	Proc.10_1
110	Data12	Proc.1_11	Proc.2_10	………	Proc.10_2
⋮	⋮	⋮	⋮	⋮	⋮
190	Data20	Proc.1_19	Proc.2_18	………	Proc.10_10

Latency: 100 ns.

Fig. 10.2 Processing order of pipelining

for the output to manifest at Reg. 10. This delay is referred to as the latency. If each clock pulse takes 10 ns to arrive, then the output is available after 100 ns, which is the same as in the traditional method. Once the pipeline is full, we get a stream of processed results every 10 ns. Thus, the advantage in pipelining is that we can have a throughput of (and also a clock of) 100 MHz instead of 10 MHz in the traditional approach. That means ten-fold processing speed when compared to the traditional method. However, we need to apply the input(s) every 10 ns, the same as the output rate. The foregoing treatment of pipelining is shown in Figure 10.2, which is self-explanatory. It may be noted that Proc.10_1, Proc.10_2, up to Proc.10_10 are the results corresponding to the inputs Data1, Data2 up to Data10. The effect of pipelining may be summarized as follows:

- Throughput increases considerably
- Latency comes into effect
- Chip area increases marginally

10.2 Partitioning of a Design

In order to incorporate pipelining in the design, we need to break a sequence of operations or a complex algorithm into convenient small steps in terms of the following:

- Partition of data width
- Partition of functionality

The following sub-sections discuss the methodology of partitioning.

10.2.1 Partition of Data Width

Let us consider a process of adding two 16-bit numbers. This will be a time consuming process if addition is carried out on 16 bits since bit-wise carry out generated need to propagate through all the 16 bits. A better way of doing this is to bifurcate it into two 8-bit numbers and add only 8 bits at a time. That will be faster than adding 16 bits at one go. This can be effectively carried out by introducing pipelining. The LSBs of the two numbers are added first and stored in a pipeline register along with the generated carry at the rising edge of the system clock. In the next rising edge of the clock, MSBs of the two numbers are added along with the carry generated while adding the LSBs. In this fashion, we can divide and conquer the entire data width, no matter how wide it is. There are no hard and fast rules for this division of width. One has to experiment with it and choose the best possible bifurcation applicable for a particular application. We will illustrate the partitioning of data width by an example, a signed adder with the following specifications:

1. Eight signed input numbers, each of width 12 bits
2. Sum of these numbers are required

Conventional approach of addition/subtraction uses all the 12 bits together. Since full adders are used for implementation, the result is delayed owing to the propagation of carry rippling through all the 12 bits. Even the usage of 'carry look ahead' circuit does not help in speeding up the computation since a large number of gates and inputs are required in this case. The answer for this problem is to divide the data widths into smaller and equal chunks, and introduce pipelining. In the data width partitioning approach, all sub-blocks do the same function, namely addition. Before we take this problem for Verilog implementation, we will also see what partitioning of functionality is in the next section.

10.2.2 Partition of Functionality

Functionality is any process such as addition, subtraction, multiplication, or division. We need to group similar functions such as multiplication together. Also, the functional block is divided into smaller sub-blocks, if this is feasible. In this type of partitioning, each sub-block does a different function, in general. This can be clearly understood by considering an example. Let us say, we wish to compute a sum of products: a1*b1 + a2*b2 + a3*b3 + a4*b4, where a1, b1, etc., are each of size 16 bits. We can group multiplication functions, a1*b1, a2*b2, and a3*b3 together and do all these computations simultaneously and register the partial products. Similarly in the subsequent pipeline stage, we can perform additions A = (a1*b1 + a2*b2) and B = (a3*b3 + a4*b4) concurrently. In a next pipeline stage, the final addition, result = A + B, which is the desired sum of products, is performed. It may be noted that products such as a1*b1, etc., can be broken down into smaller sub-blocks, namely, shift operations and additions as illustrated in multiplier design in a later section. In the signed adder example cited earlier, LSBs (7 bits) of the eight numbers are added concurrently followed by the addition of MSBs (5 bits along with carry from LSB addition) in subsequent pipeline stages. This example is simpler than the sum of products example.

10.3 Signed Adder Design

We will take the signed adder example cited in the previous sections, wherein eight signed 12-bit numbers are added together. This can be realized in two different ways:

- Feeding each input serially or sequentially
- Feeding inputs concurrently

In a serial adder realization, we apply the eight inputs one after another serially at the positive edge of the clock. In contrast to this, in parallel adder, all the eight inputs are applied simultaneously. If chip area is of great concern, we go for the serial adder. However, for high speed processing, concurrent adder is preferred.

10.3.1 Signed Serial Adder

We may regard the add/subtract circuit as a simple adder if the numbers are in twos complement. We are going to add eight numbers, n0 to n7, of width 12 bits each and, all of them are fed through a single input, n[11:0]. The accumulated result will be available in a register, sum [14:0], as shown in Figure 10.3. Note that

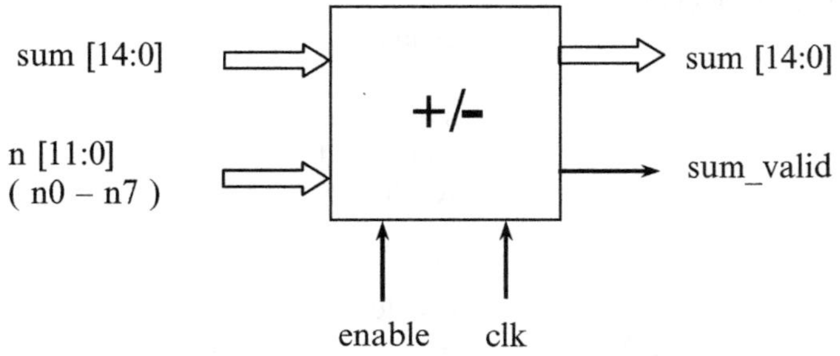

Fig. 10.3 Serial signed adder design

the output is fed back as one of the two inputs of the serial adder as it is an accumulator. Since the input is of size 12 bits, the output should be of size 15 bits. The difference of 3 bits between the input and output sizes is owing to the number of inputs being 2^3. The validity of the output is announced by a signal "sum_valid". All these transactions take place only when the "enable" is high. Although this is a pipelined approach, we are going to use only a single register, the accumulator.

Pipelined Serial Adder Design

The code for addition of eight, 12 bit, twos complement numbers is shown in Verilog Code_10.1. The inputs are fed serially at pins marked "n". The design module is declared as "serial_adder12s", listing all the inputs/outputs. The inputs are the system clock, enable, and n. The sum and result are the outputs. The signal, sum_valid, goes high when the added sum is valid. The "result" is the same as the "sum" except for the difference that the added result is prolonged at the "result" output till it is overwritten by a new result. A 3-bit counter, cnt [2:0], keeps track of the number of inputs accumulated. The first assign statement computes the "sum" in advance (sum_next [14:0]) if "enable" is high. Otherwise, it is cleared. Note that the sum is sign extended by 3 bits since the result is 3 bits more than the input number(s). Also, note carefully the number of flower brackets used. Otherwise, compiler tool will complain. The counter, cnt, is pre-advanced if enabled. The sum is valid after inputting the eighth number. An advanced valid signal, sum_val, is switched on only when "cnt" equals 7. The first "always" block registers the advance sum computed earlier when the clock strikes. Also, the "cnt" is incremented, every time an input is accumulated. The "sum_valid" is set high if all the eight input numbers are exhausted. The last "always" block registers the "result" whatever was in "sum" if "sum_valid" is active. Otherwise, the result is not disturbed.

Verilog Code_10.1

```
// Place the design in a file named "serial_adder12s.v".

module serial_adder12s (        clk,
                                enable,
                                n,
                                sum,
                                sum_valid,
                                result
                        );
input                           clk ;
input                           enable ;
input           [11:0]          n ;
output          [14:0]          sum ;
output                          sum_valid ;
```

```
output          [14:0]          result ;
                        // Prolong the result till it is overwritten by a new result.

wire            [14:0]          sum_next ;      // Declare nets in the design.
wire            [2:0]           cnt_next ;
wire                            sum_val ;

reg             [14:0]          sum;
reg             [2:0]           cnt ;
reg                             sum_valid ;
reg             [14:0]          result ;

assign  sum_next [14:0]  =  enable ? ({{3{n[11]}}, n[11:0]} + sum[14:0]) : 0 ;
                                        // Sign extend & accumulate.
assign  cnt_next [2:0]   =   enable ? (cnt + 1) : 0 ;
                                        // Pre-advance the counter.
assign  sum_val   =   (cnt == 7) ? 1 : 0 ;   // Pre-determine the validity of the sum.

always @ (posedge clk)                  // Pipeline – Register the sum.
begin
        sum [14:0]      <=      sum_next [14:0] ; // Register the sum.
        cnt [2:0]       <=      cnt_next [2:0] ;   // Advance the count.
        sum_valid       <=      sum_val ;          // Register the signal.
end

always @ (posedge clk)
                // Prolong the result till it is overwritten by the new result.
        result[14:0]    <=      sum_valid ? sum[14:0] : result[14:0] ;
                                                   // Register the sum.
endmodule
```

Test Bench for Signed Serial Adder Design

The functional verification of serial adder design can be carried out by writing a test bench as shown in Verilog Code_10.2. The simulation is carried out at 50 MHz by defining half clock period as 10 ns. The design included in the test bench is "serial_adder12s.v". The test bench module is declared as "serial_adder12s_test" and only the required outputs are listed. All the inputs are declared as usual as "reg". This is followed by calling the design, serial_adder12s, and ports by name. In the "initial" block, we apply two sets of eight inputs sequentially, each spaced by 20 ns. Enable is de-asserted for one clock cycle before applying the second set of inputs so that the accumulated "sum" is cleared.

Verilog Code_10.2

```
// Place the following test bench in a file named "serial_adder12s_test.v".
`define clkperiodby2 10
`include "serial_adder12s.v"

module serial_adder12s_test (              sum,
                                           sum_valid,
                                           result
                        );

        output          [14:0]             sum;
        output                             sum_valid ;
        output          [14:0]             result;

        reg                                clk ;
        reg                                enable ;
        reg             [11:0]             n ;

serial_adder12s   u1(   .clk(clk),
                        .enable(enable),
                        .n(n),
                        .sum(sum),
                        .sum_valid(sum_valid),
                        .result(result)
                  );
initial
begin
        clk             =       0 ;

// Apply first set of inputs sequentially every 20 ns.
                n               =       12'h0 ;         // n0 @ 0 ns.
                enable          =       0 ;
        #20     enable          =       1 ;
        #17     n               =       12'hfff ;       // n1 @ 37 ns.
        #20     n               =       12'h7ff ;       // n2 @ 57 ns, etc.
        #20     n               =       12'h800 ;
        #20     n               =       12'h001 ;
        #20     n               =       12'h001 ;
        #20     n               =       12'h7ff ;
        #20     n               =       12'haaa ;       // n7 @ 157 ns.
        #20     n               =       12'h0 ;
                enable          =       0 ;
                                // Disable before applying the next set of inputs
                                // so that the accumulated "sum" is cleared.
```

```
        #20   enable   =     1 ;  // Apply the next set of inputs.
              n        =   100 ;  // n0
        #20   n        =   200 ;
        #20   n        =   300 ;
        #20   n        =   400 ;
        #20   n        =   500 ;
        #20   n        =   100 ;
        #20   n        =   200 ;
        #20   n        =   247 ;  // n7
        #20   enable   =     0 ;
        #100
  $stop ;
end
      always
              #`clkperiodby2 clk <= ~clk ;      // Run the clock at 50 MHz.
endmodule
```

Simulation Results for Serial Adder Design

The Modelsim results for the serial adder design are shown in Figure 10.4. From the waveforms, it is seen that the clock period is 20 ns since we specified 50 MHz for the

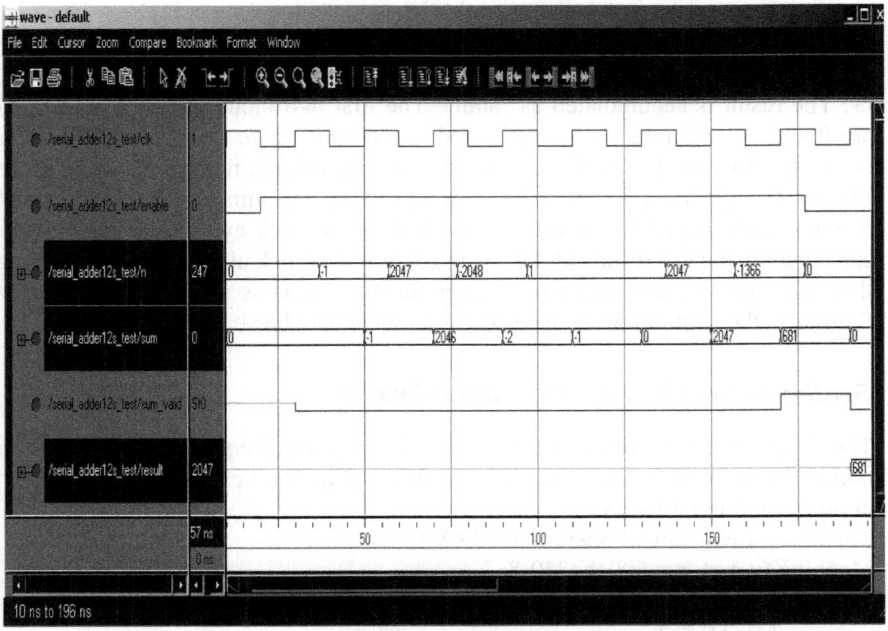

Fig. 10.4 Timing diagram of serial adder – first set of inputs (Continued)

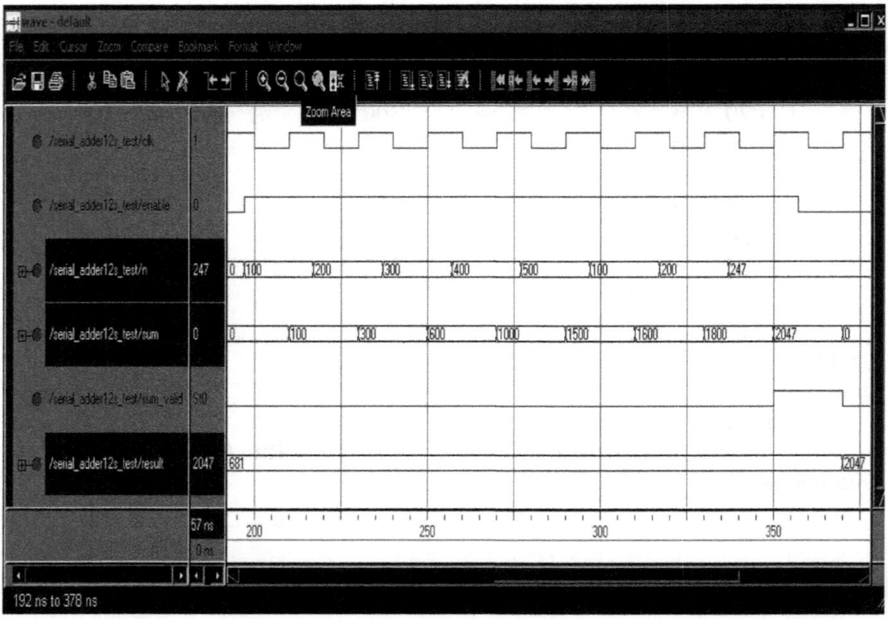

Fig. 10.4 Timing diagram of serial adder design – second set of inputs

"clk" signal in the test bench. The adder "enable" signal goes high at 20 ns and goes low at 177 ns, during which interval eight inputs, one every clock cycle, are applied at the input "n". Note that the data FFF H or –1 in decimal is applied at 37 ns. The result is accumulated as "sum". The first two input numbers, n0 and n1, are 0 and –1 adding up to –1 registered at the rising edge of "clk" at 50 ns. The next data 2047 applied at 57 ns changes the accumulated result to 2046 at 70 ns. Finally, the last input number n7 = –1366 produces the sum 681 at 170 ns. Simultaneously, the signal "sum_valid" goes high for a clock cycle. The accumulated sum is available as "result" at the rising edge of the clock at 190 ns and remains at this value till the next result overwrites this at 370 ns as shown in Figure 10.4. Similarly, the addition of the next set of numbers may be verified.

Synthesis Results for Serial Adder Design

The Synplify results are tabulated in the following. The frequency of operation reported for XCV 600 device of Xilinx is 138 MHz. The design consumes just 18 numbers of 4 input LUTs.
Maximum frequency of operation: 138 MHz.
Mapping to part: xcv600ehq240-8
Cell usage:

MUXCY_L	14	uses
XORCY	14	uses

FDR	19	uses
FDE	15	uses
GND	1	use
I/O primitives:		
IBUF	13	uses
OBUF	31	uses
BUFGP	1	use
I/O Register bits:	15	
Register bits not including I/Os:	19 (0%)	
Global Clock Buffers:	1 of 4 (25%)	
Total LUTs:	18 (0%)	

Place and Route Results

The Xilinx P&R results for serial adder design is as follows. The design consumes very little hardware, namely, 11 slices or 464 gates. The maximum frequency of operation reported by Xilinx navigator is 174 MHz, higher than that reported by the Synplify tool. However, this must be used only as a rough guidance since the maximum frequency of operation goes down when a complete system design is routed.

Design Summary:

Number of errors:	0			
Number of warnings:	0			
Number of slices:	11	out of	6,912	1%
Number of slices containing unrelated logic:	0	out of	11	0%
Number of slice flip flops:	19	out of	13,824	1%
Number of four input LUTs:	18	out of	13,824	1%
Number of bonded IOBs:	44	out of	158	27%
IOB flip flops:	15			
Number of GCLKs:	1	out of	4	25%
Number of GCLKIOBs:	1	out of	4	25%

Total equivalent gate count for design: 464
Additional JTAG gate count for IOBs: 2,160
Maximum frequency: 174.307MHz

10.3.2 Parallel Signed Adder Design

In the serial adder design, we added eight numbers, n0 to n7, and we got a sum whose size is 3 bits more than the input. The last bit is the sign bit. The design was pipelined and partitioned for the data width as well as the functionality. It is also true for the parallel adder design considered in this section. The block diagram for this design is shown in Figure 10.5. We have three stages of pipelining and five pipelined registers in this design. It may be mentioned that the last "sum"

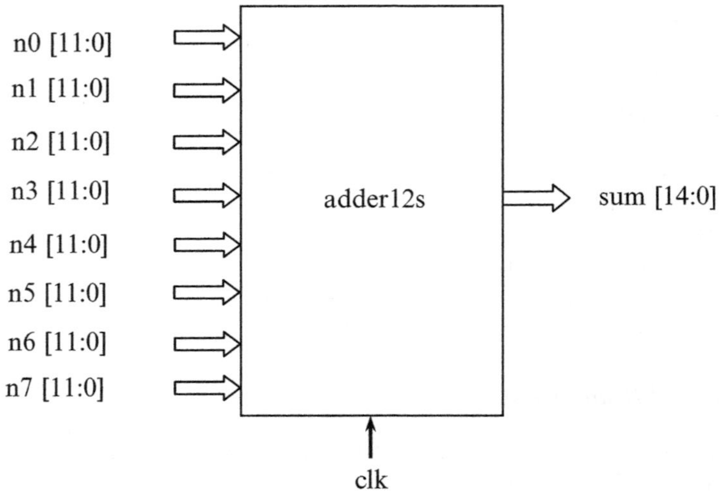

Fig. 10.5 Parallel signed adder design

is not registered. Before we consider the design, let us see how to evaluate twos complement quickly. It can be done in just two steps as follows.

Twos Complement Evaluation (Shortcut)

Let us say that we have an eight bit data 11110000, whose twos complement is required. This can be evaluated as follows. We may have to sign extend the number by 1 bit, i.e., duplicate the MSB, if we wish to add another number as shown. In the first step (other than sign extension), we scan the number from LSB till we encounter the first "1" and retain all the bits from LSB up to "1". In this example, we retain 10000. In the second and final step, we invert all other bits (1111) to get the desired result, 000010000. Once you get used to this, you will be able to compute the twos complement at one shot. When we add two numbers, the result will be 1 bit more than the precision of each number. Hence, we need to extend the sign bit of each number by one.

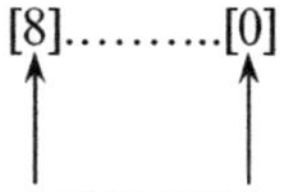

	111110000	Sign extended data.
Step		
1	10000	Retain first 1 from LSB, followed by 0s.
2	000010000	Invert other bits.

- Sign can be extended by any number of bits without affecting the actual value.
- Sign extend means duplicate MSB ([8]<=[7]).

- Without the sign extension, the MSB [7] will be mistaken as a negative number for high positive values such as +254.

Extend Sign

111111111	–1	111111111	–1
111111111	–1	000000001	+1
111111110	–2	000000000	0

Ignore Carry generated

001111111	+127	110000000	–128
001111111	+127	110000000	–128
011111110	+254	100000000	–256

Several examples are shown for the addition of two, twos complement numbers. Addition must be carried out in the same way we add two unsigned binary numbers, ignoring the carry generated.

Pipelined Design of Parallel Twos Complement Adder

The parallel signed adder shown in Figure 10.5 has a simple algorithm. This was evolved for use in the DCTQ application, where speed of processing has the top most priority, and the method is shown in Figure 10.6. The signed addition can be realized with seven two input adders and five pipeline stages. In the first stage, we have four numbers of 12 bits, twos complement adders to add all the eight numbers. They work concurrently, thereby speeding up the process. They have pipelined registers internally. The clock input is marked as (1), (2), etc., and correspond to internal pipeline registers. We will add the LSBs at the first clock pulse (1) and the MSBs at the next clock pulse (2) along with the carry generated at the LSB. In the second stage, we will add the four outputs, each of size 13 bits, generated at the first stage. Two numbers of two input adders are used at this stage. LSBs and MSBs are added with the arrival of the clock pulse (3) and clock pulse (4) respectively. In the third stage, with the arrival of the clock pulse (5), we will add the LSBs of the two inputs of size, 14 bits. Subsequently, the MSBs are added along with carry generated while adding the LSBs to produce 15 bits final result.

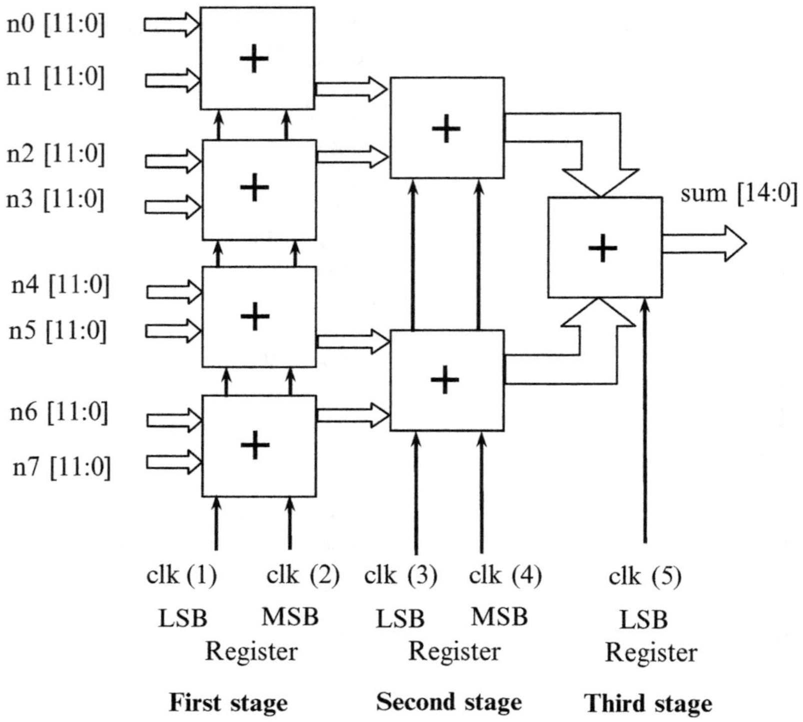

Fig. 10.6 Pipelined design partition of parallel adder

Verilog Code for the Parallel Signed Adder Design

Now, let us consider the Verilog code for this parallel, signed adder design. We will see how to add eight 12 bit, twos complement numbers n0 to n7 with 5 pipeline stages registered at positive clock. The result "sum" is a 15 bits in twos complement and the output is not registered. We have to first declare the module with the appropriate module name and declare the input clk, the input numbers n0 to n7 and the output sum. During the course of actual arithmetic operations, we will encounter many intermediate signals. Some of them may be used in assign statements and they are declared as wire along with their width. We also have some numbers, which are not used in the computation, but propagated at a particular stage. For example, the msb addition is not calculated at the beginning and so they have to be registered and propagated for use later on when it is required. The msb and lsb for the next stage are also declared as registers. This completes the "reg", "wire" declarations.

In the first stage, we add two numbers at a time, say, n0 and n1 and we add only the lsbs of the two numbers. Parallel to this, we add the others numbers n2 and n3, n4 and n5, and n6 and n7. This is same as that of using four adders concurrently and the results are stored using the assign statements. We add only the lsbs and register when the first clock arrives. Since we are not adding the msbs at this stage, we need to register it separately and propagate it through and use when the next clock arrives. Before the next clock arrives, we also preserve the sum. We have four sum results at this stage. Before we add the msbs, the sign should be extended. The msb 11 is the sign bit. The sign bit is first copied to another signal and then concatenated with the original value. This is done for both n0 and n1 and then added. We should also add the carry resulting from the msb addition. Since this is a time consuming operation, we preserve the results before the next clock pulse arrives. In the next clock, we preserve the entire msb sum in registers for use in the subsequent stages. We should also continue to preserve the lsb sum, as we need it for the final results. This completes the first stage of computation.

In the second stage, we add the 4-lsb sums we got in the first stage in two steps s00, s01 and s02, s03. The carry resulting here will be added with the msb later on. At the third clock pulse, the msbs are registered to continue addition later on. So we preserve the msbs and the lsb sum found at this stage. After the clock 4 edge rises, the added msbs of the second stage and carry generated in lsb addition are stored. At this stage, we have two msb and lsb sums.

At clk (5) rising edge, msbs and lsbs are registered to continue addition of msb. At the third stage, the two msbs are added and concatenated with LSB result to get the final result, 15 bits sum. This completes the design of the parallel signed adder.

Verilog Code_10.3

```
/*          Verilog Code for Signed Adder Design

// Adds eight numbers, n0 to n7, each of size 12 bits in 2's complement.
// Has five pipeline stages registered at positive edge of clock.
// Result, sum, is in 15 bits, 2's complement form (not registered).

module adder12s  (        clk,
                           n0,
                           n1,
                           n2,
                           n3,
                           n4,
                           n5,
                           n6,
                           n7,
                          sum

                  );

input                           clk ;
input           [11:0]          n0 ;
```

```
input    [11:0]   n1 ;
input    [11:0]   n2 ;
input    [11:0]   n3 ;
input    [11:0]   n4 ;
input    [11:0]   n5 ;
input    [11:0]   n6 ;
input    [11:0]   n7 ;
output   [14:0]   sum ;

wire     [7:0]    s00_lsb ;
wire     [7:0]    s01_lsb ;
wire     [7:0]    s02_lsb ;
wire     [7:0]    s03_lsb ;
wire     [5:0]    s00_msb ;
wire     [5:0]    s01_msb ;
wire     [5:0]    s02_msb ;
wire     [5:0]    s03_msb ;
wire     [7:0]    s10_lsb ;
wire     [7:0]    s11_lsb ;
wire     [6:0]    s10_msb ;
wire     [6:0]    s11_msb ;
wire     [7:0]    s20_lsb ;

reg      [11:7]   n0_reg1 ;
reg      [11:7]   n1_reg1 ;
reg      [11:7]   n2_reg1 ;
reg      [11:7]   n3_reg1 ;
reg      [11:7]   n4_reg1 ;
reg      [11:7]   n5_reg1 ;
reg      [11:7]   n6_reg1 ;
reg      [11:7]   n7_reg1 ;
reg      [7:0]    s00_lsbreg1 ;
reg      [7:0]    s01_lsbreg1 ;
reg      [7:0]    s02_lsbreg1 ;
reg      [7:0]    s03_lsbreg1 ;
reg      [5:0]    s00_msbreg2 ;
reg      [5:0]    s01_msbreg2 ;
reg      [5:0]    s02_msbreg2 ;
reg      [5:0]    s03_msbreg2 ;
reg      [6:0]    s00_lsbreg2 ;
reg      [6:0]    s01_lsbreg2 ;
reg      [6:0]    s02_lsbreg2 ;
reg      [6:0]    s03_lsbreg2 ;
reg      [7:0]    s10_lsbreg3 ;
reg      [7:0]    s11_lsbreg3 ;
reg      [5:0]    s00_msbreg3 ;
```

```
reg         [5:0]       s01_msbreg3 ;
reg         [5:0]       s02_msbreg3 ;
reg         [5:0]       s03_msbreg3 ;
reg         [6:0]       s10_lsbreg4 ;
reg         [6:0]       s11_lsbreg4 ;
reg         [6:0]       s10_msbreg4 ;
reg         [6:0]       s11_msbreg4 ;
reg         [6:0]       s10_msbreg5 ;
reg         [6:0]       s11_msbreg5 ;
reg                     s20_lsbreg5cy ;
reg         [6:0]       s20_lsbreg5 ;

 //     First Stage Addition
assign  s00_lsb[7:0]            =       n0[6:0]+n1[6:0] ;
                                // Add lsb first - s00_lsb[7] is the carry
assign  s01_lsb[7:0]            =       n2[6:0]+n3[6:0] ;
// n0-n7 lsb need not be registered since addition is already carried out here.
assign  s02_lsb[7:0]            =       n4[6:0]+n5[6:0] ;
assign  s03_lsb[7:0]            =       n6[6:0]+n7[6:0] ;

always @ (posedge clk)
                        // Pipeline 1: clk (1). Register msb to continue
                        // addition of msb.
begin
        n0_reg1[11:7]   <=      n0[11:7] ;
                        // Preserve all inputs for msb addition during the clk(2).
        n1_reg1[11:7]   <=      n1[11:7] ;
        n2_reg1[11:7]   <=      n2[11:7] ;
        n3_reg1[11:7]   <=      n3[11:7] ;
        n4_reg1[11:7]   <=      n4[11:7] ;
        n5_reg1[11:7]   <=      n5[11:7] ;
        n6_reg1[11:7]   <=      n6[11:7] ;
        n7_reg1[11:7]   <=      n7[11:7] ;
        s00_lsbreg1[7:0]        <=      s00_lsb[7:0] ;
// Preserve all lsb sum.  s00_lsbreg1[7] is the registered carry from lsb addition.
        s01_lsbreg1[7:0]        <=      s01_lsb[7:0] ;
        s02_lsbreg1[7:0]        <=      s02_lsb[7:0] ;
        s03_lsbreg1[7:0]        <=      s03_lsb[7:0] ;
end
                        // Sign extended & msb added with carry.
assign  s00_msb[5:0]    =       {n0_reg1[11], n0_reg1[11:7]}+
                                {n1_reg1[11], n1_reg1[11:7]}+s00_lsbreg1[7];
                                                //s00_msb[6] is ignored.
assign  s01_msb[5:0]    =       {n2_reg1[11], n2_reg1[11:7]}+
                                {n3_reg1[11], n3_reg1[11:7]}+s01_lsbreg1[7];
```

```
assign  s02_msb[5:0]    =        {n4_reg1[11], n4_reg1[11:7]}+
                                 {n5_reg1[11], n5_reg1[11:7]}+s02_lsbreg1[7];
assign  s03_msb[5:0]    =        {n6_reg1[11], n6_reg1[11:7]}+
                                 {n7_reg1[11], n7_reg1[11:7]}+s03_lsbreg1[7];

always @ (posedge clk)
// Pipeline 2: clk (2). Register msb to continue addition of msb.
begin
        s00_msbreg2[5:0]        <= s00_msb[5:0] ;    // Preserve all msb sum.
        s01_msbreg2[5:0]        <= s01_msb[5:0] ;
        s02_msbreg2[5:0]        <= s02_msb[5:0] ;
        s03_msbreg2[5:0]        <= s03_msb[5:0] ;
        s00_lsbreg2[6:0]        <= s00_lsbreg1[6:0] ; // Preserve all lsb sum.
        s01_lsbreg2[6:0]        <= s01_lsbreg1[6:0] ;
        s02_lsbreg2[6:0]        <= s02_lsbreg1[6:0] ;
        s03_lsbreg2[6:0]        <= s03_lsbreg1[6:0] ;
end

//      Second Stage Addition
assign  s10_lsb[7:0]            =       s00_lsbreg2[6:0]+s01_lsbreg2[6:0] ;
                                       //Add lsb first : s10_lsb[7] is the carry.

assign  s11_lsb[7:0]            =       s02_lsbreg2[6:0] +s03_lsbreg2[6:0] ;
                                        //s00, s01 lsbs need not be registered
                                   //since addition is already carried out here.
always @ (posedge clk)
      // Pipeline 3: clk (3). Register msb to continue addition of msb.
begin
        s10_lsbreg3[7:0]        <=      s10_lsb[7:0] ; // Preserve all lsb sum.
        s11_lsbreg3[7:0]        <=      s11_lsb[7:0] ;
        s00_msbreg3[5:0]        <=      s00_msbreg2[5:0]
                                                      // Preserve all msb sum.
        s01_msbreg3[5:0]        <=      s01_msbreg2[5:0] ;
        s02_msbreg3[5:0]        <=      s02_msbreg2[5:0] ;
        s03_msbreg3[5:0]        <=      s03_msbreg2[5:0] ;
end

assign  s10_msb[6:0]            =       {s00_msbreg3[5],
                                 s00_msbreg3[5:0]}+{s01_msbreg3[5],
                                 s01_msbreg3[5:0]}+s10_lsbreg3[7] ;
      // Add MSB of second stage with sign extension and carry in from LSB.
      // s10_msb[7] is ignored.
assign  s11_msb[6:0]            = {s02_msbreg3[5], s02_msbreg3[5:0]}+
                                  {s03_msbreg3[5], s03_msbreg3[5:0]}+
                                                      s11_lsbreg3[7] ;

always @ (posedge clk)
```

```
// Pipeline 4: clk (4). Register msb to continue addition of msb.
begin
        s10_lsbreg4[6:0]      <= s10_lsbreg3[6:0] ; // Preserve all lsb sum.
        s11_lsbreg4[6:0]      <= s11_lsbreg3[6:0] ;
        s10_msbreg4[6:0]      <= s10_msb[6:0] ;    // Preserve all msb sum.
        s11_msbreg4[6:0]      <= s11_msb[6:0] ;
end

//      Third Stage Addition
assign  s20_lsb[7:0]          = s10_lsbreg4[6:0]+ s11_lsbreg4[6:0] ;
                                  //Add lsb first : s20_lsb[7] is the carry.
always @ (posedge clk)
// Pipeline 5: clk (5). Register msb to continue addition of msb.
begin
        s10_msbreg5[6:0]      <= s10_msbreg4[6:0]; //Preserve all msb sum.
        s11_msbreg5[6:0]      <= s11_msbreg4[6:0] ;
        s20_lsbreg5cy         <= s20_lsb[7];        // Preserve all lsb sum.
        s20_lsbreg5[6:0]      <= s20_lsb[6:0];
end
// Add third stage MSB results and concatenate
// with LSB result to get the final result.
assign  sum[14:0]             = {({s10_msbreg5[6], s10_msbreg5[6:0]}+
                                 {s11_msbreg5[6], s11_msbreg5[6:0]}+
                                    s20_lsbreg5cy), s20_lsbreg5[6:0]};
endmodule
```

Test Bench for Parallel Signed Adder

As usual, we will use 50 MHz clock and, therefore, we define a clock period by 2 as 10 ns. We will use the back annotated design file and declare the test module as "adder12s_test". The final sum is not registered. It takes 5 clock cycles to produce the final result. However, if we want to register the sum, then we need one more clock cycle to produce the result. To start with, we declare the output, "sum" and all the inputs n0 to n7 and the clock as "reg". We then invoke the actual design "adder12s" and instantiate it as "u1". Using "initial block", different sets of input numbers n0 to n7 are applied every 20 ns except for the second set of inputs, which is applied at 17 ns. Finally, we invert the clock signal to create a free running clock. This ends the test module. Verilog_code_10.4 presents the test bench, which may be put in a file named "adder12s_test.v".

Verilog_Code_10.4

```
//      Test Bench for Parallel Adder/Subtractor Design
```

```
`define clkperiodby2 10           // Frequency of operation is 50 MHz.
`include "adder12s_banno.v"       // Use back annotated source code.

module adder12s_test(sum          // Declare the test bench.
                    );
output          [14:0]            sum;

reg                               clk ;
reg             [11:0]            n0 ;
reg             [11:0]            n1 ;
reg             [11:0]            n2 ;
reg             [11:0]            n3 ;
reg             [11:0]            n4 ;
reg             [11:0]            n5 ;
reg             [11:0]            n6 ;
reg             [11:0]            n7 ;

adder12s   u1(            .clk(clk), // Call the adder design.
                          .n0(n0),
                          .n1(n1),
                          .n2(n2),
                          .n3(n3),
                          .n4(n4),
                          .n5(n5),
                          .n6(n6),
                          .n7(n7),
                          .sum(sum)
           );
initial
begin
           clk    =    1'b0 ;             // Initialize the clock.
           n0     =       12'h0 ;         // Apply the first set of inputs.
           n1     =       12'h0 ;
           n2     =       12'h0 ;
           n3     =       12'h0 ;
           n4     =       12'h0 ;
           n5     =       12'h0 ;
           n6     =       12'h0 ;
           n7     =       12'h0 ;
 #17       n0     =       12'hfff ;       // Apply the second set of inputs.
           n1     =       12'hfff ;
           n2     =       12'hfff ;
           n3     =       12'hfff ;
           n4     =       12'hfff ;
           n5     =       12'hfff ;
           n6     =       12'hfff ;
```

```
        n7  =  12'hfff ;
#20     n0  =  12'h7ff ;   // Apply the third set of inputs.
        n1  =  12'h7ff ;
        n2  =  12'h7ff ;
        n3  =  12'h7ff ;
        n4  =  12'h7ff ;
        n5  =  12'h7ff ;
        n6  =  12'h7ff ;
        n7  =  12'h7ff ;
#20     n0  =  12'h800 ;   // Apply the fourth set of inputs.
        n1  =  12'h800 ;
        n2  =  12'h800 ;
        n3  =  12'h800 ;
        n4  =  12'h800 ;
        n5  =  12'h800 ;
        n6  =  12'h800 ;
        n7  =  12'h800 ;
  #20   n0  =  12'h001 ;   // Apply the fifth set of inputs.
        n1  =  12'h001 ;
        n2  =  12'h001 ;
        n3  =  12'h001 ;
        n4  =  12'h001 ;
        n5  =  12'h001 ;
        n6  =  12'h001 ;
        n7  =  12'h001 ;
  #20   n0  =  12'h001 ;   // Apply the sixth set of inputs.
        n1  =  12'hfff ;
        n2  =  12'h001 ;
        n3  =  12'hfff ;
        n4  =  12'h001 ;
        n5  =  12'hfff ;
        n6  =  12'h001 ;
        n7  =  12'hfff ;
 #20    n0  =  12'h7ff ;   // Apply the seventh set of inputs.
        n1  =  12'h7ff ;
        n2  =  12'h7ff ;
        n3  =  12'h7ff ;
        n4  =  12'h801 ;
        n5  =  12'h801 ;
        n6  =  12'h801 ;
        n7  =  12'h801 ;
  #20   n0  =  12'haaa ;   // Apply the eighth set of inputs.
        n1  =  12'h555 ;
        n2  =  12'haaa ;
        n3  =  12'h555 ;
        n4  =  12'haaa ;
```

```
        n5      =       12'h555 ;
        n6      =       12'haaa ;
        n7      =       12'h555 ;
  #20   n0      =       12'h0 ;         // Apply one more set of inputs.
        n1      =       12'h0 ;
        n2      =       12'h0 ;
        n3      =       12'h0 ;
        n4      =       12'h0 ;
        n5      =       12'h0 ;
        n6      =       12'h0 ;
        n7      =       12'h0 ;
  #400                                  // Wait for some time before stopping.
  $stop ;
end

always
        #`clkperiodby2 clk <= ~clk ;    // Toggle the clock.
endmodule
```

Simulation Results of Parallel Signed Adder

The Modelsim results are shown in Figure 10.7. As seen in the waveforms, nine sets of input numbers, n0 to n7, are 8 × 0; 8 × –1; 8 × 2047; 8 × –2048; 8 × 1; 4 × 1-1 × 4;

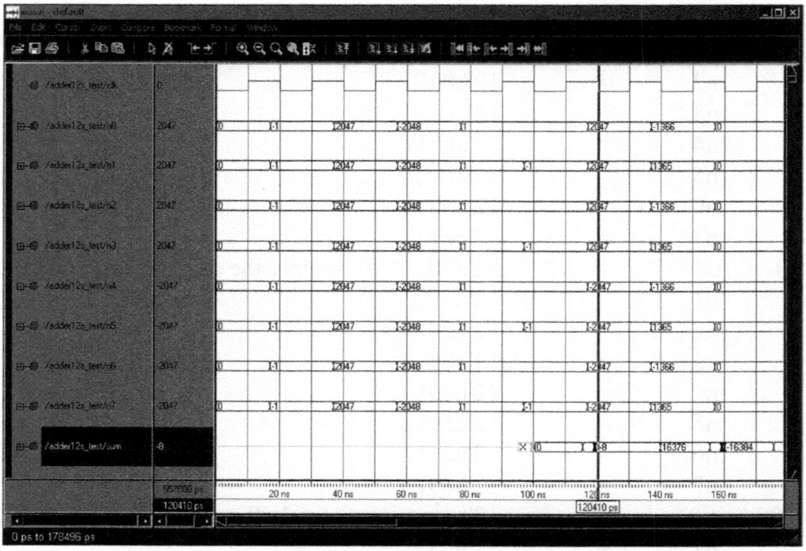

Fig. 10.7 Simulation result of back annotated design, adder12s (Continued)

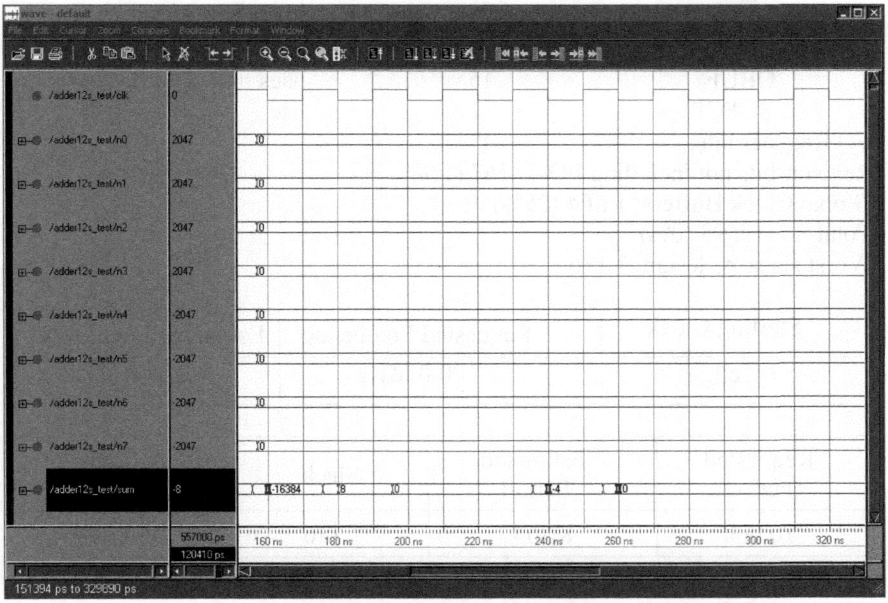

Fig. 10.7 Simulation result of back annotated design, adder12s

4 × 2047–2047 × 4; 4 × 1365 – 1366 × 4 and 8 × 0 resulting in the "sum" of 0; –8; 16376; –16384; 8; 0; 0; –4 and 0. This proves that the addition is working properly. The first result "0" occurs at 100 ns, five clock cycles after the first set of inputs are applied. Note that the gate delays are about 10 ns since we have used the back annotated file for simulation.

Synthesis Results of the Parallel Signed Adder

The Synplify results for the parallel signed adder design are as follows. The maximum clock frequency reported by the synthesis tool is 112 MHz. As in other designs we have covered so far, we map onto the xcv600ehq240-8 device. The number of LUTs consumed by the design is 95.

Performance Summary

Resource Usage Report for adder12s

Mapping to part: xcv600ehq240-8

Cell usage:

MUXCY_L	81	uses
XORCY	88	uses
MUXCY	7	uses
FD	214	uses
GND	1	use

I/O Primitives:

IBUF	96	uses
OBUF	15	uses
BUFGP	1	use

I/O Register bits: 47
Register bits not including I/Os: 167 (1%)
Global Clock Buffers: 1 of 4 (25%)
Total LUTs: 95 (0%)
Worst slack in design: 1.136

Starting Clock	Requested Frequency	Estimated Frequency
clk	100.0 MHz	112.8 MHz

Requested Period	Estimated Period	Slack	Clock Type
10.000	8.864	1.136	inferred

Place and Route Results

The Xilinx place and route results are as follows. We are primarily interested in the gate count and the maximum frequency of operation for our design. The gate count reported by the tool is about 2800 for the parallel, signed adder that adds eight, twos complement numbers, each of size: 12 bits. Surprisingly, Xilinx reports a higher frequency of operation, about 152 MHz for this design. The tool generates the bit stream "adder12s.bit", which is used for downloading onto the target FPGA.

Design Summary:

Number of errors:	0			
Number of warnings:	0			
Number of slices:	97	out of	6,912	1%
Number of slices containing unrelated logic:	0	out of	97	0%
Number of slice flip flops:	167	out of	13,824	1%
Number of four input LUTs:	95	out of	13,824	1%
Number of bonded IOBs:	111	out of	158	70%
IOB flip flops:	47			
Number of GCLKs:	1	out of	4	25%
Number of GCLKIOBs:	1	out of	4	25%

Total equivalent gate count for design: 2,810
Additional JTAG gate count for IOBs: 5,376

Timing Summary:
Design statistics:
Minimum period: 6.563 ns (maximum frequency: 152.369 MHz)

Minimum input arrival time before clock: 4.259 ns
Minimum output required time after clock: 11.083 ns
Saving bit stream in "adder12s.bit".

Comparison of Serial and Parallel Adders with Eight Numbers of Inputs

The serial and parallel adders we designed earlier, add eight numbers of inputs, each of width, 12 bits, where the MSB is the sign bit. They are basically adder cum subtractor since they perform signed addition. The output width is 15 bits. The performance of these two types of designs, which serve the same purpose of adding eight signed numbers are presented in Table 10.1. The parallel adder is nine times faster than the serial adder and may be used if speed of processing is of top most concern as it is in real time applications such as the DCTQ. However, if the chip area is vital and the speed of processing is adequate for the application, then the serial adder is a better choice. The chip area requirement for serial adder is about six times less than the parallel adder. Also, the Verilog code is shorter.

Table 10.1 Comparison of performance of eight inputs serial and parallel adders

Type of Adder	Serial	Parallel
No. of i/p clk cycles	8	1
No. of o/p clk cycles	9	1
Gate count	464	2,810
JTAG gate	2,160	5,376
Maximum frequency of operation in MHz	174	152

10.4 Multiplier Design

This is a new algorithm developed for the sake of implementing DCTQ on the FPGA or as an ASIC with an eye on achieving as high a throughput as possible. The DCTQ and other video processing applications demand very high throughputs. There is need to process very high-resolution motion pictures such as 1024 × 768 pixels or higher at a real time frame rate of 30 frames per second. High processing speeds can be achieved only by heavy pipelining and massively parallel circuits. It should be mentioned that FPGA/ASIC based designs incorporate massively parallel circuits and are highly pipelined when compared to microprocessors and DSPs. The multiplier design presented here incorporates a high degree of

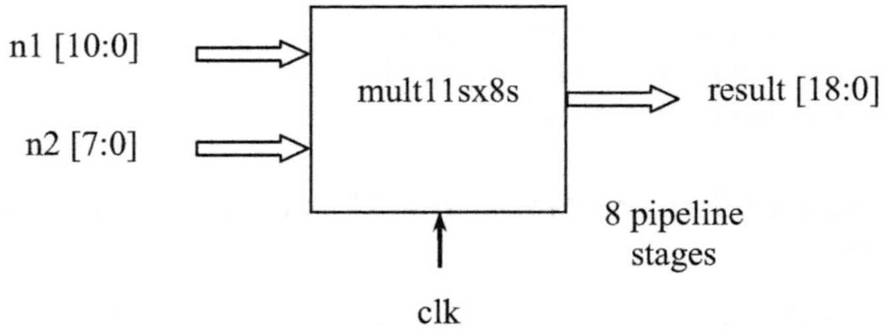

Fig. 10.8 Multiplier block

parallel circuits and a pipelining of eight levels. The multiplier, shown in Figure 10.8, performs a multiplication of two signed numbers n0 and n1, one of 11 bits and the other of 8 bits, as an example. The result is of size 19 bits in twos complement. The multiplication is done primarily on magnitudes of the two numbers, and, therefore, we will first separate out the sign and the magnitude and process only the magnitude. The sign can be dealt separately by using an exclusive or gate. On similar lines, the reader can develop HDL codes for any other size of the multiplicand and the multiplier, be it signed or unsigned. Before we go into the details of the algorithm, let us take an example.

Example:
Consider the evaluation of products of two signed numbers:

1023 × –128 = – 130944

The twos complement representation of the above product is as follows:

01111111111 × 10000000 = 1000000000010000000

Stripping the signs of the numbers, we have:

n1 (magnitude) × n2 (magnitude)
01111111111 × 10000000

00000000000	P1
00000000000	P2
00000000000	P3
00000000000	P4
00000000000	P5
00000000000	P6
00000000000	P7
01111111111	P8
011111111110000000	(magnitude)

The example mentioned earlier shows exactly the same way we multiply manually. The multiplier algorithm reflects the same pattern as the hand computing and is represented pictorially in Figure 10.9. We have used data partitioning

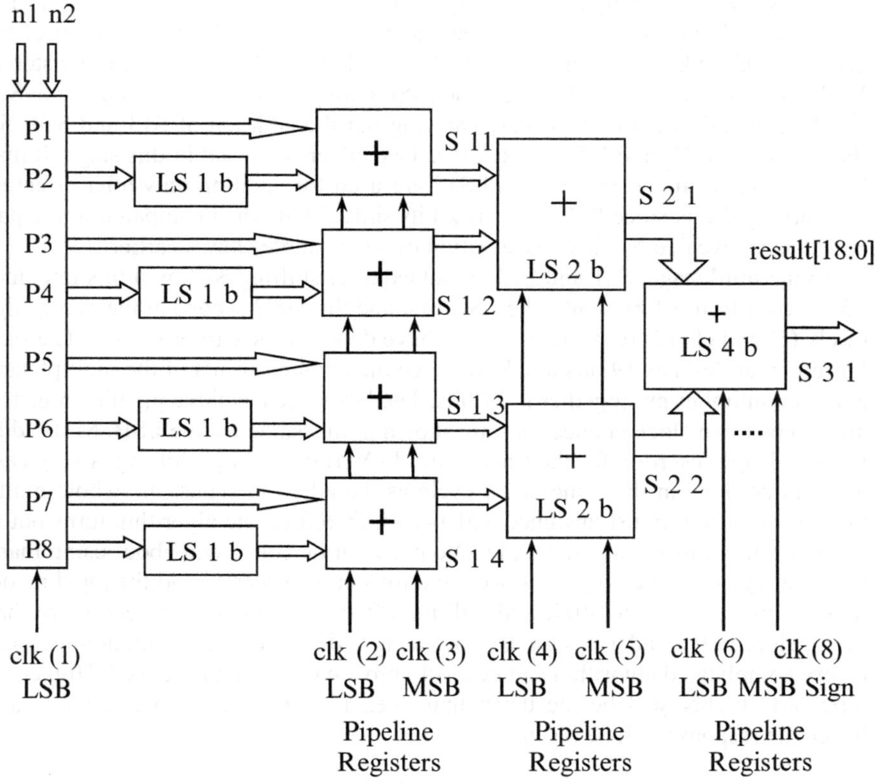

Fig. 10. 9 Pipelined multiplier design

and functionality partitioning in the design. This is once again in three stages as in the adder design we saw before, and the basic functionality in the multiplier is only addition. Extra operation we need to perform is left shifting by 1 bit, 2 bits, and 4 bits in the first, second, and third stage respectively. In the example we considered earlier, P1 through P8 are the partial products generated by hand computing. In the first stage, we will add four sets of two numbers at one stroke, and we will get four partial products S11 to S14. Note that we add P1 and P2 after left shifting P2 by 1 bit. Similarly, we add other pairs P3/P4, P5/P6, and P7/P8 in the first stage, all of them concurrently. Also internally, there are many pipelined registers. For example, two registers are hidden in the first stage and that is the reason why the clock is fed for each stage. The clock signal is marked clk (1), clk (2), etc., corresponding to pipeline registers shown in different stages. These are the reasons why we said before that FPGA/ASIC designs have massively parallel and highly pipelined circuits that offer high speed performance not feasible with DSPs.

The partial products, P1 to P8, are obtained by bit-wise "anding" the multiplicand and the multiplier corresponding to the respective bits of the multiplier. For instances, P1 = n1 [10:0] & {11{n2[0]}}, and P8 = n1 [10:0] & {11{n2[7]}}. The first clock, clk (1), will register all the "anded" P1 to P8. With the arrival of the second clock, clk (2), all the added LSBs of P1/P2 to P7/P8 pairs are registered. With the arrival of the third clk (3), the MSBs are added. In the second stage, S11 and S12 are added, simultaneously carrying out the addition of S13 and S14, for the LSBs at clk (4) and MSBs at clk (5). Two adders are used in this stage. It may be noted that at this stage, S12 and S14 are used for addition only after effecting left shift by 2 bits since P3/P4 pair is 2 bits shifted left when compared to the pair P1/P2 as is revealed by close observation of the example shown earlier.

In the third stage, S21 and S22 are added after shifting S22 by 4 bits left since P5 is offset from P1 by 4 bits, the LSB first and the MSB second at the rising edge of clk (6) and clk (7) respectively as we have done for the earlier stages. The final "result" manifests as 19 bits at clk (8) since the number of bits of the multiplicand and the multiplier put together is 19 bits. The last stage has three pipeline registers and hence three clock connections are shown in this stage for the LSB, MSB additions and sign insertion for the final result. In Verilog coding, shifting is very easy and can be done in the same clock cycle as the addition. Thus, the whole multiplier has reduced to just unsigned adders and, therefore, the algorithm turns out to be very straightforward and simple. We need not be concerned about using many pipeline registers since registers occupy only small chip area, and the pipeline delays are just a one-time affair and will not affect the processing speed as we had seen before in the adder design. By following a similar design methodology, even a very complicated algorithm can be easily broken down and pipelined. Higher the pipelining higher will be the throughput, i.e., higher the system clock rate and hence the response will be faster.

10.4.1 Verilog Code for Multiplier Design

We have two signed numbers n1 (11 bit) and n2 (8 bit), whose product is required. Since it is easier to perform the multiplication, if the numbers involved are unsigned, we will remove the signs temporarily while developing the code. We will apply the algorithm only on the magnitude and finally evaluate and combine the result with the sign. In the inputs, n1 and n2, 11th and 8th bits (msbs) are respectively the sign bits. In DCTQ application, for which we are developing the code for multiplier, we need to multiply a partial product of size 11 bits with a 8-bits "cosine" term, both of which are in twos complement. These two inputs require a multiplier with the above specification. The "cosine" values are stored in a dual address ROM, "romc", the design of which we discussed earlier in Section 9.1. The result is in twos complement. The multiplier module has eight pipelined stages and the input is not registered. In all the arithmetic calculations that we have seen so far, we have used only fixed-point arithmetic and not floating point arithmetic, as floating point arithmetic is more complex and will take more chip area.

The Verilog code for the pipelined multiplier design is presented in Verilog_Code_10.5. We start the code by declaring the design module, "mult11s × 8s" and the inputs/outputs. This is followed by declaring all the output signals in "assign" statements as "wire" and all output signals of "always" blocks as "reg". The first two "always" blocks compute the magnitudes of the two numbers, n1 and n2. The statement that assigns "n1orn2z" is to check whether n1 or n2 is zero so that we may assign the final "result" as zero. Next, we will see how to get the partial products p1 through p8. To get the partial products of the two numbers, we check whether the bit 0 of n2, which is the multiplier, is 1 or 0. If it is 0, then the partial product p1 is 0, otherwise it is the same as n1. In order to accomplish this, we perform bit-wise logical "and" of the two numbers n1 and bit 0 of n2 and assign to p1 using 'assign' statements. Similarly, assignments for other products p2 through p8 are carried out.

At the positive edge of the clock, clk (1), we will save p1 to p8 in pipeline registers for use in the next clock. We also preserve the sign bits of n1 and n2 and whether the result is 0 or not. Using "assign" statements, we compute the lsb sums, s11a, s12a, s13a, and s14a for the partial product pairs p1 and p2, etc. In the next clock, clk (2), we store the lsb partial sums in the pipelined registers, which we have already evaluated. These results will be used in the subsequent clock pulse only. Until then, they have to be propagated. We also store and propagate the unprocessed msb, the zero bits, sign bits, and the zero status indicators. All these pipeline registers will occupy more chip area. Although, manual optimization can be done to overcome this, we will leave the task to the synthesis tool to do the optimization. In the following "assign" statements, we will add the msb along with the carry got from lsb addition. Then the msbs and lsbs are concatenated. In the final result of the first stage, we will concatenate the msb sum, the lsb sum, and the least significant bit. We will have four such outputs in the first stage.

In the next clock, clk (3), the first stage results, s11 to s14, as well as the sign and the zero result are stored for further processing. Before the arrival of the next clock pulse, using assign statements, we calculate the second stage sum of the lsbs, shifting it by 2 bits. At the positive edge of the clock, clk (4), we will store the lsbs computed and propagate all the results not yet processed. Then the msb, lsb, and the last 2 bits are concatenated as the final result of the second stage. In the next clock, clk (5), the fifth pipelined stage, the second stage results as well as the sign and the zero result are stored for further processing. Before the next clock pulse, we calculate the third stage sum of the lsbs after shifting by four bits. At the positive edge of the clock, clk (6), we will propagate all the results not yet processed. Then we concatenate the msb, lsb and the last 2 bits as the final result of the present stage in clk (7). In the last clock, clk (8), we perform exclusive or of the sign bits of n1 and n2 to compute the sign bit of the final result, which we ignored in the first stage. To get the final result, result [18:0], of the multiplier, we set the sign bit along with the result and we register the output to suit the requirement of the DCTQ application. While writing this code, we haven't strained much to write an optimized code. For instance, instead of propagating the two sign bits of n1 and n2, we could have evaluated the sign bit right at the start and propagated just the sign of the final result. However, we shall delegate the optimization work to the synthesis tool.

Verilog_Code_10.5

```
/*                Verilog Code for Two Input Multiplier

Place this code in a file named "mult11s × 8s.v".
Signed multiplication of two numbers, n1 (11-bit) and n2 (8-bit).
Inputs are not registered.
Result is in twos complement.
This module has eight pipeline stages to increase the speed of processing.
*/

module mult11s × 8s (     clk,              // Declare the design module and
                          n1,
                          n2,
                          result
                         ) ;
input       clk ;                           // The inputs/outputs.
input       [10:0]     n1 ;
input       [7:0]      n2 ;
output      [18:0]     result ;

wire        n1orn2z ;                       // Declare combinational
wire        [10:0]     p1 ;                 // circuit signals.
wire        [10:0]     p2 ;
wire        [10:0]     p3 ;
wire        [10:0]     p4 ;
wire        [10:0]     p5 ;
wire        [10:0]     p6 ;
wire        [10:0]     p7 ;
wire        [10:0]     p8 ;
wire        [6:0]      s11a ;
wire        [6:0]      s12a ;
wire        [6:0]      s13a ;
wire        [6:0]      s14a ;
wire        [5:0]      s11b ;
wire        [5:0]      s12b ;
wire        [5:0]      s13b ;
wire        [5:0]      s14b ;
wire        [12:0]     s11 ;
wire        [12:0]     s12 ;
wire        [12:0]     s13 ;
wire        [12:0]     s14 ;
wire        [7:0]      s21a ;
wire        [7:0]      s22a ;
wire        [6:0]      s21b ;
wire        [6:0]      s22b ;
```

```
wire    [14:0]    s21 ;
wire    [14:0]    s22 ;
wire    [8:0]     s31a ;
wire    [7:0]     s31b ;
wire    [17:0]    s31 ;
wire              res_sign ;
wire    [18:0]    res ;

reg     [10:0]    n1_mag ;     // Declare all registers.
reg     [7:0]     n2_mag ;
reg     [10:0]    p1_reg1 ;
reg     [10:0]    p2_reg1 ;
reg     [10:0]    p3_reg1 ;
reg     [10:0]    p4_reg1 ;
reg     [10:0]    p5_reg1 ;
reg     [10:0]    p6_reg1 ;
reg     [10:0]    p7_reg1 ;
reg     [10:0]    p8_reg1 ;
reg     [6:0]     s11a_reg2 ;
reg     [6:0]     s12a_reg2 ;
reg     [6:0]     s13a_reg2 ;
reg     [6:0]     s14a_reg2 ;
reg               n1_reg1;
reg               n1_reg2;
reg               n1_reg3;
reg               n1_reg4;
reg               n1_reg5;
reg               n1_reg6;
reg               n1_reg7;
reg               n2_reg1;
reg               n2_reg2;
reg               n2_reg3;
reg               n2_reg4;
reg               n2_reg5;
reg               n2_reg6;
reg               n2_reg7;
reg               n1orn2z_reg1 ;
reg               n1orn2z_reg2 ;
reg               n1orn2z_reg3 ;
reg               n1orn2z_reg4 ;
reg               n1orn2z_reg5 ;
reg               n1orn2z_reg6 ;
reg               n1orn2z_reg7 ;
reg     [10:0]    p1_reg2 ;
reg     [10:0]    p2_reg2 ;
reg     [10:0]    p3_reg2 ;
```

```
reg         [10:0]          p4_reg2 ;
reg         [10:0]          p5_reg2 ;
reg         [10:0]          p6_reg2 ;
reg         [10:0]          p7_reg2 ;
reg         [10:0]          p8_reg2 ;
reg         [12:0]          s11_reg3 ;
reg         [12:0]          s12_reg3 ;
reg         [12:0]          s13_reg3 ;
reg         [12:0]          s14_reg3 ;
reg         [12:0]          s11_reg4 ;
reg         [12:0]          s12_reg4 ;
reg         [12:0]          s13_reg4 ;
reg         [12:0]          s14_reg4 ;
reg         [7:0]           s21a_reg4 ;
reg         [7:0]           s22a_reg4 ;
reg         [14:0]          s21_reg5 ;
reg         [14:0]          s22_reg5 ;
reg         [14:0]          s21_reg6 ;
reg         [14:0]          s22_reg6 ;
reg         [8:0]           s31a_reg6 ;
reg         [17:0]          s31_reg7 ;
reg         [18:0]          result ;

always @(n1)
begin
        if(n1[10] == 1'b0)
                n1_mag = n1[10:0] ;
        else
                n1_mag = ~n1[10:0] + 1 ;        // Evaluate twos complement.
end

always @(n2)
begin
        if(n2[7] == 1'b0)
                n2_mag = n2[7:0] ;
        else
                n2_mag = ~n2[7:0] + 1 ;     // Evaluate twos complement.
end

assign n1orn2z      =    ((n1 == 11'b0)||(n2 == 7'b0)) ?1'b1:1'b0 ;
                                        // If n1 or n2 is zero, make final result +0.
assign p1           =    n1_mag[10:0] & {11{n2_mag[0]}};
                  // Compute the partial products. Multiply n1 by n2 bit '0', etc.
assign p2           =    n1_mag[10:0] & {11{n2_mag[1]}} ;
assign p3           =    n1_mag[10:0] & {11{n2_mag[2]}} ;
assign p4           =    n1_mag[10:0] & {11{n2_mag[3]}} ;
```

```
assign p5          =    n1_mag[10:0] & {11{n2_mag[4]}} ;
assign p6          =    n1_mag[10:0] & {11{n2_mag[5]}} ;
assign p7          =    n1_mag[10:0] & {11{n2_mag[6]}} ;
assign p8          =    n1_mag[10:0] & {11{n2_mag[7]}} ;

always @ (posedge clk)    // These are the first pipeline registers at clk (1) stage.
begin
        p1_reg1         <=      p1 ;
        p2_reg1         <=      p2 ;
        p3_reg1         <=      p3 ;
        p4_reg1         <=      p4 ;
        p5_reg1         <=      p5 ;
        p6_reg1         <=      p6 ;
        p7_reg1         <=      p7 ;
        p8_reg1         <=      p8 ;
        n1_reg1         <=      n1[10] ;     // Preserve sign bits and the status
        n2_reg1         <=      n2[7] ;
        n1orn2z_reg1    <=      n1orn2z ;  // whether result is zero or not
end

//p1_reg1, etc. means p1, etc. are registered at positive edge of clk (1), clk (2), etc.

assign  s11a[6:0]       =       p1_reg1[6:1] + p2_reg1[5:0] ;
                                        // LSBs are added here after left shifting
                                        // p1_reg1 by one bit.
assign  s12a[6:0]       =       p3_reg1[6:1] + p4_reg1[5:0] ;
assign  s13a[6:0]       =       p5_reg1[6:1] + p6_reg1[5:0] ;

assign  s14a[6:0]       =       p7_reg1[6:1] + p8_reg1[5:0] ;
                                // Note: the left shifts are taken care of
                                // for p1, p3, p5 and p7.
                            // p1_reg1[0], etc. will be processed at the clk (2).
                            // s11a[6], etc. are the carry bits.
always @ (posedge clk)    // These are the second pipeline registers @ clk (2).
begin
        s11a_reg2       <=      s11a ;    // Store LSB partial sums.
        s12a_reg2       <=      s12a ;
        s13a_reg2       <=      s13a ;
        s14a_reg2       <=      s14a ;
        p1_reg2[10:7]   <= p1_reg1[10:7] ; // Store MSB of partial products.
        p2_reg2[10:6]   <= p2_reg1[10:6] ;
        p3_reg2[10:7]   <= p3_reg1[10:7] ;
        p4_reg2[10:6]   <= p4_reg1[10:6] ;
        p5_reg2[10:7]   <= p5_reg1[10:7] ;
        p6_reg2[10:6]   <= p6_reg1[10:6] ;
        p7_reg2[10:7]   <= p7_reg1[10:7] ;
```

```
        p8_reg2[10:6]  <= p8_reg1[10:6] ;
        p1_reg2[0]     <= p1_reg1[0] ;     // Store '0' th bit since
        p3_reg2[0]     <= p3_reg1[0] ;     //it is not yet processed.
        p5_reg2[0]     <= p5_reg1[0] ;
        p7_reg2[0]     <= p7_reg1[0] ;
        n1_reg2        <= n1_reg1 ;        // Also store sign bits and zero status.
        n2_reg2        <= n2_reg1 ;
        n1orn2z_reg2   <= n1orn2z_reg1 ;
end

// MSB is added here along with carry.
assign s11b[5:0]  =  {1'b0, p1_reg2[10:7]} + p2_reg2[10:6] + s11a_reg2[6] ;
assign s12b[5:0]  =  {1'b0, p3_reg2[10:7]} + p4_reg2[10:6] + s12a_reg2[6] ;
assign s13b[5:0]  =  {1'b0, p5_reg2[10:7]} + p6_reg2[10:6] + s13a_reg2[6] ;
assign s14b[5:0]  =  {1'b0, p7_reg2[10:7]} + p8_reg2[10:6] + s14a_reg2[6] ;
                                    // MSBs & LSBs are concatenated here.
assign s11[12:0]  =  {s11b, s11a_reg2[5:0], p1_reg2[0]} ;
                              // Concatenate MSB, LSB, '0' th bit respectively.
assign s12[12:0]  =  {s12b, s12a_reg2[5:0], p3_reg2[0]} ;
assign s13[12:0]  =  {s13b, s13a_reg2[5:0], p5_reg2[0]} ;
assign s14[12:0]  =  {s14b, s14a_reg2[5:0], p7_reg2[0]} ;

always @ (posedge clk)
        // These are the third pipeline registers @ clk (3). First stage results.
begin
        s11_reg3       <=    s11;        // Store for further processing.
        s12_reg3       <=    s12;
        s13_reg3       <=    s13;
        s14_reg3       <=    s14;
        n1_reg3        <=    n1_reg2;
        n2_reg3        <=    n2_reg2;
        n1orn2z_reg3   <=    n1orn2z_reg2;
end

assign s21a[7:0]  =    s11_reg3[8:2] + s12_reg3[6:0] ;  // s21a[7] is the carry.
assign s22a[7:0]  =    s13_reg3[8:2] + s14_reg3[6:0] ;  // LSB sum, 2nd stage.

always @ (posedge clk)    // These are the fourth pipeline registers @ clk (4).
begin
        s11_reg4[12:9]  <= s11_reg3[12:9] ; // Store bits not yet processed.
        s11_reg4[1:0]   <= s11_reg3[1:0] ;
        s12_reg4[12:7]  <= s12_reg3[12:7] ;
        s13_reg4[12:9]  <= s13_reg3[12:9] ;
        s13_reg4[1:0]   <= s13_reg3[1:0] ;
        s14_reg4[12:7]  <= s14_reg3[12:7] ;
        s21a_reg4       <= s21a ;
```

```
                                        // Store LSB, second stage partial sums.
        s22a_reg4       <= s22a ;
        n1_reg4         <= n1_reg3 ;
        n2_reg4         <= n2_reg3 ;
        n1orn2z_reg4    <= n1orn2z_reg3 ;
end

// Add second stage MSBs with carry.
assign s21b[6:0]  =  {2'b0, s11_reg4[12:9]} +s12_reg4[12:7] + s21a_reg4[7];
assign s22b[6:0]  =  {2'b0, s13_reg4[12:9]} +s14_reg4[12:7] + s22a_reg4[7];
assign s21[14:0]  =  {s21b[5:0], s21a_reg4[6:0], s11_reg4[1:0]} ;
                                                    // {MSB, LSB, [1:0]}
                            // Result will never effect s21b[6], which is always 0.
assign s22[14:0]  =  {s22b[5:0], s22a_reg4[6:0], s13_reg4[1:0]} ;

always @ (posedge clk)          // These are the fifth pipeline registers @ clk (5).
begin
        s21_reg5        <=      s21 ;     // Store for further processing.
        s22_reg5        <=      s22 ;
        n1_reg5         <=      n1_reg4 ;
        n2_reg5         <=      n2_reg4 ;
        n1orn2z_reg5    <=      n1orn2z_reg4 ;
end

assign s31a[8:0] =          s21_reg5[11:4] + s22_reg5[7:0] ;
                                            // Third stage LSB is computed here.
always @ (posedge clk)          // These are the sixth pipeline registers @ clk (6).
begin
        s21_reg6    [14:12]  <=   s21_reg5[14:12] ;    // Preserve MSBs.
        s22_reg6    [14:8]   <=   s22_reg5[14:8] ;
        s21_reg6    [3:0]    <=   s21_reg5[3:0] ;
        s31a_reg6   <=         s31a ;  //Third stage LSB is registered here.
        n1_reg6     <=         n1_reg5 ;
        n2_reg6     <=         n2_reg5 ;
        n1orn2z_reg6<=         n1orn2z_reg5 ;
end

assign s31b[7:0]  = {4'b0, s21_reg6[14:12]} + s22_reg6[14:8] + s31a_reg6[8] ;
                                            // Third stage MSB is computed
here.
assign s31[17:0]  = {s31b[5:0], s31a_reg6[7:0], s21_reg6[3:0]} ;
                                        // Put MSB, LSB and [3:0] bits together.
// Note that the third stage result will never effect s31b[6:5], which is always 0.

always @ (posedge clk) // These are the seventh pipeline registers @ clk (7).
begin
```

```
        n1_reg7       <=     n1_reg6 ;          // Store intermediate results.
        n2_reg7       <=     n2_reg6 ;
        s31_reg7      <=     s31 ;
        n1orn2z_reg7  <=     n1orn2z_reg6 ;
end

assign res_sign   =     n1_reg7^n2_reg7 ;  // "1" means a -ve no.
assign res[18:0] = (res_sign ) ? {1'b1, (~s31_reg7 + 1'b1)}:{1'b0, s31_reg7} ;

always @ (posedge clk) // This is the eighth pipeline register registered @ clk (8).
begin
        if (n1orn2z_reg7 == 1'b1)
                result[18:0] <= 19'b0 ;
        else
                result[18:0] <= res ;    // This is the final result (product of two
                                         // numbers) in twos complement form.
end
endmodule
```

Test Bench for the Multiplier Design

As usual, we define clkperiodby2 as 10 ns so that we may run the simulation at 50 MHz. We include the back annotated source file, "mult11s × 8s_banno.v", in the test bench to check that the design works at the maximum frequency of operation for the target FPGA. In order to run at the maximum frequency of operation reported by P&R tool, the "clkperiodby2" will have to be changed suitably. After including the design, we declare the test bench module "mult11s × 8s_test" and its inputs/output. This is followed by invoking the design module "mult11s × 8s", calling ports by name. Using initial block, we will apply different pairs of test patterns as inputs, once every 20 ns. We stagger the data and clock by a few nanoseconds so that we may apply the clock only after the data stabilizes. Initially, we force the clock and the two numbers n1 and n2 to zero, and later on for every 20 ns, we change the data. Before ending the test module, we will toggle the clock as usual. The Verilog code for the test bench follows:

Verilog_Code_10.6

```
// Place this test bench in a file named "mult11s × 8s_test.v".

`define clkperiodby2 10                    // Simulate at 50 MHz.
`include "mult11s × 8s_banno.v"            // Back annotated design file.

module mult11s × 8s_test (result           // Declare the test bench,
                         );
```

```
        output   [18:0]   result ;            // output and

        reg                                   clk ;     // input stimulants.
        reg              [10:0]               n1 ;
        reg              [7:0]                n2 ;

mult11s × 8s   u1(               .clk(clk),           // Invoke the design.
                 .n1(n1),
                 .n2(n2),
                 .result(result)
             );
initial
        begin                                 // Apply several sets of inputs.
            clk      =      1'b0 ;
             n1      =      11'h0 ;
             n2      =      8'h0 ;
        #17 n1       =      11'h555 ; // Not that the inputs are applied
             n2      =      8'h55;    // before the rising edge of "clk".
        #20  n1      =      11'h2aa ;
             n2      =      8'haa;
        #20 n1       =      11'h7ff ;
             n2      =      8'h80;
        #20 n1       =      11'h555 ;
             n2      =      8'hff;
        #20 n1       =      11'h7ff ;
             n2      =      8'h81;
        #20 n1       =      11'h555 ;
             n2      =      8'h81;
        #20 n1       =      11'h2aa ;
             n2      =      8'h81;
        #20 n1       =      11'h7ff ;
             n2      =      8'h00;
        #20 n1       =      11'h7ff ;
             n2      =      8'h7f;
        #20 n1       =      11'h000 ;
             n2      =      8'hff;
        #20 n1       =      11'h000 ;
             n2      =      8'h7f;
        #400
 $stop ;
end

always
        #`clkperiodby2 clk <= ~clk ;
endmodule
```

Simulation Results of the Multiplier Design

The Modelsim results of the multiplier are shown in Figure 10.10. As seen in the waveform, the clock rises at intervals of 20 ns starting from 10 ns. The inputs are shown in both decimal number formats as well as in hex formats so that we may correlate the results with the test bench. Input pairs applied are respectively 0, 0; –683, 85; 682, –86; –1, –128; –683, –1; –1, –127; –683, –127; 682, –127; –1, 0; –1, 127; 0, –1 and 0, 127. The multiplied results are respectively 0; –58055; –58652; 128; 683; 127; 86741; –86614; 0; –127; 0 and 0. The calculator on your computer may be used to verify the hex to decimal conversion as well as the results. It may be noted that the second pair of inputs is applied at 17 ns, 13 ns before the clock strikes. Thus, the inputs are stable when the clock arrives. The first output arrives at 155.6 ns since there are eight pipeline registers and gate delays in the data path. The subsequent results appear periodically at 20 ns interval. If, instead of back annotated design, we had used the actual design source, the first result would have manifested at the eighth clock pulse, i.e., at 150 ns. Clearly, the associated gate delays are 5.6 ns.

Synthesis Results of Multiplier Design

The Synplify results for the signed multiplier are as follows. The report reveals whatever signals are optimized. For example, it reports that the signal "s14a_reg2[6]" is always 0. Therefore, we need not spend too much time on optimizing while

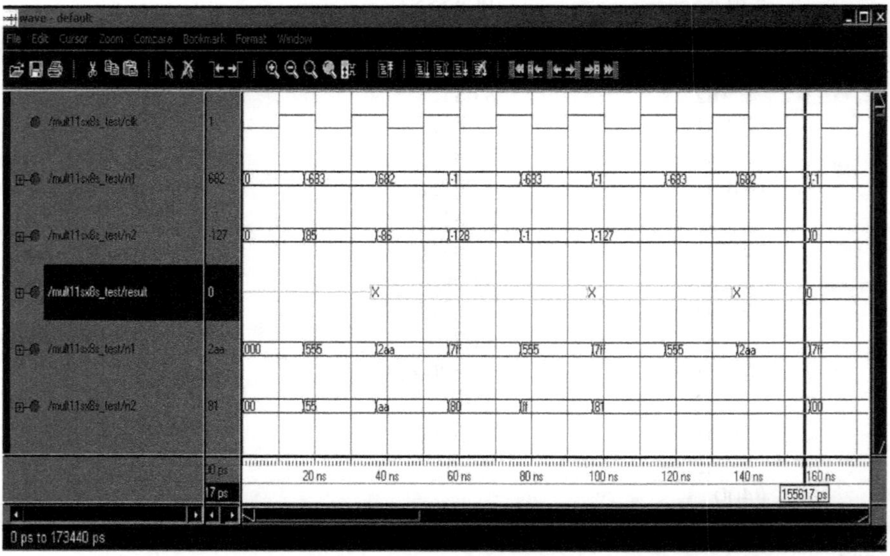

Fig. 10.10 Simulation results of the multiplier design "mult11s × 8s" (Continued)

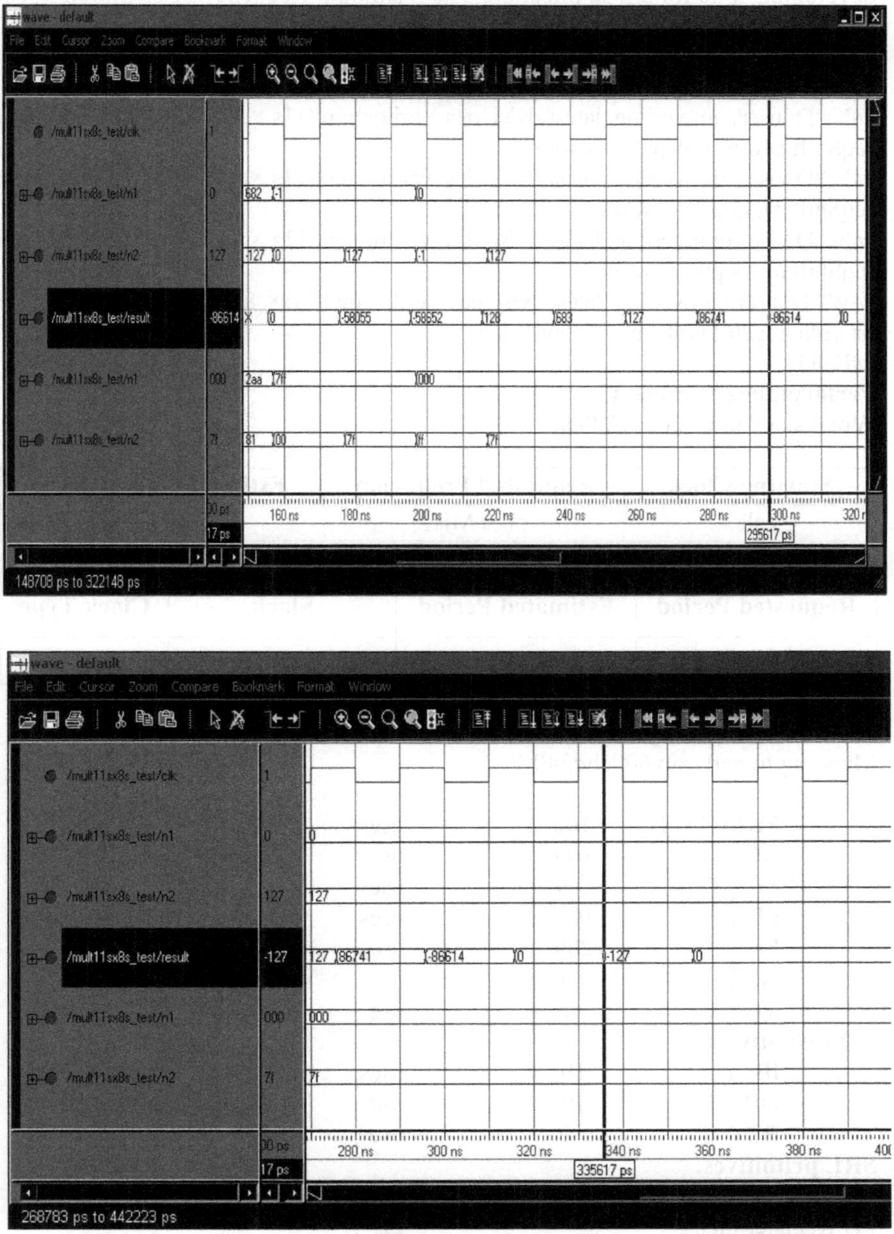

Fig. 10.10 Simulation results of the multiplier design "mult11s × 8s"

developing code for any application. The maximum frequency of operation is 125 MHz and number of LUTs is 181.

Synplify Report

Synthesizing module mult11s × 8s
@N:"D:\user\ram\verilog_latest\dvlsi_des_verilog\mult11s × 8s.v":346:0:346:5|Found seqShift n1orn2z, depth=7, width=1
@N:"D:\user\ram\verilog_latest\dvlsi_des_verilog\mult11s × 8s.v":346:0:346:5|Found seqShift n1, depth=6, width=1
@N:"D:\user\ram\verilog_latest\dvlsi_des_verilog\mult11s × 8s.v":346:0:346:5|Found seqShift n2, depth=6, width=1
@W:"D:\user\ram\verilog_latest\dvlsi_des_verilog\mult11s× 8s.v":202:0:202:5|Register bit s14a_reg2[6] is always 0, optimizing ...
@END

Performance Summary

Worst slack in design: 12.009

Starting Clock	Requested Frequency	Estimated Frequency
clk	50.0 MHz	125.1 MHz

Requested Period	Estimated Period	Slack	Clock Type
20.000	7.991	12.009	Inferred

Resource Usage Report for mult11s × 8s

Mapping to part: xcv600ehq240-8
Cell usage:

MUXCY_L	100	uses
XORCY	109	uses
MUXCY	9	uses
FDR	105	uses
FD	209	uses
GND	1	use
VCC	1	use

I/O primitives:

IBUF	19	uses
OBUF	19	uses
BUFGP	1	use

SRL primitives:

SRL16	9	uses

I/O Register bits: 22
Register bits not including I/Os: 292 (2%)
Global Clock Buffers: 1 of 4 (25%)
Total LUTs: 181 (1%)

Place and Route Results of Multiplier Design

The Xilinx place and route results are as follows. From the report listed below, the number of slices consumed is seen to be 201, and the total gate count for the design is 5,284. The maximum frequency of operation is 82 MHz although the Synplify tool has reported much higher frequency. Anyway, what matters is the maximum frequency of operation for the entire application such as DCTQ. The bit stream is saved as "mult11s × 8s.bit".

Design Summary:

Number of errors:	0			
Number of warnings:	0			
Number of slices:	201	out of	6,912	2%
Number of slices containing unrelated logic:	0	out of	201	0%
Number of slice flip flops:	292	out of	13,824	2%
Total Number 4 input LUTs:	178	out of	3,824	1%
Number used as LUTs:	161			
Number used as a route-thru:	8			
Number used as shift registers:	9			
Number of bonded IOBs:	38	out of	158	24%
IOB flip flops:	22			
Number of GCLKs:	1	out of	4	25%
Number of GCLKIOBs:	1	out of	4	25%

Total equivalent gate count for design: 5,284
Additional JTAG gate count for IOBs: 1,872

Timing summary:

Minimum period: 12.132 ns (maximum frequency: 82.427 MHz)
Minimum input arrival time before clock: 10.150 ns
Minimum output required time after clock: 5.617 ns

Saving bit stream in "mult11s × 8s.bit".

Summary

Arithmetic circuits such as add, subtract, multiply, etc., are computationally intensive and, therefore, conventional methods are not sufficient for real time implementations on FPGA or ASIC. In order to speed up the processing considerably, we will have to base our designs on massively parallel circuits and heavy pipelining. This chapter presented arithmetic circuit designs such as signed adders and multiplier with a high degree of parallelism and pipelining for computationally intensive applications such as video compression. The next chapter shows the importance of developing algorithms so that they may be effectively implemented on an FPGA or as an ASIC.

Assignments

10.1 Code and test a design for adding two signed numbers of width 16 bits each in three different manner:

(i) Add all the 16 bits at a time without any pipelining.

(ii) Add only 8 bits at every pipeline stage.

(iii) Repeat (ii) for 4 bits.

Analyze and use the optimum number of pipeline stages in order to get the best possible speed of implementation. Which of these three designs yields the best performance in terms of speed/chip area?

10.2 Write a Verilog code for subtracting two 12-bit numbers in twos complement. Pipeline your design. Test your design.

10.3 Code and test your design for adding eight signed numbers, each of width 14 bits on similar lines as the 12 bits signed adder design shown in the text. Name the adder design as "adder14sr.v" for use in DCTQ application. Use six stages of pipelining. The final output "sum" must be registered.

10.4 Code and test a design for adding eight signed numbers, each of width 12 bits on similar lines as the 12 bits signed parallel adder design shown in the text. Name the adder design as "adder12sr.v" for use in IQIDCT application. Use six stages of pipelining. The final output "sum" must be registered.

10.5 Code and test a design for adding eight signed numbers, each of width 14 bits on similar lines as the 14 bits signed adder design shown in the assignment 10.3. Name the adder design as "adder14s.v" for use in IQIDCT Processor application. Use five stages of pipelining. The final output "sum" must not be registered.

10.6 Write Verilog codes for multiplying two numbers for the following specifications:

(i) n1 is unsigned 8 bit, and n2 is signed 8 bit.

(ii) n1 is signed 12 bit, and n2 is unsigned 8 bit.

Use the multiplier algorithm presented in the text. Design each of the above with eight pipeline stages to increase the speed. Inputs are not registered. Both these designs are required for use in DCTQ application. Give apt names. Test your codes.

10.7 Write Verilog code for multiplying two numbers for the following specification:

(i) "n1" is signed 9 bit, and "n2" is unsigned 8 bit.

(ii) "n1" is signed 12 bit, and "n2" is signed 8 bit.

Use the multiplier algorithm presented in the text. Design with eight pipeline stages to increase the speed. Inputs are not registered. These designs are required for use in IQIDCT application. Give apt names. Test your codes.

10.8 Another new algorithm can be developed for multiplier based on decimal weights of a multiplier or for that matter a multiplicand. For example, consider two unsigned 4-bit numbers, n1[3:0] and n2[3:0]. The multiplied result may be simply expressed as follows:

Result = (if n2[3] = 1 then 8 (n1[3:0]); else 0) +
(if n2[2] = 1 then 4 (n1[3:0]); else 0) +
(if n2[1] = 1 then 2 (n1[3:0]); else 0) +
(if n2[0] = 1 then (n1[3:0]); else 0)

Develop this algorithm so that it may be easily coded in Verilog and prove its working for two examples:

1. 1111 × 1111
2. 1111 × 1010

10.9 Write RTL Verilog code for multiplying two unsigned 4-bit numbers for the new algorithm you have developed for the assignment 10.8.

10.10 Write a test bench for the Verilog code developed by you for the assignment 10.9 and present the waveform to prove its working.

10.11 Write RTL Verilog code for multiplying two unsigned 8-bit numbers after modifying the new algorithm you have developed for the assignment 10.8. Incorporate four stages of pipelining.

10.12 Write a test bench for the Verilog code developed by you for the assignment 10.11 and present the waveform to prove its working.

10.13 Modify the algorithm of the assignment 10.11 to multiply two 8-bit numbers, one of them being unsigned and the other signed in twos complement. Code it in Verilog with five pipeline stages. Compare the synthesis and Xilinx P&R results with that of the corresponding design you have solved for the assignment 10.6 (i).

10.14 Write a test bench for the Verilog code developed by you for the assignment 10.13 and present the waveform to prove its working.

10.15 In the assignment 10.13, change the number of pipeline stages from five to one and test it. Compare the synthesis and Xilinx P&R results with that design.

10.16 Write a test bench for the Verilog code developed by you for the assignment 10.15, if necessary, and present the waveform to prove its working.

10.17 A multiplier and accumulator (MAC) is useful in digital signal processing applications, where a sum of products is required. Develop the RTL Verilog code for a MAC using the 8- bit multiplier you have designed in the assignment 10.13. Present the synthesis and Xilinx P&R results.

10.18 Write a test bench for the MAC you have designed for the assignment 10.17. Establish its working by presenting the simulated waveform(s).

Chapter 11

Development of Algorithms and Verification Using High Level Languages

Simple applications such as a traffic light controller, etc., may be directly coded without a need for an algorithm. However, more complex applications such as video codecs, demodulators, etc., have involved algorithms at their core, which need to be adapted or developed depending upon how we wish to implement the system. The design methodology or strategy would depend upon whether we need to implement the system using software such as C or by a HDL such as Verilog. While developing algorithms for hardware implementation, we need to keep the actual hardware such as registers, counters, combination circuits, etc., in mind and, subsequently, design the architecture. Only then, we will be in a position to meet stringent specifications when the algorithm is converted into an actual working product. This is especially true for computationally intensive applications such as the discrete cosine transform (DCT), modulation/demodulation, etc., where we need to process the algorithms at real time rates. For instance, in video codecs conforming to MPEG 2 standards, the computationally intensive DCT algorithm needs to be computed at the rate of one coefficient per clock cycle running at the rate of 100 MHz or more if we are to meet the real time processing rate of 30 frames per second for a color picture of size: 1024×768 pixels or higher.

In this chapter, we will learn how to develop algorithms and verify using a high level language such as Matlab for various applications such as DCTQ, automatic quality control while speeding up processing of DCT, and a block matching algorithm for motion estimation in a motion picture. Although C codes have been developed for verifying the motion estimation algorithm, they are not presented in this book as the codes run into over 80 pages. Prior to designing architectures based on actual hardware components, we need to check whether the concepts and algorithms we have developed are really working. This is a vital step which should not be bypassed for medium to large designs, where the developments of algorithms are involved. Otherwise, rest of the processes such as the development of architectures, Verilog coding, simulation, synthesis, place and route, hardware development, etc., will go waste or the designers may end up in reworking the entire chain of processes starting from the algorithms. Handy tools for verifying the developed algorithms or concepts are evidently one of the high level languages such as Matlab and C. This chapter presents the verification of algorithms and concepts using Matlab. Matlab is generally more preferable since it has many built-in functions and hence codes are much shorter than the C codes and, therefore, they are close to the algorithms, thus serving as a standard reference for us to verify our Verilog codes later on. These algorithms will be converted into architectures that can be coded in HDL. Detailed RTL compliant Verilog codes will be presented for DCT and Quantization (DCTQ), as one of the examples of project design in a later chapter.

11.1 2D-Discrete Cosine Transform and Quantization

The discrete cosine transform closely approximates the Karhunen Loeve Transform (KLT) [21], which is known to be optimal in the sense of de-correlating the data and maximizing the energy packed into the lowest order coefficients. However, unlike the KLT, the DCT involves much less computational complexity in implementation, and is, therefore, preferred in image and video compression work. A comprehensive treatment of DCT algorithms and applications can be found in reference [22]. The DCT, which exploits the spatial redundancy to prepare the ground for effective compression, has played a key role in video data compression standards such as JPEG [23], MPEG 1 [24], MPEG 2 [25], and H.26X [26]

Over the years, considerable amount of research work have been carried out in proposing new algorithms for the DCT [27–31] and implementing them on general-purpose computers, DSPs, and ASICs. Direct 2-D approach [32] results in less parallelism, whereas separable row–column 1-D approach [33, 34] yields a faster algorithm. The authors of reference [34] have implemented 8 × 8 DCT by software on Pentium operating at 200 MHz. The fast algorithms [35–39] with minimum numbers of multiplication are often realized by flexible software approaches on the DSPs [40–43]. The speed requirement can be met by a high-speed DSP but it still needs to pay high hardware cost due to its inherent complexity of multipliers.

ICs have been fabricated [44, 45] for still image compression and decompression conforming to JPEG standard. In order to meet the real-time requirements, DCT and IDCT implementations use efficient and dedicated hardware [46–54]. In the architecture proposed in reference [46], all the multipliers are replaced by ROMs, and the number of ROMs required is large. For example, for an 8 × 8 DCT, more than 40 ROMs of size 1 K × 10 bits are required. Furthermore, a very stringent utilization of VLSI technology is required for the design to meet the required speed criteria. Reference [55] presents direct mapping of fast cosine transform (FCT) algorithm using SIMD architecture. That implementation uses a large number of switching networks and processing elements to achieve real time speeds. The authors of reference [38] have implemented 2D-DCT using only 6-bit precision for cosine coefficients. With increase in precision, the processing speed decreases drastically.

A linear, highly pipelined, parallel algorithm and architecture have been proposed and implemented by the author [56, 57] for 2D-DCT and Quantization on FPGAs. This architecture eliminates or minimizes the limitations cited in the earlier references. The scheme is further improved and incorporates dual-redundant input image memory, 45 stages of pipelining, and an optimized controller design yielding a throughput of one coefficient per clock cycle at 100 MHz. The use of dual input memory eliminates the input loading time of the host processor. The following section describes this DCTQ algorithm for fast implementation on an FPGA or an ASIC.

11.1.1 Algorithm for Parallel Matrix Multiplication for DCTQ

DCT is an orthogonal transform consisting of a set of vectors that are sampled cosine functions [22]. 2D-DCT of a block of size 8 × 8 pixels of an image is defined as

$$\mathrm{DCT}(u,v) = \frac{1}{4} c(u)\, c(v) \sum_{x=0}^{7} \sum_{y=0}^{7} f(x,y) \left[\cos\frac{(2x+1)u\pi}{16}\right] \left[\cos\frac{(2y+1)v\pi}{16}\right] \tag{11.1}$$

where f(x, y) is the pixel intensity and

$c(u) = c(v) = 1/\sqrt{2}$ for $u = v = 0$ and
$= 1$ for u, v = 1 to 7.

The DCT can be expressed conveniently in a matrix form:

$$\mathrm{DCT} = C\,X\,C^T \tag{11.2}$$

where X is the input image matrix, C the cosine coefficient matrix, and C^T, its transpose with constants (1/2)c(u) and (1/2)c(v) absorbed in C and C^T matrices respectively. For a clearer understanding, the DCT may be expressed in an expanded form:

$$\mathrm{DCT} = \begin{bmatrix} c_{00} & c_{01} & \cdots & c_{07} \\ c_{10} & c_{11} & \cdots & c_{17} \\ \vdots & \vdots & & \vdots \\ c_{70} & c_{71} & \cdots & c_{77} \end{bmatrix} \begin{bmatrix} x_{00} & x_{01} & \cdots & x_{07} \\ x_{10} & x_{11} & \cdots & x_{17} \\ \vdots & \vdots & & \vdots \\ x_{70} & x_{71} & \cdots & x_{77} \end{bmatrix} \begin{bmatrix} c_{00} & c_{10} & \cdots & c_{70} \\ c_{01} & c_{11} & \cdots & c_{71} \\ \vdots & \vdots & & \vdots \\ c_{07} & c_{17} & \cdots & c_{77} \end{bmatrix}$$

$$= \begin{bmatrix} p_{00} & p_{01} & \cdots & p_{07} \\ p_{10} & p_{11} & \cdots & p_{17} \\ \vdots & \vdots & & \vdots \\ p_{70} & p_{71} & \cdots & p_{77} \end{bmatrix} \begin{bmatrix} c_{00} & c_{10} & \cdots & c_{70} \\ c_{01} & c_{11} & \cdots & c_{71} \\ \vdots & \vdots & & \vdots \\ c_{07} & c_{17} & \cdots & c_{77} \end{bmatrix}$$

$$= \begin{bmatrix} \sum_{i=0}^{7} p_{0i} c_{0i} & \sum_{i=0}^{7} p_{0i} c_{1i} & \cdots & \sum_{i=0}^{7} p_{0i} c_{7i} \\ \sum_{i=0}^{7} p_{1i} c_{0i} & \sum_{i=0}^{7} p_{1i} c_{1i} & \cdots & \sum_{i=0}^{7} p_{1i} c_{7i} \\ \vdots & \vdots & & \vdots \\ \sum_{i=0}^{7} p_{7i} c_{0i} & \sum_{i=0}^{7} p_{7i} c_{1i} & & \sum_{i=0}^{7} p_{7i} c_{7i} \end{bmatrix} \qquad (11.3)$$

where

$$p_{jk} = \sum_{i=0}^{7} c_{ji} \; x_{ik} \qquad (11.4)$$

The two-stage matrix multiplication shown in Eq. 11.3 can be implemented by parallel architecture, wherein eight partial products, which are the row vectors of CX generated in the first stage, are fed to the second stage. Subsequently, multiplying row vector of CX by the C^T matrix generates eight DCT coefficients, corresponding to a row of $C\ X\ C^T$. While computing the $(i + 1)$th partial products of CX, the ith row DCT coefficients can also be computed simultaneously since the ith partial products of CX are already available. Application of DCT on an 8×8 pixel block, thus, generates 64 coefficients in a raster scan order.

Quantized outputs can be obtained by dividing each of the 64 DCT coefficients by the corresponding quantization table values given in the standards [25] as per the expression:

$$\text{DCTQ}(u,v) = \text{DCT}(u,v) \,/\, q(u,v); \quad u, v = 0 \text{ to } 7 \qquad (11.5)$$

These stages can be pipelined in such a way that one DCTQ output can be generated every clock cycle. Pipelining is detailed in the next chapter on the design of architecture. Similarly, inverse quantization can be computed by multiplying each of the 64 DCTQ coefficients by the corresponding quantization table values as per the expression:

$$\text{DCT}(u,v) = \text{DCTQ}(u,v) \times q(u,v); \quad u, v = 0 \text{ to } 7 \qquad (11.6)$$

The image can be reconstructed from the DCT (u,v) by evaluating the (inverse DCT) product of the matrix:

$$\text{IDCT} = C^T\,(\text{DCT})\,C \qquad (11.7)$$

The algorithm for the evaluation of IQIDCT is similar to that of DCTQ and, therefore, not presented here.

11.1.2 Verification of DCTQ–IQIDCT Processes with Fixed Pruning Level Control Using Matlab

Pruning levels indicate the stage at which the computation of DCT coefficients is stopped owing to insignificant contribution of subsequent coefficients towards the quality of the image. Pruning level (PL) simply relates to the number of DCT coefficients that is processed in every block of image. Details of pruning levels will be explained in Section 11.2. We will now develop Matlab codes to verify the DCTQ–IQIDCT algorithms incorporating a fixed pruning level that can be user selected. Matlab_code_11.1 shows the top level (main) DCTQ–IQIDCT code that is menu driven. It is named, 'fixedplcmain.m' to mean it is based on fixed pruning level control. This code is capable of processing any number of video frames of any size. To start with, the user enters the video disk file name such as 'car', followed by the starting and ending frame numbers as well as the pruning level up to which the DCTQ is to be processed. If the user enters 14 for PL input, then all the 64 coefficients in every block of a frame is processed. This, naturally, offers the highest possible quality of the reconstructed image. On the other extreme, a PL value of "0" processes only the DC coefficients, resulting in the least possible (may be, browse) quality. Intermediate qualities may be obtained by processing up to other pruning levels.

At the beginning of the code, two 'if' loops are used to check the pruning level category, inferred by a variable, k: –7 to –2; –1, to 6, and 14. The next set of instructions is a 'for' loop to process frame after frame, commencing from the start frame (f2) desired by the user. The end frame is in f3. The variable, 'frameno', gives the current frame number being processed. The current frame is read into 'i'. The cosine matrix 'c' is an 8 × 8 matrix with coefficients having a high resolution so that the Matlab reconstructed image may serve as a reference for verifying the Verilog simulated image output. The quantization matrix 'q' is as per the MPEG-1/MPEG-2 standards.

Next, the DCTQ, IQIDCT, and the compression are computed followed by the image quality using the statement:

i4 = 10*log 10(((255^2)*m*n)/(sum(sum((i–i2).^2))));

where 'i2' is the reconstructed image. The original 'i' and the reconstructed 'i2' images are displayed using 'imshow' function of Matlab. The reconstructed image is also saved as a disk file using the Matlab function, 'imwrite'. The above process is repeated (using a variable called 'counter') for all other frames as input by the user. The reconstructed image (i3) in TIFF format can also be displayed using any standard software such as Paintshop, XnView, Irfan View, etc. A number of Matlab statements are presently commented and the reader may uncomment them if required.

Matlab_code_11.1

```
%       Main program for the computation of DCTQ–IQIDCT
%       File: fixedplcmain.m
% Incorporates Fixed Pruning Level Control.
```

```
% Compression effected is expressed as bits per pixel and quality as PSNR
% in dB.
% Accepts input video frames of any size.
global exectime                                  % Declare execution time for use
                                                 % in other modules.
f1=input('Enter the ".tif"image file  : ');      % For example: 'car'
f2=input('Enter the start frame no.   : ');
f3=input('Enter the end frame no.     : ');
% INPUT PRUNING LEVEL : 14 means highest possible quality, i.e.,
% processes all the 64 coefficients in a block. '0' for processing DC coefficients
% alone.

P = input('Enter pruning level (0–14) : ');
if p < 14
  if p > 7
    k = (p–6)*(–1);                              % p range : 8–13, k range: –2 to –7
  elseif p <= 7                                  % p range : 0–7
    k = 6–p;                                     % k range: 6 to –1
  end
elseif p >= 14                                   % p = k = 14
  k = p;
end

counter  = 1 ;                                   % Initialize variables.
Psnr              = [];                          % Image quality in dB.

Bitspp            = [];                          % Compression.
Frame             = [];
for frameno = f2:1: f3
frameno = num2str(frameno);                      % Change to string format.
Fidopen = cat(2,f1,frameno,'.tif');              % Video frames must be in TIFF format.
i          = double(imread(fidopen));            % Read the video frame.
Exectime   = 0;
% C matrix:
c = [ 0.3536  0.3536  0.3536  0.3536  0.3536  0.3536  0.3536  0.3536 ;
      0.4904  0.4158  0.2778  0.0976 –0.0976 –0.2778 –0.4158 –0.4904 ;
      0.4620  0.1914 –0.1914 –0.4620 –0.4620 –0.1914  0.1914  0.4620 ;
      0.4156 –0.0976 –0.4904 –0.2778  0.2778  0.4904  0.0976 –0.4158 ;
      0.3536 –0.3536 –0.3536  0.3536  0.3536 –0.3536 –0.3536  0.3536 ;
      0.2778 –0.4904  0.0976  0.4156 –0.4158 –0.0976  0.4904 –0.2778 ;
      0.1914 –0.4620  0.4620 –0.1914 –0.1914  0.4620 –0.4620  0.1914 ;
      0.0976 –0.2778  0.4156 –0.4904  0.4904 –0.4158  0.2778 –0.0976 ];
% Quantization matrix:
q = [8 16 19 22 26 27 29 34 ;
     16 16 22 24 27 29 34 37 ;
     19 22 26 27 29 34 34 38 ;
```

```
    22 22 26 27 29 34 37 40 ;
    22 26 27 29 32 35 40 48 ;
    26 27 29 32 35 40 48 58 ;
    26 27 29 34 38 46 56 69 ;
    27 29 35 38 46 56 69 83];
 [m,n]         = size(i);                    % Get the picture size.
% Calculate  DCTQ  &  IQIDCT
  i2 = blkproc(i,[8 8],'fixedplc',c,q,k);

% Processes the image 'i' by applying the function 'fixedplc' to each distinct
% 8 by 8 block of A, with cosine, quantization matrices passed on as
% parameters. "i2" is the reconstructed image after applying DCTQ_IQIDCT.

% Calculate the compression expressed as bits per pixel
dci2        = blkproc(i,[8 8], 'fplc',c,q,k);
% Computes the DC coefficients of each 8 × 8 pixel block of the input image.
DCB         = dcbits(dci2); % Compute the number of DC bits.
aci2        = blkproc(i,[8 8], 'countac',c,q,k);
% Computes the AC coefficients of each 8 × 8 pixel block of the input image.
ACB         = sum(sum(aci2));        % Compute the number of AC bits in an image
BITS        = DCB + ACB;             % and the total number of compressed bits.
BPP         = BITS/(m*n);            % Calculate the compression effected
                                     % in terms of bits per pixel.
% Calculate Quality (PSNR in dB)
i4 = 10*log10(((255^2)*m*n)/(sum(sum((i–i2).^2))));
% clc
% disp('AC BITS')
% disp(ACB)
disp('PSNR value (in dB) is : ')
disp(i4)
disp('BITS PER PIXEL : ')
disp(BPP)
% Display the original and the reconstructed frames.
i5  = uint8(i); figure,imshow(i5),title('ORIGINAL IMAGE')
i3  = uint8(i2); figure,imshow(i3),title('RECONSTRUCTED IMAGE')
 [d] = sprintf('%2.4f  %2.4f',i4,BPP);
psnr(counter)   =  i4;
bitspp(counter) =  BPP;
frame(counter) =  (str2num(frameno));
counter = counter + 1;
e = 'r';
imwrite(i3,(cat(2,e,f1,frameno,'.tif')));       % Save the reconstructed frame.
end      % for frameno =  loop
% plot(frame,psnr),title('PSNR Vs FRAME NO.')
% figure,plot(frame,bitspp),title('BITS PER PIXEL Vs FRAME NO.')
```

In the main program, the code, 'fixedplc' was called in order to compute the DCTQ–IQIDCT processes. This code is presented in Matlab_code_11.1.1. The DCTQ is computed on an image input matrix 'x' pruning it at a given fixed level 'p' by the simple expression: (c*x*c′)./q, where c′ is the transpose of 'c'. The DCTQ is pruned by calling the pruning function 'prune', which is presented in Matlab_code_11.1.2. Finally, the reconstructed image is got by applying the IQIDCT process, c′*(y.*q)*c.

Matlab_code_11.1.1

```
%       File: "fixedplc.m"

function y = fixedplc(x,c,q,k)      % Declare the function.
% Compute DCTQ on an image input block matrix 'x' pruning it at a given
% fixed level p.
y = (fix((c*x*c')./q));             % Compute DCTQ.
If k<14
        y = prune(y,k);             % Call the pruning function to prune DCTQ.
end
y = fix(c'*(y.*q)*c);               % Compute IQIDCT to reconstruct the frame.
```

The Matlab code for the pruning DCTQ is presented in Matlab_code_11.1.2. The Matlab function, zeros(8,8) creates an 8 × 8 matrix of zeros. All other Matlab functions such as 'fliplr', 'spdiags' and 'full' along with 'zeros' are manipulations used to retain all DCTQ coefficients of an image block lying on the diagonal PL, i.e., 'k' and left of the diagonal, while clearing all other coefficients on the right hand side of the diagonal.

Matlab_code_11.1.2

```
% Pruning
% File:    prune.m
function y = prune(y,k)             % Declare the pruning function.
B = zeros(8,8);                     % Make all elements zeros.
Y = fliplr(y);
% FLIPLR(y) returns y with row preserved and columns flipped
%  in the left/right direction.
        for j=k:-1:-8
                y2 = spdiags(b,j,y);
% y2 = SPDIAGS(b,j,y) replaces the diagonals of 'y' specified by 'j' with
% the columns of 'b'. The output is sparse. Sparse form is obtained by
% squeezing out any zero elements.
                y  = full(y2);
```

```
% y = full(y2) converts a sparse matrix y2 to full storage
% organization.  If y2 is a full matrix, it is left unchanged.
        end
y = fliplr(y);                          % Flipped again.
```

The Matlab code for getting the DC coefficient in the DCTQ output of a block of image is presented in Matlab_code_11.1.3. The very first coefficient of the 'y' block matrix is the DC coefficient.

Matlab_code_11.1.3

```
% File:    fplc.m
function dccoeff = fplc(x, c, q, k)  % Declare the DC coefficient function.
% Obtain the DC coefficient of an 8 × 8 pixel block after computing DCTQ
% on input matrix 'x', pruned at a given fixed level, p.
y = (fix((c*x*c')./q));
dccoeff = y(1,1); % Returns only the DC coefficient of a block.
```

Matlab_code_11.1.4 gives the function code for computing the number of bits in DC coefficients after applying variable length coding (VLC) scheme, which is the last module in a video encoder. VLC brings about compression by assigning minimum size of variable length codes for the DCTQ coefficients. The reader may refer the VLC scheme of MPEG 1/MPEG 2 standards to get a complete picture. The first 'for' loop is for computing DC coefficients difference between two adjacent blocks, whereas the second 'for' loop is for computing the corresponding variable code from the differential DC size and VLC and additional code tables given in the standards.

Matlab_code_11.1.4

```
%        "dcbits.m file"
function dcb = dcbits(dci2)                % Declare the function.
% This counts the number of bits in DC coefficients after applying VLC
%  scheme on a block of DCTQ coefficients.
[u,v]   = size(dci2);                      % Gets the size.
dci2    = dci2';
dci5       = reshape(dci2,1,u*v);          % Returns the u × v matrix, whose
                                    % elements are taken column-wise from dci2.

dci4     = [];
dci4(1) = dci5(1);                  % First block DC coefficient is taken as it is,
  for i=1: ((u*v)-1)                % –1 is applied since dci4(1) is already got.
```

```
            first       = dci5(i);
            second      = dci5(i+1);
            diff        = second-first;  % Subsequent DC coefficients are
            dci4(i+1) = diff;            % taken as the difference.
  end
% disp(dci4)
bits = 0;
  for i=1: (u*v)
   ii = abs(dci4(i));                 % Drop the sign of the DC coefficients difference.
        if    ii<2  ;
                   bits = bits+3;          % Includes differential dc + additional code.
         elseif ii<4  ;
                   bits = bits+4;
        elseif ii<8  ;
                   bits = bits+6;
        elseif ii<16 ;
                   bits = bits+7;
        elseif ii<32 ;
                   bits = bits+9;
        elseif ii<64 ;
                   bits = bits+11;
        elseif ii<128;
                   bits = bits+13;
        elseif ii<256;
                   bits = bits+15;
        end
end
disp('DC BITS:')
disp(bits)
dcb = bits;
```

The function 'countac.m file' computes DCTQ coefficients in a block applying pruning and calls the function 'acbits' that counts the number of bits in AC coefficients. This is shown in Matlab_code_11.1.5.

Matlab_code_11.1.5

```
%    "countac.m file"
function accoeff = countac(x,c,q,k)
% This function computes DCTQ coefficients in a block applying pruning and
% calls the function "acbits" that counts the number of bits in AC coefficients.
y = (fix((c*x*c')./q));
        if k < 14
```

```
            y = prune1(y,k);
      end
accoeff = acbits(y);
```

Matlab_code_11.1.6 is a function to compute the number of bits in all the AC coefficients of a block in the VLC coding scheme. Two arrays called ‘runlength’ (RL) and ‘level’ (which means the value of DCTQ coefficients in a block) are required. The first ‘for’ loop is for computing the runlength in a block. DCTQ of a block is processed in a zig-zag order for computing the VLC. The number of zeros preceding a non-zero coefficient in the zig-zag order is known as the runlength. The first ‘for’ loop sequences through 1 to ‘L’ in steps of 1 or L to 1 in steps of –1, where L is the number of coefficients in a diagonal of a matrix block. The run length is computed here. The last ‘for’ loop, for x = 1:1:len, accumulates the actual number of bits in all the AC coefficients in a block. This function does not compute the actual variable length codes, but only the number of bits in all the AC coefficients, for we are interested only in assessing the compression effected. The reader is urged to study the AC VLC code table of MPEG 1/MPEG 2 to convince himself or herself the correct working of the Matlab codes presented in this section.

Matlab_code_11.1.6

```
%                  " acbits.m file "
function acb = acbits(y)
% This computes the number of bits in AC coefficients in VLC coding scheme.
% Two arrays called 'RL' (runlength) & 'level' (value of coefficients) are
% calculated.
y           = fliplr(y);
runlength   = 0;
r           = 1;
RL          = [];
level       = [];
for j=6:–1:–7            % Topmost diagonal is the DC coefficient and hence skip it.
  y3 = diag(y,j);                    % Select elements of a diagonal.
  L  = length(y3);
  k  = abs(j);
        if k == 6 | k == 4 | k == 2 | k == 0;
              a = 1;
              b = L;
              c = 1;
        else
              a = L;
              b = 1;
              c = -1;
```

```
        end
        for h = a : c : b          % 1 to 'L' in steps of 1 or L to 1 in steps of -1.
          check = y3(h);
            if check == 0;         % If the AC coefficient is "0",
                 runlength         = runlength + 1;  % advance the counter.
            else level(r)          = check;
                 RL(r)             = runlength;
                 r                 = r + 1;
                 runlength         = 0;
            end                    % Done for this coefficient.
        end               % Process the next coefficient in diagonal (for 'h' loop).
end                        % Process the next coefficient in diagonal (for 'j' loop).
%RL
%level
% Based on the two arrays 'RL' and 'level', the number of bits
% (for encoding AC coefficients) are counted.
len   = length(RL);
acbit = 0;
if runlength~=63;                  % 63 means all AC coefficients are 0.
 for x=1:1:len
          level(x)          = abs(level(x));
          k                 = 0;
            if RL(x)==0  & level(x)<41
                 if    level(x) < 2  ;
                       acbit = acbit + 3  ;
                 elseif level(x) < 3  ;
                       acbit = acbit + 5  ;
                 elseif level(x) < 4  ;
                       acbit = acbit + 6  ;
                 elseif level(x) < 5  ;
                       acbit = acbit + 8  ;
                 elseif level(x) < 7  ;
                       acbit = acbit + 9  ;
                 elseif level(x) < 8  ;
                       acbit = acbit + 11 ;
                 elseif level(x) < 12 ;
                       acbit = acbit + 13 ;
                 elseif level(x) < 16 ;
                       acbit = acbit + 14 ;
                 elseif level(x) < 32 ;

                       acbit = acbit + 15 ;
                 elseif level(x) < 41 ;
                       acbit = acbit + 16 ;
                 end
           elseif RL(x)==1 & level(x)<19
```

```
        if    level(x) < 2  ;
              acbit = acbit + 4  ;
        elseif level(x) < 3  ;
              acbit = acbit + 7  ;
        elseif level(x) < 4  ;
               acbit = acbit + 9  ;
        elseif level(x) < 5  ;
              acbit = acbit + 11 ;
        elseif level(x) < 6  ;
              acbit = acbit + 13 ;
        elseif level(x) < 8  ;
              acbit = acbit + 14 ;
        elseif level(x) < 15 ;
              acbit = acbit + 16 ;
        elseif level(x) < 19 ;
              acbit = acbit + 17 ;
        end
 elseif RL(x)==2 & level(x)<6
        if    level(x) == 1 ;
              acbit = acbit + 5  ;
        elseif level(x) == 2 ;
              acbit = acbit + 8  ;
        elseif level(x) == 3 ;
              acbit = acbit + 11 ;
        elseif level(x) == 4 ;
              acbit = acbit + 13 ;
        elseif level(x) == 5 ;
              acbit = acbit + 14 ;
        end
elseif RL(x)==3 & level(x)<5
        if    level(x) == 1 ;
              acbit = acbit + 6  ;
        elseif level(x) == 2 ;
              acbit = acbit + 9  ;
        elseif level(x) == 3 ;
              acbit = acbit + 13 ;
        elseif level(x) == 4 ;
              acbit = acbit + 14 ;
        end
elseif RL(x)==4 & level(x)<4
        if    level(x) == 1 ;
              acbit = acbit + 6  ;
        elseif level(x) == 2 ;
              acbit = acbit + 11 ;
        elseif level(x) == 3 ;
              acbit = acbit + 13 ;
```

```
            end
        elseif RL(x)==5 & level(x)<4
            if   level(x) == 1 ;
                acbit = acbit + 7  ;
            elseif level(x) == 2 ;
                acbit = acbit + 11 ;
            elseif level(x) == 3 ;
                acbit = acbit + 14 ;
            end
        elseif RL(x)==6 & level(x)<4
            if   level(x) == 1 ;
                acbit = acbit + 7  ;
            elseif level(x) == 2 ;
                acbit = acbit + 13 ;
            elseif level(x) == 3 ;
                acbit = acbit + 17 ;
            end
        elseif RL(x)==7 & level(x)<3
            if   level(x) == 1 ;
                acbit = acbit + 7  ;
            elseif level(x) == 2 ;
                acbit = acbit + 13 ;
            end
        elseif RL(x)< 10 & level(x)<3
            if   level(x) == 1 ;
                acbit = acbit + 8  ;
            elseif level(x) == 2 ;
                acbit = acbit + 13 ;
            end
        elseif RL(x)==10 & level(x)<3
            if   level(x) == 1 ;
                acbit = acbit + 9  ;
            elseif level(x) == 2 ;
                acbit = acbit + 14 ;
            end
        elseif RL(x)< 14 & level(x)<3
            if   level(x) == 1 ;
                acbit = acbit + 9  ;
            elseif level(x) == 2 ;
                acbit = acbit + 17 ;
            end
        elseif RL(x)< 17 & level(x)<3
            if   level(x) == 1 ;
                acbit = acbit + 11 ;
            elseif level(x) == 2 ;
                acbit = acbit + 17 ;
```

```
                    end
                elseif RL(x)< 22  & RL(x)>=17    ;
                        acbit = acbit + 13 ;
                elseif RL(x)< 27  & RL(x)>=22    ;
                        acbit = acbit + 14 ;
                elseif RL(x)< 32  & RL(x)>=27    ;
                        acbit = acbit + 17 ;
                else
                        k = 1;
            end
        if k == 1
          if  level(x)< 128  ;
                    acbit = acbit + 20;
                        % Extra bits required when ESCAPE is encountered.
          elseif level(x)>128;
                    acbit = acbit + 28;
          end
        end        % End of calculation of bits for level(x); proceed to level(x+1).
      %disp(acbit)
  end                          % End of calculation for all elements in 'level'.
end                                   % End for runlength~=63 loop.
acbit = acbit + 2 ;                   % EOB (end of block) indication bits.
acb  = acbit;
%disp(acb)
```

11.2 Automatic Quality Control Scheme for Image Compression

A parallel matrix multiplication algorithm has been presented for a fast implementation of DCT in Section 11.1.1. A new algorithm for assessing image quality on the fly [58] using a concept called pruning will be presented in this section. As a result of applying this algorithm, the processing speed of DCTQ can be doubled when compared to the implementation speed of the DCTQ without pruning. This can also speed up the next pipeline module called the variable length coder (VLC) of a video encoder. The algorithm can be used for effecting rate control, i.e., maintaining a constant bit rate while a compressed bit stream is transmitted over a serial channel. This algorithm can be used for both hardware and software implementations.

DCT applied on an 8 × 8 pixel block of image results in the generation of 64 coefficients in the raster scan order, assuming it to be arranged as an 8 × 8 matrix as shown in Figure 11.1. The first coefficient is known as DC coefficient, whereas other coefficients are known as AC coefficients. The diagonals, numbered from 0 to 14 from top-left to bottom-right in that order, are referred to as pruning levels

(PL) and indicate the stage at which computation of DCT coefficients is stopped owing to insignificant contribution of subsequent coefficients towards the quality of the image. Quantized DCT coefficients beyond pruning levels of about two are zero for most of the image blocks. Therefore, computation time is wasted in processing beyond these pruning levels. This problem is solved in the present method by computing the sum of energy of AC coefficients lying on every diagonal commencing from PL1 up to PL14. At each step, the computed energy is compared with threshold energy. If it is less than the threshold energy, the computation for the current image block is immediately terminated, and the processing for the next block commences, thus speeding the processing by over two times when compared to conventional approaches of computing up to PL14. More details are presented in Section 11.2.1.

Pruning has been applied to the Fast Fourier Transform (FFT) for applications such as filtering, transformation of zero-padded sequence, etc. Examples of the applications of the pruned FFT are found in references [59, 60]. Pruned DCT algorithms have been proposed in reference [61] and applied to fast image compression in references [62] and [63]. In order to reduce the processing time of still pictures, adaptive pruning techniques [64, 65] were introduced at the DCT stage itself. These methods, however, are not suitable for implementation on FPGAs both from the quality and exacting speed considerations encountered in motion pictures. The scheme presented [58] here considerably reduces these limitations by evaluating the image quality dynamically as DCT coefficients are being computed. The processing for the current image block is stopped as soon as the desired quality is met, thus speeding up the system considerably, while retaining the quality of the picture.

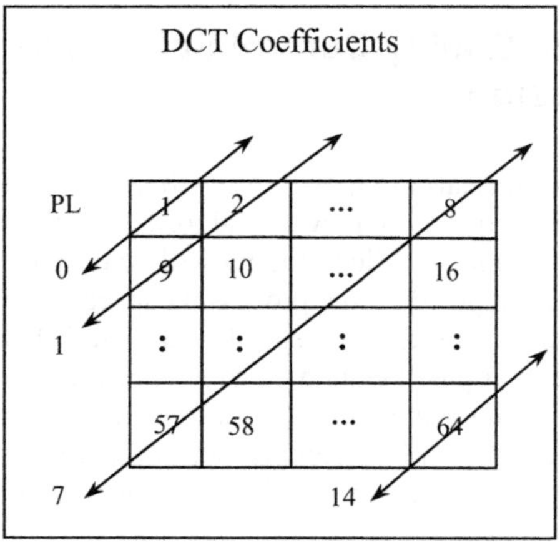

Fig. 11.1 Pruning levels in an image block

11.2.1 Algorithm for Assessing Image Quality Dynamically

In transform coding, a signal is mapped from one domain, usually spatial or temporal, into the transform domain. The signal can be one-dimensional or multi-dimensional. DCT is an orthogonal transform consisting of a set of vectors that are sampled cosine functions. The mapping is, therefore, unique and reversible. In this case, the energy is preserved in the transform domain, and the signal can be recovered completely by the inverse transform. Since the DCT transform is orthogonal, implementing the inverse transform is essentially the same as implementing the forward transform. Therefore, the properties such as fast algorithm and recursive structure are preserved in the inverse transform. In fact, the hardware, e.g., a VLSI chip, designed for the forward transform can be used with minor modifications for implementing the inverse transform. In image and video coding standards such as JPEG, MPEG, H.263, etc., 2D-DCT is the primary factor in achieving compression.

In order to achieve a regular and efficient method of implementation of DCT, a parallel matrix multiplication algorithm has been presented earlier. In general, from a visual perception viewpoint, the low frequency coefficients are much more sensitive than the high frequency coefficients. The energy is invariant to orthogonal transformation. The sum of the squares of all the DCT coefficients or the spatial data values is the energy of the block, i.e.,

$$\sum_{u=0}^{7}\sum_{v=0}^{7}(DCT_{u,v})^2=\sum_{n=0}^{7}\sum_{m=0}^{7}(x_{n,m})^2, \qquad u, v, n, m = 0\text{–}7 \tag{11.8}$$

$$\begin{bmatrix} 1260 & -1 & -12 & -5 & 2 & -2 & -3 & 1 \\ -23 & -18 & -6 & -3 & -3 & 0 & 0 & -1 \\ -11 & -9 & -2 & 2 & 0 & -1 & -1 & 0 \\ -7 & -2 & 0 & 2 & 1 & 0 & 0 & 0 \\ -1 & -1 & 2 & 2 & 0 & -1 & 1 & 1 \\ 2 & 0 & 2 & 0 & -1 & 2 & 1 & -1 \\ -1 & 0 & 0 & -2 & -1 & 2 & 1 & -1 \\ -3 & 2 & -4 & -2 & 2 & 1 & -1 & 0 \end{bmatrix}$$

a

$$\begin{bmatrix} 157 & 0 & 0 & 0 & 0 & 0 & 0 & 0 \\ -1 & -1 & 0 & 0 & 0 & 0 & 0 & 0 \\ 0 & 0 & 0 & 0 & 0 & 0 & 0 & 0 \\ 0 & 0 & 0 & 0 & 0 & 0 & 0 & 0 \\ 0 & 0 & 0 & 0 & 0 & 0 & 0 & 0 \\ 0 & 0 & 0 & 0 & 0 & 0 & 0 & 0 \\ 0 & 0 & 0 & 0 & 0 & 0 & 0 & 0 \\ 0 & 0 & 0 & 0 & 0 & 0 & 0 & 0 \end{bmatrix}$$

b

Fig. 11.2 A representative sample image block: (a) DCT coefficients; (b) Quantized DCT coefficients

1. $e_{PL1} = (DCT2)^2 + (DCT9)^2$;
   if $e_{PL1} < e_{THR}$, PLN = 0 & go to step 15 ;
   else
2. $e_{PL2} = (DCT3)^2 + (DCT10)^2 + (DCT17)^2$;
   if $e_{PL2} < e_{THR}$, PLN=1 & go to step 15 ;
   else

   ⋮

13. $e_{PL13} = (DCT56)^2 + (DCT63)^2$;
    if $e_{PL13} < e_{THR}$, PLN = 13 ;
    else
14. PLN = 14;
15. Stop current block & start processing the next image block.

Fig. 11.3 Algorithm for the computation of energy (quality) of an image

A sample 8 × 8 DCT coefficient block is shown in Figure 11.2 a. Quantized DCT coefficients beyond pruning levels of about two are zero for most of the image blocks as shown in Figure 11.2b for the same block of DCT coefficients. Therefore, as mentioned before, the computation time is wasted in processing beyond these pruning levels which is the case in all the earlier methods cited including the parallel algorithm presented in Section 11.1.1. This problem is solved by computing the sum of energy of AC coefficients lying on every diagonal commencing from PL1 up to PL14 shown in Figure 11.1.

Energy has a direct bearing on the quality of the image. PL0 is not taken into account for the energy computation since considerable amount of energy is packed in the DC coefficient and, therefore, needs to be processed without fail. At each step, the computed energy is compared with threshold energy, e_{THR}. If it is less than e_{THR}, the computation for the current image block is immediately terminated, and the processing for the next block commences, thus speeding the processing by over two times when compared to conventional approaches of computing up to PL14. The algorithm for this method is shown in Figure 11.3.

Applying this method for the DCT coefficients cited in the example, it is seen that the energies of the DCT coefficients for PL1 to PL3 are 530, 589, and 191 units respectively while those for PL4 to PL14 are each less than 23 units. By applying the energy threshold of 200 units, the final pruning level for this image block is evident to be two, as can also be readily verified from Figure 11.2b. With a few exceptions, exactly similar results were obtained for a large number of image blocks for a number of images that have been experimented. Therefore, the energy threshold of 200 units has been made as the default value, although the user can change the same. The image quality obtained by this method is quite good and is very close to that obtained by processing up to the full quality level of PL14 as presented in Section 1.1. This method of applying threshold energy has been arrived at after conducting exhaustive experiments with a number of images.

11.2.2 Verification of DCTQ–IQIDCT Processes with Automatic Pruning Level Control Incorporated Using Matlab

In the previous section, the automatic pruning level control algorithm was presented. We will now develop Matlab codes to verify the DCTQ–IQIDCT incorporating the automatic pruning level algorithm, in which the threshold of energy, that determines the quality of the image processed as well as the processing speed, is user selected.

Matlab_code_11.2.1 shows the top level (main) DCTQ–IQIDCT code that is menu driven. It is named, 'autoplcmain.m' to mean that it is based on automatic pruning level control method to meet the desired quality of the image. This code is capable of processing any number of video frames. The input file, the starting and the ending frame numbers as well as the energy threshold are user selected to start with. The first 'for' loop processes many video files depending upon the setting, 'manyframes'. The second 'for' loop:
for frameno = f2:1:f3
processes all the frames commencing from the start frame number to the end frame number. These are followed by the C and Q matrices. Calculation of DCTQ, IQIDCT, compression (bits per pixel), and quality (PSNR) are similar to the fixed PL method presented in the previous section except that 'ethreshold' is extra in "i2". Reconstructed image is in "i2" after applying auto PL, while "i" matrix contains the original image, both of which are displayed using the Matlab function, 'imshow'. The average pruning level and execution time are also computed and displayed. This is followed by displaying the original (i5) and the reconstructed image (i3) after applying the auto PL method using the Matlab command, 'imshow'. Using 'imwrite', the reconstructed image is also saved as a disk file. The PSNR, the bits per pixel, average pruning level, and execution times are computed and plotted for a number of frames. There are numerous commented codes such as processing only the desired blocks in a frame, graph plots of variables, etc. These may be selectively uncommented to activate these commands, if the reader wishes to study their behavior.

Matlab_code_11.2.1

```
% Main program for the computation of DCTQ–IQIDCT using
% Automatic Pruning Level Control

%        File: "autoplcmain.m"

% Incorporates Automatic Pruning Level Control.
% Compression effected is expressed as bits per pixel and quality (PSNR)
% in dB.
% Accepts input video frames of any size.
global k exectime c3                        % Declare variables for use
```

```
global dci3 acbit avgpl bno                    % in other modules.
% Main Program for DCT (i/p an image file of any size).
% This includes Auto PLC / bpp / psnr.
manyframes  = 2;                    % Input ('enter the number of types of files').
counter     = 0;
noframes    = [];
counter2    = 0;
psnr        = [];
bitspp      = [];
avgpln      = [];
etime       = [];
frame       = [];
for iii=1:1:manyframes
        f1 = input('Enter the "tif" image file  : ');
        f2 = input('Enter the start frame no. : ');
        f3 = input('Enter the end frame no.   : ');
        ethreshold = input('Enter energy threshold : ');
        countfig  = 0;
        for frameno=f2:1:f3
                counter2            = counter2 + 1;
                countfig            = countfig + 1;
                frame(counter2) = frameno;
                frameno = num2str(frameno);
                fidopen = cat(2,f1,frameno,'.tif');
                % im      = fread(fid,[str2num(f3),inf]);
                % rawim = im';
                % imshow(uint8(rawim));
                % imsize = size(rawim)
                i           = double(imread(fidopen));
                k           = 0;
                exectime = 0;
                c3          = 0;
                dci3        = [];
                acbit     = 0;
                avgpl     = 0;
                bno         = 0;
                %imshow(i)
% C matrix :
c = [ 0.3536  0.3536  0.3536  0.3536  0.3536  0.3536  0.3536  0.3536 ;
      0.4904  0.4158  0.2778  0.0976 -0.0976 -0.2778 -0.4158 -0.4904 ;
      0.4620  0.1914 -0.1914 -0.4620 -0.4620 -0.1914  0.1914  0.4620 ;
      0.4156 -0.0976 -0.4904 -0.2778  0.2778  0.4904  0.0976 -0.4158 ;
      0.3536 -0.3536 -0.3536  0.3536  0.3536 -0.3536 -0.3536  0.3536 ;
      0.2778 -0.4904  0.0976  0.4156 -0.4158 -0.0976  0.4904 -0.2778 ;
      0.1914 -0.4620  0.4620 -0.1914 -0.1914  0.4620 -0.4620  0.1914 ;
      0.0976 -0.2778  0.4156 -0.4904  0.4904 -0.4158  0.2778 -0.0976 ];
```

```
% Quantization matrix : As per MPEG 1/MPEG 2 standards.
        q =[8 16 19 22 26 27 29 34 ;
            16 16 22 24 27 29 34 37;
            19 22 26 27 29 34 34 38;
            22 22 26 27 29 34 37 40;
            22 26 27 29 32 35 40 48;
            26 27 29 32 35 40 48 58;
            26 27 29 34 38 46 56 69;
            27 29 35 38 46 56 69 83];
            [m,n]     = size(i);
            m1        = m/8;
            n1        = n/8;
            blocks1   = m1*n1;
            % [Q1] = sprintf('Total nos. of blocks of 8x8 pixels
            % in image are %d',blocks1);
            % disp(Q1)
            % x = input('Enter the start block no      : ');
            % a = input('Enter the end block no        : ');
            % ethreshold = input('Enter energy threshold    : ');
            % Find starting pixel location.
            % if x<= n1
            %       x1 = (x*8)–7; y1 = 1;
            % elseif x>n1 & ((floor(x/n1))*n1) == x
            %       x1 = ((8*n1)–7);y1 = ((floor(x/n1))*8)–7;
            % elseif x>n1
            %      x1 = ( (x–((floor(x/n1))*n1))*8 )–7;
            %      y1 = (floor(x/n1)*8)+1;
            % end
            % Find ending pixel location.
            % if a<= n1
            %       x2 = (a*8);
            %        y2 = 8;
            % elseif a>n1 & ((floor(a/n1))*n1) == a
            %       x2 = (8*n1);
            %       y2 = (floor(a/n1))*8;
            % else
            %       x2 = ( (a-((floor(a/n1))*n1))*8 );
            %       y2 = (floor(a/n1)+1)*8;

            % end
            % i = i(y1:y2,x1:x2);
            % [m,n] = size(i);

% Calculation of DCTQ  &  IQIDCT
            i2 = blkproc(i,[8 8], 'autoeplc',c,q,ethreshold);
```

```
                                    % 'autoplc' is a called function.
            % "i2" is the reconstructed image after applying Auto PLC.

% Calculation of BITS/PIXEL
            DCB   = dcbitsauto(dci3);
            % ACB = sum(sum(aci2));
            BITS  = DCB + acbit;
            BPP   = BITS/(m*n);
            %disp('AC BITS:')
            %disp(acbit)

% Calculation of PSNR
            i4 = 10*log10(((255^2)*m*n)/(sum(sum((i-i2).^2))));
            % disp(' PSNR value is :- ')
            % disp(i4)
            % disp(' BITS PER PIXEL :- ')
            % disp(BPP)
            % exectime = (exectime);
            % disp('AVG PLN :')
            AVGPL = avgpl/bno;              % Compute the average PL.
            % disp(AVGPL)
            % disp('EXECUTION TIME :-')
            % disp(exectime)
        [d] = sprintf('%2.4f %2.4f  %2.4f  %i',i4,BPP,AVGPL, exectime);
            cfig              = num2str(counter);
            psnr(counter2)    = i4;
            bitspp(counter2)  = BPP;
            avgpln(counter2) = AVGPL;
            etime(counter2)   = exectime;
        i5 = uint8(i);figure,imshow(i5),title('ORIGINAL IMAGE')
        i3 = uint8(i2);figure,imshow(i3),title('RECONSTRUCTED IMAGE')
            e  = 'r';
            imwrite(i3,(cat(2,e,f1,frameno,'.tif')));
      end      % for frameno =  loop
  counter = counter + 1;
  noframes(counter) = countfig;
 end      % for iii=     loop
psnr1   = psnr(1:noframes(1));
psnr2   = psnr((noframes(1)+ 1):noframes(1)+ noframes(2)));
bitspp1 = bitspp(1:noframes(1));
bitspp2 = bitspp((noframes(1)+ 1):noframes(1)+ noframes(2)));
avgpln1 = avgpln(1:noframes(1));
avgpln2 = avgpln((noframes(1)+ 1):noframes(1)+ noframes(2)));
etime1  = etime(1:noframes(1));
etime2  = etime((noframes(1)+ 1):noframes(1)+ noframes(2)));
frame1  = frame(1:noframes(1));
```

```
frame2  = frame((noframes(1)+ 1):noframes(1)+ noframes(2)));
etime1  = etime1/1000000;          % Converts the execution times to ms from ns.
etime2  = etime2/1000000;
ae1 = (sum(etime1))/(length(etime1));
ae2 = (sum(etime2))/(length(etime2));
ap1 = (sum(psnr1))/(length(psnr1));
ap2 = (sum(psnr2))/(length(psnr2));
ab1 = (sum(bitspp1))/(length(bitspp1));
ab2 = (sum(bitspp2))/(length(bitspp2));
av1 = (sum(avgpln1))/(length(avgpln1));
av2 = (sum(avgpln2))/(length(avgpln2));
disp('AVG PSNR1 value is :- ')
disp(ap1)
disp('AVG PSNR2 value is :- ')
disp(ap2)
disp('AVG BITS PER PIXEL1 :- ')
disp(ab1)
disp('AVG BITS PER PIXEL2 :- ')
disp(ab2)
disp('AVG PLN1 :')
disp(av1)
disp('AVG PLN2 :')
disp(av2)
disp('AVG EXECUTION TIME1 IN ms :')
disp(ae1)
disp('AVG EXECUTION TIME2 IN ms :')
disp(ae2)

%figure,plot(frame1,psnr1),xlabel('frame number'),ylabel('PSNR');
%figure,plot(frame2,psnr2),xlabel('frame number'),ylabel('PSNR');
%figure,plot(frame1,bitspp1),xlabel('frame number'),ylabel('bits per pixel');
%figure,plot(frame2,bitspp);xlabel('frame number'),ylabel('bits per pixel');
%figure,plot(frame1,avgpln1),xlabel('frame number'),ylabel('average pruning
%                                                                    level');

% figure,plot(frame2,avgpln2);xlabel('frame number'),ylabel('average pruning
%                                                                    level');
%figure,plot(frame1,etime1),xlabel('frame number'),ylabel('execution time, ms');
%figure,plot(frame2,etime2)xlabel('frame number'),ylabel('execution time, ms');
% subplot(8,1,1);plot(frame1,psnr1),title('PSNR Vs FRAME NO.');
% subplot(8,1,3);plot(frame1,bitspp1),title('BITS PER PIXEL Vs FRAME NO.');
% subplot(8,1,5);plot(frame1,avgpln1),title('AVG PLN Vs FRAME NO.');
% subplot(8,1,7);plot(frame1,etime1),title('EXECUTION TIME(in ms)
%                                                         Vs FRAME NO.');
%subplot(8,1,2);plot(frame2,psnr2);
%subplot(8,1,4);plot(frame2,bitspp2);
```

```
%subplot(8,1,6);plot(frame2,avgpln2);
%subplot(8,1,8);plot(frame2,etime2);
%subplot(4,2,1);plot(frame1,psnr1),title('PSNR Vs FRAME NO.');
%subplot(4,2,2);plot(frame1,bitspp1),title('BITS PER PIXEL Vs FRAME NO.');
%subplot(4,2,5);plot(frame1,avgpln1),title('AVG PLN Vs FRAME NO.');
%subplot(4,2,6);plot(frame1,etime1),title('EXECUTION TIME(m s) Vs
%                                                  FRAME NO.');
%subplot(4,2,3);plot(frame2,psnr2);
%subplot(4,2,4);plot(frame2,bitspp2);
%subplot(4,2,7);plot(frame2,avgpln2);
%subplot(4,2,8);plot(frame2,etime2);
subplot(2,2,1);plot(frame1,psnr1),xlabel('a'),ylabel('PSNR');
subplot(2,2,3);plot(frame1,bitspp1),xlabel('c'),ylabel('bits per pixel');
subplot(2,2,2);plot(frame2,psnr2),xlabel('b');
subplot(2,2,4);plot(frame2,bitspp2),xlabel('d');
figure,subplot(2,2,1);plot(frame1,avgpln1),xlabel('a'),ylabel('average pl');
subplot(2,2,3);plot(frame1,etime1),xlabel('c'),ylabel('execution time, ms');
subplot(2,2,2);plot(frame2,avgpln2),xlabel('b');
subplot(2,2,4);plot(frame2,etime2),xlabel('d');
%SUBPLOT('position',[.1 .08 .385 .15]);plot(frame1,etime1),xlabel('g'),
%                          ylabel('exec.Time, ms');set(gca,'FontSize',8);
%SUBPLOT('position',[.1 .31 .385 .15]);plot(frame1,avgpln1),
%                       xlabel('e'),ylabel('avg. pl');set(gca,'FontSize',8);
%SUBPLOT('position',[.1 .54 .385 .15]);plot(frame1,bitspp1),
%                  xlabel('c'),ylabel('bits per pixel');set(gca,'FontSize',8);
%SUBPLOT('position',[.1 .77 .385 .15]);plot(frame1,psnr1),
                       xlabel('a'),ylabel('PSNR, dB');set(gca,'FontSize',8);
%SUBPLOT('position',[.565 .77 .385 .15]);plot(frame2,psnr2),
%                                       xlabel('b');set(gca,'FontSize',8);
%SUBPLOT('position',[.565 .54 .385 .15]);plot(frame2,bitspp2),
%                                       xlabel('d');set(gca,'FontSize',8);
%SUBPLOT('position',[.565 .31 .385 .15]);plot(frame2,avgpln2),
%                                       xlabel('f');set(gca,'FontSize',8);
%SUBPLOT('position',[.565 .08 .385 .15]);plot(frame2,etime2),
%                                       xlabel('h');set(gca,'FontSize',8);
```

The Matlab_code_11.2.1.1 presents the function called by the main program presented earlier, which brings about the automatic pruning level control based on the algorithm developed in the previous chapter. This requires the energy threshold, 'ethreshold', selected by the user as a parameter to be passed by the calling function. The processing time is computed as presented in Table 11.1 in Section 11.2.3. More details are presented in the next section. This code is profusely commented for the reader to understand easily. The rest of the codes are similar to Matlab_code_11.1.6 (acbits.m) presented earlier for computing the number of AC bits in a block.

Matlab_code_11.2.1.1

% File: "autoeplc.m"

```
function yp = autoeplc(x,c,q,ethreshold)
global exectime              % Declare the variables used by other modules.
global c3 k
global dci3
global acbit avgpl bno
bno = bno + 1;
c3  = c3 + 1;
yyy = (c*x*c');              % Compute the DCT on a block.
qqq = q;                     % Get the quantization matrix.
dci3(c3) = fix(yyy(1,1)/qqq(1,1));  % Compute the quantized DC coefficient.
y   = fix(yyy);
pln = [ ];
k   = [ ];
pl  = [ ];
y   = fliplr(y);                         % Flip left to right to access
q   = fliplr(q);                         % the leading diagonals.
ch  = 1;
for m=6:-1:-7                            % Select the diagonals except DC
                                         % coefficient diagonal.
  v1 = sum((diag(y,m)).^2);            % Compute the energy of the AC diagonal
                                         % as per the Auto PLC algorithm.
    if v1<ethreshold & ch==1           % If the energy is less than the threshold,
            pln = m ;                    % return the pruning level number.
            ch  = 0;
    elseif v1>=ethreshold;               % Otherwise, clear v1 for a fresh
           v1  = 0;                      % computation.
    end
end
k = pln;
y = y./q;                                % Compute the DCTQ of the block.
b = zeros(8,8);                          % Clear all elements.
  for j=k:-1:-7                          % PL number to the last diagonal.
      y2 = spdiags(b,j,y);         % Replaces the diagonals of DCTQ specified by j
                                   % with the columns of b.
      y  = full(y2);               % Fill all elements of DCTQ matrix from
                                   % PL number to the last diagonal with zeros.
  end
        if k>=0 ;
                pl = 7 - k - 1;                  % Pruning level = 0 to 6.
        elseif k<0 ;
```

```
                    pl = 7 + abs(k) – 1;          % Pruning level = 7 to 13.
          else
                    pl = 14 ;
          end
pl;
avgpl = avgpl+pl;
          if pl>7 ;                         % Compute execution time – refer Table 11.1.
                                            % pl = 8 to 14.
                    exectime1 = 3040 + (40)*(pl–7);
          else
                    exectime1 = 800 + pl*320;          % pl = 0 to 7.
end
exectime = exectime + exectime1;
%dci3(c3)= fix(yyy(1,1)/qqq(1,1));
% acbits here
runlength = 0;
r         = 1;
RL        = [];
level     = [];
for j=6:-1:-7
  y3 = diag(y,j);
  L  = length(y3);
  k  = abs(j);
  if k==6 | k==4 | k==2 | k==0;
          a  = 1;
          b  = L;
          c1 = 1;
  else
          a  = L;
          b  = 1;
          c1 = -1;
  end
  for h=a:c1:b                              % 1 to 'L' in steps of 1 or L to 1 in steps of -1.
    check = fix(abs(y3(h)));
    if check==0;                            % If the AC coefficient is "0",
                    runlength = runlength + 1;
    else level(r) = check;
                    RL(r)      = runlength;
                    r          = r + 1;
                    runlength  = 0;
     end
  end
end
len = length(RL);
%level
%RL
```

```
if runlength~=63;                  % 63 means all AC coefficients are 0.
for z=1:1:len
  level(z) = round(abs(level(z)));
  k        = 0;
  if RL(z) ==0  & level(z)<41 & level(z)>0
    if    level(z) < 2  ;
                  acbit = acbit + 3  ;
    elseif level(z) < 3  ;
                  acbit = acbit + 5  ;
    elseif level(z) < 4  ;
                  acbit = acbit + 6  ;
    elseif level(z) < 5  ;
                  acbit = acbit + 8  ;
    elseif level(z) < 7  ;
                  acbit = acbit + 9  ;
    elseif level(z) < 8  ;
                  acbit = acbit + 11 ;
    elseif level(z) < 12 ;
                  acbit = acbit + 13 ;
    elseif level(z) < 16 ;
                  acbit = acbit + 14 ;
    elseif level(z) < 32 ;
                  acbit = acbit + 15 ;
    elseif level(z) < 41 ;
                  acbit = acbit + 16 ;
    end
  elseif RL(z) ==1 & level(z)<19 & level(z)>0
    if    level(z) < 2  ;
                  acbit = acbit + 4  ;
    elseif level(z) < 3  ;
                  acbit = acbit + 7  ;
    elseif level(z) < 4  ;
                  acbit = acbit + 9  ;
    elseif level(z) < 5  ;
                  acbit = acbit + 11 ;
    elseif level(z) < 6  ;
                  acbit = acbit + 13 ;
    elseif level(z) < 8  ;
                  acbit = acbit + 14 ;
    elseif level(z) < 15 ;
                  acbit = acbit + 16 ;
    elseif level(z) < 19 ;
                  acbit = acbit + 17 ;
    end
  elseif RL(z)==2 & level(z)<6 & level(z)>0
    if    level(z) == 1 ;
```

```
                 acbit = acbit + 5  ;
  elseif level(z) == 2 ;
                 acbit = acbit + 8  ;
  elseif level(z) == 3 ;
                 acbit = acbit + 11 ;
  elseif level(z) == 4 ;
                 acbit = acbit + 13 ;
  elseif level(z) == 5 ;
                 acbit = acbit + 14 ;
  end
elseif RL(z)==3 & level(z)<5 & level(z)>0
  if   level(z) == 1 ;
                 acbit = acbit + 6  ;
  elseif level(z) == 2 ;
                 acbit = acbit + 9  ;
  elseif level(z) == 3 ;
                 acbit = acbit + 13 ;
  elseif level(z) == 4 ;
                 acbit = acbit + 14 ;
  end
elseif RL(z)==4 & level(z)<4 & level(z)>0
  if   level(z) == 1 ;
                 acbit = acbit + 6  ;
  elseif level(z) == 2 ;
                 acbit = acbit + 11 ;
  elseif level(z) == 3 ;
                 acbit = acbit + 13 ;
  end
elseif RL(z)==5 & level(z)<4 & level(z)>0
  if   level(z) == 1 ;
                 acbit = acbit + 7  ;
  elseif level(z) == 2 ;
                 acbit = acbit + 11 ;
  elseif level(z) == 3 ;
                 acbit = acbit + 14 ;
  end
elseif RL(z)==6 & level(z)<4 & level(z)>0
  if   level(z) == 1 ;
                 acbit = acbit + 7  ;
  elseif level(z) == 2 ;
                 acbit = acbit + 13 ;
  elseif level(z) == 3 ;
                 acbit = acbit + 17 ;
  end
elseif RL(z)==7 & level(z)<3 & level(z)>0
  if   level(z) == 1 ;
```

```
                acbit = acbit + 7  ;
    elseif level(z) == 2 ;
                acbit = acbit + 13 ;
    end
  elseif RL(z)< 10 & level(z)<3 & level(z)>0
    if   level(z) == 1 ;
                acbit = acbit + 8  ;
    elseif level(z) == 2 ;
                acbit = acbit + 13 ;
    end
  elseif RL(z)==10 & level(z)<3 & level(z)>0
    if   level(z) == 1 ;
                acbit = acbit + 9  ;
    elseif level(z) == 2 ;
                acbit = acbit + 14 ;
    end
  elseif RL(z)< 14 & level(z)<3 & level(z)>0
    if   level(z) == 1 ;
                acbit = acbit + 9  ;
    elseif level(z) == 2 ;
                acbit = acbit + 17 ;
    end
  elseif RL(z)< 17 & level(z)<3 & level(z)>0
    if   level(z) == 1 ;
                acbit = acbit + 11 ;
    elseif level(z) == 2 ;
                acbit = acbit + 17 ;
    end
  elseif RL(z)< 22  & RL(z)>=17    ;
                acbit = acbit + 13 ;
  elseif RL(z)< 27  & RL(z)>=22    ;
                acbit = acbit + 14 ;
  elseif RL(z)< 32  & RL(z)>=27    ;
                acbit = acbit + 17 ;
  else k=1;
  end
  if k==1
    if level(z)< 128 & level(z)>0  ;
                acbit = acbit + 20;      % Extra bits required when ESCAPE is
                                         % encountered.
    elseif level(z)>128 & level(z)>0;
                acbit = acbit + 28;
    end
  end           % End of calculation of bits for level(x); proceed to level(z+1).
end             % End of calculation of for all elements in 'level'.
end             % End if runlength~=63 loop.
```

```
acbit = acbit + 2 ;                    % EOB indication bits.
y  = fix(fliplr(y));
q  = fliplr(q);
s  = y.*q;
y  = s;
yp = fix(c'*y*c); % Compute IQIDCT, i.e., reconstruct the image.
y  = yp;
```

Matlab_code_11.2.1.2 presents the computation of the number of DC coefficients and is similar to the Matlab_code_11.1.4 presented earlier for the computation of the corresponding DC coefficients in the fixed pruning control.

Matlab_code_11.2.1.2

% File: "dcbitsauto.m"

```
function dcb = dcbitsauto(dci3)     % Declare the function.
% This counts the number of bits in DC coefficients after applying VLC
% scheme on a block of DCTQ coefficients.
 [u,v]   = size(dci3);                 % Get the size .
dci2    = dci3';
dci5    = reshape(dci2,1,u*v);         % Returns the u x v matrix, whose
                                       % elements are taken columnwise from dci2.
dci4    = [];
dci4(1) = dci5(1);                     % First block DC coefficient is taken as it is,
for i=1: ((u*v)-1)                     % -1 is applied since dci4(1) is already got.
   first        = dci5(i);
   second       = dci5(i + 1);
   diff         = second - first;      % Subsequent DC coefficients are
   dci4(i + 1) = diff;                 % taken as the difference.
end
%disp(dci4)
bits = 0;
for i=1: (u*v)
   ii = abs(dci4(i));             % Drop the sign of the DC coefficients difference.
   if   ii<2 ;
        bits = bits + 3;          % Includes differential dc + additional code.
   elseif ii<4  ;
        bits = bits + 4;
   elseif ii<8  ;
        bits = bits + 6;
   elseif ii<16 ;
        bits = bits + 7;
   elseif ii<32 ;
```

```
        bits = bits + 9;
    elseif ii<64 ;
        bits = bits + 11;
    elseif ii<128;
        bits = bits + 13;
    elseif ii<256;
        bits = bits + 15;
    end
end
disp('DC BITS: ')
disp(bits)
dcb = bits;
```

11.2.3 Results and Discussions for the Fixed and Automatic Pruning Level Controls

Table 11.1 shows the processing time of DCTQ for various pruning levels [58]. It may be noted that these timings are different from the DCTQ timings, which will be presented in Chapter 13, which does not have either the fixed or auto pruning level control. In the present treatment, the frequency of operation is assumed to be 25 MHz. DCTQ coefficients are processed in a raster scan order, left to right and top to bottom, assuming that they are arranged as an 8 × 8 matrix. The number of coefficients processed for the pruning levels from 0 to 14 is shown in Table 11.1. The processing time of DCTQ is 800 ns for PL0 because the first DC coefficient is issued at the 20th clock cycle owing to high pipelining inherent in the design. The processing times for subsequent pruning levels are given by the expression:

$$t_T = t_{PL(N-1)} + [C_{PL(N)} - C_{PL(N-1)}] \times 40 \text{ ns},$$

where t_T is the total processing time for the current pruning level N, $t_{PL(N-1)}$ is the processing time of the previous pruning level, $C_{PL(N)}$ and $C_{PL(N-1)}$ are the number of DCTQ coefficients processed for the current and the previous pruning levels respectively. One coefficient is generated in every clock cycle with a time period of 40 ns. From the table, it is clear that the execution times for DCTQ increases steeply for every additional pruning level up to level 7 and thereafter only gradually. Further, there is no appreciable improvement of visual quality for pruning levels beyond 2 or 3 for most of the images. Therefore, by terminating the computation of DCTQ at that level, where good quality of image is already obtained, the processing speed can be stepped up. The savings in DCTQ computation, thus obtained, are also extended to VLC processing.

The encoder can be operated in two different modes – the fixed pruning level mode and the automatic pruning level mode. The DCTQ computation for an 8 × 8 pixel block of image requires 800 ns for a PL of 0. Hence, it is capable of processing monochrome images of size 1600 × 1200 pixels at the rate of 40 frames per second. However, the picture quality for PL0 will be the lowest and can be used only for fast browsing of images in which one is not interested.

By conducting experiments on various images, it is found that image quality is acceptable for a PL of 3 or 4, for which image sizes of up to 1200 × 1024 pixels for

Table 11.1 Processing time of DCTQ for various pruning levels

Pruning level	Number of DCTQ coefficients processed	DCTQ processing time in ns
0	1	800
1	9	1120
2	17	1440
3	25	1760
4	33	2080
5	41	2400
6	49	2720
7	57	3040
8	58	3080
9	59	3120
10	60	3160
11	61	3200
12	62	3240
13	63	3280
14	64	3320

monochrome and 1024 × 768 pixels for color can be processed at 25 frames per second. For color images in 4:2:0 format, 4 blocks of Y and 2 blocks of Cb and Cr will have to be processed for every macroblock of the picture. This is 50% more execution time than that for the monochrome pictures, which needs processing of only 4 blocks per macroblock. Therefore, the maximum size of color image that can be processed will be only 67% of the size of the monochrome picture. Normally, about 130 bits of header information are required to be processed per frame, which requires 2920 ns for execution in the present design. If DCTQ is processed as slices of row blocks, then this overhead goes up to 300 μs. These values are small fractions when compared to the processing speed of 40 ms for a full frame at 25 frames per second and, therefore, does not affect the overall execution time. The above results are applicable for user programed fixed pruning levels. Higher frame rate of 30 is possible with the improved DCTQ implementation presented in Chapter 13.

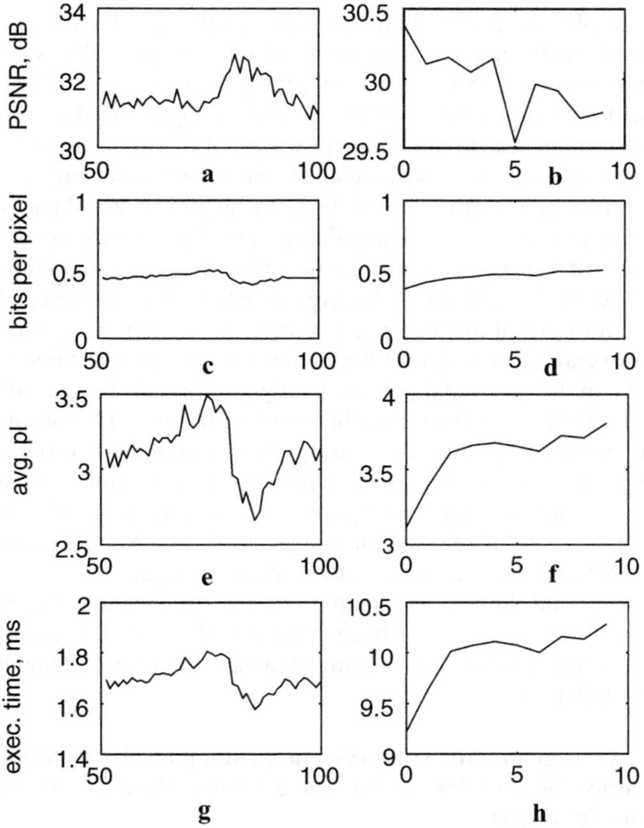

Fig. 11.4 (a/b) PSNR (c/d) bits per pixel (e/f) average PL (g/h) execution time per frame plots for Car and Rugby image sequences respectively using automatic pruning level control

The automatic, quality controlled encoder has been presented in Matlab and verified. The results for automatic pruning level control are portrayed graphically in Figure 11.4. In this figure, the picture quality obtained as PSNR, the compression effected by the scheme in terms of bits per pixel (bpp), the average pruning level attained, and finally the execution time achieved are presented for various frames for two image sequences, Car and Rugby. The quality of the image is computed by the expression:

$$\mathrm{PSNR} = 10 \log_{10} \left[(255^2 \times \mathrm{M} \times \mathrm{N}) / \sum_{j=0}^{M} \sum_{k=0}^{N} (i_{jk} - i'_{jk})^2 \right],$$

where M × N is the picture size in pixels, i_{jk} and i'_{jk} are pixel intensities of the original and the reconstructed image frames respectively. Similar results were obtained for other images such as table tennis, Susie, etc., though not presented here.

From these graphs, the following points may be inferred: Picture quality is good with PSNR of 30 dB and a maximum variation of only 1 dB. Compression is fairly constant around 0.45 bits per pixel giving a compression ratio of 18:1. Average PL per frame is between 3 and 5 for all the images tested even though quality achieved is quite close to the full quality level of PL14. Average PL and execution time graph patterns are almost identical since they are mutually proportional.

Energy threshold is user programable in the range between 0 and 4095 since it determines the end results. For example, the threshold zero gives full quality at PL14 for each of the blocks, but execution time doubles when compared to the automatic PL control at 200 units of energy threshold. This figure can be used as a default value for most of the pictures, in general, to yield good quality. Table 11.2 presents the overall average for all the frames for images with three different picture sizes. It may be noted that picture quality got for auto PL control is closer to full quality level of PL14 than that obtained for the fixed PL for all the images. Further, the execution speed attained using auto PL is about two times faster than that for the fixed PL of 14. Extrapolating the average execution times to a standard picture size, we can conclude that the auto PL controlled encoder is capable of processing 1024 × 768 pixels size of images at 42 frames per second for monochrome and 28 frames per second for color on the average.

Figures 11.5 and 11.6 show samples of the original and the reconstructed frames obtained for monochrome images using both fixed and automatic PL control modes. Processed color images using automatic PL control, though not shown, also exhibit similar results.

Table 11.2 Average quality, compression, pruning level, execution times, and speed up ratios obtained for various images using fixed as well as automatic PL-based quality controls

Per frame	**Car 51–100** (256 × 256 pixels)		**Rugby 0–9** (688 × 480 pixels)		**TT 0–15** (720 × 480 pixels)	
	Fixed PL	Auto PL	Fixed PL	Auto PL	Fixed PL	Auto PL
Avg. PSNR in dB	27.4 32.1	31.5	27.3 33.1	30.0	26.9 29.0	28.1
Avg. bits per pixel	0.33 0.48	0.45	0.39 0.53	0.46	0.4 0.55	0.46
Avg. PLN	3 14	3.12	4 14	3.60	5 14	4.70
Avg. exec. Time, ms	1.80 3.40	1.70	10.7 17.1	9.97	13.0 17.9	10.6
Speed up ratio	1.9 1	2	1.6 1	1.7	1.4 1	1.7

Fig. 11.5 (a) Original Car 70 image, size: 256 × 256 pixels. (b) Reconstructed Car 70 image by automatic PL-based quality control method for e_{THR} = 200 units. PSNR = 31.3 dB, Average PLN = 3.30, execution time = 1.75 ms/F and bpp = 0.475, speed up factor = 1.94. **(c) Reconstructed Car 70 image by fixed PL-based quality control method for FPLN = 3.** PSNR = 27.3 dB and bpp = 0.346. **(d) Original Eiffeltp 0 image, size: 520 × 732 pixels. (e) Reconstructed Eiffeltp 0 image by automatic PL-based quality control method for e_{THR} = 200 units.** PSNR = 35.1 dB, Average PLN = 1.83, execution time = 8.2 ms/F and bpp = 0.276, speed up factor = 2.43. **(f) Reconstructed Eiffeltp 0 image by fixed PL-based quality control method for FPLN = 2.** PSNR = 32.5 dB and bpp = 0.223

a b c d

Fig. 11.6 (a) Original Titanic image (352 × 288 pixels). (b) Reconstructed Titanic image by auto quality method. (c) Original Titlis/Switzerland image (908 × 600 pixels). (d) Reconstructed Titlis/Switzerland image by auto quality method

11.3 Fast Motion Estimation Algorithm for Real-Time Video Compression

11.3.1 Introduction

In multimedia applications, the key requirements are speed of processing and compression of image without sacrificing the quality. In order to process motion pictures with high resolution, one needs a highly efficient motion estimation algorithm in terms of processing speed. Several block matching algorithms are available in the literature. Full search algorithm [66–69] is a straightforward scheme, which requires a large number of searches for finding a correct match for the image block being processed. This scheme requires $(2w + 1)^2$ number of search points, where w is the maximum pixel displacement, which is usually taken as 8. One-dimensional full search [70] is another method that requires $(4w + 3)$ number of search points. Hierarchical method [71], which requires ($1 + 8\log_2 w$) number of search points, is faster than the two methods mentioned earlier.

There are scores of other fast algorithms available, references [72–83] to name a few. Even faster is the one-at-a-time step search (OSS) algorithm [84] requiring only ($2w + 3$) number of search points. In this section, a novel, fast one-at-a-time step search (FOSS) algorithm proposed by the authors [85] is presented. In this method, the maximum number of search points is ($2w + 1$). Although this number is close to that of the OSS method, the FOSS is faster by up to 40% than the OSS method when the motion of image blocks is small of the order of 1 or 2 pixels, which is usually the case in most of the commonly encountered video frame sequences. The speed up distribution for various motions is covered in detail in Section 11.3.2.

A number of software and hardware implementations [86–89] have been reported for some of the motion estimation algorithms described earlier. Although software implementations are easy to realize on general-purpose microprocessors or digital signal processors, their instruction sets are not well suited for fast processing of high-resolution moving pictures. In addition, the instructions are executed sequentially, thus slowing down the processing further. In contrast to this, the hardware implementations based on FPGAs and ASICs can exploit pipelined and massively parallel processing, resulting in faster and cost effective motion estimation implementation.

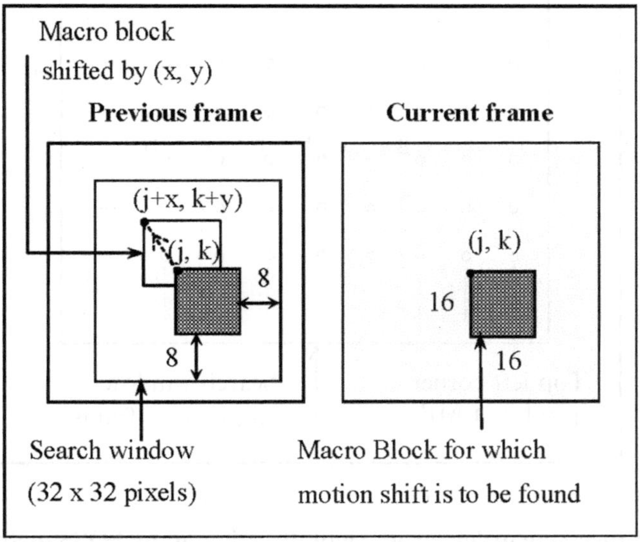

Fig. 11.7 Basic principle of motion estimation

11.3.2 The Fast One-at-a-time Step Search Algorithm

The basic principle involved in motion estimation is depicted in Figure 11.7. The current frame of the image is processed macroblock (MB) by macroblock. The macroblock (j,k) currently being processed is identified in the previous frame and a search window surrounding it is defined for conducting the search. A minimum sum of absolute pixel intensity differences (A_d) is required to be met in order to locate the shifted macroblock. A_d is defined as

$$A_d(x,y) = \sum_{j=0}^{15} \sum_{k=0}^{15} \left| i_{jk} - i'_{(j+x)(k+y)} \right| ; \qquad -8 \le x, y \le 8, \qquad (11.9)$$

where i_{jk} is the pixel intensity in the macroblock processed in the current frame, $i'_{(j+x)(k+y)}$, its corresponding intensity in the search window of the previous frame, and (x,y), referred to as the motion vector, is the shift undergone by the macroblock. For the sake of convenience, the computation of one A_d is referred to as a search point.

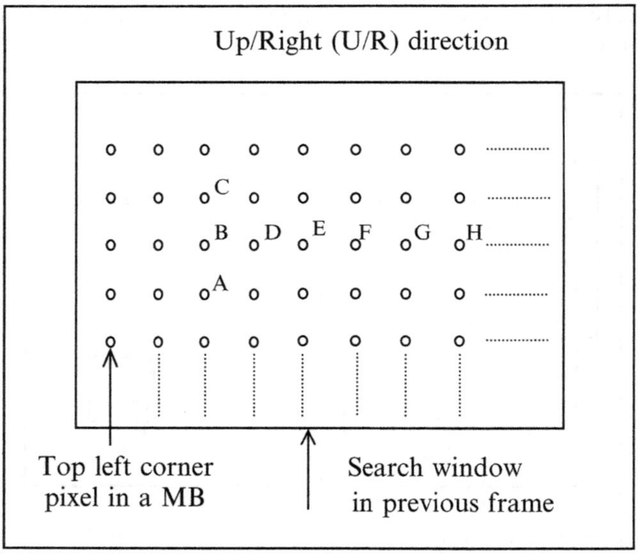

Fig. 11.8 Processing order for up/right direction in the FOSS method

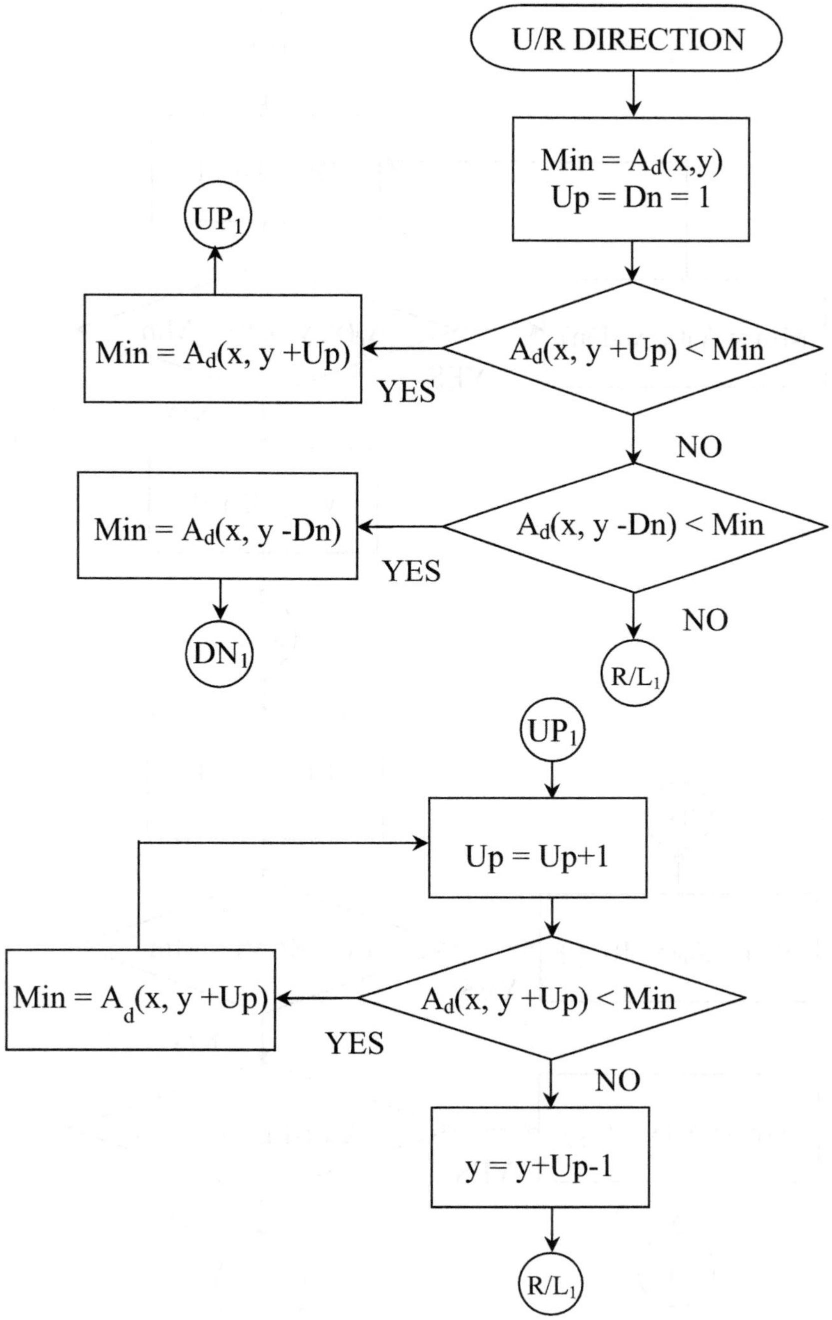

Fig. 11.9 FOSS algorithm for motion estimation for U/R direction (Continued)

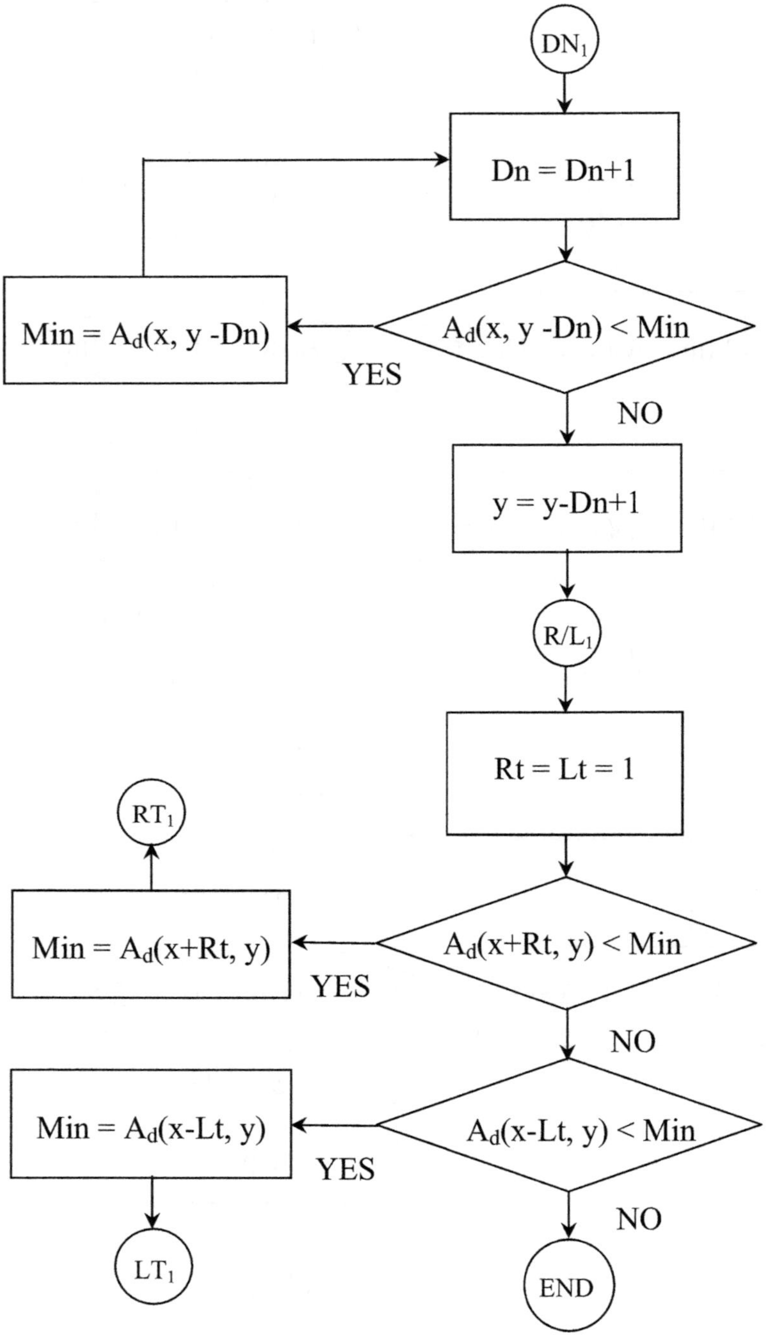

Fig. 11.9 FOSS algorithm for motion estimation for U/R direction (Continued)

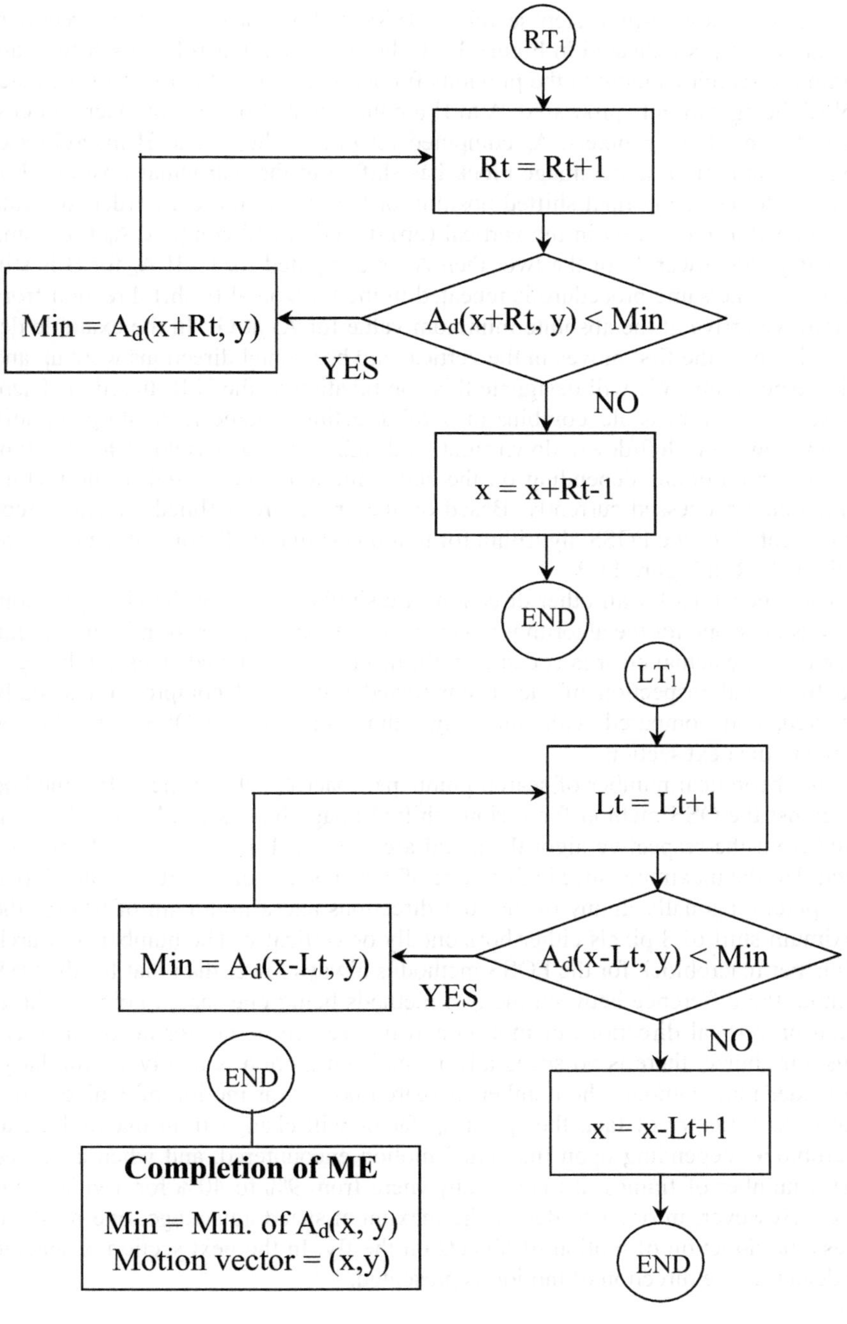

Fig. 11.9 FOSS algorithm for motion estimation for U/R direction

The fast one-at-a-time step search (FOSS) method for one of the directions, up/right (U/R), is indicated in Figure 11.8. For example, macroblocks A to H are within the search window in the previous frame, where A is the origin of the macroblock being currently processed. A to H are the origins (top-left) of macro blocks, each of size, 16 × 16 pixels. A_d computed for one of these A to H macroblocks would be minimum if the image block has shifted to that particular position. For instance, let G be the final shifted position of the image block. In order to locate G, we need to move first in the vertical (up) direction and compute A_d for A and B. If B yields lower A_d of the two, then A_d is computed for C. If A_d for B is still the lowest, the same procedure is repeated in the horizontal (right) direction from D until we arrive at the absolute minimum value for A_d at G. In the example described above, the first moves in the vertical and horizontal directions were up and right respectively. We will designate this combination as the U/R direction. There are seven other possible combinations of directions, namely, right/up, up/left, left/up, down/left, left/down, down/right, and right/down that yield different numbers of search points depending on the actual motion encountered in the picture frame being processed currently. Based on the procedure outlined, detailed steps are presented for the FOSS algorithm for motion estimation for one of the eight directions, U/R in Figure 11.9.

The algorithms for all other directions are similar to that of the U/R direction. There is no scope for the algorithm to get caught in local minima or missing motion since only the actual error is coded and ultimately reconstructed. This can be verified by visual inspection of the reconstructed image and compression actually achieved, and compared with other algorithms such as the OSS algorithm as shown in the next section.

The theoretical number of search points per macroblock for the FOSS method as against the OSS method for various shifted image blocks can be readily computed from the respective algorithms and are presented in Table 11.3. It may be noted that the maximum speed advantage of 40% results for a shift of image block by 1 pixel diagonally in any of the four directions and a minimum of 9% for the maximum shift of 8 pixels either horizontally or vertically. The number of search points per macroblock for the FOSS method is always lower than that for the OSS method, the difference between the two methods being one search point for horizontal or vertical directions of motion and two search points for all other directions. Of course, there is no speed advantage if no motion is involved. Similarly, the reader may compute the number of search points for motion of 3 pixels to 7 pixels. As a matter of fact, the speed up factor will change from macroblock to macroblock, depending upon the actual motion encountered, and when averaged over a number of frames, it may be anywhere from 9% to 40% for a video sequence. However, in order to derive the maximum speed advantage, one needs to assess the direction of motion of objects on the fly. In the next section, a scheme for detecting the direction of motion is presented.

11.3.3 Assessment of Direction of Motion of Image Blocks

The method of finding the direction of motion of image blocks in a picture is as follows. This assessment is made only for the first P frame after the I frame in every group of pictures (GOP), which can be user defined. In the present scheme, a GOP consists of only an I frame followed by P frames and does not contain B frames. An I frame is the reconstructed picture frame by processing IQIDCT without involving any motion estimation, whereas a P frame is the reconstructed picture after motion estimation such as the FOSS motion estimation described earlier. Since the computation of motion estimation is a time consuming process, motion estimation for all the eight directions as explained in the FOSS algorithm is carried out only for representative samples of five macroblocks. The total number of search points for all the five macroblocks for each of the eight directions is computed first. The direction for which the total number of search points is a minimum is reckoned as the optimum direction of motion of image blocks in the proposed method. The optimum direction, thus found, is applied to all the P frames in the current GOP. These macroblocks are located at (M/4, N/4), (3M/4, N/4), (M/2, N/2), (M/4, 3N/4), and (3M/4, 3N/4) co-ordinates as shown in Figure11.10, where M × N is the picture size in pixels.

The processing time overhead involved in motion estimation for these macroblocks is under 0.4% of the overall processing time for the entire GOP of 10 frames. The placement of macroblocks has been arrived at after conducting elaborate experiments on a number of images, trying various locations and numbers of samples in a frame. This placement yielded the minimum number of search points of all the combinations tried out for various images of sizes of up to 720 × 480 pixels. Beyond this picture size, if required, one can use nine macroblocks instead of five yielding marginally better performance.

11.3.4 Detection of Scene Change

The FOSS algorithm is robust and can adapt seamlessly even in the event of a radical scene change. The algorithm detects scene changes by keeping track of the total number of search points for every frame and comparing it with that for the previous frame. If the total number of search points for the current frame exceeds that for the previous frame by more than 25%, a scene change is deemed to have occurred. This figure of 25% has been arrived at after testing with a number of images. In such an event, the frame following the scene change frame is taken as the reference frame for a fresh group of pictures. Quality of the picture does not degrade for the scene change frame since only the actual error is processed and appropriate correction effected. Also, no motion of image blocks is lost track of owing to the same reason. However, compression falls only for the scene change frame, and normalcy is restored immediately with succeeding frame.

Table 11.3 The number of search points and the speed up factors in a macroblock for the FOSS method over the OSS method

Shift in image block position	Number of search points in a MB by		Speed up factor for FOSS method
	FOSS method	OSS method	
1 pixel diagonally	5	7	1.40
2 pixels diagonally	7	9	1.29
8 pixels diagonally	17	19	1.12
1 pixel horizontally or vertically	5	6	1.20
2 pixels horizontally or vertically	6	7	1.17
8 pixels horizontally or vertically	11	12	1.09
No motion	5	5	1.00

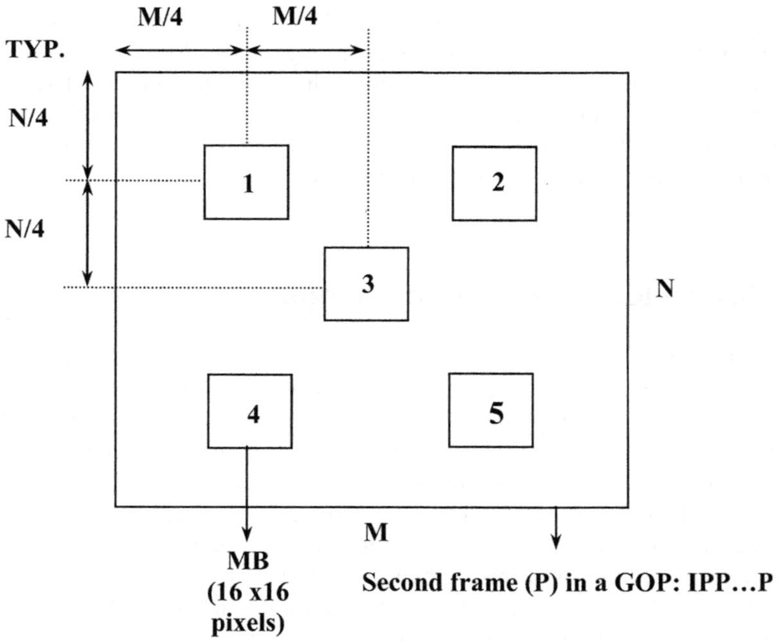

Fig. 11.10 Assessment of the direction of motion of objects for sample macroblocks in the second (P) frame of a GOP being processed

11.3.5 Results and Discussions of FOSS Motion Estimation Algorithm

The FOSS algorithm for all the eight directions coded in C, although not presented since it runs to over 75 pages, is successfully tested using a number of images of different sizes. Before we go into these details, let us see how this algorithm fares with the block matching algorithms of other researchers. Table 11.4 presents the relative speeds of various algorithms including the FOSS algorithm we developed. The FOSS algorithm is faster than other algorithms as can be seen from the table. The next best algorithm in terms of processing speed is the OSS algorithm. The savings effected in computations in the FOSS method compared to the OSS method are given in Table 11.5. It is clear from the table that, irrespective of the direction chosen initially and applied to all the frames of a group of picture (GOP), the FOSS method tracks the shifted image block faster than the OSS method. The direction for which the number of search points is minimum is selected as the optimum direction of motion for each image sequence.

Table 11.4 Comparison of speeds of various algorithms with FOSS algorithm

Algorithm	Number of search points per macroblock	Speed factor
Full search [68]	289	1.0
One-dimensional full search [70]	35	8.3
Hierarchical search [71]	25	11.6
Three-step search [90]	33	8.8
New three-step search [77]	31	9.3
Four-step search [80]	27	10.7
Center-biased diamond search [82]	25	11.6
Normalized partial distortion search [91]	22	13.1
Minima-bound area search [83]	36	8.0
Fast BMME [81]	35 to 66	8.2 to 4.4
Alternating subsampling search [76]	74	3.9
One-at-a-time step search (OSS) [84]	7 to 19	41.3 to 15.2
FOSS [85]	5 to 17	57.8 to 17.0

Table 11.5 Savings effected in the number of search points for various directions

Image sequence	Percentage savings effected in number of search points of a GOP							
	FOSS method over OSS method							
	Direction							
	L/U	R/U	R/D	L/D	U/L	U/R	D/R	D/L
Rugby	10.8	11.1	11.3	10.6	14.8	15.3	15.0	14.7
Table tennis	12.6	12.4	12.1	12.2	13.5	13.6	13.3	13.4

Plots for the number of search points versus the frame number are presented for one of the video sequences, viz., the Car in Figure 11.11 for both the FOSS and OSS methods. The first frame is an I frame and all others are P frames. As can be seen from the plots, the number of search points in the FOSS method is lower compared to that in OSS method for all the frames. Similar results have been obtained for all the video sequences tested, although not presented here.

The quality measure, the peak signal-to-noise ratio (PSNR) and the compression effected in bits per pixel for one of the video sequences, Car, are presented in Figures 11.12 and 11.13 respectively. These results show that the FOSS method offers marginally better performance in terms of the quality of the reconstructed image or on the compression effected. The PSNR and bits per pixel averaged over all the

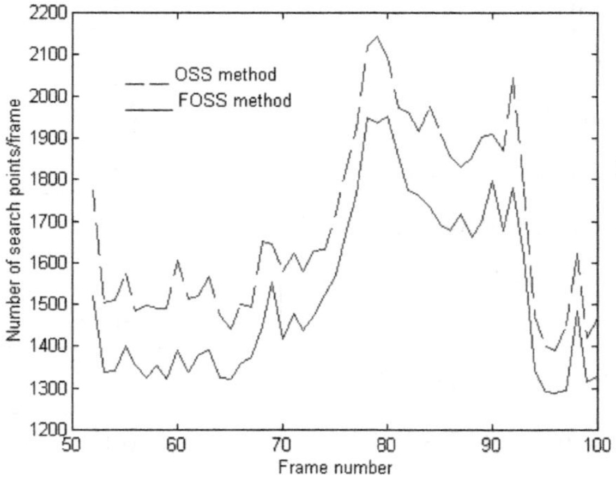

Fig. 11.11 The number of search points versus the frame number for the Car sequence for the FOSS and the OSS methods

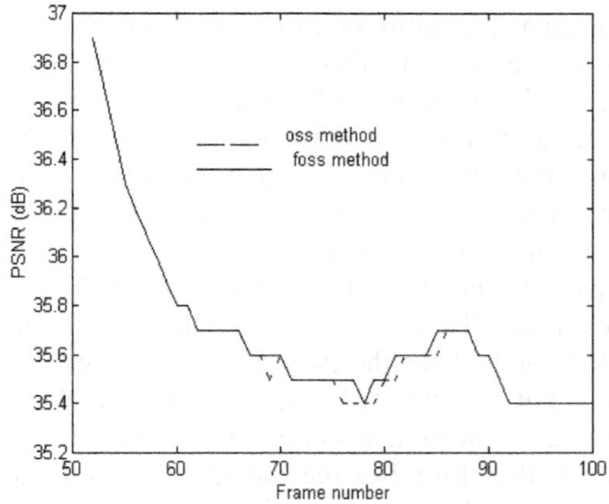

Fig. 11.12 The image quality versus the frame number for the Car sequence for the FOSS and the OSS methods

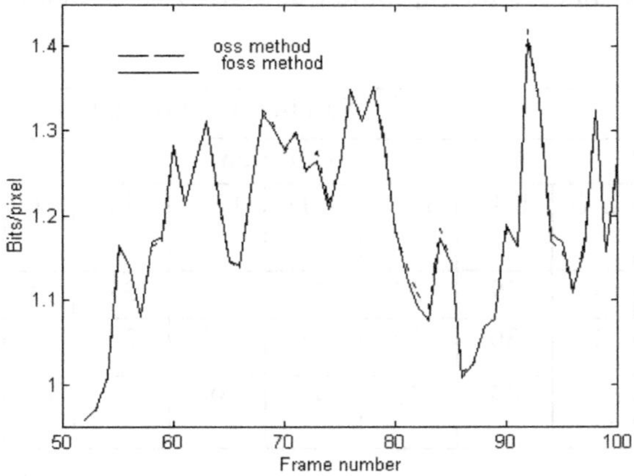

Fig. 11.13 The compression effected versus the frame number for the Car sequence for the FOSS and the OSS methods

frames are 33.9 dB and 0.536 respectively for the FOSS method, whereas for the OSS method they are 33.5 dB and 0.54 respectively. This indicates that the FOSS method does not compromise on the quality of the reconstructed image and in effecting compression, while improving the speed of execution. Similar results, though not presented here, have been obtained for other video sequences as well.

The total number of search points for various directions for the five sample macroblocks in a frame as explained earlier is given in Table 11.6. The optimum directions of motion found by the FOSS algorithm for various image sequences are also presented. In case there are more than one minima, the first occurrence from the left is taken as the optimum direction. For instance, the Rugby sequence yields a minimum of 35 search points for two directions D/R and D/L and, therefore, D/R is recognized as the optimum direction in this case. Table 11.7 presents the speed up ratios for the FOSS method over the OSS method for various image sequences. The overhead time required for detecting the direction of motion is included in the total number of search points found by the FOSS method. The proposed algorithm preserves the visual quality of the picture as can be seen from Figure 11.14, which presents the original and the reconstructed images using the FOSS method, for one of the video frames.

Two video sequences, namely, the 'bmw_tram' and the 'Car_susie' were tested for the scene change. Table 11.8 presents the speed up ratios for the FOSS method over the OSS method for the two video sequences. In spite of a sudden scene change, there is not only a speed advantage in the FOSS method, but reconstruction of a good quality image is also possible as is evident from Figure 11.15. The FOSS algorithm is, therefore, flexible enough to accommodate scene changes, while preserving the speed advantage as well as the quality of the processed image over the OSS method.

Table 11.6 The total number of search points for various directions for five sample macroblocks and optimum directions of motion found by the FOSS algorithm

Image	**Number of search points**								
	Direction								Optimum direction found
	L/U	R/U	R/D	L/D	U/L	U/R	D/R	D/L	
Rugby 1	42	39	39	41	37	37	35	35	D/R
TT 1	29	30	29	28	25	25	29	29	U/L
Bmw 161	45	45	47	47	38	41	43	40	U/L

Table 11.7 The speed up ratio in a GOP for the FOSS method over the OSS method

Image sequence	**Frame numbers**	**Avg. search points per MB**	**Speed up ratio**
Rugby Image size: 480 × 688 pixels	0–9	8.1	1.146
TT Image size: 480 × 720 pixels	0–15	7.0	1.134
bmw Image size: 352 × 288 pixels	90–110	8.1	1.114

a b

Fig. 11.14 Simulation image using FOSS motion estimation algorithm. (a) Original bmw image (frame number: 91, picture size: 352 × 288 pixels) (b) Reconstructed bmw image by the FOSS method (PSNR: 35.6 dB)

Table 11.8 The speed up ratio for the FOSS method over the OSS method for video sequences with scene changes

Image sequence	Frame numbers	Scene change at	Avg. search points per MB	Speed up ratio
bmw_tram Image size: 352 × 288 pixels Dir.: R/U	90–110 170–180	170 Tram	7.9	1.096
Car_susie Image size: 256 × 256 pixels Dir.: U/R	51–70 0–19	0 Susie	5.3	1.090

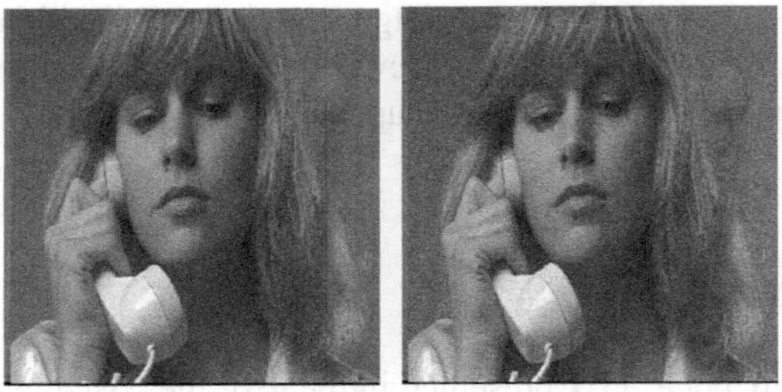

a b

Fig. 11.15 Simulation image with scene change from Car to Susie (a) Original Susie image (frame number: 0, picture size: 256 × 256 pixels). (b) Reconstructed Susie image by the FOSS method (PSNR: 35.9 dB)

Summary

Complex applications such as video codecs, etc. have involved algorithms at their core, which need to be adapted or developed depending upon how we wish to implement the system. This chapter dealt with the development of algorithms for these applications so that they are suitable for implementation on FPGAs/ASICs. Before going for the architectural and Verilog designs, it is of paramount importance to code them in a high level language such as Matlab or C and test them to ascertain their feasibility. The following algorithms and their verifications using Matlab were presented.

A parallel algorithm for the computation of DCTQ for achieving high throughput was presented. This was followed by a novel, automatic, pruning level-based quality control scheme. By incorporating dynamic control based on assessing the quality of a picture on-the-fly, more than twofold speed advantage was demonstrated over the conventional approach of processing without pruning and without sacrificing the quality. The processing power of the video encoder can be further enhanced, at the cost of additional hardware, by designing a reconfigurable video encoder system that caters to a wide variety of applications conforming to JPEG, MPEG, and H.263 standards.

A new, fast, one step search method for motion estimation in video frame sequences along with automatic assessment of direction of motion of image blocks was also presented. The simulation results show that this method is faster than the OSS method without compromising either on the quality of picture or the compression effected. Although the present implementation is for processing only I and P frames, the algorithm can be easily extended to cover B frames as well.

While developing algorithms for hardware implementation, we need to keep the actual hardware such as registers, counters, combination circuits, etc. in mind, and subsequently design the architecture. Only then, we will be in a position to meet stringent specifications when the algorithm is converted into an actual working product. The next in the chain of developments is the hardware architectural design, which is presented in the next chapter.

Assignments

11.1 In Matlab_code_11.1 presented in the text, the cosine matrix, C, was presented without details. Show how these values were arrived at. You may refer the DCT algorithm presented in this chapter and also the MPEG 2 standard.

11.2 A parallel algorithm was presented in the text for the discrete cosine transform and quantization, which operations are required for effecting compression of an image. The DCTQ processor is used in encoder, be it for still image compression conforming to JPEG standard or for video compression conforming to MPEG standards. On similar lines to the algorithm developed in the text for DCTQ, inverse quantization can be computed by multiplying each of the 64 DCTQ coefficients by the corresponding quantization table values as per the expression:

DCT (u,v) = DCTQ (u,v) x q (u,v); u, v = 0 to 7.

The image can be reconstructed from the DCT (u,v) by evaluating the (inverse DCT) product of the matrices as follows:

IDCT = C^T (DCT) C

Develop the algorithm for the evaluation of IQIDCT on similar lines to that of DCTQ, bearing in mind that the throughput of FPGA/ASIC implementation must match the DCTQ throughput, namely, one reconstructed pixel per system clock.

11.3 A new algorithm for assessing image quality on the fly using a concept called pruning was presented in the text. As a result of applying the algorithm, the processing speed of DCTQ can be increased by over 150% when compared to the implementation speed of the DCTQ without pruning. The sum of the squares of all the DCT coefficients or the spatial data values, ($x_{n,m}$), is the energy of the block, i.e.,

$$\sum_{u=0}^{7} \sum_{v=0}^{7} (DCT_{u,v})^2 = \sum_{n=0}^{7} \sum_{m=0}^{7} (x_{n,m})^2, \qquad u, v, n, m = 0\text{–}7$$

This algorithm was based on computing the sum of energy of AC coefficients lying on every diagonal commencing from PL1 up to PL14 shown in Figure 11.1. This algorithm used DCT for the computation of energy. Instead of the DCT, apply DCTQ in the algorithm. Work out details with the representative sample image block provided in the text and check the feasibility of its implementation as an MPEG encoder.

11.4 The algorithm for assessing image quality presented in the text can also speed up the next pipeline module called the variable length coder (VLC) of a video encoder. Discuss how this can be done. Details of VLC can be found in the last chapter.

11.5 The algorithm for assessing image quality presented in the text can be used for effecting rate control, i.e., maintain a constant bit rate while a compressed bit stream is transmitted over a serial channel. This algorithm can be used for both hardware and software implementations. There are three possible ways to bring about rate control. Discuss how these can be brought about.

11.6 In the text, it was mentioned that the execution speed that can be achieved using auto PL is about two times than that for the fixed PL of 14. Extrapolating the average execution times to a standard picture size, we can conclude that the auto PL controlled encoder is capable of processing 1024 × 768 pixels size of images at 42 frames per second for monochrome and 28 frames per second for color in 4:2:0 format on the average. Work out these details to verify the above statements. Also, work out the frame rates for other formats such as 4:2:2 and 4:4:4.

11.7 Full search block matching algorithm used in motion estimation of a video sequence is a straightforward scheme, which requires a large number of searches for finding a correct match for the image block being processed. This scheme requires $(2w + 1)^2$ number of search points, where w is the maximum pixel displacement, which is usually taken as 8. Derive this number of search points.

11.8 A much faster algorithm than Full search block matching algorithm is the one-at-a-time step search (OSS) algorithm requiring only (2w + 3) number of search points. Discuss how this may be derived.

11.9 In the text, a novel, fast one-at-a-time step search (FOSS) algorithm was presented. In this method, the maximum number of search points is (2w + 1). How was this arrived at?

11.10 The theoretical number of search points per macroblock for the FOSS method as against the OSS method for various shifted image blocks were presented in Table 11.3. The maximum speed advantage of 40% results for a shift of image block by 1 pixel diagonally in any of the four directions and a minimum of 9% for the maximum shift of 8 pixels either horizontally or vertically. Explain how these results were obtained. Similarly, compute the number of search points for motion of 3 pixels to 7 pixels.

11.11 There are eight possible combinations of directions, namely, up/right (U/R), right/up, up/left, left/up, down/left, left/down, down/right, and right/down that yield different numbers of search points depending upon the actual motion encountered in a picture frame. Detailed steps were presented for the FOSS algorithm for motion estimation for one of the eight directions, U/R in Figure 11.9. Draw flow charts for any other direction other than the U/R and R/U directions.

11.12 In color image/video processing applications, a popular format is the Y, Cb, Cr format governed by the following expressions:

$$Y = 0.299\,R + 0.587\,G + 0.114\,B$$

$$Cb = -0.169\,R - 0.331\,G + 0.500\,B$$

$$Cr = 0.500\,R - 0.419\,G - 0.081\,B$$

where R, G, and B are the picture element (pixel) values of three fundamental colors: red, green, and blue. Each of these color components is of size, 8 bits. Develop an apt algorithm without using multipliers for the computation of Y, Cb, Cr so that it may be efficiently mapped on to an FPGA/ASIC. Suggest the right pipeline stages. Hint: Scale the coefficients up in the above expressions by 128, retain only integers, replace multiplication operations by addition of relevant decimal weights as was suggested in the assignment 10.8 in Chapter 10 and, finally, scale down the result by 128.

11.13 Write a Verilog RTL code for the algorithm you have developed for the assignment 11.12. Retain only 9 bits precision for each of the components, Y, Cb, and Cr. Apply multi-pipeline stages in your design.

11.14 Write a test bench for the Verilog code you have designed for effecting the color format conversion from R, G, B to Y, Cb, Cr. Demonstrate the working of your design by presenting the simulation waveforms. For an alternative test bench, see the assignment 11.17.

11.15 On similar lines to the algorithm you have developed for the assignment 11.12, develop the algorithm for the inverse format conversion from Y, Cb, Cr to R, G, B.

11.16 Write a Verilog RTL code for the algorithm you have developed for the assignment 11.15. Retain the 9 bits precision for each of the components, Y, Cb, and Cr. The R, G, B outputs are each 8 bits. Apply multi-pipeline stages in your design.

11.17 Write a test bench for the Verilog code you have developed for effecting the conversion from Y, Cb, Cr to R, G, B. Alternatively, you can write an integrated test bench for testing the two color format conversions at one shot. Demonstrate the working of your designs by presenting the simulation waveforms.

11.18 Develop a simple algorithm for converting three digit BCD number to a binary number.

11.19 Write a Verilog RTL code for the algorithm you have developed for the assignment 11.18. Use pipelining if necessary.

11.20 Write a test bench for testing the BCD to binary conversion. Demonstrate the working of your design by presenting the simulation waveforms.

Chapter 12

Architectural Design

In the previous chapter, we learnt how to develop algorithms for a number of applications and verify the same in the field of video processing as examples. These algorithms were developed in such a way that the applications may be mapped onto an FPGA or an ASIC. The next logical step is to work out a detailed architecture keeping the actual hardware in mind. In the present chapter, we will consider the architectures for the same applications that we had undertaken in the previous chapter. Verilog coding and results of implementations for a couple of applications will be presented in detail in the chapter on project design.

12.1 Architecture of Discrete Cosine Transform and Quantization Processor

Detailed architecture of DCTQ as implemented on an FPGA is shown in Figure 12.1. As shown therein, one row (8 pixels) of an image block (8 × 8 pixels) is input via the data bus, 'di[63:0]'. 'di[7:0]' receives the first pixel and 'di[63:56]', the last pixel in the row. A pixel is of size 8 bits, unsigned for each of the color components, Y, Cb, and Cr. 'be[7:0]' is the input data byte enable signal. 'be[0]' selects di[7:0] and so on. 'wa[2:0]' furnishes the row address of an image block, where 'wa[0]' is the first row. Eight rows need to be written for inputting a block of image. 'pci_clk' is the image input synchronous clock. 'di[63:0]' is written into the core with 'wa[2:0]' serving as the address synchronous to positive edge of this clock. 'din_valid' input signals when the input data, 'di[63:0]' is valid. Active high at 'start' commences and maintains the DCTQ processing. If de-asserted and re-applied, latency will come into effect once again.

DCTQ processing can be frozen by the host, which inputs image data, or the variable length coder (VLC) processor, which is a subsequent process to effect image compression by asserting a 'hold' signal. De-asserting hold resumes the processing without any latency coming into play again. This signal together with the corresponding signals in VLC processor provides a convenient handshake for the two processors to work concurrently. 'clk' is the DCTQ system clock, which can be the same as the 'pci_clk'. 'ready' signal indicates that the DCTQ core is ready to accept an image input block. DCTQ output in twos complement is issued out of 'dctq[8:0]' pins, valid at the positive edge of 'clk'. 'addr[63:0]' is the DCTQ coefficient address. 'addr[0]' is the DC coefficient address and all other addresses are for AC coefficients. 'dctq_valid' signal indicates the validity of

DCTQ coefficient and its address. The DCTQ processor can be reset at any point of time by asserting the asynchronous, active low signal, 'reset_n'.

The module designated 'dualram' contains RAM storage for two image blocks. When one block of RAM is written into, the DCTQ processing takes place concurrently by reading from the other block of RAM. One complete block can be written in eight 'pci_clk' cycles, whereas reading for processing of DCTQ takes 64 'clk' cycles. The writing time is much faster than the reading time, thus freeing the host computer to other domestic chores in addition to fetching the image input periodically. 'cnt1_reg [2:0]' generated by the 'dctq_controller' serves as the read address for getting the 'dualram' content. The signal 'rnw' selects the appropriate RAM bank. While reading, the RAM is accessed column-wise, since in the computation of C*X of the DCT algorithm we need to multiply a row of C matrix with the column of X (image input) matrix as explained in Chapter 11 on the development of algorithms. Eight multipliers, mult8ux8s, accomplish this, where X is the unsigned input and C is the signed cosine term for evaluating the DCT. The resulting products (result1–result8) are summed in the next module, 'adder12s'. Each of the inputs of this adder is of size, 12 bits. The result, 'sum1' is stored in dctreg2x8xn registers, which contains eight numbers of 11-bit registers, qr0–qr7. These registers store one row of partial products of CX in eight 'clk' cycles and will be preserved for the next eight 'clk' cycles so that (CX) * C^T may be computed. 'cnt2_reg [2:0]' selects one of the eight registers at a time.

In the second stage multiplication, eight numbers of 11 × 8 bit precision multiplier, mult11sx8s, is made use of to generate res1-res8. These are summed using 14-bit precision, 'adder14sr' module, to get the DCT of precision, 12 bits. The final stage divides the DCT output by the corresponding quantization value, as per MPEG 2 standard, to get the desired DCTQ output of precision, 9 bits. The division has been replaced by a 12 × 8 bit multiplier, mult12sx8s, taking the inverse of the quantization value, qout, as one of the inputs. The bit precisions for various stages of processing have been arrived at after conducting a number of experiments on several video sequences and by the computation of PSNR value, a popular measure of the reconstructed image quality, defined in Chapter 11. A PSNR value of 30 dB or more is generally reckoned as good quality images. This criterion has been adopted for arriving at the requisite precision.

The cosine matrix C and its transpose C^T are stored in on-chip read only memory, 'romc', and their values are retrieved for processing simultaneously by accessing C and C^T matrices using 'cnt1_reg [5:3]' and 'cnt3_reg [2:0]' respectively as addresses generated by the controller. Similarly, the inverse quantization values stored in 'romq' module are accessed using 'cnt4_reg [5:0]' as the address. Various time-bound activities of DCTQ processor are finely orchestrated by the 'dctq_controller'. Actual input/output pins of the processor are shown in bold in order to easily distinguish them from intermediate signals. The algorithms for the multipliers and adders used in the design have been developed in order to speed up the computation by introducing high level of pipelining. These designs were covered in depth in the chapter on arithmetic circuits. The adders: 'adder12s' and 'adder14sr' compute the sum of eight, 12 bits, twos complement numbers and have

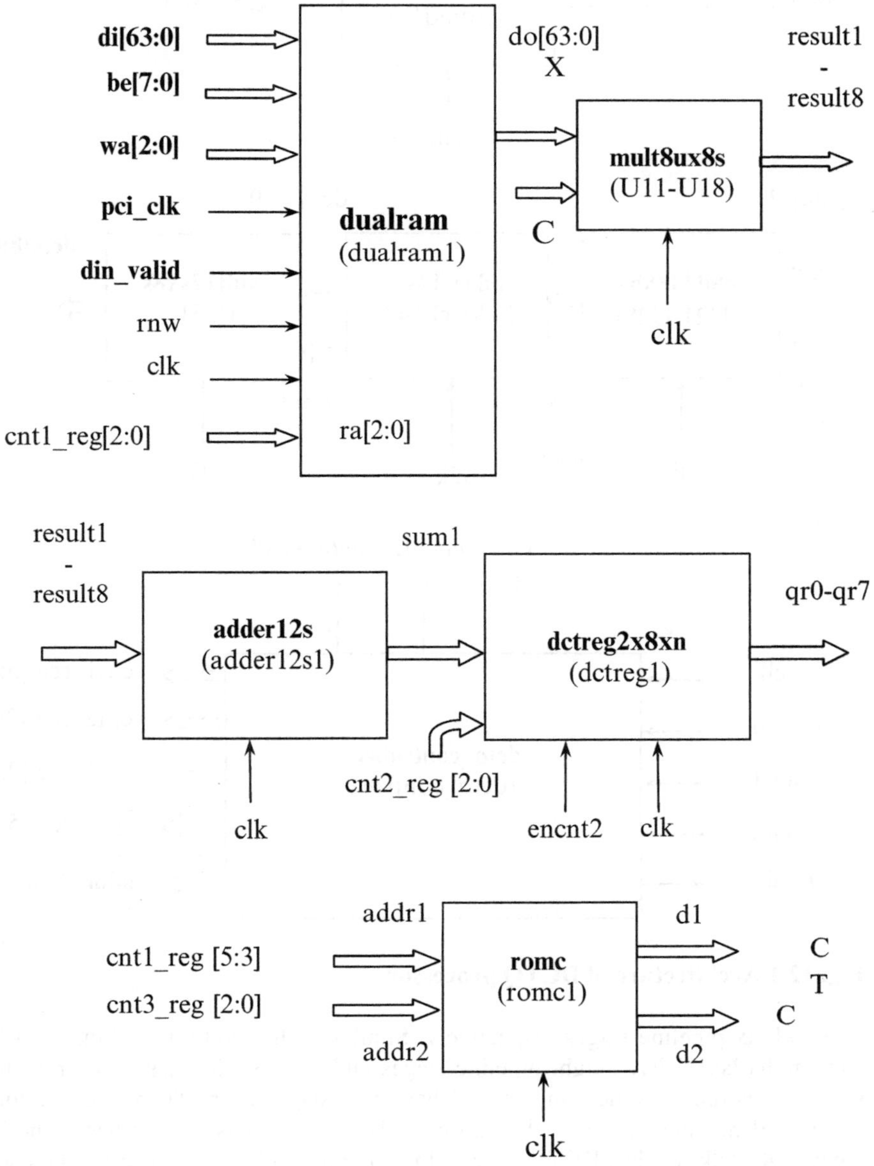

Fig. 12.1 Architecture of DCTQ processor (Continued)

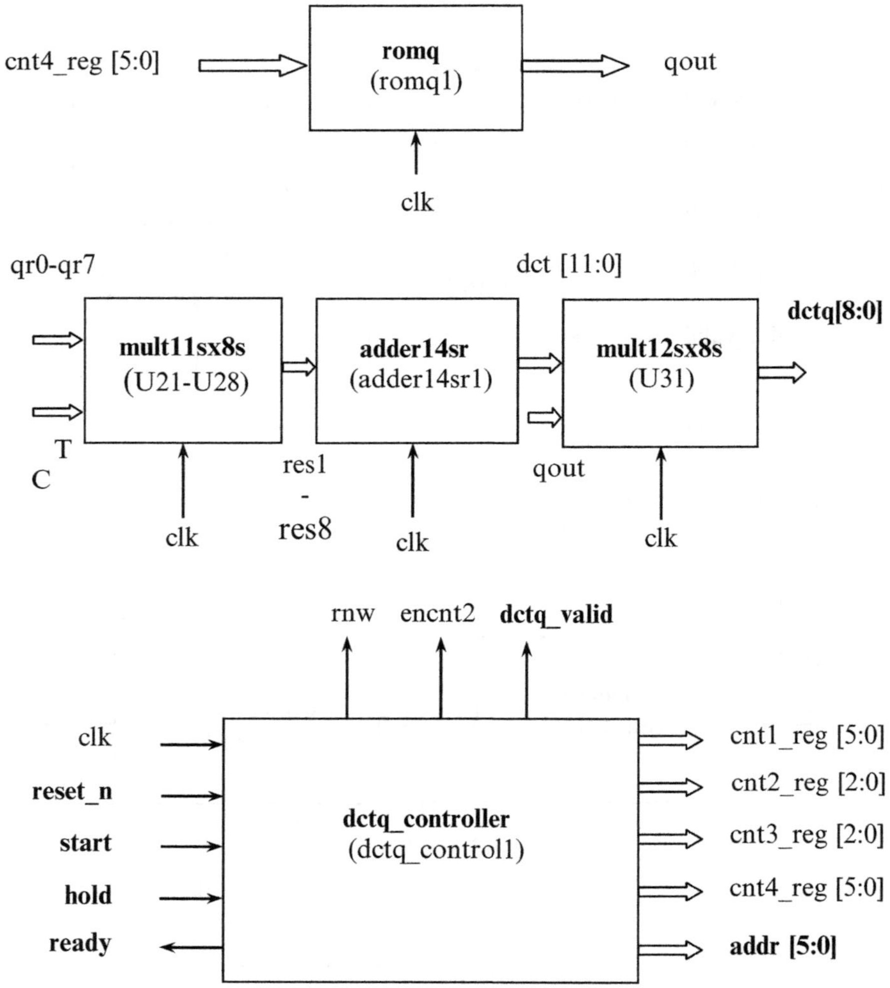

Fig. 12.1 Architecture of DCTQ processor

five and six pipeline stages respectively. Similarly, the multipliers: 'mult11sx8s' and 'mult12sx8s' have eight pipeline stages and 'mult8ux8s' has seven pipeline stages. 'dualram', 'romc', and 'romq' have two stages each. Therefore, the total number of pipeline stages in the entire DCTQ processor is 44 and hence the latency is 45 'clk' cycles. DCTQ is issued from 46th 'clk' pulse onwards, one coefficient every 'clk' cycle. This latency is applicable only for the very first block and not for subsequent blocks. In other words, the processing time is 64 clock cycles per block of image once the pipeline is full. It is also true for the inverse processor: IQIDCT, although not presented in this book. The Verilog code for the DCTQ design is presented in the next chapter.

12.2 Architecture of a Video Encoder Using Automatic Quality Control Scheme and DCTQ Processor

The basic architecture of a video encoder for processing Intra (I) frames is depicted in Figure 12.2. The image to be processed is input block by block, by a host computer such as a Pentium Processor, into the DCTQ processor, where the discrete cosine transform is performed followed by quantization. The optimum pruning level PLN up to which the DCTQ is to be computed is processed in the automatic quality controller circuit. The algorithm for this method was presented in the chapter on the development of algorithms. The computed PLN, which is communicated to both DCTQ and VLC processors, changes dynamically from block to block depending upon the picture content and energy (which has a direct bearing on quality) computed. The resulting quantized coefficients are applied to the next stage, VLC, where they are assigned variable length codes and buffered by FIFO before they are sent out onto a serial channel as a bit stream. The color information Y, Cb, and Cr are input once per macroblock. The energy threshold e_{THR}, which is a measure of image quality is user programmable. The encoder is also capable of processing up to a fixed pruning level.

12.2.1 The Automatic Quality Controller

The automatic quality controller is shown in Figure 12.3. It basically consists of a squaring circuit to evaluate $(DCT)^2$, adders/registers to accumulate 14 energy levels, e_{PL1} through e_{PL14}, registered comparators, and a controller to evaluate different steps of the algorithm given in the previous chapter. Although the system clock is not shown in the figure, all the blocks are connected to the clock. There are two modes of operations possible: fixed pruning level up to which the processing is required and automatic control of pruning level in order to get the desired quality level. The DCT coefficients generated in a raster scan order together with its address 'DCTCA' are input to the controller one by one. Since squaring and additions are time consuming operations, they are pipelined using signals, 'WS1', 'WS2', and 'WA', derived from the DCTQ processor.

When the DC coefficient is processed, the DCTQ controller issues 'RESET' signal to clear all the 14 e_{PL} registers. While the subsequent AC coefficients are processed, the DCTQ processor generates write signals, 'We_{PL1}' through 'We_{PL14}' to store energies at various levels from 1 to 14. The comparator compares the accumulated energy for every pruning level with the threshold energy programmed and if the accumulated energy is less than the threshold energy, then the controller outputs the pruning level number PLN as presented in the previous chapter. Both DCTQ and VLC processors process the quantized DCT coefficients only up to this PLN level and not beyond, thus speeding up the entire system without compromising on the image quality. The pulse 'PLNV' signals the validity of the 'PLN'. The signals: 'DCTCA', 'WS1', 'WS2', 'WA', and 'RESET' were not shown in the DCTQ

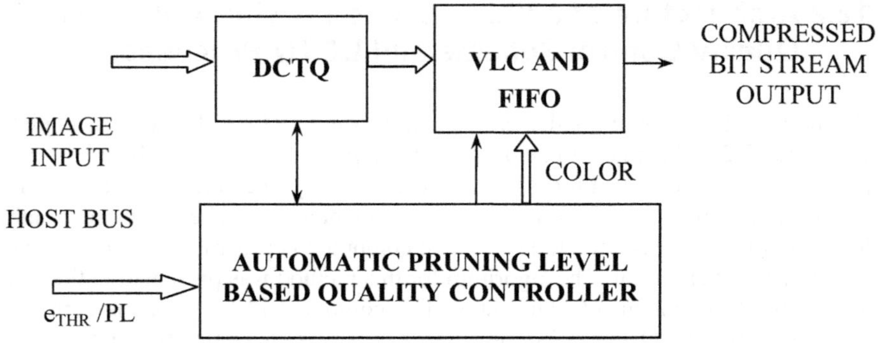

Fig. 12.2 The Basic architecture of the video encoder using pruning level based control

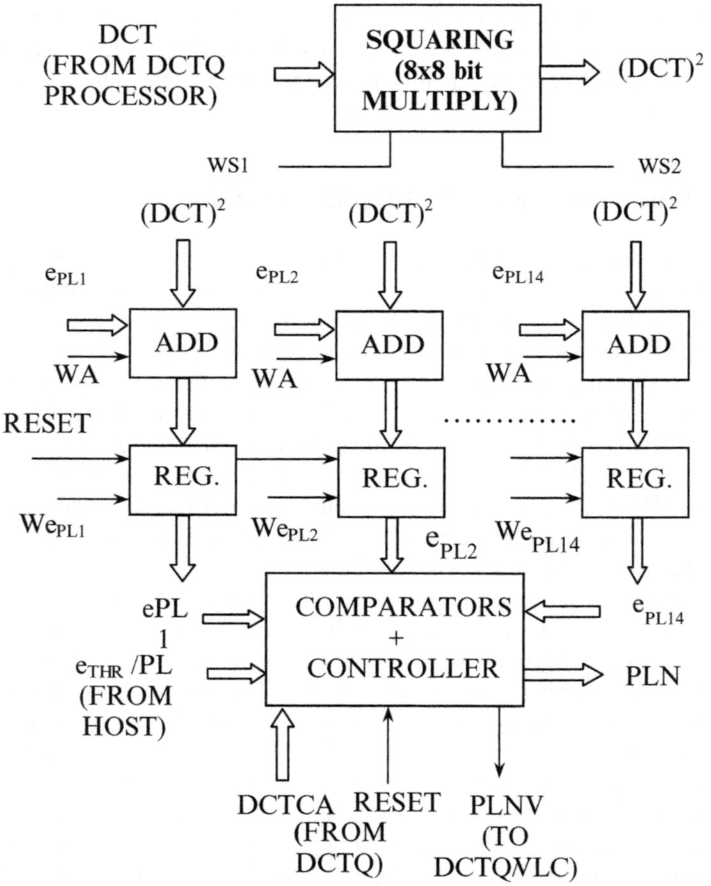

Fig. 12.3 The Automatic quality controller

architecture presented earlier in order to keep the treatment simple. See Chapter 15 for details of VLC processor.

12.3 Architecture for the FOSS Motion Estimation Processor

Figure 12.4 shows the architecture for the FOSS motion estimation processor. It consists of a motion estimation controller, which contains the circuit for executing the FOSS algorithm presented in the previous chapter, a dual redundant current MB RAM, a module for evaluating A_d and external RAMs to hold the processed I and P frames. To start with, the host processor communicates the picture size, the luminance or the color information and the macroblock number to be processed to the motion estimation (ME) controller. After ensuring that the EDATA signal is set, the host writes the image macroblock information into one of the two current

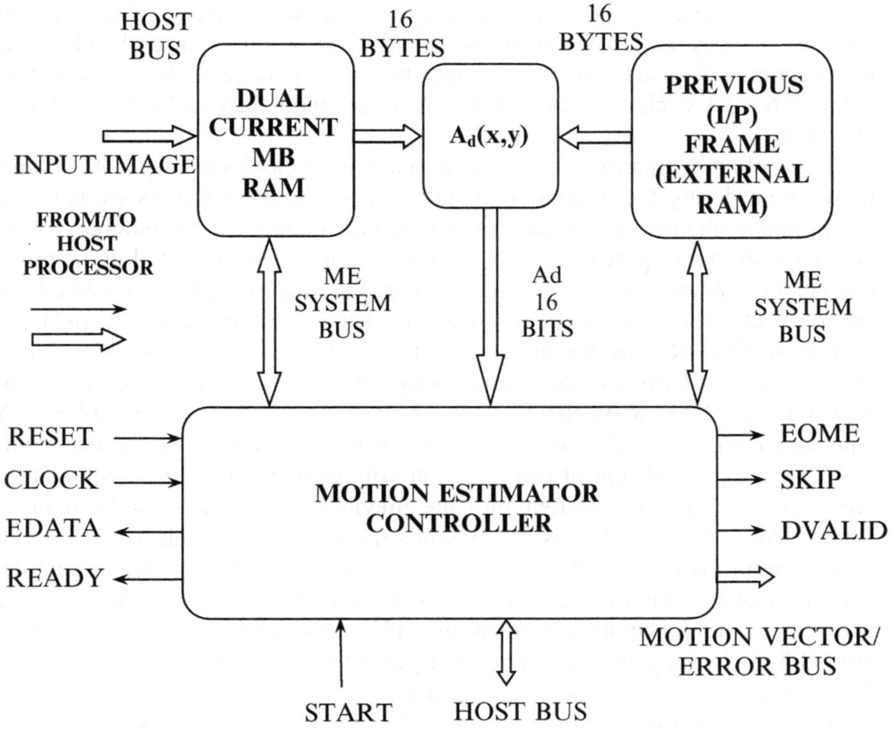

Fig. 12.4 Architecture of the FOSS motion estimation processor

macroblock RAMs. When the ME controller is ready to begin the motion estimation processing, the READY signal is set, receiving which the host asserts START signal, thus initiating the processing. EDATA signal is immediately asserted so that the host can enter the next image input concurrently with the processing of motion estimation.

Before processing the P frame, the I frame is processed by the DCTQ processor followed by the pipelined inverse quantization and inverse DCT by the IQIDCT processor. The processed I frame is stored in the external RAM designated as the previous frame RAM and serves as the reference frame for carrying out the motion estimation. The ME controller contains the FOSS algorithm in the form of a sequential circuit, executing each step of the algorithm as explained in Chapter 11. The minimum value of A_d is stored in an internal 16-bit register. The motion vector variables x and y are cleared at the start of motion estimation for every macroblock. A_d is computed using the $A_d(x,y)$ module. This is cleared before starting every A_d computation. The controller converts the x, y variables into appropriate addresses for the current macroblock and the previous frame RAMs. At one time, one row of a macroblock containing 16 pixels of data is fetched, each from the current and the previous frame RAMs and the sum of absolute differences for all these 16 pixel pairs is computed and accumulated. The controller takes 16 clock cycles to accumulate the sum for 16 rows of the macroblock. Since these computations are time consuming, they are pipelined with an inherent latency of 6 clock cycles. As a result, one A_d computation takes 22 clock cycles for execution.

The motion estimation is completed in about 30 clock cycles per A_d computation, considering the internal steps involved in the algorithm as explained in the previous chapter. For a macroblock, a maximum of eight numbers of A_d computations are required as can be seen from the results presented in the previous chapter. When the motion estimation for one macroblock is completed, the motion vector followed by the row-wise intensity errors are output at MOTION VECTOR/ERROR pins for use as inputs for the subsequent DCTQ, Inverse Quantization and Inverse DCT processing, thus reconstructing the error. The MOTION VECTOR and ERROR codes are generated in a sequential order for Y and color components Cb and Cr as per the MPEG 2 format. The corresponding motion-compensated sum of row-wise intensity of previous macroblock and the reconstructed error are written into the previous frame RAM to form the P frame. This P frame serves as the previous frame for processing the next frame of a group of pictures (GOP). These operations require 25 clock cycles each for execution of one luminance and two color components. As a result, the total execution time for motion estimation and compensation per macroblock is around 315 clock cycles for a true color picture in 4:4:4 format. This can come down for other formats such as 4:2:2 and 4:2:0.

Motion estimation is applied only on the luminance part, Y. A synchronous signal DVALID is asserted for writing the MOTION VECTOR/ERROR. If no motion is detected (motion vectors, x = y = 0), SKIP signal is issued. End of motion estimation signal EOME is generated after completing the motion estimation for the current macroblock. This process is repeated for all the macroblocks in the frame.

Summary

This chapter presented the development of architectural designs. In the last chapter, we learnt how to develop algorithms for a number of applications in the field of video processing as examples and verified the same. These algorithms were developed in such a way that the applications may be mapped onto an FPGA or an ASIC. The next logical step is to work out a detailed architecture keeping the actual hardware in mind. In this chapter, the architectural designs were developed for the same applications that we had undertaken in the previous chapter. The next chapter presents a very detailed description of project design and complete Verilog codes, test benches and results for a couple of applications.

Assignments

12.1 Detailed architecture of DCTQ processor as implemented on an FPGA was presented in Figure 12.1. On similar lines, design and describe a detailed

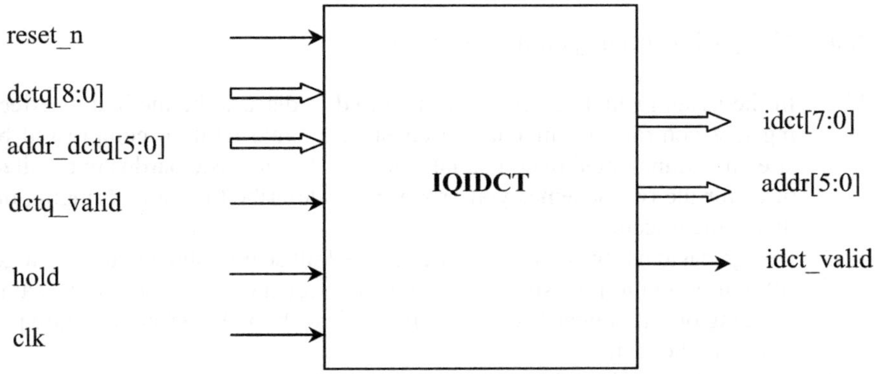

Fig. A12.1 Block diagram of IQIDCT

architecture of IQIDCT Processor such that one pixel may be processed every clock cycle. Use the algorithm of IQIDCT you have developed for the assignment 2 in Chapter 11. Block diagram of IQIDCT is shown in Figure A12.1. The signal descriptions are as follows.

Signal description

Signal	Input/ Output	Description
reset_n	Input	Asynchronous, active low.
dctq[8:0]	Input	Quantized DCT input.
addr_dctq[5:0]	Input	This furnishes the 'dctq' address of an image block. 'addr_dctq[0]' is the first (DC) coefficient. 'dctq' and its address are valid at positive edge of 'clk'. The coefficients are issued out in a raster scan order.
dctq_valid	Input	This input signals when the input data, 'dctq[8:0]', is valid.
hold	Input	IQIDCT processing can be kept on hold by this signal. De-asserting this signal resumes the processing without any latency coming into play.
clk	Input	System clock.
idct[7:0]	Output	IDCT output (unsigned), valid at positive edge of 'clk'.
addr[5:0]	Output	IDCT address. 'addr[0]' is the first reconstructed pixel and 'addr[63]' is the last pixel value in a block of image.
idct_valid	Output	This signal indicates the validity of IDCT and its address.

Note: All signals excepting reset_n are active high.

12.2 In the assignment 11.5, you were required to discuss the method of effecting rate control, i.e., maintain a constant bit rate while a compressed bit stream is transmitted over a serial channel. Design basic hardware architectures for the two schemes you propose and describe the same, bringing out their salient features.

12.3 Design basic hardware architecture for the Full search block matching algorithm used in motion estimation of a video sequence and describe the same, bringing out its salient features. Compare it with the FOSS architecture presented in the text.

12.4 A much faster algorithm than Full search block matching algorithm is the one-at-a-time step search (OSS) algorithm. Design and describe the hardware architecture for this algorithm. Compare it with the FOSS architecture presented in the text.

12.5 Programmable Logic Controller (PLC) is a digital equipment which is used in a number of industries/plants for bringing in automation. It consists of a processor which solves user programmed logic fetched from user memory. Each instruction is arranged as a 16-bit word in a typical PLC. MSB 5 bits of the memory word contain the instruction operation code and the balance

– a parameter, which can be an input/output (I/O) number, a Timer/Counter number or set time/count value, etc. Brief Specifications of the PLC is as follows:

128 Discrete inputs (parameter: 0–127)
128 Discrete outputs (parameter: 128–255)
256 Discrete flags (parameter: 256–512)
64 Programmable timers/counters (parameter: 0–63 for number and 0–2047 for timing/count value)
2K words user instruction memory (instruction pointer/parameter: 0–2047)
20 PLC instructions

The following is the instruction set along with its function, where IP is the instruction pointer and SP is the stack pointer.

Instruction	***Parameter***	***Operation***
NOP		No operation
READ	0 to 511	RR <= (I/O)
NOT		RR <= !RR
AND	0 to 511	RR <= RR & (I/O)
OR	0 to 511	RR <= RR | (I/O)
XOR	0 to 511	RR <= RR ^ (I/O)
STO	256–511	(Parameter) <= RR
JMP	0 to 2047	IP <= Parameter
JMPC	0 to 2047	IP <= Parameter if RR = 1
JSR	0 to 2047	IP <= Parameter; SP <= Parameter + 1
RET		IP <= SP
STRTC	0 to 63	Start Timer/Counter if RR = 1
TT		Set time base as 0.1 Sec.
TS		Set time base as 1 Sec.
TON	0–2047	Switch on the ON delay timer
TOF	0–2047	Switch on the OFF delay timer
UC	0–2047	Up Counter Set Value
DC	0–2047	Down Counter Set Value
TCEN		Enable Timer/Counter if RR = 1
STC	256–511	Output T/C Result

RR is a single bit accumulator called result register, JMP is an unconditional jump, JMPC is a conditional jump (if RR = 1), JSR is an unconditional subroutine and RET is a return instruction used by JSR instruction. No nesting of subroutines is permitted. Design the architecture of this PLC and describe them in detail. Note that the Timer/Counter routine and PLC instruction execution must take place concurrently. How will you debounce inputs (parameter = 0 to 127)? State your assumptions clearly.

12.6 Suggest more PLC instructions for different applications. Explain their operations.

12.7 In order to enter user programs for the PLC mentioned in the assignment 12.5, we need to design a programming unit. Describe how such a unit may be designed. Work out the basic architecture and describe the same.

12.8 Arithmetic Logic Units (ALU) are used for realizing arithmetic functions such as addition, subtraction and logic functions such as AND, OR, complement, NAND, NOR, EX-OR, EX-NOR, increment, decrement, etc. 74xx181 of Texas Instruments, whose logic symbol is shown in Figure A12.2. A 4-bit select code (S3–S0) and a mode bit (M) are used to decide the operation to be performed on data inputs. For more details, refer Chapter 2. Develop architecture for this ALU so that it may be implemented using a HDL.

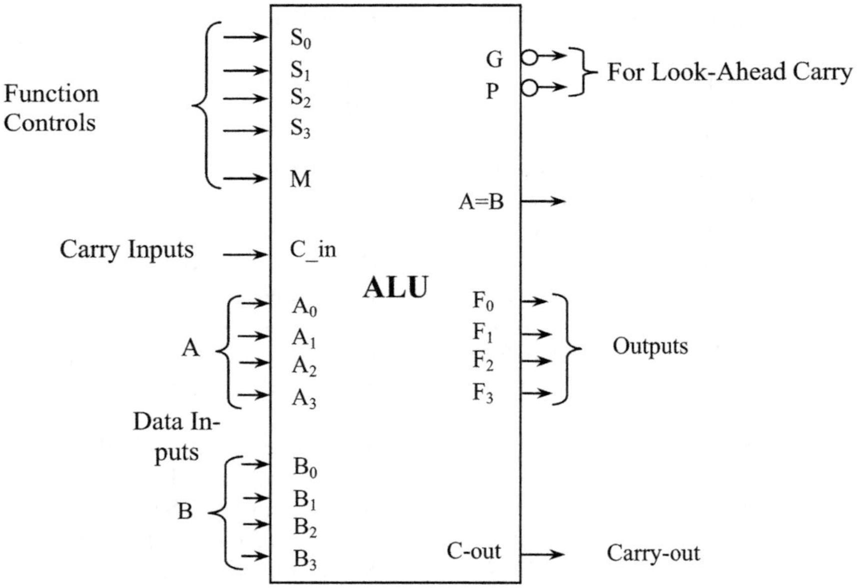

Fig. A12.2 ALU logic symbol (Courtesy of Texas Instruments Inc.)

Chapter 13

Project Design

Most challenging design applications involve the development of moderate to complex algorithms and hardware architectures that configure these algorithms efficiently. In Chapter 11, we learnt how to develop algorithms for a number of projects, namely, the discrete cosine transform and quantization, the automatic quality control system based on pruning level control and the fast one at a time step search for motion estimation for image/video compression systems. The next logical step is to check the concepts involved in the algorithm without worrying about the hardware that is required for implementation. Once the concepts are proven to be correct by coding and testing in a high level language such as the Matlab or C, as presented in Chapter 11, one can confidently embark on more involved design of architecture with actual hardware in mind. These architectural design aspects were elaborated in Chapter 12.

The sequel to architecture is to code the design using a HDL such as Verilog or VHDL. This chapter presents a couple of applications, namely, PCI Bus Arbiter and DCTQ as examples, using Verilog. The reader may develop codes for other applications such as Video encoder with automatic quality control and motion estimation presented in the previous chapters on similar lines as shown in the following sections. Before we take up the involved DCTQ design, we will start with a simple application, namely, a PCI Bus Arbiter, which uses ASM chart to aid in the design. For simple design applications, one need not develop any algorithm as it is mandatory for involved designs.

13.1 PCI Bus Arbiter

In a multiprocessor environment, several processors share the same system bus such as a PCI bus. A system based on multiprocessors can work in a coordinated manner only if bus arbitration is in force. The PCI Bus Arbiter design we are going to cover is for arbitrating four processors, some of which can be configured as the masters and others as targets. As an illustration, we will consider an application such as a video compression system for bus arbitration, although the design can be modified or extended to any other multiprocessor application and bus. Sharing the PCI bus are four masters, namely:

- Video Grabber, which will input a raw video data. It can be NTSC, PAL or SECAM sequence or may be in XGA, SVGA or in any other format. Any color motion picture can be processed, say, at 30 frames per second or 25 frames per second. Using the PCI, we can input the raw data into the Video Codec.
- Video Codec brings about the compression and reconstruction. We have an encoder and a decoder in the Codec, which brings about respectively the compression and decompression. This has to be designed in Verilog and implemented on either FPGA or ASIC.
- Fire Wire is a serial bus, which can be connected up to 64K nodes. It serializes the compressed data and broadcasts the compressed bit stream. Concurrently, it can receive a compressed bit stream from external source and send it to the decoder in Video Codec for effecting decompression.
- CPU (PC), which configures and coordinates the system activities via a north bridge.

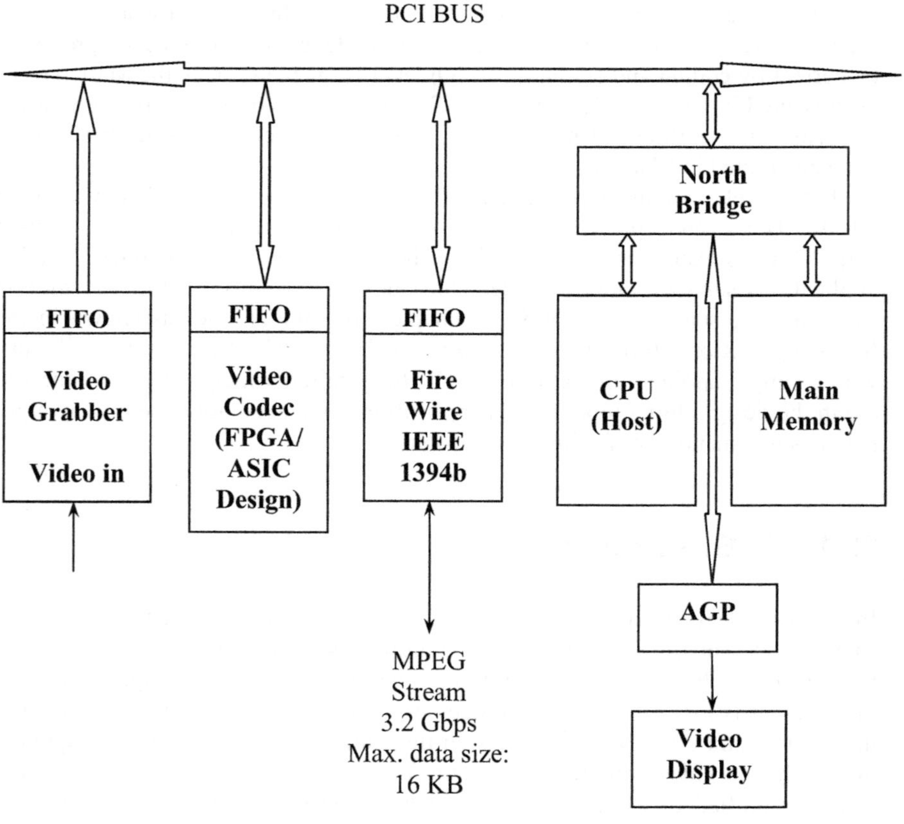

Fig. 13.1 PCI_bus arbiter for video codec (Continued)

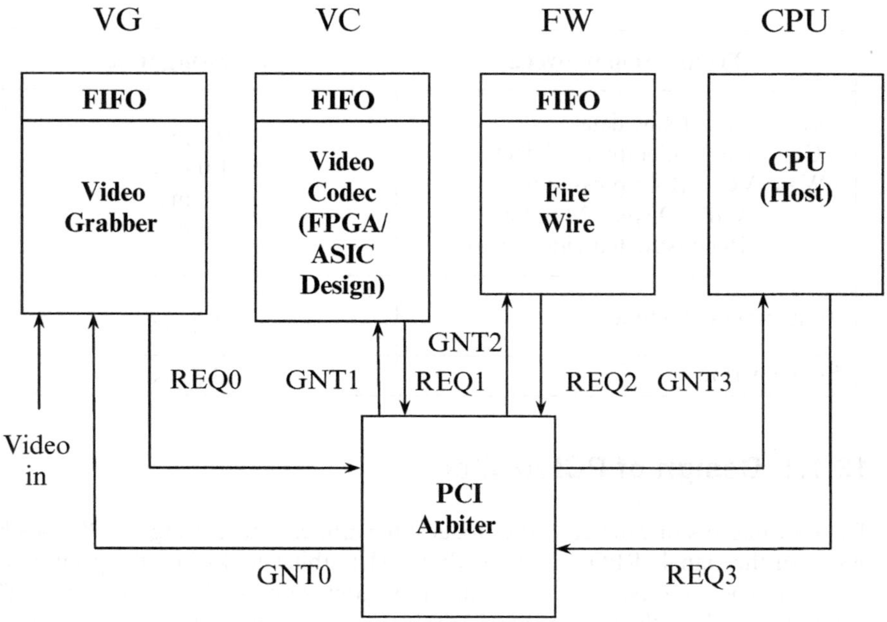

Fig. 13.1 PCI_bus arbiter for video codec

All the masters have built-in FIFO memory connected to the PCI bus in order to buffer data and maintain the frame rate. The PCI Bus Arbiter for the application, Video Codec, is shown in Figure 13.1. If we want to communicate the raw video data to the Codec in order to compress it, we make a request to the bus arbitrator, which grants the request as per a priority protocol among different masters. We divide the four processors into two groups, Video Grabber and Codec in the high priority group and the Fire Wire and the CPU in the second group. Among these, the Video Grabber and the Fire Wire has the highest priority in each group. We can compress the data in the Codec and send it through the PCI bus to the Fire Wire to serialize and transmit it over the channel. Likewise, compressed data can be obtained from any other system and serial to parallel conversion done in the Fire Wire and decompressed in the Codec and display through an interface called advanced graphics port (AGP) in the PC. As far as Fire Wire is concerned, its throughput is very high and, therefore, it can easily handle the MPEG 2, the fastest stream of the MPEG group (MPEG 1, MPEG 2, MPEG 4) or still picture JPEG and JPEG 2000 or any upcoming standards. The maximum data that the Fire Wire can handle at a time is 16 KB. The CPU and the main memory are connected to the PCI bus via the north bridge. These are all the standard PCI architecture. So is the case with AGP and the Display.

Table 13.1 Projected Processing Time for Video Codec Application

Transaction between	Processing time
VG => VC (Raw data) VC => FW (Compressed data) FW => VC (Compressed data) VC => AGP (Display Monitor) (Reconstructed video data)	9.6 ms 1.0 ms 1.0 ms 9.6 ms
Total processing time	21.2 ms
Frame period	33.3 ms

13.1.1 Design of PCI Arbiter

The four masters mentioned earlier request the arbiter before using the PCI bus by asserting the signals REQ0 through REQ3. The arbiter looks into the priority assigned for each request and asserts one of the grant signals, GNT0 through GNT3. The order in which the masters would receive access to the PCI bus is as follows:

1. Video Grabber (VG)
2. Video Codec (VC)
3. Fire Wire (FW)
4. Video Grabber
5. Video Codec
6. CPU (Host)

After the CPU accesses the PCI bus, the priority sequence repeats from step 1. The Video Grabber VG and Codec VC access the bus more frequently than the Fire Wire and the host processor since the raw data and compression/decompression have to be processed immediately and are time consuming operations as can be seen from Table 13.1.

The Verilog coding for the design will be easier if an ASM chart is drawn. Accordingly, we present the chart in Figure 13.2. Initially, the arbiter will be in a wait state. We use decimal values for the state representation as it is easy to code in Verilog. In the wait state, the arbiter will check for the REQ signal as per their priority order one after another. If a request is true, the arbiter will enable the grant signal GNT to that device. On the other hand, if the request is not made, the arbiter will check the next priority device request and so on. In wait state "0", for instance, if REQ0 to REQ3 are asserted simultaneously, then the arbiter grants the signal GNT0 in the next state "1" since the Video Grabber VG gets the top most priority for using the PCI bus. If none of the masters make any request for the use of bus, the arbiter continues to remain in the same wait state. In the VG state (1), if REQ0 is still asserted, the arbiter continues in the same state.

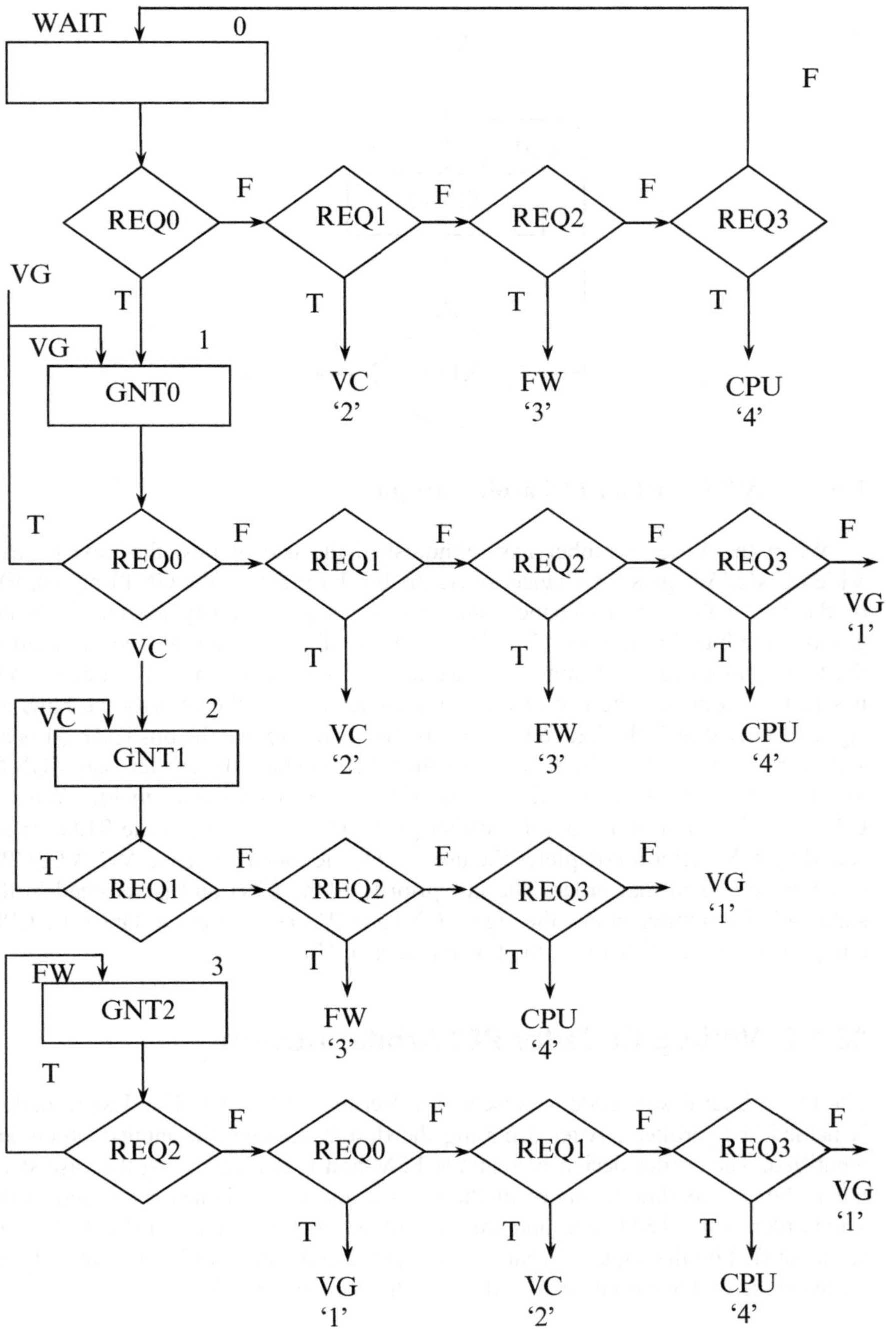

Fig. 13.2 ASM chart for PCI arbiter design (Continued)

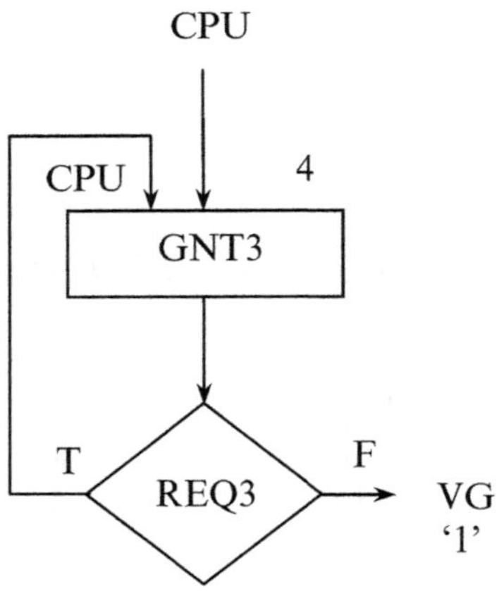

Fig. 13.2 ASM chart for PCI arbiter design

When the Video Grabber has relinquished the use of bus, the next priority Video Codec VC gets the chance to use the bus by granting the GNT1 signal. FW or the host CPU gets its chance only if other higher priority masters have not grabbed the bus. In the state "1", if VC, FW or CPU have not availed the chance, the token passes to VG again. Here, we have given chance to CPU since the VG has just relinquished the use of bus. In accordance with the priority we have assigned as per step 2, the Video Codec gets the chance to use the bus if REQ1 is asserted in the state "2" or in the earlier states. In this state, the arbiter issues GNT1 to VC. When VC is done with the use of bus, the token passes to Fire Wire or CPU or VG. In FW state "3", the arbiter grants GNT2 so long as the REQ2 is asserted by FW. After it completes the use of bus, the token passes to VG, VC, CPU or again to VG in that order. The last priority is the CPU and is serviced in the state "4". The arbiter grants the signal GNT3 if REQ3 is asserted. Once the CPU completes the use of PCI bus, the token passes to VG.

13.1.2 Verilog Code for PCI Arbiter Design

The PCI arbiter design code is presented in Verilog_code_13.1. The design module is named "pci_arbiter". After declaring the design module, the inputs/outputs are identified. The arbiter design is a simple FSM and is realized using the case statement. All the conditional states of the request and grant signals are coded in the same order as the ASM chart and are self-explanatory. The states of the ASM chart are identified by the signal, "arbiter_state", in the code. Priority is automatically assigned since we have used "if–else if–else" structure in the code.

Verilog_code_13.1

```
module pci_arbiter (                                    // Declare the design module.
                            clk,                        // List I/Os.
                            reset_n,
                            REQ0,
                            REQ1,
                            REQ2,
                            REQ3,
                            GNT0,
                            GNT1,
                            GNT2,
                            GNT3
                       ) ;

        input               clk ;                       // Declare the inputs
        input               reset_n ;                   // and outputs of the
        input               REQ0 ;                      // module.
        input               REQ1 ;
        input               REQ2 ;
        input               REQ3 ;
        output              GNT0 ;
        output              GNT1 ;
        output              GNT2 ;
        output              GNT3 ;

        reg                 GNT0 ;                      // Declare outputs as registers.
        reg                 GNT1 ;
        reg                 GNT2 ;
        reg                 GNT3 ;
        reg      [2:0]      arbiter_state ;             // State declaration.

always @ (posedge clk or negedge reset_n)

begin
  if (reset_n == 0)
                begin
                            // Switch OFF all grant signals to start with.
                    GNT0           <=      0 ;
                    GNT1           <=      0 ;
                    GNT2           <=      0 ;
                    GNT3           <=      0 ;
                    arbiter_state  <=      0 ;
                            // Initialize the state when the system is reset.
                end
```

```
else
    case (arbiter_state)
      0:
            begin                          // Wait state.
                                           // Switch OFF all grants signals.
                        GNT0          <=       0 ;
                        GNT1          <=       0 ;
                        GNT2          <=       0 ;
                        GNT3          <=       0 ;
                        if (REQ0 == 1)
                                // If Video Grabber request is asserted,
                                // go to the Video Grabber state "1".
                              arbiter_state <=     1 ;
                                // Otherwise, go to the Video Codec, state "2".
                        else if (REQ1 == 1)
                                arbiter_state <=   2 ;
                                // Otherwise, go to the Fire Wire, state "3".
                        else if (REQ2 == 1)
                                arbiter_state <=   3 ;
                                // Otherwise, go to the Host (CPU), state "4".
                        else if (REQ3 == 1)
                                arbiter_state <=   4 ;
                                // Otherwise, go to the WAIT, state "0".
                        else
                                arbiter_state <=   0 ;
             end
      1:            begin                  // Switch OFF all grant signals
                                           // except that of Video Grabber.
                                GNT0          <=       1 ;
                                GNT1          <=       0 ;
                                GNT2          <=       0 ;
                                GNT3          <=       0 ;
                        if (REQ0 == 1)
                                  // If Video Grabber request is still asserted,
                                  // remain in  the Video Grabber state "1".
                                  arbiter_state <=          1 ;
                                // Otherwise, go to the Video Codec, state "2".
                        else if (REQ1 == 1)
                                  arbiter_state <=          2 ;
                                    // Otherwise, go to the Fire Wire, state "3".
                        else if (REQ2 == 1)
                                  arbiter_state <=          3 ;
                                  // Otherwise, go to the Host (CPU), state "4".
                        else if (REQ3 == 1)
                                  arbiter_state <=          4 ;
                                        // Otherwise, go to the VG, state "1".
```

```
                        else
                                arbiter_state <=          1 ;
                      end
          2:  begin                          // Switch OFF all grant signals
                                             // except that of Video Codec.
                              GNT0          <=      0 ;
                              GNT1          <=      1 ;
                              GNT2          <=      0 ;
                              GNT3          <=      0 ;
                              if (REQ1 == 1)
// If Video Codec request is still asserted,  remain in the Video Codec state "2".
                                arbiter_state <=          2 ;
                                  // Otherwise, go to the Fire Wire state "3".
                              else if (REQ2 == 1)
                                arbiter_state <=          3 ;
                                     // Otherwise, go to the CPU state "4".
                              else if (REQ3 == 1)
                                arbiter_state <=          4 ;
                                     // Otherwise, go to the VG state "1".
                              else
                                arbiter_state <=          1 ;
                      end
          3:          begin   // Switch OFF all grant signals except Fire Wire.
                              GNT0          <=      0 ;
                              GNT1          <=      0 ;
                              GNT2          <=      1 ;
                              GNT3          <=      0 ;
                              if (REQ2 == 1)
      // If Fire Wire request is still asserted, remain in the Fire Wire state "3".
                                arbiter_state   <=      3 ;
                            // Otherwise, go to the Video Grabber, state "1".
                              else if (REQ0 == 1)
                               arbiter_state   <=       1 ;
                               // Otherwise, go to the Video Codec, state "2".
                              else if (REQ1 == 1)
                                arbiter_state   <=      2 ;
                               // Otherwise, go to the Host (CPU), state "4".
                              else if (REQ3 == 1)
                                arbiter_state   <=      4 ;
                                     // Otherwise, go to the VG state "1".
                             else
                                arbiter_state    <=      1 ;
                        end
          4:            begin           // Switch OFF all grant signals except
                                        // that for the Host.
                              GNT0          <=      0 ;
```

```
                                  GNT1          <=      0 ;
                                  GNT2          <=      0 ;
                                  GNT3          <=      1 ;
                                  if (REQ3 == 1)
                    // If CPU request is still asserted, remain in the CPU state "4".
                                   arbiter_state  <=        4 ;
                                       // Otherwise, go to the VG state "1".
                                  else
                                   arbiter_state  <=        1 ;
                                  end
                    default:       arbiter_state  <=        0 ;
                                       // Otherwise, remain in the WAIT state.
      endcase
end
endmodule
```

13.1.3 Test Bench for the Functional Testing of PCI Arbiter

Verilog _code_13.2 presents the test bench for the PCI arbiter design presented in Verilog _code_13.1. As usual, we will run the simulation at 50 MHz. The back annotated design, "pci_arbiter_banno.v", is included and is followed by declaring the test bench as "pci_arbiter_test". All the test bench stimulus are declared as "reg". It may be noted that the grant signals are declared as nets or wires in order to interconnect the output signals wherever necessary. Next, we shall instantiate the design of the arbiter. The inputs and outputs can be in any order, calling ports by name. When we initialize, the timing is zero and, initially, let us make all the bus requests active. Also initialize the clock and the active low reset signals. The reset signal is applied at 60 ns for 20 ns, after which the normal working of the arbiter commences. We can clear the request signals at regular intervals in order to study the corresponding effect of the output waveforms during simulation.

Verilog _code_13.2

```
`define clkperiodby2 10             // 10 ns is the half time period (50 MHz).
`include "pci_arbiter_banno.v"      // This is the back annotated design file.

module pci_arbiter_test ;           // Declare the test module.

reg                                 REQ0 ;  // Declare all inputs of
reg                                 REQ1 ;  // the design as  registers.
reg                                 REQ2 ;
reg                                 REQ3 ;
```

```
reg                         clk ;
reg                         reset_n ;  // Declare Bus Grant outputs as nets.

wire                        GNT0 ;
wire                        GNT1 ;
wire                        GNT2 ;
wire                        GNT3 ;

pci_arbiter     u1(     // Instantiate the design module, calling ports by name.
                            .REQ0(REQ0) ,           // Inputs.
                            .REQ1(REQ1) ,
                            .REQ2(REQ2) ,
                            .REQ3(REQ3) ,
                            .GNT0(GNT0) ,           // Outputs.
                            .GNT1(GNT1) ,
                            .GNT2(GNT2) ,
                            .GNT3(GNT3) ,
                            .clk(clk) ,             // Inputs.
                            .reset_n(reset_n)
                        );
initial
        begin
                REQ0    = 1 ;   // At time zero, let the request inputs
                REQ1    = 1 ;   // be active.
                REQ2    = 1 ;
                REQ3    = 1 ;
                clk     = 0 ;   // Initialize clk, and reset_n.
                reset_n = 1 ;
        #60     reset_n = 0 ;   // At 60 ns, apply reset.
        #20     reset_n = 1 ;   // At 80 ns, let the reset be withdrawn.
        #400    REQ0    = 0 ;   // At time 480 ns, let the request input be 0.
        #80     REQ1    = 0 ;   // At time 560 ns, let the request input be 0.
        #80     REQ2    = 0 ;   // At time 640 ns, let the request input be 0.
        #160    REQ0    = 1 ;   // At time 800 ns, let the request input be
                                // asserted again.
        #200    REQ3    = 0 ;   // At time 1000 ns, let the request input be 0.

        #1200                   // Run long enough to complete the test
        $stop ;                 // and stop.

        end
always
        #`clkperiodby2 clk <= !clk ;        // Toggle to get a free running clk.

endmodule
```

13.1.4 Simulation Results

The back annotated design has been used to run the simulation using Modelsim. The simulated results for the PCI arbiter are shown in Figures 13.3.1 to 13.3.4. Inspecting the waveform of reset_n, we see that the active low reset is applied at 60 ns and withdrawn at 80 ns, which is in agreement with the test bench we wrote before. At 0 ns, all the request signals from REQ0 to REQ3 are asserted. Although in the design, reset is applied at 60 ns, the grant outputs, GNT0 to GNT3, and the "arbiter_state" are cleared only after gate delays of about 5 ns. The first rising edge of the "clk", after the reset is withdrawn, occurs at 90 ns. After a delay of about 2 ns, the state changes to "1". When the "clk" strikes again at 110 ns, GNT0 is asserted at about 115 ns owing to gate delays. This is because REQ0 is the highest priority although all other requests are also asserted.

In state "1", the REQ0 is withdrawn at 480 ns. In state "2", VC request, REQ1, is recognized and its grant signal GNT1 is asserted at 515.4 ns, simultaneously withdrawing the VG grant signal. The gate delays are marked at the time axis as can be seen clearly in Figure 13.3.2. It may be observed in the figures that arbiter states are different from what we have keyed in in the design. This is because the synthesis tool has changed their assignments. The grant signals GNT2 and GNT3 are respectively asserted in states 4 (corresponds to 3 of the design) and 8 (corresponds to 4 of the design). Carrying out the waveform analysis in the foresaid manner, the reader can easily correlate the waveforms and the design.

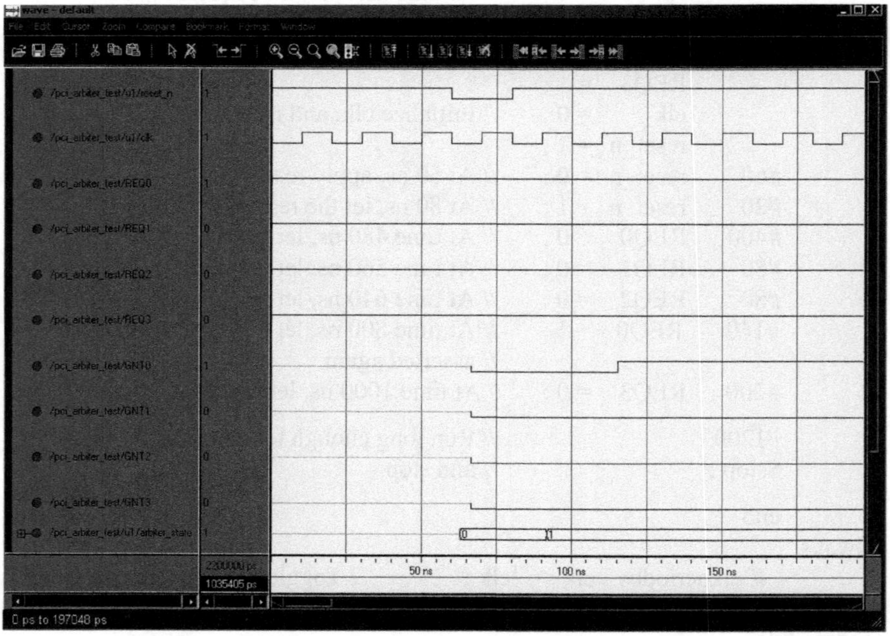

Fig. 13.3.1 Simulation results of back annotated PCI arbiter (Continued)

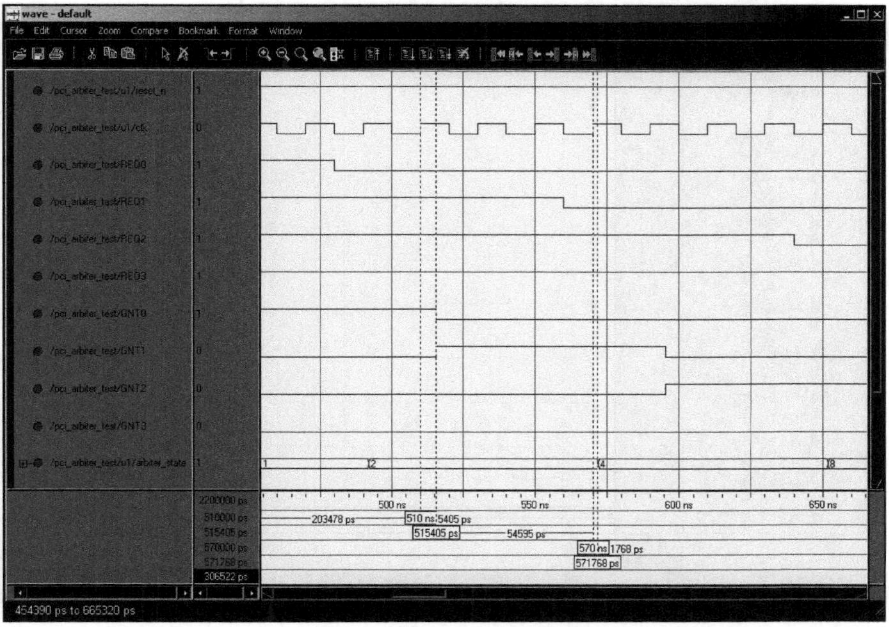

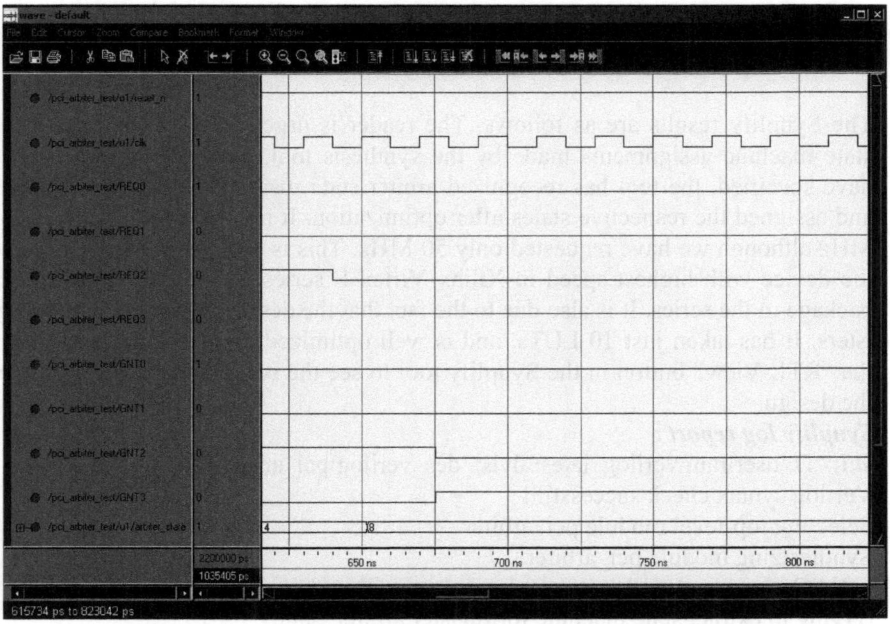

Fig. 13.3.2 and 13.3.3 Simulation results of back annotated PCI arbiter (Continued)

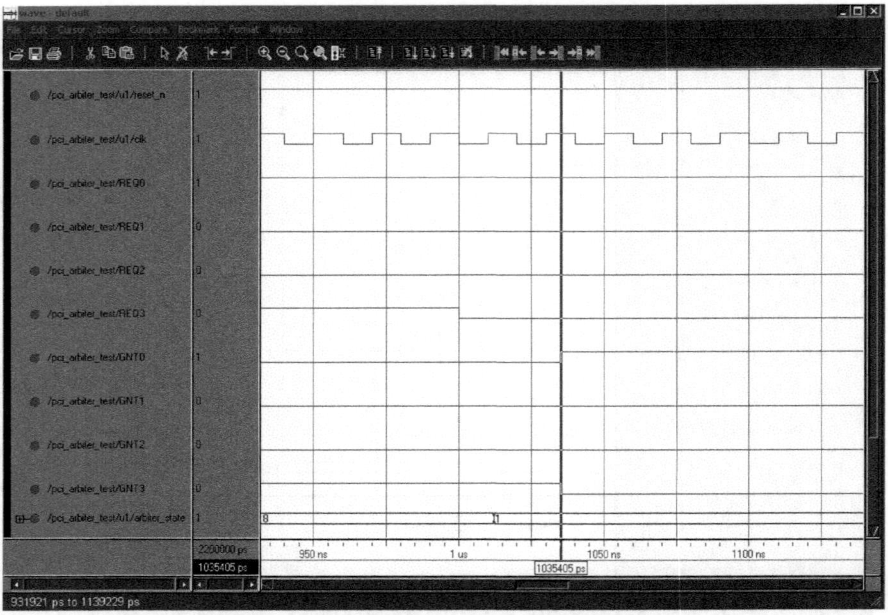

Fig. 13.3.4 Simulation results of back annotated PCI arbiter

13.1.5 Synthesis Results for PCI Arbiter

The Synplify results are as follows. The reader is urged to examine closely the state machine assignments made by the synthesis tool. Irrespective of what we have specified, the tool has recognized arbiter state machine as one hot machine and assigned the respective states after optimization. It reports a frequency of 242 MHz although we have requested only 50 MHz. This is because we have selected the device with highest speed in Xilinx Virtex E series and lowest capacity and package in the series. It is also due to the fact that the design is predominantly registers. It has taken just 10 LUTs, and is well optimized. The reader can click on the "RTL View" button in the Synplify tool to see the schematic circuit design of the design.

Synplify log report :

@I::"D:\user\ram\verilog_latest\dvlsi_des_verilog\pci_arbiter.v"
Verilog syntax check successful!
Selecting top-level module pci_arbiter
Synthesizing module pci_arbiter
@N:"D:\user\ram\verilog_latest\dvlsi_des_verilog\pci_arbiter.v":53:0:53:5
|Trying to extract state machine for register arbiter_state
Extracted state machine for register arbiter_state
State machine has 5 reachable states with original encodings of:

```
      000
      001
      010
      011
      100
    @END
```

Encoding state machine work.pci_arbiter (verilog)-arbiter_state_h.arbiter_state[4:0]

Original code	->	New code
000	->	00000
001	->	00001
010	->	00010
011	->	00100
100	->	01000

Worst slack in design: 15.864 (ns)

Starting clock	**Requested frequency**	**Estimated frequency**	**Requested period**	**Estimated period**	**Slack**
Clk	50.0 MHz	241.8 MHz	20.000	4.136	15.864

Resource usage report for pci_arbiter

Mapping to part: xcv50ecs144-8

Cell usage:

FDC	8	uses
FDP	1	use

I/O primitives:

IBUF	5	uses
OBUF	4	uses
BUFGP	1	use

I/O register bits: 4

Register bits not including I/Os:	5 (0%)
Global Clock Buffers:	1 of 4 (25%)
Total LUTs:	10 (0%)

13.1.6 Xilinx Place and Route Results for PCI Arbiter

The Xilinx P&R tool also gives a good result as summarized in the following report. It shows the number of gates used for the design as 132. The frequency reported is (294 MHz) higher than that reported by the Synplify tool. But this frequency information may be misleading because this is not the total design. The total design is when we design the entire Video Codec and map all modules on a single chip or multiple chips. In that case, one may expect the frequency to drop somewhere between 50 to 100 MHz for the Virtex E series FPGAs. However, it is advisable to run the P&R tool to part of a design such as the present design since each of the design parts contribute towards the overall processing speed. If each of the submodules of a design is taken due care of, then the overall design will take care of itself. It is; therefore, better to get the best possible processing speed for smaller design modules individually. The adage that prevention is better than cure is equally valid in the realms of digital designs. The tool generates the bit stream, which can be used to download into the mapped FPGA.

Design Summary:

Number of slices:	6	out of	768	1%
Number of slices containing unrelated logic:	0	out of	6	0%
Number of slice flip flops:	5	out of	1,536	1%
Number of four input LUTs:	10	out of	1,536	1%
Number of bonded IOBs:	9	out of	94	9%
IOB flip flops:	4			
Number of GCLKs:	1	out of	4	25%
Number of GCLKIOBs:	1	out of	4	25%

Total equivalent gate count for design: 132
Additional JTAG gate count for IOBs: 480

Device, speed: xcv50e,-8 (PRELIMINARY 1.65 2001-12-19)

Timing Summary:

Minimum period: 3.401ns (maximum frequency: 294.031MHz)
Minimum input arrival time before clock: 2.671 ns
Minimum output required time after clock: 5.419 ns

Saving bit stream in "pci_arbiter.bit".

13.2 Design of the DCTQ Processor

We need to have an overall bird's eye view before we design any system. If we wish to design a chip for an application such as a video compression system using the DCTQ, we must first define the application clearly. For instance, the application is to receive a burst of image/video data and apply a transform such as the DCT followed by quantization in order to effect compression on a picture. We may view the DCTQ processor as a black box with inputs and outputs defined to suit the application requirements. Based on the emerging details, we formulate the specifications. Now, let us examine the specification of the DCTQ design. To start with, we will see what signals are required to communicate the pixel information of a picture. For example, we can use a host processor such as the personnel computer in order to communicate the pixel data. Since we need a very high throughput, especially for compressing high resolution pictures, we can think of using a parallel bus such as a PCI bus, which we have already discussed in the design of PCI Bus Arbiter.

13.2.1 Specification of DCTQ Processor

In an earlier chapter, we developed an algorithm for processing DCTQ, which requires the application of an image data in 8 × 8 pixel blocks. Any block of image can be input as one row of a block, i.e., 8 pixels at a time. A 64-bit PCI bus will be handy here since a block of data can be input in just 8 clock cycles. It may be recalled that the DCTQ takes 64 clock cycles for processing one block of image. This aspect was detailed in a previous chapter on architectural design. Assuming that PCI bus clock, "pci_clk", and the DCTQ clock, "clk", are the same, the host has plenty of free time (56 clock cycles) to attend to other processing cores such as getting image related information from a Video Grabber card, send data to display, etc. Since we have established that we need 64 bits of data input, let us label this data bus as "di[63:0]". Further, we need a signal to identify which bytes in "di" are enabled. "be[7:0]" serves this purpose since the data bus is 8 bytes wide. Also, we need 3 bits of write address, "wa[2:0]", so that we may write eight rows of a block of data. All these activities must be synchronous to the clock signal, "pci_clk". Further, we need to tell the DCTQ engine when the data input, "di", is valid. Let us designate such a signal as "din_valid". If we have a hardware core such as the DCTQ, we need to reset it at any point of time. We have an asynchronous, active low signal, "reset_n" for the same. Once we have these signals, we can communicate the image information from the host processor or any other processor such as a Video Grabber, which we have discussed earlier. Consolidating all that we have discussed so far, we can draw a block diagram for the DCTQ processor as shown in Figure 13.4.

The DCTQ processing can be commenced by asserting the "start" signal. We can suspend the processing by activating the "hold" pin. Before we input image data, we need to check whether the DCTQ is "ready" to receive the input. The output "dctq" is of width, 9 bits to comply with the MPEG 1/MPEG 2 standards. The validity of this output is indicated by the signal, "dctq_valid". The coefficients are identified by an address signal, "addr[5:0]". Address "0" means the DCTQ output is the DC coefficient, while other addresses are for the AC coefficients. The address width is 6 bits since there are 64 coefficients for a block of image. Table 13.2 presents the signal descriptions of the DCTQ processor.

After we have written one block of information, we can start the DCTQ process. While this DCTQ process is going on, we can input the next block of image concurrently. As there are 64 coefficients for a block, the DCTQ processor is so

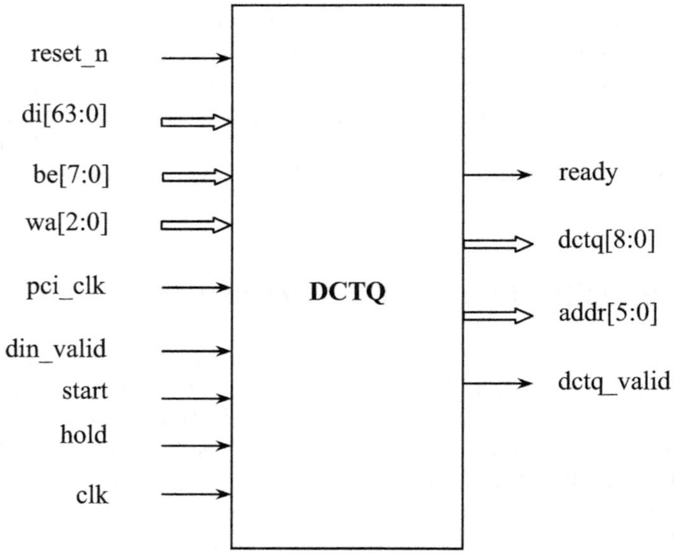

Fig. 13.4 Block diagram of DCTQ processor

designed that coefficients are issued, one every clock cycle, as described in the chapter on architecture. Thereby, we will require 64 clock cycles for processing one block of DCTQ, while we need only 8 clock pulses in order to input a block of raw image. This way, the burden of the host processor, which inputs the image, is relieved. The DCTQ is really a time consuming operation, which takes $2N^3$ number of computations per block, where N is 8 for an 8×8 pixel block. Multiplications and additions are involved in matrix manipulations as was shown in the chapter on algorithms.

The DCTQ output is 9 bits wide and in twos complement form as per the requirements of JPEG and MPEG standards. All the specifications must conform to the standards in order to maintain a healthy communication between the product being designed and compatible products of other vendors. Having formulated the standards-compliant specifications, we can very well design the DCTQ processor now.

13.2.2 Sequence of Operations of the Host and the DCTQ Processors

Before we take the Verilog coding of DCTQ, let us see how the DCTQ processor communicates with a host processor. The step-by-step operation sequence of the host and the DCTQ processors are as follows:

Table 13.2 DCTQ Core Signal Description

Signal	Input/ Output	Description
reset_n	Input	Asynchronous, active low.
di[63:0]	Input	One row (8 pixels) of an image block (8 × 8 pixels) is input. di[7:0] is the first pixel and di[63:56] is the last in the row. A pixel is of size 8 bits, unsigned.
be[7:0]	Input	Byte enable signal. Active low be[0] selects di[7:0] and so on.
wa[2:0]	Input	This furnishes the row address of an image block. wa[0] is the first row.
pci_clk	Input	Image input synchronous clock. di[63:0] is written into the core, wa[2:0] serving as the address synchronous to positive edge of this clock.
din_valid	Input	This input signals when the input data, di[63:0] is valid.
start	Input	Active high starts and maintains the DCTQ processing. If de-asserted and re-applied, latency will come into effect again.
hold	Input	DCTQ processing can be kept on hold. De-asserting resumes the processing without any latency coming into play.
clk	Input	DCTQ system clock.
ready	Output	This signal indicates that the core is ready to accept image input block.
dctq[8:0]	Output	DCTQ output in twos complement, valid at positive edge of "clk".
addr[5:0]	Output	DCTQ coefficient address. "addr[0]" is the DC coefficient.
dctq_valid	Output	This signal indicates the validity of DCTQ coefficient and its address.

Note: All signals excepting "reset_n" and "be" are active high.

By Host:

1. Assert "reset_n" signal to initialize the DCTQ core.
2. Write an 8 × 8 pixel block of image data (64 bytes) into one block of the 64-byte Dual RAM in the DCTQ processor, 8 bytes at a time, via the data bus "di[63:0]" after ascertaining that "ready" signal is set by the DCTQ processor. All the "be" bits may be simultaneously asserted (PCI compatible with "pci_clk" as clock). Also assert "din_valid" signal while inputting the image data.

3. Issue "start" signal to begin DCTQ processing. The start signal must be continuously asserted for continuous processing without latency. If "start" is withdrawn and re-applied, the latency comes into play once again for the first block.
4. If the "ready" signal is set now, write the next block of image data into another block of 64-byte Dual RAM in the DCTQ engine. Otherwise wait.

Notes:

1. Repeat step 4 for processing subsequent blocks in succession.
2. The host asserts "reset_n" signal once at the beginning of processing normally. However, it may apply "reset_n" at any point of time for terminating the processing.

By DCTQ Core:

1. If "reset_n" is active, initialize all the internal registers and terminate the current DCTQ processing. Select one RAM bank for the host to write the image block. One bank is in read-only mode for DCTQ processing, while the other bank is in write-only mode for the host to write the image block concurrently. Set the "ready" signal.
2. If "start" is asserted, begin processing. Otherwise, wait. Assert "ready" signal for the host to write into the other RAM bank concurrently. Compute DCTQ. The DCTQ coefficients are issued at dctq[8:0] pins, valid at the positive edge of "clk" with "dctq_valid" signal asserted after a latency of 45 clk cycles. "dctq_valid" signal is continuously asserted as long as the processing continues without a break. "addr[5:0]" is valid when the DCTQ is valid. DCTQ is issued every "clk" cycle from 46 th "clk" cycle onwards. 64 coefficients are issued per image block. This implies that one image block is processed in 64 clk cycles. The first coefficient (addr[5:0] = 0) is called the DC coefficient and the other 63 are known as AC coefficients.

Note:

1. Step 2 is continuously processed as long as "start" is kept asserted and "reset_n" or "hold" is not active.

13.2.3 Verilog Code for the DCTQ Design

In the chapter on architectural design, we presented the architecture for the DCTQ processor. Various blocks therein reflect the corresponding Verilog modules, which we are presently in the process of coding in this section. In the chapter on simulation, we presented a model for a design hierarchy. The same is presented again in Figure 13.5 in order to show all the Verilog modules of DCTQ processor, wherein the top design module is "dctq". This module, in turn, calls various submodules as depicted in Figure 13.5. In the chapter on memories, we presented the designs of "dualram", "ram_rc", "romc" and "romq". Similarly, we presented the designs of "adder12s" and "mult11sx8s" in the chapter on arithmetic circuit designs. The following designs: "adder14sr", "mult8ux8s", and "mult12sx8u" were left as assignments in the chapter on arithmetic circuits. Therefore, these designs

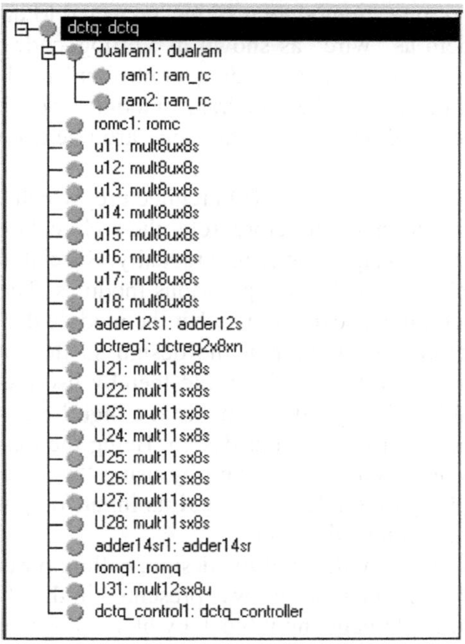

Fig. 13. 5 Verilog design modules of DCTQ processor

will not be presented in this book. The remaining Verilog modules, namely, the “dctq”, “dctreg2x8xn”, and “dctq_controller”, will be presented in this section.

We will see how to code the top design module, “dctq”. Verilog _code_13.3 presents the “dctq” module. The design file is named “dctq.v”. We identify all the submodules in the design using the “include” statements. This is followed by the declaration of “dctq”. 2D-DCT is a two-stage multiplication of three 8 × 8 matrices: C, X, and C^T. The algorithm and architecture of DCTQ were presented in earlier chapters. Input X of DCT is a block (8 × 8 pixels) of image information. DCT produces 64 coefficients per block. The first coefficient is called the DC coefficient, and others are known as AC coefficients. The resulting DCT is divided by the corresponding quantization value stored in a 64B ROM, “romq.v”. The DCTQ coefficients are identified by the address, “addr”, and are issued keeping in step with the corresponding “dctq” coefficients. Valid range for the DCTQ address is 0 to 63. This address will prove to be handy in the design of subsequent module, variable length coder (VLC). Detailed specifications and basic architecture for VLC will be presented in the last chapter.

It is a good design practice not to include any logic in the main design, which is referred to as the top module. We merely call all the submodules in the top design. Any logic required at the top design can be pushed into one or more of the existing submodules or a new module created exclusively for this purpose. Each of the submodules may be viewed as a block of circuits for accomplishing certain functionalities. Naturally, these modules need to be interconnected to form the final application. Signals of one submodule may be connected to one or more submodules by declaring them as "wire" as shown in Verilog_code_13.3. For example, "cnt1_reg[2:0]" signals issued out of "dctq_controller" module is connected to the read address input "ra" of the "dualram" module and, therefore, the signal "cnt1_reg[2:0] is declared as "wire". The same argument holds good for other signals.

The design closely follows the DCTQ architecture, which was presented in an earlier chapter. The reader may, therefore, refer the architecture quite frequently in order to get a feel of the design flowering gradually. Special attention may be bestowed on the handshake signals among various modules. The DCTQ code starts with a good deal of comments explaining what the module deals with. In fact, one must include apt comments throughout the design so that other designers may readily understand what we have written and thereby use them in their own designs, where applicable. Also, good comments at a high level will be highly beneficial for the customers or users of your design. Being the top design module, we identify all the modules used in the design by using "include" statements. Next, we identify the top design module "dctq" and its inputs/outputs. This is followed by declaring "wire" as explained earlier.

Next, we invoke all the modules of the design, calling ports by name. The first module is the "dualram", which design was presented in the chapter on memories. Dual RAM is used to read image input block by block (8 × 8 pixels). This is the X matrix in the algorithm. The signal "rnw", derived from the "dctq_controller" is to configure one set of RAM in the "dualram" in "read-only" mode, while the other set of RAM is automatically configured in "write-only" mode. The next module is a ROM, "romc", storing 2C and $2C^T$. Two times C/C^T are used in order to improve the accuracy. Later on, we will divide the final result by two in order to get the correct value for DCTQ. C and C^T are accessed simultaneously. Both require row accesses for the computation of DCT. "addr1" and "addr2" inputs are once again derived from the counters in "dctq_controller" module. They are respectively "cnt1_reg[5:3]" and "cnt3_reg[2:0]". "addr1" and "addr2" are for fetching C and C^T matrices. "do" is the image input, "X" and "d1" is the C output. "do" is unsigned, while "d1" is in twos complement. C is used in the first stage multiplication, u11–u18, while C^T is used in the second stage multiplication, u21–u28.

CX is computed using eight multipliers, "mult8ux8s". "result" are the products of C and X and are in twos complement. These results correspond to one row of cosine coefficients (in 8 × 8 matrix) and one column of X (also of size 8 × 8). Partial products of CX (result1 to result8) are added by the next module, "adder12s" in order to get 12 bits of result. Excess bits are discarded since the chip area will be less and the quality of reconstructed image does not suffer much. The added result is stored in partial product registers, p0–p7, using the next module, "dctreg2x8xn". Only 11-bit signed (integer) is retained after dropping 3 bits after decimal point. It may be recalled that C was actually taken as 2C in the ROM and hence, 1 more bit is dropped. Similarly for C^T. "cnt2_reg[2:0]" is used as the address pointer for "p0" to "p7". This counter and its enable signal, "encnt2", are generated by the controller.

In the second stage, we take the partial products, p0–p7, and multiply them with the corresponding column elements of C^T, as presented in the algorithm and architecture, to produce independent multiplied results indicated as "res1" through "res8", each of precision, 19 bits. The eight multipliers, "mult11sx8s", of this stage are u21 to u28. Note that the bit precision has increased. "d2" is the C^T output, fed as one of the inputs to these multipliers. These results, "res1" through "res8" are added together using the module, "adder14sr". Note that we retain only the most significant 14 bits for each input of this adder. The added result, naturally, is 3 bits more. Once again, we truncate the result and retain only 12 bits. This is the DCT output. 12 bits precision for DCT complies with the standards.

The precision required at various stages can be arrived at based on the quality of the reconstructed image. We will, however, have high precision computation for DCTQ and reconstruction in Matlab in order to serve as a standard reference for verifying the performance of hardware coded in Verilog. The Matlab codes for this application were presented in a previous chapter on verification of algorithms using high level languages. If the hardware result is close to the Matlab result, say, within 0.5 dB, then we accept the precisions as such at various stages of the DCTQ computation.

Next process is the quantization, in which each of the 64 DCT coefficients is divided by a corresponding quantization value such as 8, 16, etc. In the present design, the default values recommended in the standards is adopted. Instead of using a divider, we can use a multiplier, taking inverse quantization values instead of quantization values. We multiply the inverse quantization values by 16 so that precision of the resulting values increases. Later on, the final result can be divided by 16 to get the correct value for the DCTQ. 16/Quantization value, "qout" is fetched from a ROM, "romq" as shown in the code. The address for this ROM is fed from a 6-bit counter, "cnt4_reg" generated by the "dctq_controller". The final stage is a multiplier, "mult12sx8u" u31 to multiply the "dct" (12 bits in twos complement) and "qout" (unsigned) to get the final output, DCTQ. The result, which is 20 bits is divided by 16 by dropping 4 bits. The 16-bit DCTQ obtained as a result is further truncated in order to get a 9-bit integer. The 9-bit DCTQ, "dctq[8:0]" in twos complement, conforms to the image/video standards, JPEG/MPEG-1/MPEG-2 standards, etc. "dctq_valid" is asserted whenever DCTQ is valid and "addr" provides the address of the DCTQ coefficient. "addr" is 0 for the DC coefficient, and 1 to 63 for AC coefficients.

Verilog_code_13.3

```
/*                         DCTQ RTL Code

This is the top-level design module for the computation of DCTQ. The design file
is "dctq.v". DCTQ prepares the ground for effective compression of data, espe-
cially that from images, be it still or motion pictures (also referred to as Video).

2D-DCT is a simple two-stage multiplication of three 8 × 8 matrices: C, X, and C^T.
```

```
  Input of DCT is a block (8 × 8 pixels) of image information.
  DCT produces 64 coefficients
              - The first coefficient is referred to as the DC coefficient,
                while others are known as AC coefficients.
  The resulting DCT is divided by the corresponding quantization value stored
  in 64B ROM, "romq.v".
  "addr" is the address of the issued DCTQ coefficient. Valid range: 0–63.
  See the DCTQ document for details of signals used.
*/

`include "dualram.v"            // These files are the submodules
`include "adder12s.v"           // used in the design.
`include "adder14sr.v"
`include "dctreg2x8xn.v"
`include "mult8ux8s.v"
`include "mult11sx8s.v"
`include "mult12sx8u.v"
`include "romc.v"
`include "romq.v"
`include "dctq_controller.v"

module dctq (                                     // Declare the design module
                        pci_clk,                  // and its inputs/outputs.
                        clk,
                        reset_n,
                        start,
                        di,
                        din_valid,
                        wa,
                        be,
                        hold,
                        ready,
                        dctq,
                        dctq_valid,
                        addr
            );
      input             pci_clk ;
      input             clk ;
      input             reset_n ;
      input             start ;
      input             din_valid ;
      input             hold ;
      input   [63:0]    di ;
      input   [2:0]     wa ;
      input   [7:0]     be ;
```

```
output              ready ;
output    [8:0]     dctq ;
output              dctq_valid ;
output    [5:0]     addr ;

wire                ready ;  // Declare the nets of the design.
wire      [8:0]     dctq ;
wire                dctq_valid ;
wire      [5:0]     addr ;
wire                rnw ;
wire                encnt2 ;
wire      [15:0]    result1 ;
wire      [15:0]    result2 ;
wire      [15:0]    result3 ;
wire      [15:0]    result4 ;
wire      [15:0]    result5 ;
wire      [15:0]    result6 ;
wire      [15:0]    result7 ;
wire      [15:0]    result8 ;
wire      [14:0]    sum1 ;
wire      [11:0]    dct ;
wire      [63:0]    do ;
wire      [63:0]    d1 ;
wire      [63:0]    d2 ;
wire      [10:0]    qr0 ;
wire      [10:0]    qr1 ;
wire      [10:0]    qr2 ;
wire      [10:0]    qr3 ;
wire      [10:0]    qr4 ;
wire      [10:0]    qr5 ;
wire      [10:0]    qr6 ;
wire      [10:0]    qr7 ;
wire      [18:0]    res1 ;
wire      [18:0]    res2 ;
wire      [18:0]    res3 ;
wire      [18:0]    res4 ;
wire      [18:0]    res5 ;
wire      [18:0]    res6 ;
wire      [18:0]    res7 ;
wire      [18:0]    res8 ;
wire      [5:0]     cnt1_reg ;
wire      [2:0]     cnt2_reg ;
wire      [2:0]     cnt3_reg ;
wire      [5:0]     cnt4_reg ;
wire      [7:0]     qout ;
```

```
// Dual RAM to read image input block (8 × 8 pixels) by block.
// This is the X matrix in the algorithm.

dualram dualram1 (          .clk(clk),
                            .pci_clk(pci_clk),
                            .rnw(rnw),
                            .be(be),
                            .ra(cnt1_reg[2:0]),
                            .wa(wa),
                            .di(di),
                            .din_valid(din_valid),
                            .do(do)
          );
// Dual RAM has two pipeline registers, one after "ram_rc.v" and the other at
// the output of dualram.v.
/*
The following module is a ROM storing 2C and 2C^T (two times C/C^T in order to
improve the accuracy). C and C^T are accessed simultaneously. Both require row
accesses for the computation of DCT.
*/
romc  romc1 (               .clk(clk),
                            .addr1(cnt1_reg[5:3]),
                            .addr2(cnt3_reg[2:0]),
                            .dout1(d1),
                            .dout2(d2)
                  ) ;
/*
 ROM (romc.v) also has two pipeline registers, to keep pace with "dualram".
"addr1" and "addr2" are for fetching C and C^T matrices. C is used in the first stage
multiplication, u11–u18, while C^T is used in second stage multiplication, u21–u28.
CX is computed using the following eight multipliers. "do" is the image input,
"X" and "d1", is the C input. "do" is unsigned, while "d1" is in twos complement.
"result" is in twos complement.
*/

mult8ux8s  u11(             .clk(clk),
                            .n1(do[63:56]),
                            .n2(d1[63:56]),
                            .result(result1)          //16-bit signed
                  ) ;
mult8ux8s  u12(             .clk(clk),
                            .n1(do[55:48]),
                            .n2(d1[55:48]),
                            .result(result2)          //16-bit signed
                  ) ;
```

```
mult8ux8s  u13(          .clk(clk),
                         .n1(do[47:40]),
                         .n2(d1[47:40]),
                         .result(result3)          //16-bit signed
              );
mult8ux8s  u14(          .clk(clk),
                         .n1(do[39:32]),
                         .n2(d1[39:32]),
                         .result(result4)          //16-bit signed
              );
mult8ux8s  u15(          .clk(clk),
                         .n1(do[31:24]),
                         .n2(d1[31:24]),
                         .result(result5)          //16-bit signed
              );
mult8ux8s  u16(          .clk(clk),
                         .n1(do[23:16]),
                         .n2(d1[23:16]),
                         .result(result6)          //16-bit signed
              );
mult8ux8s  u17(          .clk(clk),
                         .n1(do[15:8]),
                         .n2(d1[15:8]),
                         .result(result7)          //16-bit signed
              );

mult8ux8s  u18(          .clk(clk),
                         .n1(do[7:0]),
                         .n2(d1[7:0]),
                         .result(result8)          //16-bit signed
              );
// Partial products of CX are added here.
adder12s adder12s1(
                        .clk(clk),
                        .n0(result1[15:4]),  // 12-bit signed  Ex.: (156).0010
                        .n1(result2[15:4]),  // Five pipeline stages – output
                                             // not registered
                        .n2(result3[15:4]),  // since it is registered in the
                                             // following dctreg1.
                        .n3(result4[15:4]),
                        .n4(result5[15:4]),
                        .n5(result6[15:4]),
                        .n6(result7[15:4]),
                        .n7(result8[15:4]),
                        .sum(sum1)
          );
```

```
                // Three stage addition means 3 bits more.
                // Therefore, sum1[14:0] is 15-bit signed.

dctreg2x8xn #(11) dctreg1(                  // Partial product registers, p0–p7.
                .clk(clk),
                .din(sum1[14:4]),
                // 11-bit signed (integer) – dropping 3 bits after decimal point.
                // C was actually taken as 2C in the ROM and
                // hence 1 more bit is dropped. Similarly for CT.
                .wa(cnt2_reg[2:0]),
                .enreg(encnt2),
                .qr0(qr0),                  // This is "p0",
                .qr1(qr1),                  // "p1", etc.
                .qr2(qr2),
                .qr3(qr3),
                .qr4(qr4),
                .qr5(qr5),
                .qr6(qr6),
                .qr7(qr7)                   // "p7", 11-bit signed
                );
mult11sx8s u21( .clk(clk),
                .n1(qr0),
                        // 11-bit signed – Partial sum of product, "p0" of CX
                .n2(d2[63:56]),  // 8-bit signed CT
                .result(res1)    // 19-bit signed, [18:0]
               );
mult11sx8s u22( .clk(clk),
                .n1(qr1),        // 11-bit signed – Partial sum of  product of CX
                .n2(d2[55:48]),
                .result(res2)
                );
mult11sx8s u23( .clk(clk),
                .n1(qr2),        // 11-bit signed – Partial sum of  product of CX
                .n2(d2[47:40]),
                .result(res3)
                );
mult11sx8s u24( .clk(clk),
                .n1(qr3),        // 11-bit signed – Partial sum of product of CX
                .n2(d2[39:32]),
                .result   (res4)
                );
mult11sx8s u25( .clk(clk),
                .n1(qr4),        // 11-bit signed – Partial sum of product of CX
                .n2(d2[31:24]),
                .result(res5)
                );
```

```
mult11sx8s u26( .clk(clk),
                .n1(qr5),         // 11-bit signed – Partial sum of product of CX
                .n2(d2[23:16]),
                .result(res6)
                );
mult11sx8s u27( .clk(clk),
                .n1(qr6),         // 11-bit signed – Partial sum of product of CX
                .n2(d2[15:8]),
                .result(res7)
                );
mult11sx8s u28( .clk(clk),
                .n1(qr7),         // 11-bit signed – Partial sum of product, CX
                .n2(d2[7:0]),     // 8-bit signed
                .result(res8)     // 19-bit signed
                );
// Above multiplied results are added as follows:
adder14sr adder14sr1(      .clk(clk),
                           .n0(res1[18:5]), // 14-bit signed
                           .n1(res2[18:5]),
                           .n2(res3[18:5]),
                           .n3(res4[18:5]),
                           .n4(res5[18:5]),
                           .n5(res6[18:5]),
                           .n6(res7[18:5]),
                           .n7(res8[18:5]),
                           .dct(dct[11:0])
                                  // 12-bit signed, [11:0] – This is the DCT output.
                           );
// This module (adder14sr.v) has six pipeline stages – output is registered.

// Quantization stage – 64B ROM, decimal point before msbs of the
// quantization values.
romq romq1  (              // 16/Quantization value is fetched from ROM.
                .clk(clk),
                .a(cnt4_reg),
                .d(qout)
              );
mult12sx8u u31  (          .clk(clk),
                           .n1(dct[11:0]),
                           .n2(qout),
                           .dctq(dctq)
                );
/*
16/quantization value is multiplied with the DCT output above to get the final
9-bit DCTQ output.
```

```
"n1" is DCT, signed, 12 bits integer.
"n2" is unsigned, 8 bits, decimal point before msb.
"dctq[8:0]" conforms to JPEG/MPEG-1/MPEG-2 standards, etc.
"dctq_valid" is asserted whenever DCTQ is valid and "addr" provides the address
of the DCTQ coefficient.
addr = 0 for the DC coefficient, addr = 1 to 63 for AC  coefficients.
*/

// Following is the DCTQ Controller.
dctq_controller dctq_control1 (   .clk(clk),
                                  .reset_n(reset_n),
                                  .start(start),
                                  .hold(hold),
                                  .ready(ready),
                                  .rnw(rnw),
                                  .dctq_valid(dctq_valid),
                                  .encnt2(encnt2),
                                  .cnt1_reg(cnt1_reg),
                                  .cnt2_reg(cnt2_reg),
                                  .cnt3_reg(cnt3_reg),
                                  .cnt4_reg(cnt4_reg),
                                  .addr(addr) // This is essentially cnt5_reg[5:0].
                                  );
endmodule
```

Code for Partial Products, p0–p7, of CX Matrix

Verilog _code_13.4 presents the "dctreg2x8xn" module. This code is put in a file named "dctreg2x8xn.v". A set of eight numbers of 11-bit registers qr0 to qr7 are used to store the partial products of p0–p7 of CX. As usual, in the module declaration, we list all the inputs and outputs. "din" is the input of this module to receive partial products, which appear sequentially starting from p0. Every clock, a new data in the order, p0, p1, …, p7, will come into "din", which will have to be stored in the registers qr0 to qr7 in the same order. The partial products p0, p1, …, p7 are respectively addressed by the signal "wa" as 0, 1, …, 7. These are stored only when the enable signal "enreg" is set. The "controller" module takes care to activate these signals "wa" and "enreg" at the appropriate time.

As in "C" language, the Verilog permits declaring a variable as a parameter. We can assign different values to the parameter and use the variable elsewhere in the code. In the present code, the "WIDTH" of partial product registers is declared as 11 (bits). The "always" block is used to store the partial products at the positive edge of the clock, "clk". In this block, we simply assign the "din" value to the appropriate registers depending upon the write address, "wa". The address itself is one of the inputs. All these are processed only if the enable signal is high. These partial products must be stable until the end of next clock pulse. The reader may wonder, why extra registers, q0 through q6 are used. The explanation for this is as follows. With the arrival of the first clock, p0 (appearing at "din" pin) is stored in "q0" register. In subsequent six clocks, p1 to p6 are stored in "q1" to "q6" registers.

With the arrival of the eighth clock, the values in q0 to q6 are transferred to the registers qr0 to qr6. Simultaneously, the last partial product, p7, is also stored into "qr7". When the ninth clock arrives, "q0" will be overwritten by a new data. Similarly, other registers "q1" to "q6" will be overwritten in subsequent clocks. But we need the partial products p0–p7 to be stable for 8 clock cycles in order to compute the multiplications of p0–p7 with all the eight columns of C^T, without which DCT cannot be computed. This problem can be solved if we provide extra registers, "q0" to "q6". Although "q1" to "q6" registers are overwritten, the registers qr0 to qr7 will be stable for 9th to 16th clock cycles. They will be overwritten only with the arrival of sixteenth clock pulse.

Verilog _code_13.4

```
/*
Place this code in a file named "dctreg2x8xn.v".
A set of eight 11-bit registers (qr0–qr7) to store the partial products, p0–p7, of CX
is created in this module.
*/

module dctreg2x8xn (            clk,
                                wa,      // Pointer to p0–p7.
                                din,
                                enreg,
                                qr0,
                                qr1,
                                qr2,
                                qr3,
                                qr4,
                                qr5,
                                qr6,
                                qr7
                                );
        parameter          WIDTH = 11 ;   // Change this for any other size.

        output             [(WIDTH-1):0]  qr0;
        output             [(WIDTH-1):0]  qr1;
        output             [(WIDTH-1):0]  qr2;
        output             [(WIDTH-1):0]  qr3;
        output             [(WIDTH-1):0]  qr4;
        output             [(WIDTH-1):0]  qr5;
        output             [(WIDTH-1):0]  qr6;
        output             [(WIDTH-1):0]  qr7;
```

```
        input        [(WIDTH-1):0]  din;
        input        [2:0]          wa;
        input        enreg,
        input        clk;

        reg          [(WIDTH-1):0]  qr0;
        reg          [(WIDTH-1):0]  qr1;
        reg          [(WIDTH-1):0]  qr2;
        reg          [(WIDTH-1):0]  qr3;
        reg          [(WIDTH-1):0]  qr4;
        reg          [(WIDTH-1):0]  qr5;
        reg          [(WIDTH-1):0]  qr6;
        reg          [(WIDTH-1):0]  qr7;

        reg          [(WIDTH-1):0]  q0;
        reg          [(WIDTH-1):0]  q1;
        reg          [(WIDTH-1):0]  q2;
        reg          [(WIDTH-1):0]  q3;
        reg          [(WIDTH-1):0]  q4;
        reg          [(WIDTH-1):0]  q5;
        reg          [(WIDTH-1):0]  q6;

always @ (posedge clk)
begin
        if (enreg)
            begin
                case (wa)
                    3'b000: q0  <=   din;    // Register "p0" in clk(1)
                    3'b001: q1  <=   din;    // Register "p1" in clk(2),
                    3'b010: q2  <=   din;    // etc.
                    3'b011: q3  <=   din;
                    3'b100: q4  <=   din;
                    3'b101: q5  <=   din;
                    3'b110: q6  <=   din;    // Register "p6" in clk(7)
                    3'b111:
                        begin
                            qr0    <=    q0;   // Register "p0–p7" in
                            qr1    <=    q1;   // clk(8).
                            qr2    <=    q2;
                            qr3    <=    q3;
                            qr4    <=    q4;
                            qr5    <=    q5;
                            qr6    <=    q6;
                            qr7    <=    din;
                        end
                endcase
```

```
            end
    end
endmodule
```

DCTQ Controller Code

The controller is the last module in the DCTQ design, named "dctq_controller" as presented in Verilog_code_13.5. Put the module in a file called "dctq_controller.v". This module generates all the control and handshake signals required for various other modules for effective computation of DCTQ coefficients. We have five counters, "cnt1_reg" to "cnt4_reg" and "addr" in the controller design. We need pre-incremented signals for these counters in order to increase the processing speed of the DCTQ Processor, and they are realized in the first five "assign" statements. The first four counters are the pipeline registers and counter "addr" serves as the address of the DCTQ coefficients. Each of the five counters mentioned earlier has enable signals. These signals, "encnt1" to "encnt5" are realized by five "always" sequential blocks. The following two statements are for generating the advanced enable and disable signals respectively for the first counter:

assign encnt1_next = ((start_reg1 == 1'b1)&&(cnt1_reg == 0)) ? 1'b1 : 1'b0;
assign discnt1_next = ((start_reg1 == 1'b0)&&(cnt1_reg == 6'd63)) ? 1'b1:1'b0;

When the "start_reg1", which is derived from the "start" input of the top design "dctq", is high and the cnt1_reg is 0, then the first counter is enabled. This condition occurs right at the beginning when the user asserts the "start" signal (after reset signal is withdrawn) to commence the DCTQ computation. When the start_reg1 is low (signifying that the input "start" is withdrawn to terminate the DCTQ processing) and cnt1_reg is 63, that is, when the DCTQ engine completes processing the last coefficient of a block, then the counter "cnt1_reg" is disabled.

It is always a good practice to have only one signal in one "always" block. This practice is in vogue in industries. If we mix a number of signals in one always block, we are likely to end up in a mess. Only on rare occasions, we may have two or three closely related signals in an always block. Note that all the "always" sequential blocks have their respective signals initialized when system reset (reset_n) is applied, and the signals' value are frozen so long as "hold" signal is asserted. The enable signals, "encnt2" to "encnt5", are activated when the first counter "cnt1_reg" is respectively 14, 20, 35, and 44 in order to allow the counters "cnt2_reg" to "addr" to start running at the appropriate time. As described earlier, "discnt1_next" disables all the enable signals, "encnt1" to "encnt5". The enable signals are followed by "always" blocks for counters "cnt2_reg" to "addr", which are self-explanatory. "cnt1_reg" to "cnt4_reg" serve as the read addresses for the modules, "dualram", "romc", "dctreg2x8xn" and "romq", and "addr" serves as the address of the DCTQ coefficients as was described earlier in the top design, "dctq".

We will now see how the "rnw" signal works. This signal is for configuring one block (64B) of RAM in the "dualram" module in write-only mode and the other block of RAM in read-only mode. In the "rnw" always block, toggle "rnw" after the first block of RAM is written, which is kept track of by signals, “swrnw1” and “cnt_0 = 1”. “cnt_0” is cleared if “swrnw1” is asserted. Also, toggle “rnw” after every DCTQ block is processed. This is based on “swrnw2”. The next always block is the “ready” block, which informs the host processor that the DCTQ processor is ready to accept image input block. This signal reacts within 2 clock cycles (owing to cnt1_reg = 1 after a block of DCTQ is processed).

The next always block is the DCTQ valid signal, “dctq_valid”. It is de-asserted when the system reset or hold signal is asserted. When the “cnt1_reg” is 44, then the DCTQ valid is set, and another signal, “dctq_prev signal” is also set. This signal keeps track of where the processing stopped after the hold signal was asserted. In order to continue from where DCTQ processing stopped earlier, the hold signal is de-asserted. This can be understood by studying the “dctq_valid” block carefully. The last always block is for the “start_reg1” signal, which we considered earlier. When the “start” input is asserted and “cnt1_reg” is ‘0’, then “start_reg1” will be set with the following clock pulse. If the “start” input is withdrawn and “cnt1_reg” is 62, then “start_reg1” is reset. This completes the controller module.

The DCTQ design has a total of 17 multipliers, two adders, Dual RAM, two ROMs, etc. and, all of them work concurrently with fresh data being input at every clock cycle. Once the pipeline (45 depth) is full, the DCTQ coefficients are issued one every clock cycle without a break. Two sets of eight multipliers and two adders work in parallel, thus providing a high throughput. Hence the design becomes massively parallel and highly pipelined, which features make it ably suited for implementation on an FPGA or as an ASIC.

Verilog _code_13.5

```
// Place this code in a file named “dctq_controller.v” file.
// This submodule generates all the control and handshake signals required for
// effective computation of DCTQ coefficients.

module dctq_controller (                              // Declare the submodule
                              clk,                    // and I/Os.
                              reset_n,
                              start,
                              hold,
                              ready,
                              rnw,
                              dctq_valid,
                              encnt2,
                              cnt1_reg,
                              cnt2_reg,
                              cnt3_reg,
                              cnt4_reg,
                              addr
                         );
```

```
        input                   clk ;
        input                   reset_n ;
        input                   start ;
        input                   hold ;
        output                  ready ;
        output                  rnw ;
        output                  dctq_valid ;
        output                  encnt2 ;
        output      [5:0]       cnt1_reg ;
        output      [2:0]       cnt2_reg ;
        output      [2:0]       cnt3_reg ;
        output      [5:0]       cnt4_reg ;
        output      [5:0]       addr ;

        reg                     ready ;
        reg                     rnw ;
        reg                     dctq_valid ;
        reg                     dctq_valid_prev ;
        reg                     start_reg1 ;
        reg                     cnt_0 ;
        reg         [5:0]       cnt1_reg ;
        reg         [2:0]       cnt2_reg ;
        reg         [2:0]       cnt3_reg ;
        reg         [5:0]       cnt4_reg ;
        reg         [5:0]       addr ;
        reg                     encnt1 ;
        reg                     encnt2 ;
        reg                     encnt3 ;
        reg                     encnt4 ;
        reg                     encnt5 ;

        wire                    start_next1 ;
        wire                    encnt1_next ;
        wire                    discnt1_next;
        wire                    swrnw1 ;
        wire                    swrnw2 ;
        wire                    swon_ready ;
        wire        [5:0]       cnt1_next ;
        wire        [2:0]       cnt2_next ;
        wire        [2:0]       cnt3_next ;
        wire        [5:0]       cnt4_next ;
        wire        [5:0]       cnt5_next ;

assign  cnt1_next       = cnt1_reg + 1 ;  // Increment counters in advance.
assign  cnt2_next       = cnt2_reg + 1 ;
assign  cnt3_next       = cnt3_reg + 1 ;
```

```
assign   cnt4_next          = cnt4_reg + 1 ;
assign   cnt5_next          = addr + 1 ;
assign encnt1_next  = ((start_reg1 == 1'b1)&&(cnt1_reg == 0)) ? 1'b1 : 1'b0 ;
assign discnt1_next = ((start_reg1 == 1'b0)&&(cnt1_reg == 6'd63)) ? 1'b1:1'b0 ;
                     // Conditions for enabling/disabling counter, "cnt1_reg".
always @ (posedge clk or negedge reset_n)
        begin                               // Generate enable for cnt1.
                if (reset_n == 1'b0)
                        encnt1  <=      1'b0 ;
                else if (hold == 1'b1)
                        encnt1  <=      encnt1 ;
                else if (encnt1_next == 1'b1)
                        encnt1  <=      1'b1 ;
                else if (discnt1_next == 1'b1)
                        encnt1  <=      1'b0 ;
                else
                        encnt1  <=      encnt1 ;
        end

always @ (posedge clk or negedge reset_n)
        begin                               // Generate enable for cnt2.
                if (reset_n == 1'b0)
                        encnt2  <=      1'b0 ;
                else if (hold == 1'b1)
                        encnt2  <=      encnt2 ;
                else if (discnt1_next == 1'b1)
                        encnt2  <=      1'b0 ;
                else if (cnt1_reg == 6'd14)
                                // cnt2 is enabled when cnt1_reg = 14 dec.
                        encnt2  <=      1'b1 ;
                else
                        encnt2  <=      encnt2 ;
        end

always @ (posedge clk or negedge reset_n)
        begin                               // Generate enable for cnt3.
                if (reset_n == 1'b0)
                        encnt3  <=      1'b0 ;
                else if (hold == 1'b1)
                        encnt3  <=      encnt3 ;
                else if (discnt1_next == 1'b1 )
                        encnt3  <=      1'b0 ;
                else if (cnt1_reg == 6'd20)
                                // cnt3 is enabled when cnt1_reg = 20 dec
                                // since ROM C^T has two pipeline stages.
                        encnt3  <=      1'b1 ;
                else
```

```
                    encnt3   <=        encnt3 ;
        end

always @ (posedge clk or negedge reset_n)
        begin                              // Generate enable for cnt4.
                if (reset_n == 1'b0)
                    encnt4   <=        1'b0 ;
                else if (hold == 1'b1)
                    encnt4   <=        encnt4 ;
                else if (discnt1_next == 1'b1)
                    encnt4   <=        1'b0 ;
                else if (cnt1_reg == 6'd35)
                            // cnt4 is enabled when cnt1_reg = 35 dec.
                    encnt4   <=        1'b1 ;
                else
                    encnt4   <=        encnt4 ;
        end

always @ (posedge clk or negedge reset_n)
        begin                              // Generate enable for cnt5.
                if (reset_n == 1'b0)
                    encnt5   <=        1'b0 ;
                else if (hold == 1'b1)
                    encnt5   <=        encnt5 ;
                else if (discnt1_next == 1'b1)
                    encnt5   <=        1'b0 ;
                else if (cnt1_reg == 6'd44)
                            // cnt5 is enabled when cnt1_reg = 44 dec.
                    encnt5   <=        1'b1 ;
                else
                    encnt5   <=        encnt5 ;
        end

always @ (posedge clk or negedge reset_n)

        begin                                    // Realize cnt1.
                if (reset_n == 1'b0)
                    cnt1_reg <=        6'd00 ;
                else if (hold == 1'b1)
                    cnt1_reg <=        cnt1_reg ;
                else if (encnt1 == 1'b1)
                    cnt1_reg <=        cnt1_next ;
                else
                    cnt1_reg <=        cnt1_reg ;
        end

always @ (posedge clk or negedge reset_n)
        begin                                    // Realize cnt2.
```

```
                if (reset_n == 1'b0)
                        cnt2_reg <=       6'd00 ;
                else if (hold == 1'b1)
                        cnt2_reg <=       cnt2_reg ;
                else if (encnt2 == 1'b1)
                        cnt2_reg <=       cnt2_next ;
                else
                        cnt2_reg <=       cnt2_reg ;
        end

always @ (posedge clk or negedge reset_n)
        begin                                   // Realize cnt3.
                if (reset_n == 1'b0)
                        cnt3_reg <=       6'd00 ;
                else if (hold == 1'b1)
                        cnt3_reg <=       cnt3_reg ;
                else if (encnt3 == 1'b1)
                        cnt3_reg <=       cnt3_next ;
                else
                        cnt3_reg <=       cnt3_reg ;
        end

always @ (posedge clk or negedge reset_n)
        begin                                   // Realize cnt4.
                if (reset_n == 1'b0)
                        cnt4_reg <=       6'd00 ;
                else if (hold == 1'b1)
                        cnt4_reg <=       cnt4_reg ;
                else if (encnt4 == 1'b1)
                        cnt4_reg <=       cnt4_next ;
                else
                        cnt4_reg <=       cnt4_reg ;
        end

always @ (posedge clk or negedge reset_n)
        begin                                   // Realize "cnt5" or "addr".
                if (reset_n == 1'b0)
                        addr     <=       6'd00 ;
                else if (hold == 1'b1)
                        addr     <=       addr ;
                else if (encnt5 == 1'b1)
                        addr              <=      cnt5_next ;
                else
                        addr              <=      addr ;
        end

assign swrnw1 = ((start_reg1 == 1'b1)&&(cnt1_reg == 0)&&(cnt_0 == 1'b1))
                                                        ? 1'b1 : 1'b0 ;
```

```
assign swrnw2 = ((start_reg1 == 1'b1)&&(cnt1_reg == 63)) ?1'b1 : 1'b0 ;

always @ (posedge clk or negedge reset_n)
        begin
            if (reset_n == 1'b0)
                begin
                         cnt_0   <=      1'b1 ;
                        rnw      <=      1'b1 ;
                end
            else if (hold == 1'b1)
                        rnw      <=      rnw ;
            else if (swrnw1)
                begin
                         cnt_0   <=      1'b0 ;
                         rnw     <=      !rnw ;    // Toggle after the first
                                                   // block of RAM is written.
                end
            else if (swrnw2)
                         rnw     <=      !rnw ;    // Toggle after every DCTQ
                                                   // block is processed.
            else
                         rnw     <=      rnw ;
        end

assign swon_ready = ((start_reg1 == 1'b1)&&(cnt1_reg ==6'd01)) ? 1'b1 : 1'b0 ;

always @ (posedge clk or negedge reset_n)
          begin
                    if (reset_n == 1'b0)
                             ready  <=      1'b1;
                    else if (hold == 1'b1)
                             ready  <=      ready ;
                    else if (swon_ready)
                              ready <=      1'b1;
                    else
                             ready  <=      !start_reg1 ;
          end
always @ (posedge clk or negedge reset_n)

begin
           if (reset_n == 1'b0)
               begin
                    dctq_valid_prev <=      1'b0 ;
                    dctq_valid      <=      1'b0 ;
               end
           else if (hold == 1'b1)           // Asserting hold clears
               begin
```

```
                dctq_valid       <=    1'b0 ;    // valid signal.
            end
        else if (cnt1_reg == 6'd44)            // DCTQ is valid from cnt1_reg = 44
                                               // onwards.
            begin
                dctq_valid       <=    1'b1 ;
                dctq_valid_prev  <=    1'b1 ;
            end
        else if (hold == 1'b0)
                dctq_valid       <=    dctq_valid_prev ;
        else
                dctq_valid       <=    dctq_valid ;
end
assign start_next1 = (start == 1'b1)&&(cnt1_reg == 0) ;
always @ (posedge clk or negedge reset_n)
begin
        if (reset_n == 1'b0)
                start_reg1       <=    1'b0 ;
        else if (hold == 1'b1)
                start_reg1       <=    start_reg1 ;
        else if (start_next1)
                start_reg1       <=    1'b1 ;
        else if ((start == 1'b0)&&(cnt1_reg == 6'd62))
                start_reg1       <=    1'b0 ;
        else
                start_reg1       <=    start_reg1 ;
        end
endmodule
```

13.2.4 Test Bench for the DCTQ Design

Verilog_code_13.6 presents the test bench for the DCTQ Design. As usual, we will run the simulation at 100 MHz and, therefore, we have a "clock period divided by two" variable assigned the value of 5 ns. We define a variable "NUM_BLKS" as 1024 in order to indicate the number of blocks in a picture such as "Lena" of size, 256 × 256 pixels. We also include the design module, "dctq". The test bench is declared as "dctq_test". All inputs are declared as "Reg" and outputs as "wire". We declare "i", the current number of blocks processed and "fp1", the handle of the output file as integers. The following statement defines the register buffer to accommodate an image or one frame of a video sequence:

```
        reg        [63:0]        mem  [8191:0] ;
```

The explanation of this statement is offered in the comments of the test bench.

Next, we invoke the design, "dctq", calling ports in the design by name. This is followed by an "initial" block. The statement, $readmemh ("lena.txt", mem); reads the disk file of the image, "lena.txt", in hex format into the buffer named "mem", which was described earlier. In a later section, we will see how to get an image text file such as "lena.txt" using Matlab. We also identify the output disk file, "dctq.txt", which will contain the DCTQ output after running the test bench of the design in the simulator. After initializing various signals, we apply the reset followed by the start data input (start_din) signal. After running the simulation for a specified time, the DCTQ output file is closed and processing stopped. By that time, the simulation would have processed one frame or an image. The next two always statements are for running the two clocks.

The always block, after the two assign statements, is processed only when any of the inputs listed therein changes and not otherwise. With the arrival of "pci_clk" after "start_din" is set and the running counter "i", which keeps track of the number of blocks of image remaining to be processed, is not "0", then the data input valid signal "din_valid" is asserted, the write address "wa" is initialized to point to the first row of an image block and read them into "di" input of the "dctq" design. "wa" is also incremented to address the next row of the image block. The above process is repeated seven more times to complete writing a whole image block in 8 clock cycles. With the arrival of the next clock, "din_valid" is de-asserted and waits for DCTQ Processor to be "ready". When the next clock arrives, "start" signal is asserted for the DCTQ engine to commence the processing and the block counter "i" is decremented. Since "i" has changed, control reverts back to start of the "always" block to repeat the above process till all the blocks are processed. When the last block is processed, "i" reduces to "0" and the control branches to wait for the "eobcnt_reg" to equal the `NUM_BLKS. When this condition is met, the DCTQ output file "dctq.txt" is closed and the process stops.

"stopproc" signal indicates the condition that the last block is already processed, and the next sequential block senses this condition to disable writing of the DCTQ output file. The end of block "eob" is sensed when the DCTQ address, "addr", is 63, meaning that it has processed all the 64 coefficients in a block. This is processed in a sequential block for "eob". The next sequential block is a simple counter, "eobcnt_reg", to advance the same whenever a block (eob = 1) is processed. This completes the test bench module.

Verilog_code_13.6

```
// This is the test bench for the DCTQ Design. Input image is "lena.txt".
// Change it for processing a different image.
// dctq.txt is the DCTQ output of the image, lena.txt.
//  File name: "dctq_test.v".
`define clkperiodby2 5          // Both clocks clk & pci_clk operate at 100 MHz.
`define pci_clkperiodby2 5
`define NUM_BLKS 1024     // Defines number of blocks in a frame. A 256 × 256
                          // pixel picture contains 1024 blocks.
```

```
                                    // Change this for a different image size.
`include "dctq.v"                   // Design module.

module dctq_test ;                  // Declare the test bench and inputs.

    reg                             pci_clk ;
    reg                             clk ;
    reg                             reset_n ;
    reg                             start ;
    reg         [63:0]              di ;
    reg                             din_valid ;
    reg         [2:0]               wa ;
    reg         [7:0]               be ;
    reg                             hold ;

    wire                            ready ;
    wire        [8:0]               dctq ;          // DCTQ output.
    wire                            dctq_valid ;
    wire        [5:0]               addr ;
    wire                            stopproc ;
    reg                             eob ;
    wire        [10:0]              eobcnt_next ;
    reg         [10:0]              eobcnt_reg ;
    reg                             start_din ;
// Change the above two "eobcnt" statements for a different image size,
// sufficient to accommodate the total number of blocks in a frame.
integer   i ;           // Keeps track of the current number of blocks processed.
integer   fp1 ;         // Points the DCTQ output file.

reg    [63:0]   mem  [8191:0] ; // Buffer to accommodate one frame.
reg    [12:0]   mem_addr ;      // 13 bits address to accommodate
                                // up to 8191.
// reg  [63:0] contains one row (8 pixels) of an image block – eight such rows make
// one block; 1024 such blocks mean 8192 rows. Change mem [8191:0] and
// reg[12:0] for a different image size.

dctq    dctq1(       // Invoke DCTQ design module to get the DCTQ output.
                .pci_clk(pci_clk),
                .clk(clk),
                .reset_n(reset_n),
                .start(start),
                .di(di),
                .din_valid(din_valid),
                .wa(wa),
                .be(be),
                .hold(hold),
```

```
                    .ready(ready),
                    .dctq(dctq),
                    .dctq_valid(dctq_valid),
                    .addr(addr)
                  ) ;
initial
begin
  $readmemh ("lena.txt", mem) ;
                          // "mem" receives the input image frame, lena.txt.
                          // Change the name for a different image frame.
  fp1 = $fopen ("dctq.txt") ;
                        // dctq.txt is the DCTQ output of the image frame, lena.txt.
          pci_clk        =       0 ;
          clk            =       0 ;
          reset_n        =       1 ;
          start          =       0 ;
          di             =       0 ;
          din_valid      =       0 ;
          wa             =       0 ;
          be             =       8'h00 ;          // Enable bytes to be written.
          hold           =       0 ;
          mem_addr       =       0 ;
          start_din      =       1'b0 ;
          i              =       `NUM_BLKS ;  // i = 1024
          #20 reset_n    =       1'b0 ;
          #40 reset_n    =       1'b1 ;
          start_din      =       1'b1;
         #700000                 // Run long enough to process the entire frame.
         $fclose(fp1) ;          // Close the output file and
         $stop ;                 // stop the simulation.
end

always
          #`clkperiodby2 clk          <= ~clk ;       // Run the two clocks.
always
         #`pci_clkperiodby2 pci_clk <=  ~pci_clk ;

always @ (start_din or i or clk or pci_clk or reset_n or wa or mem_addr)
begin
         if (start_din == 1'b1)
          begin
            @(posedge pci_clk)
                 if(i != 0)                              // Image block counter.
                   begin
                        @(posedge pci_clk) ;
                        #1 ;
```

```
                        din_valid = 1 ;
                        wa = 0 ;
                        di = mem[mem_addr] ;        // Inputs first row of an
                                                    // image  block.
                        mem_addr = mem_addr + 1 ;
                    end

                  repeat(7)
                    begin
                        @(posedge pci_clk) ;
                        #1 ;
                        din_valid = 1 ;
                        wa = wa + 1 ;
                        di = mem[mem_addr] ;   // Inputs second to eight rows
                                               // of the image block.
                        mem_addr = mem_addr + 1 ;
                    end
                        @(posedge pci_clk) ;
                        #1 ;
                        din_valid = 0 ;
                        wait (ready) ;         // Wait for ready to go high.
                        @(posedge clk) ;
                        #1 start = 1'b1 ;      // Start the DCTQ process after
                                               // inputting the image block and
                                               // when ready signal is high.
                         i = i -1 ;            //Address the next image block.
            end
          else
            begin
              wait(eobcnt_reg==`NUM_BLKS);
                                        // Completion of all the image blocks.
              $fclose(fp1) ;
              $stop ;
            end
end

assign stopproc =((eobcnt_reg==`NUM_BLKS-1)&&(eob== 'b1)) ? 1'b1 : 1'b0 ;
                                  // Condition to stop DCTQ processing.
always @ (posedge clk)
begin
          if(dctq_valid == 1'b1)
            begin
                  if (stopproc == 1'b0)       // Means, the process has not stopped.
                  $fdisplay(fp1,"%h", dctq) ;
                                        // DCTQ coefficients are written into
```

```
                                        // the "dctq" output file every time the DCTQ is
                                        // valid. Don't write into "dctq.txt" file when
                                        // all the coefficients are already written.
        end
end

always @ (posedge clk or negedge reset_n)
begin
        if (reset_n == 1'b0)
                eob             <=      1'b0 ;
        else if (addr == 6'd63)
                eob             <=      1'b1 ;  // End of block is issued
                                                // when the last coefficient of
                                                // a block is processed.
        else
                eob             <=      1'b0 ;
end

assign eobcnt_next = eobcnt_reg + 1 ;   // Count the number of blocks
                                        // processed.
always @ (posedge clk or negedge reset_n)
        begin
         if (reset_n == 1'b0)
                eobcnt_reg      <= 0 ;
         else if (eob == 1'b1)
                eobcnt_reg      <= eobcnt_next ;
        end

endmodule
```

13.2.5 Simulation Results for DCTQ Design

The simulation results of DCTQ design are presented in Figures 13.6.1 to 13.6.7. Observing Figure 13.6.1, we see that the first block of data is applied soon after the reset is withdrawn. The "di" input is of width 8 bytes and changes at every positive edge of "pci_clk". So also the corresponding write address, "wa". Thus in eight clock pulses, one block of image data is written. Note that during the application of this input, "din_valid" signal is asserted. "be" signals are active (low) in order to write all the 8 bytes in a row of image. Soon after one block of image is applied, the "start" signal is asserted to commence the DCTQ operation. At 215 ns, the "rnw" toggles and the next block of data is written into a second set of "dualram". This is a concurrent operation to the DCTQ operation. All the counters are cleared when "reset_n" pulse is applied. The counter, "cnt1_reg", starts operation immediately after "rnw" goes low.

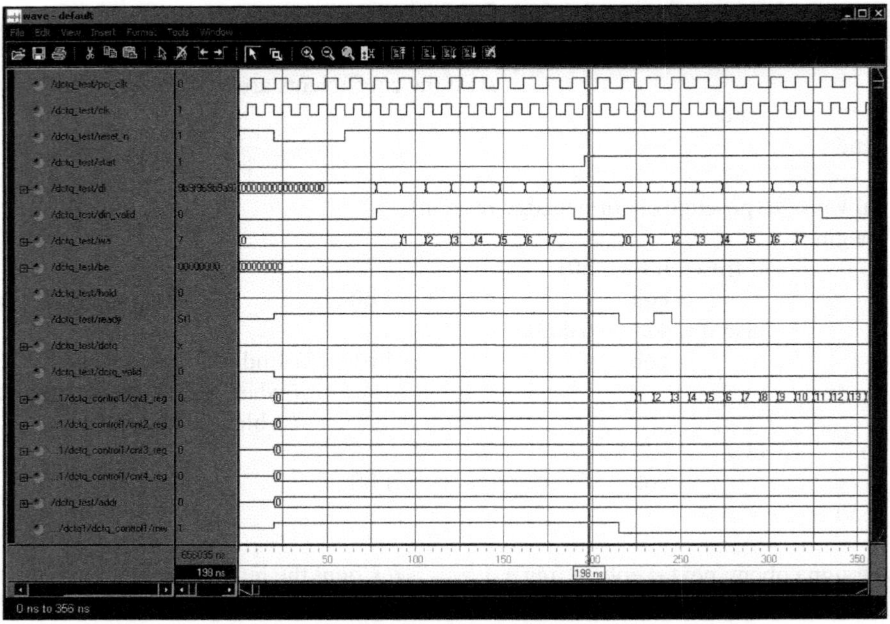

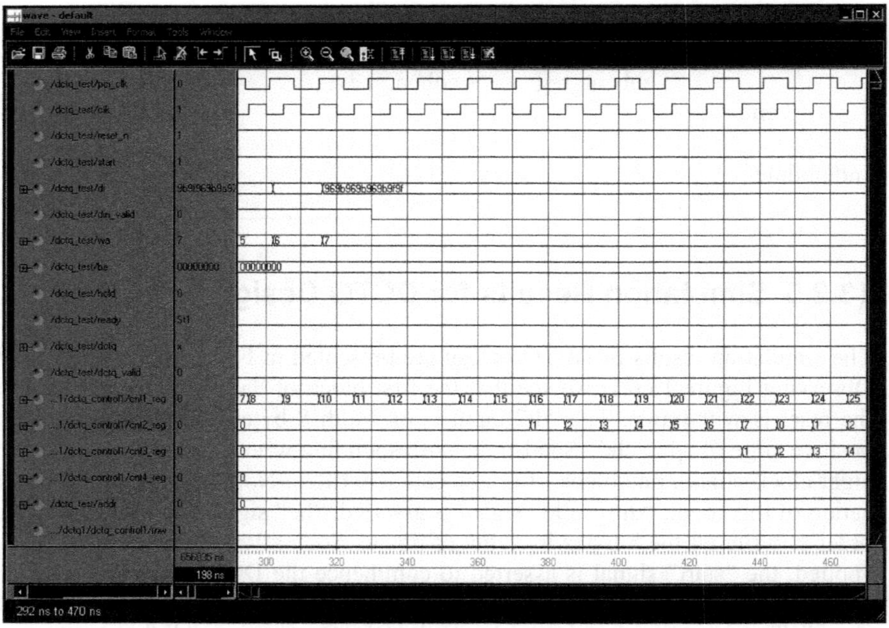

Fig. 13.6.1 and 13.6.2 Simulation results of DCTQ (Continued)

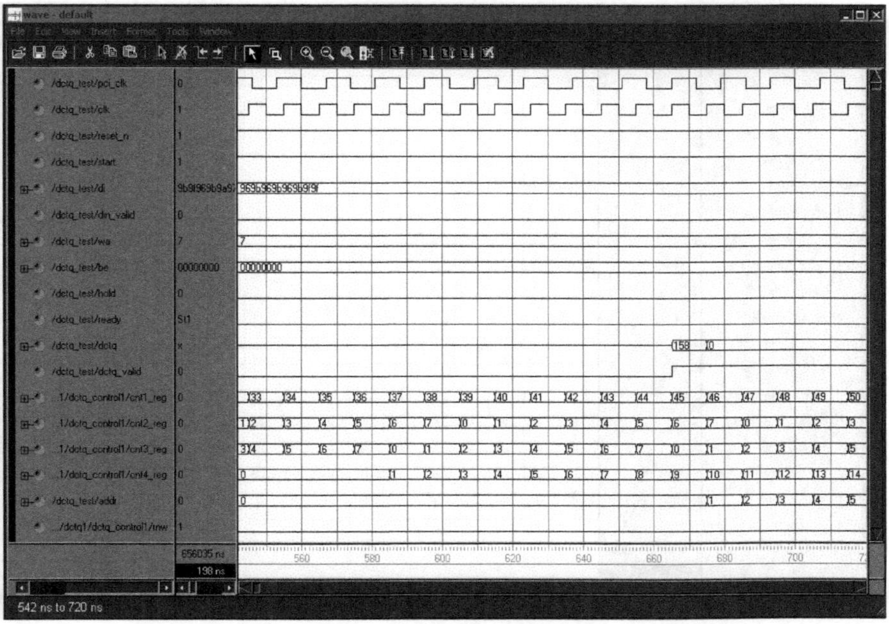

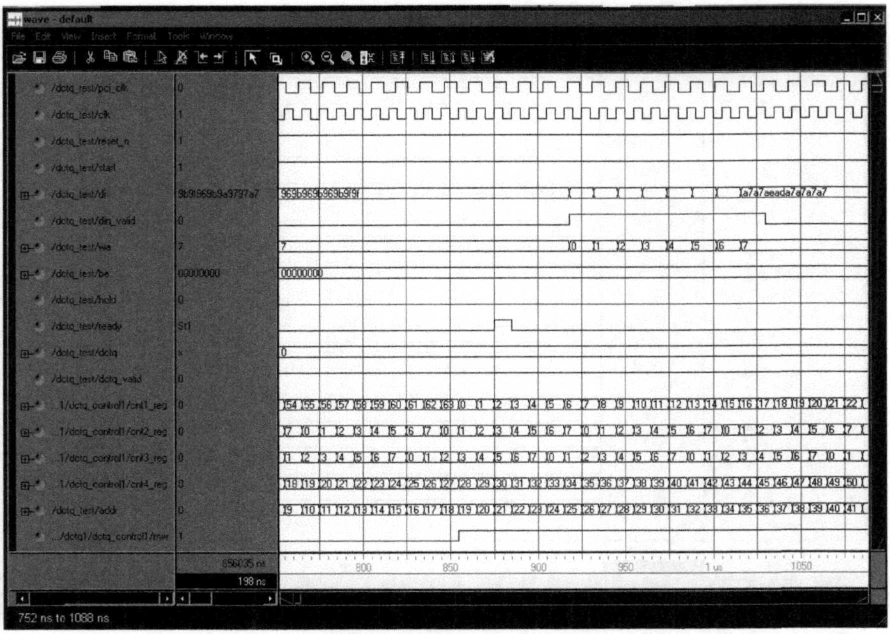

Fig. 13.6.3 and 13.6.4 Simulation results of DCTQ (Continued)

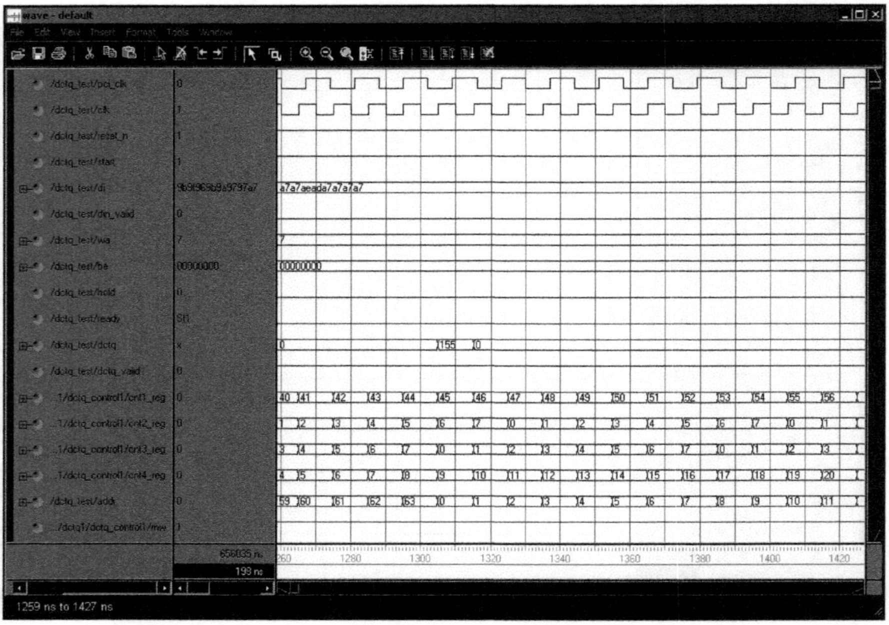

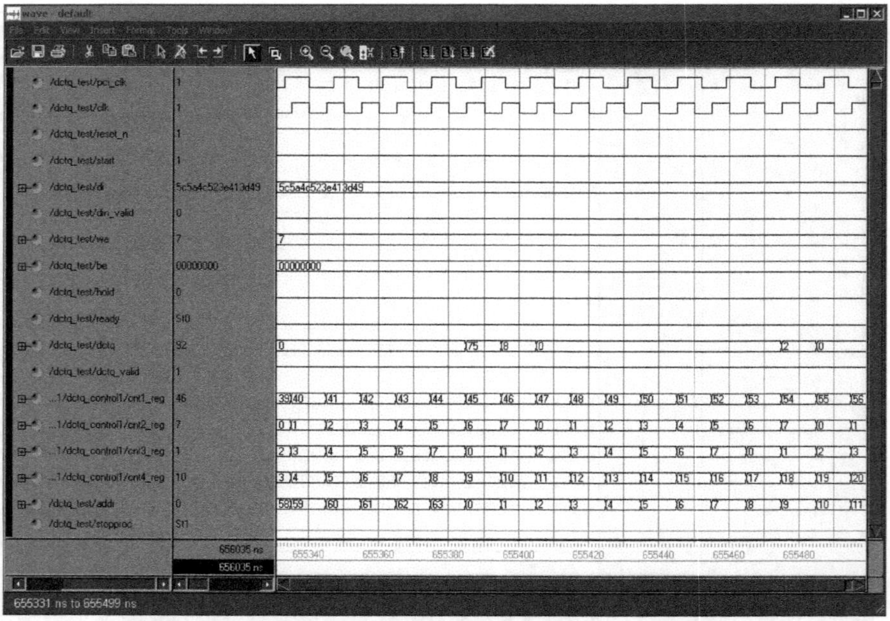

Fig. 13.6.5 and 13.6.6 Simulation results of DCTQ (Continued)

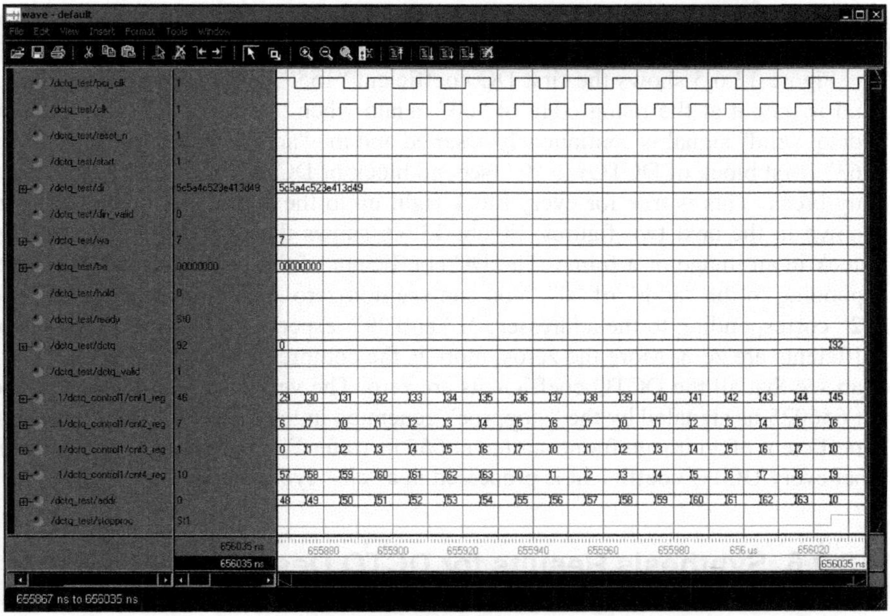

Fig. 13.6.7 Simulation results of DCTQ

Figure 13.6.2 shows that the counters "cnt2_reg" and "cnt3_reg" start working when "cnt1_reg" becomes 14 and 20 respectively, effective at the positive edge of "clk". Note that "rnw" continues to be low since the DCTQ for the first block of data is not yet computed. Similarly, "cnt4_reg" and "addr" start working when "cnt1_reg" becomes 35 and 44 respectively, effective at the positive edge of "clk" as can be seen in Figure 13.6.3. The first DCTQ coefficient (DC) "158" is output at the rising edge of "clk" when "cnt1_reg" is 44. Simultaneously, the "dctq_valid" signal goes high. The coefficient address is provided by "addr" signal. The latency is, therefore, 45 "clk" cycles. Note that the DCTQ is issued one coefficient per "clk" cycle thereafter, without any break.

Figure 13.6.4 shows that all AC coefficients are "0", which is not coded in the next functional module, "VLC", of the video encoder, thus bringing about good deal of compression. When "cnt1_reg" (which serves as the read address of "dual-ram" module) is 63, it resets and toggles the "rnw" signal immediately, thus reading the second block of image data without any break. Note the continuing progress of other counters including the DCTQ address counter, all of which, except "addr", serve as address pointers for various modules of the DCTQ engine. This reminds us of a perfectly synchronized orchestra, where every musician maintains perfect harmony with other performers. Evidently, counter based design proves invaluable in a "controller" of a system, which acts like the "music conductor" in an orchestra performance. Incidentally, the counter based design takes the least possible chip area. When "cnt1_reg" is "1", the DCTQ processor asserts the "ready" signal and the host processor sends in the next (third) block of image data.

Thus, the host processor has plenty of spare time to attend to other domestic chores.

Figure 13.6.5 shows the first DC coefficient "155" of the second image block and is output at the rising edge of "clk" again when "cnt1_reg" is 44. Note that "dctq_valid" signal is continuously asserted and the "addr" smoothly transits from "63" (first block of DCTQ) to "0" (second block of DCTQ) and continues without any break. This is true for every block right up to the end of the image/frame as shown in the next two figures. Figure 13.6.6 shows the DCTQ result of the last block in an image or a frame. The DC coefficient of the last block is "75" corresponding to the "addr" of "0". You can see non-zero AC coefficients of "8" and "2" corresponding to the addresses, "1" and "9" respectively. Majority of AC coefficients are zero. More the zeros, more is the compression. In Figure 13.6.7, we can see that all the DCTQ coefficients are zero. The very last coefficient is issued at 656025 ns signaled by the "stopproc" generated in the test bench, while the first coefficient of an image/frame started at 665 ns (refer Figure 13.6.3). This confirms that every DCTQ coefficient is processed in a "clk" cycle.

13.2.6 Synthesis Results for DCTQ Design

The device we have mapped is XCV600EHQ240-8, the same device we used for various modules presented in earlier chapters. The frequency of operation reported is around 100 MHz. The number of 16×1 RAMs consumed in the design is 128 and the number of LUTs is 3728, which is around 25% of the total capacity. Synplify generates "dctq.edf" output, which is input to the P & R tool. The Synplify results are as follows.

Worst slack in design: 9.829

Clock starting	Requested frequency	Estimated frequency	Requested period	Estimated period
clk	50.0 MHz	98.3 MHz	20.000	10.171
pci_clk	50.0 MHz	101.1 MHz	20.000	9.890

Resource usage report for dctq
Mapping to part: xcv600ehq240-8
Cell usage:

VCC	21	uses
GND	23	uses
MUXCY_L	1616	uses
XORCY	1706	uses
FDCE	33	uses
FDP	3	uses

FDPE	2	uses
FDC	7	uses
MUXCY	168	uses
FDR	1459	uses
FD	3805	uses
MUXF5	266	uses
MUXF6	131	uses
FDE	1317	uses
FDS	38	uses
FDRS	26	uses

I/O primitives:

IBUF	79	uses
OBUF	17	uses
BUFGP	2	uses

SRL primitives:

SRL16	144	uses

I/O Register bits: 11

Register bits not including I/Os: 6679 (48%)

RAM/ROM usage summary

Single Port Rams (RAM16X1S): 128

Global Clock Buffers: 2 of 4 (50%)

Total LUTs: 3728 (26%)

13.2.7 Place and Route Results for DCTQ Design

The Xilinx place and route results are as follows. In Xilinx P & R, we will get more optimized results. The input for Xilinx P & R tool is the "dctq.edf" that was generated by Synplify tool. It lists the slices, etc. The reader may note that the number of LUTs have come down to 3247. The number of gates consumed by the design is around 120,000 and maximum frequency of operation over 100 MHz.

"D:\USER\RAM\VERILOG_LATEST\DVLSI_DES_VERILOG\dctq_iqidct\rev_1\dctq.edf "

Using target part "v600ehq240-8".

Design Summary:

Number of errors:	0			
Number of warnings:	0			
Number of slices:	4,410	out of	6,912	63%
Number of slices containing unrelated logic:	0	out of	4,410	0%
Number of slice flip flops:	6,677	out of	13,824	48%
Total number of four input LUTs:	3,676	out of	13,824	26%
Number used as LUTs:	3,247			
Number used as a route-thru:	157			
Number used as 16 × 1 RAMs:	128			

Number used as shift registers	144			
Number of bonded IOBs:	96	out of	158	60%
IOB flip flops:	11			
Number of GCLKs:	2	out of	4	50%
Number of GCLKIOBs:	2	out of	4	50%
Total equivalent gate count for design:	119,451			
Additional JTAG gate count for IOBs:	4,704			

Timing summary:

Minimum period: 9.766 ns (Maximum frequency: 102.396 MHz)
Minimum input arrival time before clock: 12.156 ns
Minimum output required time after clock: 8.911 ns

13.2.8 Matlab Codes for Pre-processing and Post-processing an Image

An image or a frame of a video sequence is in raw format and is organized as a contiguous location of bytes (see Figure 13.7a) if it is a monochrome image. These appear in raster scan order, i.e., the same way we read a text: left to right; top to bottom (Figure 13.7b) when displayed. But we need an image in the block format (c) since the DCTQ processing is based on block, consisting of 8 pixels from eight different lines as depicted in Figure 13.7c for 256 × 256 pixels image, as an example. This format conversion from raster scan order to block order is presented in Matlab_code_13.1. This is put into a file named "read_image.m". The first statement is a function call for reading an image. The file name, number of rows, number of columns and the block size are specified, an example of which is shown in the following commented (%) statement. The input file and the output file are specified in the variables, file_in and file_out. "fp1" and "fp2" are handles

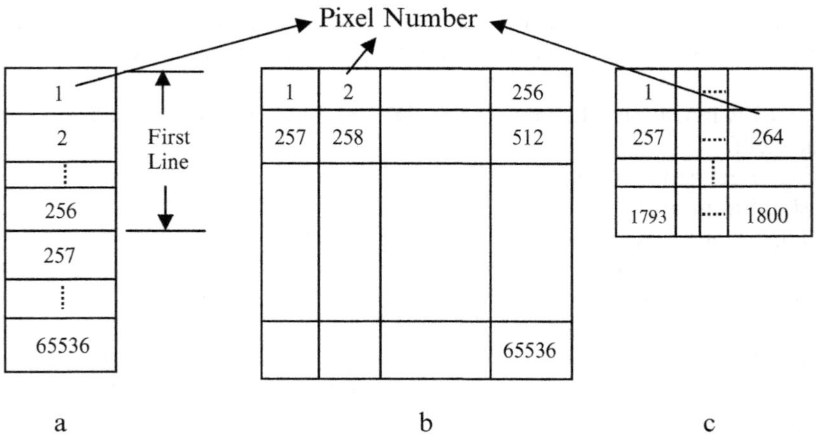

Fig. 13.7 (a) Contiguous raw image. (b) An image. (c) Block (8 × 8 pixels)

for the same, identifying the files as read-only or writeable types as shown. The input file, which is in raster scan order is read into the variable called “blks”. This is in a 1-D vector form. The image read function is then called in order to organize it as blocks. Refer “imageread.m” file for further details. The three “for” loops extract the data from an image or a frame block after block in “oneblk(i,j)” variable. Finally, call the function “mem” to write the block into the output file. “k” is the block counter. For example, if “lena.raw” is the input file, then processing this Matlab code produces an output file, “lena.txt” in block order.

Matlab_code_13.1

```
% This is the top module to read an image file in raw format.
% Put this in a file named "read_image.m".

function read_image (filename, rows, cols, blksize) ;
% Execute "read_image ('lena', 256,256,8)" for reading the input image, Lena.
file_in  =        [filename,'.raw']; % Specify the input image file in raw format.
file_out =        [filename,'.txt'];  % Output image file in block (text) format.
        fp1            =      fopen(file_in,'r') ;
        fp2            =      fopen(file_out,'w');
        image   =      fscanf(fp1,'%c') ;
                              % Read the input file which is in raster scan order.
        blks    =   imageread(image, rows, cols, blksize) ; % 1-D vector.
% Call the image read function to organize it into blocks. Refer "imageread.m"
% file for details.
        blkn = [];
        n = 1 ;                                % Number of pixels counter.
        for k = 1:1024,                        % 1024 blocks in an image frame.
                for i = 1:8,                   % Process one block along height after

                  for j = 1:8,                 % processing along width first.

                    oneblk(i,j) = blks(n) ; % Rearrange it as one block at a time.
                    n           = n + 1 ;   % Advance the pixel counter.
                  end
                end
                 mem(oneblk,0,fp2) ;
        % Call this function to write the block into the output file. "0" for
        % writing row-wise or "1" for column-wise writing. "fp2" is the output
        % image file in text format.
        end
fclose('all') ;
```

The function, "imageread.m", presented in Matlab_code_13.2, is for reading the image frame block by block from left to right and top to bottom. This function is called by "read_image.m" module presented in Matlab_code_13.1. This function comprises four "for" loops with the outer loop, "i", which selects succeeding block rows. One block row consists of eight lines (8 × 256 pixels) of the image as shown in Figure 13.7b. The next loop "j" covers all blocks (numbering 32) in a block row. The next two inner loops are for processing a block. "k" is for row addressing pixels and "l" is for column addressing pixels within a block. "x" is the output of this submodule, which is a 1-D vector.

Matlab_code_13.2

```
% Function "imageread.m" to read the image frame block by block from left to
% right and top to bottom. This function is called by "read_image.m" module.

function [x] = imageread(image, rows, cols, size)
x = [];
        for i = 0 : size*cols : rows*cols - size*cols ,
              % Ex.: 0:(8*256):(256*256-8*256) select succeeding block rows.
          for j = i + 1 : size : I + cols-size + 1,
                                   % Ex.: 1:8:249 – this covers all blocks in a row.
        % j is the starting element of a block => size = 8 for an 8 × 8 pixels block.
            for k = 0 : 1 : size-1,
                          % 0:1:7 => Block processing – row address within a block.
              for m = 0 : 1 : size-1,      % 0:1:7 – column address within a block.
                x = [x , image(j + k*cols + m) ];          % 1-D vector, appended.
  % (249 + 7 × 256 + 7 = 8 × 256) for the last pixel, as an example. In order to process
  % the next row of a block, we will have to skip 8*256 pixels.
              end
            end
          end
        end
```

Matlab_code_13.3 presents the function "mem.m" which is used to write 8 × 8 pixels image block as a hex ASCII string. This function is called by "read_image.m" module. This generates an output file of an image, "lena.txt" for instance, arranged in block order is to be used as input by Modelsim for processing DCTQ. "for" loops involving "i" and "j" fetch each pixel (in "num"), convert the pixel value in decimal into hex format (temp) and append it into the final 1-D output file "lena.txt", for example.

Matlab_code_13.3

```
% Function (mem.m) to write 8x8 pixels image block as a hex ASCII string.
```

```
% This function is called by "read_image.m" module.
% The output file is to be used by Modelsim for processing DCTQ.

function mem (blk8, col, fp) ;
                    % col is '0' for writing row-wise or "1" for column-wise writing.
for i = 1:8,
                    hexstr = [] ;
          for j = 1:8,
                    if (col == 1)
                        num = blk8(j,i) ;          % Select column-wise writing.
                    else
                        num = blk8(i,j) ;          % Otherwise, select row-wise writing.
                    end
            temp = dec2hex(num, 2);                % Character string – 2 digits.
           hexstr = [hexstr,temp] ;                % Append into the final 1-D output.
          end
fprintf(fp,'%s\n',hexstr) ;     % Write into the output file, each pixel in a new line.
end
%fclose(fp) ;
```

The DCTQ design (actually its test bench, "dctq_test.v") is run in Modelsim with "lena.txt" as an input generated by executing the above Matlab codes. The simulation output, "dctq.txt" is taken as the input for the inverse design (IQIDCT) of DCTQ design and simulation run to get a reconstructed image output, which is in block format. This output needs to be converted into raster scan order or "raw" format. Matlab_code_13.4 presents the function "write_image.m", which takes the reconstructed image output (produced by IQIDCT Verilog design in Modelsim) and converts the block format into raw format for display using "showim.m" file. It may be noted that IQIDCT Verilog design is not presented in this book and, therefore, the reader is expected to design the same before proceeding any further. However, "showim.m" Matlab code is presented in this section.

"file_in" is a variable that points to the "iqidct.txt" file produced by the IQIDCT design in simulator environment and "file_out" is the output file in the raw format. "iqidct.txt" is read into a vector named "blks" in decimal block format and the function "imagewrite.m" is called to finally store row-wise (raster scan order) into the output file. The function "imagewrite.m" is similar to the function "imageread.m" presented in Matlab_code_13.2 except that writing is carried out instead of reading. This is presented in Matlab_code_13.5.

Matlab_code_13.4

```
% This code (write_image.m) takes the reconstructed image output (produced
% by Modelsim and converts the block format into raw format for display using
% "showim.m" file.
```

```
        function [image] =        write_image(filename,rows,cols,blksize)
        file_in          =        [filename,'.txt'] ;
        file_out         =        [filename,'.raw'] ;
        fp1              =        fopen(file_in,'r') ;
        fp2              =        fopen(file_out,'w');
        blks             =        fscanf(fp1,'%d');
image = imagewrite(blks,rows,cols,blksize);      % Call function "imagewrite.m".
fprintf(fp2,'%c',image) ;                        % Store row-wise into the output file.
fclose('all');
```

Matlab_code_13.5

```
% This function (imagewrite.m) returns a vector that can be directly written
% into an image file. This is similar to "imageread.m" file.
function [image] =        imagewrite(blks,rows,cols,size)

n = 1 ;
for i = 0 : size*cols : rows*cols - size*cols ,
                                             % Ex.: 0:(8*256):(256*256-8*256).
                                             % Select succeeding block rows.
 for j = i+1 : size : i+cols-size+1          % j is the starting element of  a block.
                                             % 1:8:249 covers all blocks in a row.
  for k = 0 : 1 : size-1                     % 0:1:7 => Block processing
                                             % Row address within a block.
   for m = 0 : 1 : size-1                    % 0:1:7 – column address within a block.
          image(j + k*cols + m) = blks(n);         % 1-D vector, appended.
    n = n + 1;
   end

  end

 end

end
```

Running the Matlab_code_13.6, we can display an image, be it original or reconstructed. "file1" and "file2" are two such files. "file1" is read as characters into a variable, "im1". The first two "for" loops get the original stored image (im1) pixel by pixel, convert it into 8-bit unsigned integers and is stored in "image(i,j)". "n" keeps track of the number of pixels processed in the loops. Similarly, next two "for" loops get the reconstructed image (im2) pixel by pixel, convert it into 8-bit unsigned integers and is stored in "recon(i,j)". "figure(1)" identifies the first figure we are about to display. The statement: title("Original Image"); displays the title, "Original image". This is followed by displaying the actual original image, which was processed and stored in "image". Similar explanation holds good for displaying the reconstructed image.

Matlab_code_13.6

```
% This module (show_image.m) displays both the original and the
% reconstructed images.
% Run "psnr.m" file for the PSNR computation to get the image quality.

    function showim(file1, file2);
    %file1       =       'lena.raw';
    %file2       =       'iqidct.raw';
    fp1          =       fopen(file1,'r');
    im1          =       fscanf(fp1,'%c');
    n            =       1;
    for i = 1:256,
        for j = 1:256,
            image(i,j) = uint8(double(im1(n)));
                        % Convert it into 8-bit unsigned integers.
            n = n + 1;
        end
    end

    fp2          =       fopen(file2,'r');
    im2          =       fscanf(fp2,'%c');
    n            =       1;
    for i = 1:256,
        for j = 1:256,
            recon(i,j) = uint8(double(im2(n)));
            n = n + 1;
        end
    end
    figure(1) ;
    title("Original Image");          % Display the original image.
    imshow(image,256);
    figure(2);
    title("Reconstructed Image");     % Display the reconstructed image.
    imshow(recon,256);
    fclose("all");
```

Although visual comparison of the original and the reconstructed image can reveal unto us how the quality of the latter image is, it is always better to have a quantitative measure of the quality of the reconstructed image with reference to the original image. A popular expression for such a measure is "PSNR" and is as follows. The double summation is computed for 1 to M and 1 to N corresponding to the two dimensions of the image. M and N are respectively the width and the height of the image.

PSNR = 10*log10{((255^2) × M × N)/(ΣΣ ((orig-recon)^2)} (13.1)

(orig-recon) is the pixel-wise error. To start with, the original image and the reconstructed image are read in TIFF format. If they are in raw format, they can be changed to TIFF format using softwares such as Paintshop, XnView, etc. After converting the read files into double precision, obtain the size of the image, m and n. The next statement for "PSNR" is the core of this Matlab code. "." after the error, (orig-recon), implies element by element computation. Typical PSNR values are presented for an image. Normally, a very good quality, indistinguishable from the original image yields a PSNR value of 35 dB or higher, whereas a PSNR value of 30 dB or higher is reckoned as a good quality image. Reconstructed image with a PSNR value of 25 dB or lower is taken as a poor quality image. The Matlab code for the PSNR computation is presented in Matlab_code_13.7.

Matlab_code_13.7

```
% Computation of image quality (PSNR).
% "psnr.m" file.
clear
x = imread('./lena.tif' );              % Read the original image file in TIFF format.
y = imread('./iqidct.tif');             % Read the Verilog reconstructed image file.
                    % Change iq_idct to idct to read the Verilog DCT–IDCT output.
orig =  double(x);                      % Convert to double precision.
recon = double(y);
[m,n]=size(orig);               % Get the size of the image.
psnr = 10*log10(((255^2)*m*n)/(sum(sum((orig-recon).^2)))) % Compute PSNR
                    % (orig-recon). => dot means element by element computation.
% For Lena image,  Verilog dct–idct PSNR =  38.9 dB, and Verilog dctq–iqidct
% PSNR =  29.4 dB. Matlab dctq–iqidct PSNR =  29.9 dB. Compression
% expressed as bits per pixel = 0.6514 for Lena image. Compression effected in
% this case is 12.28.
```

13.2.9 Verification of Verilog DCTQ – IQIDCT Cores

While the DCTQ design was presented in the book, the IQIDCT core is left as an assignment to the reader. The following steps verify the functioning of the DCTQ and the IQIDCT Cores:

1. Invoke Matlab and run the following file for reading an image disk file, "lena.raw", for example:
   read_image('lena', 256, 256, 8)

Execution of this file creates another disk output file, "lena.txt", which is in block format and, can be input into a simulator such as Modelsim for DCTQ computation.

2. Invoke Modelsim.
   Compile DCTQ test bench using the command:
   vlog dctq_test.v
   Load design using the command:
   vsim work.dctq_test
   Make sure that all the relevant source files of this design are in the same folder. Invoke the waveform and signals (select all) from View menu and "run all" command. The DCTQ output "dctq.txt" is created as a disk file in about 5 minutes.
3. In Modelsim, compile IQIDCT test bench using the command:
   vlog iqidct_test.v
   Load design using the command:
   vsim work.iqidct_test
   Make sure that all the relevant source files of this design are in the same folder. Invoke the waveform and signals (select all) from View menu and "run all" command. "dctq.txt" file created in step 2 is taken as the input and the IQIDCT output is created as a disk file "iqidct.txt" in about 5 min. This is in block order.
4. Invoke Matlab and run the following file for reading the "iqidct.txt" disk file created in step 3:
   write_image('iqidct', 256, 256, 8)
   Execution of this file creates another disk output file, "iqidct.raw", which is in raster scan order and can be input into the following file for displaying the reconstructed image as well as the original "lena.raw" images.
   show_image("lena.raw", "iqidct.raw")
5. Run "psnr.m" file to compute the image quality (PSNR) between the original image file (say, "lena.tif") and the Verilog reconstructed image file, "iqidct.tif". You can use software such as "XnView" (http://www.tucows.com/preview/290806.html) to effect conversion from "raw" to "tif" format, resizing, cropping and so on. Similarly, the DCT–IDCT cores (subsets of DCTQ–IQIDCT cores) can be verified.

13.2.10 Simulation Results

Figure 13.8 shows the final results along with the original image, Lena. The original image is shown in Figure 13.8a and the reconstructed images are shown in b to d. The DCTQ–IQIDCT coded in Matlab as presented in chapter on verification of algorithms and concepts yielded the result shown in Figure 13.8b (PSNR : 29.9 dB). The corresponding result of the DCTQ–IQIDCT cores coded in Verilog is shown

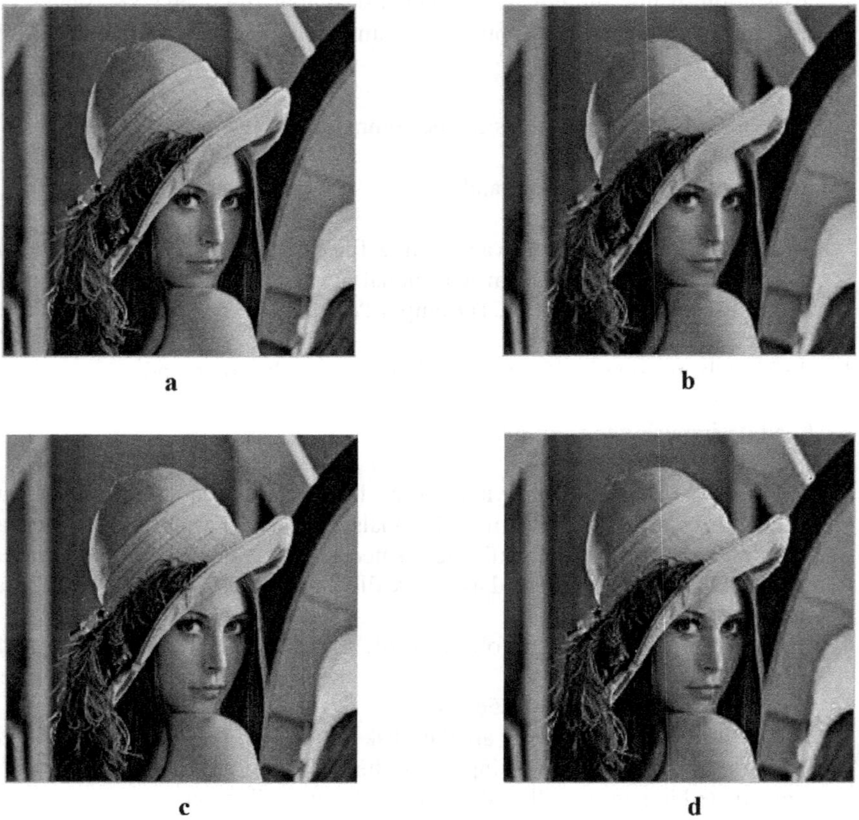

Fig. 13.8 Reconstructed Image, Lena. (a) Original Lena, 256 × 256 pixels. (b) Reconstructed by (DCTQ–IQIDCT) Matlab (29.9 dB). (c) Verilog (DCT–IDCT) (38.9 dB). (d) Reconstructed by (DCTQ–IQIDCT) Verilog (29.4 dB)

in Figure 13.8 d. Note that the PSNR value of 29.4 dB is very close to the Matlab result, which serves as a standard reference since the code is short and the precision high when compared to the Verilog codes. The reason for lower PSNR value is due to the loss introduced by quantization and inverse quantization. This can be easily cross checked by observing the result of simulating DCT–IDCT Verilog codes, which yields a very high PSNR value of about 39 dB as shown in Figure 13.8c. It may be noted that there are no blocking artifacts in this case in spite of the fact that there are no correlations among neighboring blocks. In fact, this reconstructed image is indistinguishable from the original image. This also proves that the precision we have adopted at various stages of Verilog coding of DCTQ as well as IQIDCT are perfectly alright. Of course, there is little room for increasing the bit precision for the "romq" module, which stores the inverse quantization value. Such increase can only improve the PSNR value from 29.4 dB to less than

29.9 dB. Since this improvement will not make any discernible change visually, attempts were not made to code to such precisions.

Close inspection of reconstructed images, shown in Figure 13.8b and d, reveals the blocking artifacts, which are typical characteristics of DCT based transform. This is owing to the fact that the DCT/IDCT are block based transforms and, therefore, there are no correlations between adjacent blocks. This is accentuated by the loss introduced by quantization and inverse quantization. The blocking artifacts can be minimized by applying filter on pixels across the border of adjacent blocks. A number of research papers are available [92–97] to minimize the blocking artifacts. The reader may undertake developing new algorithms and implementation on FPGA/ASIC without affecting the high processing speed we have already achieved.

The upcoming standard, MPEG 4, Part 10 (also called H.264), based on integer transform derived from DCT, recommends the deblocking filters and offers promise of better quality of reconstructed image. Similarly, JPEG 2000 and Motion JPEG 2K standards, based on the discrete wavelet transform (DWT), may offer better quality. However, they are computationally intensive and chip area is very high. Memory store requirements are also very high. The blocking artifacts are only minimized and not eliminated in toto in any of these transform based codecs. The reader may make an attempt to implement the H.264 based codecs, wherein the design methodology and part of Verilog codes presented for the DCTQ design may be made use of.

13.2.11 Implementation of DCTQ/IQIDCT IP Cores

The implementation of DCTQ Processor, which design we presented earlier, is summarized in Table 13.3. In addition to the Xilinx place and route tool we used, the Synopsys Design Manager (for ASIC implementation) was also run with TSMC Vendor Library for 0.13 μm technology. The ASIC implementation is nearly 2.7 times faster than the FPGA implementation and consumes only about 57% of the gate count. The ASIC tool reports only in terms of chip area and, therefore, the gate count is extrapolated by getting the chip area for a two input NAND gate on Synopsys platform. This has been done with the intention of getting an idea of the gate count in ASIC relative to FPGA implementation. Similar results were obtained for other IP Cores: IQIDCT, DCT, and IDCT and is presented in Table 13.4. It takes about one and half hour for each core to complete the synthesis with the Synopsys tool. By increasing the frequency of operation, say, to 300 MHz or more, the synthesis takes even more time.

Higher speeds of implementation can be obtained by using higher technology that would be available in future. The Verilog codes presented in this book can be readily used without any change, no matter what the technology is. Therefore, stay tuned to the latest technology. It may be noted that the development costs for ASIC based designs are very high when compared to FPGA based designs. For IP

Table 13.3 Implementation of DCTQ

ASIC/ FPGA	Vendor	Technology/ Device	Logic Gates	Performance (Mega Samples/Sec.)*
ASIC	TSMC	0.13 μm	68,000	270
FPGA	Xilinx	Virtex-E	120,000	102

Table 13.4 Implementations of IQIDCT, DCT, and IDCT

Design	ASIC/ FPGA	Logic Gates	Performance (Mega Samples/Sec.)*
IQIDCT	ASIC	80,000	270
	FPGA	114,000	81
DCT	ASIC	64,500	270
	FPGA	115,000	102
IDCT	ASIC	77,500	270
	FPGA	110,000	103

*Operating frequency in MHz

Core development, R&D activity and low quantity of market demand, the FPGA implementation is the right choice and for bulk requirements, ASIC is the best choice in terms of cost as well as manufacturing and marketing lead time. We need to work out the development time required and overhead costs involved in order to arrive at the viable quantities for the FPGA and ASIC implementations. Even if the market demands are seemingly high, manufacturers, as a rule, start the development on FPGA platform, release few chips or systems to assess the performance as well as the real market potential before embarking on the development of ambitious ASIC products.

13.2.12 Capabilities of the IP Cores

For processing a motion picture at a real time rate of 30 frames/second, each of the processors, DCTQ, VLC with rate control, VLD, and IQIDCT need to execute one picture frame in 33.33 ms or less. Therefore, the maximum picture size that can be processed will be as follows:

$$M \times N < 10^5 \times (f/3),$$

where M and N are picture width and height in pixels respectively, and f is the worst case frequency of operation of the cores in MHz.

The above expression is valid for both monochrome and 24 bits true color pictures, assuming that all the three-color components, Y, Cb, and Cr are processed concurrently. The latter case demands three times the chip area than the former. Presently, full color processing is not covered by the MPEG 2 standards. What is covered in the standards is four blocks and two blocks respectively for Y and Cb/Cr color components processed sequentially, one after another. It is known as 4:2:0 format, offering the highest possible compression. Full coding of this type of video encoder conforming to MPEG 2 standards developed by the author, yielded a compression of 20 for a tennis video sequence in QCIF format. Chip area, however, will remain the same as that for monochrome pictures in accordance with the tables already presented. In this case, the maximum picture size that can be processed will be as follows.

$$M \times N < 10^5 \times (2f/9)$$

The standard also permits other formats such as 4:2:2 and 4:2:1, with reduced compression. Table 13.5 presents the maximum picture size that can be processed using the Cores listed in Tables 13.3 and 13.4. A conservative 80 MHz is taken as the frequency of operation for FPGA and 270 MHz for ASIC implementation in order to arrive at the maximum picture size that can be processed at 30 frames/second. Trading off the picture size, one can multiplex many channels to implement a cable digital TV. Assuming a picture size of 640 × 480 pixels, one can have 20 channels or more with the 0.09 μm technology. In order to make this idea a reality, one must add the audio compression core based on modified DCT or filters. Interested readers may work on this project.

Table 13.5 Maximum Picture Size that can be Processed using the Cores

ASIC/ FPGA	Frequency of Operation (MHz)	Maximum Picture Size in Pixels *	Type
FPGA	80	1600 × 1600	Monochrome/ Full color
FPGA	80	1600 × 1100	4:2:0 format (color)
ASIC	270	2900 × 2900	Monochrome/ Full color
ASIC	270	2300 × 2300	4:2:0 format (color)

* Standard sizes of pictures are 1600 × 1200, 1024 × 768 (XGA), 800 × 600 pixels (SVGA), etc.

Summary

VLSI System Design examples were presented for a couple of projects, namely, PCI Arbiter, the Discrete Cosine Transform and Quantization Processor for Video compression applications. While presenting these designs, emphasis were on systematic design. The systematic design comprises identification of a project based on need, formulating detailed specifications, verifying or proving an algorithm or concept using a high level language such as Matlab or C to establish its feasibility, development of detailed architecture based on actual hardware components, Verilog RTL coding, simulation, synthesis, place and route and back annotation. The design methodologies adopted in these designs are building highly parallel, heavily pipelined circuits in order to increase the throughput and, platform independent, be it FPGA or ASIC implementation. No vendor specific modules are used and hence these designs are universal and can work on any FPGA or ASIC. The design methodologies presented in this book are equally applicable to other HDLs such as VHDL. In the next chapter, a couple of design applications based on actual FPGA and input/output boards will be presented so that the serious reader may have hands on experience in product development.

Assignments

13.1 In the PCI arbiter design presented, add a 4-bit counter to monitor the bus activity after a master is granted the bus. If the master fails to avail the bus within 16 clock cycles, withdraw its grant, and allocate the grant to the next priority master that is waiting.

13.2 To the PCI arbiter design, add a 16-bit counter that keeps track of the maximum allocated time elapsed (which is different for different masters) since the time a master is granted the use of bus. After the elapse of time programmed for the master using the bus currently, withdraw the grant after 2 clock cycles of warning and, allocate the bus to the next priority master that is waiting.

13.3 Video Codec (actually the decoder) to the AGP (display monitor) communication is not incorporated in the arbiter design. Include the same into your PCI arbiter design.

13.4 Amend the ASM chart, the design and the test bench to incorporate all the above changes. The test bench described earlier is by no means exhaustive. Therefore, include more possible combination of inputs to test the design.

13.5 Run the simulation of DCTQ design and explain why these events take place in the waveforms.

   a. In Figure 13.6.1, “ready” signal goes low/high at 215 ns/235 ns. Why?
   b. Why is the second block of image data applied so early?

c. Why does "cntl_reg" start its operation when "rnw" goes low?

13.6 In the section on capabilities of the IP Cores for the DCTQ design and the like, it was mentioned that the maximum picture size that can be processed is governed by the expression:

$M \times N < 10^5 \times (2f/9)$

Derive this expression. Will it be valid, if audio is added to the video transmission? State your assumptions clearly.

13.7 In the assignment 1 of Chapter 12, you were asked to design the architecture of IQIDCT Processor. The block diagram of IQIDCT was presented in Figure A12.1 along with the description of signals. Write Verilog design codes for the IQIDCT Processor on similar lines to the DCTQ design presented in the text.

13.8 Write a test bench for IQIDCT Processor you have coded in response to the assignment A13.7 and run all the three tools: simulation, synthesis, and place and route and report the results of your design.

13.9 Write the step-by-step operation sequence of the host/IQIDCT processors.

Chapter 14

Hardware Implementations Using FPGA and I/O Boards

In the previous chapter, we have been discussing the design applications using FPGA, namely, the PCI Arbiter, Discrete Cosine Transform and Quantization Processor for video compression application. The next step is to design a printed circuit board, which will house the target FPGA and other necessary components that make up the system, populate and test the board using oscilloscope, logic analyzer, etc. When the FPGA board is ready, we need to download the bit stream generated by the place and route tool for the particular application. The frequency of operation reported by the place and route tool is to be taken into account when we actually design the system. The maximum frequency reported by the place and route tool shall not be exceeded on the FPGA board. The hardware development work for the applications we have covered in the previous chapter are quite involved and, therefore, a couple of simpler design applications, namely, a traffic light controller and a real time clock will be presented as examples in this chapter. These applications will be based on ready made boards available such as an FPGA board and a digital input/output board.

There are many vendors for the populated and tested FPGA boards: XESS, Avnet, Nu Horizons, Digilent Inc. [98–101], Xilinx [20], to name a few. The FPGA boards from these vendors have many features that can be used readily in our development work. In general, you may use any board available with you since the Verilog code we are going to develop for the applications mentioned earlier are platform independent. However, you need to be careful to choose a board that provides at least 50 input/output pins that can be connected to an external input/output card, if your application demands these, as is the case with the applications that we are going to cover in detail. We will be using XSV800 board of XESS Corporation. You can get the details of the board from the vendor's website, www.xess.com. In case this board is not available, you can get other boards such as XSA-200 (Spartan-2 FPGA, 200 K gates, 72 free I/O pins), XSA-3S 1000 (Spartan-3 FPGA, 1 M gates, 65 free I/O pins), XSB-300E (Spartan-2E FPGA, 300 K gates, 75 free I/O pins). Similar boards such as XCS3S200 of Digilent Inc. are also available. Check the availability of the boards in case you want to procure them.

14.1 FPGA Board Features

The XSV800 board of XESS has two programable logic chips; one is a Xilinx Virtex FPGA for downloading the application that we develop and the other is a complex programable logic device (CPLD), XC95108, for board configuration and other house-keeping activities. The XSV800 board is based on Xilinx Virtex FPGA device, XCV800 as shown in Figure 14.1. The salient features of the board are as follows:

- Xilinx Virtex XCV50 (50 K gates) to XCV 800 (800 K gates) into which our application bit stream can be downloaded.

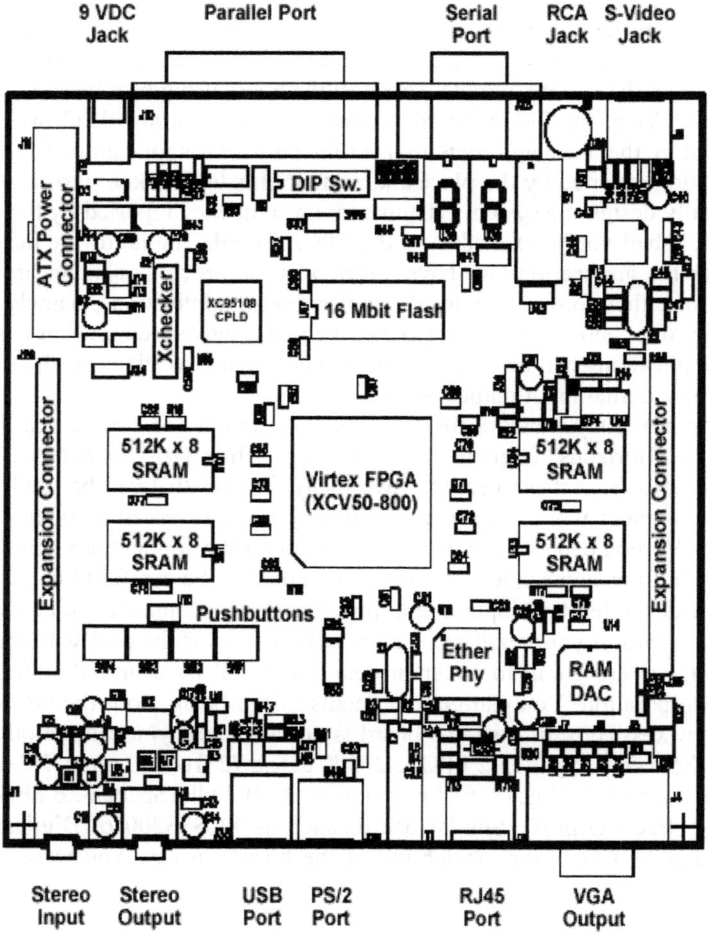

Fig. 14.1 XSV800 board (Courtesy of XESS Corporation)

- Xilinx CPLD, XC95108, is used to manage the configuration of the Virtex FPGA.
- 100 MHz Programable Clock for FPGA and CPLD.
- It has 16 M bit Flash RAM (non-volatile), which can store multiple configurations or general purpose data for FPGA.
- Two independent 512 K × 16 SRAM banks to store data and used by the FPGA.
- Two expansion headers interface connected to 76 I/O ports of the FPGA.
- Four push buttons and one eight position DIP switch connected to FPGA and CPLD.
- Two seven-segment LEDs and one bar graph LED connected to FPGA and CPLD.
- Parallel/serial ports connected to CPLD.
- Cable interface for downloading and read back of the FPGA configuration.
- Video decoder to accept NTSC/PAL/SECAM signals.
- RAMDAC with 256 entry, 24-bit color map used by the FPGA to output video to a VGA monitor.
- Stereo codec, 0–50 KHz, 20-bit resolution.
- Ethernet PHY to access a LAN at 100 Mbps.
- Mouse/Keyboard port.
- USB port with bandwidths up to 12 Mbps.
- Power requirement: 9 V/1.5 A power supply.

The onboard parallel port can be connected to the LPT 1 or LPT 2 port of the host processor, namely, the Pentium PC. In addition to this, we have a serial port that can be connected to RS232 serial link of the PC and can be used for any application needing it. There is an 8-bit DIP switch, which can be used for any of the applications like setting a timer value or any other engineering parameter required for the particular application. We also have 2 digits, seven-segment LEDs. In addition to this, we have a 10 LED bar display, which can be used for displaying histograms. We have XCV800 FPGA board, capable of holding about 900,000 gates, which means that we can house a huge circuitry in a single FPGA chip. Our application is going to reside right in the FPGA. We will download a ".bit" file via the parallel port. For example, if the application is going to be a traffic light controller, we will generate the "traffic_controller.bit" using Xilinx place and route. This particular bit stream will be downloaded via the parallel port. In order to download bit streams, we need to install XESS software supplied along with the FPGA board.

Besides a 16 M bit Flash RAM, the board has 2 MB of static RAM. They are arranged as two pairs of 512K × 16 bits. Further, the board has expansion connectors, which is used either for accessing external memory or external inputs/outputs. In the case of using an external memory, care should be taken to disable the chip enable signal of internal memory. In the two examples we are going to consider, we will be connecting these expansion connectors to an external input/output board. The next section describes the digital input/output board.

14.2 Features of Digital Input/Output Board

This board consists of forty eight digital inputs/outputs with a maximum of 32 inputs and balance can be outputs. Alternatively, all the 48 I/Os can be outputs connected to six numbers of seven-segment LEDs. The board features are summarized as follows:

- 48 inputs/outputs – user selected.
- 4 push button inputs, H/W debounced.
- 8, 4-bit binary switches.
- 8 BCD switches.
- 6 seven-segment LEDs.
- 16 discrete LEDs.
- Single supply operation.

The digital input/output board is shown in Figure 14.2. As shown in the figure, PB1–PB4 are four push button switches that can be selected by installing jumpers on the left of X1 to X4. Installing all the four jumpers on right, the push button switches are deselected, while selecting SW1/SW5. The push button switches are

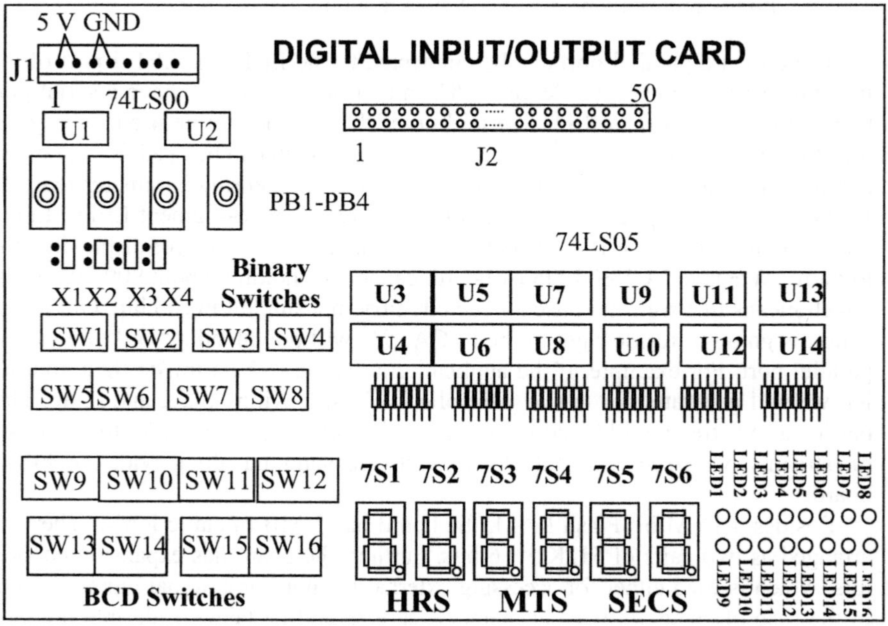

Fig. 14.2 Digital input/output board

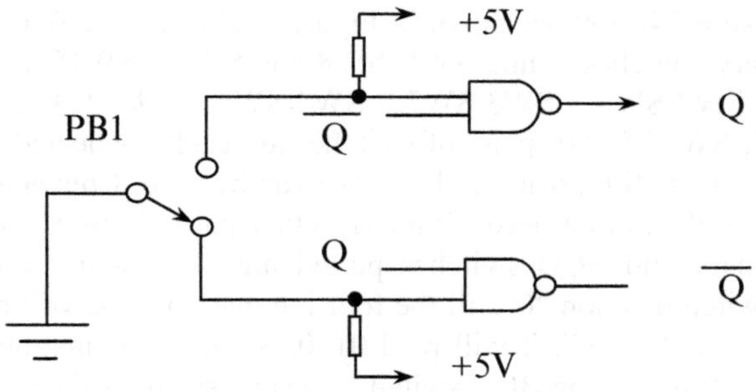

a Typical Push Button Debouncing Circuit

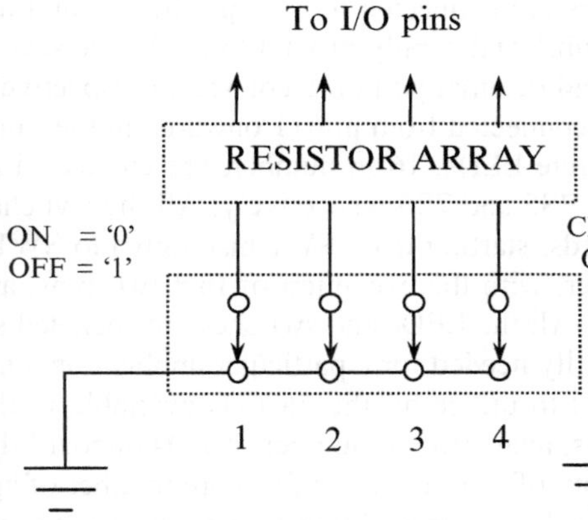

b Binary Switch

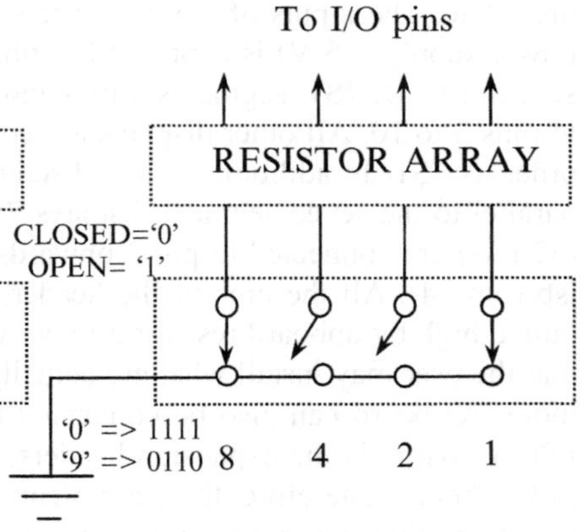

c BCD Switch

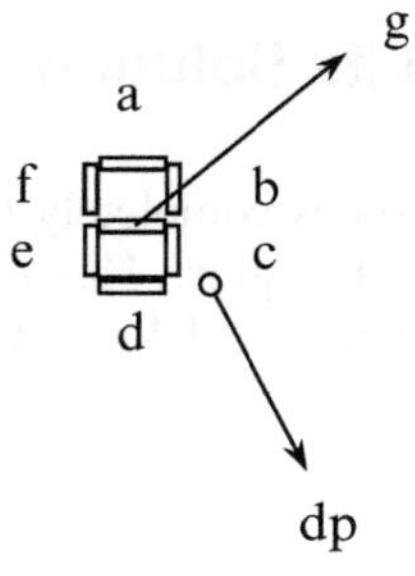

d Seven segment LED

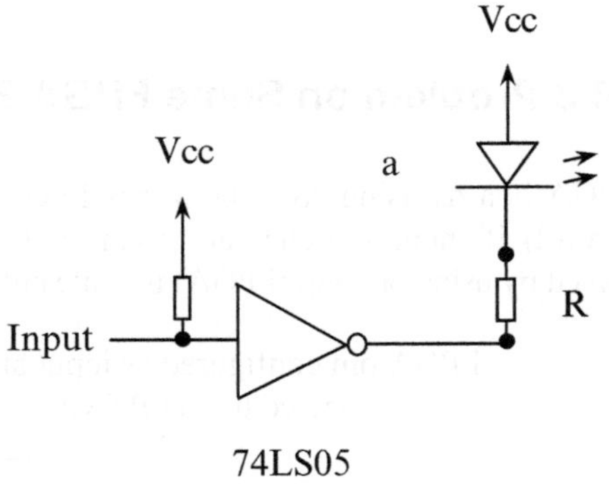

e Typ. Driving Ckt. (Common Anode)

Fig. 14.3 Interface circuits used in the digital input/output board

debounced using 74LS00 as shown in Figure 14.3a. SW1–SW4 and SW9–SW12 are 4-bit binary switches, while SW5–SW8 and SW13–SW16 are BCD switches. SW1/SW5, SW2/SW6, SW3/SW7, SW4/SW8, SW9/SW13, SW10/SW14, SW11/SW15, SW12/SW16 pairs of switches are each connected in parallel. As a result, if SW1 is in OFF position, then SW5 can be used. Conversely, if SW5 is in "0" position, SW1 can be used. Similarly other pairs. Figures 14.3b and c show the wired binary and BCD switches pulled high by resistor arrays. In a BCD switch, for switch position "0", all the four internal contacts will be off and hence it will read 1111. For "9", it will read 0110. Similar explanation holds good for other "digit" settings of the BCD switch. We can use either the push button switch or the DIP switch for the binary setting or the BCD switch for the decimal setting depending upon the needs of an application.

A seven-segment LED display is shown in Figure 14.3d. Seven-segment LEDs, 7S1-7S6, are driven by open collector inverters, 74LS05 as shown in Figure 14.3e. The inputs of these inverters are connected to a 50 pin header and the power supply (+5 V) is connected to pin1 and supply ground to pin2. The seven-segment LED, 7S1, segments a to g and decimal point are connected respectively to pins 3 to 10. All other displays are connected from pin 11 onwards in the same order as 7S1. In addition to this, discrete LEDs, 16 in number, are connected in parallel to the seven segment displays 7S5 and 7S6 respectively. All the switches (32 bits) are connected to pin 3 onwards, starting from SW1 msb down to SW12 lsb (pin 34). All the pins of the header, with the exception of first two pins, are pulled high by onboard resistor arrays. All the LEDs and switches are socketed so that the user may install what are actually needed for a particular application. One more I/O board can also be connected to utilize all the 76 I/Os available in the FPGA board via the expansion headers, apart from a number of I/Os onboard the FPGA board. Therefore, there are plenty of I/Os available for more number of applications that the reader may want to design, in addition to catering to the two applications we are going to discuss in detail.

14.3 Problem on Some FPGA Boards and Its Solution

FPGA boards using XC4000 series FPGAs malfunction (goes completely out of control) if input switches are connected directly to their I/O pins. This can be solved by using on-chip (FPGA) tri-state buffers as shown in Figure 14.4 for a typical

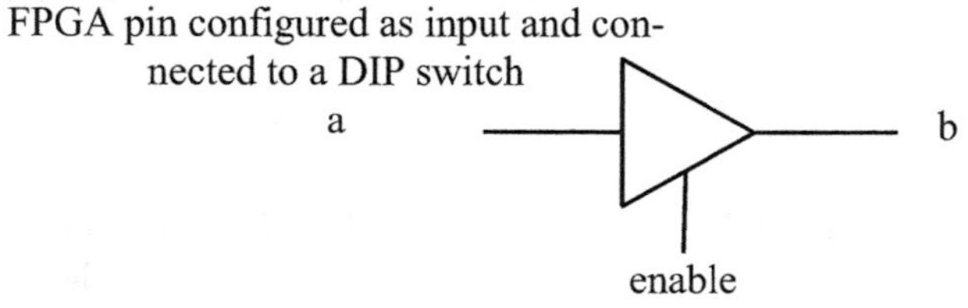

Fig. 14.4 Typical on-chip tri-state buffer

buffer. Although external buffers such as 74LS244 or 74LS240 can be used, on-chip buffers are more cost-effective. In the case of external buffers, the enable pin and pin “b” must be connected to separate FPGA pins, while “a” is connected to the DIP switch. During FPGA configuration, the “enable” goes low, thus isolating the DIP switch and the FPGA pin.

14.3.1 Verilog Code to Solve the Malfunctioning of System Using XC4000 Series FPGA Boards

The code for overcoming the malfunction mentioned above is presented in Verilog_code_14.1. The width of input switches is parameterized by the variable, “WIDTH”. Typical sizes can be 4 or 8, although any other size can be used as per needs of the application. The signal “a” is connected to the FPGA pin, which in turn is connected to a DIP switch, whereas the buffered signal “b” is used for further processing within the FPGA. At the configuration time, the enable signal “en” is not activated, thus tri-stating the buffer and, therefore, isolating the external switch, which is the cause of malfunctioning of the board. After the configuration is over, the two “assign” statements in the present code helps to set the “en” signal and thereby connect the switches to the FPGA pins gracefully, thus preventing the malfunction. You need to embed these simple codes in your application codes to overcome the problem mentioned. The XSV800 board does not have this problem and, therefore, these codes are not required for the same.

Verilog_code_14.1

```
// Tri-state input buffers to correct the malfunctioning of an FPGA based board,
// where FPGA pins configured as inputs are connected directly to switches.
Parameter        WIDTH 4 ;

module inbuf_fpga   (   a,
                        en,
                        b
                       ) ;
        input           [(WIDTH-1):0]  a ;
        output                         en ;
        output          [(WIDTH-1):0]  b ;

        wire                           en ;
        wire            [(WIDTH-1):0]  b;

        assign          en             = 1 ;
        assign          b              = en  ?  a : 0 ;
endmodule
```

We will consider two examples of design applications implemented on FPGA board and digital I/O board starting from the next section. The first example is the traffic light controller. This was developed after observing one of the busiest traffic junctions in a Metropolitan city. This application will be followed by the next application, the real time clock.

14.4 Traffic Light Controller Design

A traffic junction with a main road and two side roads is shown in Figure 14.5. The traffic flows in all possible directions. The straight traffic on the main road is timed for 45 s and 25 s for all other traffic. Yellow lights are activated for 5 s. The traffic allows free right. For convenience of identifying, the main road is bifurcated as "Main Road 1" and "Main Road 2". Similarly, the side roads are named "Side Road 1" and "Side Road 2". Other notations employed in the design are indicated in the figure.

The traffic sequences through 12 states, S0 to S11, primarily. There is one more state, "S12", where blinking of yellow lights is processed. Initially in S0 state, there will be green signal for the straight traffic on main roads in both the directions and the other two cross roads have red signals. In the second sequence S1, both the green lights changes to yellow. In the next sequence S2, the straight main road traffic is blocked and the left turn is allowed from Main Road 1 to the Side Road 2. In S3 state, the green signal changes to yellow for the left turn from main road to side road. In S4 state, the Main Road 2 traffic is allowed to turn left on the Side Road 1. In S5 state, the green signal changes to yellow for the left turn from Main Road 2 to Side Road 1. In S6 state, the two side roads are given green signals to move straight in both the directions. In S7 state, the green changes to yellow. In S8 state, the Side Road 1 traffic is allowed to turn left onto the Main Road 1. In S9 state, the green signal changes to yellow. In S10 state, the Side Road 2

M	Main
S	Side
G	Green
R	Red
Y	Yellow
LT	Left

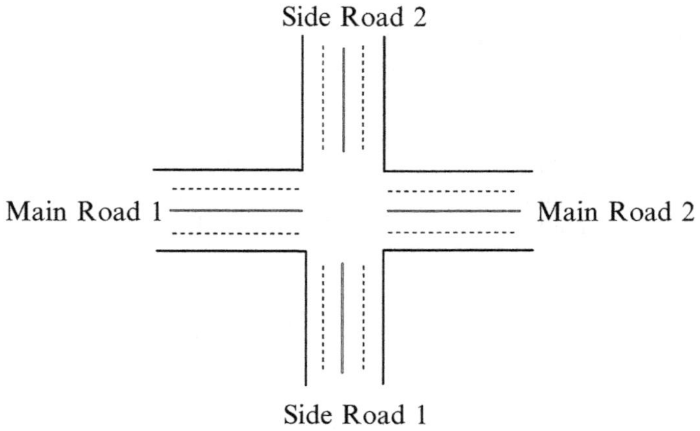

Fig. 14.5 Notation used for the traffic lights control

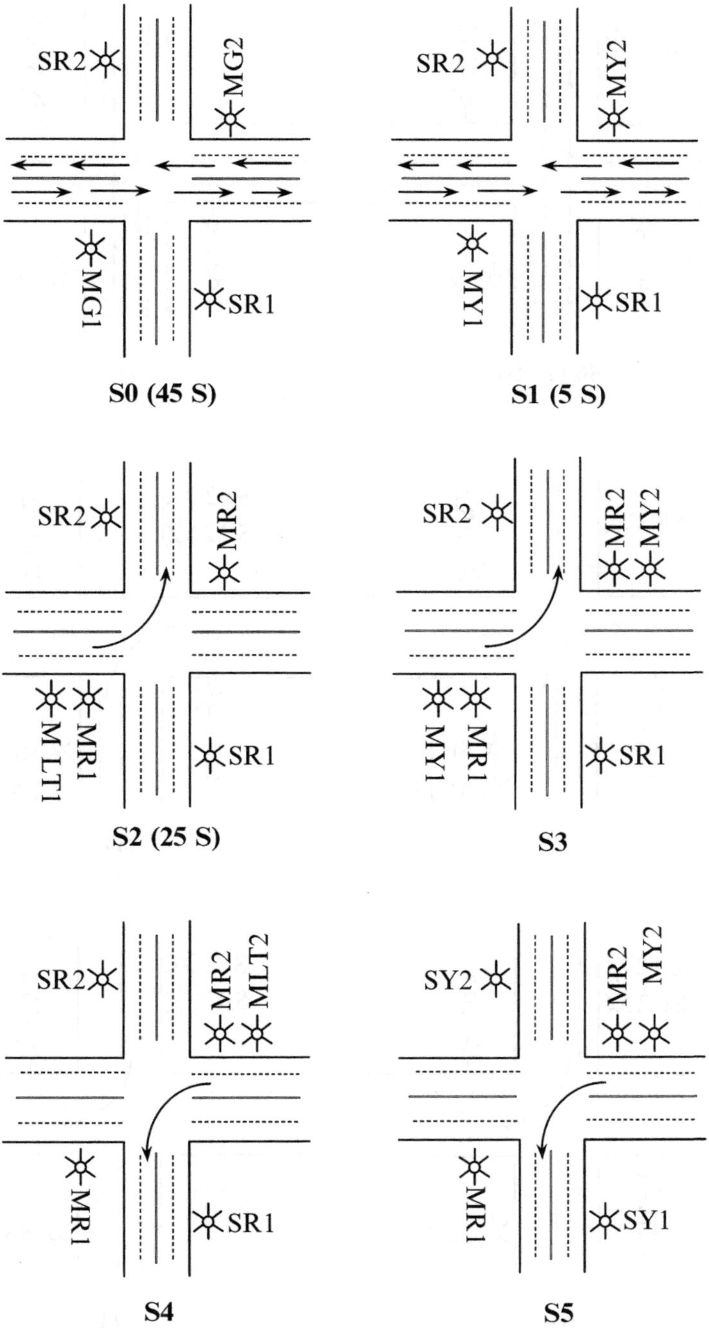

Fig. 14.6.1 Traffic lights control sequence (Continued)

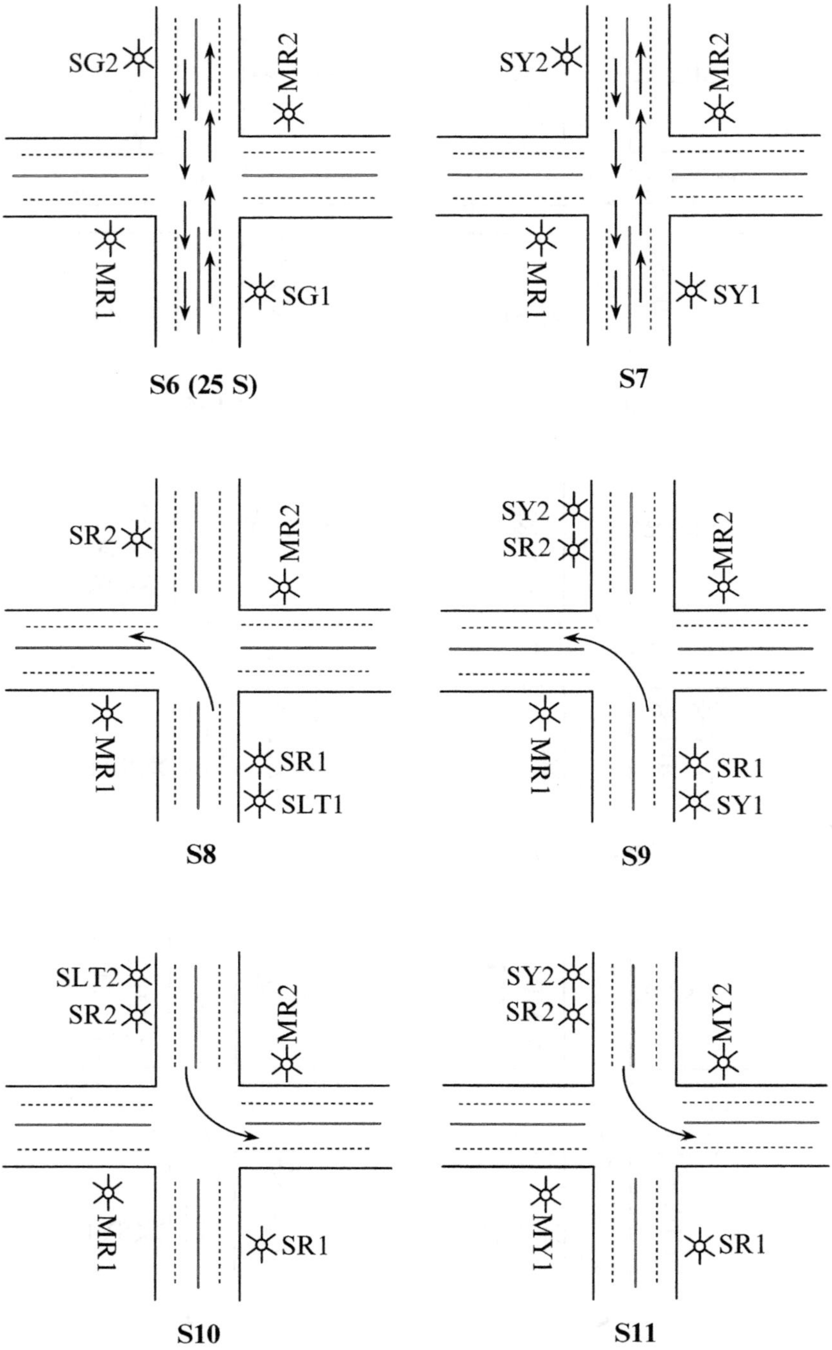

Fig. 14.6.2 Traffic lights control sequence (Continued)

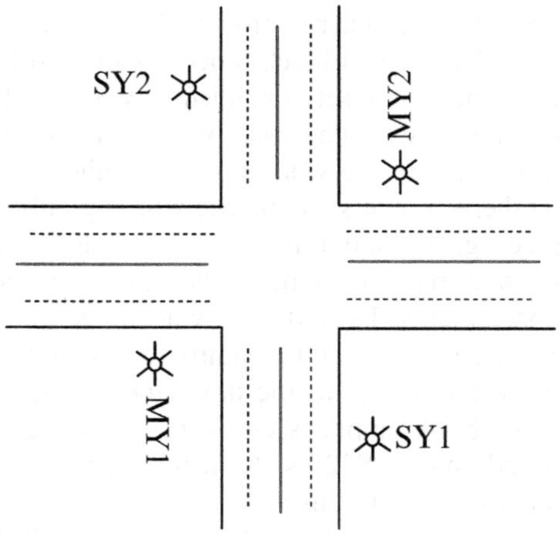

Fig. 14.6.3 Traffic lights control sequence

traffic turns left onto the Main Road 2. In S11 state, the green signal changes to yellow for the left turn from the Side Road 2. Beyond the traffic hours, when a blink input is switched on, the traffic controller enters the S12 state and all the yellow lights start flashing at the rate of 1 Hz. Normal operation is restored from "S0" state onwards when the blink input is switched off.

14.4.1 Verilog RTL Code for Traffic Light Controller

The code for this design is presented in Verilog_code_14.2. The design file is "traffic_controller.v" and the test bench is "traffic_controller_test.v". This is basically an FSM (state machine) realization. Each of the states is indicated from S0 to S12. The timing for the main road traffic is assumed to be 45 s. For the side road, it is 25 s. Yellow lights will be on for 5 s when they are switched on. Beyond normal traffic hours, all yellow lights flash at 1 Hz. We need 45 s, 25 s, 5 s, and 0.5 s timers for this application, which can be realized using counters. We also need a time base for keeping track of passing time. Therefore, a "time_base" is defined as 22'd4999999, suitable for a 50 MHz operation resulting in a convenient time base of 0.1 s. However for simulation, we shall change this parameter to 9. Similarly, we have load values, "load_cnt2" to "load_cnt5", of 449 for 45 s, 49 for 5 s, 249 for 25 s, and 4 for 0.5 s. These timings can be changed to suit any other field requirements. After the above definitions, the design module is declared, identifying all the I/Os. All of them are single bits and, therefore, width is not specified. All the wires and registers in the design are also declared.

"cnt1_reg" is a free-running counter to provide the time base of 0.1 s for timers 1 thru' 4. The first "always" sequential block processes the "cnt1_reg" counter at the positive edge of clock. When the reset is applied or when "cnt1_reg" reaches the "time_base", the counter is reset. Otherwise, we will assign the pre-incremented value of the counter. Similar are the workings of all other counters "cnt2_reg" to "cnt5_reg" except that these counters commence working only when the respective timer is started using the signal, "start_timer_1" = 1, for instance.

The traffic lights' state machine starts at the next "always" sequential block, realized using "case" statements. To start with, when power on reset is applied and the first "clk" arrives, we switch off all the lamps and the timers, and the "state" is initialized to "S0". With the arrival of the next clock pulse, the FSM enters the "S0" state, wherein the "blink" input is checked. If it is set, then the controller skips all other states and enters "S12" state in the following clock pulse. Otherwise, the "S0" state is processed. In this state, we will start the timer 1 and remain in the same "S0" state, turning on main road green (MG1, MG2) and side road red (SR1, SR2) lights. When the set time has elapsed, then the timer is reset and the controller goes to the next state, "S1". A similar explanation holds good for all other states. The reader may refer to Figures 14.6.1 to 14.6.3, and check the Verilog codes for all other states. Finally in the blink mode, we just toggle all the yellow lamps every 0.5 s. If the blink control is switched off, then the controller would go to the first state and repeat the normal sequence thereafter. The blink input is checked in every state.

Verilog_code_14.2

```
/*            Verilog RTL Code for Traffic Light Controller

This is the top design module.
Design file: traffic_controller.v
This controls the traffic lights of a four-road junction.
The timing for the main road traffic is assumed to be 45 s. For the side road, it is
25 s. Yellow lights will be on for 5 s when they are switched on.
Beyond normal traffic hours, all yellow lights flash at 1 Hz.
See traffic light controller document file for specifications.
Test bench for this design is traffic_controller_test.v.
*/
//      Define the states of the controller.
        `define  S0          4'd0
        `define  S1          4'd1
        `define  S2          4'd2
        `define  S3          4'd3
        `define  S4          4'd4
        `define  S5          4'd5
        `define  S6          4'd6
        `define  S7          4'd7
```

```
        `define  S8           4'd8
        `define  S9           4'd9
        `define  S10          4'd10
        `define  S11          4'd11
        `define  S12          4'd12
        `define  time_base    22'd4999999
/*
  For 50 MHz operation, the time base is 0.1 s. Use 22'd9 for simulation purposes.
*/
        `define  load_cnt2      9'd449
  // This is the Timer 1 count value in units of  0.1 s providing 45 s delay.
        `define  load_cnt3      6'd49
  // This is the Timer 2 count value in units of  0.1 s providing 5 s delay.
        `define  load_cnt4      8'd249
  // This is the Timer 3 count value in units of  0.1 s providing 25 s delay.
        `define  load_cnt5      8'd4
  // This is the Timer 4 count value in units of 0.1 s providing 0.5 s delay.
  //  Change these if you desire different timings.

module traffic_controller (          clk,
                                     reset_n,
                                     MR1,         // Main Red 1
                                     MR2,         // Main Red 2
                                     MY1,         // Main Yellow 1
                                     MY2,         // Main Yellow 2
                                     MG1,         // Main Green 1
                                     MG2,         // Main Green 2
                                     MLT1,        // Main Left 1
                                     MLT2,        // Main Left 2
                                     SR1,         // Side Red 1
                                     SR2,         // Side Red 2
                                     SY1,         // Side Yellow 1
                                     SY2,         // Side Yellow 2
                                     SG1,         // Side Green 1
                                     SG2,         // Side Green 2
                                     SLT1,        // Side Left 1
                                     SLT2,        // Side Left 2
                                     blink
                                  );
// Declare inputs/outputs.
        input                     clk ;
        input                     reset_n ;
        input                     blink ;
        output                    MG1 ;  // Traffic lights, main green,
        output                    MG2 ;    // etc. are declared as outputs.
        output                    MY1 ;
```

```
output                  MY2 ;
output                  MR1 ;
output                  MR2 ;
output                  MLT1 ; //Outputs for Main road left turn.
output                  MLT2 ;
output                  SR1 ;
output                  SR2 ;
output                  SY1 ;
output                  SY2 ;
output                  SG1 ;
output                  SG2 ;
output                  SLT1 ;  //Outputs for Side road left turn.
output                  SLT2 ;

// Declare nets (combinational circuit outputs).
wire                            adv_cnt2 ;
wire                            adv_cnt3 ;
wire                            adv_cnt4 ;
wire                            adv_cnt5 ;
wire                            res_cnt2 ;
wire                            res_cnt3 ;
wire                            res_cnt4 ;
wire                            res_cnt5 ;
wire            [21:0]          cnt1_next ;
wire            [8:0]           cnt2_next ;
wire            [5:0]           cnt3_next ;
wire            [7:0]           cnt4_next ;
wire            [7:0]           cnt5_next ;

// Declare registered signals.
reg             [21:0]          cnt1_reg ;
reg             [8:0]           cnt2_reg ;
reg             [5:0]           cnt3_reg ;
reg             [7:0]           cnt4_reg ;
reg             [7:0]           cnt5_reg ;
reg                             start_timer_1 ;
reg                             start_timer_2 ;
reg                             start_timer_3 ;
reg             [3:0]           state ;
reg                             MR1 ;
reg                             MR2 ;
reg                             MY1 ;
reg                             MY2 ;
reg                             MG1 ;
reg                             MG2 ;
reg                             MLT1 ;
```

```
        reg                                MLT2 ;
        reg                                SR1 ;
        reg                                SR2 ;
        reg                                SY1 ;
        reg                                SY2 ;
        reg                                SG1 ;
        reg                                SG2 ;
        reg                                SLT1 ;
        reg                                SLT2 ;

//                      Timer implementation
/*
        cnt1_reg is a free-running counter to provide the time base of 0.1 s for
        timers 1 thru' 4.
*/
assign cnt1_next  =    cnt1_reg + 1 ;          // Pre-increment the counter.

always @ (posedge clk or negedge reset_n)
begin
        if  (reset_n == 1'b0)
                cnt1_reg <= 22'd0 ;            // Initialize when the system is reset.
        else if (cnt1_reg  == `time_base)
                cnt1_reg <= 22'd0 ;            // Reset if terminal count is reached.
        else
                cnt1_reg <= cnt1_next ;        // Otherwise, advance the count once.
end
/*
  This is the Timer 1, programmed for 45 s in order to facilitate the smooth run-
  ning of the main road traffic.
*/
assign adv_cnt2  = (start_timer_1 == 1'b1) &(cnt1_reg  == `time_base) ;
                                        // Condition for Pre-incrementing the counter.
assign res_cnt2  = (cnt1_reg  == `time_base)&(cnt2_reg  == `load_cnt2) ;
                                        // Condition for resetting the counter.
assign cnt2_next =         cnt2_reg + 1 ;      // Pre-increment the counter.

always @ (posedge clk or negedge reset_n)
begin
        if (reset_n == 1'b0)
                cnt2_reg <= 9'd0 ;             // Initialize when the system is reset.
        else if (res_cnt2 == 1'b1)
                cnt2_reg <= 9'd0 ;             // Reset if terminal count is reached.
        else if (adv_cnt2 == 1'b1)
```

```
                cnt2_reg <= cnt2_next ;
                                        // 45 s timer – advance the count once if
                                        // the timer is  still running.
        else
                cnt2_reg            <= cnt2_reg ;      // Otherwise, don't disturb.
end
/*
  This is the Timer 2, programed for 5 s (activating yellow lights) for the
  smooth  transition while switching from one traffic to another.
*/
assign adv_cnt3 = (start_timer_2 == 1'b1)&(cnt1_reg  == `time_base) ;
                                        // Condition for Pre-incrementing the counter.
assign res_cnt3 = (cnt1_reg  == `time_base)&(cnt3_reg  == `load_cnt3) ;
                                              // Condition for resetting the counter.
assign cnt3_next = cnt3_reg + 1 ;             // Pre-increment the counter.

always @ (posedge clk or negedge reset_n)
begin
        if (reset_n == 1'b0)
                cnt3_reg <= 6'd0 ;            // Initialize when the system is reset.
        else if (res_cnt3 == 1'b1)
                cnt3_reg <= 6'd0 ;            // Reset if terminal count is reached.
        else if (adv_cnt3 == 1'b1)
                cnt3_reg <= cnt3_next ;
          // 5 s timer – advance the count once if the timer is still running.
        else
                cnt3_reg <= cnt3_reg  ;    // Otherwise, don't disturb.
end
                // This is the Timer 3, programed for 25 s delay, used for        //
        the side road traffic.
assign adv_cnt4 = (start_timer_3 == 1'b1)&(cnt1_reg  == `time_base) ;
                                        // Condition for Pre-incrementing the counter.
assign res_cnt4 = (cnt1_reg  == `time_base)&(cnt4_reg  == `load_cnt4) ;
                                              // Condition for resetting the counter.
assign cnt4_next = cnt4_reg + 1 ;             // Pre-increment the counter.

always @ (posedge clk or negedge reset_n)
begin
        if (reset_n == 1'b0)
                cnt4_reg <= 8'd0 ;            // Initialize when the system is reset.
        else if  (res_cnt4 == 1'b1)
                cnt4_reg <= 8'd0 ;            // Reset if terminal count is reached.
        else if (adv_cnt4 == 1'b1)
                cnt4_reg <= cnt4_next ;
          // 25 s timer – advance the count once if the timer is still running.
        else
```

```
                cnt4_reg <=  cnt4_reg ;    // Otherwise, don't disturb.
        end
// This is the Timer 4, programed for 0.5 s delay, used for blinking of all
// the yellow lights after the normal traffic hours.
assign adv_cnt5 = (blink  == 1'b1)&(cnt1_reg  == `time_base) ;
                                    // Condition for Pre-incrementing the counter.
assign res_cnt5 = (cnt1_reg  == `time_base)&(cnt5_reg  == `load_cnt5) ;
                                           // Condition for resetting the counter.
assign cnt5_next = cnt5_reg + 1 ;          // Pre-increment the counter.

always @ (posedge clk or negedge reset_n)
begin
        if (reset_n == 1'b0)
                cnt5_reg <= 8'd0 ;         // Initialize when the system is reset.
        else if (res_cnt5 == 1'b1)
                cnt5_reg <= 8'd0 ;         // Reset if terminal count is reached.
        else if (adv_cnt5 == 1'b1)
                cnt5_reg <= cnt5_next ;
                        // Advance the count once if the timer is still running.
        else
                cnt5_reg <= cnt5_reg ;     // Otherwise, don't disturb.
end
//                              Traffic lights state machine
always @ (posedge clk or negedge reset_n)
        begin
                if (reset_n == 1'b0)
                        begin              // Switch OFF all lights to start with.
                                MR1             <=      1'b0 ;
                                MR2             <=      1'b0 ;
                                MG1             <=      1'b0 ;
                                MG2             <=      1'b0 ;
                                MY1             <=      1'b0 ;
                                MY2             <=      1'b0 ;
                                MLT1            <=      1'b0 ;
                                MLT2            <=      1'b0 ;
                                SR1             <=      1'b0 ;
                                SR2             <=      1'b0 ;
                                SY1             <=      1'b0;
                                SY2             <=      1'b0;
                                SG1             <=      1'b0 ;
                                SG2             <=      1'b0 ;
                                SLT1            <=      1'b0 ;
                                SLT2            <=      1'b0 ;
                                // Also, switch OFF the timers.
                                start_timer_1   <=      1'b0 ;
                                start_timer_2   <=      1'b0 ;
```

```
                    start_timer_3        <=      1'b0 ;
                    state                <=      `S0 ;
                end     // Initialize the state when the system is reset.
        else

        case (state)
`S0:
        if (blink == 1'b1)
                state    <=     `S12 ;   // Change to the blink state.
        else
            begin
                MG1    <=     1'b1 ;  // Switch ON main
                MG2    <=     1'b1 ;  // green lights and
                SR1    <=     1'b1 ;  // side red lights.
                SR2    <=     1'b1 ;
                            // Switch OFF all other lights not wanted.
                MR1     <=    1'b0 ;
                MR2     <=    1'b0 ;
                MY1     <=    1'b0 ;
                MY2     <=    1'b0 ;
                MLT1    <=    1'b0;
                MLT2    <=    1'b0;
                SY1     <=    1'b0 ;
                SY2     <=    1'b0 ;
                SG1     <=    1'b0 ;
                SG2     <=    1'b0 ;
                SLT1    <=    1'b0 ;
                SLT2    <=    1'b0 ;
                if (res_cnt2 == 1'b1)
                      // This corresponds to 45 s timing of timer 1.
                begin
                  start_timer_1   <=      1'b0 ;
                       // Stop the timer if the terminal count is reached.
                  state           <=      `S1 ;      // Change the state.
                end
                else
                begin
                  start_timer_1   <=      1'b1;  // Otherwise, let it run.
                  state            <= `S0 ;
        // Remain in the same state until the terminal count is reached.
                end
             end
`S1:
        if (blink == 1'b1)
                state               <= `S12 ;  //Change to the blink state.
        else
```

```
            begin
                // Switch ON main yellow lights and side red lights.
                        MY1         <=      1'b1 ;
                        MY2         <=      1'b1 ;
                        SR1         <=      1'b1 ;
                        SR2         <=      1'b1 ;
                        MG1         <=      1'b0 ;
                        MG2         <=      1'b0 ;
                        MR1         <=      1'b0 ;
                        MR2         <=      1'b0 ;
                        MLT1        <=      1'b0 ;
                        MLT2        <=      1'b0 ;
                        SY1         <=      1'b0 ;
                        SY2         <=      1'b0 ;
                        SG1         <=      1'b0 ;
                        SG2         <=      1'b0 ;
                        SLT1        <=      1'b0 ;
                        SLT2        <=      1'b0 ;
                if (res_cnt3 == 1'b1)
                    // This corresponds to 5 s timing of  timer 2.
                    begin
                        start_timer_2   <=   1'b0 ;
                      // Stop the timer if the terminal count is reached.
                        state   <=  `S2 ;          // Change the state.
                    end
                else
                    begin
                        start_timer_2 <=  1'b1 ; // Otherwise, let it run.
                        state          <=  `S1 ;
          // Remain in the same state until the terminal count is reached.
                    end
                end
`S2:
            if (blink == 1'b1)
                        state     <=  `S12 ; // Change to the blink state.
            else
                begin     // Switch ON main red lights, main road1 right,
                          // main road1  left and side red  lights
                        MR1         <=      1'b1 ;
                        MR2         <=      1'b1 ;
                        MLT1        <=      1'b1 ;
                        SR1         <=      1'b1 ;
                        SR2         <=      1'b1 ;
                            // Switch OFF all other lights not wanted.
                        MY1         <=      1'b0 ;
                        MY2         <=      1'b0 ;
```

```
                        MG1         <=     1'b0 ;
                        MG2         <=     1'b0 ;
                        MLT2        <=     1'b0 ;
                        SY1         <=     1'b0 ;
                        SY2         <=     1'b0 ;
                        SG1         <=     1'b0 ;
                        SG2         <=     1'b0 ;
                        SLT1        <=     1'b0 ;
                        SLT2        <=     1'b0 ;
                if (res_cnt4 == 1'b1)
                      // This corresponds to 25 s timing of  timer 3.
                    begin
                        start_timer_3    <=     1'b0 ;
                       // Stop the timer if the terminal count is reached.
                        state    <=     `S3 ; // Change the state.
                    end
                else
                        begin
                        start_timer_3    <=     1'b1 ;
                                         // Otherwise, let it run.
                        state            <=     `S2 ;
            // Remain in the same state until the terminal count is reached.
                        end
              end
`S3:
              if (blink == 1'b1)
                        state    <= `S12 ;  // Change to the blink state.
              else
                  begin
      // Switch ON main red, main road1  yellow light and side red lights.
                        MR1         <=     1'b1 ;
                        MR2         <=     1'b1 ;
                        MY1         <=     1'b1 ;
                        MY2         <=     1'b1 ;
                        SR1         <=     1'b1 ;
                SR2            <=     1'b1 ;
                         // Switch OFF all other lights not wanted.
                        MG1         <=     1'b0 ;
                        MG2         <=     1'b0 ;
                        MLT1        <=     1'b0 ;
                        MLT2        <=     1'b0 ;
                        SG1         <=     1'b0 ;
                        SG2         <=     1'b0 ;
                        SY1         <=     1'b0 ;
                        SY2         <=     1'b0 ;
                        SLT1        <=     1'b0 ;
```

```
                    SLT2              <=        1'b0 ;
            if (res_cnt3 == 1'b1)
                   // This corresponds to 5 s timing of  timer 2.
                begin
                    start_timer_2     <=        1'b0 ;
                  // Stop the timer if the terminal count is reached.
                    state     <=         `S4 ;     // Change the state.
               end
            else
               begin
                    start_timer_2 <= 1'b1 ; // Otherwise, let it run.
                    state            <= `S3 ;
    // Remain in the same state until the terminal  count is reached.
               end
            end
`S4:
     if (blink == 1'b1)
            state       <=         `S12 ;   // Change to the blink state.
     else
            begin
   // Main red lights continue to be ON. Switch ON Main road 2 left,
   // and  also side red lights.
                    MR1               <=        1'b1 ;
                    MR2               <=        1'b1 ;
                    MLT2              <=        1'b1 ;
                    SR1               <=        1'b1 ;
                    SR2               <=        1'b1 ;
                        // Switch OFF all other lights not wanted.
                    MY1               <=        1'b0 ;
                    MY2               <=        1'b0 ;
            MG1              <=        1'b0 ;
                    MG2               <=        1'b0 ;
                    MLT1              <=        1'b0 ;
                    SY1               <=        1'b0 ;
                    SY2               <=        1'b0 ;
                    SG1               <=        1'b0 ;
                    SG2               <=        1'b0 ;
                    SLT1              <=        1'b0 ;
                    SLT2              <=        1'b0 ;
             if (res_cnt4 == 1'b1)
                  // This corresponds to 25 s timing of timer 3.
             begin
                    start_timer_3     <=        1'b0 ;
                  // Stop the timer if the terminal count is reached.
                    state             <=        `S5 ;
                                                  // Change the state.
```

```
                end
                else
                    begin
                    start_timer_3 <= 1'b1 ;  // Otherwise, let it run.
                    state         <= `S4 ;
          // Remain in the same state until the terminal  count is reached.
                    end
               end
    `S5:
          if (blink == 1'b1)
               state     <=      `S12 ;    // Change to the blink state.
          else
               begin
            // Switch ON main red lights, MY2, and side yellow lights.
                    MR1       <=    1'b1 ;
                    MR2       <=    1'b1 ;
                    MY2       <=    1'b1 ;
                    SY1       <=    1'b1 ;
                    SY2       <=    1'b1 ;
                    // Switch OFF all other lights not wanted.
                    MY1       <=    1'b0 ;
                    MG1       <=    1'b0 ;
                    MG2       <=    1'b0 ;
                    MLT1      <=    1'b0 ;
                    MLT2      <=    1'b0 ;
                    SR1       <=    1'b0 ;
                    SR2       <=    1'b0 ;
                    SG1       <=    1'b0 ;
                    SG2       <=    1'b0 ;
                    SLT1      <=    1'b0 ;
                    SLT2      <=    1'b0 ;
                if (res_cnt3 == 1'b1)
                   // This corresponds to 5 s timing of timer 2.
                  begin
                    start_timer_2    <=    1'b0 ;
                  // Stop the timer if the terminal count is reached.
                    state     <=     `S6 ;    // Change the state.
                  end
                else
                  begin

                    start_timer_2 <= 1'b1 ;  // Otherwise, let it run.
                    state         <= `S5  ;
          // Remain in the same state until the terminal count is reached.
                  end
               end
```

```
`S6:
        if (blink == 1'b1)
                state    <=       `S12 ; // Change to the blink state.
        else
            begin
                    // Switch ON main red lights and side green lights.
                MR1           <=       1'b1 ;
                MR2           <=       1'b1 ;
                SG1           <=       1'b1 ;
                SG2           <=       1'b1 ;
                            // Switch OFF all other lights not wanted.
                MY1           <=       1'b0 ;
                MY2           <=       1'b0 ;
                MG1           <=       1'b0 ;
                MG2           <=       1'b0 ;
                MLT1          <=       1'b0 ;
                MLT2          <=       1'b0 ;
                SR1           <=       1'b0 ;
                SR2           <=       1'b0 ;
                SY1           <=       1'b0 ;
                SY2           <=       1'b0 ;
                SLT1          <=       1'b0 ;
                SLT2          <=       1'b0 ;
                if (res_cnt2 == 1'b1)
                      // This corresponds to 45 s timing of timer 1.
                 begin
                  start_timer_1  <=       1'b0 ;
                     // Stop the timer if the terminal count is  reached.
                  state          <=       `S7 ;    // Change the state.
                 end
                else
                   begin
                       start_timer_1   <= 1'b1 ; // Otherwise, let it run.
                       state      <=       `S6 ;
    // Remain in the same state until the terminal count is reached.
                   end
            end
`S7:
        if (blink == 1'b1)
                state    <=       `S12 ;    // Change to the blink state.
        else
            begin                                // Let Main roads red be ON,
                MR1    <=      1'b1 ;
                MR2    <=      1'b1 ; // side roads yellow ON and
                SY1    <=      1'b1 ;
                SY2    <=      1'b1 ;
```

```
                    MY1    <=      1'b0 ;   // switch OFF all unwanted
                    MY2    <=      1'b0 ;   // lights.
                    MG1    <=      1'b0 ;
                    MG2    <=      1'b0 ;
                    MLT1   <=      1'b0 ;
                    MLT2   <=      1'b0 ;
                    SR1    <=      1'b0 ;
                    SR2    <=      1'b0 ;
                    SG1    <=      1'b0 ;
                    SG2    <=      1'b0 ;
                    SLT1   <=      1'b0 ;
                    SLT2   <=      1'b0 ;
                    if (res_cnt3 == 1'b1)
                          // This corresponds to 5 s timing of timer 2.
                      begin
                        start_timer_2  <=        1'b0 ;
                            // Stop the timer if the terminal count is reached.
                        state              <=        `S8 ;
                                // Change the state to the eighth sequence.
                      end
                    else
                    begin
                      start_timer_2    <= 1'b1 ;        // Otherwise, let it run.
                      state                <=        `S7 ;
        // Remain in the same state until the terminal count is reached.
                      end
                end
`S8:
            if (blink == 1'b1)
                state              <=      `S12 ;   // Change to the blink state.
            else
                    begin
                    MR1    <=      1'b1 ;
                    MR2    <=      1'b1 ;  // Let Main roads red be ON and
                    SR1    <=      1'b1 ;
                    SR2    <=      1'b1 ;  // switch ON side roads red.
                    SLT1   <=      1'b1 ;  // Also switch ON side road1 left.
                    MY1    <=      1'b0 ;  // Switch OFF all unwanted lights.
                    MY2    <=      1'b0 ;
                    MG1    <=      1'b0 ;
                    MG2    <=      1'b0 ;
                    MLT1   <=      1'b0 ;
                    MLT2   <=      1'b0 ;
                    SY1    <=      1'b0 ;
                    SY2    <=      1'b0 ;
                    SG1    <=      1'b0 ;
```

```
                SG2    <=      1'b0 ;
                SLT2   <=      1'b0 ;
                if (res_cnt4 == 1'b1)
                      // This  corresponds to 10 s timing of timer 3.
                begin
                  start_timer_3  <=       1'b0 ;
                       // Stop the timer if the terminal count is reached.
                  state          <=       `S9 ;      // Change the state.
                end
                else
                begin
                  start_timer_3  <=       1'b1 ; // Otherwise, let it run.
                  state          <=       `S8 ;
        // Remain in the same state until the terminal count is reached.
                end
                end
`S9:
        if (blink == 1'b1)
                state   <=      `S12 ;   // Change to the blink state.
        else
                begin
                MR1    <=      1'b1 ;
                MR2    <=      1'b1 ;    // Retain Main roads red and
                SR1    <=      1'b1 ;
                SR2    <=      1'b1 ;    // switch ON side roads red.
                SY2    <=      1'b1 ;
                SY1    <=      1'b1 ; // Switch ON side roads yellow.
                                      // Switch OFF all unwanted lights.
                MY1    <=      1'b0 ;
                MY2    <=      1'b0 ;
                MG1    <=      1'b0 ;
                MG2    <=      1'b0 ;
                MLT1   <=      1'b0 ;
                MLT2   <=      1'b0 ;
                SY1    <=      1'b0 ;
                SG1    <=      1'b0 ;
                SG2    <=      1'b0 ;
                SLT1   <=      1'b0 ;
                SLT2   <=      1'b0 ;
                if (res_cnt3 == 1'b1)
                       // This corresponds to 5 s timing of timer 2.
                  begin
                      start_timer_2     <=       1'b0 ;
                    // Stop the timer if the terminal count is reached.
                      state             <=       `S10 ;
                             // Change the state to the first sequence.
```

```
                    end
                    else
                         begin
                           start_timer_2 <= 1'b1 ;  // Otherwise, let it run.
                           state           <= `S9 ;
             // Remain in the same state until the terminal count is reached.
                           end
                   end
`S10:
           if (blink == 1'b1)
                           state      <= `S12 ; // Change to the blink state.
           else
               begin
                   MR1     <=        1'b1 ;    // Retain Main roads red and
                   MR2     <=        1'b1 ;    // switch ON side roads red.
                   SR1     <=        1'b1 ;
                   SR2     <=        1'b1 ;    // Switch ON side road 2 left.
                   SLT2    <=        1'b1 ;   //Switch OFF all other lights.
                   MY1     <=        1'b0 ;
                   MY2     <=        1'b0 ;
                   MG1     <=        1'b0 ;
                   MG2     <=        1'b0 ;
                   MLT1    <=        1'b0 ;
                   MLT2    <=        1'b0 ;
                   SY1     <=        1'b0 ;
                   SY2     <=        1'b0 ;
                   SG1     <=        1'b0 ;
                   SG2     <=        1'b0 ;
                   SLT1    <=        1'b0 ;
                   if (res_cnt4 == 1'b1)
                        // This corresponds to 10 s timing of timer 3.
                     begin
                           start_timer_3 <= 1'b0 ;
                         // Stop the timer if the terminal count is reached.
                           state      <= `S11 ;          // Change the state.
                     end
                   else
                     begin
                           start_timer_3 <= 1'b1 ; // Otherwise, let it run.
                           state                 <=         `S10 ;
             // Remain in the same state until the terminal count is reached.
                     end
               end
`S11:
           if (blink == 1'b1)
                           state      <= `S12 ; // Change to the blink state.
```

```
        else
          begin
               MY1     <=       1'b1 ;  // Switch ON Main roads yellow,
               MY2     <=       1'b1 ;  // side roads red, SY2 and
               SR1     <=       1'b1 ;
               SR2     <=       1'b1 ;
               SY2     <=       1'b1 ;
               SY1     <=       1'b0 ;  // all other lights OFF.
               MR1     <=       1'b0 ;
               MR2     <=       1'b0 ;
               MG1     <=       1'b0 ;
               MG2     <=       1'b0 ;
               MLT1    <=       1'b0 ;
               MLT2    <=       1'b0 ;
               SG1     <=       1'b0 ;
               SG2     <=       1'b0 ;
               SLT1    <=       1'b0 ;
               SLT2    <=       1'b0 ;
               if (res_cnt3 == 1'b1)
                     // This corresponds to 5 s timing of timer 2.
                 begin
                        start_timer_2      <=        0 ;
                      // Stop the timer if the terminal count is reached.
                        state              <=        S0 ;
                              // Change the state to the first sequence.
                 end
               else
                 begin
                        start_timer_2 <= 1'b1 ; // Otherwise, let it run.
                        state         <= `S11 ;
         // Remain in the same state until the terminal count is reached.
                 end
          end
`S12:
        if (blink == 1'b1)
          begin
              begin
                 MR1    <=       1'b0 ;    // Switch OFF all lights
                 MR2    <=       1'b0 ;    // except yellow.
                 MG1    <=       1'b0 ;
                 MG2    <=       1'b0 ;
                 MLT1   <=       1'b0 ;
                 MLT2   <=       1'b0 ;
                 SR1    <=       1'b0 ;
                 SR2    <=       1'b0 ;
                 SG1    <=       1'b0 ;
```

```
                    SG2    <=     1'b0 ;
                    SLT1   <=     1'b0 ;
                    SLT2   <=     1'b0 ;
                  end
              if ((cnt1_reg == `time_base)&&(cnt5_reg == `load_cnt5))
                  begin                           // Blink all yellow lights.
                    SY1    <=     ~SY1 ;
                    SY2    <=     ~SY2 ;
                    MY1    <=     ~MY1 ;
                    MY2    <=     ~MY2 ;
                    state  <=     `S12 ;
         // Remain in the same state until the terminal count is reached.
                  end
              else
                    state   <=     `S12 ;
            end
           else
                    state <= `S0 ;  // Change the state to the first sequence.
            default: state <= `S0 ;
            endcase
    end
endmodule
```

14.4.2 Test Bench for the Traffic Light Controller

Verilog_code_14.3 presents the test bench of the traffic light controller. We include the design file, "traffic_controller.v" in the test bench. The frequency of operation is set at 50 MHz. The test bench is declared as "traffic_controller_test" and the inputs as registers. This is followed by instantiating the design, "traffic_controller". To start with, we initialize all the inputs before the normal operation starts. Once the reset pulse is applied and withdrawn, the traffic light controller starts sequencing starting from "S0" state. We will allow it to run long enough to capture one full sequence or more. After it runs for about 300,000 ns, we apply the blink control for about 50,000 ns. During this period, all the yellow lights will flash at a rate of 1 Hz. After this period, the normal traffic lights operation is resumed.

Verilog_code_14.3

```
/*                Test Bench for Traffic Light Controller
Put this in a file named, "traffic_controller_test.v"  */
`define clkperiodby2 10   // 10 ns is the half time period –
                          // Frequency of operation: 50 MHz.
```

```
`include "traffic_controller.v"              // This is the design file.
`timescale 1ns/100ps
module traffic_controller_test ;
reg          clk ;                           // Declare input signals.
reg          reset_n ;
reg          blink ;
traffic_controller tc1(                 // Instantiate the traffic controller design module.
                                          .clk(clk),
                                          .reset_n(reset_n),
                                          .MG1(MG1),
                                          .MG2(MG2),
                                          .SR1(SR1),
                                          .SR2(SR2),
                                          .MY1(MY1),
                                          .MY2(MY2),
                                          .MR1(MR1),
                                          .MR2(MR2),
                                          .SG1(SG1),
                                          .SG2(SG2),
                                          .SY1(SY1),
                                          .SY2(SY2),
                                          .MLT1(MLT1),
                                          .MLT2(MLT2),
                                          .SLT1(SLT1),
                                          .SLT2(SLT2),
                                          .blink(blink)
                 );
initial
begin
        clk          = 1'b0 ;             // Initialize input signals.
        reset_n      = 1'b1 ;
        blink        = 0 ;
 #20    reset_n      = 1'b0 ;             // Pulse low.
 #20    reset_n      = 1'b1 ;
                        // Run long enough to capture one full sequence or more.
 #600000 blink = 1 ;                       // Blink all yellow lights.
 #50000  blink = 0 ;                      // Resume normal traffic lights operation
 #50000
 $stop ;                                  // and stop.
end
always
 #`clkperiodby2 clk <= !clk ;             // Toggle to get a free running clock.
endmodule
```

14.4.3 Simulation of Traffic Light Controller

The simulation results are shown in Figures 14.7.1 to 14.7.3. Looking at the first waveform, we see that the reset is applied at 20 ns and withdrawn at 40 ns, after which time the normal sequence commences. With the application of the reset pulse, the FSM state is initialized to “S0” state. The traffic lights are, however, turned on only with the arrival of the first clock pulse after the reset is withdrawn. Accordingly, MG1, MG2, SR1, and SR2 lights are turned on at 50 ns with the rising edge of the “clk” signal. Figure 14.7.2 shows the results of all other states, “S1” to “S11”. The lights that are turned on for various states as per the waveforms are as follows:

S1 MY1, MY2, SR1, and SR2
S2 MR1, MR2, SR1, SR2, and MLT1
S3 MR1, MR2, MY1, MY2, SR1, and SR2
S4 MR1, MR2, SR1, SR2, and MLT2
S5 MR1, MR2, MY2, SY1, and SY2
S6 MR1, MR2, SG1, and SG2
S7 MR1, MR2, SY1, and SY2
S8 MR1, MR2, SR1, SR2, and SLT1
S9 MR1, MR2, SR1, SR2, SY1, and SY2
S10 MR1, MR2, SR1, SR2, and SLT2
S11 MY1, MY2, SR1, SR2, and SY2

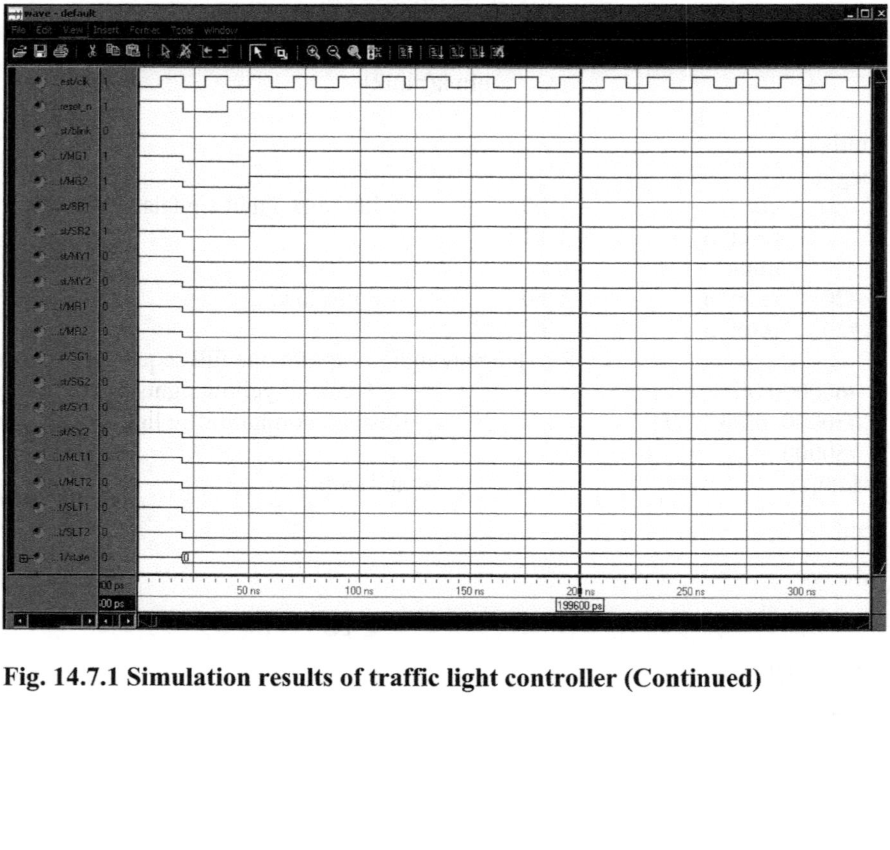

Fig. 14.7.1 Simulation results of traffic light controller (Continued)

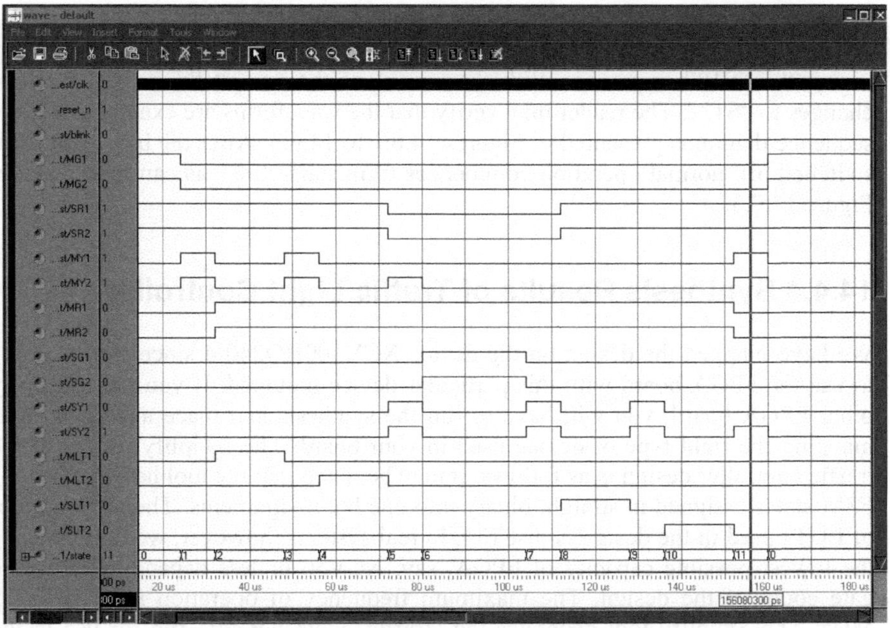

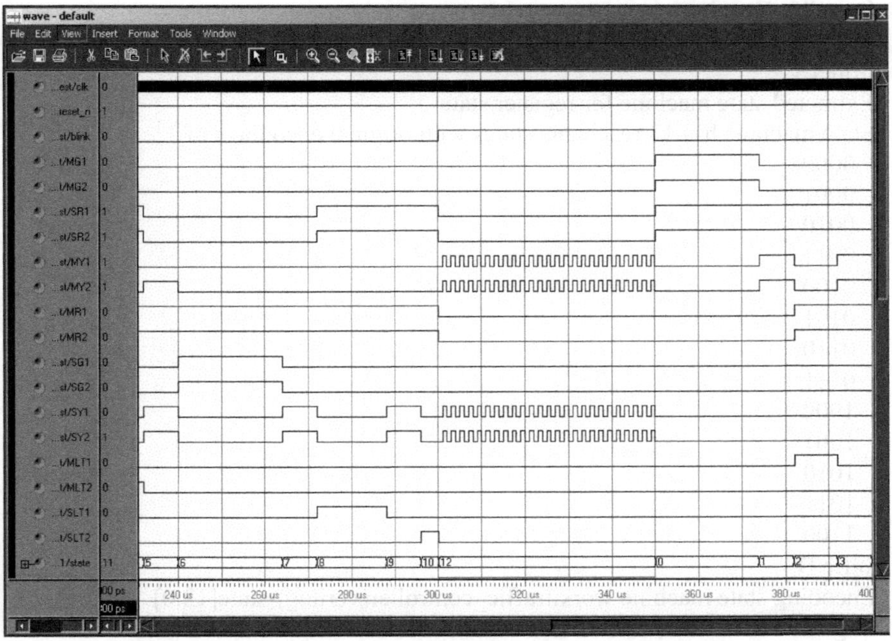

Fig. 14.7.2 and 14.7.3 Simulation results of traffic light controller

Figure 14.7.3 shows the flashing of all yellow lights, MY1, MY2, SY1, and SY2 during 300 to 350 μs when blink input is switched on while the controller was servicing the "S10" sequence. With the arrival of the next "clk" pulse, the state changes to "S12". The reader may verify that the waveforms are exactly as per the sequence diagrams presented in Figures 14.6.1 to 14.6.3. After the blink control is switched off, normal operation commences from state, "S0", as can be seen from Figure 14.7.3.

14.4.4 Synthesis Results of Traffic Light Controller

We have mapped the design on the device XCV800HQ240-4 since we are going to use the FPGA board with this particular device mounted. If you are using any other FPGA board, you will have to run the synthesis and place and route tools mapping the right type of device used in your board. The Synplify results of the traffic controller design is as follows. It may be noted that the tool has changed the FSM states assigned in straight binary into one hot assignments. The total number of LUTs used in the design is just 134. In real systems, however, we will be using the lowest possible capacity of FPGA, say, XCV50 or less depending upon the gate count of the design. The maximum frequency of operation reported is 75 MHz. The Synplify tool generates the "traffic_controller.edf" file for use in the P&R tool.

Synplify Results:

@I::"D:\RAM\book\Traffic_controller_seq12\Traffic_controller_Right\traffic_controller.v"
Extracted state machine for register state
State machine has 13 reachable states with original encodings of:
  0000
  0001
  0010
  0011
  0100
  0101
  0110
  0111
  1000
  1001
  1010
  1011
  1100
@END
Encoding state machine work.traffic_controller(verilog)-state[12:0]
original code -> new code
  0000 -> 0000000000001
  0001 -> 0000000000010
  0010 -> 0000000000100

0011 -> 0000000001000
0100 -> 0000000010000
0101 -> 0000000100000
0110 -> 0000001000000
0111 -> 0000010000000
1000 -> 0000100000000
1001 -> 0001000000000
1010 -> 0010000000000
1011 -> 0100000000000
1100 -> 1000000000000

Worst slack in design: 6.693

Clock Starting	Requested Frequency	Estimated Frequency	Requested Period	Estimated Period
clk	50.0 MHz	75.1 MHz	20.000	13.307

Resource Usage Report for traffic_controller
Mapping to part: xcv800hq240-4
Cell usage:

FDC	65 uses
FDCE	15 uses
FDP	1 use
GND	1 use
MUXCY_L	48 uses
VCC	1 use
XORCY	53 uses
I/O primitives:	18
IBUF	2 uses
OBUF	16 uses
BUFGP	1 use

I/O Register bits:	0
Register bits not including I/Os:	81 (0%)
Global Clock Buffers:	1 of 4 (25%)
Total LUTs:	134 (0%)

14.4.5 Place and Route Results of Traffic Controller

The "traffic_controller.edf" file generated by the synthesis tool is input into the Xilinx place and route tool for creating the bit stream. The "traffic_controller.ucf" file, explained in the next sub-section, is also used while running the P&R tool. The Xilinx P&R tool report is as follows. The gate count of the design is 1437. The maximum frequency of operation reported is 60 MHz (16.672 ns).

Command Line: C:/Xilinx/bin/nt/map.exe -intstyle ise -p xcv800-hq240-4 -cm area -pr b -k 4 -c 100 -tx off -o traffic_controller_map.ncd traffic_controller.ngd traffic_controller.pcf
Target Device: xv800
Target Package: hq240
Target Speed: -4

Logic Utilization:				
Number of slice flip flops:	81	out of	18,816	1%
Number of four input LUTs:	81	out of	18,816	1%
Logic Distribution:				
Number of occupied slices:	79	out of	9,408	1%
Number of slices containing only related logic:	79	out of	79	100%
Number of slices containing unrelated logic:	0	out of	79	0%
Total number four input LUTs:	134	out of	18,816	1%
Number used as logic:	81			
Number used as a route-thru:	53			
Number of bonded IOBs:	18	out of	166	10%
Number of GCLKs:	1	out of	4	25%
Number of GCLKIOBs:	1	out of	4	25%

Total equivalent gate count for design: 1,437
Additional JTAG gate count for IOBs: 912

User Constraint File for Traffic Light Controller

The place and route tool of Xilinx assigns default pin numbers to the design signals if no user constraint file, ".ucf ", is specified while running the tool. However, the user can assign them as per actual hardware connections by specifying the desired pin configuration in a ".ucf " file. This should be a separate file, say, "traffic_controller.ucf " and located, preferably, in the same folder where the "traffic_controller.edf " is located. The FPGA pins to be included are as follows. They have to be declared as NET, specifying the signals and their corresponding pins.

User Constraint File, "traffic_controller.ucf"

```
NET  "clk"      LOC   =   P89 ;
NET  "blink"    LOC   =   P108 ;
NET "reset_n"   LOC   =   P188 ;
NET  "MR1"      LOC   =   P94 ;
NET  "MR2"      LOC   =   P223;
NET  "MY1"      LOC   =   P224;
NET  "MY2"      LOC   =   P228;
NET  "MG1"      LOC   =   P93;
NET  "MG2"      LOC   =   P229;
NET  "MLT1"     LOC   =   P230;
```

```
NET "MLT2"  LOC   =   P231;
NET "SR1"   LOC   =   P87;
NET "SR2"   LOC   =   P232;
NET "SY1"   LOC   =   P234;
NET "SY2"   LOC   =   P235;
NET "SG1"   LOC   =   P86;
NET "SG2"   LOC   =   P236;
NET "SLT1"  LOC   =   P237;
NET "SLT2"  LOC   =   P222;
```

14.4.6 Hardware Setup of Traffic Light Controller

Figure 14.8.1 shows the capture of one of the live demo sequences (S8) display of the traffic light controller. The 16 lamps on the display board are connected to LED1 to LED16 of the digital I/O board, which we discussed in an earlier section. Figure 14.8.2 shows the hardware setup of the traffic light controller. The demo setup comprises the computer (PC), the FPGA board, the digital I/O board, the traffic display board, and power supplies. The power supplies deliver 9 V DC, 1.5 A for the FPGA board and 5 V DC, 1A to the digital I/O board. As shown in Figure 14.8.2, the LEDs in the display board are connected to the digital I/O board, whose I/Os in turn are connected to the expansion headers of the FPGA board. Actual connections established were presented in the user constraint file, "traffic_controller.ucf ". The close-up views of the FPGA and the digital I/O boards are shown in Figures 14.8.3 and 14.8.4 respectively. The downloading of the bit stream is shown in Figure 14.8.5. The last push button switch, PB4, on the digital I/O board is used as system reset and is enabled by installing the jumper X4 on the left hand side. Make sure that the jumper X1 is installed on the right hand side and the BCD switch SW5 (refer Figure 14.8.2) is in "0" position in order to use the binary switch SW1, first bit position from left, as the "Blink" control. Note that the "ON" position switches off the blink. For normal sequence, this switch must be in the ON position and for blinking of yellow lights, it must be in the OFF position (which corresponds to a logical high as explained in an earlier section).

In the demo setup described earlier, only single LED displays were used for the traffic lights. However, in the actual traffic lights, we need to use a group of LEDs in lieu of single LED. Rest of the hardware we have developed would work without any changes. Conventional traffic bulb signals may be replaced with a group of discrete LEDs. These offer brighter illumination and consume less power. Also, the life of LEDs are much longer (over 10 years). In the conventional traffic signals, once the bulbs get fused, it is a major problem for road users. Everyday, the police spend a lot of time changing bulbs at various signal points. In the discrete LED displays, there are about 200 small LEDs and not all of them get fused at the

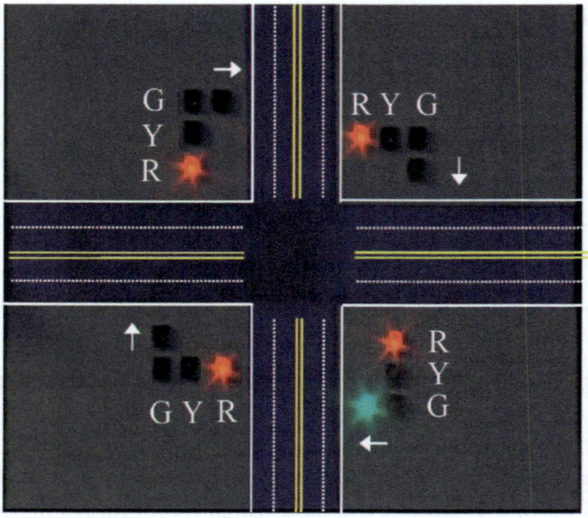

Fig. 14.8.1 Traffic light display board of the demo setup

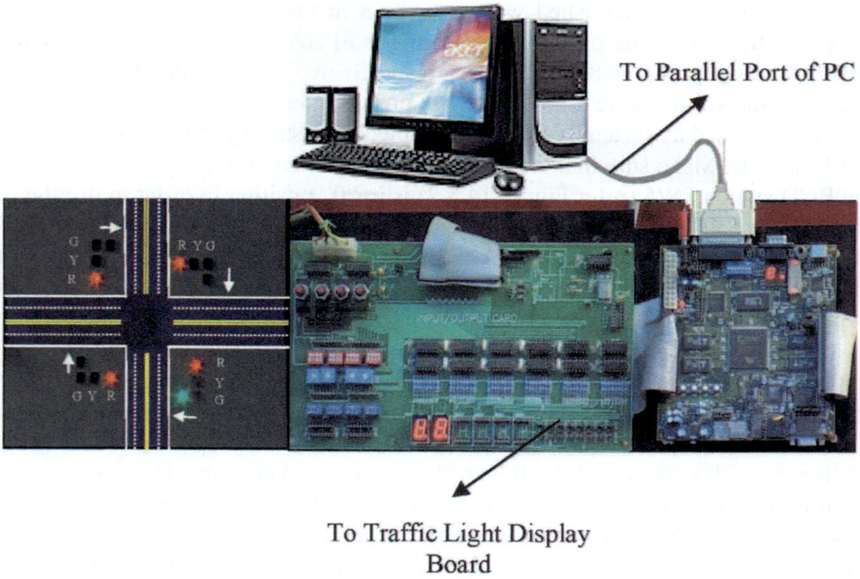

Fig. 14.8.2 Hardware setup for the traffic light controller

Fig. 14.8.3 Close-up view of the FPGA board

Fig. 14.8.4 Close-up view of the digital input/output board

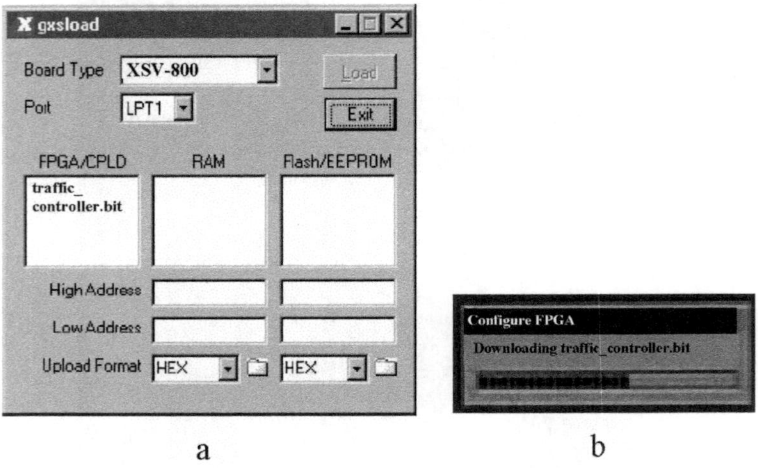

Fig. 14.8.5 Down loading of traffic light controller bit stream

same time. Moreover, even if a batch of them got fused, the signal would be still visible from a distance. Maintaining a plug-in, modular display design, replacement of LED display boards will be quick and economical in the long run.

14.5 Real Time Clock Design

This is another design example suitable for implementation on the FPGA board and digital I/O board we have used in the earlier design. This design is more involved than the traffic controller design. The I/O board needs to be populated with all the six, seven-segment LEDs for the real time clock application. It has many applications and features as presented in the following sub-sections.

14.5.1 Applications

The following lists the applications for the real time clock (RTC). The first application is to display the real time on 24 h basis. The RTC can display in hours, minutes, and seconds. It has a counter which can be configured as an up counter or a down counter. We can use the same as a stopwatch, an industrial timer or a photographic timer. We can also use the RTC with three different alarm settings for time-bound medical treatment, as an example:

- Real time display
- Stop watch
- Industrial timer

- Photographic timer
- Medical application using three alarm settings

14.5.2 Features

The main features of the real time clock are as follows:

- 24-h clock
  - Hrs/mts/secs push button settings
- Stopwatch
  - Up counter
  - Down counter
  - Count setting by push buttons
  - Timer out if the running counter matches the set time
- Three independent alarms
  - Alarm settings by push buttons
  - Common audio alarm

We will be designing a 24-h clock, which can be easily converted to accommodate a 12-h clock as well. In this system, hours, minutes, and seconds can be set by push buttons. We need three independent push button switches to set them. We have a stopwatch, which can count up or down, and we can set this count by using the same push button switches once again. There is also a "timer out", which is a single bit signal. It will be turned on when the running counter matches the set time. For example, in firing a rocket, we can use a down counter (or an up counter, whichever is preferred), which counts from a preset value. The current time remaining over is indicated in the display. When the display touches "00 00 00", the rocket is fired. When this happens, an audio alarm is also activated for 30 s. We can also use the real time counter in up counting mode. Once started, the counting commences from "00 00 00" progressing upwards. When it touches the set value, it stops, simultaneously sounding the alarm. We will also include three independent alarms that can be set using DIP switches and the alarm settings by push buttons.

When we design a system, the layout of the product is very important. From the perspective of a user, the outer look and ease of operation of the product matters a lot, in addition to correct and reliable functioning. It may be noted that the present hardware setup is only for the R&D phase to prove our design. Once the design is completed and is working satisfactorily, a compact hardware will have to be fabricated retaining only those components that are absolutely required for the particular application. For instance, the entire real time clock can be mapped on a single ASIC and having a small LCD for display and miniature switches for operation, looking very much like a wrist watch or a small table top equipment. In the case of a wrist watch, all switches can be push buttons. We can configure the FPGA/digital I/O boards for any number of applications subject to the limitations of the hardware on the boards available with us. Depending upon the actual needs, the Verilog code may require modifications. So also the user constraints file.

Therefore, before we start coding, it is better to make sure which hardware we need ultimately. For the present implementation, we will use the same hardware we used for the earlier design.

14.5.3 Hardware Requirements for the Real Time Clock

We will use the same hardware we used for the traffic light controller application, namely, the XSV 800 FPGA board and the digital input/output card for the real time clock application. If you have different boards, you may amend the design, Verilog codes and ".ucf " file accordingly to suit the specific boards you have. The design methodology and most of the Verilog codes presented in this section will hold good for any other board. If I/O board is not available, you may try to fabricate it yourselves using the hardware details presented earlier. We need six numbers of seven-segment LEDs for displaying hours, minutes, and seconds. Therefore, we will install the same in the sockets on the digital I/O card. Figure 14.9.1 depicts the real time clock front fascia that is desirable. However, we need to remain content with the controls and displays spread over the two boards. With the exception of seven-segment LED displays, buzzer and Alarm ON/OFF switch, all other controls and display for "Timer" output (the bar LED display) are available

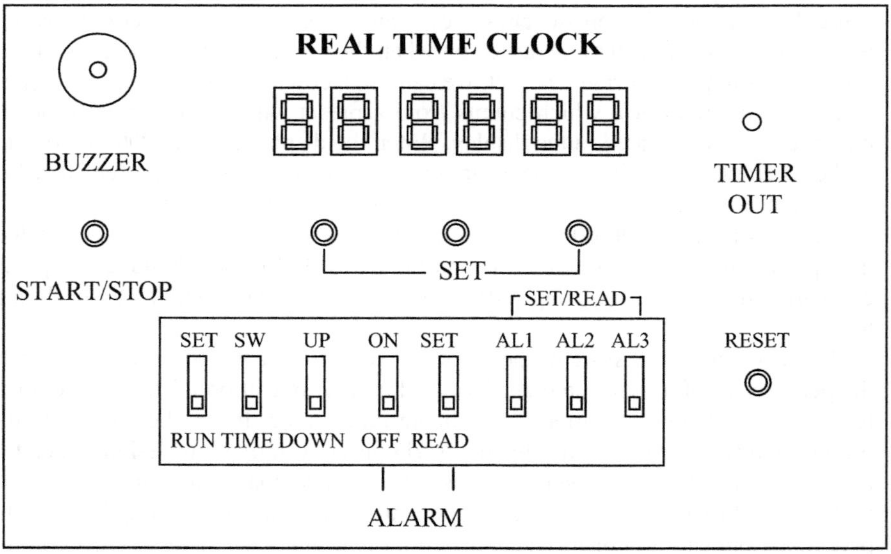

Fig. 14.9.1 Real time clock front fascia desired

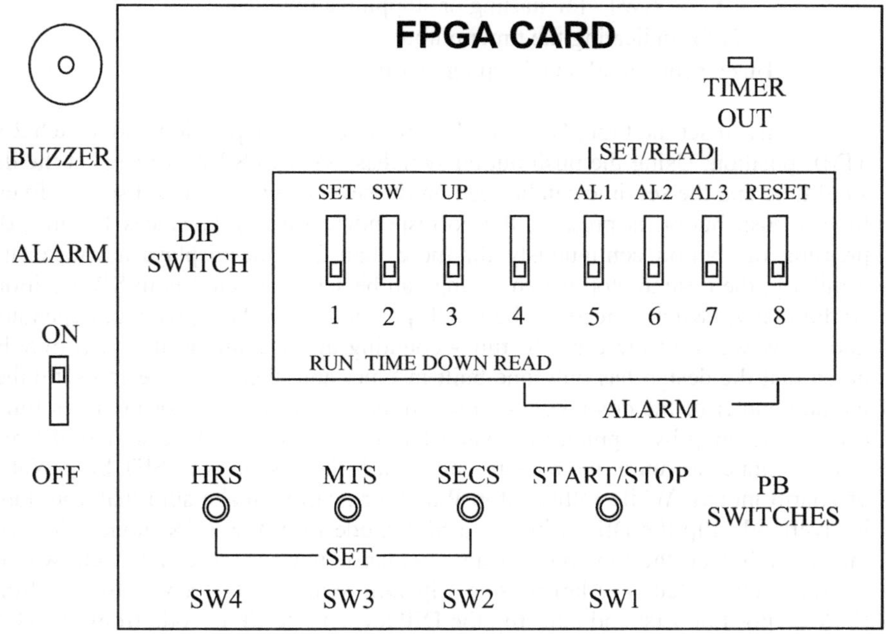

Fig. 14.9.2 Real time clock controls on FPGA board

on the FPGA board. This "Timer" output LED is switched on after the set timing is complete. The seven-segment LED displays are housed in the digital I/O card, while the buzzer and alarm ON/OFF switch are located externally.

Figure 14.9.2 presents the changed scenario to suit the FPGA board available. It shows the real time clock controls on FPGA board as well as those that are external to it shown on the left hand side. The usage of various controls and displays are as follows:

- DIP switch marked 1 to 8 in that order
  - "1": SET/RUN for setting the time or stop watch/running the clock or stop watch
  - "2": TIME/SW for configuring the equipment in the time or stop watch mode
  - "3": DOWN/UP for configuring the equipment in the up or down counter mode
  - "4": ALARM READ/SET for reading either what is set or setting what alarm time is required
  - "5" to "7": AL1 to AL3 for setting (top position) or reading (bottom position) what is set by the corresponding switch
  - "8": RESET for resetting or disabling the set alarms
- ALARM ON/OFF for switching on or off the sound alarm
- Push button switches marked 4 to 1 in that order
  - "SW4" to "SW2" for setting HRS/MTS/SECS respectively

    - "SW1" for starting or stopping a timer
- LED for indicating the timer output
- Buzzer for sounding a beeping alarm.

In order to set the time, DIP switch 1 must be in SET position and switch 2 in TIME position. Using the push button switches, SW4 to SW2, we can set the desired time and the set time can be seen on the seven-segment LED displays. Every time a push button is pressed, the corresponding display advances by one. By pressing the button continuously for more than 2 s, the displays advance fast. Similarly, the desired stop watch timing can be set if switch 2 is in SW position. Additionally, switch 3 needs to be in UP position or DOWN position, depending upon how we want the timer to run – counting up or counting down. It may be noted that the design has only one built-in timer and, therefore, we can set either the up counter or the down counter and not both. The real time or up/down timer can start running by flipping the switch 1 to RUN position. Once started, the real time continues to run in the background even if the system is in SET Stop Watch or Alarm modes. While setting Stop Watch or Alarms, make sure that you don't inadvertently flip the DIP switch 1 to SET mode from the RUN mode. This will immediately stop the running real time, which is not desirable. The right way to set the Stop Watch or Alarms is to flip DIP switch 2 to "SW" position from "TIME" position first and later flip the DIP switch 1 to SET mode from the RUN mode. In short, ensure that the system is in SW mode (DIP switch 2) before you switch to SET (DIP switch 1) mode. To remove this problem, in real dedicated equipment, you can use a thumbwheel switch, preferably a press +/– button.

In SET Stop Watch mode, you can reuse SW4 to SW2 to set the HRS/MTS/SECS just as you did for the real time. This setting holds good for UP or DOWN counting timer depending upon the DIP switch 2 setting. The push button switch SW1 (START/STOP) is used to start a timer or stop the running timer, be it UP or DOWN. This switch toggles between the two functions, start and stop. If the running timer is in UP mode, the display advances every second from "00 00 00" until it reaches the set time. At this moment, the display freezes at the set point turning on the "TIMER" output LED and sounds the beeping alarm in the buzzer for 30 s. On the other hand, if the timer is in DOWN mode, the display decrements every second from the set time value until it reaches "00 00 00". Here too, the display freezes although at zero, turning on the "TIMER" output LED and sounds the beeping alarm in the buzzer for 30 s. The "TIMER" output may be optionally connected to a relay, a contactor or a solid state relay to fire a rocket, switch on an equipment, a heater, or any other device as per user application. The timer can be started or stopped at any point of time. If it is stopped, the display freezes at the current running time. If START is pressed once again, it continues from where it stopped previously. This feature will be convenient if this timer is used in a dark room for developing photographs and the user wants to hold the processing temporarily.

There are three alarms, AL1 to AL3, available in the present design and can be extended to more numbers by adding Verilog codes similar to that shown for the three alarms. These alarms can be set by DIP switch 4 in SET position, followed by flipping one of the switches, 5 to 7, to SET position (Up), one at a time. In this mode of alarm setting, you can reuse SW4 to SW2 to set the HRS/MTS/SECS just as you did for the real time or the stop watch. You can read back the alarm setting by flipping the DIP switch 4 to READ position. If all the alarm switches 5 to 7 are in UP position, then AL1 alone will be displayed. AL1 has the top most priority and AL3 has the least priority while reading. Therefore, in order to read AL3, you will have to switch down the other priority alarms AL1 and AL2. The DIP switch 8, RESET, is used (in Up position) to clear all the alarm settings. All the three alarms can be disabled (OFF position) by a switch external to the FPGA board. This only switches off the beeping alarm for real time, and not its setting or stop watch/timer functioning.

14.5.4 Detailed Specification of the Real Time Clock

In the previous section, we defined the problem, identified possible applications and formulated its features. This was followed by more details such as how the front fascia must look like, and take stock of what hardware we have on hand in order to develop the product based on these hardware. Once the product design is proven, we may take up building the minimum possible hardware, be it industrial model mounted on a control panel, a table top model, or a mobile model. The last model needs power reduction techniques in the design and its implementation in order to consume least possible power, that is typical of any battery operated device. Before we start coding this application in Verilog, we need to formulate the detailed specification. While coding in Verilog, we will use the same nomenclature

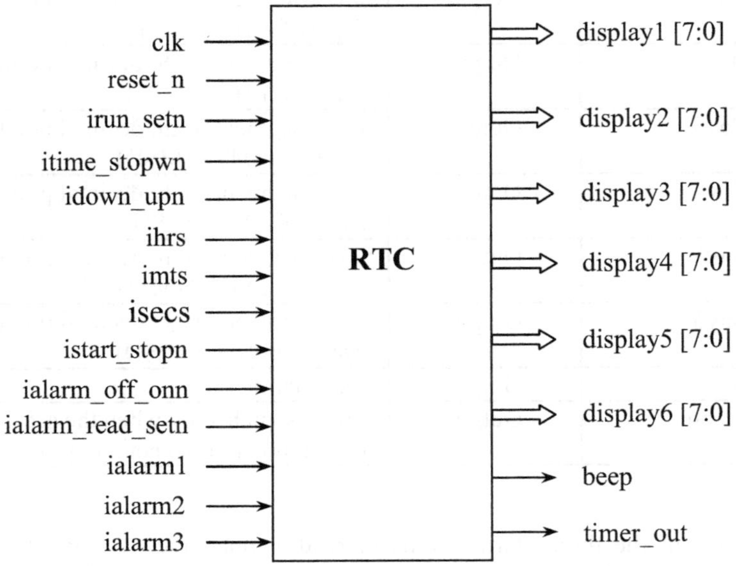

Fig. 14.10 Block diagram of real time clock

Table 14.1 Signal Description of Real Time Clock

Signal	Input/Output	Description
clk	Input	System clock
reset_n	Input	Asynchronous, active low sys. reset
irun_setn	Input	RUN/SET mode switch
itime_stopwn	Input	TIME/STOP WATCH mode switch
idown_upn	Input	UP/DOWN mode switch
ihrs	Input	Push button switch for setting "hrs"
imts	Input	Push button switch for setting "mts"
isecs	Input	Push button switch for setting "secs"
istart_stopn	Input	Start/Stop push button switch
ialarm_off_onn	Input	Sound alarm ON/OFF switch (common to all the three alarms)
ialarm_read_setn	Input	This switch enables reading or setting of the alarms
ialarm1	Input	Alarm SET or READ/OFF switch for Alarm 1
ialarm2	Input	Alarm SET or READ/OFF switch for Alarm 2
ialarm3	Input	Alarm SET or READ/OFF switch for Alarm 3
display1	Output	Seven segment, right decimal point LED display for HRS (MSD)
display2	Output	Seven segment, right decimal point LED display for HRS (LSD)
display3	Output	Seven segment, right decimal point LED display for MTS (MSD)
display4	Output	Seven segment, right decimal point LED display for MTS (LSD)
display5	Output	Seven segment, right decimal point LED display for SECS (MSD)
display6	Output	Seven segment, right decimal point LED display for SECS (LSD)
beep	Output	Beep alarm
timer_out	Output	This is switched on when the set time expires in up or down counter mode

Note:
Only one of the three alarms is set or read at one time. If more than one is set, the highest priority alarm will alone be actually read and the others are ignored. Alarm 1 has the highest priority and Alarm 3 has the lowest priority.

for various signals as shown in the block diagram of the real time clock in Figure 14.10. Various signals used in the design are presented in Table 14.1. Except for the system clock, "clk", the input signals have one to one correspondence with the switches presented in the front fascia earlier. For examples, "reset_n" signal is the same as "RESET" on the front fascia, "irun_setn" is the same as the "SET/RUN" switch, and "ihrs", "imts", "isecs" are inputs corresponding to "HRS", "MTS", and "SECS" push button switches. In these signals, "I" denotes "input" and "n" denotes negative, rather meaning an active low signal. For example, the single bit input signal, "itime_stopwn" indicates that the system is in "Time" mode or in "Stop Watch" mode if its digital value is "High" and "Low" respectively. If "ialarm1" to "ialarm3" signals are high (UP position), then the corresponding alarms can be set or read. If a switch is in DOWN position, the corresponding alarm can neither be set, nor read. The outputs of the RTC Core are "display1" to "display6" to drive the "HRS MTS SECS" display, "beep" for sounding the buzzer and "time_out" for switching on a LED/Relay output after the set time has expired.

14.5.5 Simplified Architecture of RTC

The simplified architecture of the real time clock is shown in Figure 14.11. All the input signals are exactly as shown in the block diagram in Figure 14.10. So also are the two outputs, "beep" and "timer_out". Instead of display1 to display6,

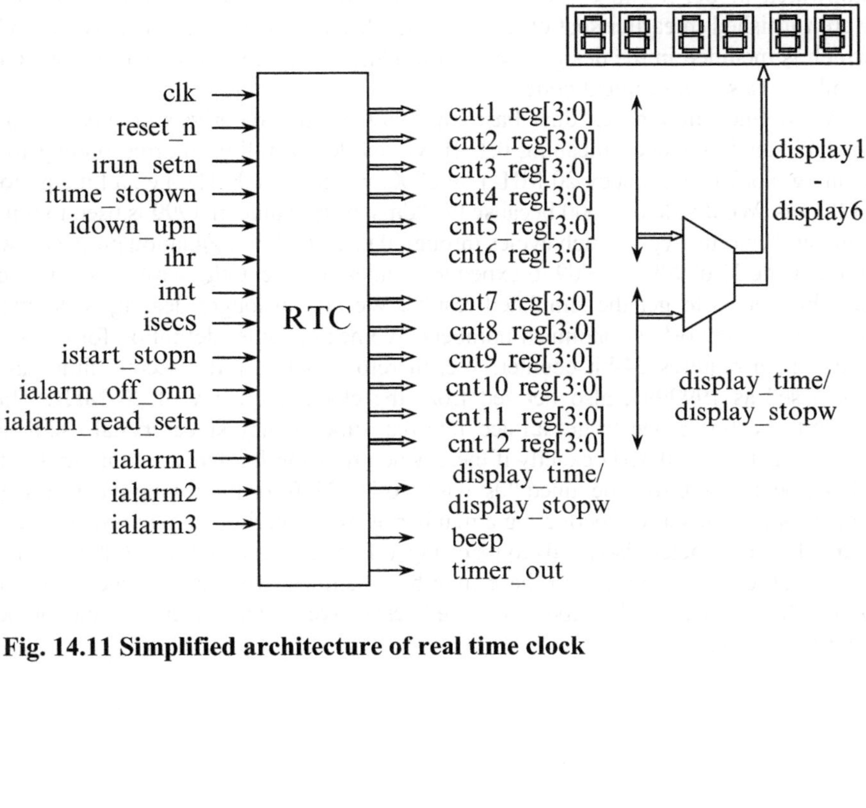

Fig. 14.11 Simplified architecture of real time clock

we have two pairs of outputs, cnt1 to cnt6 and cnt7 to cnt12. The counters, "cnt1 to cnt6", are the cores of running real time and the counters, "cnt7 to cnt12" are for running stop watch counter/timer. While using a stop watch, we should not disturb the smooth running of the real time. In other words, both the real time and the stop watch must run concurrently. This will be possible only if two pairs of counters are used, instead of one set. The counters, "cnt1 (MSD), cnt2" and "cnt7 (MSD), cnt8" are for display of 'HRS'; the counters, "cnt3 (MSD), cnt4" and "cnt9 (MSD), cnt10" are for display of 'MTS'; and the counters, "cnt5 (MSD), cnt6" and "cnt11 (MSD), cnt12" are for display of "SECS". When the system is in "TIME" mode, "cnt1 to cnt6" contents are output to the seven-segment LED displays via "display1–display6", whereas in the "SW" mode, the counters "cnt7 to cnt12" are displayed. This is done by a MUX as shown in the figure, whose inputs are selected by the control, "display_time/display_stopw".

14.5.6 Verilog Code for Real Time Clock

Verilog_code_14.4 presents the code of the real time design, "rtc_alarm.v". The test bench for this design is "rtc_alarm_test.v" and will be presented in a later section. As presented in earlier sections, this design of a real time clock is for displaying time as well as to serve as a stopwatch or a timer. Timing range is 00 00 00 to 23 59 59 (HRS MTS SECS). Up to three different sound alarms can be set. In the stopwatch mode, it can count either Up or Down. More detailed specifications were furnished in earlier sections. The design has a sub-module, "display_rom.v", which is included in the design. The sub-module is required for converting a BCD number to a seven segment code.

We define a time base, "dms_base", as 1999 in order to derive a time base of 0.1 ms "d" stands for deci, meaning 0.1. If we divide 20 million (corresponding to a running clock of frequency 20 MHz) by 2000, we get 10,000 Hz, i.e., a time period of 0.1 ms. We divide by 2000 because the "dms_base" value of 1999 is used to run a counter, "cntdms_reg", which cycles through 0 to 1999. For simulation purposes, we will change 13'd1999 to 13'd9 to expedite simulation time. Otherwise, it will be impossible for us to get the simulated results. We are no longer dealing with small numbers but almost astronomical numbers. We need another definition for decisecond, which requires 24-bit counter. We, therefore, define a decisecond time base, "ds_base" as 1999999, also derived from the clock frequency of 20 MHz. This needs to be fine tuned to get an accurate real time display since crystal oscillator may not generate 20 MHz exactly if there is no provision for hardware tuning. In the FPGA board used, we fine tuned "ds_base" to 2003340 for improving the accuracy. This was obtained after conducting a number of trials running the real time on hardware. This parameter also needs to be changed from 23'd2003340 to 23'd9 for simulation purposes. In addition to these time base settings, we need 1 s and 3 m time bases, "time_base" and "debounce_time" respectively. The latter is used for debouncing switches.

We now declare the design module, "rtc_alarm", listing all inputs/outputs. This is followed by declarations of I/Os, nets, and registers in the design. The I/Os are precisely the same as presented in previous sections. Remember that all the combinational signals assigned are "wire" and registers (in sequential always blocks) are "reg". These signals will be made clear at the appropriate time when we discuss the codes using them.

The real time clock implementation starts with the coding of counters, whose time base values were discussed earlier. These counters are "cntdms_reg", "cntds_reg", "cntb_reg", and "deb_cnt_reg", and are reset with power on/system reset conditions or when the respective running counter equals the corresponding time base mentioned earlier. Otherwise, these counters are advanced by one at the rising edge of the "clk" if the set conditions are satisfied. The first two counters are free of conditions, while "cntb_reg" and "deb_cnt_reg" have the following conditions, "tbsec" and "dmsec" respectively:
tbsec = (cntb_reg == `time_base)&(cntds_reg == `ds_base), which means time base in seconds and
dmsec = (deb_cnt_reg == `debounce_time)&(cntdms_reg == `dms_base), which gives the debounce time of 3 ms.
In all the cases of increments/decrements or presets, we use the "assign" statements to realize these in advance, exactly the same way we did in other designs.

The next sequential block is for reading all the input switches such as RUN/SET (signal "irun_setn"), TIME/SW (signal "itime_stopwn"), etc., where "n" implies that particular signal is low. For example, if RUN/SET is in SET mode, then irun_setn is low. At the start, before debouncing, all the inputs are registered. This is the first input sampling, preparing for debouncing next. After 3 ms, all these inputs are debounced. The method of debouncing switches depends upon whether the signal is active high or active low. For examples, the RUN/SET DIP switch and HRS push button switch are debounced differently, the first by "AND" gate and the second by "NOR" gate as per the following condition:

```
run_setn   <=   irun_setn && rrun_setn ;
hrs        <=   !(ihrs || rhrs) ;
```

Similarly, other switches.

"cnt1_reg" is the Time watch's most significant HOUR digit. This is reset or advanced only in the RUN and TIME mode of operation, which can be detected by the signal, "run_time", realized using the statement:

```
assign run_time = (run_setn == 1)&(time_stopwn == 1) ;
```

Similarly, other modes are identified by a number of "assign" statements that follow the above statement. The next three assign statements after the above statement sense when the HRS/MTS/SECS push buttons are pressed or sense 0.1 s clock tick if the push buttons are kept pressed for 2 s or more. The signal, "hrs_d", etc. goes high only after 2 s of ON delay. The condition for resetting the "cnt1_reg" counter is "cnt1 - cnt6" = 23 59 59 in TIME RUN mode or "cnt1–cnt2" = 23 in HRS TIME SET mode. Similarly, "cnt1–cnt2" = 09 or 19, "cnt3–cnt6" = 59 59 in TIME RUN mode or "cnt–cnt2" = 09 or 19 in HRS TIME SET mode are the conditions for advancing this counter. The next sequential block is the realization of "cnt1_reg" counter. The above treatment holds good for all other counters, "cnt2_reg" to "cnt6_reg".

Code for stop watch implementation is presented next. Run and stop watch DOWN mode is coded first. Counters, "cnt7_reg" to "cnt12_reg", are used to hold the running value of stopwatch, where "cnt7" is HRS (MSD) and so on, as is the case with "cnt1" to "cnt6". "cnt7_reg" is the stop watch's most significant HOUR digit. This is reset or advanced only in the RUN and STOP WATCH mode of operation. For presetting in down counting mode, "cnt7_reg" to "cnt12_reg" are used, whereas for up counting, "term_count_reg1" to "term_count_reg6" are used for presetting the user desired terminal values. "cnt7_reg" to "cnt12_reg" are the running counters for both up and down counting. Reset, advance (for up counter), decrement (for down counter) and preset signals for these counters are generated by "assign" statements and are self-explanatory. Reset, advance conditions and counter realizations for "term_count_reg1" to "term_count_reg6" are similar to other counters we have covered already. The signal, "timer out", is set when the terminal count for Up or Down is reached. For the up counter, the preset values are contained in "term_count_reg1" to "term_count_reg6". The terminal count for down counter is "00 00 00".

When "timer out" is set, the buzzer is activated using the signal, "timer_out_alarm" and the "timer_out_alarm_counter". This signal is high for 30 s after the terminal count is reached, i.e., when the "timer_out" is set. The signal, "start_stopn", is the debounced status of START/STOP push button switch and "start_stopn_reg" stores the detected depression of the push button switch. This signal toggles between START and STOP using the same push button switch. The next three sequential blocks are for "hrs2s_reg", "mts2s_reg" and "secs2s_reg", to generate 2 s delays when the respective push button switches marked HRS, MTS, and SECS are pressed. The set display advances by one with every key depression. If setting needs to be advanced fast automatically, then the push button switches are depressed and held for more than 2 s. When the desired setting closes in, the button may be pressed repeatedly a few times till the display shows what is required. These codes are self-explanatory since they are profusely commented.

Next group of codes present the alarm implementation, wherein "temp_alarm_reg1" to "temp_alarm_reg6" are a set of 4-bit temporary registers, which hold the alarm time when it is being set. The conditions for advancing, resetting the alarm settings are similar to the counters described earlier. So also is the case for the functioning of the counters, "temp_alarm_reg1" to "temp_alarm_reg6". "set_alarm1" is a signal which indicates that Alarm 1 is being set. When this signal is high, the contents of "temp_alarm_reg1", etc. are copied into "alarm1_reg1" and so on. Similarly, for the other two alarms, Alarm 2 and Alarm 3. These are realized by simple sequential blocks. Alarm 1 has the top most priority while reading. "read_alarm_reg" is a 2-bit register, which stores the number of the alarm to be read. If no alarm is on, it stores "0". If more than one alarm is on, the one displayed is the top priority alarm. Alarm 1 has the top most priority while reading. Alarm 3 is the least priority. The next combinational "always" block accomplishes this feat.

The next sequential block displays real time or stopwatch or alarm on the seven-segment LEDs. The registers, "data1" to "data6", derive their input data from "cnt1_reg" to "cnt6_reg" if the signal, "display_time", is asserted or from "term_count_reg1" to "term_count_reg6" if "display_stopw" is asserted while counting "Up" or from "cnt7_reg" to "cnt12_reg" if the signal, "display_stopw" is asserted while counting "Down". Otherwise, if "set_alarm = 1", which indicates that alarm is set, then "temp_alarm_reg1" to temp_alarm_reg6 will be written into "data1" to "data6". On the other hand, if "display_alarm = 1", which means display the Alarm set via "data1" to "data6", then "alarm1_reg1" to "alarm1_reg6" or "alarm2_reg1" to "alarm2_reg6" or "alarm3_reg1" to "alarm3_reg6" are displayed, depending upon the value of "read_alarm_reg" described earlier.

The next three sequential blocks are for running a 30 s timer to sound the audio alarm corresponding to the three alarms, Alarm 1, Alarm 2, and Alarm 3 if the corresponding set points are reached by the real time clock. For example, "alarm1_30sec_delay" is a single bit which becomes "1" when "alarm1_match = 1". It stays high for 30 s and then goes low. alarm1_30_sec is a counter which counts till 30. It counts so long as alarm1_30sec_delay is high. It is incremented every 1 s, i.e., when "tbsec = 1". The reader is urged to figure this out from the codes presented.

The signal, "ring" indicates that one or more alarms is/are active. "beep" is the signal (square wave) which is actually output to the speaker if ring is high. "beep_counter" counts till 2 (means 0.2 s). When it is 2, "beep" signal is toggled repeatedly producing 2.5 Hz beeping tone if alarm OFF/ON switch is in the ON position. Otherwise, the sound alarm is OFF.

Before we wind up the description of the Verilog code for the real time clock, we need to call the "display_rom" module, which converts the BCD code to seven segment code for eventual display on the LED. This is called six times corresponding to six numbers of seven segment displays. The inputs to these displays are the registers, "data1" to "data6", which was described earlier. "display1" (HRS _ MSD) through "display6" (SECS _ LSD) are the final outputs of the real time clock, which drives the seven-segment LED displays. It may be noted that all decimal points are turned off.

Verilog_code_14.4

```
/*
                Verilog RTL Code for Real Time Clock

Design is "rtc_alarm.v".
Test bench for this design is "rtc_alarm_test.v".
This is the design for a real time clock to display time as well as function as a
stopwatch. In the latter mode, it can count Up or Down. Up to three different
sound alarms can be set. Timing range: 00 00 00 to 23 59 59 (HRS MTS SECS).
For more details, see specification sheet.
*/
```

```
`include "display_rom.v"            // Sub-module for converting a BCD number
                                    // to drive seven segment LED outputs.
`define  dms_base          13'd1999
                           // For 20 MHz operation, the time base is 0.1 ms.
                           // Change 13'd1999 to 13'd9 for simulation purposes.
`define  ds_base 23'd2003340
         // For 20 MHz operation, the time base is 0.1 s for 23'd1999999
         // setting. Fine tuned to 23'd2003340 for improving the accuracy.
         // Change 23'd2003340 to 23'd9 for simulation purposes.
`define  time_base         4'd9     // This is the 1 s time base.
`define  debounce_time     5'd29    // Switch debounce time is 3 ms.

module rtc_alarm (
                           clk,
                           reset_n,
                           irun_setn,                // "i" stands for input.
                           itime_stopwn,
                           idown_upn,
                           ihrs,
                           imts,
                           isecs,
                           istart_stopn,
                           ialarm_off_onn,
                                  // "0" switches ON alarm, otherwise OFF.
                           ialarm_read_setn,   // "0" is set mode, otherwise read.
                           ialarm1,          // "0" sets or reads the
                           ialarm2,          // corresponding alarm.
                           ialarm3,          // Otherwise, off.
                           display1,         // seven-segment LED outputs –
                           display2,         // display1 (MSD), 2 are HRS,
                           display3,         // display3 (MSD), 4 are MTS,
                           display4,         // display5 (MSD), 6 are SECS.
                           display5,
                           display6,
                           beep,  // Beeping alarm => use a piezo-electric buzzer.
                           timer_out         // Signals when the set time is over.
                    );
input          clk ;               // Declare inputs/outputs.
input          reset_n ;           // Asynchronous, active low.
input          irun_setn ;         // RUN/SET mode switch.
input          itime_stopwn ;      // TIME/STOP WATCH mode switch.
input          idown_upn ;         // UP/DOWN mode switch.
input          ihrs ;              // Push button switches for setting "hrs",
input          imts ;              // "mts" and
input          isecs ;             // "secs".
input          istart_stopn ;      // Start/Stop PB.
```

```
input          ialarm_off_onn;        // Sound alarm ON/OFF switch
                                      // (common to all the three alarms).
input          ialarm_read_setn;      // This switch enables reading
                                      // or setting of the alarms.
input          ialarm1;          // Alarm SET or READ/OFF switch for Alarm 1,
input          ialarm2;               // Alarm 2 and
input          ialarm3;               // Alarm 3.
/*
Note: Only one of the three alarms can be read at one time. If more than one is
attempted to be read, the highest priority alarm alone will be actually read, and
others are ignored. Alarm 1 is the highest priority and Alarm 3 is the lowest.
*/
output         [7:0]          display1 ;
output         [7:0]          display2 ;
output         [7:0]          display3 ;
output         [7:0]          display4 ;
output         [7:0]          display5 ;
output         [7:0]          display6 ;
output                        beep;
output                        timer_out;

wire           [7:0]          display1 ;   // Declare outputs as nets
                                           // (combinational circuit outputs).
wire           [7:0]          display2 ;
wire           [7:0]          display3 ;
wire           [7:0]          display4 ;
wire           [7:0]          display5 ;
wire           [7:0]          display6 ;
wire           [12:0]         cntdms_next ; // Declare other combinational
                                            // circuit  signals as nets.
wire           [22:0]         cntds_next ;
wire           [3:0]          cntb_next ;
wire           [4:0]          deb_cnt_next ;
wire                          dmsec ;
wire                          tbsec ;
wire                          run_time ;
wire                          set_time ;
wire                          set_stopw ;
wire                          run_stopw;
wire                          adv_hrs ;
wire                          adv_mts  ;
wire                          adv_secs ;
wire                          adv_hrs_time ;
wire                          adv_hrs_sw ;
wire                          adv_mts_time ;
wire                          adv_mts_sw ;
```

```
wire                     adv_secs_time ;
wire                     adv_secs_sw ;
wire                     res_cnt1 ;
wire                     res_cnt2 ;
wire                     res_cnt3 ;
wire                     res_cnt4 ;
wire                     res_cnt5 ;
wire                     res_cnt6 ;
wire                     res_cnt7 ;
wire                     res_cnt8_sw ;
wire                     res_cnt8_set ;
wire                     res_cnt9 ;
wire                     res_cnt10 ;
wire                     res_cnt11 ;
wire                     res_cnt12 ;
wire                     pres_cnt8 ;
                     // Signal to indicate preset condition for the counter.
wire                     pres_cnt9 ;
wire                     pres_cnt10 ;
wire                     pres_cnt11 ;
wire                     pres_cnt12 ;
wire                     adv_cnt1 ;
wire                     adv_cnt2 ;
wire                     adv_cnt3 ;
wire                     adv_cnt4 ;
wire                     adv_cnt5 ;
wire                     adv_cnt6 ;
wire                     adv_cnt7 ;
wire                     adv_cnt8 ;
wire                     adv_cnt8_set ;
wire                     adv_cnt8_sw ;
wire                     adv_cnt9 ;
wire                     adv_cnt10 ;
wire                     adv_cnt11 ;
wire                     adv_cnt12 ;
wire         [3:0]       cnt1_next ;
wire         [3:0]       cnt2_next ;
wire         [3:0]       cnt3_next ;
wire         [3:0]       cnt4_next ;
wire         [3:0]       cnt5_next ;
wire         [3:0]       cnt6_next ;
wire         [3:0]       cnt7_next ;
wire         [3:0]       cnt8_next ;
wire         [3:0]       cnt9_next ;
wire         [3:0]       cnt10_next ;
wire         [3:0]       cnt11_next ;
```

```
wire    [3:0]    cnt12_next ;
wire             rsd ;  // RUN, STOPWATCH, DOWN mode.
wire             rsd_cnt8_res ;
wire             cnt8_res ;
wire             decr_cnt7 ;      // Signal to indicate decrement
                                  // condition for the stopwatch counter.
wire             decr_cnt8 ;
wire             decr_cnt9 ;
wire             decr_cnt10 ;
wire             decr_cnt11 ;
wire             decr_cnt12 ;
wire    [3:0]    cnt7_nextd ;     // "d" stands for decrement.
wire    [3:0]    cnt8_nextd ;
wire    [3:0]    cnt9_nextd ;
wire    [3:0]    cnt10_nextd ;
wire    [3:0]    cnt11_nextd ;
wire    [3:0]    cnt12_nextd ;
wire    [4:0]    hrs2s_next ;
wire    [4:0]    mts2s_next ;
wire    [4:0]    secs2s_next ;

reg     [3:0]    data1 ;          // Declare registers.
reg     [3:0]    data2 ;
reg     [3:0]    data3 ;
reg     [3:0]    data4 ;
reg     [3:0]    data5 ;
reg     [3:0]    data6 ;
reg     [12:0]   cntdms_reg ;
reg     [22:0]   cntds_reg ;
reg     [3:0]    cntb_reg ;
reg     [4:0]    deb_cnt_reg ;
reg              rrun_setn ;      // " r" for registered values of
                                  // RUN/SET switch, etc.
reg              rtime_stopwn ;
reg              rdown_upn ;
reg              rhrs ;
reg              rmts ;
reg              rsecs ;
reg              rstart_stopn ;
reg              ralarm_off_onn;
reg              ralarm1;
reg              ralarm2;
reg              ralarm3;
reg              run_setn ;
reg              time_stopwn ;
reg              down_upn ;
```

```
reg                        hrs ;
reg                        mts ;
reg                        secs ;
reg                        start_stopn ;
reg                        alarm_off_onn;
reg                        alarm_read_setn;
reg                        alarm1;
reg                        alarm2;
reg                        alarm3;
reg          [3:0]         cnt1_reg ;   // cnt1 cnt2  cnt3 cnt4  cnt5 cnt6
                                 // stores running time in
reg          [3:0]         cnt2_reg ;   // HRS MTS SECS
reg          [3:0]         cnt3_reg ;
reg          [3:0]         cnt4_reg ;
reg          [3:0]         cnt5_reg ;
reg          [3:0]         cnt6_reg ; // cnt7 cnt8  cnt9 cnt10  cnt11 cnt12
                             // stores the time count (counter) in
                             // HRS MTS SECS
reg          [3:0]         cnt7_reg ;
reg          [3:0]         cnt8_reg ;
reg          [3:0]         cnt9_reg ;
reg          [3:0]         cnt10_reg ;
reg          [3:0]         cnt11_reg ;
reg          [3:0]         cnt12_reg ;
reg                        start_stopn_reg ;
               // START/STOP mode register => start_stopn_reg = 1 means
                                                   // START, otherwise STOP.
reg                  start_stopnp_reg ;        // Previous value of start/stop.
reg                        hrsp_reg ;          // Previous value of HRS PB.
reg                        hrs_d ;             // ON delay output of HRS PB.
reg          [4:0]         hrs2s_reg ;
                            // Counter to keep track of 2 s ON delay.
reg                        mtsp_reg ;          // Similar signals for MTS and
reg                        mts_d ;
reg          [4:0]         mts2s_reg ;
reg                        secsp_reg ;         // SECS.
reg                        secs_d ;
reg          [4:0]         secs2s_reg ;
wire                       adv_res_cnt2 ;
                // "adv", "res" mean advance and reset counter respectively.
wire                       res_cnt2_time ;
wire                       res_cnt2_set ;
reg          [3:0]         temp_alarm_reg1 ;
                      // Individual alarms are set via temporary registers.
reg          [3:0]         temp_alarm_reg2 ;
reg          [3:0]         temp_alarm_reg3 ;
```

```
reg      [3:0]    temp_alarm_reg4 ;
reg      [3:0]    temp_alarm_reg5 ;
reg      [3:0]    temp_alarm_reg6 ;
wire     [3:0]    temp_alarm_reg1_next ;
wire     [3:0]    temp_alarm_reg2_next ;
wire     [3:0]    temp_alarm_reg3_next ;
wire     [3:0]    temp_alarm_reg4_next ;
wire     [3:0]    temp_alarm_reg5_next ;
wire     [3:0]    temp_alarm_reg6_next ;
wire              adv_temp_alarm_reg1 ;
wire              adv_temp_alarm_reg2 ;
wire              adv_temp_alarm_reg3 ;
wire              adv_temp_alarm_reg4 ;
wire              adv_temp_alarm_reg5 ;
wire              adv_temp_alarm_reg6 ;
wire              res_temp_alarm_reg1 ;
wire              res_temp_alarm_reg2 ;
wire              res_temp_alarm_reg3 ;
wire              res_temp_alarm_reg4 ;
wire              res_temp_alarm_reg5 ;
wire              res_temp_alarm_reg6 ;
reg      [3:0]    alarm1_reg1 ;          // Alarm set registers
reg      [3:0]    alarm1_reg2 ;
reg      [3:0]    alarm1_reg3 ;
reg      [3:0]    alarm1_reg4 ;
reg      [3:0]    alarm1_reg5 ;
reg      [3:0]    alarm1_reg6 ;
reg      [3:0]    alarm2_reg1 ;
reg      [3:0]    alarm2_reg2 ;
reg      [3:0]    alarm2_reg3 ;
reg      [3:0]    alarm2_reg4 ;
reg      [3:0]    alarm2_reg5 ;
reg      [3:0]    alarm2_reg6 ;
reg      [3:0]    alarm3_reg1 ;
reg      [3:0]    alarm3_reg2 ;
reg      [3:0]    alarm3_reg3 ;
reg      [3:0]    alarm3_reg4 ;
reg      [3:0]    alarm3_reg5 ;
reg      [3:0]    alarm3_reg6 ;
wire              adv_hrs_temp_alarm ;
wire              adv_mts_temp_alarm ;
wire              adv_secs_temp_alarm ;
wire              set_alarm ;
reg               ralarm_read_setn ;
wire              set_alarm1 ;
wire              set_alarm2 ;
```

```
wire                     set_alarm3 ;
reg          [1:0]       read_alarm_reg ; // "0" – No alarm display,
                                    // "1" – Alarm1 display, etc.
wire                     display_time ;
wire                     display_stopw ;
wire                     display_alarm ;
reg          [4:0]       alarm1_30sec_counter ;
reg                      alarm1_30sec_delay ; // 30 s alarm1 timer
                                                  // output, active high.
wire                     alarm1_match ; // Goes high when the run time
                                  // matches the alarm1 set point.
wire         [4:0]       alarm1_30sec_counter_next ;
wire                     adv_alarm1_30sec_counter ;
reg          [4:0]       alarm2_30sec_counter ;
reg                      alarm2_30sec_delay ;
wire                     alarm2_match ;
wire         [4:0]       alarm2_30sec_counter_next ;
wire                     adv_alarm2_30sec_counter ;
reg          [4:0]       alarm3_30sec_counter ;
reg                      alarm3_30sec_delay ;
wire                     alarm3_match ;
wire         [4:0]       alarm3_30sec_counter_next ;
wire                     adv_alarm3_30sec_counter ;
reg                      beep ;    // Generates square pulse for beeping
wire                     ring ;    // of a buzzer if this signal is active.
reg          [2:0]       beep_counter ;
                         // Counter for generating square pulse and its
wire         [2:0]       beep_counter_next ;        // advanced counter.
reg          [3:0]       term_count_reg1;  // For use with up counter.
reg          [3:0]       term_count_reg2 ;
reg          [3:0]       term_count_reg3 ;
reg          [3:0]       term_count_reg4 ;
reg          [3:0]       term_count_reg5 ;
reg          [3:0]       term_count_reg6 ;
wire         [3:0]       term_count_reg1_next ;
wire         [3:0]       term_count_reg2_next ;
wire         [3:0]       term_count_reg3_next ;
wire         [3:0]       term_count_reg4_next ;
wire         [3:0]       term_count_reg5_next ;
wire         [3:0]       term_count_reg6_next ;
wire                     adv_term_count_reg1;
wire                     res_term_count_reg1;
wire                     adv_term_count_reg2 ;
wire                     res_term_count_reg2 ;
wire                     adv_term_count_reg3 ;
wire                     res_term_count_reg3 ;
```

```
wire                        adv_term_count_reg4 ;
wire                        res_term_count_reg4 ;
wire                        adv_term_count_reg5 ;
wire                        res_term_count_reg5 ;
wire                        adv_term_count_reg6 ;
wire                        res_term_count_reg6 ;
wire                   adv_hrs_tcr ;    // "tcr" means terminal count register.
wire                        adv_mts_tcr ;    // Applicable while setting.
wire                        adv_secs_tcr ;
wire                        term_count_reached_up ;
wire                        term_count_reached_down ;
reg          [4:0]          timer_out_alarm_counter ;
                                   // 30 s counter for audio alarm.
wire         [4:0]          timer_out_alarm_counter_next ;
wire                        timer_out_alarm;
```

// **Real time clock implementation**

```
// cntdms_reg is a free-running counter to provide the time base of 0.1 ms.
assign cntdms_next = cntdms_reg + 1 ;    // Pre-increment the 0.1 ms counter.

always @ (posedge clk or negedge reset_n)
begin
        if (reset_n == 1'b0)
                cntdms_reg <= 13'd0 ; // Initialize when the system is reset.
        else if (cntdms_reg  == `dms_base )    // Also reset if terminal count is
                cntdms_reg <= 13'd0 ;          // reached. Otherwise,
        else
                cntdms_reg <= cntdms_next ; // advance the count once.
end

// cntds_reg is a free-running counter to provide the time base of
// decisecond (0.1 s) for the time watch or the stopwatch.
assign cntds_next = cntds_reg + 1 ;          // Pre-increment the counter.

always @ (posedge clk or negedge reset_n)
begin
        if (reset_n == 1'b0)
                cntds_reg <= 23'd0 ;       // Initialize when the system is reset.
        else if (cntds_reg   == `ds_base )  // Also reset if terminal count is
                cntds_reg <= 23'd0 ;       // reached.
        else
                cntds_reg <= cntds_next ; // Otherwise, advance the count once.
end

assign tbsec = (cntb_reg  == `time_base)&(cntds_reg  == `ds_base) ;
                                                // Time base in seconds.
```

```
assign cntb_next = cntb_reg + 1 ;              // Pre-increment the counter.

always @ (posedge clk or negedge reset_n)
begin
        if (reset_n == 1'b0)
                cntb_reg <= 4'd0 ;         // Initialize when the system is reset.
        else if (tbsec == 1)
                cntb_reg <= 4'd0 ;  // Reset if terminal count (1 s) is  reached.
        else if (cntds_reg  == `ds_base)
                cntb_reg <= cntb_next ; // Advance the count once every  0.1 s.
        else                         ;         // Otherwise, ignore.
end

assign dmsec = (deb_cnt_reg  == `debounce_time)&(cntdms_reg == `dms_base) ;
                                               // Debounce time in milli seconds.
assign deb_cnt_next = deb_cnt_reg + 1 ;        // Pre-increment the counter.

always @ (posedge clk or negedge reset_n)
begin
        if (reset_n == 1'b0)
                deb_cnt_reg <= 5'd0 ; // Initialize when the system is reset.
        else if (dmsec == 1)
                deb_cnt_reg <= 5'd0 ;             // Reset after 3 ms delay.
        else if (cntdms_reg  == `dms_base)
                deb_cnt_reg <= deb_cnt_next ;  // Advance the count
                                               // once every 0.1 ms.
        else                         ;         // Otherwise, ignore.
end

always @ (posedge clk or negedge reset_n)
    begin
        if (reset_n == 1'b0)
          begin
                rrun_setn        <= 1 ;  // Initialize when the system is reset.
                rtime_stopwn     <= 1 ;  // "r" stands for register or store.
                rdown_upn        <= 0 ;
                rhrs             <= 0 ;
                rmts             <= 0 ;
                rsecs            <= 0 ;
                rstart_stopn     <= 0 ;
                ralarm_off_onn   <= 1 ;
                ralarm_read_setn <= 1 ;
                ralarm1          <= 1 ;
                ralarm2          <= 1 ;
                ralarm3          <= 1 ;
                run_setn         <= 1 ;
```

```
            time_stopwn      <= 1 ;
            down_upn         <= 0 ;
            hrs              <= 0 ;
            mts              <= 0 ;
            secs             <= 0 ;
            start_stopn      <= 0 ;
            alarm_off_onn    <= 1 ;
            alarm_read_setn  <= 1 ;
            alarm1           <= 1 ;
            alarm2           <= 1 ;
            alarm3           <= 1 ;
    end
  else if ((deb_cnt_reg == 0)&&(cntdms_reg == 0))
                                       // At the start before debouncing.
    begin
            rrun_setn        <= irun_setn  ; // Read the status of all inputs.
            rtime_stopwn     <= itime_stopwn  ;
            rdown_upn        <= idown_upn ;
            rhrs             <= ihrs ;
            rmts             <= imts ;
            rsecs            <= isecs ;
            rstart_stopn     <= istart_stopn ;     // istart_stopn = 0 means
            ralarm_off_onn    <= ialarm_off_onn;          // button pressed.
            ralarm_read_setn <= ialarm_read_setn;
            ralarm1          <= ialarm1;
            ralarm2          <= ialarm2;
            ralarm3          <= ialarm3;
    end
  else if ((deb_cnt_reg == `debounce_time)&&
                                            (cntdms_reg == `dms_base))
    begin
            run_setn         <= irun_setn && rrun_setn ;
                              // Read the status of all inputs and debounce.
            time_stopwn      <= itime_stopwn && rtime_stopwn ;
            down_upn         <= idown_upn && rdown_upn ;
            hrs              <= !(ihrs || rhrs) ;
            mts              <= !(imts || rmts) ;
            secs             <= !(isecs || rsecs) ;
            start_stopn      <= !(istart_stopn || rstart_stopn) ;
                                // start_stopn == 1 means button pressed.
            alarm_off_onn    <= (ialarm_off_onn && ralarm_off_onn) ;
            alarm_read_setn  <= (ialarm_read_setn && ralarm_read_setn) ;
            alarm1           <= !(ialarm1 || ralarm1);
            alarm2           <= !(ialarm2 || ralarm2);
            alarm3           <= !(ialarm3 || ralarm3);
    end
```

```
        else                          ;          // Otherwise, ignore.
end

    // cnt1_reg is the Time watch's most significant HOUR digit. This is reset or
    // advanced only in the RUN and TIME mode of operation.
assign run_time   =   (run_setn == 1)&(time_stopwn == 1) ;
                                      // These signals identify RUN TIME,
assign set_time   =   (run_setn == 0)&(time_stopwn == 1)&(set_alarm == 0) ;
                                      // SET TIME &
assign set_stopw  =   (run_setn == 0)&(time_stopwn == 0)&(set_alarm == 0) ;
                                      // SET STOPWATCH modes respectively.
              // Note that if alarm is being set, we can't set time or stop watch.
assign run_stopw = (run_setn == 1)&(time_stopwn == 0) ;
                                                    // Run stopwatch mode.
assign set_alarm  = (alarm_read_setn == 0)&(set_stopw == 1);
                                        // Set or Read alarm in set stopwatch mode.
/*
The following three statements sense when the HRS/MTS/SECS  push buttons are
pressed or sense 0.1 s clock tick if the push buttons are kept pressed for 2 s or
more. hrs_d etc. goes high only after 2 s ON delay.
*/
assign adv_hrs  = ((hrs == 1)&(hrsp_reg == 0))|
                                    ((hrs_d == 1)&(cntds_reg  == `ds_base )) ;
                                       // Advance every 0.1 s after 2 s delay.
assign adv_mts  = ((mts == 1)&(mtsp_reg == 0))|
                                    ((mts_d == 1)&(cntds_reg  == `ds_base )) ;
assign adv_secs = ((secs == 1)&(secsp_reg == 0))|
                                    ((secs_d == 1)&(cntds_reg  == `ds_base )) ;
assign adv_hrs_time = (adv_hrs == 1)&(set_time == 1) ;
assign adv_hrs_sw   = (adv_hrs == 1)&(set_stopw == 1)&(down_upn == 1) ;
assign adv_hrs_tcr  = (adv_hrs == 1)&(set_stopw == 1)&(down_upn == 0) ;
                                      // tcr => terminal count reg. Up count mode.
assign adv_mts_time  = (adv_mts == 1)&(set_time == 1) ;
assign adv_mts_sw    = (adv_mts == 1)&(set_stopw == 1)&(down_upn == 1) ;
assign adv_mts_tcr   = (adv_mts == 1)&(set_stopw == 1)&(down_upn == 0) ;
assign adv_secs_time = (adv_secs == 1)&(set_time == 1) ;
assign adv_secs_sw   = (adv_secs == 1)&(set_stopw == 1) (down_upn == 1) ;
assign adv_secs_tcr  = (adv_secs == 1)&(set_stopw == 1)&(down_upn == 0) ;
assign adv_hrs_temp_alarm       = (adv_hrs == 1)&(set_alarm == 1) ;
assign adv_mts_temp_alarm       = (adv_mts == 1)&(set_alarm == 1) ;
assign adv_secs_temp_alarm      = (adv_secs == 1)&(set_alarm == 1) ;
assign display_alarm            = ((alarm1 == 1)|(alarm2 == 1)|
                                     (alarm3 == 1))&(alarm_read_setn == 1) ;
assign display_time        = (time_stopwn == 1)&(display_alarm == 0)&
                                                          (set_alarm == 0) ;
assign display_stopw = (time_stopwn == 0)&(display_alarm == 0)&
```

```
                                        (set_alarm == 0) ;
// Only one of the three displays: Alarm/Time/Stopwatch is possible at one time.
assign res_cnt1 = ((set_time != 1)&(tbsec == 1)&(cnt1_reg == 2)&
                   (cnt2_reg == 3)&(cnt3_reg == 5)&(cnt4_reg == 9)&
                   (cnt5_reg == 5)&(cnt6_reg == 9))|((adv_hrs_time == 1)&
                   (cnt1_reg == 2)&(cnt2_reg == 3)) ;
                        // cnt1 - cnt6 = 23 59 59 in TIME RUN mode
                        // or cnt1 - cnt2 = 23 in HRS TIME SET mode is the
                        // condition for resetting this counter.
assign adv_cnt1 = ((set_time != 1)&(tbsec == 1)&(cnt1_reg < 2)&
                   (cnt2_reg == 9)&(cnt3_reg == 5)&(cnt4_reg == 9)&
                      (cnt5_reg == 5)&(cnt6_reg == 9))|((adv_hrs_time== 1)
                                    &(cnt1_reg < 2)&(cnt2_reg == 9)) ;
      // cnt1 - cnt2 = 09 or 19, cnt3 - cnt6 = 59 59 in TIME RUN mode or
      // cnt1 - cnt2 = 09 or 19 in HRS TIME SET mode is the condition for
      // advancing this counter.
assign cnt1_next = cnt1_reg + 1 ;          // Pre-increment the counter.

always @ (posedge clk or negedge reset_n)
begin
       if (reset_n == 1'b0)
            cnt1_reg <= 4'd0 ;             // Initialize when the system is reset.
       else if (res_cnt1 == 1'b1)          // Reset if terminal count is reached.
            cnt1_reg <= 4'd0 ;
       else if (adv_cnt1 == 1'b1)
            cnt1_reg <= cnt1_next ;        // Advance the count once if the
                                           // time watch is still running.
       else                    ;           // Otherwise, don't disturb.
end

      // cnt2_reg is the Time watch's least significant HOUR digit. This is
      // reset or advanced only in the RUN mode of operation.
assign adv_res_cnt2 =   (set_time != 1)&(tbsec == 1)&(cnt3_reg == 5)&
                        (cnt4_reg == 9)&(cnt5_reg == 5)&(cnt6_reg == 9) ;
assign res_cnt2_time =  (adv_res_cnt2 == 1) & ( (cnt1_reg < 2)&
                        (cnt2_reg == 9) | ((cnt1_reg == 2)&(cnt2_reg == 3))) ;
assign res_cnt2_set  =  (adv_hrs_time == 1)&( (cnt1_reg < 2)&
                        (cnt2_reg == 9) |  (cnt1_reg == 2)&(cnt2_reg == 3)) ;
assign res_cnt2      =  res_cnt2_time | res_cnt2_set ;
                        // cnt1 cnt2 = 23 or 09 or 19 and cnt3 - cnt6 = 59 59
                        // are the conditions for resetting this counter.
assign adv_cnt2      =  (adv_res_cnt2 == 1)|(adv_hrs_time == 1) ;
      // Other conditions are implied since res_cnt2 has a higher priority than
      // adv_cnt2. cnt1 cnt2 = 00 to 18 (except 09) or 20 to 22 & cnt3 - cnt6 =
      // 59 59 are the conditions for pre-incrementing this counter.
assign cnt2_next = cnt2_reg + 1 ;          // Pre-increment the counter.
```

```
always @ (posedge clk or negedge reset_n)
begin
            if (reset_n == 1'b0)
                    cnt2_reg <= 4'd0 ;              // Initialize when the system is reset.
            else if (res_cnt2 == 1'b1)              // Reset if terminal count is reached.
                    cnt2_reg <= 4'd0 ;
            else if (adv_cnt2 == 1'b1)
                    cnt2_reg <= cnt2_next ;         // Advance the count once if the
                                                    // time watch is still running.
            else                          ;         // Otherwise, don't disturb.
end

     // cnt3_reg is the Time watch's most significant MINUTES digit. This is reset
     // or advanced only in the RUN and TIME mode of operation every 1 s.
assign res_cnt3 = ((set_time !=  1)&(tbsec == 1)&(cnt3_reg == 5)&
                    (cnt4_reg == 9)&(cnt5_reg == 5)&(cnt6_reg == 9))|
                    ((adv_mts_time == 1)&(cnt3_reg == 5)&(cnt4_reg == 9)) ;
                // cnt3 - cnt6 = 59 59 are the conditions for resetting this counter.
assign adv_cnt3 = ((set_time != 1)&( tbsec == 1)&(cnt3_reg  <  5)&
                    (cnt4_reg == 9)&(cnt5_reg == 5)&(cnt6_reg == 9))|
                    ((adv_mts_time == 1)&(cnt3_reg  <  5)&(cnt4_reg == 9)) ;
                        // cnt3 = 0-4 and cnt4 - cnt6 = 9 59 are the conditions
                        // for pre-incrementing this counter.
assign cnt3_next = cnt3_reg + 1 ;   // Pre-increment the counter.

always @ (posedge clk or negedge reset_n)
begin
            if (reset_n == 1'b0)
                    cnt3_reg <= 4'd0 ;              // Initialize when the system is reset.
            else if (res_cnt3 == 1'b1)              // Reset if terminal count is reached.
                    cnt3_reg <= 4'd0 ;
            else if (adv_cnt3 == 1'b1)
                    cnt3_reg <= cnt3_next ;         // Advance the count once if the
                                                    // time watch is still running.
            else                          ;         // Otherwise, don't disturb.
end

     // cnt4_reg is the Time watch's least significant MINUTES digit. This is reset
     // or advanced only in the RUN and TIME mode of operation every 1 s.
assign res_cnt4 = ((set_time != 1)&(tbsec  == 1)&(cnt4_reg == 9)&
                    (cnt5_reg == 5)&(cnt6_reg == 9)) |((adv_mts_time == 1)&
                                                        (cnt4_reg == 9)) ;
                // cnt4 - cnt6 = 9 59 are the conditions for resetting this counter.
assign adv_cnt4 =((set_time != 1) &(tbsec == 1)&(cnt4_reg < 9)&
                    (cnt5_reg == 5)&(cnt6_reg == 9)) |
                                    ((adv_mts_time == 1)&(cnt4_reg  <  9)) ;
```

```
                                 // cnt4 = 0 to 8 & cnt5 - cnt6 = 59 are the
                                 // conditions for pre-incrementing this counter.
assign cnt4_next = cnt4_reg + 1 ;  // Pre-increment the counter.

always @ (posedge clk or negedge reset_n)
begin
        if (reset_n == 1'b0)
                cnt4_reg <= 4'd0 ;           // Initialize when the system is reset.
        else if (res_cnt4  == 1'b1)          // Reset if terminal count is reached.
                cnt4_reg <= 4'd0 ;
        else if (adv_cnt4 == 1'b1)
                cnt4_reg <= cnt4_next ;  // Advance the count once if the
                                         // time watch is still running.
        else                      ;      // Otherwise, don't disturb.
end

   // cnt5_reg is the Time watch's most significant SECONDS digit. This is reset
   // or advanced only in the RUN and TIME mode of operation every 1 s.
assign res_cnt5 = ((set_time != 1)&(tbsec == 1)&(cnt5_reg == 5)
                              &(cnt6_reg == 9))|((adv_secs_time ==1)&
                                     (cnt5_reg == 5)&(cnt6_reg == 9)) ;
              // cnt5 - cnt6 = 59 are the conditions for resetting this counter.
assign adv_cnt5 = ((set_time != 1) &(tbsec  == 1)&(cnt5_reg < 5)&
              (cnt6_reg == 9))|((adv_secs_time == 1)&(cnt5_reg < 5)&
                                                  (cnt6_reg == 9)) ;

  // cnt5 = 0 to 4 & cnt6 = 9 are the conditions for pre-incrementing this counter.
assign cnt5_next = cnt5_reg + 1 ;          // Pre-increment the counter.

always @ (posedge clk or negedge reset_n)
begin
        if (reset_n == 1'b0)
                cnt5_reg <= 4'd0 ;           // Initialize when the system is reset.
        else if (res_cnt5 == 1'b1)           // Reset if terminal count is reached.
                cnt5_reg <= 4'd0 ;
        else if (adv_cnt5 == 1'b1)
                cnt5_reg <= cnt5_next ;  // Advance the count once if the
                                         // time watch is still running.
        else                      ;      // Otherwise, don't disturb.
end

// cnt6_reg is the Time watch's least significant SECONDS digit. This is reset
// or advanced only in the RUN and TIME mode of operation every 1 s.
assign res_cnt6 = ((set_time != 1'b1)&( tbsec  == 1'b1)&(cnt6_reg == 9))|
                                  ((adv_secs_time == 1)&(cnt6_reg == 9)) ;
                        // cnt6 = 9 is the condition for resetting this counter.
```

```
assign adv_cnt6 = ((set_time != 1'b1)&( tbsec  == 1'b1)&
                      (cnt6_reg < 9))|((adv_secs_time == 1)&(cnt6_reg < 9)) ;
          // cnt6 = 0 to 8 are the conditions for pre-incrementing this counter.
assign cnt6_next = cnt6_reg + 1 ;              // Pre-increment the counter.

always @ (posedge clk or negedge reset_n)
begin
          if (reset_n == 1'b0)
                    cnt6_reg <= 4'd0 ;         // Initialize when the system is reset.
          else if (res_cnt6 == 1'b1)           // Reset if terminal count is reached.
                    cnt6_reg <= 4'd0 ;
          else if (adv_cnt6 == 1'b1)
                    cnt6_reg <= cnt6_next ;    // Advance the count once if the
                                               // time watch is still running.
          else                        ;        // Otherwise, don't disturb.
end
//                         Stop watch implementation
// RUN, STOP WATCH mode
assign term_count_reached_up = (set_stopw != 1)&(down_upn == 0)&
                         (start_stopn_reg == 1)&(cnt7_reg == term_count_reg1)&
              (cnt8_reg == term_count_reg2)&(cnt9_reg == term_count_reg3)&
          (cnt10_reg == term_count_reg4)&(cnt11_reg == term_count_reg5)&
                                            (cnt12_reg == term_count_reg6) ;
assign term_count_reached_down = rsd&(cnt7_reg == 0)&(cnt8_reg == 0)&
          (cnt9_reg == 0)&(cnt10_reg == 0)&(cnt11_reg == 0)&(cnt12_reg == 0) ;
assign timer_out = (term_count_reached_up == 1)|
                                            (term_count_reached_down == 1) ;
assign rsd = (set_stopw != 1)&(down_upn == 1)&(start_stopn_reg == 1) ;
                              // "rsd" means run stopwatch in down counter mode.
                              // start_stopn_reg = 1 means START, otherwise STOP.

// cnt7_reg is the Stop watch's most significant HOUR digit. This is reset or
// advanced only in the RUN and STOP WATCH mode of operation. For Down
// counting, cnt7_reg to cnt12_reg are used, whereas for Up counting,
// term_count_reg1 to term_count_reg6 are used for  presetting. cnt7_reg to
// cnt12_reg are the Running counters for both UP & Down counting.
assign res_cnt7 = (((set_stopw == 1)&(down_upn == 0))|
                         ((adv_hrs_sw == 1)&(cnt7_reg == 2)&(cnt8_reg == 3)));
          // Counter must be reset in SET UP mode. cnt7 – cnt8 = 23 are the
          // conditions for resetting the counter in SET DOWN COUNTER mode.
assign adv_cnt7 = ((set_stopw != 1)&(down_upn == 1'b0)&
                    (term_count_reached_up == 0)&(tbsec == 1)&(cnt7_reg < 2)&
                              (cnt8_reg == 9)&(cnt9_reg == 5)&(cnt10_reg == 9)&
               (cnt11_reg == 5)&(cnt12_reg == 9)&(start_stopn_reg == 1'b1)) |
                         ((adv_hrs_sw == 1)&(cnt7_reg < 2)&(cnt8_reg == 9)) ;
                                   // cnt7 - cnt12 = 09 59 59 or 19 59 59 are the
```

```
                                        // conditions for pre-incrementing the counter.
assign cnt7_next = cnt7_reg + 1 ;   // Pre-increment the counter.
assign decr_cnt7 = rsd&(cnt8_reg == 0)&(cnt9_reg == 0)&(cnt10_reg == 0)&
                    (cnt11_reg == 0)&cnt12_reg == 0)&(cnt7_reg > 0)&
                                                  (cnt7_reg <= 2)&(tbsec == 1) ;
assign cnt7_nextd = cnt7_reg – 1 ;              // Pre-decrement the counter.

always @ (posedge clk or negedge reset_n)
begin
          if (reset_n == 1'b0)
                 cnt7_reg <= 4'd0 ;         // Initialize when the system is reset.
          else if (res_cnt7 == 1'b1)        // Reset if terminal count is reached.
                 cnt7_reg <= 4'd0 ;
          else if (adv_cnt7 == 1'b1)
                 cnt7_reg <= cnt7_next ;   // Advance the count once if the
                                           // time watch is still running.
          else if (decr_cnt7 == 1'b1)
                 cnt7_reg <= cnt7_nextd ; // Decrement the count once if
                                           // the stop watch is still running.
          else              ;              // Otherwise, don't disturb.
end

// cnt8_reg is the Stop watch's least significant HOUR digit. This is reset or
// advanced/decremented only in the RUN and STOPWATCH mode of operation.
assign rsd_cnt8_res = rsd&((cnt7_reg > 2)|((cnt7_reg == 2)&
                   (cnt8_reg > 3))|((cnt7_reg < 2)&(cnt8_reg > 9))) ;
                                          // These are illegal values.
assign res_cnt8_set = ((adv_hrs_sw == 1)&(cnt7_reg < 2)&(cnt8_reg == 9))|
                    ((adv_hrs_sw == 1)&(cnt7_reg == 2)&(cnt8_reg == 3)) ;
assign res_cnt8_sw = (set_stopw != 1)&(down_upn == 1'b0)&
                    (term_count_reached_up == 0)&(tbsec == 1)&
                    (((cnt7_reg == 2)&(cnt8_reg == 3))|
                     ((cnt7_reg  < 2)&(cnt8_reg == 9)))&(cnt9_reg == 5)&
                      (cnt10_reg == 9)&(cnt11_reg == 5)&(cnt12_reg == 9) ;
                       // cnt7 cnt8 = 23 or 09 or 19 and cnt9 - cnt12 = 59 59
                       // are the conditions for resetting this counter.
assign cnt8_res = (rsd_cnt8_res | res_cnt8_sw | res_cnt8_set) |
                                    ((set_stopw == 1)&(down_upn == 0)) ;
assign adv_cnt8_set = ((adv_hrs_sw == 1)&(cnt7_reg < 2)&(cnt8_reg  < 9))|
                    ((adv_hrs_sw == 1)&(cnt7_reg == 2)&(cnt8_reg < 3)) ;
assign adv_cnt8_sw =  (set_stopw != 1)&(down_upn == 1'b0)&
                     (term_count_reached_up == 0)&(tbsec == 1)&
                     (((cnt7_reg < 2)&(cnt8_reg < 9))|((cnt7_reg  == 2)&
                      (cnt8_reg < 3)))&(cnt9_reg   == 5) & (cnt10_reg == 9) &
              (cnt11_reg == 5)&(cnt12_reg == 9)&(start_stopn_reg == 1'b1) ;
assign adv_cnt8 = adv_cnt8_set | adv_cnt8_sw ;
```

```
                              // cnt7 cnt8 = 00 to 18 or 20 to 22 & cnt9-cnt12 = 59 59
                              // are the conditions for pre-incrementing this counter.
assign cnt8_next = cnt8_reg + 1 ;  // Pre-increment the counter.
assign decr_cnt8 = rsd&(cnt9_reg == 0)&(cnt10_reg == 0)&
                   (cnt11_reg == 0)&(cnt12_reg == 0)&(((cnt7_reg == 0)&
                   (cnt8_reg > 0)&(cnt8_reg <= 9))|((cnt7_reg == 1)&
                      (cnt8_reg > 0)&(cnt8_reg <= 9))|((cnt7_reg == 2)&
                       (cnt8_reg > 0)&(cnt8_reg <= 3))) & (tbsec == 1) ;
// Decrement if cnt7 cnt8 = 01-09 or 11-19 or 21-23 & cnt9-cnt12 = 00 00.
assign cnt8_nextd = cnt8_reg – 1 ;          // Pre-decrement the counter.
assign pres_cnt8 = rsd&(tbsec == 1)&(cnt8_reg == 0)&
                     (cnt9_reg == 0)&(cnt10_reg == 0)&(cnt11_reg == 0)&
                     (cnt12_reg == 0)&(cnt7_reg > 0)&(cnt7_reg <= 2) ;
                        // Preset if cnt7-cnt12 = 10 00 00 or 20 00 00.

always @ (posedge clk or negedge reset_n)
      begin
       if (reset_n == 1'b0)
             cnt8_reg <= 4'd0 ;          // Initialize when the system is reset.
       else if (cnt8_res == 1'b1)        // Reset if terminal count is reached.
             cnt8_reg <= 4'd0 ;
       else if (adv_cnt8 == 1'b1)
             cnt8_reg <= cnt8_next ;    // Advance the count once if the
                                         // stop watch is still running.
       else if (decr_cnt8 == 1'b1)
             cnt8_reg <= cnt8_nextd ; // Decrement the count once if the
                                         // stop watch is still running.
       else if (pres_cnt8 == 1'b1)       // Preset if count down terminal
            cnt8_reg <= 4'd9 ;           // count is reached.
       else                       ;      // Otherwise, don't disturb.
end

// cnt9_reg is the Stop watch's most significant MINUTES digit. This is reset or
// advanced only in the RUN and STOP WATCH mode of operation every 1 s.
assign res_cnt9 = ((set_stopw != 1)&(down_upn == 1'b0)&
                  (term_count_reached_up == 0)&(tbsec ==1)&(cnt9_reg == 5)&
                  (cnt10_reg == 9)&(cnt11_reg == 5)&(cnt12_reg == 9)&
                (start_stopn_reg == 1))|((adv_mts_sw ==1)&(cnt9_reg ==5)&
                (cnt10_reg == 9))|((set_stopw == 1)&(down_upn == 0))  ;
                // cnt9 - cnt12 = 59 59 are the conditions for resetting this counter.
assign adv_cnt9 = ((set_stopw != 1)&(down_upn == 1'b0)&
                  (term_count_reached_up == 0)&(tbsec == 1)&(cnt9_reg < 5)&
                  (cnt10_reg == 9)&(cnt11_reg == 5)&(cnt12_reg == 9)&
                (start_stopn_reg ==1'b1))| ((adv_mts_sw ==1)&(cnt9_reg <5)&
                                                          (cnt10_reg == 9)) ;
                          // cnt9 = 0-4 and cnt10 - cnt12 = 9 59 are the
```

```
                                    // conditions for pre-incrementing this counter.
assign cnt9_next = cnt9_reg + 1 ;   // Pre-increment the counter.
assign decr_cnt9 = rsd&(cnt10_reg == 0)&(cnt11_reg == 0)&(cnt12_reg == 0)
                   &(cnt9_reg > 0)&(cnt9_reg <= 5)&(tbsec == 1) ;
                                    // For cnt9 = 1-5 and cnt10-cnt12 = 0 00.

assign cnt9_nextd = cnt9_reg – 1 ; // Pre-decrement the counter.
assign pres_cnt9 = rsd&(tbsec == 1)&(cnt9_reg == 0)&
                   (cnt10_reg == 0)&(cnt11_reg == 0)&
                     (cnt12_reg == 0)&((cnt7_reg  != 0)|(cnt8_reg != 0)) ;
                     // For cnt7 or cnt8 not equal to "0" and cnt9-cnt12 = 00 00.

always @ (posedge clk or negedge reset_n)
begin
        if (reset_n == 1'b0)
                cnt9_reg <= 4'd0 ;          // Initialize when the system is reset.
        else if (res_cnt9 == 1'b1)          // Reset if terminal count is reached.
                cnt9_reg <= 4'd0 ;
        else if (adv_cnt9 == 1'b1)
                cnt9_reg <= cnt9_next ;     // Advance the count once if the
                                            // stop watch is still running.
        else if (decr_cnt9 == 1'b1)
                cnt9_reg <= cnt9_nextd ;    // Decrement the count once if the
                                            // stop watch is still running.
        else if (pres_cnt9 == 1'b1)         // Preset if count down terminal
                                            // count is reached.
                cnt9_reg <= 4'd5 ;
        else                        ;       // Otherwise, don't disturb.
end
/*
cnt10_reg is the Stop watch's least significant MINUTES digit. This is reset or
advanced only in the RUN and STOP WATCH mode of operation every 1 s.
*/
assign res_cnt10 = ((set_stopw != 1)&(down_upn == 1'b0)&
            (term_count_reached_up == 0)&(tbsec == 1)&(cnt10_reg == 9)&
            (cnt11_reg == 5)&(cnt12_reg == 9)&(start_stopn_reg == 1))|
               ((adv_mts_sw == 1)&(cnt10_reg == 9))|((set_stopw == 1)&
                                                   (down_upn == 0)) ;
            // cnt10 - cnt12 = 9 59 are the conditions for resetting this counter.
assign adv_cnt10 = ((set_stopw != 1)&(down_upn == 1'b0)&
             (term_count_reached_up == 0)&(tbsec == 1)&(cnt10_reg < 9)&
            (cnt11_reg == 5)&(cnt12_reg == 9)&(start_stopn_reg == 1'b1))|
                                   ((adv_mts_sw == 1)&(cnt10_reg < 9)) ;
                                 // cnt10 = 0 to 8 & cnt11 – cnt12 =  59 are the
                                 // conditions for pre-incrementing this counter.
assign cnt10_next = cnt10_reg + 1 ;               // Pre-increment the counter.
```

```
assign decr_cnt10 = rsd&(cnt11_reg == 0)&(cnt12_reg == 0)&
                              (cnt10_reg > 0)&(cnt10_reg <= 9)&(tbsec == 1) ;
                    // Decrement cnt10 if cnt10 = 1-9 and cnt11 - cnt12 = 00.
assign cnt10_nextd = cnt10_reg – 1 ;        // Pre-decrement the counter.
assign pres_cnt10 = rsd&(tbsec == 1)&(cnt10_reg == 0)&
                     (cnt11_reg == 0)&(cnt12_reg == 0)&
                          ((cnt9_reg != 0)|(cnt8_reg != 0)|(cnt7_reg != 0)) ;
                     // Preset cnt10 to 9 only if cnt7-cnt9 not equal to 00 0
                     // and cnt10-cnt12 = 0 00.

always @ (posedge clk or negedge reset_n)
begin
        if (reset_n == 1'b0)
               cnt10_reg <= 4'd0 ;          // Initialize when the system is reset.
        else if (res_cnt10 == 1'b1)         // Reset if terminal count is reached.
               cnt10_reg <= 4'd0 ;
        else if (adv_cnt10 == 1'b1)
               cnt10_reg <= cnt10_next ;       // Advance the count once if the
                                               // stop watch is still running.
        else if (decr_cnt10 == 1'b1)
               cnt10_reg <= cnt10_nextd ;      // Decrement the count once if
                                               // the stop watch is still running.
        else if (pres_cnt10 == 1'b1)           // Preset if count down terminal
                                               // count is reached.
               cnt10_reg <= 4'd9 ;
        else                     ;          // Otherwise, don't disturb.
end

// cnt11_reg is the Stop watch's most significant SECONDS digit. This is reset
// or advanced only in the RUN and STOP WATCH mode of operation every 1 s.
assign res_cnt11 = ((set_stopw != 1)&(down_upn == 1'b0)&
                      (term_count_reached_up == 0)&(tbsec == 1)&
                   (cnt11_reg == 5)&(cnt12_reg == 9)&(start_stopn_reg == 1))|
                   ((adv_secs_sw == 1)&(cnt11_reg == 5)&(cnt12_reg == 9))|
                                        ((set_stopw == 1)&(down_upn == 0)) ;
                   // cnt11 - cnt12 = 59 are the conditions for resetting this counter.
assign adv_cnt11 = ((set_stopw != 1)&(down_upn == 1'b0)&
                      (term_count_reached_up == 0)&(tbsec == 1)&
                   (cnt11_reg < 5)&(cnt12_reg == 9)&(start_stopn_reg == 1'b1))|
                      ((adv_secs_sw == 1)&(cnt11_reg < 5)&(cnt12_reg == 9)) ;

// cnt11 = 0 to 4 & cnt12 = 9 are the conditions for pre-incrementing this counter.
assign cnt11_next = cnt11_reg + 1 ;         // Pre-increment the counter.
assign decr_cnt11 = rsd&(cnt12_reg == 0)&(cnt11_reg > 0)&
                     (cnt11_reg <= 5)&(tbsec == 1) ;
                                                // For cnt11 = 1–5.
```

```
assign cnt11_nextd = cnt11_reg – 1 ;                    // Pre-decrement the counter.
assign pres_cnt11 = rsd&(tbsec == 1) & (cnt11_reg == 0)&
                    (cnt12_reg == 0)&((cnt10_reg != 0)|(cnt9_reg != 0)|
                                 (cnt8_reg != 0)|(cnt7_reg != 0)) ;
                              // Preset cnt11 to 5 only if cnt7–cnt10 not
                              // equal to 00 00 and cnt11–cnt12 = 00.

always @ (posedge clk or negedge reset_n)
begin
        if (reset_n == 1'b0)
             cnt11_reg <= 4'd0 ;         // Initialize when the system is reset.
        else if (res_cnt11 == 1'b1)      // Reset if terminal count is reached.
             cnt11_reg <= 4'd0 ;
        else if (adv_cnt11 == 1'b1)
             cnt11_reg <= cnt11_next ;     // Advance the count once if the
                                           // stop watch is still running.
        else if (decr_cnt11 == 1'b1)
             cnt11_reg <= cnt11_nextd ;    // Decrement the count once if
                                           // the stop watch is still running.
        else if (pres_cnt11 == 1'b1)       // Preset if count down terminal
                                           // count is reached.
             cnt11_reg <= 4'd5 ;
        else                    ;        // Otherwise, don't disturb.
end
/*
cnt12_reg is the Stop watch's least significant SECONDS digit. This is reset or
advanced every 1 s only in the RUN and STOP WATCH mode of operation.
*/
assign res_cnt12 = ((set_stopw != 1)&(down_upn == 1'b0)&
                    (term_count_reached_up == 0)&(tbsec == 1)&
                  (cnt12_reg == 9)&(start_stopn_reg == 1))|
                  ((adv_secs_sw == 1)&(cnt12_reg == 9))|
                  ((set_stopw == 1)&(down_upn == 0)) ;
                        // cnt12 = 9 is the condition for resetting this counter.
assign adv_cnt12 = ((set_stopw != 1)&(down_upn == 1'b0)&
                         (term_count_reached_up == 0)&(tbsec == 1)&
                         (cnt12_reg < 9)&(start_stopn_reg == 1'b1))|
                              ((adv_secs_sw == 1)&(cnt12_reg < 9));
        // cnt12 = 0 to 8 are the conditions for pre-incrementing this counter.
assign cnt12_next = cnt12_reg + 1 ;          // Pre-increment the counter.
assign decr_cnt12 = rsd&(cnt12_reg > 0)&(cnt12_reg <= 9)&(tbsec == 1) ;
                              // Decrement cnt12 every second if cnt12 = 1–9.
assign cnt12_nextd = cnt12_reg – 1 ;         // Pre-decrement the counter.
assign pres_cnt12 = rsd & (tbsec == 1)&(cnt12_reg == 0)&
                  ((cnt11_reg != 0)|(cnt10_reg != 0)|(cnt9_reg != 0)|
                                      (cnt8_reg != 0)|(cnt7_reg != 0)) ;
```

```
        // Preset cnt12 to 9 only if cnt7-cnt11 not equal to 00 00 0 and cnt12 = 0.

always @ (posedge clk or negedge reset_n)
begin
        if (reset_n == 1'b0)
            cnt12_reg <= 4'd0 ;          // Initialize when the system is reset.
        else if (res_cnt12 == 1'b1)      // Reset if count up terminal count
            cnt12_reg <= 4'd0 ;          // is reached.
        else if (adv_cnt12 == 1'b1)      // Advance the count once if the
            cnt12_reg <= cnt12_next ; // stop watch is still running.
        else if (decr_cnt12 == 1'b1)
            cnt12_reg <= cnt12_nextd ;  // Decrement the count once if
                                         // the stop watch is still running.
        else if (pres_cnt12 == 1'b1)     // Preset if count down terminal
            cnt12_reg <= 4'd9 ;          // count is reached.
        else                  ;          // Otherwise, don't disturb.
end

assign res_term_count_reg1 = (adv_hrs_tcr == 1)&(term_count_reg1 == 2)&
                                             (term_count_reg2 == 3) ;
                             // Reset Terminal count register for Up counter.
assign adv_term_count_reg1 = (adv_hrs_tcr == 1)&(term_count_reg1 < 2)&
                                             (term_count_reg2 == 9) ;
assign term_count_reg1_next = term_count_reg1 + 1;

always @ (posedge clk or negedge reset_n)
begin
      if (reset_n == 0)
            term_count_reg1 <= 0 ;
      else if (res_term_count_reg1 == 1)
            term_count_reg1 <= 0 ;
      else if (adv_term_count_reg1 == 1)
            term_count_reg1 <= term_count_reg1_next ;
      else                  ;
end

assign res_term_count_reg2 = (adv_hrs_tcr == 1)&
                  (((term_count_reg1 == 2)&(term_count_reg2 == 3))
                   ((term_count_reg1 < 2)&(term_count_reg2 == 9))) ;
assign adv_term_count_reg2 = (adv_hrs_tcr == 1)&
                  (((term_count_reg1 < 2)&(term_count_reg2 < 9))|
                   ((term_count_reg1 == 2)&(term_count_reg2 < 3))) ;
assign term_count_reg2_next = term_count_reg2 + 1;

always @ (posedge clk or negedge reset_n)
begin
      if (reset_n == 0)
            term_count_reg2 <= 0 ;
```

```
        else if (res_term_count_reg2 == 1)
                term_count_reg2 <= 0 ;
        else if (adv_term_count_reg2 == 1)
                term_count_reg2 <= term_count_reg2_next ;
        else                            ;
end

assign res_term_count_reg3 = (adv_mts_tcr == 1)&(term_count_reg3 == 5)&
                                                (term_count_reg4 == 9) ;
assign adv_term_count_reg3 = (adv_mts_tcr == 1)&(term_count_reg3 < 5)&
                                                (term_count_reg4 == 9) ;
assign term_count_reg3_next = term_count_reg3 + 1;

always @ (posedge clk or negedge reset_n)
begin
        if (reset_n == 0)
                term_count_reg3 <= 0 ;
        else if (res_term_count_reg3 == 1)
                term_count_reg3 <= 0 ;
        else if (adv_term_count_reg3 == 1)
                term_count_reg3 <= term_count_reg3_next ;
        else                            ;
end

assign res_term_count_reg4  = (adv_mts_tcr == 1)&(term_count_reg4 == 9) ;
assign adv_term_count_reg4  = (adv_mts_tcr == 1)&(term_count_reg4 < 9) ;
assign term_count_reg4_next = term_count_reg4 + 1;

always @ (posedge clk or negedge reset_n)
begin
        if (reset_n == 0)
                term_count_reg4 <= 0 ;
        else if (res_term_count_reg4 == 1)
                term_count_reg4 <= 0 ;
        else if (adv_term_count_reg4 == 1)
                term_count_reg4 <= term_count_reg4_next ;
        else                    ;
end

assign res_term_count_reg5 = (adv_secs_tcr == 1)&(term_count_reg5 == 5)&
                                                (term_count_reg6 == 9) ;
assign adv_term_count_reg5 = (adv_secs_tcr == 1)&(term_count_reg5 < 5)&
                                                (term_count_reg6 == 9) ;
assign term_count_reg5_next = term_count_reg5 + 1 ;

always @ (posedge clk or negedge reset_n)
begin
        if (reset_n == 0)
```

```
                term_count_reg5 <= 0 ;
        else if (res_term_count_reg5 == 1)
                term_count_reg5 <= 0 ;
        else if (adv_term_count_reg5 == 1)
                term_count_reg5 <= term_count_reg5_next ;
        else                    ;
end

assign res_term_count_reg6  = (adv_secs_tcr == 1)&(term_count_reg6 == 9) ;
assign adv_term_count_reg6  = (adv_secs_tcr == 1)&(term_count_reg6 < 9) ;
assign term_count_reg6_next = term_count_reg6 + 1 ;

always @ (posedge clk or negedge reset_n)
begin
        if (reset_n == 0)
                term_count_reg6 <= 0 ;
        else if (res_term_count_reg6 == 1)
                term_count_reg6 <= 0 ;
        else if (adv_term_count_reg6 == 1)
                term_count_reg6 <= term_count_reg6_next ;
        else                        ;
end

// Timer out  is set when the terminal count (Up/Down) is reached.
assign timer_out_alarm_counter_next = timer_out_alarm_counter + 1 ;
                                                // 30 s audio alarm counter.
assign timer_out_alarm = (timer_out)&(timer_out_alarm_counter != 31) ;
                            // This signal is high for 30 s after terminal count
                            // is reached, i.e., timer_out =1.

always @ (posedge clk or negedge reset_n)
begin
        if (reset_n == 0)
                timer_out_alarm_counter  <= 0 ;
        else if (timer_out == 0)
                timer_out_alarm_counter  <= 0 ;
        else if ((timer_out == 1)&(timer_out_alarm_counter != 31)&
                                                                (tbsec == 1))
                timer_out_alarm_counter  <= timer_out_alarm_counter_next ;
        else                        ;
end

always @ (posedge clk or negedge reset_n)
begin
        if (reset_n == 1'b0)
        begin
            start_stopn_reg        <= 1'b0 ;          // Initialize to STOP mode
                                                      // when the system is reset.
```

```
                                              // This stores the start/stop value.
        start_stopnp_reg        <= 1'b0       // Previous start/stop value.
      end
      else if (set_stopw == 1)
        start_stopn_reg         <= 0;
      else if ((start_stopn == 1'b1)&&(start_stopnp_reg == 0)&&
                                                    (run_stopw == 1))
                      // start_stopn is the debounced START/STOP
                      // PB input. Look for rising edge (Depression of the
                      // push button switch).
      begin
        start_stopn_reg         <= !start_stopn_reg ;
                                          // Toggle between START & STOP.
        start_stopnp_reg        <= start_stopn ;
                                    // Preserve as the previous start/stop value.
      end
      else
        start_stopnp_reg        <= start_stopn ;
                                    // Preserve as the previous start/stop value.
end

assign hrs2s_next = hrs2s_reg + 1 ;          // Pre-increment the counter.

always @ (posedge clk or negedge reset_n)
begin
      if (reset_n == 1'b0)
            begin                             // Initialize when the system is reset.
                  hrsp_reg   <= 0 ;           // Initialize previous "hrs" value
                  hrs_d      <= 0 ;           // and clear ON delay output.
                  hrs2s_reg  <= 0 ;           // Clear 2 s counter.
            end
      else if (hrs == 0)                      // If "hrs" PB is released,
            begin
                  hrs_d      <= 0 ;           // clear ON delay output.
                  hrsp_reg   <= hrs ;         // Preserve as the previous "hrs" value.
                                              // "hrs" is the HRS PB input.
                  hrs2s_reg  <= 0 ;           // Clear 2 s counter.
            end
      else if (hrs2s_reg == 20)               // After 2 s delay (or greater),
            begin
                   hrs_d     <= 1 ;           // switch ON delay output.
                   hrsp_reg <= hrs ; // Preserve as the previous "hrs" value.
                                              // "hrs" is the HRS PB input.
                    // Note: hrs2s_reg is not reset here. It is reset when
                    // HRS PB is released as above.
            end
```

```
        else if ((cntds_reg  == `ds_base)&(hrs == 1))
                begin
                        hrs2s_reg <= hrs2s_next ;
                                // Advance the count once every 0.1 s.
                                // so long as "HRS" PB is kept pressed.
                                // Otherwise, ignore.
                        hrs_d     <= 0 ;   // Clear ON delay output.
                        hrsp_reg <= hrs ; // Preserve as the previous "hrs" value.
                end                       // "hrs" is the HRS PB input.
        else            hrsp_reg <= hrs ; // Preserve as the previous "hrs" value.
                                          // "hrs" is the HRS PB input.
end

assign mts2s_next = mts2s_reg + 1 ;           // Pre-increment the counter.
always @ (posedge clk or negedge reset_n)
begin
        if (reset_n == 1'b0)
                begin                         // Initialize when the system is reset.
                    mtsp_reg    <= 1'b0 ; // Initialize previous "mts" value
                    mts_d       <= 0 ;    // and clear ON delay output.
                    mts2s_reg   <= 0 ;    // Clear 2 s counter.
                end
        else if (mts == 0)                    // If MTS PB is released,
                begin
                     mts_d      <= 0 ;    // clear ON delay output and
                     mtsp_reg   <= mts ; // preserve the previous "mts" value.
                                          // "mts" is the MTS PB input.
                     mts2s_reg<= 0 ;      // Clear 2 s counter.
                end
        else if (mts2s_reg == 20)             // After 2 s delay or greater,
                begin
                    mts_d       <= 1 ;    // switch ON delay output.
                    mtsp_reg    <= mts ; // Preserve the previous "mts" value.
                                          // "mts" is the MTS PB input.
// Note: mts2s_reg is not reset here. It is reset when MTS PB is released as above.
                end
        else if ( (cntds_reg  == `ds_base)&(mts == 1)) // Advance the count once
                begin   // every 0.1 sec. so long as "MTS" PB is kept pressed.
                    mts2s_reg   <= mts2s_next ;   // Otherwise, ignore.
                    mts_d       <= 0 ;              // Clear ON delay output.
                    mtsp_reg    <= mts ; // Preserve the previous "mts" value.
                end                                 // "mts" is the MTS PB input.
        else        mtsp_reg    <= mts ; // Preserve the previous "mts" value.
                                                    // "mts" is the MTS PB input.
end
assign secs2s_next = secs2s_reg + 1 ;         // Pre-increment the counter.
```

```
always @ (posedge clk or negedge reset_n)
begin
        if (reset_n == 1'b0)
                begin                          // Initialize when the system is reset.
                  secsp_reg       <= 1'b0 ; // Initialize previous "secs." value
                  secs_d          <= 0 ;     // and clear ON delay output.
                  secs2s_reg      <= 0 ;     // Clear 2 s counter.
                end
        else if (secs == 0)                    // If secs. PB is released,
                begin
                  secs_d      <= 0 ;          // Clear ON delay output.
                  secsp_reg   <= secs ;       // Preserve the previous "secs" value.
                                              // "secs" is the SECS PB input.
                  secs2s_reg <= 0 ;           // Clear 2 s counter.
                end
        else if (secs2s_reg == 20)            // After 2 s delay or greater,
                begin
                  secs_d      <= 1 ;          // switch ON delay output.
                  secsp_reg   <= secs ;       // Preserve the previous secs. value.
                                              // secs is the SECS PB input.
//Note: secs2s_reg is not reset here. It is reset when SECS PB is released asabove.
                end
        else if ((cntds_reg   == `ds_base)&(secs == 1))
                begin
                 secs2s_reg   <= secs2s_next ;      // Advance the count once
         // every 0.1 sec. so long as "SECS" PB is kept pressed. Otherwise, ignore.
                 secs_d       <= 0 ;          // Clear ON delay output.
                 secsp_reg    <= secs ;       // Preserve the previous "secs" value.
                end                           // "secs" is the SECS PB input.
        else     secsp_reg    <= secs ;       // Preserve the previous 'secs' value.
                                              // "secs" is the SECS PB input.
end
//                          Alarm implementation
// temp_alarm_reg1 to 6 are a set of 4-bit temporary registers which hold the
// alarm time when it is being set.
assign adv_temp_alarm_reg1 = (adv_hrs_temp_alarm == 1)&
                          (temp_alarm_reg1 < 2)&(temp_alarm_reg2 == 9) ;
                                              // Means 09 or 19
assign res_temp_alarm_reg1 = (adv_hrs_temp_alarm == 1)&
                          (temp_alarm_reg1 == 2)&(temp_alarm_reg2 == 3) ;
assign temp_alarm_reg1_next = temp_alarm_reg1 + 1 ;
          // Common alarm setting counter, one each for the 6 digits display.
always @ (posedge clk or negedge reset_n)
        begin
                if (reset_n == 0)
```

```
                        temp_alarm_reg1 <= 0 ;
                else if (res_temp_alarm_reg1 == 1)
                        temp_alarm_reg1 <= 0 ;
                else if (adv_temp_alarm_reg1 == 1)
                        temp_alarm_reg1 <= temp_alarm_reg1_next;
                else            ;
        end

assign adv_temp_alarm_reg2 = (adv_hrs_temp_alarm == 1)&
                        (((temp_alarm_reg1 < 2)&(temp_alarm_reg2 < 9))|
                      ((temp_alarm_reg1 == 2)&(temp_alarm_reg2 < 3))) ;
                                // Advance for 00-18 (except 09) & 20-22.
assign res_temp_alarm_reg2 = (adv_hrs_temp_alarm == 1)&
                        (((temp_alarm_reg1 < 2)&(temp_alarm_reg2 == 9))|
                        ((temp_alarm_reg1 == 2)&(temp_alarm_reg2 == 3))) ;
                                                // Reset for 09, 19 & 23.
assign temp_alarm_reg2_next = temp_alarm_reg2 + 1 ;

always @ (posedge clk or negedge reset_n)
        begin
                if (reset_n == 0)
                        temp_alarm_reg2 <= 0 ;
                else if (res_temp_alarm_reg2 == 1)
                        temp_alarm_reg2 <= 0 ;
                else if (adv_temp_alarm_reg2 == 1)
                        temp_alarm_reg2 <= temp_alarm_reg2_next ;
                else                    ;
        end

assign adv_temp_alarm_reg3 = (adv_mts_temp_alarm == 1)&
                        (temp_alarm_reg3 < 5)&(temp_alarm_reg4 == 9) ;
                                        // Advance for 09, 19, 29, 39, 49.
assign res_temp_alarm_reg3 = (adv_mts_temp_alarm == 1)&
                        (temp_alarm_reg3 == 5)&(temp_alarm_reg4 == 9) ;
                                        // Reset for 59.
assign temp_alarm_reg3_next = temp_alarm_reg3 + 1 ;
always @ (posedge clk or negedge reset_n)
        begin
                if (reset_n == 0)
                        temp_alarm_reg3 <= 0 ;
                else if (res_temp_alarm_reg3 == 1)
                        temp_alarm_reg3 <= 0 ;
                else if (adv_temp_alarm_reg3 == 1)
                        temp_alarm_reg3 <= temp_alarm_reg3_next ;
                else                    ;
        end
assign adv_temp_alarm_reg4 = (adv_mts_temp_alarm == 1)&
```

```
                                  (temp_alarm_reg4 < 9) ;  // Advance for  0–8.
assign res_temp_alarm_reg4 =      (adv_mts_temp_alarm == 1)&
                                  (temp_alarm_reg4 == 9) ; // Reset for 9.
assign temp_alarm_reg4_next = temp_alarm_reg4 + 1 ;

always @ (posedge clk or negedge reset_n)
      begin
            if (reset_n == 0)
                  temp_alarm_reg4 <= 0 ;
            else if (res_temp_alarm_reg4 == 1)
                  temp_alarm_reg4 <= 0 ;
            else if (adv_temp_alarm_reg4 == 1)
                  temp_alarm_reg4 <= temp_alarm_reg4_next ;
            else                  ;
      end

assign adv_temp_alarm_reg5 = (adv_secs_temp_alarm == 1)&
                              (temp_alarm_reg5 < 5)&(temp_alarm_reg6 == 9) ;
                                          // Advance for 09, 19, 29, 39, 49.
assign res_temp_alarm_reg5 = (adv_secs_temp_alarm == 1)&
                              (temp_alarm_reg5 == 5)&(temp_alarm_reg6 == 9) ;
                                          // Reset for 59.
assign temp_alarm_reg5_next = temp_alarm_reg5 + 1 ;

always @ (posedge clk or negedge reset_n)
begin
            if (reset_n == 0)
                  temp_alarm_reg5 <= 0 ;
            else if (res_temp_alarm_reg5 == 1)
                  temp_alarm_reg5 <= 0 ;
            else if (adv_temp_alarm_reg5 == 1)
                  temp_alarm_reg5 <= temp_alarm_reg5_next ;
            else                  ;
end

assign adv_temp_alarm_reg6 = (adv_secs_temp_alarm == 1)&
                                                (temp_alarm_reg6 < 9) ;
                                          // Advance for  0–8.
assign res_temp_alarm_reg6  = (adv_secs_temp_alarm == 1)&
                                                (temp_alarm_reg6 == 9) ;
                                          // Reset for 9.
assign temp_alarm_reg6_next = temp_alarm_reg6 + 1 ;

always @ (posedge clk or negedge reset_n)
      begin
            if (reset_n == 0)
```

```
                temp_alarm_reg6 <= 0 ;
            else if (res_temp_alarm_reg6 == 1)
                temp_alarm_reg6 <= 0 ;
            else if (adv_temp_alarm_reg6 == 1)
                temp_alarm_reg6 <= temp_alarm_reg6_next ;
            else        ;
    end

// set_alarm1 is a signal which indicates that alarm1 is being set. When this
// signal is 1, the contents of temp_alarm_reg 1, etc. are copied into
// alarm1_reg1 and so on.
assign set_alarm1 = (set_alarm == 1)&(alarm1 == 1) ;
            // Set stop watch and Alarm Read/Set in set mode with Alarm1 set.
always @ (posedge clk or negedge reset_n)
    begin
            if (reset_n == 0)
                begin
                alarm1_reg1 <=  0 ;
                alarm1_reg2 <=  0 ;
                alarm1_reg3 <=  0 ;
                alarm1_reg4 <=  0 ;
                alarm1_reg5 <=  0 ;
                alarm1_reg6 <=  0 ;
                end
            else if (set_alarm1 == 1)
                begin
                alarm1_reg1 <= temp_alarm_reg1 ;
                        // Copy alarm setting from common set register
                        // into the particular Alarm register.
                alarm1_reg2 <=  temp_alarm_reg2 ;
                alarm1_reg3 <=  temp_alarm_reg3 ;
                alarm1_reg4 <=  temp_alarm_reg4 ;
                alarm1_reg5 <=  temp_alarm_reg5 ;
                alarm1_reg6 <=  temp_alarm_reg6 ;
                end
            else            ;
end

assign set_alarm2 = (set_alarm == 1)&(alarm2 == 1) ;
always @ (posedge clk or negedge reset_n)
begin
            if (reset_n == 0)
                begin
                alarm2_reg1 <=  0 ;
                alarm2_reg2 <=  0 ;
                alarm2_reg3 <=  0 ;
```

```
                    alarm2_reg4 <=  0 ;
                    alarm2_reg5 <=  0 ;
                    alarm2_reg6 <=  0 ;
                    end
            else if (set_alarm2 == 1)
                    begin
                    alarm2_reg1 <=  temp_alarm_reg1 ;
                    alarm2_reg2 <=  temp_alarm_reg2 ;
                    alarm2_reg3 <=  temp_alarm_reg3 ;
                    alarm2_reg4 <=  temp_alarm_reg4 ;
                    alarm2_reg5 <=  temp_alarm_reg5 ;
                    alarm2_reg6 <=  temp_alarm_reg6 ;
                    end
            else            ;
    end

assign set_alarm3 = (set_alarm == 1)&(alarm3 == 1) ;
always @ (posedge clk or negedge reset_n)
    begin
            if (reset_n == 0)
                    begin
                    alarm3_reg1 <=  0 ;
                    alarm3_reg2 <=  0 ;
                    alarm3_reg3 <=  0 ;
                    alarm3_reg4 <=  0 ;
                    alarm3_reg5 <=  0 ;
                    alarm3_reg6 <=  0 ;
                    end
            else if (set_alarm3 == 1)
                    begin
                    alarm3_reg1 <=  temp_alarm_reg1 ;
                    alarm3_reg2 <=  temp_alarm_reg2 ;
                    alarm3_reg3 <=  temp_alarm_reg3 ;
                    alarm3_reg4 <=  temp_alarm_reg4 ;
                    alarm3_reg5 <=  temp_alarm_reg5 ;
                    alarm3_reg6 <=  temp_alarm_reg6 ;
                    end
            else            ;
    end

always @ (alarm1 or alarm2 or alarm3 or reset_n)
    begin
            if (reset_n == 0)
                    read_alarm_reg  <=      0 ;
              // "read_alarm_reg" is a 2 bit register which stores the number
              // of the alarm to be read. If no alarm is on, it stores "0".
```

```
                    else if (alarm1 == 1)
                     // If more than one alarm is on, the one displayed (read) is
                     // the top priority alarm. alarm1 is the top most priority.
                              read_alarm_reg  <= 1 ;
                    else if (alarm2 == 1)
                              read_alarm_reg  <= 2 ;
                    else if (alarm3 == 1)
                              read_alarm_reg  <= 3 ;  // Least priority.
                    else
                              read_alarm_reg  <= 0 ;
          end

// Display real time or stopwatch or alarm on the seven-segment LEDs.
always @ (posedge clk or negedge reset_n)
          begin
                    if (reset_n == 0)
                              begin
                              data1     <=        0 ;
                              data2     <=        0 ;
                              data3     <=        0 ;
                              data4     <=        0 ;
                              data5     <=        0 ;
                              data6     <=        0 ;
                              end
                    else if (display_time == 1)// Display the Time.
                              begin
                              data1     <=        cnt1_reg ;
                              data2     <=        cnt2_reg ;
                              data3     <=        cnt3_reg ;
                              data4     <=        cnt4_reg ;
                              data5     <=        cnt5_reg ;
                              data6     <=        cnt6_reg ;
                              end
                    else if (display_stopw == 1)          // Display the Stopwatch.
                         begin
                         if ((set_stopw == 1)&(down_upn == 0))
                              begin
                              data1     <=        term_count_reg1 ;
                              data2     <=        term_count_reg2 ;
                              data3     <=        term_count_reg3 ;
                              data4     <=        term_count_reg4 ;
                              data5     <=        term_count_reg5 ;
                              data6     <=        term_count_reg6 ;
                              end
                         else
                              begin
```

```
            data1    <=    cnt7_reg ;
            data2    <=    cnt8_reg ;
            data3    <=    cnt9_reg ;
            data4    <=    cnt10_reg ;
            data5    <=    cnt11_reg ;
            data6    <=    cnt12_reg ;
            end
        end
else if (set_alarm == 1)
// Indicates that alarm is set and, therefore, temp_alarm_reg1 –
//  temp_alarm_reg6 will be displayed.
begin
            data1    <=    temp_alarm_reg1 ;
            data2    <=    temp_alarm_reg2 ;
            data3    <=    temp_alarm_reg3 ;
            data4    <=    temp_alarm_reg4 ;
            data5    <=    temp_alarm_reg5 ;
            data6    <=    temp_alarm_reg6 ;
end
else if (display_alarm == 1)          // Display the Alarm set.
begin
case (read_alarm_reg)
1:                                    // Display the Alarm 1.
begin
            data1    <=    alarm1_reg1 ;
            data2    <=    alarm1_reg2 ;
            data3    <=    alarm1_reg3 ;
            data4    <=    alarm1_reg4 ;
            data5    <=    alarm1_reg5 ;
            data6    <=    alarm1_reg6 ;
end
2:                                    // Display the Alarm 2.
begin
            data1    <=    alarm2_reg1 ;
            data2    <=    alarm2_reg2 ;
            data3    <=    alarm2_reg3 ;
            data4    <=    alarm2_reg4 ;
            data5    <=    alarm2_reg5 ;
            data6    <=    alarm2_reg6 ;
end
3:                                    // Display the Alarm 3.
begin
            data1    <=    alarm3_reg1 ;
            data2    <=    alarm3_reg2 ;
            data3    <=    alarm3_reg3 ;
            data4    <=    alarm3_reg4 ;
```

```
                        data5   <=      alarm3_reg5 ;
                        data6   <=      alarm3_reg6;
                end
                default:        ;
                endcase
                        end
                else            ;
        end

assign alarm1_match = (alarm1_reg1 == cnt1_reg)&(alarm1_reg2 == cnt2_reg)&
                      (alarm1_reg3 == cnt3_reg)&(alarm1_reg4 == cnt4_reg)&
                      (alarm1_reg5 == cnt5_reg)&(alarm1_reg6 == cnt6_reg) ;
                                // Set if present time = alarm1 set time.
// alarm1_30sec_delay is a bit which becomes "1" when alarm1_match = 1. It
// stays high for 30 s and then goes low. alarm1_30_sec is a counter which
// counts till 30. It counts so long as alarm1_30sec_delay is high. It is
// incremented every 1 s, i.e., when tbsec = 1.
assign adv_alarm1_30sec_counter = (tbsec == 1)&(alarm1_30sec_delay == 1) ;
assign alarm1_30sec_counter_next = alarm1_30sec_counter + 1 ;

always @ (posedge clk or negedge reset_n)
        begin
            if (reset_n == 0)
                begin
                alarm1_30sec_counter    <=      0 ;
                alarm1_30sec_delay      <=      0 ;
                end
            else if (alarm1_match == 1)
                                // When the present running time equals the
                                // Alarm 1 set time, turn the 30 s delay ON
                alarm1_30sec_delay      <=      1 ;
            else if (alarm1_30sec_counter == 5'd30)
                                // and turn it Off when the delay is complete.
                begin           // Also reset the counter.
                alarm1_30sec_counter    <=      0 ;
                alarm1_30sec_delay      <=      0 ;
                end
            else if (adv_alarm1_30sec_counter == 1)
                alarm1_30sec_counter <= alarm1_30sec_counter_next ;
            else                ;
        end

        // Alarm 2, Alarm 3 delays and counters work similar to that of Alarm 1.
assign alarm2_match = (alarm2_reg1==cnt1_reg)&(alarm2_reg2 == cnt2_reg)&
                      (alarm2_reg3==cnt3_reg)&(alarm2_reg4 == cnt4_reg)&
                      (alarm2_reg5== cnt5_reg)&(alarm2_reg6 == cnt6_reg) ;
```

```
                                        // Set if present time = alarm 2 time.
assign adv_alarm2_30sec_counter = (tbsec == 1)&(alarm2_30sec_delay == 1) ;
assign alarm2_30sec_counter_next = alarm2_30sec_counter + 1 ;
always @ (posedge clk or negedge reset_n)
        begin
            if (reset_n == 0)
                begin
                alarm2_30sec_counter    <=      0 ;
                alarm2_30sec_delay      <=      0 ;
                end
            else if (alarm2_match == 1)
                alarm2_30sec_delay      <=      1 ;
            else if (alarm2_30sec_counter == 5'd30)
                begin                   // 30 s complete.
                alarm2_30sec_counter    <=      0 ;
                alarm2_30sec_delay      <=      0 ;
                end
            else if (adv_alarm2_30sec_counter == 1)
                alarm2_30sec_counter  <=  alarm2_30sec_counter_next ;
            else                    ;
        end

assign alarm3_match =(alarm3_reg1==cnt1_reg)&(alarm3_reg2 == cnt2_reg)&
                     (alarm3_reg3==cnt3_reg)&(alarm3_reg4 == cnt4_reg)&
                     (alarm3_reg5==cnt5_reg)&(alarm3_reg6 == cnt6_reg) ;
                                        // Present time = alarm3 time.
assign adv_alarm3_30sec_counter = (tbsec == 1)&(alarm3_30sec_delay == 1) ;
assign alarm3_30sec_counter_next = alarm3_30sec_counter + 1 ;

always @ (posedge clk or negedge reset_n)
        begin
            if (reset_n == 0)
                begin
                alarm3_30sec_counter    <=      0 ;
                alarm3_30sec_delay      <=      0 ;
                end
            else if (alarm3_match == 1)
                alarm3_30sec_delay      <=      1 ;
            else if (alarm3_30sec_counter == 5'd30)  // 30 s complete.
                begin
                alarm3_30sec_counter    <=      0 ;
                alarm3_30sec_delay      <=      0 ;
                end
            else if (adv_alarm3_30sec_counter == 1)
                alarm3_30sec_counter  <=  alarm3_30sec_counter_next ;
            else                    ;
```

```
        end
/*
"ring" is a signal that indicates that one or more alarms is/are active. "beep" is the
signal (square wave) which is actually output to the speaker if ring is high.
"beep_counter" counts till 2 (means 0.2 s). When it is 2, beep is toggled repeat-
edly producing 2.5 Hz beeping tone if alarm OFF/ON switch is in ON position.
Otherwise, the sound alarm is OFF.
*/
assign ring = ((alarm1_30sec_delay == 1)|(alarm2_30sec_delay == 1)|
              (alarm3_30sec_delay == 1))|(timer_out_alarm      == 1) ;
/*
"timer_out_alarm" signal is high for 30 s (to sound the audio alarm) after the ter-
minal count, Up or Down, is reached, i.e., timer_out =1.
*/
assign beep_counter_next = beep_counter + 1 ;

always @ (posedge clk or negedge reset_n)
begin
        if (reset_n == 0)
                begin
                        beep_counter    <=      0 ;
                        beep            <=      0 ;
                end
        else if (ring == 0)                     // This means no alarm is active.
                begin
                        beep_counter    <=      0 ;
                        beep            <=      0 ;
                        end
        else if (beep_counter == 2)
                begin
                        beep    <= (~beep)&((alarm_off_onn == 0)|
                                        (timer_out_alarm == 1)) ;
                                // Toggle if alarm switch is in ON position or if
                                // "timer_out_alarm" is high.
                        beep_counter    <=      0 ;
                end
        else if ((ring == 1)&(cntds_reg == `ds_base))
                        beep_counter <= beep_counter_next ;
        else                    ;
end

// Call the BCD to seven-segment display conversion ROM.
// Turn off all display decimal points.
display_rom disp1       (       .addr(data1),
                                .dec_pt(1'b0),
                                .out(display1)
```

```
                              );
display_rom disp2    (    .addr(data2),
                          .dec_pt(1'b0),
                          .out(display2)
                              );
display_rom disp3    (    .addr(data3),
                          .dec_pt(1'b0),
                          .out(display3)
                              );
display_rom disp4    (    .addr(data4),
                          .dec_pt(1'b0),
                          .out(display4)
                              );
display_rom disp5    (    .addr(data5),
                          .dec_pt(1'b0),
                          .out(display5)
                              );
display_rom disp6    (    .addr(data6),
                          .dec_pt(1'b0),
                          .out(display6)
                              );
endmodule
```

We have used a display ROM sub-module for converting a BCD number to a seven-segment LED display code. This ROM stores the seven segment display values including a right decimal point. Logic 1 lights the LED configured in the common anode mode. Put this sub-module in a separate file called "display_rom.v". The code for "display_rom" is presented in Verilog_code_14.4.1. This module has two inputs named, "addr" and "dec_pt" and an output, "out". The BCD number, say, any one of "data1" to "data6", that needs to be converted is input into "addr". The seven segments and decimal point are : a b c d e f g dp, where "a" is the msb and decimal point, "dp", is the lsb. The Verilog code for display ROM is a simple realization using "case" statements in "always" block. For example, the statement:

```
9 :     out =    {7'b1111_011, dec_pt} ;
```

means if "addr", i.e., BCD input number, is "9", then the display output, "out", is 1111 0110. Note that the segments "e" and "dp" are both off for a display of "9".

Verilog_code_14.4.1

```
/*
                              display_rom
"display_rom.v" is the sub-module for converting a BCD number to seven-
segment LED display code. This ROM stores the seven segment display values
including a right decimal point.
```

```
Logic "1" lights the LED configured in common anode mode.
*/
module display_rom (addr, dec_pt, out) ;
input      [3:0] addr ;           // 0000 displays "0", 1001 displays "9" & so on.
input                  dec_pt ;   // Input logic "1" if you wish to turn it on.
output     [7:0]       out ;      // Segments: a b c d e f g dp, where "a" is the
                                  // msb and decimal point, "dp", is the lsb.
                                  // "a" is the top segment, "b" is the next segment
                                  // clockwise and "g" is the center segment.
reg        [7:0]       out ;

always @ (addr or dec_pt)
begin
  case (addr)
            0 :    out = {7'b1111_110, dec_pt} ;
                                  // dec_pt = 1 turns ON the decimal point.
            1 :    out = {7'b0110_000, dec_pt} ;  // Order:  abcd_efg_dp
            2 :    out = {7'b1101_101, dec_pt} ;
            3 :    out = {7'b1111_001, dec_pt} ;
            4 :    out = {7'b0110_011, dec_pt} ;
            5 :    out = {7'b1011_011, dec_pt} ;
            6 :    out = {7'b1011_111, dec_pt} ;
            7 :    out = {7'b1110_000, dec_pt} ;
            8 :    out = {7'b1111_111, dec_pt} ;
            9 :    out = {7'b1111_011, dec_pt} ;
      default :    out = 8'b0 ;        // Blank the display for illegal values.
  endcase
end
endmodule
```

14.5.7 Test Bench for Real Time Clock Design

This test is not going to be elaborate. We shall only test the running of the real time since other tests are going to be conducted on the actual hardware. The testing of other modes can be done on your own, if you wish. To start with, we need to declare the timescale. Let us say that we need to run simulation at 1GHz, for which we need high precision. Therefore, we will have the time base as 100 ps and the accuracy as 10 ps. We also have a clkperiodby2 declared as 5. The basic unit is 100 ps and hence this value of 5 denotes 500 ps each for the ON time and OFF time. This sums up to 1000 ps or 1ns time period. This means that the frequency is 1 GHz. Then we include the actual design, "rtc_alarm.v" and declare the module for the test bench. This is followed by the declaration of the inputs in the test bench. As usual, we have used "reg" for the inputs in the test bench and wire for the outputs. Note that this is different from what we have been using in the design.

Next step is to instantiate the "rtc_alarm" design module. Here, we call all ports by name. Thereafter, we shall apply stimulants in the initial block. Various inputs like "clk", etc. are initialized and the system is reset. After 100 units of time, i.e., 10 ns, the reset is withdrawn to start the normal functioning of real time clock. The RUN, TIME mode is also selected so that the real time clock may run from "00 00 00" onwards. It will be cumbersome to test time/alarm settings, etc., while simulating. Therefore, we will defer elaborate testing until setting up the Demo unit with FPGA and digital I/O boards. The simulation is allowed to run for 10 ms so as to get one complete 24 h cycle of real time starting from zero. In the next "always" statement, we toggle "clk" every 0.5 ns to get a free running clock. This ends the test bench, which is as follows.

Verilog_Code_14.5

```
//                    Test Bench for Real Time Clock Design
`timescale 100ps/10ps
`define clkperiodby2 5                     // Run at 1 GHz for simulation.
`include "rtc_alarm.v"                     // This is the design file.

module rtc_alarm_test ;

    reg       clk ;
    reg       reset_n ;
    reg       irun_setn ;                  // RUN/SET mode switch.
    reg       itime_stopwn ;               // TIME/STOP WATCH mode switch.
    reg       idown_upn ;                  // UP/DOWN mode switch.
    reg       ihrs ;                       // Push button switches for setting "hrs",
    reg       imts ;                       // "mts"
    reg       isecs ;                      // "secs" and
    reg       istart_stopn ;               // "Start/Stop" .
    reg       ialarm_read_setn;
    reg       ialarm_off_onn;
    reg       ialarm1;
    reg       ialarm2;
    reg       ialarm3;
    wire      [7:0]     display1 ;
    wire      [7:0]     display2 ;
    wire      [7:0]     display3 ;
    wire      [7:0]     display4 ;
    wire      [7:0]     display5 ;
    wire      [7:0]     display6 ;
    wire                beep ;
    wire                timer_out ;
```

```
// Instantiate the "rtc_alarm" design module.
   rtc_alarm u1 (           .clk(clk),
                            .reset_n(reset_n),
                            .irun_setn(irun_setn),
                            .itime_stopwn(itime_stopwn),
                            .idown_upn(idown_upn),
                            .ihrs(ihrs),
                            .imts(imts),
                            .isecs(isecs),
                            .istart_stopn(istart_stopn),
                            .ialarm_read_setn(ialarm_read_setn),
                            .ialarm_off_onn(ialarm_off_onn),
                            .ialarm1(ialarm1),
                            .ialarm2(ialarm2),
                            .ialarm3(ialarm3),
                            .display1(display1),  // seven-segment LED outputs –
                            .display2(display2),  // display1 (MSD), 2 are HRS,
                            .display3(display3),  // display3 (MSD), 4 are MTS,
                            .display4(display4),
                            .display5(display5),  // display5 (MSD), 6 are SECS.
                            .display6(display6),
                            .beep(beep),
                            .timer_out(timer_out)
                        );
initial
    begin
        clk                 <=    0 ;
        reset_n             <=    0 ;
        irun_setn           <=    0 ;     // SET
        itime_stopwn        <=    0 ;     // STOPWATCH
        idown_upn           <=    0 ;     // UP count mode.
        ihrs                <=    1 ;
        imts                <=    1 ;
        isecs               <=    1 ;
        istart_stopn        <=    1 ;     // istart_stopn = 1 means that
                                        // START/STOP button is pressed.
        ialarm_read_setn    <=    1 ;     // Don't set alarm mode.
        ialarm_off_onn      <=    1 ;
        ialarm1             <=    1 ;
        ialarm2             <=    1 ;
        ialarm3             <=    1 ;
      #100
        reset_n             <=    1 ;
        irun_setn           <=    1 ;     // RUN
        itime_stopwn        <=    1 ;     // TIME mode.
```

```
        #100000000      $stop ;                  // Stop after 10 ms.
end
always
        #`clkperiodby2 clk <= ~clk ;       // Toggle to get a free running clock.
endmodule
```

14.5.8 Simulation Results of Real Time Clock

The simulation results are presented in Figures 14.12.1 to 14.12.3. From Figure 14.12.1, we see that reset is withdrawn, RUN/SET switch is set to RUN position and TIME/SW switch is set to TIME position at 10 ns, conforming to the conditions set in the test bench. Note that the running time counters, "cnt1_reg" to "cnt6_reg", are cleared when reset is applied at the beginning. The six digit seven segment LED display outputs, "display1" to "display6", which correspond to the counters, "cnt1_reg" to "cnt6_reg" are also cleared (segments "g" and "dp" are off) when reset is applied. Figure 14.12.2 shows one complete 24 h cycle. The real time is progressing systematically from "00 00 00" to "23 59 59" and raps round to "00 00 00" to repeat the time. This can be observed in the running time counters, "cnt1_reg" to "cnt6_reg". Note the one to one correspondence between "cnt1_reg" to "cnt6_reg" and "display1" to "display6". Figure 14.12.3 shows the zoomed in view, centered around midnight, real time.

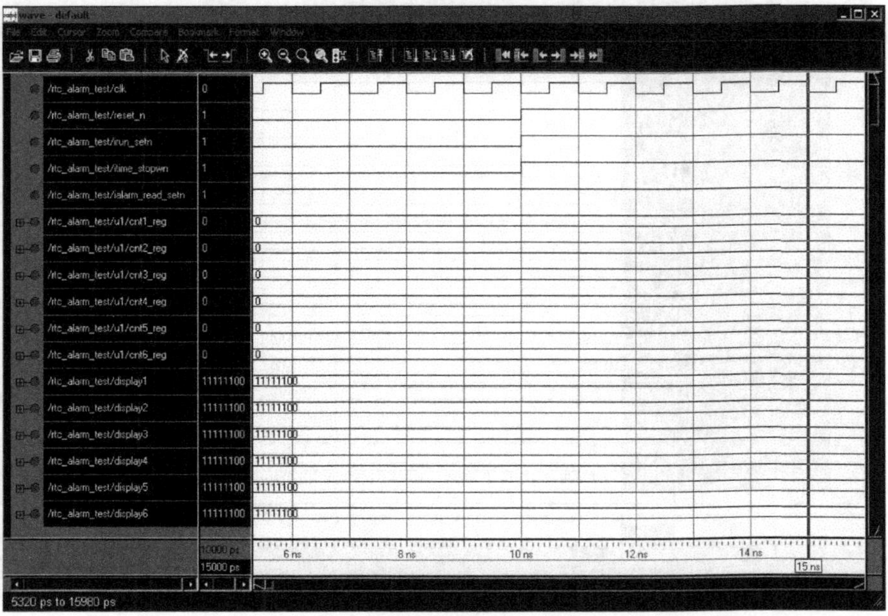

Fig. 14.12.1 Simulation results of real time clock (Continued)

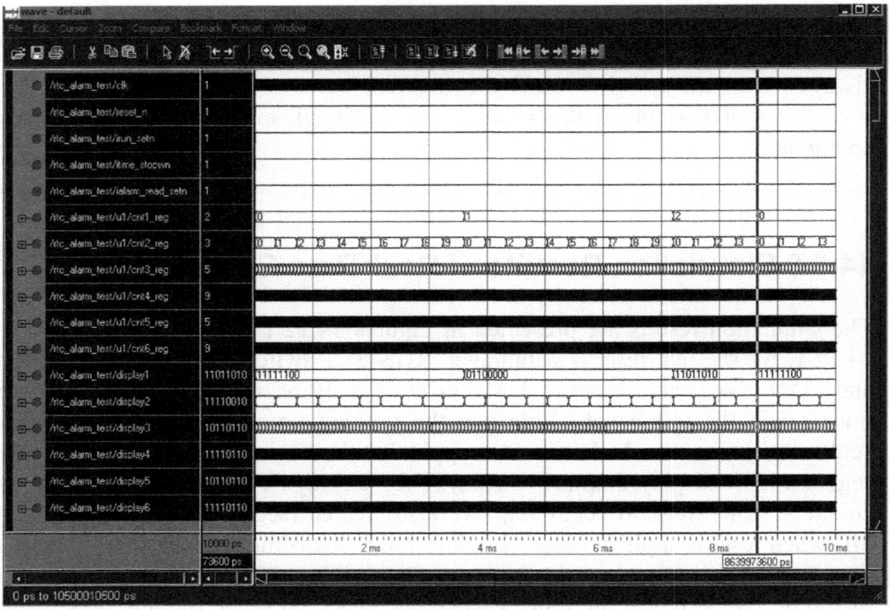

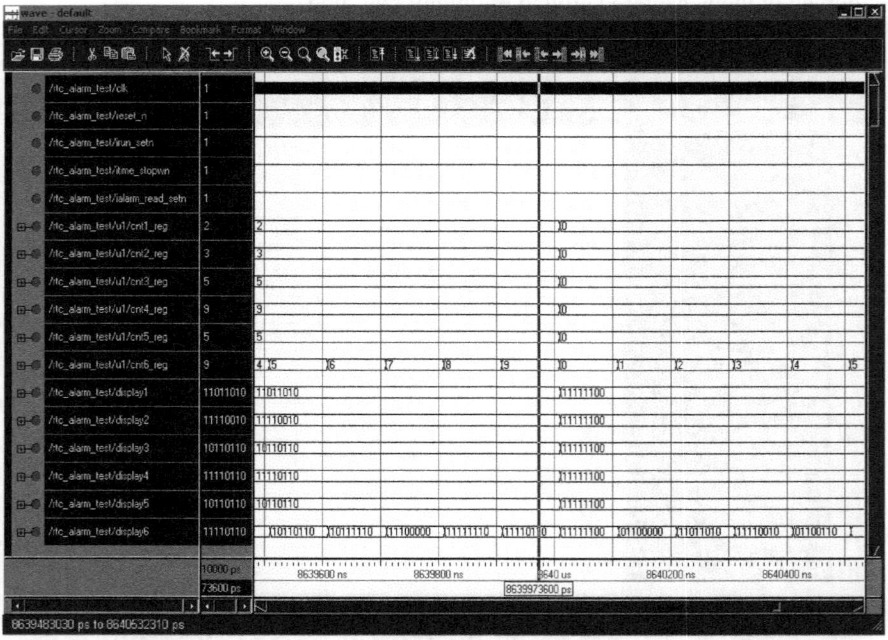

Fig. 14.12.2 and 14.12.3 Simulation results of real time clock

14.5.9 Synthesis Results of Real Time Clock

Synplify results are summarized in this section. The device used is XCV800hq240-4. During simulation, we have deliberately chosen an impractical frequency of operation of 1 GHz. Modelsim permits up to 2.5 GHz. Even in previous chapters, these details were discussed. As discussed before, simulation can be run at any GHz, but after place and route, the frequency of operation will reduce drastically. For example, in the present real time clock design, it has come down to 42.1 MHz as per Synplify tool. Even the synthesis will not reveal the actual frequency of operation. Only after place and route, we will get the true picture. Of course, it would be enough if we run at 20 MHz. A positive "Slack" reported by the tool means that there is more free time available so that we can improve the speed of the design by introducing additional pipeline stages. Watch out for negative slack time. Guard against it. The tool reports the primitives that we have used in the FPGA. The total number of LUTs consumed by the design is 681. "rtc_alarm.edf" file is finally created by the synplify tool, which needs to be exported to the Xilinx Place & Route tool for the creation of bit file.

Synplify Report

Worst slack in design: 26.244

Clock Starting	Requested Frequency	Estimated Frequency	Requested Period	Estimated Period
clk	20.0 MHz	42.1 MHz	50.000	23.756

Mapping to part: xcv800hq240-4
Cell usage:

MUXCY_L	78	uses
MULT_AND	12	uses
XORCY	87	uses
MUXF5	20	uses
FDCE	114	uses
FDC	159	uses
FDPE	14	uses
GND	1	use

I/O primitives:

IBUF	13	uses
OBUF	50	uses
BUFGP	1	use
I/O Register bits:	12	
Register bits not including I/Os:	275 (1%)	
Global Clock Buffers:	1 of 4 (25%)	
Total LUTs:	681 (3%)	

14.5.10 Xilinx P&R Results

The input for this tool is the "rtc_alarm.edf " file created by the synthesis tool. The number of LUTs reported by this tool is 675, whereas synthesis tool reported 681. The place and route tool gives the exact number. The total gate count for the design is 6721. The maximum frequency of operation is also reduced to about 39 MHz. Therefore, the system can work at 20 MHz without any problem.

User Constraint File for Real Time Clock

While running the place and route tool, we must have a separate file named user constraint file, ".ucf ", which lists all the signals in the design as well as the exact pin numbers of the signals. Put the following constraints in a separate file and name it as "rtc.ucf ". Place it in the same folder, where the "rtc_alarm.edf " is located. This is not mandatory, but preferable. It may be noted that the ".edf" file was created using the Synplify tool. We will declare the inputs and outputs as "NET" and assign the pins as per the actual connections in the Demo setup, which we will describe in the next section. The bussed outputs such as the six numbers of seven segment displays will have to be allotted pins, bit-wise. The user constraint file, "rtc.ucf ", is as follows:

```
NET    "clk"                LOC    = P89 ;

NET    "reset_n"            LOC    = P140 ;
NET    "irun_setn"          LOC    = P161 ;
NET    "itime_stopwn"       LOC    = P159 ;
NET    "idown_upn"          LOC    = P155 ;
NET    "ihrs"               LOC    = P185 ;
NET    "imts"               LOC    = P176 ;
NET    "isecs"              LOC    = P175 ;
NET    "istart_stopn"       LOC    = P174 ;
NET    "ialarm_read_setn"   LOC    = P153 ;
NET    "ialarm_off_onn"     LOC    = P53 ;

NET    "ialarm1"            LOC    = P149 ;
NET    "ialarm2"            LOC    = P146 ;
NET    "ialarm3"            LOC    = P142 ;

NET    "display1[7]"        LOC    = P108 ;
NET    "display1[6]"        LOC    = P188 ;
NET    "display1[5]"        LOC    = P189 ;
NET    "display1[4]"        LOC    = P191 ;
NET    "display1[3]"        LOC    = P107 ;
NET    "display1[2]"        LOC    = P192 ;
NET    "display1[1]"        LOC    = P193 ;
```

```
NET    "display1[0]"    LOC    = P194;

NET    "display2[7]"    LOC    = P103;
NET    "display2[6]"    LOC    = P195;
NET    "display2[5]"    LOC    = P199;
NET    "display2[4]"    LOC    = P200;
NET    "display2[3]"    LOC    = P102;
NET    "display2[2]"    LOC    = P101;
NET    "display2[1]"    LOC    = P202;
NET    "display2[0]"    LOC    = P203;

NET    "display3[7]"    LOC    = P100;
NET    "display3[6]"    LOC    = P205;
NET    "display3[5]"    LOC    = P206;
NET    "display3[4]"    LOC    = P207;
NET    "display3[3]"    LOC    = P99;
NET    "display3[2]"    LOC    = P208;
NET    "display3[1]"    LOC    = P209;
NET    "display3[0]"    LOC    = P215;

NET    "display4[7]"    LOC    = P97;
NET    "display4[6]"    LOC    = P216;
NET    "display4[5]"    LOC    = P217;
NET    "display4[4]"    LOC    = P218;
NET    "display4[3]"    LOC    = P96;
NET    "display4[2]"    LOC    = P220;
NET    "display4[1]"    LOC    = P221;
NET    "display4[0]"    LOC    = P222;

NET    "display5[7]"    LOC    = P94;
NET    "display5[6]"    LOC    = P223;
NET    "display5[5]"    LOC    = P224;
NET    "display5[4]"    LOC    = P228;
NET    "display5[3]"    LOC    = P93;
NET    "display5[2]"    LOC    = P229;
NET    "display5[1]"    LOC    = P230;
NET    "display5[0]"    LOC    = P231;

NET    "display6[7]"    LOC    = P87;
NET    "display6[6]"    LOC    = P232;
NET    "display6[5]"    LOC    = P234;
NET    "display6[4]"    LOC    = P235;
NET    "display6[3]"    LOC    = P86;
NET    "display6[2]"    LOC    = P236;
NET    "display6[1]"    LOC    = P237;
NET    "display6[0]"    LOC    = P238;
```

NET	"beep"	LOC	= P109;
NET	"timer_out"	LOC	= P162;

You may need to change these pin assignments if you are using any other FPGA or I/O board for realizing the real time clock. The Xilinx P&R tool creates a bit file, "rtc_alarm.bit", which is used for downloading into the FPGA. The P&R report is as follows:

Target Device : xv800		
Target Package : hq240		
Target Speed : –4		
Logic Utilization:		
Number of slice flip flops:	275 out of 18,816	1%
Number of four input LUTs:	641 out of 18,816	3%
Number of occupied slice:	398 out of 9,408	4%
Number of slice containing only related logic:	398 out of 398	100%
Number of slices containing unrelated logic:	0 out of 398	0%
Total number 4 input LUTs:	675 out of 18,816	3%
Number used as logic:	641	
Number used as a route-thru:	34	
Number of bonded IOBs:	63 out of 166	37%
IOB flip flops:	12	
Number of GCLKs:	1 out of 4	25%
Number of GCLKIOBs:	1 out of 4	25%

Total equivalent gate count for design: 6,721
Additional JTAG gate count for IOBs: 3,072
Timing summary:
Minimum period: 25.194 ns (maximum frequency: 39.692 MHz)
Minimum input required time before clock: 6.604 ns
Minimum output required time after clock: 25.551 ns
Saving bit stream in "rtc_alarm.bit"

14.5.11 Hardware Setup of Real Time Clock

The hardware setup for this design is basically the same as that used in the traffic light controller design except that this design does not have the traffic light display board. Further, we need all the six numbers of seven-segment LED displays mounted on the digital input/output board. Accordingly, we have met these hardware requirements, and the overall setup for the real time clock is shown in Figure 14.13. As shown therein, the two expansion headers on the FPGA board are connected to the digital I/O board using a flat cable. The FPGA board is also connected to a parallel port of a PC. One of the output ports of the FPGA is connected to a piezoelectric buzzer to sound the audio alarm. The FPGA board is connected to a DC power supply, 8.5 V. It consumes only 0.71 A as can be seen on the right side of the figure. The I/O board is powered by another supply, 5V DC, placed behind the two boards. Since the display on the I/O board is not clearly visible in the figure, a zoomed version is also shown below the setup. The snapshot shows the working of the real time clock in the RUN TIME mode. Various input/output signals used in the design are connected as per the pin definition in the user constraint file presented in the previous section.

Fig. 14.13 Hardware setup for the real time clock

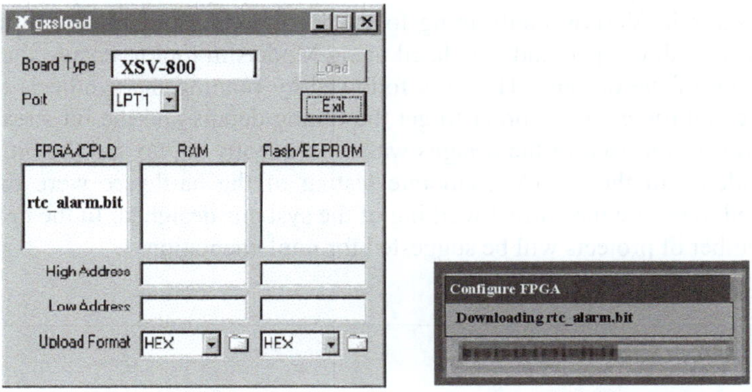

a **b**

Fig. 14.14 Downloading the bit stream into the FPGA

Figure 14.14 shows the downloading of the bit stream, "rtc_alarm.bit" into the FPGA. It may be recalled that this bit stream was generated using the Xilinx P&R tool, also presented in the previous section. Testing the real time working takes 24 hours in the normal time mode. The testing can be expedited by reducing the definition, "ds_base", declared at the beginning of the design, "Verilog_code_14.4". By decreasing it 60-fold, we can advance the minutes display every second. Similarly, by decreasing it further by 60 times, we can advance the hours display every second. Thus, the entire real time test can be carried out in about 3 min. time, instead of 24 hours in the normal mode. However, we need to run the synthesis and P&R tools two more times to get these bit stream files. This has been done and the two bit streams are "rtc_alarm_mts_fast.bit" and "rtc_alarm_hrs_fast.bit" respectively for advancing minutes and hours display fast. These two bit streams are downloaded into the FPGA in turn for fast display to complete the test as mentioned earlier. Other modes such as time and alarm settings, stop watch settings and functioning of the three alarms, and Up and Down counter/timer were also checked and found to work satisfactorily. These tests are left as exercise for the readers.

Summary

A couple of complete hardware implementations, namely, a traffic light controller and a real time clock were presented as examples in this chapter. These applications were based on ready made boards available such as an FPGA board and a digital input/output board. The design methodology adopted in these designs may be extended to any other project design or any other FPGA and I/O boards. These system designs were presented in a systematic manner, starting from detailed specification. The need for formulating the right type of architecture was emphasized and designed with actual hardware components in mind. The signal nomenclature adopted in the architectures were actually used as it is in realizing their designs in Verilog conforming to the RTL coding guidelines. Simple test benches were developed and simulated using Modelsim tool to ensure the correct functioning of the designs. This was followed by running the synthesis tool and the place and route tool in order to get the timing details and the bit stream files. The hardware for each of the designs was subsequently set up and the bit streams downloaded into the FPGA. Elaborate testing of the hardware were thereafter conducted to ensure the correct working of the systems designed. In the next chapter, a number of projects will be suggested for implementation.

Assignments

14.1 A traffic light controller design for right flowing traffic was presented in the text. Redesign for left flowing traffic as is in vogue in eastern countries.

14.2 The traffic light controller design covered in the text does not have "Pedestrian crossing". Include the same and redesign for right or left flowing traffic. Do not use push button switch requests for the same. Make it automatic. Run all the three tools, Modelsim, Synplify, and Xilinx P&R and demonstrate the working of the design on the hardware depending upon the availability.

14.3 Design a traffic light controller for a T junction with free right. Assume right side flowing traffic and the side road approach to be from bottom. Run all the three tools: Modelsim, Synplify, and Xilinx P&R and demonstrate the working of the design on the hardware if feasible.

14.4 It is desirable to have large seven segment displays for displaying time remaining at every approach road of a four road traffic junction covered in the text. This will show the road users how much time is remaining for the current traffic to stop or for resuming it. Each of these displays may be formed using discrete LEDs as shown in Figure A14.1. Each digit will be a seven segment display with a decimal point in the same way a normal seven

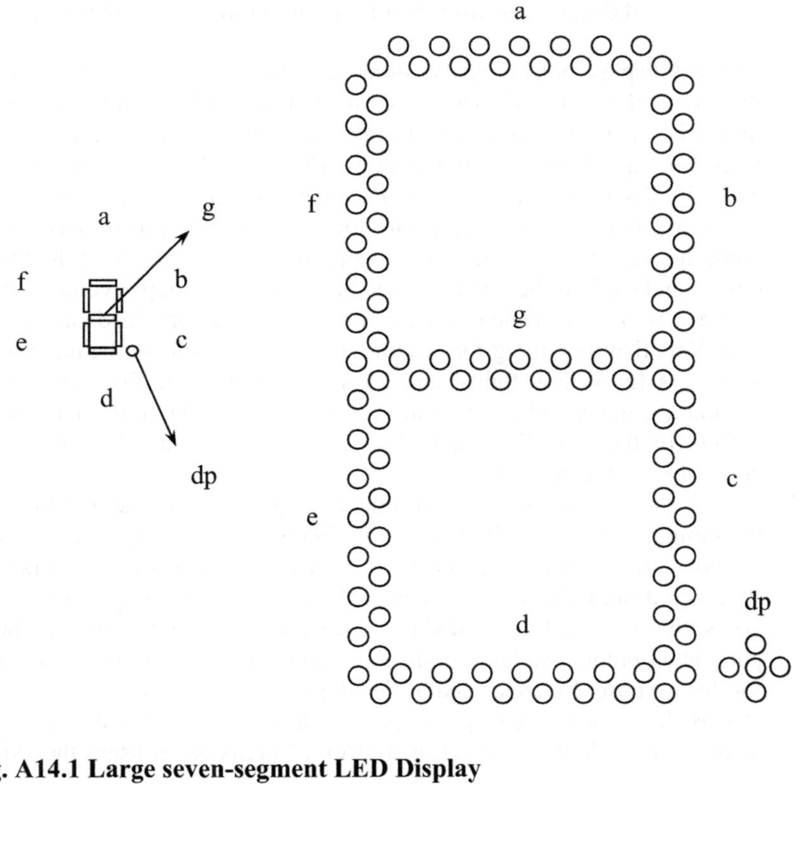

Fig. A14.1 Large seven-segment LED Display

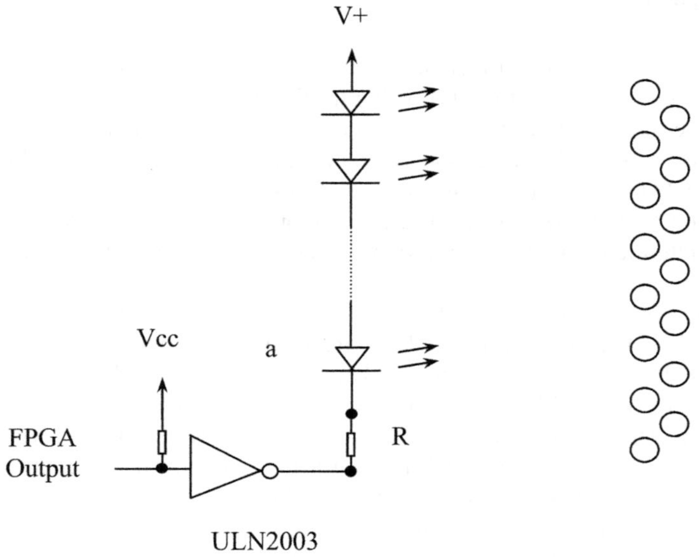

Fig. A14.2 Typical driving circuit for individual segment of a large display

segment display looks like, also shown alongside. A typical interface circuit that drives these large displays is shown in Figure A14.2. In this circuit, the driver is an open collector, Darlington pair inverter using the ULN2003 chip. The input to this inverter is an FPGA output. The ULN2003 can work up to 50 V and sink 500 mA current. Each segment is realized by connecting individual LEDs in series. Do you envisage any problem connecting them the way it is shown? If so, amend the circuit to improve the product design and explain the amendments. Select suitable supply voltage and resistors. Provide the same sequences and timings we provided earlier in the text. Note that the timing for the straight flowing traffic on the main road is 45 s, 5 s for yellow lights, and 25 s for all other traffic. Formulate an optimum scheme and incorporate it in the traffic light controller designed before for the right flowing traffic. You need only simulate the Verilog code. No demo is required.

14.5 Modify the Verilog code of real time clock presented in the text to display the timing range of 00 00 00 to 11 59 59 (HRS MTS SECS) instead of the 24 hours range, retaining other features already designed. Simulate the amended design. Discuss how you will amend the design if both 24 and 12 hours ranges are to be included in the same clock. You may use another input and an output available in the expansion connector of the FPGA board for this purpose. No code need be developed.

14.6 Discuss how you will change the present design to accommodate additional displays for YEAR, MONTH, and DAY (2 digits each) using the existing

LEDs. There is no need to write codes. Just discuss the design methodology. Will these displays present any problems? If so, how will you solve them?

14.7 If we are to use the timer designed in the text for applications demanding timing in milliseconds in addition to hours, minutes, and seconds, then we need to modify the existing design. Use appropriate numbers of additional inputs/outputs and provide user presetting to include this feature in your design change. The display shall be in the range: 0 to 999.999 s. The timing must commence when the user presses the START push button switch, simultaneously turning on the output, "TIMER OUT". After the lapse of the set delay, the output is turned off and the beeping audio alarm sounds for 30 s. Use a separate switch, preferably a push button switch, for presetting (or clearing) the timer before restarting by pressing START push button. The digital timer may be stopped or resumed at any point of time as was done in the design presented before. What are the potential problems in this aspect of the design? How will you solve any problem that may be encountered? Discuss how you will go about the design. No Verilog code need be developed.

14.8 Lots of power is being wasted in offices, banks, factories, institutions, especially during lunchtime etc., which can be easily avoided by installing a real time clock such as that we have designed in the text, of course, needing quite many changes. Include in your design change four outputs for switching on or off four different electrical circuits that power lights, fans, air conditioners, machineries, etc. as per the following schedule as an example:

DAY	Switch ON at			Switch OFF at		
	HRS	MTS	SECS	HRS	MTS	SECS
Monday thru' Friday	8 13	00 00	00 00	12 17	00 00	00 00

You must facilitate the user to program for any day of the week, any time and up to two different time settings each for switching on and switching off lights, etc., an example of which is shown in the table. Each of the power loads must be individually programed. Provide additional switches for overriding the automatic control to switch on any of the four circuits when the occasion demands. Present your detailed specification, design methodology, architecture and an algorithm to aid in the design. No Verilog code is necessary.

14.9 Design a controller that switches on lights, air conditioner(s), and other power points in a mini-concert hall as the first person enters the hall and switches off the lights and other devices as the last person leaves it. The concert hall has four entry/exit doors equipped with infra red sensors. Only one person can enter or leave through a door at a time. Realize the RTL Verilog design and test it.

14.10 A soft drink bottling plant that requires automation is shown in the Figure A14.3. Empty bottle container dispenses one bottle whenever the solenoid SOL1 is operated for 1 s. When the container has only small number of bottles left, the sensor BL sends logic high signal. Similarly, SOL2 operates for 5 s to fill soft drink into the bottle when its presence is sensed by the sensor S1, and cap sealer fixes the cap in the filled bottle by operating the solenoid SOL3 for 2 s when its arrival is sensed by S2. The sensors DL and CL generate high signals when the respective items are in short supply.

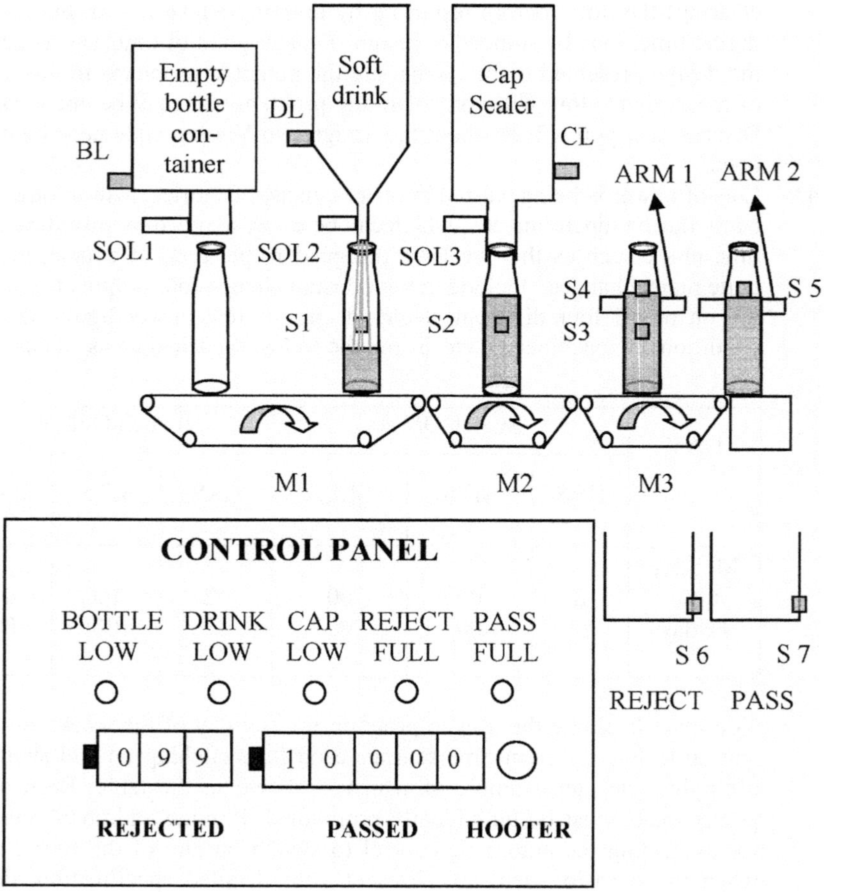

Fig. A14.3 Bottling plant

In the next position, the arrival of the capped bottle is sensed using S3. The sensor S4 inspects the filled level of the bottle and generates logic high if the level is acceptable, in which case the bottle proceeds to the next stage. Otherwise, ARM1 is activated for 2 s to grab the bottle and put it into a vacant slot in a crate designated as "REJECT". The accepted filled bottle sensed by S5 is grabbed by the ARM2 (also activated for 2 seconds) and placed in the crate below marked "PASS". S6 and S7 sense the presence or absence of the respective crates. These crates are manually handled. When they are full (each capable of accommodating up to 32 bottles), a HOOTER in the control panel is sounded continuously and, REJECT FULL or PASS FULL lamp is turned on as the case may be, and the operator replaces the filled crate by empty ones. The HOOTER generates a beeping sound if any of the sensors BL, DL, or CL is activated; turning on the appropriate lamp(s): BOTTLE LOW, DRINK LOW, or CAP LOW, as the case may be. The bottles are transported by conveyor belts driven by the motors M1, M2, and M3, which may be switched on or off as is appropriate. The total number of "REJECTED" or "PASSED" bottles are indicated by using electromagnetic counters as shown. They preserve their values even during a power failure. They may be reset at any point of time using the respective push button. Draw a detailed specification and the architecture so that the controller design may be realized using Verilog RTL. You may add any other devices or features to make the design more perfect. State your assumptions clearly.

Chapter 15

Projects Suggested for FPGA/ASIC Implementations

We have seen how to design VLSI systems using Verilog in the previous chapters. Complete system designs were presented for some projects, such as PCI Arbiter and Discrete Cosine Transform and Quantization Processor for Video compression applications. The design complexities were up to about 120,000 gates mapped on FPGAs. We have also implemented a couple of designs as examples, using FPGA and digital input/output boards. These are Traffic Light Controller and a Real Time Clock. All the codes developed work readily on any FPGA or as an ASIC. In this chapter, a number of applications are suggested for you to design on FPGA/ASIC.

The design methodologies and Verilog codes presented for a number of project designs in the earlier chapters may be readily applied to the new projects. Some of the codes may need modifications to suit the particular application. This design approach reduces the development cycle time considerably. If it may be suggested, it would be a good idea to put various commonly used Verilog modules such as the adders, multipliers, etc., and other modules you and other design team members may develop in a library folder for ready access by all others. However, extreme care must be taken while including comments in the codes, be it Verilog or higher level languages such as Matlab, C and, in the preparation of documents for users, without which the design will be useless. Also, do not forget to include aptly commented test benches. In this connection, mention may be made that there are few websites [102] catering to this need. You may contribute as well as download Verilog/VHDL codes for some applications along with documentation.

15.1 Projects for Implementation

We have presented a number of applications in this section and classified them into various categories. These categories, by no means exhaustive, are automotives, avionics, control system applications, medical applications, and video processing applications, to name a few. Brief descriptions are also presented for some of the applications. The reader may gather more ideas and information from websites, magazines, journals, conference papers, books, newspapers, TV shows, etc.; above all use fertile imagination before undertaking any serious design. All these systems need lots of intelligence embedded in the chips being designed. Some of the applications mentioned here are already available as embedded systems, implemented probably with 8051 family of microcontrollers or any other processors such as 8085, 6800, 68000, Arm processors, DSP processors, etc.

15.1.1 Automotive Electronics

The first group suggested is the automotive electronics. Brief descriptions of some of the applications falling in this category are as follows: We can design intelligent controller for anti-lock braking system, also called ABS. This is already in vogue in many countries. When the road is slippery and if you jam or apply brake continuously, then your vehicle is sure to skid and go out of control. With a gentle and effective pumping action on the brake, skidding can be eliminated. But manual application on the spur of the moment will not be effective since the road conditions are not known. Even professional drivers cannot stop as quickly without ABS as an average driver can with ABS. Hence, we need a controller that will monitor the road conditions and intelligently apply the brakes intermittently at the right time.

Next, we have automatic transmission. This will dispense with the application of the gear manually. Once we start the engine and engage the gear system, normally there is no need for manual changing of gear till we reach the destination. Even high flyovers can be negotiated easily. Cruise control maintains a constant speed of the vehicle at the press of a button, thereby freeing the right foot for the brake instead of the gas pedal. These are realized by embedded systems gathering enough intelligence of the road conditions and the weather conditions. Then we have curbside check in system that issues a ticket for parking vehicles.

Older cars that came with hydraulic power steering were powered by hydraulic pumps mounted on the engine running off the engine's crankshaft. This consumes a part of the power produced by the engine. The latest development, the electronic power steering (EPS), is powered by the battery and does not draw any power from the engine. Thus, all the power produced by the engine is used to propel the car rather than run the power steering pump. Wind resistance and rolling friction stretch a car's engine and its efficiency the most and prove a drag on the car's fuel efficiency. The electronic power steering involves the use of an electric motor and related electronics for providing directional control to the car. EPS systems draw power directly from the car's battery and are not dependent on the engine for doing their job. This directly translates into an improvement in fuel efficiency.

EPS systems are more dynamic than the traditional hydraulic power steering and capable of finer inputs for varying the time and amount of power assistance being offered. This means that EPS offers higher assistance during low speed travel or during a parking situation. On the other hand, EPS decreases the level of assistance as speed builds up, a feature that gives the driver firmer control over the car at all times. This also means that the electric motor does not draw power from the battery when there is no demand for assistance, such as during straight line travel. In contrast, hydraulic systems require the fluid to be kept at a constant pressure and hence suck up power from the engine, even when the car is idling.

Apart from the electric motor, the two other components of an EPS system are the control module and the torque sensor. All these components fit together on one compact unit, installed just below the upper steering column. The electric motor set to the side of the column transfers power through a reduction and worm gear. The torque sensor detects right or left direction and the crank speed or extent of torque that the driver applies to the steering wheel and transmits the data to the control module, which in turn decides the level of aid required. It also powers the motor to turn the wheels either to the right or to the left by reversing the applied voltage to the electric power unit. EPS systems are much more efficient than hydraulic counterpart even though they draw a bit of power from the engine indirectly as the alternator has to work overtime to compensate for the battery power spent by the electric motor.

Real time monitoring system for cars and vehicles shows close-quarter dangers that a driver might miss while driving. A nice road and a big fast car/truck blend well together. However, hazardous points of driving are the packed traffic, blind spots at the immediate sides and rear of the vehicle, and painful parking maneuvers. Vehicle manufacturers are working at eliminating these dangers. These problems can be completely solved by designing a video processing system that offers an integrated display of the roof-top view of the vehicle on the dashboard, clearly showing the surroundings to decide whether overtaking another vehicle or a lane change can be safely undertaken. This system is also helpful while parking and reversing out of tight spots. This system requires front, rear, and side mounted cameras to take care of the blind spots surrounding the vehicle. The projects described earlier in this category and more projects included in the following list may be undertaken for FPGA/ASIC implementation:

- Anti-lock brakes
- Automatic transmission
- Cruise control
- Digital speed measurement of passing vehicles on roads
- Electronic power steering
- Global positioning system for automobiles
- Real time monitoring system for cars and vehicles
- Vehicle parking check in systems
- Wireless remote control for automobile AC/door/lights/alarm control

15.1.2 Avionics

The next category we will consider is the avionic systems. They are systems in aircrafts, which can measure and record digitally the parameters like latitude, longitude, altitude, inside/outside temperatures, wind speed, cabin pressure, oxygen level, identify flying objects around the flying aircraft, etc. In the airports, we see lots of baggage moving to and fro. Often, passengers have trouble in locating their baggage, especially at the destination airport. A control system which receives these baggages and routes to a particular announced baggage collection place using a sequence of conveyor belts may be designed, saving lots of trouble for passengers. Displays at strategic points starting from the passenger arrival points must guide the passengers to the place where the passenger may collect his/her baggage without any anxiety. The list of projects for this category is as follows:

- Automated baggage clearance system in airports
- Avionic systems
  - Digital altitude meter of aircraft
  - Wind pressure display of aircraft
  - External temperature and pressure display of aircraft
- Flight simulator
- Instrument landing system (air navigation in airports – landing)
- Unmanned aircraft control

15.1.3 Cameras

In the next category, we have cameras such as auto-focus cameras, which will focus all by itself, the camcorders used as a video camera in the digital domain rather than analog domain, and the digital cameras that can record short duration compressed video sequence as per MPEG 2 or MPEG 4, Part 10 formats. These are as follows:

- Auto-focus cameras
- Digital camcorders
- Digital cameras

15.1.4 Communication Systems

Any communication system which sends data over a serial channel be it wired or wireless, is susceptible to noise and, therefore, requires error correction codes. Further, the data needs to be secure, which can be accomplished by designing encryption and decryption hardware using FPGAs or ASICs. The following list gives some of the communication systems that may be designed by the reader:

- Demodulator for satellite communication
- Encryption/decryption
- Error correction codes
- Modulator for satellite communication
- Network card
- Network switches/routers
- Quadrature amplitude modulator (QAM) and demodulator
- Radar imagery system
- Submarine detector
- Wireless LAN/WAN

15.1.5 Computers and Peripherals

Low cost computer is a low cost PC costing under $200, which is Linux-based or Windows-based. However, windows operating system based PC may cost more than the Linux-based machine, unless the windows operating system prices are slashed to compete with the Linux counterpart. In order to make this project viable, an open source for manufacturing such low cost PCs will have to be created on the net, the idea being that prices of PCs should come down benefiting people. The PC may have a one GHZ processor, 128 MB RAM, 40 GB hard disk, 15-in. color monitor, 52X optical drive, a keyboard, and a mouse. The low cost PC shall support applications such as word processing, spreadsheet, presentation, web browsing, email clients, and audio–video playback, etc.

Scan pen and PC notes taker costing under $200 captures printed text at the stroke of a pen. It is useful for researchers, journalists, doctors, lawyers, and students. It can store data up to 1000 pages of text, which can be edited and stored as separate files. It captures handwriting from any paper and enables direct downloading into MS Office. It is useful for creating and saving sketches, handwritten notes, and memos in any language without requiring much knowledge of a PC. It also can send email in our handwriting and language. The above project designs along with one more are listed in the following:

- Low cost computer
- Mobile phone personal computers
- Scan pen and PC notes taker

15.1.6 Control Systems

First one in the control system category is the alarm annunciator. As the name implies, abnormal activities in industrial plants need to be announced by monitoring various engineering parameters such as low pressure, high temperature, low fuel, etc. The industries may be a power plant, a cement plant, a sugar plant, and so on. You would have seen huge control panels (at least in the TV) in various plants such as thermal and nuclear plants, which have several flashing lamps on the top of the control panels. They are annunciators. These equipments come with various flavors of 'sequences', well over 50, designed by a number of manufacturers around the globe. Most of these are based on microcontrollers. Therefore, it would be a good idea to design these equipments using FPGAs/ASICs. The market for this product is huge and hence ASIC based design will be viable. In addition to the above applications, many other applications for project design are listed:

- Alarm annunciator
- Ash level controller for Electrostatic precipitator
- Automatic packaging/sealing machines
- Electrostatic precipitator communication controller
- Data acquisition system
- Electrostatic precipitator (EP) controller

- Injection molding machine control
- Lift controller
- Medicine blend control machine
- Programable logic controllers
- PIC
- Quality control system
- Rapper controller
- Remote control for air conditioners
- Robot controller
- SCADA
- Simulator for EP controller
- Temperature controllers
- Machine vision
- Smart scales
- Unmanned railway line crossing
- Vending machines

15.1.7 Image/Video Processing Systems

Many interesting project designs are available for FPGA/ASIC implementation as listed towards the end of this section. We will discuss some of these projects. Digital cinema is a new technology that is poised to create a digital revolution. It enables the projection of movies simultaneously across several theaters using satellite communication. Currently, producers are unable to release new films in many centers due to the high variable cost of film prints. This is where the digital cinema comes in handy for the producers, distributors, and exhibitors. Incidentally, this gives a new lease of life to old theaters, crying for renovation. Once the renovated theaters are equipped with adequate facilities, screening of digital cinema will become a reality. The negative may be changed to HD 5 format and encrypted and put on a centralized server. It will then be up-linked to satellite. The theater concerned will receive the signals, which in turn will go into a local server, decrypted and then on to a digital projector for screening.

A major advantage of digital cinema would be the elimination of piracy. The theaters have very little expense in terms of print cost, film transportation, or other related charges. A single film can be viewed in hundreds of theaters simultaneously. Digital cinema systems, in another embodiment, will offer theater managers the facility to choose a movie from a catalog of films and download any film from anywhere through broadband internet and satellite. It will provide for transparency as distributors can login to the internet and monitor which film of his is playing in which theater and at what time. Digital cinema also supports MPEG 2 format. MPEG 4, Part 10 format may also be included.

Detailed specification for a new digital cinema format has been released by the Digital Cinema Initiative, a forum which represents Hollywood studios: Disney, Fox, Paramount, Sony pictures entertainment (erstwhile Columbia), Universal and

Warner, which have dominated the world's English language cinema since the dawn of the movies. The full technical document can be downloaded from the website:
http://www.dcimovies.com/DCI_Digital_Cinema_System_Spec_v1.pdf.
This document is a single standard for the entire process of making and showing films digitally, namely, mastering, compression, encryption, transport, storage, playback, and projection. The picture sizes can be 2048 × 1080 pixels known as '2K' format or 4096 × 2160 pixels ('4K' format).

Mobile film making is the making of a short video backed by a brief description of a favorite icon such as an old shop house in an alley that holds many memories, a vintage car, or even a childhood experience. Image and video processing systems that may be realized as FPGA or ASIC are listed in the following:

- Conversion of black and white movies to color motion pictures
- Digital camera interface
- Digital cinema
- Digital TV and digital cable TV
- Digitizer for analog NTSC/PAL/SECAM cameras
- Display interface
- H.264 codec
- JPEG codec
- JPEG 2000 codec
- Motion JPEG 2000 codec
- MPEG 1 codec
- MPEG 2 codec
- MPEG 4 codec
- MPEG 4, Part 10 or H.264 advance video coding (AVC) codec
- Object segmentation system
- Teleconferencing systems
- TV set-top boxes
- TV tuner card
- Video conference codec
- Video grabber card
- Video karaoke
- Videophone
- Video scalar
- Video spotlighting effect and other special effects creation system
- Video watermarking

15.1.8 Measuring Instruments

High precision measuring instruments are indispensable while developing systems, be they analog or digital systems. These instruments may be used for testing a finished product or for calibration of test equipments in quality control departments.

For example, a digital high voltage tester of capacity 100 KV can be used in the quality control of a transformer cubicle used in power stations. This equipment, used by control panel manufacturers, helps in finding the breakdown voltage between copper bus bars mounted on insulators and the cubicle. Another equipment the reader can design is a 3 GHz (or more) digital frequency meter that can measure frequencies of an oscillator and thereby carry out factory setting, say for instance, setting a real time clock quickly, whose design was presented in the previous chapter of this book.

Virtual instrumentation places the personal computer at the epicenter of the task and exploits graphical programing aids such that even a novice can drag and drop ready-made instrument panels which can look like the real multimeter, spectrum analyzer, or waveform generator. The virtual creation of measuring system ranges from the simple digital voltage, current meter to the most complex multi-sensor data acquisition system. Other creative directions of the virtual instrumentation have taken it to the embedded systems developer and the virtual electrical engineering laboratory. Standard and classical experiments on DC machines and transformers, analog and digital circuits, etc. can be virtually performed on the PC, complete with variable running speed, operation amplifiers, gates, flip-flops, counters, stunningly realistic meters and controls. Some of the measuring instruments suggested for implementation are listed in the following:

- Digital high voltage tester
- 3 GHz Digital frequency meter
- Digital LCR meter
- Digital megohmmeter
- Six digit digital multimeter
- Digital Ph meter
- Digital oscilloscope
- Embedded systems
- Virtual instrumentation using PC

15.1.9 Medical Applications

High blood pressure increases the chance of getting heart disease and kidney disease and consequent stroke. It can also result in blindness. High pressure is especially dangerous because it often has no warning signs or symptoms. Regardless of race, age, or gender, anyone can develop high blood pressure. It is estimated that one in every four American adults has high blood pressure. More or less, the same is true around the world. Once high blood pressure develops, it usually lasts a lifetime. Blood pressure is the force of blood against the walls of arteries. The pressure rises and falls during the day. If we exercise or just walk, the pressure increases even if we are normal. However, when blood pressure stays elevated over a time, it is called high blood pressure.

The medical term for high blood pressure is hypertension. A blood pressure level of 140/90 mm Hg or higher is reckoned as high. If it is within the range of

120/80 mm Hg and 139/89 mm Hg, then it means that one is likely to develop high blood pressure. The first number is called the systolic pressure and the second number is called the diastolic pressure. The systolic pressure is the force of blood in the arteries as the heartbeats, whereas the diastolic pressure is the force of blood in the arteries as the heart relaxes in between beats. Causes of high blood pressure may be due to narrowing of the arteries, a greater than normal volume of blood, heart beating faster or more forcefully than a normal beating, etc. You can prevent and control high blood pressure. The applications that may be developed are listed below along with others in the medical applications category:

- Digital acupressure
- Digital blood glucose meter
- Digital blood pressure and heart rate monitor
- Electrocardiograph
- Life-support systems
- MRI/CT scan
    - Doppler
    - Echo
    - Mamogram
    - Ultrasound

15.1.10 Miscellaneous Applications

A number of project designs for implementation are listed in this category. Some of these projects are discussed in the following. An electronic voting machine (EVM) consists of two inter-connected units: the control unit and the ballot box. The control unit is operated by the presiding electoral officer. The names and symbols of all the candidates are displayed on the top of the ballot box. There is a push button besides each name. Each machine can accommodate up to 16 names. If there are more candidates, then another machine is linked to the first unit. When a voter enters the booth where the voting machine is kept, the presiding electoral officer presses a button marked 'Ballot' on his control unit. A LED marked 'Busy' comes on in the control unit and one marked 'Ready' glows on the ballot unit. 'Ready' LED remains on till the vote is cast. When the voter presses the push button adjoining the candidate's name of his choice, a red LED switches on besides the candidate's name and a loud beeping alarm sounds, indicating that the voter has cast his/her vote. Once all the votes are polled, the presiding electoral officer closes a key operated switch marked 'Close', after which the machine automatically stops registering any votes.

An EVM is fast, with a capacity for five votes a minute. This eliminates the cost of printing ballot papers and is tamper proof. Even an illiterate voter can use it. Counting and declaration of results are quick, which means that manpower requirement is drastically cut. A single magnesium battery in the control unit powers all the linked ballot units. The machine even prevents malpractices like vote duplication. If a voter were to press more than one button at the same time, no vote is cast. On the other hand, if buttons are pressed one after another, the EVM detects which was pressed first and registers it as the only vote. The memory lasts five years even when the machine is switched off and not in use, and so it comes handy if a result is disputed much after the poll is over.

Futuristic capsule simulator is one of the entertainment systems, which offers the excitement of spectacular fantasy worlds in one of its kind outdoor simulator.

A pedometer is a pager like device, worn on the waist to record the number of steps a person takes in a day. It translates that into the distance covered and tells how many calories the person has burned. This is a simple tool for athletes, joggers, health buffs, and people out to lose weight. The pedometer works by sensing the up and down movements of the hip and thereby counting the steps. Before that, a user needs to record the length of steps he or she takes with a measuring tape. Once that is done, the user has to key in the person's weight, and then, the user is ready to go. This may be regarded as a motivating tool to remind us to walk more and be active and is an excellent recorder for tracking our activity level throughout the day. An average person walks about 6000 steps a day, and we need to hit 10000 if we need to loose weight. One can wear it all day, everyday and record the total number of steps one takes or just wear it whenever one takes walk or go for a workout. Using it also helps one set and reach daily targets, since one can sneak in 10 min or more of walking at every opportunity, whether it is taking the dog out for a run or just taking the stairs instead of the elevator. The pedometer may gain new lease of life marketed by government officials, fast food outlets, gyms, potato chip companies, etc.

A satellite view search service allows users to zoom in on any spot on earth for a dramatic satellite eye view in three-dimension. Drawing on a huge library resource of satellite imagery, merged with cartographic information from the ground, the application provides resolutions down to one meter or less. The application works on PCs that include a 3D graphics board with resolutions of 1024 × 768 pixels or 800 × 600 pixels. Many global locations are book marked and clicking on this spins the globe and zooms down to the desired place clearly identifying all the details. Pan and tilt controls allow one to rotate the view so that the buildings can be seen in sharp 3D. Entering the latitude and longitude of any place on earth sends the application zooming to that spot at one meter or less resolution. In such areas, the user can search for motels, gas stations, bus stations, etc. The application also provides for higher resolution of graphics and links the application to a global position system (GPS) position locater, if the user has one.

A treadmill is a type of fitness equipment used in gymnasiums. It is attached with a 3 HP motor that gives a speed between 1 to 12 miles per hour with five speed profiles and five intensity levels. To increase the intensity of training, the treadmill has an electric inclination system. The treadmill is designed with an integrated double fan, extra wide, and long shock absorbing running track with auto-safety key and hand rail bottoms for comfortable and safe training. Heart rate control, body fat control, system, telemetric pulse control, and contact pulse measurement system are a value addition to the equipment. The equipment may be designed with LCD display and the sequence of speed, inclination, and timings can be user programed. This equipment comes with six challenging programs targeted at runners and walkers, with a provision to make more programs by the users. The list belonging to this category of applications follows:

- Automatic teller machines
- Automatic toll systems
- Digital lockers
- Digital petrol/diesel dispenser
- Dishwashers
- Dryers
- Electronic card readers
- Electronic voting machine
- Fault location in cables
- Futuristic capsule simulator
- Hearing aids
- Intelligent cane to lead the blind/deaf persons
- Leather area measurement equipment
- Non-destructive test of ceramic bricks using ultrasonic sound
- Non-destructive test of dams/buildings using ultrasonic sound
- On-board navigation
- Point-of-sale systems
- Pedometer
- Satellite view search
- Smart ovens
- Speech recognizers
- Treadmill with heart meter for gym.
- Universal PROM/PAL/FPGA programer
- Virtual reality system
- Xpendable bathy thermograph

15.1.11 Music

Separation of human voices from orchestrated music, a challenging design task, can be used to replace the original old voices by the current singer's voices while retaining the original orchestra instruments. On the other hand, the original voices can be recast in new orchestrated music. These are useful for professionals, amateur musicians, and audio karaoke. The following lists some of the musical equipments:

- Digital filter for separation of human voices from orchestrated music
- Digital voice/music recorder cum digital camera
- Special effects generators for audio:
    - Bathroom effect
    - Cave effect
    - Echo fade in/fade out effect
- Synthesizer or music keyboard

15.1.12 Office Equipments

Some of the office equipments are as follows. Digital dictaphone is a digital recorder cum player useful for secretary in an office. FAX, Scanner, Copier, and Printer can be integrated into one machine, designed as a single ASIC. Personal digital assistant is a mobile unit to store telephone/cell numbers, addresses, email addresses, and other details of persons. It can be connected to a PC using USB port for downloading or uploading the information:

- Digital dictaphone
- FAX/scanner/copier/printer four-in-one machine
- Personal digital assistant (PDA)

15.1.13 Phones

Cell phones are currently distributed on two ASICs, analog and digital. Using mixed signal HDL such as the AMS CAD of Cadence, one can design a single chip cell phone, bringing down the cost. Accordingly, the base stations can also be designed. Low resolution, low frame rate videophone/video conferencing based on H.264 can be attempted on the cell phone. A short list of the phone based equipments is as follows:

- Cell phone, single chip
- Low resolution, low frame rate videophone/video conferencing based on H.264 on the cell phone
- Multi-channel TV reception on cell phone
- FM radio on cell phone
- Cell phone base station
- Satellite phone
- Telephone exchange

15.1.14 Security Systems

Systems known as biometrics reduce an image such as a fingerprint, facial feature, or other personal characteristics to a template of minutia points or other personal characteristics. Rather than use passwords, biometric devices identify people by behavior or physical characteristics like fingerprints. Notable features of these minutia points are loop in a fingerprint or the position of an eye. These points are converted to a numeric string by an algorithm and stored as templates. These templates can be dangerous if stolen. Altering biometric images enhances security, keeping hackers at bay. Researchers may develop ways to alter images in a defined, repeatable way so that hackers who managed to crack a biometric database would be able to steal only the distortion and not the original image. This is done by distorting the image before it is scanned by a biometric reader, and the template of the distorted image is stored in a database. Thereafter, when the same person uses the biometric reader, once again the original image is distorted and transformed, creating a match with the database. It may be noted that the original image is not stored anywhere. That means, even if hackers get the altered biometric, it would be of little use as long as organizations maintained their own formulas for transforming images before scanning.

Home security systems primarily have passive infrared motion detector, which detects infrared radiations from an intruder. It then triggers an alarm loud enough to alert the occupants of the house/office and even neighbors. The system consists of a control panel and communicatively coupled to various sensors installed in a house/office/bank and a remote control with which the system can be armed or disarmed. Apart from triggering an alarm, the system can be programed to call a pre-set telephone numbers in case of a break-in. The sensors, normally attached to the doors and windows and connected to the control panel (wired or wireless), set off an alarm immediately after they detect a movement when the intruder tries to force open the doors and windows. The client's control panel can also communicate round-the-clock with a central monitoring police station. Optionally, the system can have a closed circuit TV. Some of the security systems are as follows:

- Biometrics such as fingerprint identifiers
- Fire alarm system
- Home security systems
- Surveillance camera control system
- Theft tracking system
- Tsunami warning system

15.1.15 Toys and Games

Video games command an ever increasing huge market of over $20 billion worldwide, with a large untapped market in the east. Games such as car racing, star wars, boxing, asteroids, space traveling, etc. demand faster processors than what exists currently. ASIC based video games are better alternatives than the PC processors, especially in terms of processing speed and price. Game development is a multi-disciplinary field demanding diverse skills such as drawing, art design, painting, graphic designing, 3D graphics, story narration, screen writing, direction, etc. with a strong knowledge of digital video technology, HDL, computer programing using C++, physics, mathematics, etc. Video games require a plethora of hardware such as the sound cards, graphic cards, 3D graphic accelerators, joysticks, remote controls, CD drives, etc. These applications are listed as follows:

- Electronic toys
- Portable video games
- Toy robots
- Video game consoles

And the list goes on and on, limited only by one's imagination.

15.2 Embedded Systems Design

Computing systems have proliferated everywhere, so much so that we are conditioned to think only in terms of personal computers on our desktops, laptop computers, main frame computers, servers, etc. However, there is another class of computing or controlling system that is far more common. Yes, you guessed it right – the embedded systems/controllers. Again, when we speak of controllers, what pop up in our mind are the programable logic controllers (PLC) or the programable controllers that have invaded every conceivable industrial application. There is also a general impression among system designers that embedded systems mean only a microcontroller. Against this background, a formal definition for an embedded system is indeed hard to make.

In the recent years, there has been a spurt in embedded systems reported for wide variety of applications, which make use of microprocessors, microcontrollers, and DSPs right from 4 bits to 32 bits on one hand to FPGA/ASIC on the other. These applications include digital cameras, automobile automation, avionics, ATMs, cell phones, electronic toys/games, medical equipments, defense equipments, industrial controllers, etc. We have discussed a number of them in the previous section. If one scrutinizes these systems closely, one would infer that they have certain common features such as executing a single program repeatedly, having to meet tight constraints, i.e., they are characterized by low cost, low power, small, fast, etc., and continually reacting to changes in the system's environment and computing certain results in real time without delay. So long as these criteria are satisfied to the extent feasible, we may not have any objection to defining an embedded system as a system that is designed to perform only a dedicated application, no matter what processor is used.

An embedded system performs a dedicated function. For instance, a digital camera that can do only one function, namely, capture an image, bring about compression, store them, and upload the captured still images to a computer; nothing more, nothing less may be regarded as an embedded system. This embedded application is best realized as an ASIC since it finds a huge market. As another example, we may take the implementation of electrostatic precipitator controller used in thermal power plant for the disposal of ash. This controller is based on Intel's 8085 microprocessor by many leading vendors, rather than going for 8051 family microcontroller that came later. This may also be realized using FPGA and subsequently as an ASIC as it has good market potential and can compete with the existing versions. A detailed specification and architecture of this application for FPGA and ASIC implementation will be presented in Section 15.4. Most embedded systems need to be designed with built-in real time clock and/or watch dog timers. These designs were presented in earlier chapters.

Design metric is a measurable feature of a system's implementation. Common metrics are the functionality implemented, ease of handling the system, the processing time or throughput of the system, sale price of each system, non-recurring engineering (NRE) cost, the physical size of the system, the amount of power consumed by the system, and flexibility, i.e., the ability to change the functionality of the system without incurring heavy NRE cost. Optimizing design metrics is a key challenge that needs to be addressed while designing an embedded system.

Microprocessors are used in a variety of applications, small to medium-sized in complexity. 8085, 8086, and 68000 processors are some of the earliest general purpose microprocessors used in embedded system applications. Likewise, digital signal processors (DSP) such as TMS320C6X (Texas), ADSP 21020 (Analog Devices), DSP32C (Lucent) are used for specialized applications involving multiply-accumulator (MAC) operations and are generally costlier than microprocessor-based products. Microcontrollers such as 8051, 89C52 (Atmel), 68HC811 (Motorola), PIC 16F84 microcontroller (Microchip Technology Inc., USA) are popular for small-end applications. For medium to high-end embedded systems design, FPGAs/ASICs are the right choice. In the near future, FPGAs may be expected to be cost effective even for small end applications and can outperform the above mentioned processors.

15.3 Issues Involved in the Design of Digital VLSI Systems

Any product is saleable only if it is cost effective and competitive. These requirements can be met if we build the system with minimum of hardware: both on-chip resources and the external hardware, and conform to optimum specification. If the system design is based on FPGA and requires a large memory in the order of 16 KB or more, the system is cost effective only if the memory is located external to the FPGA. This may mean a reduction of throughput since the external memory design is slower than the on-chip memory by about two times as was shown in the chapter on design of memories. As the technology is changing rapidly, the limit of on-chip memory of 16 KB can be jacked up if found cost effective. In ASIC implementation, it may be advantageous to integrate the memory with the ASIC and bring it out as a system-on-chip (SOC). This requires vendor library for memory while using the ASIC (front end and back end) development tools such as the Synopsys, Magma, and Cadence. Of course, we can go in for ASIC implementation only if there is a huge assured market and a promise of recurring demand. Otherwise, FPGA implementation is cost effective. Similarly, the system must have a requisite number of external hardware such as integrated circuits; passive and active components such as connectors, cables, resistors, capacitors, switches, relays; transistors, drivers, zener diodes, etc. with the right specifications, nothing more, nothing less as required by the particular application.

By minimizing the hardware, the system cost is kept low, consumes less power, development cycle as well as the production times are low, reliability high and the system is compact. User controls and displays must be simple and conveniently placed and the system must be designed with aesthetics in mind. Specification must be met completely without making any compromise. Otherwise, credibility is lost. Likewise, over indulgence of specification must be curtailed since it corrodes the profitability. Codes must be optimized in order to minimize the chip area and hence reduce the cost.

Before coding in HDL, the design concepts and algorithms developed must be tested in higher languages such as Matlab or C. Their end results can also serve as references for verifying the outputs of HDL codes. Development of HDL codes must be undertaken only if Matlab or C simulation is satisfactory. If not, one must look into the possibility of compromising on the specifications and get user concurrence before proceeding further. HDL codes, be it Verilog or VHDL, must conform to RTL coding guidelines discussed at length in Chapter 5, without which FPGA or ASIC implementation cannot work. Serious designers, be they practicing engineers or students working on their projects, need to use the right tools such as Modelsim, Synplify, and Place and Route which allow large designs without any restriction, besides being easy to learn and handle subsequently. It may be noted that free downloads may have restrictions of about 750 lines, while most VLSI system designs are above 1500 lines. The Verilog/VHDL codes developed must, in general, be technology independent, device as well as vendor independent so that we are free to use any device: FPGA or ASIC. This way, we have the flexibility of migrating from one FPGA to another FPGA or ASIC when the occasion demands without needing to recode.

Once the coding is completed, printed circuit board (PCB) which houses the target FPGA will have to be fabricated. Usually, this is time consuming and development costs incurred are high. A better alternative to the PCB development and testing of the assembled board is to buy suitable, populated, pre-tested FPGA and input/output boards. Once the system is proven, one can take up the PCB development work, if the demand is high. Similarly, one can start with the FPGA implementations for small to moderate demands and graduate to ASIC implementations later on for bulk production.

The following summarizes the strategy we have already adopted in designing VLSI systems in this book:

- An efficient application involves designing with minimum of internal and external hardware in addition to developing optimized codes.
- Complex algorithms and concepts must be verified for establishing viability using high level languages such as Matlab or C.
- HDL code must conform to RTL coding guidelines.
- Right tools must be used to minimize the development cycle time.
- System development can be dramatically expedited if based on bought out, populated electronic cards.

15.4 Detailed Specifications and Basic Architectures for a Couple of Applications Suggested for FPGA/ASIC Implementations

We will formulate detailed specifications and develop basic architectures for a couple of applications, which the reader may take up for implementation subsequently:

- Electrostatic precipitator controller
- JPEG/H.263/MPEG codec

15.4.1 Electrostatic Precipitator Controller – an Embedded System

Electrostatic precipitator controllers are used in fly ash disposal in a thermal power plant. Several tons of fly ash are generated, disposal of which is quite cumbersome. For example, a 210 MW thermal power plant generates about 4000 tonnes of ash everyday. If released in the air, the entire township will be covered by ash. Water stream cannot directly wash the ash away – passage will get clogged in a short time. The solution is to apply a high DC voltage in the order of 80 KV in the EP, a large chamber with electrodes all over, where the fly ash is blown in from a boiler. Ash gets attracted to negative electrode and hence tamed. Activating special hammers frees the ash, which is promptly washed away by a water stream at the bottom of the electrostatic precipitator and finally disposed of in huge ash ponds situated about 5 miles away from the power house. The special hammers need to be activated in a specific sequence in order to dislodge the ash from the electrode. These hammers are, however, controlled by another controller called Rapper controller. The DC high voltage is generated by firing a couple of thyristors configured as full-wave rectifiers. The firing circuits for these thyristors are based on pulse transformers housed in a control panel in which the EP controller unit is mounted. A small capacity transformer (30 VA), together with opto-isolated transistors identify the zero crossover and AC positive/negative swings in order to generate firing triggers for the thyristors at appropriate time. The transformer is also used to supply power to the EP controller unit.

Front fascia of electrostatic precipitator controller, made of membrane key pad, is shown in Figure 15.1. All the LED displays seen through front fascia are mounted on a FPGA board. LEDs N1 to N7 announce the status of the EP such as high voltage transformer high temperature, coolant top/bottom float level, EP controller unit supply under voltage, thyristors overload/high voltage transformer very high temperature, when voltage peak is reached and the mode 'REMOTE' or 'LOCAL' respectively. Based on these annunciations, the operator may take corrective actions. DS1 to DS4 are seven segment displays used to display the mode in which the EP may be configured, EP voltage, current, etc. DS1 displays the mode the EP is set. Function modes are as follows:

– Precipitator Current	E Precipitator Voltage
H Sparks Per Min.	Bl Peak & Valley Voltage
0 Im Limit	5 Uv Limit
1 Is Limit	6 Charge Ratio
2 S Control	7 Pulse Current Limit
3 T Control	8 Loop Gain
4 Slopes After Spark	9 Addresses
P Base charge Set	L Base Charging Current

'Bl' stands for blank space. When the EP controller is switched on, the display is '0% –' in DS2, N8 (LED on) and DS1 respectively. Up/Down DISPLAY SELECT keys may be used for changing the modes as listed earlier. Various settings in different modes are tabulated in Table 15.1. Mode '1' is set only after setting all other modes. For normal operation, the EP controller is set to '–'.

The DC high voltage of the precipitator may be switched on by pressing 'HT ON/OFF' switch. The LED N11 lights up and the DS4–DS2 display increases from 0 to 100 and remains at 100. The display increase may be expedited by pressing the 'T/O' key. The calibration of the system is as follows: Set the unit to Uv (meaning under voltage of power supply of the EP controller unit) limit mode (5) and adjust 'Is' potmeter on the front fascia so that the EP current is 0.75 A. Switch to the current mode '–' and adjust the potmeter P3 in I/O board to display 75 for the precipitator current of 0.75 A. Similarly, adjust potmeter P6 in I/O board to read 75 in 'E' mode for the precipitator voltage of 75 KV. P8 is adjusted till continuous counting takes place in the six digit electromagnetic counter and set it to slightly lesser value when the counting just stops. This counter advances by one every time a spark occurs in the electrostatic precipitator.

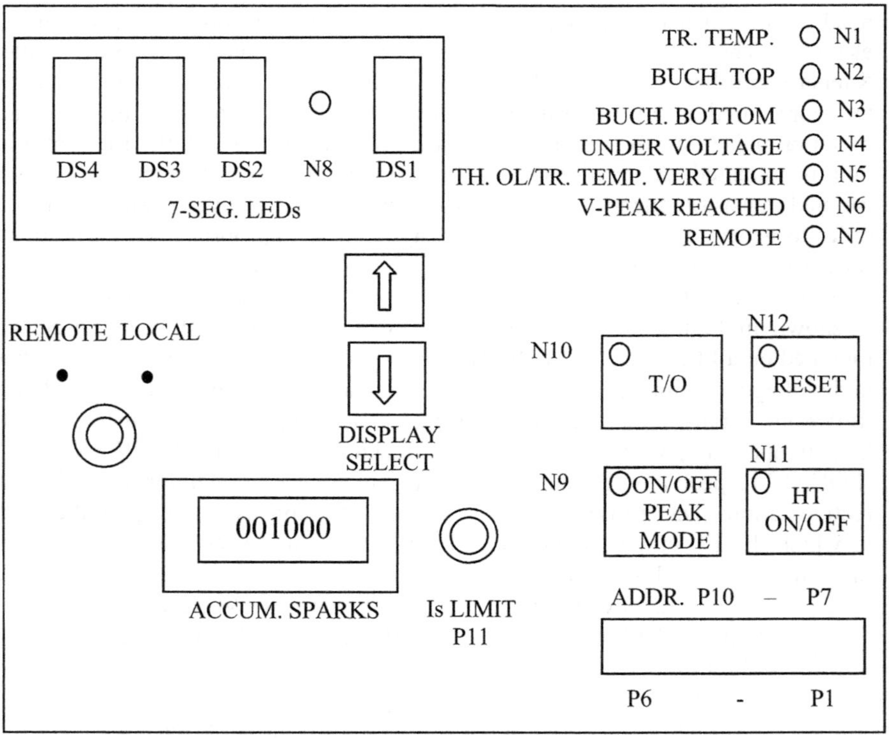

Fig. 15.1 Front panel of electrostatic precipitator controller

Table 15.1 Mode settings of the EP controller

Set mode in DS1 using Up/Down DISPLAY SELECT	Adjust potmeter (as shown in DS1) on FPGA board	Set value displayed in DS4-DS2	Check full range of potmeter setting in DS4-DS2
–	Nil	Nil	Not Applicable
E	Nil	Nil	Not Applicable
H	Nil	Nil	Not Applicable
0	0	100	0–104
2	2	5	0–25
3	3	20	0–109
4	4	30	0–99
5	5	10	0–104
6	6	1	0–31
7	7	200	0–209
8	8	20	0–99
9	9	Address in BCD switches (DS3–DS2)	
P	P	10	0–49
1	Is limit	75	0–100
L/Blank	Nil	Nil	Not applicable

Occasionally, high voltage sparks occur in the electrostatic precipitator owing to fluctuations in the flue gas flowing in the EP chamber. Peak and valley and voltage peak reached by the precipitator are required to be monitored. In the 'Peak & Valley Voltage' mode, the display DS4–DS3 shows alternately the peak and valley high voltage of the EP every 3 s. The HT voltage may be switched off by pressing 'HT ON/OFF' switch again when not required. Various potmeters and components mentioned in the foregoing description will be explained while describing the I/O and FPGA boards. Two digit BCD switches mounted on the FPGA board identify the EP controller unit. Up to 100 such controllers may be networked using serial interface circuit in the FPGA board. In the 'REMOTE' mode, only the Up/Down DISPLAY SELECT keys would be working, whereas in the 'LOCAL' mode all the keys would work.

Industrial Input/Output Board

Input/output connections and LED indicators and their partial signal conditionings are shown in Figure 15.2. Input/output board is shown in Figure 15.3. As shown in the figure, the I/O board houses signal conditioning of various analog and digital signals from the field such as sensing positive and negative AC swings, EP high voltage, EP current, sample and hold circuit to measure EP high voltage peak, watch dog timer to restart the system automatically in the event of system getting stuck and spark sensing circuit, all of which are primarily conditioned by OP. Amps. EP high voltage is measured by sensing the current while the EP current is measured by sensing the voltage as shown in Figure 15.2. These are followed by two stage differential amplifiers with P6 and P3 potmeters for adjusting the gain of the EP high voltage and the current respectively. These analog signals are fed to a 16 channel, 8 bit ADC such as ADC0816 housed in the FPGA board. +12/–12 V supply healthiness check is also fed as one of the inputs to the ADC. All the I/Os are connected to the J3 connector.

Safety line, Contactor ON information, Buch bottom/top floats, HV transformer temperature high indication, thyristors overload or HV transformer temperature very high indication, and Alarm reset are potentially free contact inputs, signal conditioned by opto-isolators (such as CNY17-2) with 2500 V isolation. The resulting digital signals are connected to FPGA input pins. Serial input and output are connected to FPGA I/O pins via a relay and a couple of opto-isolators. HT OFF, HT ON, Warning, Tripped conditions are output using four sets of line drivers followed by one change over (1 C/O) relays. Also, a couple of SCR triggers (positive and negative) derived from FPGA are output via two numbers of 1 C/O relays.

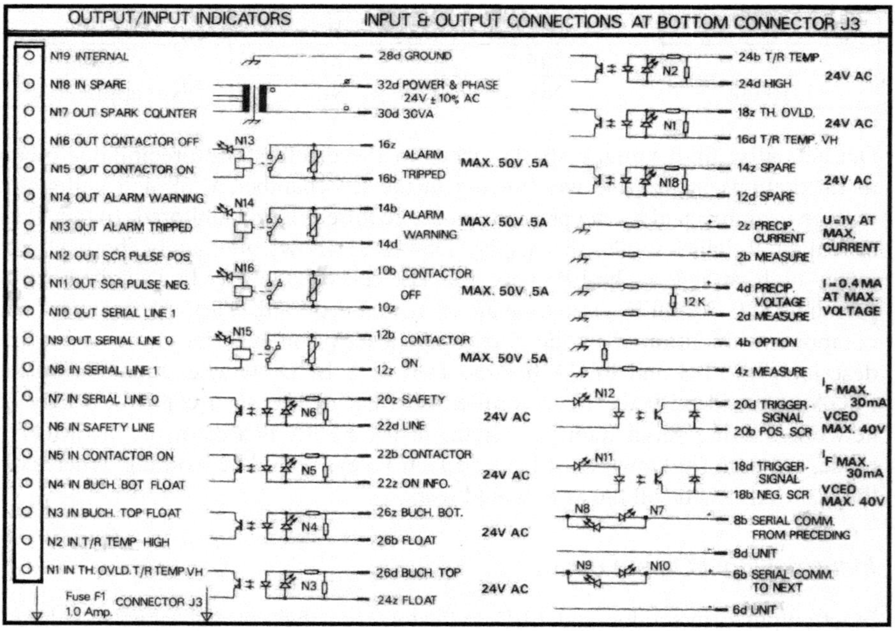

Fig. 15.2 Rear panel of electrostatic precipitator controller

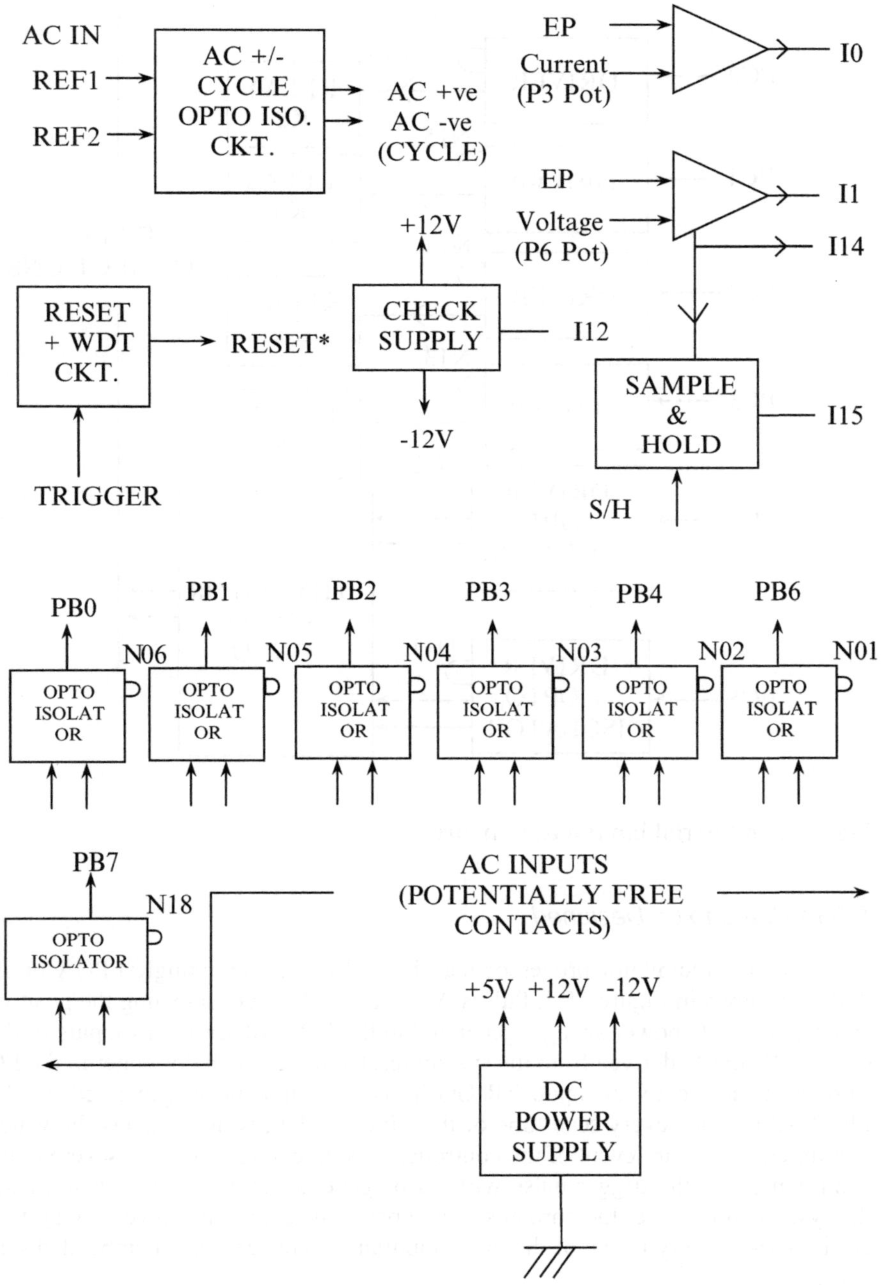

Fig. 15.3 Industrial input/output board (Continued)

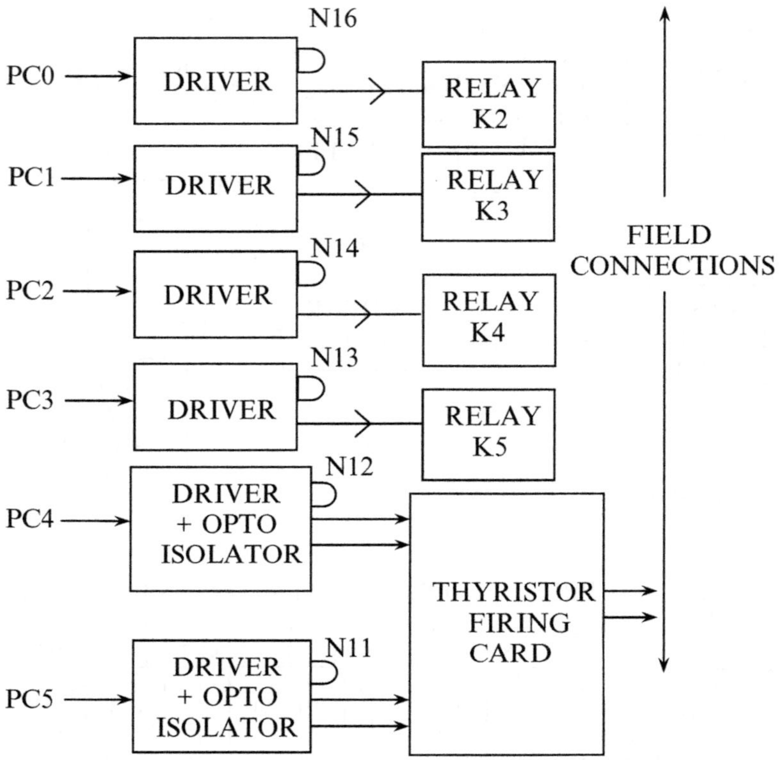

Fig. 15.3 Industrial input/output board

FPGA Board to be Designed

Electrostatic precipitator processor may be realized using a single FPGA or an ASIC as shown in Figure 15.4. Pulses AC +ve and AC –ve, signaling the positive and negative AC power swings, generated in the I/O card are fed as inputs to the device. RESET* derived from the system reset and the watch dog timer in the I/O card is connected to an input. TRIGGER is an output pulse generated by the FPGA/ASIC once every scan time of the EP controller, which triggers the watch dog timer. In the rare event of the controller loosing control owing to severe noise conditions, etc., the trigger pulse will not be generated. This in turn would reset the system and recover the normal system operation again, thus preventing system crash. Self-recovery is one of the most important characteristics in embedded systems.

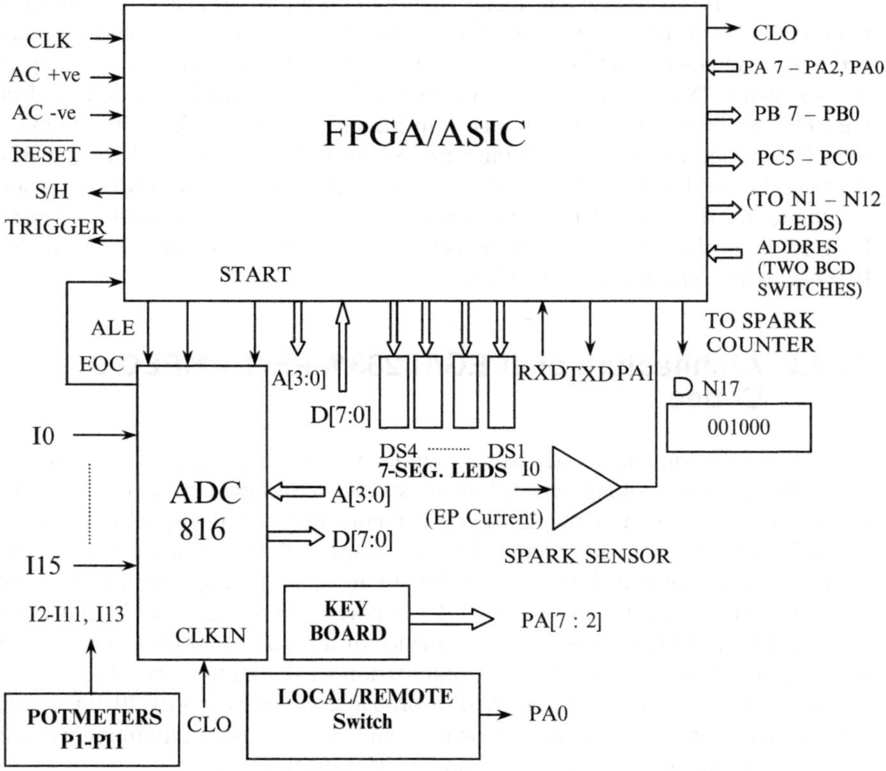

Fig. 15.4 FPGA board of the electrostatic precipitator controller

S/H output signal is asserted whenever the sample and hold is to be processed. The sampled output of I14 is fed to I15 input of an ADC 816 mounted on the FPGA/ASIC board. I1 senses the integrated EP high voltage, whereas I14 senses the dynamically changing high voltage to measure the peak or valley of the EP high voltage. I0 input of the ADC is the EP current measured by the I/O card. I2 to I11 and I13 are connected to potmeters P1 to P11 respectively shown in the front fascia. The FPGA/ASIC generates a 4-bit address, A[3:0], for the ADC to select one of the 16 analog channels I0 to I15 at a time. This address can be registered by applying 'ALE' signal. The analog to digital conversion can be initiated by asserting the 'START' signal. Once the conversion is complete, the ADC will assert 'EOC' signal. Subsequently, the FPGA/ASIC reads the converted digital channel information via D[7:0] with 'OE' asserted. The FPGA/ASIC also generates a low frequency clock, CLO, for the ADC operation.

The six key pad shown on the front fascia are connected to input ports PA[5:2] and the LOCAL/REMOTE switch to PA0. A SPARK SENSOR circuit derived from the EP current and comprising an analog comparator and a register indicates when the sparking takes place in the electrostatic precipitator through PA1 port.

DS4 to DS1 are seven segment LEDs shown in the front fascia and are connected to output ports of the FPGA/ASIC. TXD and RXD are the transmit and the receive serial data signals respectively connected to a serial network after conditioning the signals using CNY17-2 opto-isolators. PB7 to PB0 are field inputs derived from the I/O card. PC5 to PC0 are outputs from the FPGA/ASIC to drive four relays in the I/O card and the thyristors firing card. N1 to N12 are discrete LEDs shown in the front fascia driven by the output port. Another digital output advances a non-resettable six digit electromagnetic counter once every time a spark is sensed. Each EP controller unit has a unique identity by setting an 'ADDRESS' using two BCD switches. The address range is 00 to 99.

15.4.2 Architecture of JPEG/H.263/MPEG 1/MPEG 2 Codec

Video compression finds wide use in applications such as education, industries, medicine, defense, training, entertainment, sports, multimedia, desktop publishing, videophone, video conferencing, digital cameras, digital TV, digital cinema, and so on. Raw video sequences demand large storage and huge transmission channel bandwidth requirements. For example, the storage capacity required for 2 hours of raw, color motion picture of size 1024 × 768 pixels is 396 GB. Speed requirement for real time transmission of a video sequence of this size at 30 frames per second over a serial channel is 540 Mbps. Compression is, therefore, inevitable for storage and transmission of images. With a probable compression of 20 for a color motion picture in 4:2:0 format, the memory and channel speed requirements come down to manageable levels of 20 GB and 27 Mbps respectively.

High demand for these products has led to the development of various image compression techniques. Image compression methods aim at reduction in the amount of data without appreciable loss in the image quality. Design must conform to standards so that systems developed by different industries worldwide can communicate with one another. Connectivity and compatibility among different services such as videophone, video conference, MPEG 1/MPEG 2/MPEG 4 codecs are important. Standards deal with only the basic services, providing room for innovation and entrepreneurship. Several standards are available for image/video sequences:

- JPEG, JPEG 2000 for still images
- MPEG 1, MPEG 2, MPEG 4, H.264, MPEG 7 for motion pictures
- Multimedia hyper-media expert's group (MHEG)
- HDTV
- H.261/H.263 for videophone and video conferencing

Functional modules used in standards are as follows:

- ❖ JPEG: For still picture compression – DCT/Q/Huffman coding and their inverses
- ❖ JPEG 2000 is also for still picture compression but uses discrete wavelet transform (DWT) – DWT, Q, bit plane coding (BPC),

binary arithmetic coding (BAC), rate control, bit stream assembly and their inverses

- H.261/H.263: For videophone/conferencing – low bit rate (p X 64 Kbps, p = 1–30)
- MPEG 1: Audio-visual codec for digital storage – transmission rate: up to 1.5 Mbps
- MPEG 2: Consumer electronics/Telecommunications/ Broadcasting – transmission rate: 2 to 100 Mbps
- MPEG 4 Part 10 also known as H.264 Advance Video Coding for mobile and broadcasting.

DCTQ/VLC and their inverses, rate control, motion estimation, and compensation are involved in H.261/H.263/MPEG 1/MPEG 2. DCTQ and IQIDCT are common to all the above standards listed except JPEG 2000 and H.264. The basic operations that bring about image compression are the DCTQ and the VLC. Still image or I frame processing employs the DCTQ and VLC, exploiting the spatial redundancy. The motion picture processing employs motion estimation and compensation in addition to DCTQ and VLC, effecting more compression owing to the exploitation of temporal redundancy.

The basic building blocks of codecs for still image compression, conforming to JPEG standards, videophone/video conference conforming to H.261/H.263 standards, and motion pictures conforming to MPEG 1/MPEG 2 standards are DCTQ/IQIDCT and variable length coder/decoder. In earlier chapters, we presented

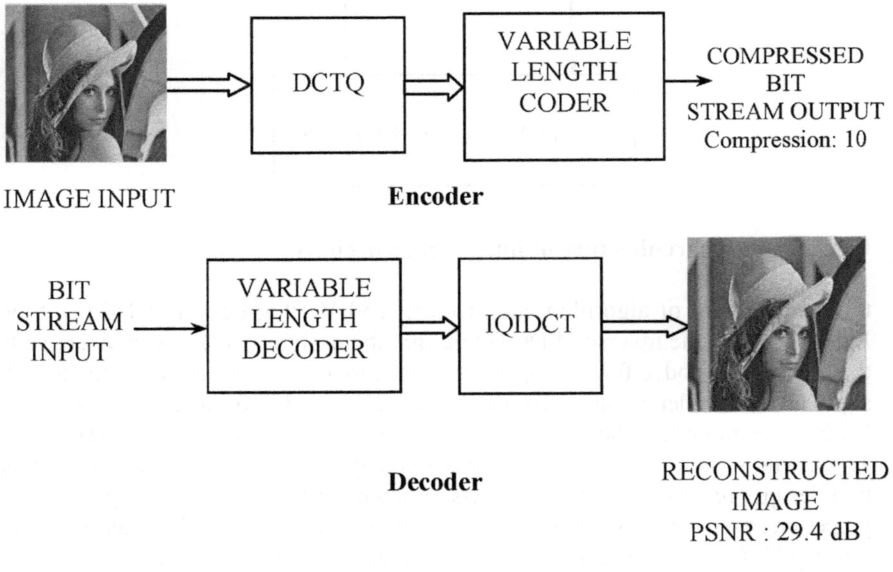

Fig. 15.5 Basic architecture of JPEG/ H.261/H.263/MPEG codec

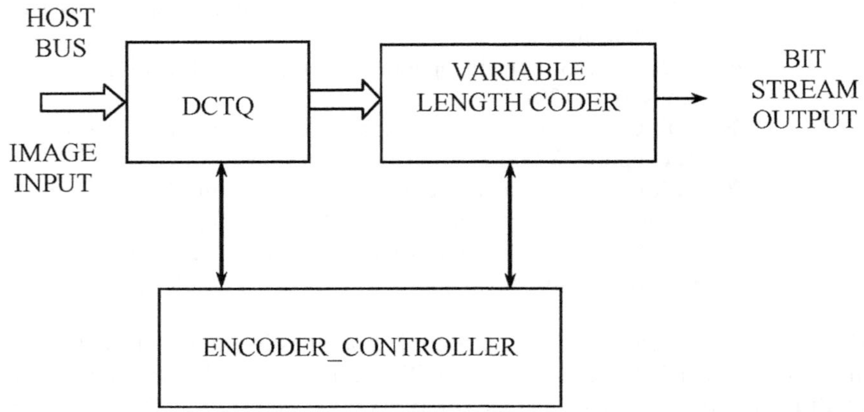

Fig. 15.6 Basic architecture of image/video encoder

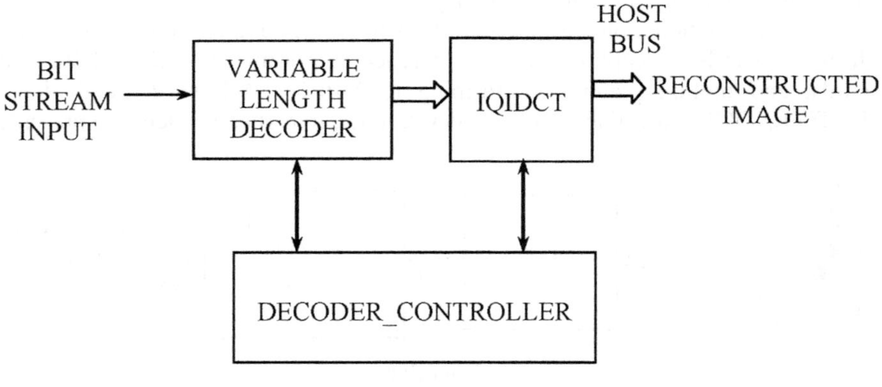

Fig. 15.7 Basic architecture of image/video decoder

the development of algorithm, architecture, and Verilog design of DCTQ. Since IQIDCT is just the inverse of DCTQ design, the design was left as an exercise for the reader. The codec for the applications mentioned earlier can be completely designed if the reader implements VLC and VLD in addition to IQIDCT modules. With these modules, the motion pictures can be processed as intra (I) frames. Those who wish to effect more compression may include the FOSS motion estimation design presented in earlier chapters or any other block matching algorithms for processing predicted (P) or bi-directionally predicted (B) frames. Detailed specification for VLC is presented in this section so that the reader may design the system without any difficulty. The reader may refer to the relevant standards [23–26] and technical papers [103, 104] before commencing the development.

The basic architecture of JPEG/H.263/MPEG codec is shown in Figure 15.5. The input image or a video sequence is applied to the DCTQ processor block-by-block resulting in quantized DCT coefficients. This is followed by the variable length coder, which assigns minimum of variable length codes, thus bringing about compression. This is at the encoder end. At the decoder end, the inverse operations take place, namely the variable length decoding, inverse quantization, and the inverse DCT. A typical compression is about 20 for a color picture in 4:2:0 format and 10 for a monochrome picture such as Lena as shown in the figure. A good quality picture can be obtained as indicated by the PSNR value of about 30 dB. The basic architectures of the video encoder and the decoder are shown in Figures 15.6 and 15.7 respectively. In addition to the modules described earlier, the encoder and the decoder modules have the encoder controller and the decoder controller respectively.

Figure 15.8 depicts the basic architecture of the implemented MPEG 2 encoder [104], capable of processing I frames. The image or the video sequence to be compressed is input block by block, by a host computer such as the Pentium, into the DCTQ processor, where the discrete cosine transform is performed followed by quantization. The input can be applied after ascertaining that READY is set. When the DCTQ processor is ready to receive the image input data, the host asserts START signal to commence the processing. The resulting quantized coefficients from DCTQ process are applied to the next stage, VLC, where they are assigned variable length codes and buffered by FIFO before they are sent out onto a serial channel as a compressed bit stream. After ensuring that VRDY is set, VSTRT may be asserted to initiate the VLC processing. EOCV indicates the completion of the process. Prior to processing the variable length codes, the header information is processed by VLC processor by writing the same into the on-chip header RAM after ensuring HRDY is set. The processing starts when the host asserts SENDH signal. The color information, Y, Cb, and Cr are input once for each macroblock.

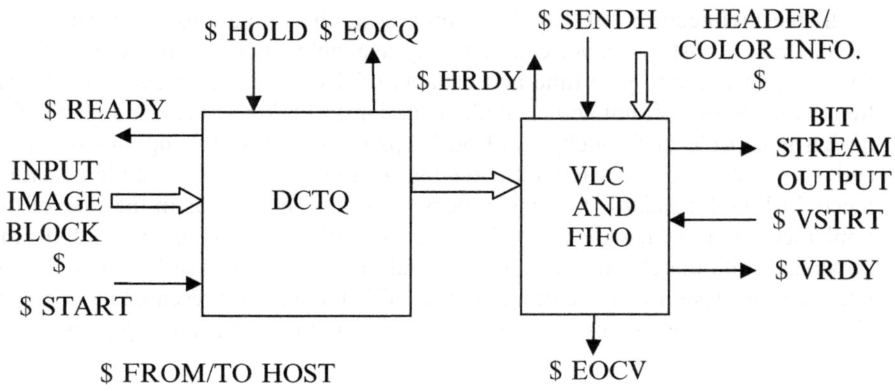

Fig. 15.8 Basic architecture of the implemented MPEG 2 encoder

On similar lines, the MPEG 2 decoder can be implemented. Details of VLC design carried out earlier are available in reference [103]. The following section describes the VLC architecture.

Variable Length Coder

A new, parallel algorithm, architecture and Verilog code for DCTQ processor were presented in Chapters 11 to 13. A video encoder that can process intra (I) frames can be developed by integrating the DCTQ processor and a variable length coder (VLC). In the following sections, architecture of a VLC featuring header information and color processing are covered.

Park and Prasanna [105] have proposed a simple and area efficient VLSI architecture for Huffman coding [106] that conforms to MPEG 1 standard. The throughput achieved with that architecture is 40 Mbps. The design implemented therein was for 8-bit symbols only, and not for a full-fledged Huffman coding whose symbol sizes can go right up to 16 bits for the DC coefficient and up to 28 bits for AC coefficients. The design is not capable of processing either the header or color information that is vital for a total working system. Being a VLSI implementation, design changes to add these features or any other modifications will be practically impossible to achieve. Naturally, this calls for a redesign.

Chang *et al.* [107] have proposed architecture for VLC encoder based on PLAs which meets the JPEG standards only. Further, the design packs the VLC code into a 24-bit constant parallel output, which eventually requires a host processor to convert the parallel information into a serial bit stream before it is sent out to the channel. As a result, the host is likely to be over burdened. Likewise, Chang and Messerschmitt [108], Lin and Messerschmitt [109], Hashermian [110], and Hsieh and Kim [111] have implemented concurrent VLC decoders. Saito [112] has implemented a real time VLC processor as a VLSI. Jeong and Jo [113] have presented an adaptive Huffman coder.

In the architecture [103] of VLC processor, the limitations cited earlier are eliminated. A cost effective, commercially available FPGA with a quick design and implementation cycle time and capable of fast design changes or modifications is made use of, unlike the implementation in Park and Prasanna [105]. The design is capable of throughputs of 50 Mbps with a 50 MHz single-phase clock and about 20:1 compression ratio on the average. The FPGA implementation meets MPEG 2 standard. The host processor is not burdened as in the case of the implementation of Chang *et al.* [107] since the implementation directly outputs the bit stream onto the channel without the need to use the host. Further, it is easy to integrate this design with the design of the DCTQ processor presented in an earlier chapter. The two processes, namely, the DCTQ and the VLC, can be pipelined.

Architecture of the VLC

DCTQ processor presented in earlier chapters prepares the ground for compression of an image or a video sequence. The next processing module is the VLC,

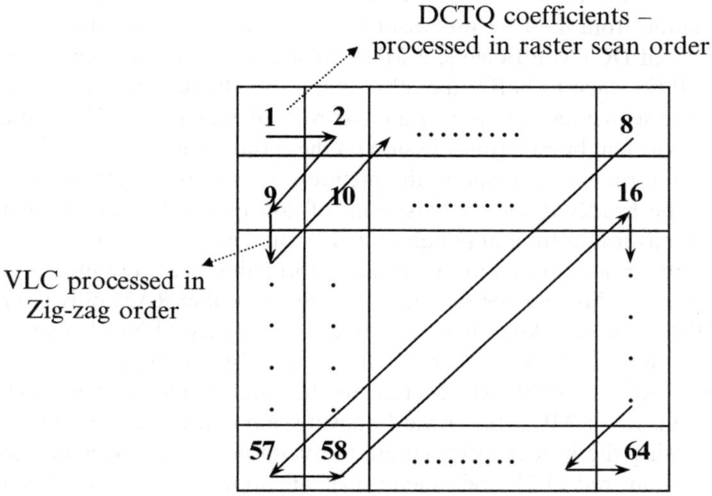

Fig. 15.9 Processing order of variable length code

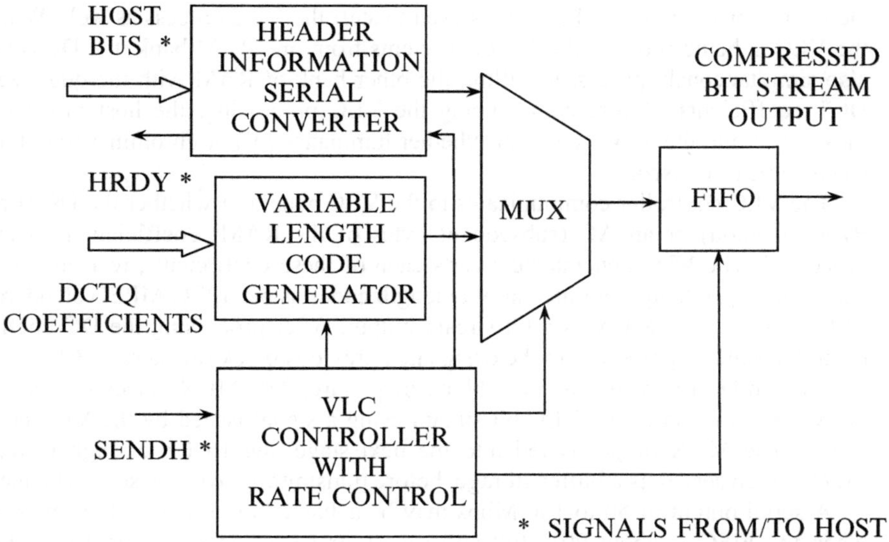

Fig. 15.10 Architecture of VLC

which assigns variable length codes to the DCTQ coefficients and transmits compressed bit stream over a serial channel. The DCTQ processor processes the DCTQ coefficients in a raster scan order, whereas the VLC processes them in a zig-zag order as shown in Figure 15.9. The basic architecture of the proposed VLC coder is shown in Figure 15.10. It essentially consists of circuitry to process header information from the host processor and to generate variable length codes from the quantized DCT coefficients, a MUX to select one of the above, a First-in–First-out (FIFO) stack to buffer the VLC output bit stream and a controller to coordinate all the sequential activities. Rate control embedded in the VLC controller maintains a constant bit rate transmission on the serial channel.

The header information containing the picture size, etc. is written by the host processor into the header RAM. A maximum of 240 bits of header information can be written into it at a time although only 129 bits per frame is required normally. The valid number of bits in the header information is also written into the header processor. The host processor can write into the header RAM only after ensuring that HRDY is set. After the host processor asserts SENDH signal, the header serial output RAM converter reads the header, byte-by-byte, using its internal counter to address the RAM. It converts the parallel information into a serial data and sends it to a MUX for onward transmission to the output FIFO.

After processing the header information, the variable length codes are generated from the quantized DCT coefficients that are input into one of the dual-redundant RAMs in the variable length code generator. RAM address is provided by the DCTQ processor. The addresses and read/write pulses required for the individual RAMs are generated or coordinated by the VLC controller. The DCTQ processor issues the end of conversion signal when it has filled all the 64 coefficients into one RAM and the same is used to start the VLC process as well. While the VLC is being processed using coefficients from one RAM bank, the DCTQ is also simultaneously processed, filling the other bank of RAM with the quantized DCT coefficients. Before commencing the VLC processing, the host processor must write into the VLC generator whether luminance (Y) or chrominance (Cb or Cr) is to be processed.

The VLC controller communicates to the VLC generator whether the DC (first RAM location) or an AC (subsequent bytes of the RAM) coefficient is being processed. The VLC generator converts each of these coefficients, read in a zig-zag sequence, into appropriate variable length codes as per JPEG/MPEG standards and sends it to the MUX as a bit stream output. After processing the DC coefficient, the same is preserved in the designated previous block registers Y, Cb, or Cr for use while processing the next block of picture. The MUX selects either the header information or the VLC bit stream using a signal issued by the VLC controller. The MUX output is fed into the next stage, the 16 Kb or higher sized FIFO, which serves as a buffer storage, before transmitting over the serial channel.

A serial output of 50 to 100 Mbps may be achieved by initiating transmission when the FIFO is about 90% full and suspending it when the content of FIFO reaches about 80% of its capacity. These limits are, however, user programable and must be experimented with actual video sequence before finalizing these set points. Rate control is incorporated in order to maintain a constant bit stream.

The VLC and the DCTQ functional modules process concurrently and have adequate interlock signals between themselves. As a result, no processing of image data will be missed. While the VLC processes image block n, the DCTQ processes (n + 1)th block. Usually, the VLC is slower than the DCTQ. Reading and writing of FIFO take place simultaneously. The VLC controller issues the end of conversion signal when the coding of the current image block is complete. If VLC is coded efficiently, it is possible to process a color motion picture of size 1024 × 768 pixels in 4:2:0 format at 30 frames per second using FPGA. A higher picture size, possibly, 1600 × 1200 pixels can be processed in ASIC implementation.

Header Serial Output Converter

This unit consists of a left shift register to convert parallel header information in RAM to a serial output and a controller to regulate various events as depicted in Figure 15.11. After making sure that HRDY is set, the SENDH signal can be asserted by the host to start the conversion. The controller addresses the header RAM using HRA[4:0] and loads the byte data into the shift register at the rising edge of CLK using the data bus, HRD[7:0] and by asserting the LD signal. An internal 4-bit counter in the controller keeps track of the number of bits to be shifted while signal LD is disabled. HRA[4:0] is incremented and the process is repeated till the entire header information in the RAM is converted into serial bits and sent out of the shift register. The total number of bits of the header information to be processed is supplied by the host. Usually, about 130 bits of header

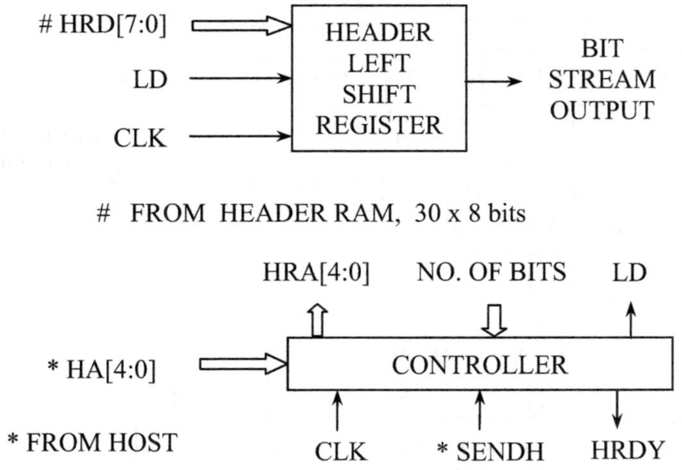

Fig. 15.11 Header serial output converter

information is transmitted per picture frame. HA[4:0] address is used by the host while writing into the header RAM and is disabled during the serial bit conversion.

VLC Generator

A simplified diagram of VLC generator is shown in Figure 15.12. It consists of a 2-bit register, C[1:0], containing the luminance or chrominance information written by the host, three 9-bit registers to store Y, Cb, and Cr DC coefficients of the previous block using the data bus, Q[8:0], a 9-bit sign-magnitude subtract circuit to get the differential DC coefficient between that of the current and the previous blocks, DC and AC VLC coders which output the variable length codes serially, a MUX to select either the DC or the AC codes, and a controller to regulate the control sequence. Which of the components: Y, Cb, or Cr is to be processed is loaded by the host before processing every macroblock.

A zig-zag counter built into the controller addresses the RAM to read the DC/AC coefficients. After processing the current DC coefficient, it is stored in one of the three previous block registers using the write signal, WR. Bit output

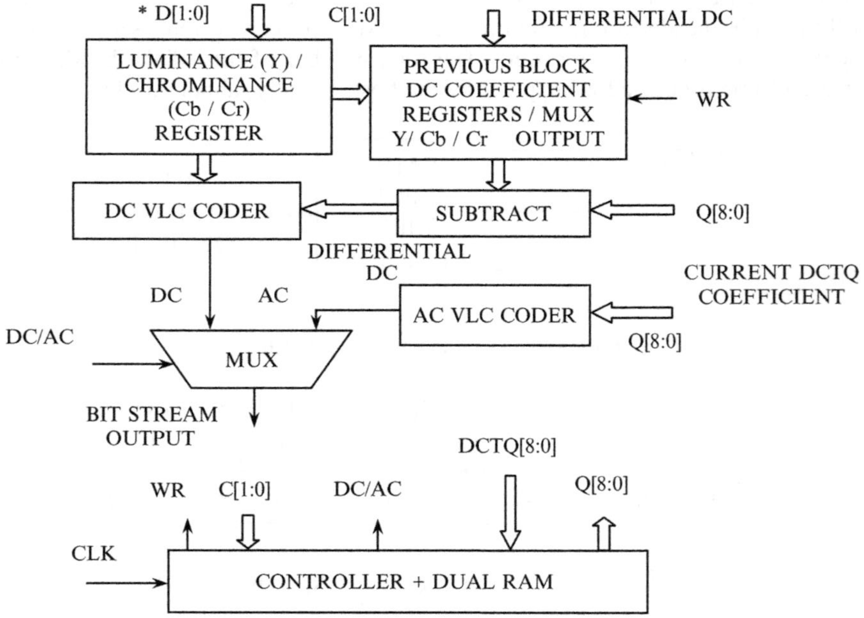

Fig. 15.12 VLC generator

from the MUX is issued once every clock cycle. The controller generates appropriate write signals, not shown in the figure, for registering the pipeline registers inside the DC and AC VLC coders.

The DCTQ or VLC, whichever process is slow, determines the overall processing time of the system since they are pipelined. It is possible to process monochrome images of size 1600 × 1200 pixels at the rate of 30 frames per second depending upon the device selected. For color images in 4:2:0 format, 50% more execution time is required than that for the monochrome picture. As a result, the maximum size of color image that can be processed will only be 67% of the size of the monochrome picture. Header processing, whose execution time is about 3000 ns is a parallel process to DCTQ and is negligible when compared to the VLC processing time of 33 ms per frame.

Oral and written presentations are very important for researchers, students, and practicing engineers. Guidelines for these presentations and a sample presentation are included in the CD. In addition, the reader is urged to develop skills in writing technical papers by studying the existing papers in the literature. A number of them can be found in the references listed in the book and in numerous websites.

Summary

Numerous project designs were suggested for implementation on FPGA or ASIC. These applications were arranged into various categories for the convenience of designers. Brief descriptions were presented for some of these projects. An introduction to embedded systems design was presented. Various issues involved in the design of Digital VLSI Systems were discussed. These were followed by the presentation of detailed specifications and architectures for a couple of projects so that the reader may straightaway start working on these projects to gain hands on experience in designing projects.

Assignments

15.1 A number of projects were suggested for FPGA/ASIC implementation in the text. Suggest some more projects for implementation for each of the following areas of applications:
Automotive electronics
Avionics
Communication
Computer products
Control engineering
Video processing

Medical applications
Miscellaneous applications
Music
Office equipments
Phones
Security systems

15.2 For each of the categories of applications you have suggested for the assignment 15.1, write a brief description.

15.3 A driver less shuttle, a light rail car, which plies between two airports at a distance of 2 miles, is to be controlled automatically. When it is waiting for the passengers at one of the stations, the two entrance/exit doors of the car must remain open. So also the corresponding doors at the station. After the car comes to a halt at a station, the car doors as well as the station doors open. In each of the stations, a push button is installed for use by the passenger(s) to request service of the car which is waiting for passengers at the other station, and has radio linked switches to detect the requests. The car leaves a station after 10 min of arrival, provided there is at least one passenger in the car at the time of departure. At the appointed time of departure, if there is no passenger in the car and, if a service request from the other station is pending, all the doors of the car and the station close, and the car departs to the other station without passengers. However, if there is no request pending, the car waits for the passengers with doors shut. When a passenger arrives, the passenger is allowed to get in and the car departs. At the time of closing, if any passenger arrives, the doors open for 5 s and close again, provided the car is not full. The entry to the car or exit from the car can be made through any of the two doors. Each of the two doors allows only one person at a time. The car can carry a maximum of 25 passengers. Draw a detailed specification of the controller and design the architecture so that the design may be coded in Verilog/VHDL RTL. State your assumptions clearly.

15.4 An alarm annunciator is a watch dog for keeping the process variables in a plant under unceasing surveillance. Usually the annunciators are mounted on the top of control panels. They keep the control engineer posted with abnormal variations in process parameters by providing visual and audible alarms, so that timely corrective action can be taken. The inputs to the alarm annunciator are potentially free, normally open (NO), or normally closed (NC) contacts. These contacts are required to be debounced. A wide variety of sequences are available depending upon the types of applications. A choice of automatic reset, manual reset, ring back, etc. are available. Three such sequences are shown in Figure A15.1. Special sequences can also be tailor-made. The visual indications are provided (to enhance the reliability) by dual-redundant, 24V, 60 mA lamps mounted on windows covered by translucent acrylic sheets. Legends are engraved on them to announce the alarm conditions prevalent at any time. Typical legends are boiler pressure high, main transformer temperature high, oil tank level low, turbine generator vibration high, excitation circuit failure, nuclear activity high, etc., depending upon the needs of industry/plant. Typical number of windows are (expandable) 4, 8, 16, 32, etc., each servicing one control contact to monitor one plant variable. Push buttons are provided in the equipment so that the alarm may be acknowledged, reset, or tested. The visual alarms are also accompanied by audio alarms, which may be an electronic horn, whose timings of operations are shown in the figure.

Auto-alarm sequences

Sequence	Normal	Abnormal	ACK	Reset	Window	Audio 1
AA1	•				Off	Off
		•			Flashing	On
			•		On	Off
	•				Flashing	On
			•		On	Off
				•	Off	Off
				•	Flashing	On
AA2	•				Off	Off
		•			Flashing	On
			•		On	Off
	•				Off	Off
	•				On	Off
				•	Off	Off
				•	On	Off

Ring back sequence

State	Window	Audio alarm	
		Audio 1	Audio 2
Normal	Off	Off	Off
Abnormal	Fast flashing (2 Hz)	On	On
Acknowledge	Steady On	Off	Off
Normal again	Slow flashing (1/2 Hz)	Off	On
Reset	Off	Off	Off
Normal before Ack.	Fast flashing (2 Hz)	On	Off
Acknowledge	Slow flashing (1/2 Hz)	Off	On
Reset	Off	Off	Off

Fig. A15.1 Alarm annunciator (Continued)

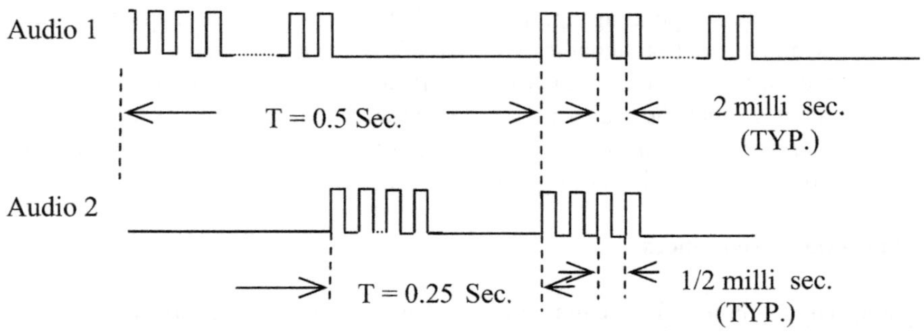

Fig. A15.1 Alarm annunciator

Debounce time of the contacts must be capable of being programed from 2 ms to 10 ms. Provide only one debounce time program control for all the points. Develop a detailed architecture for the annunciator such that RTL Verilog code may be implemented for 16 points or windows.

15.5 A music synthesizer is required to be designed. The output of the synthesizer is one of the musical instrument voices, whose waveform for one complete cycle is given in Figure A15.2a as an example. The waveform of an instrument voice can be manually digitized or captured using a real synthesizer for a finite time and stored in a ROM. The number of samples assumed or arrived at after experimentation must be adequate for getting good quality music. The output of the ROM must be in twos complement in order to accommodate both positive as well as the negative swings of the voice. This output is fed to a digital to analog converter such as DAC1000 of National Semiconductors in bipolar mode, which accepts offset binary (MSB of twos complement inverted) input, followed by a power amplifier and a speaker system (PA) to produce the music. PA may be assumed to be available. The key board of the synthesizer is shown in Figure A15.2b. So long as a key is pressed, the ROM must output complete cycles of the digital voice at a frequency which is marked in the keyboard diagram and the table in Figure A15.2c. If more than one key is pressed, then the highest frequency of the keys pressed is to be recognized. Develop a detailed RTL compliant architecture for the music synthesizer. Explain how you can reconfigure the entire keyboard one octave higher or lower at the flick of two push button switches. No Verilog code need be written.

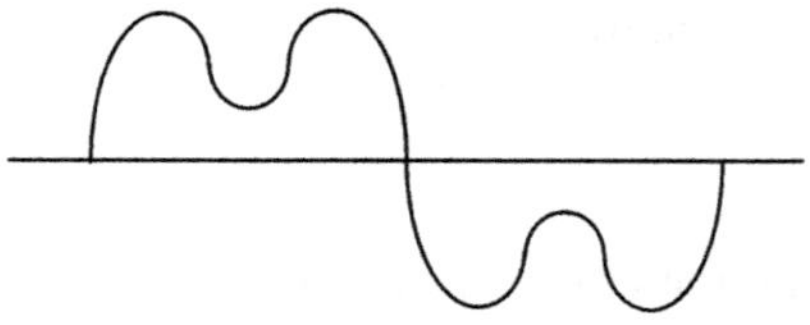

a Note C at 523 Hz

Fig. A15.2 Music Synthesizer (Continued)

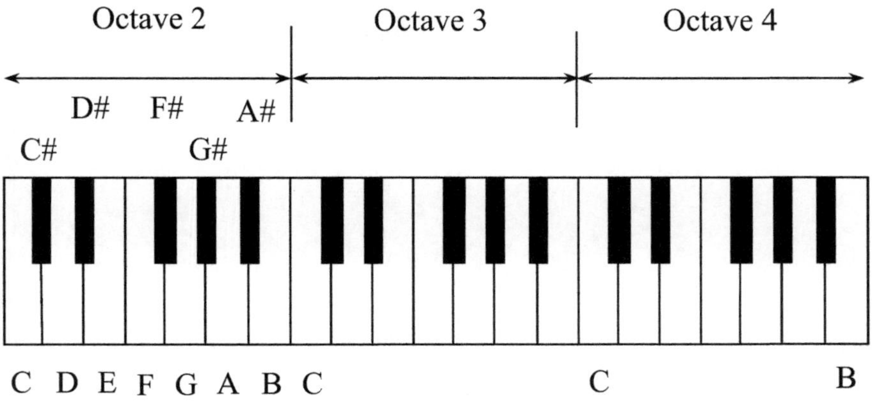

b Synthesizer Keyboard

Note	Frequency (Hz)	Note	Frequency (Hz)	Note	Frequency (Hz)
C	523	D#	660	F#	831
C#	554	E	698	G	880
D	587	F	740	G#	933
A	622	A#	784	B	988

The ratios of frequencies of same notes in Octave 2, Octave 3 and Octave 4 are 1:2:4.

c Frequencies of musical notes

Fig. A15.2 Music Synthesizer

References

[1] Frank Vahid and Tony Givargis, Embedded System Design–A Unified Hardware/Software Introduction, John Wiley and Sons, Inc., MA, 2002.
[2] Raj Kamal, Embedded Systems – Architecture, Programming and Design, Tata McGraw Hill, New Delhi, 2003.
[3] K. R. Rao and J.J. Hwang, Techniques and Standards for Image, Video and Audio Coding, Prentice Hall, Englewood Cliffs, NJ, 1996.
[4] M. Morris Mano, Digital Design, Prentice Hall, NJ, 2002.
[5] M. Morris Mano and C.R. Kime, Logic and Computer Design Fundamentals, Prentice Hall, NJ, 2000.
[6] Parag K. Lala, Digital System Design Using Programmable Logic Devices, Prentice Hall, NJ, 1990.
[7] J.F. Wakerly, Digital Design: Principles and Practices, Prentice Hall, NJ, 2000.
[8] Franklin P. Prosser and David E. Winkel, The Art of Digital Design, Prentice Hall, 1987.
[9] Charles H. Roth, Jr, Digital Systems Design Using VHDL, PWS Publishing Company, Boston, MA, 1998.
[10] PAL Programmable Array Logic Handbook, Monolithic Memories, Santa Clara, CA.
[11] Samir Palnitkar, Verilog HDL – A Guide to Digital Design and Synthesis, Prentice Hall, 2004.
[12] Michael D. Ciletti, Advanced Digital Design with the Verilog HDL, Prentice Hall, 2003.
[13] Michael John Sebastian Smith, Application-Specific Integrated Circuits, Addison-Wesley, 2000.
[14] D.E. Thomas and P.R. Moorby, The Verilog Hardware Description Language, Kluwer Academic Publishers, Boston, 1998.
[15] J. Bhaskar, A Verilog HDL Primer, Star Galaxy Publishing, PA, 1998.
[16] J. Bhaskar, Verilog HDL Synthesis, Star Galaxy Publishing, PA, 2001.
[17] Bob Zeidman, Verilog Designer's Library, Prentice Hall, 1999.
[18] Mentor Graphics, Wilsonville, OR, www.model.com.
[19] Synplicity, Inc., Sunnyvale, CA, http://www.synplicity.com.
[20] Xilinx Inc., San Jose, CA, www.xilinx.com.
[21] N. Ahmed, T. Natarajan and K.R. Rao, Discrete cosine transform, IEEE Trans. Comput., C-23, pp. 90–93, 1974.
[22] K.R. Rao and P. Yip, Discrete Cosine Transform: Algorithms, Advantages, Applications, Academic Press, New York, NY, 1990.
[23] ISO/IEC JTC1 10918-1-ITU-T Rec. T.81, Information technology – Digital compression and coding of continuous-tone still images: Requirements and guidelines, 1994.
[24] ISO/IEC 11172 Information Technology: Coding of moving pictures and associated audio for digital storage media at up to about 1.5 Mbit/s, Part 2: Video, 1993.

[25] ISO/IEC MPEG 2 standards for generic coding of moving pictures: Part 2, Video, 1998.

[26] ISO/IEC 13818-2-ITU-T Rec. H.262 & H263, Generic coding of moving pictures and associated audio information: Video, 1995.

[27] B.G. Lee, A new algorithm to compute the discrete cosine transform, IEEE Trans. Acoust., Speech and Signal Proc., ASSP-32, pp. 1243–1245, 1984.

[28] B.G. Lee, Input and output index mappings for a prime-factor-decomposed computation of discrete cosine transform, IEEE Trans. Acoust., Speech and Signal Proc., ASSP-37, pp. 237–244, 1989.

[29] A.N. Skodras, and A.G. Constantinides, Efficient input-reordering algorithms for fast DCT, Electron. Lett., 27, pp. 1973–1975, 1991.

[30] P. Lee and F.Y. Huang, An efficient prime-factor algorithm for the discrete cosine transform and its hardware implementations, IEEE Trans. Signal Process., 42, pp. 1996–2005, 1994.

[31] C.L. Wang and C.Y. Chen, High throughput VLSI architectures for the 1-D and 2-D discrete cosine transforms, IEEE Trans. Circuits Syst. Video Technol., 5, pp. 31–40, 1995.

[32] Yung-Pin Lee, Thou-Ho Chen, Liang-Gee Chen, Mei-Juan Chen and Chung-Wei Ku, A cost-effective architecture for 8 × 8 2-D DCT/IDCT using direct method, IEEE Trans. Circuits Syst. Video Technol., 7, 1997.

[33] Yukihiro ARAI, Takeshi AGUI and Masayuki NAKAJIMA, A fast DCT-SQ scheme for images, Trans. IEICE, E71, pp. 1095–1097, 1997.

[34] Yi-Shin Tung, Chia-Chiang Ho and Ja-Lung Wu, MMX-based DCT and MC Algorithms for real-time pure software MPEG decoding, IEEE Computer Society Circuits and Systems, Signal Processing, 1, Florence, Italy, pp. 357–362, 1999.

[35] H.S. Hou, A fast recursive algorithm for computing the discrete cosine transform, IEEE Trans. Acoustics, Speech: Signal Proc., ASSP-35, pp. 1455–1461, 1987.

[36] C. Loeffler, A. Ligtenberg and G.S. Moschytz, Practical fast 1-D DCT algorithms with 11 multiplications, Proceedings of IEEE ICASSP, 2, pp. 988–991, 1989.

[37] N.I. Cho and S.U. Lee, DCT algorithms and VLSI implementations, IEEE Trans. Acoust., Speech and Signal Process., ASSP-38, pp. 121–127, 1990.

[38] N.I. Cho and S.U. Lee, Fast algorithm and implementation of 2-D discrete cosine transform, IEEE Trans. Circuits Syst., 38, pp. 297–305, 1991.

[39] Y.P. Lee, T.H. Chen, L.G. Chen, M.J. Chen and C.W. Ku, A cost-effective architecture for 8 × 8 2D-DCT/IDCT using direct method, IEEE Trans. Circuits Syst. Video Technol., 7, pp. 459–467, 1997.

[40] M. Yoshida, H. Ohtomo and I. Kuroda, A new generation 16-bit general purpose programmable DSP and its video rate application, IEEE Workshop on VLSI Signal Processing, pp. 93–101, 1993.

[41] I. Kuroda, Processor architecture driven algorithm optimization for fast 2-D DCT, IEEE Workshop on VLSI Signal Processing, VIII, pp. 481–490, 1995.

[42] J. Golston, Single-chip H. 324 video conferencing, IEEE Micro., 16, pp. 21–33, 1996.

[43] W. Houl, An 8 × 8 Discrete cosine transform implementation on the TMS320C25 or TMS320C30, Texas Instruments, Application Rep. SPRA 115, 1997.

[44] M. Nakagawa, DCT-based still image compression ICs with bit-rate control, IEEE Trans. Consum. Electron., 38, pp. 711–717, 1992.

[45] K. Ogawa, A single chip compression/decompression LSI based on JPEG, IEEE Trans. Consum. Electron., 38, pp. 703–710, 1992.

[46] M.T. Sun, T.C. Chen and A.M. Gottlieb, VLSI implementation of a 16 × 16 Discrete Cosine Transform, IEEE Trans. Circuits Syst., 36, pp. 610–617, 1989.

[47] C.T. Chiu and K.J.R. Liu, Real-time parallel and fully-pipelined two-dimensional DCT lattice structures with application to HDTV systems, IEEE Trans. CSVT, 2, pp. 25–37, 1992.

[48] C.T. Chiu and K.J.R. Liu, Parallel implementation of transform based DCT filter bank for video communications, IEEE International Conference on Consumer Electronics, Chicago, II, pp. 152–153, 1994.

[49] N. Subramani and T. Ogunfunmi, VLSI design and implementation of a DCT chip for video compression using synthesis tools, 37th Midwest Symp. Circuits and Systems, Lafayette, LA, 1994.

[50] P. Pirsch, N. Demassieux and W. Gehrke, VLSI architectures for video and audio coding – A survey, Proceedings of IEEE, 83, pp. 220–246, 1995.

[51] C. Chen, T. Chang and C. Jen, The IDCT processor on the adder-based distributed arithmetic, Proceedings of Symp. VLSI Circuits, pp. 36–37, 1996.

[52] C.Y. Hung and P. Landman, Compact inverse discrete cosine transform circuit for MPEG video decoding, Proceedings of IEEE Workshop Signal Processing Systems, pp. 364–373, 1997.

[53] T. Xanthopoulos and A. Chandrakasan, A low-power IDCT macrocell for MPEG2 MP@ ML exploiting data distribution properties for minimal activity, Proceedings of Symposium VLSI Circuits, pp. 38–39, 1998.

[54] Tian-Sheuan Chang, Chin-Sheng Kung and Chein-Wei Jen, A simple processor core design for DCT/IDCT, IEEE Trans. Circuits Syst. Video Technol., 10, pp. 439–447, 2000.

[55] C.M. Wu and A. Chiou, A SIMD systolic architecture and VLSI chip for the two-dimensional DCT and IDCT, IEEE Trans. Consum. Electron. 39, pp. 859–869, 1993.

[56] D.V.R. Murthy, S. Ramachandran and S. Srinivasan, Parallel implementation of 2D-discrete cosine transform using EPLDs, International Conference on VLSI Design, Goa, January, 1999.

[57] S. Ramachandran, S. Srinivasan and R. Chen, EPLD-based Architecture of Real Time 2D-Discrete Cosine Transform and Quantization for Image Compression, IEEE International Symposium on Circuits and Systems (ISCAS '99), Orlando, Florida, May–June 1999.

[58] S. Ramachandran and S. Srinivasan, A novel, automatic quality control scheme for real time image transmission, VLSI DESIGN J., USA, 14(4), pp. 329–335, 2002.

[59] J.D. Markel, FFT pruning, IEEE Trans. Audio, Electroacoust., AU-19, pp. 305–311, 1971.

[60] K. Nagai, Pruning the decimation in time FFT algorithm with frequency shift, IEEE Trans. Acoust., Speech and Signal Proc., ASSP-34, pp. 1008–1010, 1986.

[61] Z. Wang, Pruning the fast discrete cosine transform, IEEE Trans. Commun., COM-39, pp. 640–643, 1991.

[62] A.N. Skodras, Fast discrete cosine transform pruning, IEEE Trans. Signal Proc., 42, pp. 1833–1836, 1994.

[63] N.P. Walmsley, A.N. Skodras and T.M. Curtis, A fast picture compression technique, IEEE Trans. Consum. Electron., CE-40, pp. 11–19, 1994.

[64] S.R. Rangarajan and S. Srinivasan, Fast image compression by adaptive pruning, Proceedings of the Conference on Signal Processing, Communications and Networking, Indian Institute of Science, Bangalore, India, 1997.

[65] S. Venkatesh, Design and implementation of an efficient progressive image transmission system using pruning algorithms and a parallel architecture, Thesis work for the degree of Master of Science (by research), Department of Electrical Engineering, Indian Institute of Technology, Madras, 1998.

[66] L.D. Vos and Stegherr, Parameterized VLSI architectures for the full-search block-matching algorithm, IEEE Trans. Circuits Syst. Video Technol., 36, pp. 1309–1316, 1989.

[67] L.D. Vos, M. Stegherr and T.G. Noll, VLSI architectures for the full search block matching algorithm, ICASSP'89, Glasgow, Scotland, pp. 1687–1690, 1989.

[68] S. Chang, J.H. Hwang and C.W. Jen, Scalable array architecture design for full search block matching , IEEE Trans. CSVT, 5, pp. 332–343, 1995.

[69] S.C. Cheng and H.M. Hang, A comparison of block matching algorithms mapped to systolic array implementation, IEEE Trans. Circuits Syst. Video Technol., 7, pp. 741–757, 1997.

[70] M.J. Chen, L.G. Chen and T.D. Chieuh, One-dimensional full search motion estimation algorithm for video coding, IEEE Trans. Circuits Syst. Video Technol., 4, pp. 504–509, 1994.

[71] H.M. Jong, L.G. Chen and T.D. Chiueh, Parallel architectures for three step hierarchical search block matching algorithm, IEEE Trans. Circuits Syst. Video Technol., 4, pp. 407–417, 1994.

[72] A. Puri, H.M. Hang and D.L. Schilling, An efficient block matching algorithm for motion-compensated coding, Proc. IEEE Int. Conf. Acoust., Speech and Signal Proc., pp. 1063–1066, 1987.

[73] M. Ghanbari, The cross-search algorithm for motion estimation, IEEE Trans. Commun., 38, pp. 950–953, 1990.

[74] E. Chan and S. Panchanathan, Motion estimation architecture for video compression, IEEE Trans. Consum. Electron., 39, pp. 292–297, 1993.

[75] L.W. Lee, J.F. Wang, J.Y. Lee and J.D. Shie, Dynamic search window adjustment and interlaced search for block-matching algorithm, IEEE Trans. Circuits Syst. Video Technol., 3, pp. 85–87, 1993.

[76] B. Liu and A. Zaccarin, New fast algorithms for the estimation of block motion vectors, IEEE Trans. Circuits Syst. Video Technol., 3, 1993.

[77] R. Li, B. Zeng and M.L. Liou, A new three-step search algorithm for block motion estimation, IEEE Trans. Circuits Syst. Video Technol., 4, pp. 438–442, 1994.

[78] W. Li and E. Salari, Successive estimation algorithm for motion estimation, IEEE Trans. Signal Proc., 4, pp. 105–107, 1995.

[79] K.M. Nam, A fast hierarchical motion vector estimation algorithm using mean pyramid, IEEE Trans. CSVT, 5, pp. 344–351, 1995.

[80] L.M. Po and W.C. Ma, A novel four-step search algorithm for fast block motion estimation, IEEE Trans. Circuits Syst. Video Technol., 6, pp. 313–317, 1996.

[81] Zhongli He and Ming L. Liou, A high performance fast search algorithm for block matching estimation, IEEE Trans. Circuits Syst. Video Technol., 7, pp. 826–828, 1997.

[82] J.Y. Tham, S. Ranganath, M. Ranganath and A.A. Kassim, A novel unrestricted center-biased diamond search algorithm for block motion estimation, IEEE Trans. Circuits Syst. Video Technol., 8, pp. 369–377, 1998.

[83] Vassilios Christopoulos and Jan Cornelis, A center-based adaptive search algorithm for block motion estimation, IEEE Trans. Circuits Syst. Video Technol., 10, pp. 423–426, 2000.

[84] R. Srinivasan and K. R. Rao, Predictive coding based on efficient motion estimation, IEEE Trans. Commun., COM-33, pp. 888–896, 1985.

[85] S. Ramachandran and S. Srinivasan, FPGA implementation of a novel, fast motion estimation algorithm for real-time video compression, ACM International Symposium on Field-Programmable Gate Arrays, Monterey, California, February, 2001.

[86] K.M. Yang and D.J. Le Gall, Hardware design of a motion video decoder for 1–1.5 Mbps rate applications, Signal Proc.: Image Commun., 2, pp. 117–126, 1990.

[87] Y. Kim, C. Rim and B. Min, A block matching algorithm with 16:1 sub-sampling and its hardware design, IEEE International Symposium on Circuits and Systems, 1, Seattle, WA, pp. 613–616, 1995.

[88] T.N.R. Rajesh, Optimization of fast search block matching motion estimation algorithms and their VLSI implementation, Thesis work for the degree of Master of Science (by research), Department of Electrical Engineering, Indian Institute of Technology, Madras, 1999.

[89] Yi-Shin Tung, Chia-Chiang Ho and Ja-Lung Wu, MMX-based DCT and MC Algorithms for real-time pure software MPEG decoding, IEEE Computer Society Circuits and Systems, Signal Processing, 1, Florence, Italy, pp. 357–362, 1999.

[90] T. Koga, K. Llinuma, A. Hirano, Y. Llinuma and T. Ishiguro, Motion-compensated interframe coding for video conferencing, Proceedings of NTC 81, New Orleans, LA, C9.6.1–C9.6.5, 1981.

[91] Chok-Kwan Cheung and Lai-Man Po, Normalized partial distortion search algorithm for block motion estimation, IEEE Trans. Circuits Syst. Video Technol., 10, pp. 417–422, 2000.

[92] G. Lakhani, Improved Equations for JPEG's Blocking Artifacts Reduction Approach, IEEE Trans. Circuits Syst. Video Technol., 7(6), pp. 930–934, December 1997.

[93] R. Rosenholtz and A. Zakhor, Iterative procedures for reductions of blocking effects in transform image coding, IEEE Trans. Circuits Syst. Video Technol., 2, pp. 91–95, Mar. 1992.

[94] Y.Q. Zhang, R.L. Pickholtz and M.H. Loew, A new approach to reduce the blocking effect of transform coding, IEEE Trans. Commun., 41, pp. 299–302, February 1993.

[95] Y. Yang, N.P. Galatsanos and A.K. Katsaggelos, Iterative projection algorithm for removing the blocking artifacts of block-DCT compressed images, IEEE Conf. Acoustics, Speech and Signal Processing, pp. 408–412, 1993.

[96] Zhen Li and Edward J. Delp, Block artifact reduction using a transform-domain Markov random field model, IEEE Trans. Circuits Syst. Video Technol., 15(12), pp. 1583–1593, December 2005.

[97] Y. Yang, N.P. Galatsanos and A.K. Katsaggelos, Regularized reconstruction to reduce blocking artifacts of block discrete cosine transform compressed images, IEEE Trans. Circuits Syst. Video Technol., 3, pp. 421–432, December 1993.

[98] XESS Corp., Raleigh, NC, www.xess.com.

[99] Avnet Inc., Phoenix, AZ, www.avnet.com.

[100] Nu Horizons Electronics Corp., Melville, NY, http://www.nuhorizons.com

[101] Digilent Inc., Pullman, WA, http://www.digilentinc.com/

[102] Open cores, http://www.opencores.org

[103] S. Ramachandran and S. Srinivasan, Design and implementation of an EPLD-based variable length coder for real time image compression applications. IEEE International Symposium on Circuits and Systems (ISCAS), Geneva, Switzerland, May, 2000.

[104] S. Ramachandran and S. Srinivasan, A fast, FPGA-based MPEG-2 video encoder with a novel automatic quality control scheme, Elsevier, J. Microprocessors Microsystems, UK, 25, pp. 449–457, 2002.

[105] H. Park and V.K. Prasanna, Area efficient VLSI architectures for Huffman coding, IEEE Trans. on Circuits Syst. Video Technol., 40, pp. 568–575, 1993.

[106] D.A. Huffman, A method for the construction of minimum-redundancy codes, Proceedings of IRE 40, 1952.

[107] H.C. Chang, L.G. Chen Y.C. Chang and S.C. Huang, A VLSI architecture design of VLC encoder for high data rate video/image coding, IEEE International Symposium on Circuits and Systems, Orlando, Florida, pp. iv 398–401, 1999.

[108] S.F. Chang and D.G. Messerschmitt, Designing high throughput VLC decoder part-1 concurrent VLSI architectures, IEEE Trans. Circuits Syst. Video Technol., 2, pp. 187–196, 1992.

[109] H.D. Lin and D.G. Messerschmitt, Designing high throughput VLC decoder part-2 concurrent VLSI architectures, IEEE Trans. Circuits Syst. Video Technol., 2, pp. 197–206, 1992.

[110] R. Hashermian, Design and hardware implementation of a memory efficient Huffman decoding, IEEE Trans. Consum. Electron., 40, pp. 345–352, 1994.

[111] C.T. Hsieh and S.P. Kim, A concurrent memory-efficient VLC decoder for MPEG applications, IEEE Trans. Consum. Electron., 42, pp. 439–445, 1996.

[112] R. Saito, VLSI implementation of a variable-length coding processor for real time video, Proceedings of IEEE Workshop on Visual Signal Processing and Communication, Hsinchu, Taiwan, ROC, pp. 87–90, 1991.

[113] J. Jeong and J.M. Jo, Adaptive Huffman coding of 2-D DCT coefficients for image sequence compression, Signal Proc.: Image Commun., 7, pp. 1–11, 1995.

Index

L

M

N

O

P

X

9781402058288